60984 81800

THE CONSERVATION ATLAS OF TROPICAL FORESTS

AFRICA

Contributors

Simon Anstey, WWF-International, Gland, Switzerland
E.O.A. Asibey, World Bank, Washington, D.C., USA
Serge Bahuchet, Centre Nationale de Recherche Scientifique, Paris, France
Robert Bailey, Department of Anthropology, University of California, Los Angeles, USA
Andrew Balmford, Large Animal Research Group, University of Cambridge, UK
R.K. Bamfo, Forestry Commission, Accra, Ghana
Richard Barnes, Wildlife Conservation International, University of California, San Diego, USA
Richard Barnwell, WWF, Godalming, Surrey, UK
Joseph B. Bessong, Forestry Department, Yaoundé, Cameroon
Neil Bird, ODA, Kumasi Ghana
A. Blom, WWF, Epulu, Zaïre
K.T. Boateng, Forestry Department, Accra, Ghana
Denys Bourque, Quebec, Canada
Neil Burgess, RSPB, Sandy, Bedfordshire, UK
Peter Burgess, Suffolk, UK
John Burlison, Nature Conservancy Council, Balloch, Scotland
Tom Butynski, Impenetrable Forest Conservation Project, Uganda
G. Caballé, Institut Botanique, Montpellier, France
Julian Caldecott, Cambridge, UK
Pierre Campredon, IUCN, Bissau, Guinea-Bissau
Richard Carroll, WWF, Dzanga-Sangha, Central African Republic
Javier Castroviejo, Asociacion Amigos de Doñana, Seville, Spain
Kevin Cleaver, World Bank, Washington, D.C., USA
Nonie Coulthard, RSPB, Sandy, Bedfordshire, UK
Michael Crosby, ICBP, Cambridge, UK
Alan Cross, UNEP/GRID, Geneva, Switzerland
Glyn Davies, ODA, Nairobi, Kenya
Jean-Pierre d'Huart, WWF, Brussels, Belgium
Charles Doumenge, IUCN, Gland, Switzerland
Françoise Dowsett-Lemaire, Liège, Belgium
Joseph Dudley, Department of Biology and Wildlife, University of Alaska, Fairbanks, USA
Pat Dugan, IUCN Wetlands Programme, Gland, Switzerland
Chris Elliott, WWF-International, Gland, Switzerland
John Fa, Irish Town, Gibraltar
Julia Falconer, ODA, Kumasi, Ghana
J. H. Françoise, Forestry Department, Accra, Ghana
K. Frimpong-Mensah, Institute of Renewable Resources, Kumasi, Ghana
Steve Gartlan, WWF, Douala, Cameroon
K. Ghartey, Forestry Department, Accra, Ghana
Donald Gordon, WCMC, Cambridge, UK
Arthur Green, WWF, Korup National Park, Cameroon
Glen Green, Geology Department, Macalester College, Saint Paul, Minnesota, USA
Michael Green, WCMC, Cambridge, UK
John Hall, School of Agricultural and Forest Sciences, Bangor, UK
Alan C. Hamilton, WWF, Godalming, Surrey, UK
Alexander Harcourt, Department of Anthropology, University of California, Davis, USA
John Hart, Wildlife Conservation International, Project Okapi, Epulu, Zaïre
Terese Hart, Wildlife Conservation International, Project Okapi, Zaïre
William Hawthorne, ODA, Kumasi, Ghana
Philippe Hecketsweiler, Institut Botanique, Montpellier, France
Barry Hewlett, Tulane University, New Orleans, USA
Peter Howard, Kampala, Uganda
Mark Infield, WWF-International, Gland, Switzerland
Gil Isabirye-Basuta, Kibale, Uganda
Martin Jenkins, Cambridge, UK
Andy Johns, Kibale, Uganda
Peter Jones, Department of Natural Resources and Forestry, Edinburgh University, UK
Scott Jones, Bristol, UK
Chris Justice, NASA, Goddard Space Flight Center, Greenbelt, USA
Francis Kasisi, WWF-International, Gland, Switzerland
Ronald Keay, Cobham, Surrey, UK
Jackie Kendall, NASA, Goddard Space Flight Center, Greenbelt, USA
Olivier Langrand, WWF, Antananarivo, Madagascar
Nigel Leader-Williams, Large Animal Research Group, University of Cambridge, UK
Damien Lewis, London, UK
Michel Louette, Musée Royal de l'Afrique Centrale, Belgium
Richard Lowe, Botany Department, University of Ibadan, Nigeria
Peter Lowry, Missouri Botanical Garden, St Louis, USA
H.F. Maitre, Centre Technique Forestier Tropical, Nogent-sur-Marne, France
Claude Martin, WWF-International, Gland, Switzerland
James Mayers, WWF, Dar-es-Salaam, Tanzania
Mankoto ma Mbaelele, Zaïre Institue for Nature Conservation, Kinshasa, Zaïre
Jeff McNeely, IUCN, Gland, Switzerland
Tom McShane,WWF-US, Libreville, Gabon
Erica McShane-Caluzi, WWF-US, Libreville, Gabon
Jean-Boniface Memvie, Forest Service, Libreville, Gabon
Hadelin Mertens, WWF, Kinshasa, Zaïre
Alain Monfort, Liège, Belgium
Don Moore, US Geological Survey, Eros Data Center, Sioux Falls, USA
Th. Muller, National Herbarium and National Botanic Garden, Harare, Zimbabwe
Dominique N'Sosso, Ministry of Forest Economy, Brazzaville, Congo
John Oates, Hunter College, City University of New York, USA
Katie Offert, Nyungwe Forest Conservation Project, Rwanda
Nicola O'Neill, Swansea, Wales
J.G.K. Owusu, Insitute of Renewable Resources, Kumasi, Ghana
Risto Paivinen, FINNIDA, Finland
Prince Palmer, Forestry Division, Sierra Leone
Alexander Peal, Forestry Development Authority, Liberia
Jean-Yves Pirot, IUCN Wetlands Programme, Gland, Switzerland
Roger Polhill, Royal Botanic Gardens, Kew, Surrey, UK
Derek Pomeroy, Zoology Department, Makerere University, Uganda
G. Pungese, Department of Game and Wildlife, Accra, Ghana
S.J. Quashie-Sam, Institute of Renewable Resources, Kumasi, Ghana
Simon Rietbergen, IIED, London, UK
Anne Robertson, National Museums of Kenya, Nairobi, Kenya
Alan Rodgers, Cambridge, UK
Alison Rosser, Cambridge, UK
Per Ryden, IUCN, Gland, Switzerland
Jacqueline Sawyer, IUCN, Gland, Switzerland
Gotz Schreiber, World Bank, Washington, D.C., USA
Heinrich Stoll, Bremen, Germany
David Stone, Begnins, Switzerland
Simon Stuart, IUCN, Gland, Switzerland
Robert Sussman, Anthropology Department, Washington University, St Louis, Missouri, USA
Ian Thorpe, School of Biology, University of East Anglia, UK
Raphael Tsila, Ministry of Forest Economy, Brazzaville, Congo
K. Tufour, Forestry Commission, Accra, Ghana
Caroline Tutin, Lopé Reserve, Gabon
Amy Vedder, Wildlife Conservation International, New York, USA
Fred Vooren, Forestry Department, University of Wageningen, Netherlands
John Waugh, IUCN, Washington, D.C., USA
Clive Wicks, WWF, Godalming, Surrey, UK
Roger Wilson, FFPS, London, UK
Ron Witt, UNEP/GRID, Geneva, Switzerland
Peter Wood, RSPB, Sandy, Bedfordshire, UK
Ipalaka Yobwa, Forest Inventory and Management Service, Kinshasa, Zaire

In addition authors and reveiwers are acknowledged at the end of each chapter.

THE CONSERVATION ATLAS OF TROPICAL FORESTS

AFRICA

Editors

JEFFREY A. SAYER
International Union for Conservation of Nature and Natural Resources, Gland, Switzerland

CAROLINE S. HARCOURT
World Conservation Monitoring Centre, Cambridge, UK

N. MARK COLLINS
World Conservation Monitoring Centre, Cambridge, UK

Editorial Assistant: Clare Billington · Map Editor: Mike Adam
World Conservation Monitoring Centre, Cambridge

The World Conservation Union

SIMON & SCHUSTER
A Paramount Communications Company

New York London Toronto Sydney Tokyo Singapore

ACKNOWLEDGEMENTS

This atlas was produced under the Forest Conservation Programme of IUCN, The World Conservation Union. IUCN's work in tropical forests receives financial support from the government of Sweden. Much of the research, editing and map preparation was done at the World Conservation Monitoring Centre which is supported by IUCN, the World Wide Fund for Nature (WWF) and the United Nations Environment Programme (UNEP); the Centre is also part of UNEP's Global Environment Monitoring System (GEMS) towards which this atlas is a contribution.

IUCN is especially indebted to The British Petroleum Company p.l.c. for the original idea for the atlas and for the generous funding which has enabled the research for the project to be undertaken.

Thanks also go to IBM, for providing a computer which was used for running the geographic information system (GIS) needed to compile the maps, and to the Environmental Systems Research Institute (ESRI) of California which provided the ARC/INFO software for the project. Petroconsultants Ltd of Cambridge kindly made available 'MundoCart', a world digital mapping database which proved invaluable in the preparation of this atlas.

Contributors to the atlas are listed earlier and their labours are much appreciated. A work of this nature, however, inevitably represents the labours of hundreds of people who have painstakingly documented the forests, researching their ecology and wildlife, and who have laboured over the production of the maps from field work to final printing. Heartfelt thanks are offered by the editors to all these unnamed people.

The editors would also like to thank all their colleagues at IUCN and the World Conservation Monitoring Centre, without whose dedicated work this project would not have been possible. In WCMC, particular thanks go to Harriet Gillett and Donald Gordon for information on conservation areas, to Simon Blyth and Gillian Bunting for work on the maps and to Barbara Brown, James Culverwell, Brian Groombridge and Martin Jenkins for much appreciated and varied assistance. At IUCN, invaluable help was provided by Ursula Senn, Jacqueline Sawyer and Jill Blockhus. Finally, Paul Woodman at the Royal Geographical Society gave us considerable aid with, among other things, correct spellings of ever-changing place names.

First published in the United Kingdom by Macmillan Publishers Ltd, 1992
Distributed by Globe Book Services Ltd, Brunel Road, Houndmills Basingstoke, Hants RG21 2XS

and in the USA by
Academic Reference Division
Simon & Schuster
15 Columbus Circle, New York, New York 10023
A Paramount Communications Company

Library of Congress Cataloging-in-Publication Data
The Conservation atlas of tropical forests: Africa/ the World Conservation Union: edited by Jeffrey A. Sayer, Caroline S. Harcourt, N. Mark Collins
p. cm.
First published in the United Kingdom by Macmillan Publishers Ltd — Verso t.p.
Copyright IUCN —Verso t.p.
Includes bibliographical references and index
Contents: The Issues — Country Studies
ISBN 0-13-175332-0
1. Rain forests—Africa—Maps. 2. Man—Influence on nature—Africa—Maps. 3. Conservation of natural resources—Africa—Maps.
I. Sayer, Jeffrey. II. International Union for Conservation of Nature and Natural Resources
G2446.K3C6 1991 <G&M>
333.75'096022—dc20
91-39120 CIP MAP

Acknowledgement of Sources
The sources of the country maps are given at the end of each chapter.
The sources of the illustrations and maps are given in footnotes and captions.
Designed by Robert Updegraff Map Production by Lovell Johns, Oxford
Typeset by BP Intergraphics, Bath, Avon Printed and bound in Singapore

Contents

Foreword

The loss of the world's tropical forests is one of today's most publicised, debated and least understood environmental issues. Some articles give the impression that the destruction is so rapid and catastrophic that by the end of the century there will be only scattered remnants of forest in increasingly embattled national parks. More than half the species that live on land are inhabitants of the tropical forests and a simple extrapolation leads to dire conclusions about what forest clearance means for the world's biological diversity.

But the situation is far more complicated than that. The statistics of total forest loss - 17 million hectares a year, an area considerably bigger than Switzerland - mask an intricate pattern of variation from country to country. The causes of forest loss also vary, though clearance for cultivation is generally the most important.

Deciding what policy to pursue is not easy for tropical governments who are striving to meet the needs of growing populations and to secure economic growth that will allow them to end degrading poverty and provide food security, health care, education and employment. For such nations, forest resources are vital. For many people the forests are the only homes they have ever known.

Used sustainably for meat, nuts, fruits, gums, wild rubber, fibre, medicines, rattans and carefully extracted timber, tropical forests can provide a continuous supply of materials and income to human communities and at the same time maintain local climate, regulate the run-off of rainfall and lock up some of the carbon dioxide, the accumulation of which is causing climatic change. Used destructively the forests may give Gross National Product a quick boost but often leave local communities ruined.

Governments everywhere are reviewing their policies and moving towards sustainable management. They are negotiating international conventions to conserve biological diversity and halt climate change. Conventions on Forests and Biodiversity are also being discussed. Wise use is central to all these initiatives but conventions cannot work without sound knowledge of the forests themselves: where they are, what species exists in them and what essential services they provide. It is a remarkable and disconcerting truth that we lack much of this essential knowledge today.

In 1974 Reider Persson wrote, in a ground-breaking survey of the world's forest resources, 'we know quite a lot about the moon, but we do not know how much of the earth is covered by forests and woodlands.' His words are still true. The problem is particularly acute for Africa. Although we have the capacity to use remote sensing to monitor in considerable detail what is happening in tropical forests, no forest map has ever been produced for some countries and for many the statistics available from different sources are contradictory.

This atlas is an attempt to present the facts on forest extent and loss in Africa. It addresses the issues central to forest conservation and sustainable use. What are the real causes of loss? What are the values of the forests to the people of Africa? How can these values be translated into tangible benefits for the poor rural societies who live in and around the most diverse forests?

The volume begins with an analysis of ecological history. Contrary to popular belief that tropical forests are ancient and unchanging, those of Africa have changed a great deal with the past few tens of thousands of years in response to alterations in climate and sea level. These dynamics need to be understood. The later chapters analyse the characteristics of today's forest, the ways forest-dwelling peoples use them and the implications of agricultural and social trends. The role of the timber industry as a potential force for conservation or destruction is evaluated.

The maps are the heart of the atlas. They have been much more difficult to compile than in our previous volume on the forests of the Asia/Pacific region. Those for most of West Africa and large parts of Central Africa have never before been published. These maps are based on satellite imagery obtained in the past few years and they give a new picture of the dramatic decline in the forests of these areas. The continent is losing its forests faster than any other region. Thirty per cent have already gone and the remainder are being eroded at 1 per cent per year. In Central Africa, where very large tracts of forest remain, they are being fragmented and encroached upon by small farmers. Even light disturbance makes them very vulnerable to fire. Finally, most of Africa's closed forest occur under rainfall regimes which are marginal for this type of vegetation and as a consequence they are more vulnerable to disturbance or small changes in climate than those of other regions.

This atlas is offered to all concerned with conservation and sustainable living in the forested zone of Africa. Those processes will only come about if they are a priority of the peoples of Africa. Conservation programmes that seek to impose external views are doomed to failure. There is a new emerging generation of African conservationists who are well aware of the materials and cultural value of the forests to African societies. Many of them have contributed to this atlas. We hope that the atlas will be of value to them and to their nations in ensuring that Africa's wonderful forests, and the diverse animal life they support, remain a prized asset in the 21st century.

MARTIN HOLDGATE
Director General
IUCN - The World Conservation Union

PART I

Geological Time Scale

Eon	Era	Period		Epoch	Time (Ma)
Phanerozoic	Cenozoic	Quaternary		Holocene	0.1
				Pleistocene	1.8
		Tertiary	Neogene	Pliocene	5
				Miocene	26
			Palaeogene	Oligocene	37
				Eocene	53
				Palaeocene	65
	Mesozoic	Cretaceous		late Cretaceous	100
				early Cretaceous	136
		Jurassic		late Jurassic	160
				middle Jurassic	176
				early Jurassic	190
		Triassic			225
	Palaeozoic	Permian			280
		Carboniferous	Pennsylvanian		315
			Mississippian		345
		Devonian			395
		Silurian			430
		Ordovician			500
		Cambrian			570
Proterozoic	Pre-cambrian	Vendian			650
		Riphean	late Riphean		900
			middle Riphean		1300
			early Riphean		1600
		early Proterozoic			2500
Archaean		Archaean			4550

Scale: 65– 225– 570– 1000– 2000– 2500– 3000– 4000– 4550–

1 Introduction

Africa is, essentially, a continent of woodlands and grasslands; it contains more than twice as much open woodland as closed canopy forest. Indeed, satellite images of Africa show clearly that this is the driest of the three main tropical continents (National Geographic, 1990, pp. ii–iii). Rain forests now cover only about 7 per cent of the land area. Africa's rain forests represent slightly less than one-fifth of the total remaining global resources, while Asia holds slightly more than a fifth and Latin America still contains almost three-fifths. Asia's rain and monsoon forests are depleted by half (Collins *et al.*, 1991), while those of tropical America remain more intact covering at least four-fifths of their early 20th century extent. The figures in this Atlas reveal that the forests of Africa are the most depleted of all with only one-third or so of their historical extent still remaining. (See Table 10.1, which assumes that areas classified by White (1983) as forest/savanna mosaics were once completely forested.) Furthermore, West Africa's forests are being lost faster than those of any other region.

In the 1990s Africa's forests are under severe and growing pressures. Annual deforestation rates in Africa's closed canopy forests for the years 1976–80 were estimated to be about 0.61 per cent of the total closed forest area in 1980 (FAO/UNEP, 1981). FAO has yet to finalise its statistics to a 1990 baseline but moist forest loss is likely to be about 1 per cent per year (FAO, 1990). Annual deforestation is more serious in West Africa (2.1 per cent) than in Central Africa (0.6 per cent). There are indications that unplanned deforestation, and environmental degradation in general, are close correlates of human population growth. As chapter 6 of this volume reveals, Africa's population growth rate is now running at 2.9 per cent (doubling time 24 years), an expansion that is resulting in massive demands for agricultural land, water, fuelwood and other natural products. Notwithstanding this, as chapter 2 relates, the forests of Africa are even now considerably more extensive than they were during the most recent high latitude glacial advance around 18,000 years ago.

Forests of the Region

The limits of African tropical forests shown in this Atlas are based on the recent definitive vegetation classification provided by F. White's memoir and 1:5 million scale map (1983). As in the Asia–Pacific Atlas (Collins *et al.*, 1991), only the closed canopy tropical forests are mapped. Selected forest types are shown, generalised from White (1983), in Figure 1.1.

Africa's closed canopy tropical moist forests run from the mangroves of Senegal on the west coast of the continent to the montane forests of Jebel Hantara near the eastern tip of Somalia. Most of the countries of West Africa were once clothed in forest from the coastline to deep inland, but agricultural and urban expansion have led to large-scale deforestation and fragmentation, graphically presented in Figure 1.2. The relict blocks of forest left at Gola in Sierra Leone (chapter 29), Sapo in Liberia (chapter 25) and Taï in Côte d'Ivoire (chapter 16), are now of global importance as the last significant remnants of the structurally complex, species-rich forests of the Upper Guinea zone.

In Central Africa there still remains a vast, more or less continuous expanse of rain forest. Although whittled away by fire and agriculture on its borders, and by exploitation along the banks of the great rivers of the region, large areas of little-disturbed forest remain. Indeed around 80 per cent of the rain forest on the continent is concentrated in this region, particularly in Zaïre. As Figure 1.3 indicates, this area of the continent still has the opportunity for strategic planning for conservation and economic development. To the south, the main forest block gives way to dense miombo woodlands with scattered patches of dry deciduous forest.

In East Africa the moist forest peters out as the climate gradually becomes more arid. Increasingly, forest occurs only in strips bordering rivers, along the tops of mountains, or on the wet coastal hills. These fragmented forest patches share problems of severe encroachment and exploitation, yet they harbour a high proportion of plant and animal species which are found nowhere else in the world. Many of these forests are the subject of individual conservation programmes, which are rearguard actions to save the last remnants of pristine forest lands.

Forest Classification

White's memoir and map, published by Unesco, was the result of some 15 years of cooperation between Unesco and the Association pour l'Etude Taxonomique de la Flore de l'Afrique Tropicale (AETFAT). AETFAT's Vegetation Map Committee, whose members compiled the materials from which White worked, included many distinguished authorities. Building upon earlier works such as the well-known Yangambi classification of tropical Africa (Trochain, 1957) and Keay's (1959) vegetation map of Africa, White's classification has withstood scrutiny for almost a decade, and looks set to do so for several more. The greatest threat to its boundaries, predictably, is the rapid expansion of anthropic landscapes at the expense of natural vegetation.

White's classification identifies 16 major vegetation types or formations, all based on structure and physiognomy without recourse to climatic or other environmental considerations (Table 1.1). There is a wide diversity of woody vegetation types, including forest, thicket, shrubland, Afroalpine vegetation, scrubland, mangrove and bamboo; woodland being the most widespread (White, 1983, p. 47). Africa differs from tropical Asia in the occurrence of

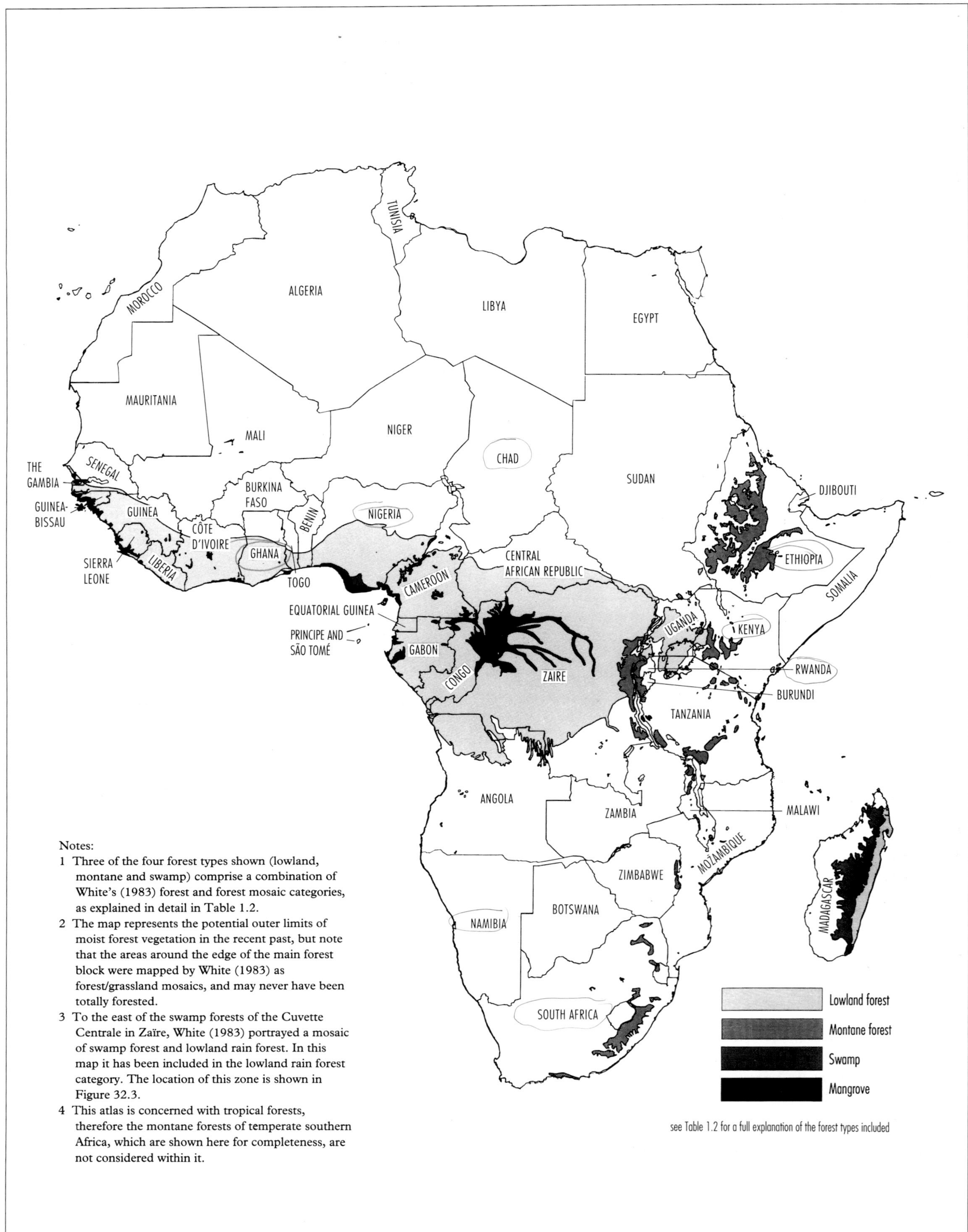

Figure 1.1 The distributional limits of the forest types from which selections have been made for this Atlas (generalised from White, 1983)

numerous closed canopy woody vegetation types which do not contain trees, but shrubs or bushes instead. For this reason the four major categories of Schimper (1903) which were broadly applied in the Asia–Pacific volume of this Atlas, are inadequate here. In the Unesco memoir the 16 formations are sub-divided into 80 mapping units (White, 1983, Table 4). Table 1.2 sets out in detail the Unesco mapping units that have been selected and combined to delimit the classification used on the maps appearing in Part II of this book and Figure 1.1 shows them in a generalised form.

Table 1.2 should be studied in detail to appreciate the context of the maps within this Atlas, but a few words of explanation are needed:

In the main, the limits of forest lands covered by this Atlas are defined by the tropical rain forests and montane forests found within the boundaries of White's 'Forest' and 'Forest Transition and Mosaics' (i.e. categories 1–5, 8, 9, and 11–19). Added to these are forests in the tropical altimontane category (category 65 of White) and mangroves (category 77). Forest types excluded from the coverage are dry forests (categories 6 and 7 of White), dry forest mosaics (categories 21 and 22) and forests outside the tropics (10, 20 and 33).

Dry evergreen forests in the Zambezian region (category 6) are separated from the rain forests by a broad swathe of miombo woodland. Although they were once probably extensive (Aubréville, 1949), they are today confined to tiny, mostly disturbed fragments in a matrix of wooded grassland. Dry forest also occurs in West Africa where it is virtually confined to deep riverine ravines in Mali, and to the coastal plain of Ghana (White, 1983, p. 46). Mali is not covered in this Atlas, and any relicts of dry forest remaining in Ghana cannot be distinguished from the rain forests. At the scale

Table 1.1 The main vegetation types of Africa, as described by White (1983, Tables 1 and 3)

Formations of Regional Extent

1 Forest
A continuous stand of trees at least 10 m tall, their crowns interlocking.

2 Woodland
An open stand of trees at least 8 m tall with a canopy cover of 40 per cent or more. The field layer is usually dominated by grasses.

3a Bushland
An open stand of bushes usually between 3 and 7 m tall with a canopy cover of 40 per cent or more.

3b Thicket
A closed stand of bushes and climbers usually between 3 and 7 m tall.

4 Shrubland
An open or closed stand of shrubs up to 2 m tall.

5 Grassland
Land covered with grasses and other herbs, either without woody plants or the latter not covering more than 10 per cent of the ground.

6 Wooded grassland
Land covered with grasses and other herbs, with woody plants covering between 10 and 40 per cent of the ground.

7 Desert
Arid landscapes with a sparse plant cover, except in depressions where water accumulates. The sandy, stony or rocky substrate contributes more to the appearance of the landscape than does the vegetation.

8 Afroalpine vegetation
Physiognomically mixed vegetation occurring on high mountains where night frosts are liable to occur throughout the year.

Transitional Formations of Local Extent

9 Scrub forest
Intermediate between forest and bushland or thicket.

10 Transition woodland
Intermediate between forest and woodland.

11 Scrub woodland
Stunted woodland less than 8 m tall or vegetation intermediate between woodland and bushland.

Edaphic Formations

12 Mangrove
Open or closed stands of trees or bushes occurring on shores between high and low water mark. Most mangrove species have pneumatophores or are viviparous.

13 Herbaceous fresh-water swamp and aquatic vegetation

14 Halophytic vegetation
(saline and brackish swamp).

Formation of Distinct Physiognomy but Restricted Distribution

15 Bamboo

Unnatural Vegetation

16 Anthropic landscapes

(*Source:* White, 1983)

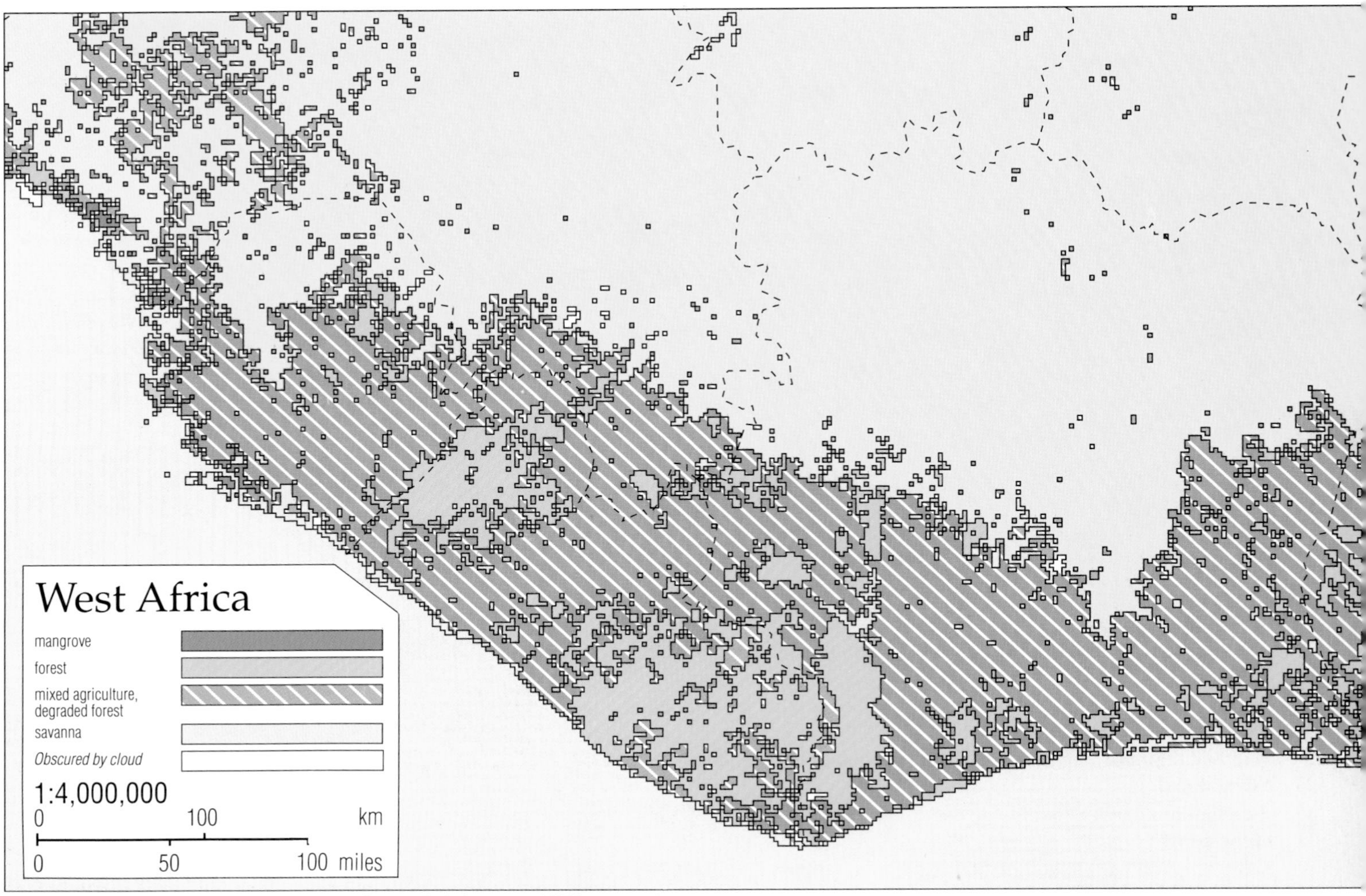

West Africa
mangrove
forest
mixed agriculture, degraded forest
savanna
Obscured by cloud
1:4,000,000
0
100
km
0
50
100 miles

Figure 1.2 Rain forests of West Africa (Päivinen and Witt, 1989)

Central Africa
Rain Forest
lowland
montane
degraded lowland
inland swamp forest
mangrove
Non Forest
1:15,500,000
0
250
km
0
250 miles
No data
No data

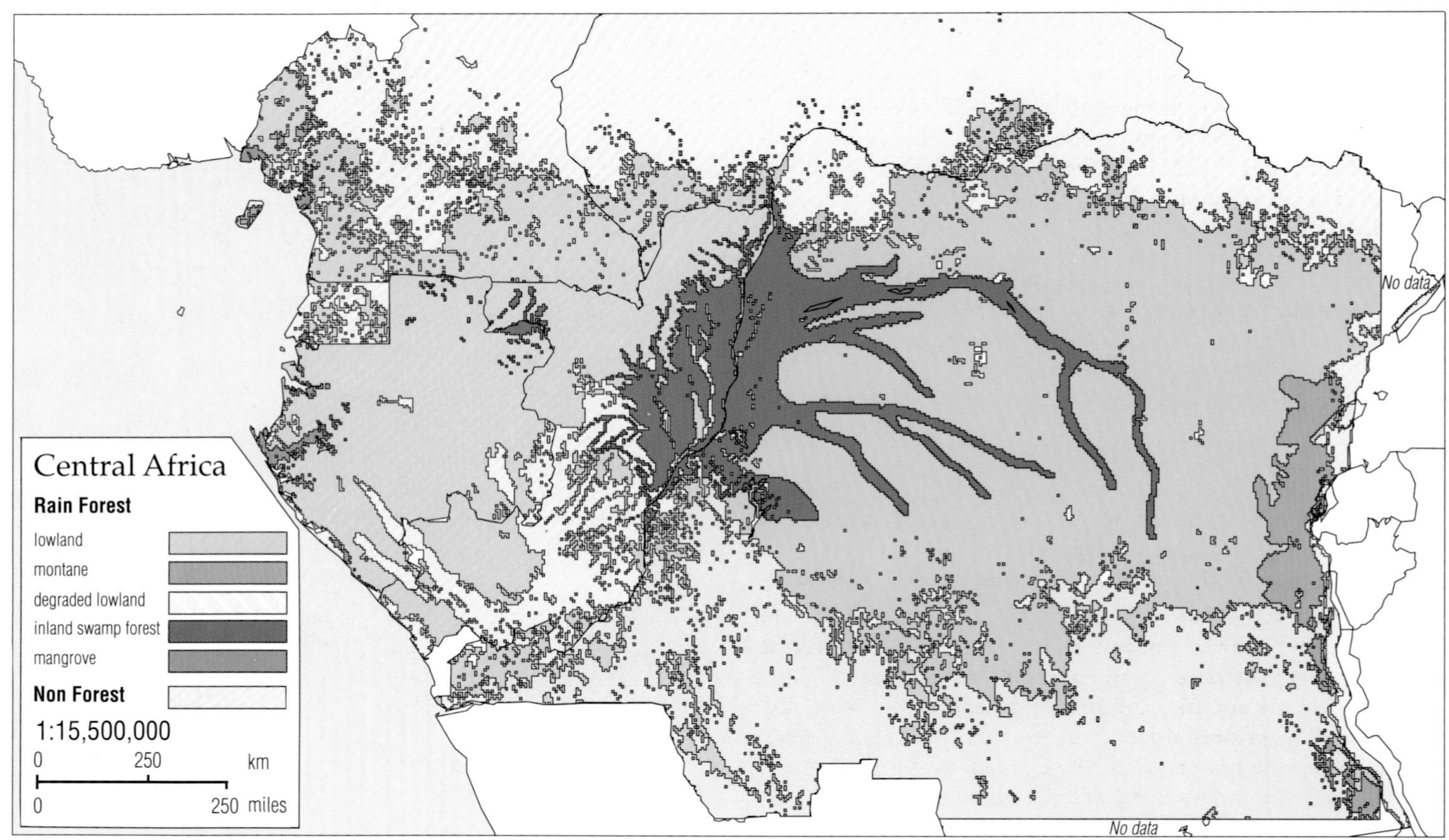

Figure 1.3 Rain forests of Central Africa

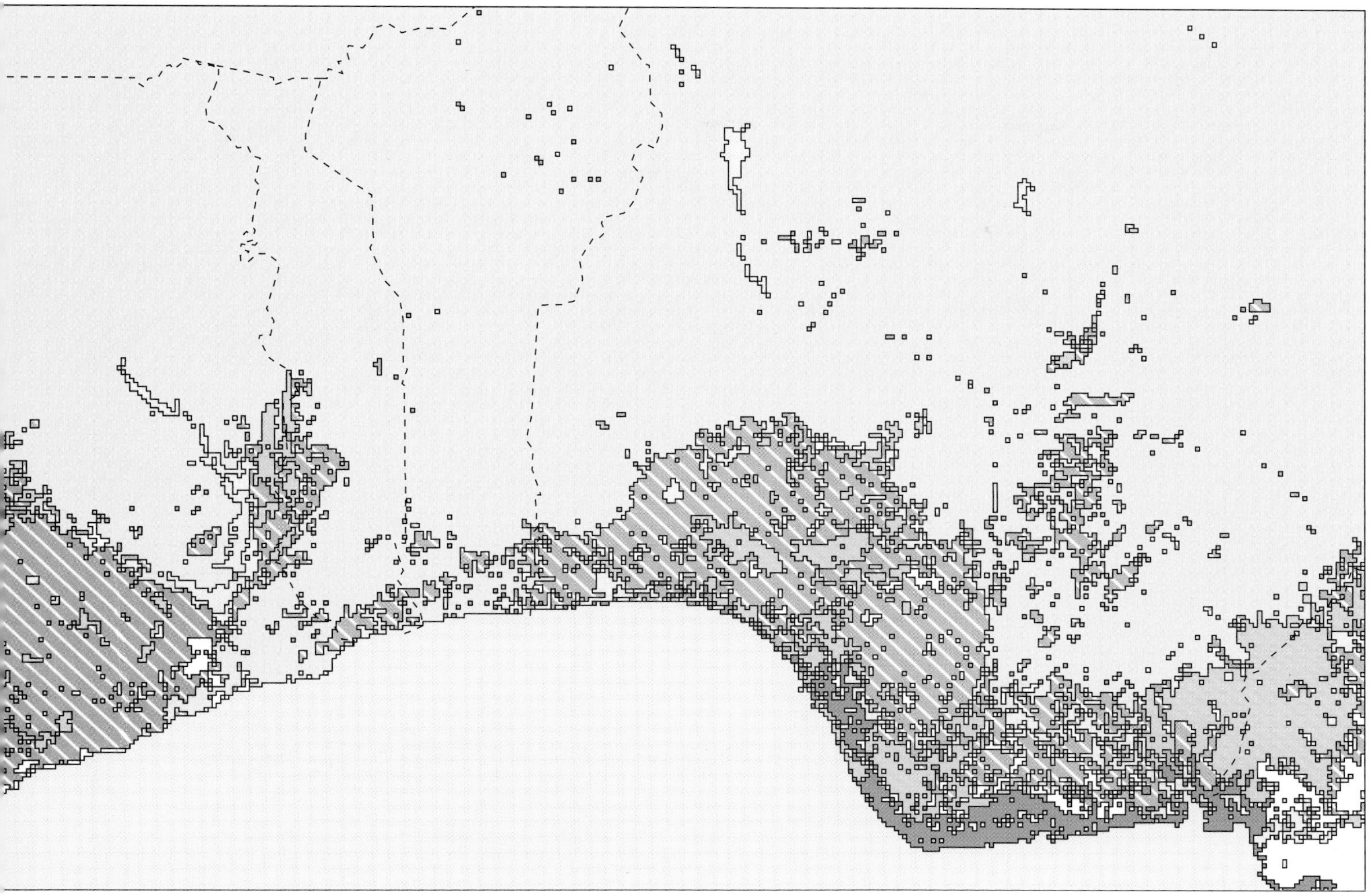

of resolution of Map 21.1, the whole of the coastal thicket and grassland zone as defined by the 'Vegetation Zones' maps published by the Survey of Ghana (1969) is, in any case, deforested.

In Madagascar the dry deciduous forest (White's category 7) was the dominant vegetation below 800 m on the western side of the island, but the forests have now been much degraded by fire and grazing livestock. As in mainland Africa, this forest type has not been mapped in the Atlas, but it is shown on a figure (26.1) in chapter 26.

At the continental scale of mapping employed by White (1983), swamp forests are mostly confined to the more or less permanently flooded areas of rain forests in the Cuvette Centrale of Zaïre. In this Atlas, where nations have been mapped individually, we have departed from White (1983) in retaining the swamp forest vegetation type where national maps have indicated this formation, even though it may fall outside White's categories 8 and 9.

In mountain areas low stature forests (*sensu lato*) are classified by Unesco as shrubland or thicket; bamboo also occurs. Details of the limitations applied to the montane forest category are given in Table 1.2. These mapping units have taken priority over other, national, definitions of montane forest (where they exist), in the interests of consistency.

Gallery forests growing on river banks in otherwise very seasonal vegetation types are not mapped in this Atlas and their contribution to total forest cover is not assessed or discussed. This is because of practical difficulties in that gallery forests (also called riverine or riparian forests) are often too narrow to map at the scales used and they are not consistently depicted on source materials. The biological importance of gallery forests in seasonal climates is well recognised in Africa (particularly along East Africa's rivers), but they fall outside the scope of this Atlas.

Africa's rain forests range from the dense evergreen forests of the lowlands to montane forests, such as that shown here in Kahuzi-Biega National Park in Zaïre, where the vegetation is dominated by tree heathers (Ericaceae). C. Doumenge

Table 1.2 Forest formations of Africa as described by White (1983) with details of the categories of White's classification used to define each forest type mapped in this Atlas.

White's (1983) Categories		*Categories used to define various rain forest formations in this Atlas*				*Notes*
Number	*Name*	*Lowland dryland rain forest*	*Wetland inland (swamp) forest*	*Montane rain forest*	*Wetland coastal (mangrove) forest*	
FOREST						
1	Lowland rain forest: wetter types					
	(a) Guineo-Congolian;	+				1a
	(b) Malagasy	+				1b
2	Guineo-Congolian rain forest:					
	drier types:	+				
3	Mosaics of 1a and 2	+				
4	Transitional rain forest			+		4
5	Malagasy moist montane forest			+		
6	Zambezian dry evergreen forest					6
7	Malagasy dry deciduous forest					7
8	Swamp forest		+			8
9	Mosaic of 8 and 1a	+	+			9
10	Mediterranean sclerophyllous forest					10
FOREST TRANSITIONS AND MOSAICS						
11	Mosaic of lowland rain forest					
	and secondary grassland					
	(a) Guineo-Congolian;	+				11a
	(b) Malagasy	+				
12	Mosaic of lowland rain forest,					
	Isoberlinia woodland and					
	secondary grassland	+				
13	Mosaic of lowland rain forest,					
	secondary grassland and					
	montane elements	+				13
14	Mosaic of lowland rain forest,					
	Zambezian dry evergreen					
	forest and secondary grassland	+				
15	West African coastal mosaic	+				15
16	East African coastal mosaic					16
	(a) Zanzibar-Inhambane;	+				
	(b) forest patches;	+				
	(c)Tongaland-Pondoland					16c
17/18	Cultivation and secondary					
	grassland replacing upland					
	and montane forest 17 African;			+		
	18 Malagasy			+		17/18
19	Undifferentiated montane					
	vegetation (a) Afromontane;			+		
	(b) Sahelomontane; (c) Malagasy			+		19a-c
20	Transition from Afromontane					
	scrub forest to Highveld					
	grassland					20
21	Mosaic of Zambezian dry					
	evergreen forest and wetter					
	miombo woodland					21
22	Mosaic of dry deciduous forest					
	and secondary grassland					
	(a) Zambezian; (b) Malagasy					22
23	Mediterranean montane forest					
	and altimontane shrubland					23
24	Mosaic of Afromontane					
	scrub forest, Zambezian scrub					
	woodland and secondary grassland					24
OTHER FOREST VEGETATION TYPES						
65	Altimontane vegetation in					
	tropical Africa			+		65
77	Mangrove				+	77

The areas of remaining forest are taken from the most reliable source for individual countries and superimposed on the areas delimited by White's (1983) map. This makes it possible to depict the existing forest patches within the extensive areas categorised by White as mosaics of forest and other vegetation (mainly anthropogenic grassland, but with some natural grassland in montane areas).

Forest Cover

As in the Asia–Pacific volume of this Atlas, the maps shown here do not include woodland, forest plantations or extensive areas of shifting cultivation where these are shown in source material. In Africa, as in Asia, it is, however, very difficult to recognise forest that has been locally disturbed by cultivation or, even extensively, by logging (chapters 7 and 8 describe the various systems of forest use and management in greater detail). The areas mapped as single units of forest are therefore frequently mosaics of relatively undisturbed, plus disturbed, forests. It is important for the reader to appreciate that even though large belts of cultivation and plantations have been excluded from the maps, the areas of forest still include enclaves, sometimes quite extensive, of disturbed and degraded vegetation. In some cases, therefore, our maps show larger areas of forest than are suggested by other sources.

Geographic Boundaries

The Atlas limits are essentially the forested countries within the tropics of Cancer and Capricorn. On the eastern side of the continent, the big mountain massif of Ethiopia contains the northernmost rain forests while in the south, the Atlas includes Mozambique and Zimbabwe, which has small patches of montane forest near its eastern border, but omits South Africa. The southern limit here is arbitrary because, as White's memoir makes clear, there is no sharp change in forest structure, physiognomy or floristics at any particular distance from the equator. On the west side, the most northerly forests to be mapped are the mangroves of Senegal and the southern limit is the rain forest in Angola. In the Indian Ocean the whole of Madagascar is included (southern limit 25.5°), as well as the Comoros, Seychelles, Mauritius and Réunion.

Notes for Table 1.2

1a This is mostly semi-evergreen forest but in West Africa includes three relatively small enclaves of evergreen 'hygrophilous coastal evergreen Guineo-Congolian rain forest' (White, 1983 p. 76); for a discussion of these two kinds of lowland rain forest see Whitmore (1990).

1b Evergreen rain forest formerly extended along the entire length of the east coast of Madagascar but much has now been replaced by secondary grassland and regrowth and destroyed for cultivation.

4 Transitional to Afromontane forest; largely destroyed.

6 Small areas of dry forest separated from the rain forests by miombo woodland. Not mapped in this Atlas.

7 Once dominant in western Madagascar, now fragmented and restricted. Not mapped in this Atlas (but see Figure 26.1).

8 In a few places, well-supported information on swamp forests falling outside White's areas 8 and 9 has been included in the swamp forest category.

9 Discussed in 8 above. Peatswamp forest has not yet been found in Africa (Whitmore, 1990).

10 Outside the tropics. Not mapped in this Atlas.

11a This occurs as a very extensive arc surrounding the central rain forest core and very little remains as forest.

13 Occurs within 11a.

15 Occurs within 11a.

16 At the scale of the Atlas 16b includes most of the remaining forest. The more extensive type 16a is extensively modified by humans and has only very tiny forest patches in some cases too small to map and 16c, outside the tropics, is not mapped.

17/18 This mapping unit, which is extensive in central Madagascar, includes three vegetation types in east Madagascar: (i) sclerophyllous montane forest (1300–2300 m), (ii) tapia forest (800–1600 m) and (iii) secondary grassland. Only the first is a rain forest; tapia forest occurs on the western slopes of the upland massif in a rain shadow and has a hot, dry climate. For the Atlas those parts of 18 are mapped which occur east of the watershed, on the rain relief slopes. When remaining forest cover is superimposed, very little of this montane forest type remains.

19a The high mountains of Africa occur as small areas in Liberia and Cameroon in the west and as a long archipelago-like chain along the eastern side running from c. 17.5°N in Eritrea to 32.5°S in the Winterberge range in South Africa. This chain includes White's mapping unit 65. The most extensive massif is in Ethiopia south of Eritrea at c. 7–15°N. The vegetation is a mosaic (White, 1983 pp. 163–9) of various sorts of forest, bamboo, evergreen bushland and thicket, shrubland, mixed Afroalpine communities (above the treeline and including the famous giant senecios and giant lobelias: mapping unit 65), and grassland which 'today is the most extensive vegetation type on the African mountains'.

19b Very small, dry forest formations. Not mapped in this Atlas.

19c Small areas of high mountain bushland and thicket set in secondary grassland.

20 Outside the tropics. Not mapped in this Atlas.

21 See 6 above. Not mapped in this Atlas.

22 See 7 above. Not mapped in this Atlas.

23 Outside the tropics. Not mapped in this Atlas.

24 Scrub forest outside the tropics. Not mapped in this Atlas.

65 See notes on 19a.

77 In a few places, well-supported information on mangrove forests falling outside White's area 77 has been included in the mangrove category.

Issues which Affect Africa's Forests

The layout used in the Asia–Pacific volume of this Atlas has been adopted here, with the first part setting out some of the important issues affecting the conservation and management of Africa's forests today. Chapter 2 gives an insight into the recent (in geological terms) history of the main forest blocks. The popular misconception that forests are primeval and unchanging is laid to rest with descriptions of how the forest boundary has fluctuated during the Pleistocene. In chapters 3 and 4 the biological richness of the forests is described, and the management problems associated with elephants and primates are selected as examples to illustrate the issues in detail. Chapters 5 and 6 bring in the human element, with a chapter on forest peoples, concentrating on the so-called pygmy tribes, and an evaluation of the links between population, environment and agriculture. The forests have always been a source of revenue from timber, and aspects of the trade are detailed in chapter 7, while the following chapter analyses current management of forests and the impact of the ever-growing demand for timber on the forests and economics of the region.

Protected areas remain the cornerstone of conservation strategies in Africa, yet the rain forests remain less well protected than other biomes on the continent. Chapter 9 examines the effectiveness of protected areas, the need for expansion of the regional system, and the limitations in training, management and knowledge that reduce its effectiveness. The final chapter in Part I looks at the future of Africa's tropical forests, teasing the issues apart and drawing them together again in an effort to inject realism into programmes to conserve forests.

Country Studies

Part II examines the situation of each country in detail. Basic statistics are provided at the head of each chapter. The land area, which excludes bodies of water in the country, is from FAO (1989) and it is this area that is used in calculations such as per cent of the country protected or per cent covered with rain forest. Actual country area (i.e. including water bodies) is frequently given in the text. Economic data, demographic statistics and predictions are from the 1990 Datasheet supplied by the Population Reference Bureau (PRB, 1990). Forest cover statistics from FAO (1988) are compared with data from the maps shown in this Atlas. However, for countries where good maps of forest cover do not exist, we have been able to produce only generalised maps based on sketch maps provided by people familiar with the country. In these cases we have not attempted to derive any statistics from our maps as these might be misleading. FAO's figures are for closed broadleaved forest, i.e. 'those which cover with their various storeys and undergrowth a high proportion of the ground and do not have a continuous dense grass layer allowing grazing and spreading of fires. They are often, but not always, multistoreyed. They may be evergreen, semi-deciduous, deciduous, wet, moist or dry' (FAO/UNEP, 1981). In other words, this definition includes the mangrove, swamp, montane and lowland forests for which the Atlas maps give separate figures but also includes areas of forest along river banks in dry country and dry forests, neither of which is considered here. Forest product information (see Table 1.3) is compiled from the 1991 FAO Yearbook (FAO, 1991).

Country chapters use a standard format as far as is practicable, with a preliminary overview followed by an introduction to the

Table 1.3 Definition of forest products

Industrial roundwood this is wood in the rough, i.e. in its natural state as felled or otherwise harvested. It includes wood removed from outside, as well as inside, forests. The commodities included are sawlogs and veneer logs, pulpwood, other industrial roundwood and, in the case of trade, chips and particles and wood residues. The statistics include recorded volumes as well as estimated unrecorded volumes. Fuelwood and charcoal is excluded from this figure, whereas it was included in the roundwood figures given in the Asian Atlas. It is a much greater component of roundwood in Africa than it is in Asia.

Fuelwood and charcoal both coniferous and non-coniferous wood are included.

Processed wood the figures given are aggregates of the figures in FAO (1991) for sawnwood and sleepers and wood-based panels. The sawnwood may be planed or unplaned and it generally exceeds 5 mm in thickness. The wood-based panels include veneer sheets, plywood, particle board and fibreboard.

In cases where countries have not reported to FAO, the information supplied in the Yearbook has been taken from national yearbooks, from reports, from unofficial publications or has had to be estimated by FAO. (*Source:* FAO, 1991)

nation and a detailed account of its tropical forests, their management and extent. Statistics on the areas of forest, their rate of deforestation and information on the map provided are discussed, with full references and sources. Floristics are not described in detail since this would take the Atlas beyond its intended size, but sources for further reference are generally noted, along with basic information on dominant tree species and species of economic significance.

Nations of eastern Africa (chapter 17), southern Africa (chapter 30) and small islands in the Indian Ocean (chapter 24) are considered together to enable their many common problems to be treated within a space consistent with the very limited area of their moist forests.

A central aim of this series of atlases is to place the resources and management of tropical forests into the context of conservation of biological and ecological diversity, i.e. to manage the forests sustainably without degrading their productivity or biological diversity. An indication of the richness of fauna and flora is presented for each nation, attention being drawn in particular to threatened species and species of economic concern. Initiatives in policy, strategy or on the ground are recorded and discussed. The representativeness of the protected area system is scrutinised and details of all the existing and officially proposed protected areas are presented in tabular form.

Those protected areas that include moist forests within their boundaries are noted, and the data compiled for consideration regionally in chapter 9, where data on Biosphere Reserves and World Heritage Sites will also be found. At the generalised level of mapping developed here, conservation areas that include relict fragments of forest may not be recorded.

Maps

Every country chapter includes a map or series of maps, mostly at a scale of 1:3 million or 1.4 million, detailing the remaining rain forests and the conservation areas in the region. These maps are drawn from a wide variety of sources. Geographic Information System (GIS) technology has been used to superimpose satellite imagery data, forest cover maps, protected area maps and vegetation maps (notably White, 1983 – see Table 1.2) to produce a series of coverages consistent in content and classification.

Each map is accompanied by a comprehensive legend that not only acknowledges in detail the sources used, but also explains what steps have been taken to harmonise the sources with the classification given in Table 1.2. All protected areas are mapped where their locations are known. However, for some of the areas precise boundaries were unobtainable, in which case the areas are represented by circles based on the centre-point of the park or reserve. In the first volume of this Atlas conservation areas smaller than 50 sq. km were not shown; in this volume they are indicated by a circle and listed in the tables.

Availability of Data

The spatial data recorded in this volume are stored in digital form at the World Conservation Monitoring Centre, Cambridge, UK. The Centre will be pleased to collaborate with organisations wishing to apply the data in the interest of nature conservation.

References

Aubréville, A. (1949) *Climats, Forêts et Désertifications de l'Afrique Tropicale.* Société d'Editions Géographiques, Maritimes et Coloniale, Paris, France.

Collins, N. M., Sayer, J. A. and Whitmore, T. C. (1991) *Conservation Atlas of Tropical Forests. Asia and the Pacific.* Macmillan, London in association with IUCN, Gland, Switzerland. 256 pp.

FAO (1988) *Second Interim Report on the State of Tropical Forests* Report to the 10th World Forestry Congress, Paris, France. Forest Resources Assessment Project. FAO, Rome, Italy. 2pp.

FAO (1989) *FAO Production Yearbook* Vol. 42. FAO, Rome, Italy.

FAO (1990) *Second Interim Report on the State of Tropical Forests.* Report to the 10th World Forestry Congress, Paris, France. Forest Resources Assessment Project. FAO, Rome, Italy. 2pp.

FAO (1991) *FAO Yearbook of Forest Products 1977–1989.* FAO Forestry Series No. 24 and FAO Statistics Series No. 97. FAO, Rome, Italy.

FAO/UNEP (1981) *Forest Resources of Tropical Africa. Part 1: Regional Synthesis.* FAO, Rome, Italy.

Keay, R. W. J. (1959) *Vegetation Map of Africa South of the Tropic of Cancer. Explanatory Notes.* Oxford University Press, Oxford, UK. 24 pp. with coloured map 1:10 million.

National Geographic (1990) *Atlas of the World.* Sixth Edition. National Geographic, Washington, DC, USA. 136 pp.

Päivinen, R. and Witt, R. (1989) *The Methodology Development Project for Tropical Forest Cover Assessment in West Africa.* Unpublished report. UNEP/GRID, Geneva, Switzerland.

PRB (1990) *1990 World Population Datasheet.* Population Reference Bureau Inc., Washington, DC, USA.

Schimper, A. F. W. (1903) *Plant-geography upon a Physiological Basis.* Fisher, W. R., Groom, P. and Balfour, I. B. (translators). Oxford University Press, Oxford, UK.

Survey of Ghana (1969) *Ghana: Vegetation Zones.* Survey of Ghana, Accra. 1 sheet map.

Trochain, J. (1957) Accord interafricain, sur la définition des types de végétation de l'Afrique tropicale. *Bulletin Institut Etudes Centrafricaine* n.s. 13–14, 55–93.

White, F. (1983) *The Vegetation of Africa: a descriptive memoir to accompany the Unesco/AETFAT/UNSO vegetation map of Africa.* Unesco, Paris, France. 356 pp.

Whitmore, T. C. (1990) *An Introduction to Tropical Rain Forests.* Clarendon Press, Oxford, UK. 226 pp.

Authorship

Mark Collins, WCMC

2 History of Forests and Climate

Introduction

Mankind has greatly altered the distribution and characteristics of tropical forests in Africa (Hall and Swaine, 1981). Forest margins have shrunk with the spread of agriculture and burning, while forest composition and structure have been influenced even in apparently remote places by past settlement or collection of forest products. In a soil survey in apparently pristine or near-pristine forest on the East Usambara Mountains, Tanzania, charcoal was found at almost all sites, pottery was encountered in two, and one soil pit even passed through house foundations (Hamilton and Bensted-Smith, 1989). This is not particularly unusual. Ecologists who have walked many kilometres into apparently natural forest often report similar signs of former human occupation (although it is noted that in some cases charcoal could have originated from natural fires).

Agriculture may be less than a few thousand years old in tropical Africa. Some of the earliest indications are increases in the abundance of oil palm pollen in sediments dating to 3500 BP (years before present) in Ghana (Talbot, 1983) and 3000 BP in Nigeria (Sowunmi, 1981). Major forest clearance believed to be associated with agriculture dates back to about 2000 BP in East Africa (Hamilton, 1982). Before agriculture, woodland burning started by people could have been a factor modifying forest margins for hundreds of thousands of years. Indeed, the oldest archaeological sites indicate that humans in China had controlled use of fire about half a million years ago (Leakey, 1981). Burning will have sharpened the forest/savanna boundary and made these two vegetation types more distinct from one another (Keay, 1959).

Human impact is superimposed on large-scale alterations in the distribution and composition of African forests caused by climatic change. Climate is always changing, but fluctuations have been particularly marked in tropical Africa during geologically recent times, notably during the 2.43 million years which have passed since the first major glaciation in the northern hemisphere (Shackleton *et al.*, 1984). The first evidence for major temperature reduction in Africa during the upper Tertiary or Quaternary Periods dates to sometime between 2.51 and 2.35 million years ago (Bonnefille, 1983), that is, probably coincident with the onset of that first glaciation. This is but one example of the numerous links which have now been found between the timing of climatic changes in northern temperate latitudes and in the African tropics; some further instances are noted later in this chapter.

Climatic and Forest Changes Before 40,000 BP

African tropical forests are known to contain fewer species of animals and plants (e.g. trees) than those of Southeast Asia and South America (Haffer, 1974; Hamilton, 1982). It has been suggested that African forests were once more diverse, but suffered extinctions at times of severe climate, especially dry phases during the Quaternary. The effects of Quaternary aridity could have been moderated in Southeast Asia by the proximity of many forests to the sea, while in America there may have been a greater number of moist forest refugia than was the case in Africa.

Although extinctions may have resulted from fluctuations in forest cover and forest types related to climatic change, climatic fluctuations must also have created opportunities for rapid evolution for some forest organisms. Climatic change will have resulted in the repeated isolation and connection of populations of some species. Speculation on patterns of speciation related to forest history (Gautier-Hion *et al.*, 1988) suggests that some groups of organisms, such as guenons, have had complicated histories and it can be difficult to describe evolutionary connections of modern taxa from their present distribution and morphological similarities alone. This is a field in which considerable progress will soon be made with the application of modern techniques for comparing the relationships of different populations using isoenzyme and DNA analyses.

A summary of evidence for climatic and forest variation in Africa for the past 8.8 million years is given in Hamilton (1988) and on Figure 2.1. There are various indications, including those from analysis of minerals and other sediment constituents in deep-sea cores collected from the North Atlantic (Stein and Sarnthein, 1984), that there has indeed been a trend towards increasing aridity between 8.8 million years ago and the present in tropical Africa. Superimposed on this general tendency, there have been short-term climatic oscillations related to earth orbital variations, known as the Croll-Milankovitch variations. These oscillations have always existed, but were relatively subdued before 2.43 million years ago; after this date, and especially since 1 million years ago, Croll-Milankovitch climatic oscillations have become more pronounced, related to the beginning and later the intensification of the Quaternary ice ages.

The 21 or so major world glaciations over the past 2–3 million years have been marked by temperature changes in Africa and, more significantly from a biological point of view, by changes in rainfall. The fact that, during the Quaternary, Africa has been subject to repeated fluctuations between more humid and arid phases is well shown by deep-sea cores from the Atlantic and Indian Oceans. These contain bands of sediment rich in desert dust alternating with layers with reduced quantities of such terrestrial sediment (Pokras and Mix, 1985). Pollen in some cores has been studied (e.g. by van Campo *et al.*, 1982) and the results compared with meteorological models of climatic change (Rossignol-Strick, 1983; Rossignol-Strick *et al.*, 1982). The comparisons show that most,

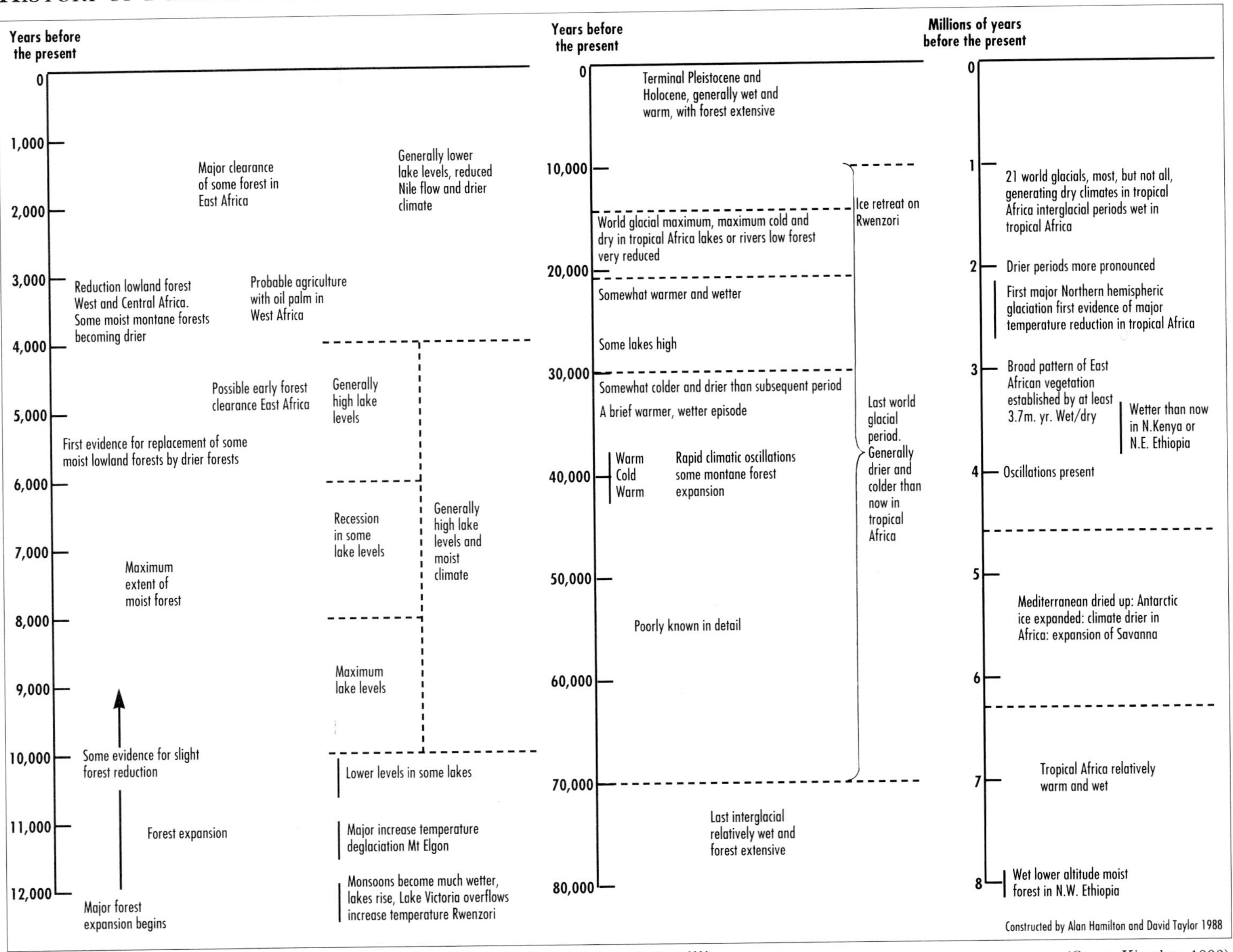

Figure 2.1 Changes in climate and forest in Africa during the last 8.6 million years (*Source:* Kingdon, 1990)

though not all, ice ages have been dry times in tropical Africa, marked by forest reduction, and that interglacial periods have been wetter, with forest expansion. The severity of the arid periods has increased during the Quaternary (Stein and Sarnthein, 1984).

The fossil record yields very little palaeobotanical evidence for continent-wide extinctions of African forest plants which can be attributed to a Miocene-to-Quaternary drying trend in the climate. Most interesting are fossils of dipterocarp trees at several pre-Quaternary localities in East Africa (Ashton, 1982). Dipterocarps are a major component of modern moist tropical lowland forests in Southeast Asia (see Collins *et al.*, 1991), but are totally absent today from continental Africa.

In contrast to a lack of evidence for extinctions (very probably due to a shortage of studies), there is considerable macrofossil and pollen evidence from East Africa showing a decline in the extent of both forest as a whole and of mesic forest types in particular over the past 20 million years (Bonnefille and Letouzey, 1976; Yemane *et al.*, 1985). A lower Miocene macrofossil assemblage from Rusinga Island, Lake Victoria, is noteworthy for its abundance of Annonaceae seeds (Chesters, 1957). This plant family is commonest today in the wettest African forests and is poorly represented in the modern, relatively dry semi-deciduous forests which grow near Lake Victoria. The previous abundance of Annonaceae at Rusinga provides some evidence of former moister conditions.

A pollen diagram from the plateau of Ethiopia shows moist lowland or submontane forest existing 8 million years ago. There is a complication here because the site has been tectonically uplifted, but, when this is discounted, there is no doubt that the climate was wetter than it is today. Some plants were present which no longer grow in Ethiopia, but survive in the main Guineo-Congolian forests (Yemane *et al.*, 1985). A remarkable absentee is *Podocarpus* pollen, which is always present in sites younger than 3 million years in Ethiopia (Bonnefille, 1987). Considering the good representation of this pollen type in the modern pollen rain in East Africa, it seems certain that the genus must have been absent from Ethiopia at the time. It is presumed to have been present in southern Africa, from which it subsequently migrated northward (Bonnefille, 1987).

Fossil fruits of the lowland forest tree *Antrocaryon*, as well as associated mollusc and mammal fossils, provide evidence of a more diverse forest flora and fauna and a wetter climate near Lake Turkana 3.4–3.3 million years ago than exist at present (Bonnefille and Letouzey, 1976; Williamson, 1985). Fossil wood from the same area, derived from riverine forest trees and dating to 4–1.5 million years (mostly 2–1.8 million years ago), is rather rich in species, some of which no longer occur in Kenya or Ethiopia, but survive in the Guineo-Congolian forests (Bonnefille, 1984; Deschamps and Maes, 1985).

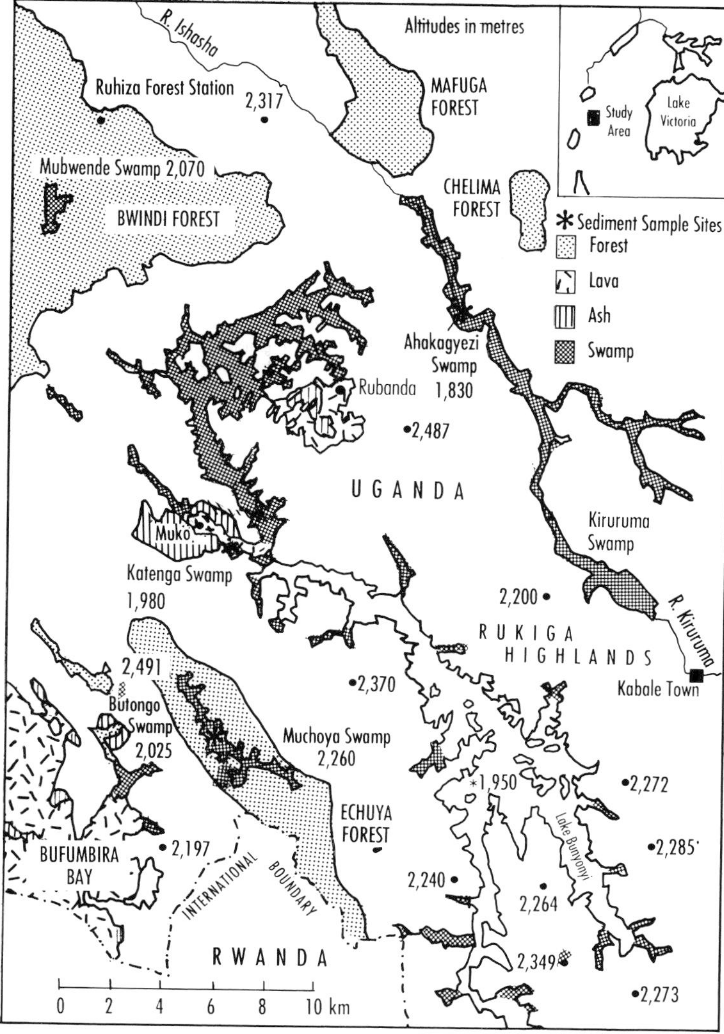

Figure 2.2 South-west Uganda (*Source:* Taylor, 1988)

An unresolved issue related to climatic and forest history concerns the time (or times) of connection of the isolated and endemic-rich forests near the East African coast (fragmented lowland forest patches occurring from Somalia to Mozambique and submontane forest on the Eastern Arc Mountains of Kenya and Tanzania) and the main Guineo-Congolian forests further west. This is of great interest from the evolutionary point of view

Forty per cent of forest tree species on the East Usambara Mountains (one of the Eastern Arc forests in Tanzania) are not found in the main Guineo-Congolian forests (Hamilton and Bensted-Smith, 1989), while for other groups of organisms such as millipedes and tree-frogs, levels of endemism are even higher. These high levels of endemism indicate long isolation. Certainly there must have been little biotic connection, so far as trees, tree-frogs and millipedes are concerned, between east and west forest zones for millions of years. However, the mammals of the East African coastal forests are less distinctive and have probably been able to move more freely between west and east, perhaps making use of intermittent connections through riverine forest.

Climatic and Forest Changes during the Past 40,000 Years

The environmental history of tropical Africa is comparatively well known for the past 40,000 years and it is possible to give a more continuous and detailed account of climatic and forest changes. This is a useful time-band, because it includes part of the last world glaciation (about 70,000 to 12,500/10,000 BP), containing the particularly severe Würm II glaciation, centred on 18,000 BP and an interglacial stage, extending from the end of the glacial up to the present.

Some of the best evidence comes from analysis of the fossils, especially pollen and other constituents of sediments beneath modern lakes, swamps and bogs. The account here is woven around the pollen records contained within the sediments under two swamps in the Rukiga Highlands, Kigezi, south-western Uganda (Figures 2.2–2.4). Many valleys in these highlands contain organic-rich sediments, offering some of the best opportunities in

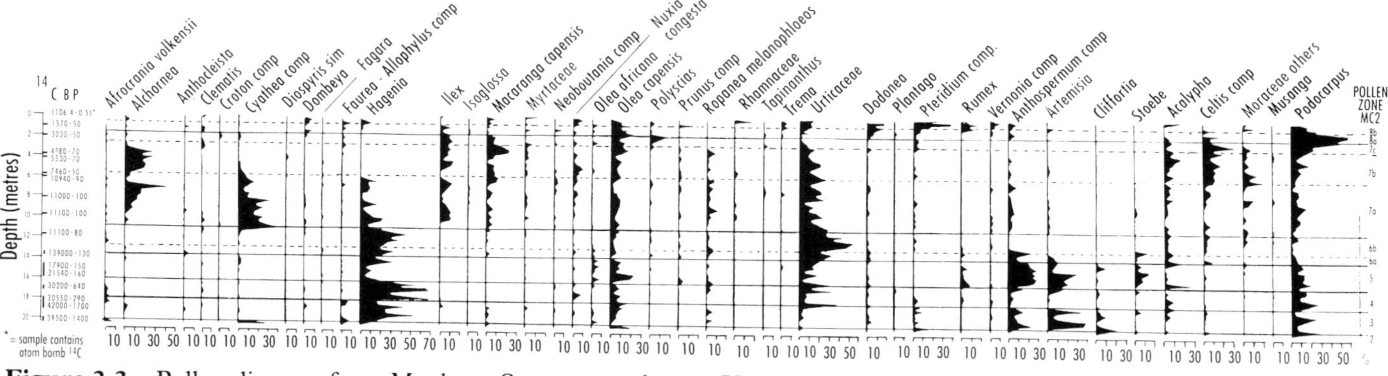

Figure 2.3 Pollen diagram from Muchoya Swamp, south-west Uganda (altitude 2260 m)

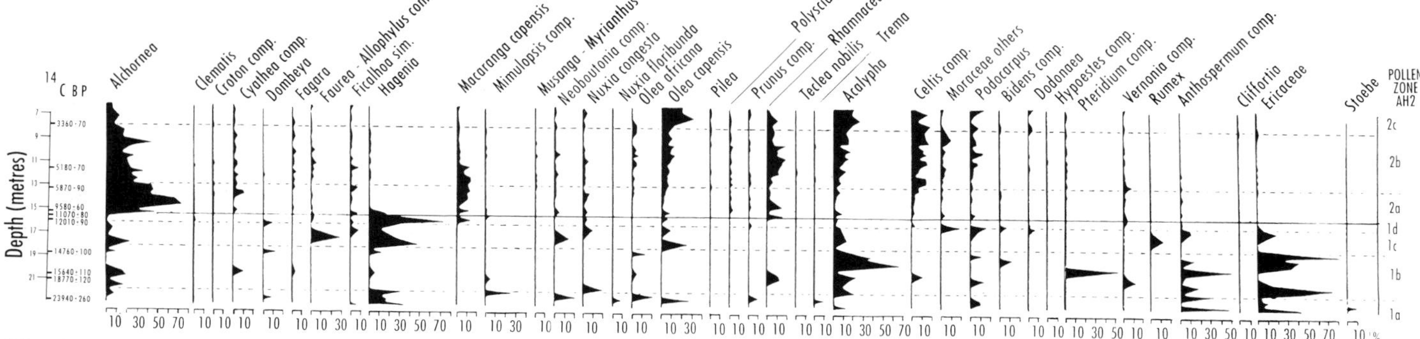

Figure 2.4 Pollen diagram from Ahakagyezi Swamp, south-west Uganda (altitude 1830 m)

Only pollen types believed to be derived from non-wetland species are included (thus Gramineae is excluded). Abundances are expressed as percentages of dryland pollen. (*Source:* Taylor, 1988)

tropical Africa for studies of environmental evolution. A more detailed account of the environmental history of the Rukiga Highlands is contained in Taylor (1988), which is fully acknowledged as a source of much of the information given here.

One of the mires, Muchoya Swamp (2260 m altitude), is today surrounded by mountain bamboo *Arundinaria alpina* of probable anthropogenic origin and the other, Ahakagyezi Swamp (1830 m), by cultivated hillsides. Before the arrival of agriculture, the vegetation around both sites was moist montane forest, a remnant of which still survives at Bwindi (see case study on the Impenetrable Forest, chapter 31) about 10 km distant from each of the two sites (Figure 2.2). Bwindi is the richest forest in East Africa in terms of numbers of species of mammals, birds and possibly trees and contains, among other rare species, a population of mountain gorilla *Gorilla gorilla berengei*.

Both pollen diagrams show a major division at about 11,000 BP, with higher altitude types of vegetation being present on the hills around the swamp before this date (back to about 40,000 BP) and lower altitude vegetation thereafter. Evidence for this includes the abundance at one or both of the sites of such pollen types as *Anthospermum comp.* (i.e. the pollen is similar to (compares with) pollen from the genus *Anthospermum*), *Artemisia, Cliffortia, Ericaceae, Hagenia* and *Stoebe* (all from characteristically higher altitude plants) before about 11,000 BP and of such pollen types as *Alchornea, Cyathea comp., Ilex* and *Macaranga capensis* later (all from characteristically lower altitude forest plants). This significant finding confirms that temperature depression was a feature of central Africa during much of the last ice age, as it was in many other parts of the world.

Summarising the evidence for temperature reduction in tropical Africa at the height of the last ice age (18,000 BP), it is estimated from the altitudes of past glaciers that temperatures in East Africa were reduced by 6.7–9.5°C compared with today; these calculations make allowance for likely changes in precipitation (Hurni, 1981; Livingstone, 1980). Pollen diagrams from altitudes between 1830 and 4000 m in East Africa show a lowering of vegetation zones by about 1000 m, equivalent to a fall in temperature of about 6°C (Hamilton, 1972; van Zinderen Bakker and Coetzee, 1972). The lowest place for which there is fossil evidence for temperature depression lies below an altitude of 1000 m in Ghana, where the presence of the olive tree, *Olea europaea capensis*, and pollen and cuticles of pooid grasses point to temperature reduction of several degrees centigrade (Talbot *et al.*, 1984).

The effects of this and presumed earlier phases of major temperature depression on lowland forests in Africa need to be further explored. Past phases of cold climate may have exerted a lasting influence on the characteristics of modern altitudinal distribution of African forests, through causing the extinction of species intolerant of relatively low temperatures. Assuming that vegetation at all altitudes was depressed by 1000 m during the more severe Quaternary cold periods, it seems that many species restricted during pre-Quaternary warmer times to the lowermost 850 m altitudinal band of forest may have become extinct (sea-levels were lowered by about 150 m during the ice ages). If such extinctions did occur, then it can be predicted that the basal 850 m of forest vegetation in tropical Africa at the present time will show a different pattern of altitudinal change to that at higher altitudes. The 850 m of forest should represent a single floristic zone, occupied after

Fire has had profound impacts upon Africa's forests for millennia: an annual grass fire in savanna habitat in Ghana. D. and I. Gordon

increase in temperature at the end of the last, and earlier, ice ages by the upward altitudinal colonisation of near sea-level species that survived. In contrast, higher altitude forest vegetation may be expected to show a more continuous pattern of floristic change with increasing altitude.

Turning now to consider the earliest period recorded in these Kigezi pollen diagrams, there is evidence from Muchoya Swamp (from another pollen diagram, not that shown on Figure 2.3) and from a nearby swamp in Rwanda, Kamiranzovu (Hamilton, 1982), that there was at least one phase at about 40,000 BP when the climate was as warm as it is today. Temperatures as high as this do not recur in central Africa until after 11,000 BP. Interestingly, a very similar phenomenon is known from England, where temperatures were at least as high as they are today for a few hundred years at about 40,000 BP (Lowe and Walker, 1984); this was the warmest episode known in England between the end of the last interglacial period about 70,000 years ago and the modern post-glacial period, after about 13,000 BP. Surely, this inter-continental correlation cannot be coincidental: very widespread climatic change is indicated. Whatever the cause of the climatic events at about 40,000 BP, they may have been of considerable biological significance. Neanderthal man was replaced by modern man at about this time in Europe; was this triggered in some way by sudden adverse climatic change? Did the sudden reversion to cold conditions at the end of the warm interval place populations of Neanderthal man under stress, facilitating the invasion of Europe by modern man?

Soon after 40,000 BP, the evidence from Ahakagyezi and Muchoya indicates a period colder than now, lasting up to 11,000 BP. However, this long period was not climatically uniform. The first phase, up to about 32,000 BP, was cold and dry, temperatures being as much as 6°C colder than now; the next phase between about 32,000 and 21,000 BP was warmer (only about 3°C colder than at present), and wetter. The phase from about 21,000 to 14,000 BP shows a return to very cold conditions (6°C colder than now) and aridity was at a maximum. The climate became somewhat warmer and wetter at about 14,000 BP and again at about 12,000 BP.

The period between about 21,000 and 14,000 BP was a critical time for forest survival in tropical Africa. Indeed, modern patterns of distribution of many plants and animals in African forests still reflect the restrictions in range imposed during this unfavourable climatic period. This was the time of the last major world glaciation (centred on 18,000 BP). Evidence for major temperature depression is very apparent in the Muchoya pollen diagram (Figure 2.3), where there is an abundance of pollen of shrub and herb taxa such as *Anthospermum, Artemisia* and *Stoebe* (now typical of high altitude ericaceous thicket), as well as grass pollen (the latter not shown on Figure 2.3). The Ahakagyezi diagram (Figure 2.4) contains relatively large quantities of *Anthospermum comp.*, Ericaceae (heathers) and grass pollen, as might be expected, and also, more surprisingly, *Acalypha* (a well dispersed pollen type probably originating from distant vegetation and over-represented due to low local pollen production).

Environmental conditions in tropical Africa at about 18,000 BP are quite well known. Figure 2.5 shows sites from which there is pollen or plant macrofossil evidence of more arid vegetation than is present today. Geomorphological studies suggest that much of tropical Africa was caught in the grip of aridity during the Würm II glaciation. The dune front in north Africa lay about 500 km south of its present position, many lakes were very low (including Lake Victoria) or even dried up completely, and river flows were everywhere reduced. The rivers Senegal, Niger and Nile were much less active than they are today.

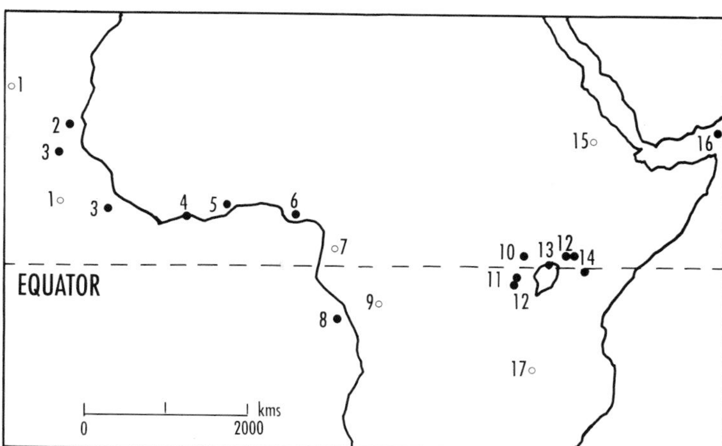

Figure 2.5 Sites (closed circles) in tropical Africa or neighbouring oceans where there is pollen or plant macrofossil evidence of more arid vegetation during the last glacial maximum (18,000 BP). The open circles indicate some other important sites revealing evidence of Quaternary environmental history.

References: (1) Stein and Sarnthein, 1984; (2) Rossignol-Strick and Duzer, 1979; (3) Agwu and Beung, 1984; (4) Assemien *et al.*, 1970; (5) Talbot *et al.*, 1984; (6) Sowunmi, 1981; (7) Kadomura, 1982; (8) Giresse and Lanfranchi, 1984; (9) de Ploey, 1968; (10) Livingstone, 1967; (11) Taylor, 1988; (12) Hamilton, 1982; (13) Kendall, 1969; (14) Coetzee, 1967; (15) Hurni, 1981; (16) Van Campo *et al.*, 1982; (17) Livingstone, 1971. (*Source:* Hamilton, 1988)

Forest was undoubtedly greatly reduced in extent at 18,000 BP, but the exact scale of this reduction is unknown for most areas, notably in the little-investigated central Zaïre forests. The best documented country palynologically is Uganda, where there is fossil evidence from Mt Elgon (Hamilton, 1987), Lake Victoria (Kendall, 1969), Rwenzori (Livingstone, 1968; Hamilton, 1982) and Kigezi (Taylor, 1988) to show that very little, if any, lowland forest remained in the country. This spectacular reduction in forest cover compared to the present, is in fact no greater than is known for many other parts of the world during the last glacial maximum, for example Europe and North America.

A more detailed reconstruction of forest extent in tropical Africa at 18,000 BP is directly relevant to studies of patterns of evolution in forest organisms and for devising strategies for conservation of genetic resources. The sites of forest refugia at 18,000 BP are likely to be not only relatively rich in number of species and endemics but also centres of genetic diversity for species which occur both within these areas and elsewhere (Hamilton, 1981).

Modern patterns of distribution of forest organisms are believed to provide clues to past forest history. Some important patterns of distribution of plants and animals in African forests are shown in Figure 2.6 which is based on a wide survey of the literature. Notable features are centres of biotic diversity (core areas) and intervening gradients of declining numbers of species. The two principal core areas are in Cameroon/Gabon and eastern Zaïre, with other, less diverse core areas in West Africa and near the East African coast. The core areas are not only rich in numbers of species and endemics but also the centres of distribution of disjunct species. It is believed that the core areas were the main centres of forest survival during the severe arid period around 18,000 BP.

There have been criticisms of this refuge theory. Undoubtedly, the pattern shown in Figure 2.6 is only a broad generalisation and some forest, albeit impoverished, did survive outside the core areas. Figure 2.7 shows a more detailed pattern for forest refugia

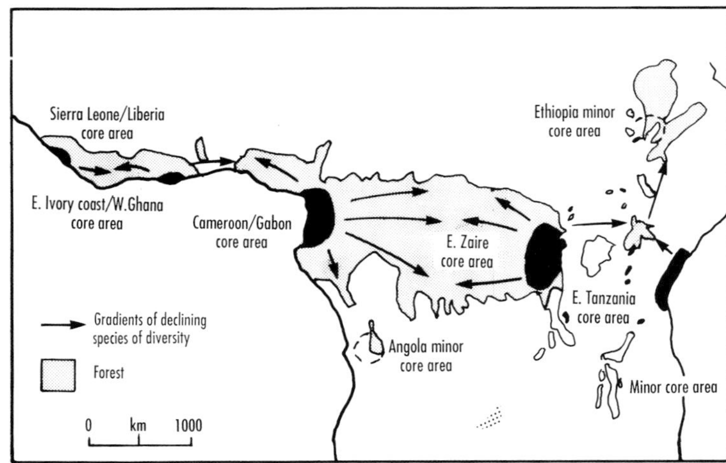

Figure 2.6 Distribution of forest, core areas and gradients of decreasing biotic diversity in tropical Africa

The core areas are believed to approximate to sites of forest refugia at the time of the last world glacial maximum, at 18,000 BP.
(*Source:* Hamilton, 1988)

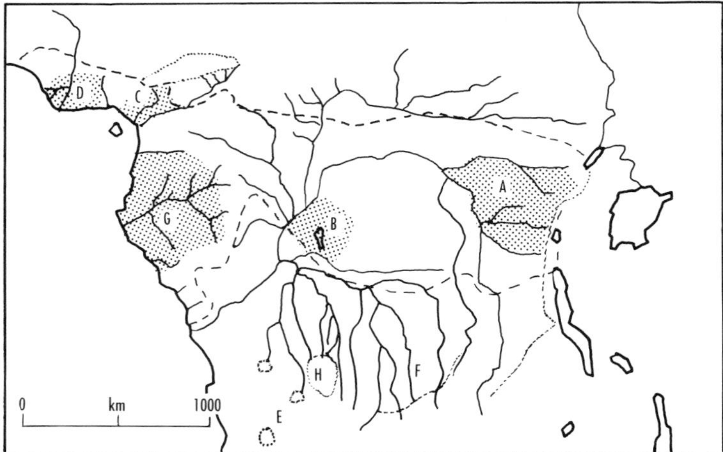

Figure 2.7 Forest refugia during arid periods in Central Africa

A Central refuge. **D, C, G** Cameroon/Gabon refuge (**D**: Niger section; **C**: Cameroon section; **G**: Gabon or Ogoowe basin section). **B** Southern Zaïre basin refuge. **E** North Angola refuge. **F** Southern scarps of Zaïre basin. **H** Lunda Plateau.
(*Source:* Kingdon, 1980)

in central Africa, largely based on the distribution of mammals. A more intricate pattern is indicated, with fragmentation of the Cameroon/Gabon refuge and minor refugia in the centre of the Zaïre basin, sustained by groundwater, and on escarpments and hills to its south.

A more serious argument against using modern patterns of distribution of forest organisms to indicate the sites of past refugia is that these patterns are a reflection of modern environmental conditions and tell us little about the past. In support of this argument may be cited the correlation between high species numbers and high precipitation which seems to be the normal pattern in African forests (e.g. in Ghana; see Hall and Swaine, 1981), and the fact that the core areas shown in Figure 2.6 are places of relatively modern high precipitation.

A number of responses can be made to this criticism. First, it can be argued that it is not surprising that places which receive high precipitation today were sites of forest survival in the past. There is evidence that patterns of atmospheric circulation over Africa at 18,000 BP were broadly similar to those of today, though with stronger and equatorially closer subtropical high pressure cells (Rognon and William, 1977); given this, it is reasonable to assume that the topographic patterns which contribute so greatly to determining modern patterns of precipitation would have ensured that places which are relatively wet today are likely to have been so in the past.

Second, it should be noted that the core areas are not only rich in total numbers of species and endemics, but that they are also the centres of distribution of the isolated populations of many species which show disjunct distributions. Some of these species are unlikely to be able to disperse from core area to core area without a continuous forest cover and some explanation is needed as to how their ranges became fragmented. Consider, for example, the gorilla, which is disjunctly distributed across the Zaïre basin (Figure 2.8); the forests between the two populations seem suitable for the species. Many other species of plants and animals show similar disjunct distributions across the Zaïre basin. Surely the most likely explanation of this for the gorilla and other obligate forest species is that their ranges have become fragmented due to forest retraction at times of aridity and that, subsequently, the species have been slow to expand their range to include all potentially suitable habitat.

Finally, if modern environmental conditions alone are responsible for restricting many species to core areas, then, by inference, environmental conditions outside the core areas are different and it might be expected that there would be a substantial number of species restricted to these areas and absent from the core areas themselves. However, this is not the case. For example, there is not a single species of forest passerine restricted to the central part of the Zaïre basin (Figure 2.9; Diamond and Hamilton, 1980).

Coming back from this digression concerning the distribution of forest cover in tropical Africa at 18,000 BP, we return to the Muchoya and Ahakagyezi pollen diagrams and the evidence for climatic amelioration at the end of the ice age maximum. The time of about 14,000 BP shows warmer and wetter conditions at both sites, notably with expansion of *Hagenia*, a tree characteristic of upper montane forest, and a decline in the representation of ericaceous belt plants. Elsewhere in central Africa there are other signs of slight climatic improvement at 15,000–14,000 BP, with retreats of glaciers from their maxima on Rwenzori and Mt Kenya (Livingstone, 1962; Mahaney, 1982). On the broader world stage, analysis of deep-sea sediments indicates that the first substantial easing of the grip of the last ice age dates to about 14,500 BP (Berger *et al.*, 1985).

The time of about 12,500–12,000 BP shows further movement towards greater warmth and wetness in Kigezi, though temperatures remained about 3°C lower than now. The pollen diagrams from Ahakagyezi and Muchoya show ever higher values of the montane tree *Hagenia* and also much Urticaceae pollen at Muchoya (often a sign of very wet climate in African pollen diagrams). On a wider geographical scale, the date of about 12,500 BP appears to be most significant for forest expansion right across tropical Africa. Marine cores off Senegal and the mouth of the Zaïre River register strong increases in lowland forest pollen after 12,500 BP (Rossignol-Strick and Duzer, 1979; Giresse and Lanfranchi, 1984), while in Uganda at this time forest expanded in the west and south (Livingstone, 1967; Kendall, 1969). Simultaneously, the levels of many tropical African lakes increased greatly, some overflowing and contributing to exceptionally strong river flows. All these changes can be attributed to a great increase in the water content of air masses moving on to Africa from the ocean; the monsoons seem to have become suddenly very active, perhaps due to the attainment of critical threshold sea-surface temperatures. Positive water balances on the continent would have been encouraged by the still depressed temperatures.

The time around 11,000 BP is very significant at Muchoya, where a series of radiocarbon studies firmly dates a major change in vegetation, with replacement of *Hagenia* forest by moist lower montane forest. A similar drastic alteration in vegetation occurred at Ahakagyezi, but is less exactly dated. A rather sudden temperature rise of 3°C is postulated as the main cause of these events. Evidence from elsewhere in Africa provides further support for this temperature increase; for example, this was the time of final deglaciation of Mt Elgon (Hamilton and Perrott, 1978). Perhaps as a result of higher evaporation under increased temperatures, there is some evidence for drier conditions in East Africa for some centuries after 11,000 BP. There was a major reduction in the level of Lake Kivu between 11,000 and 10,000 BP (Hecky and Degens, 1973) and Lake Victoria was reduced in level briefly at around 10,000 BP (Kendall, 1969). Lake Victoria sediments contain reduced quantities of forest pollen at this time, another indication of climatic dryness (Kendall, 1969).

It is intriguing that the major rise in temperature recorded at 11,000 BP in central Africa corresponds to a major decline in temperatures in north-west Europe. Earlier, after the end of the last ice age, temperatures had risen in Britain to values as high as the present, but between about 11,000 and 10,000 BP, severe weather returned, with formation and advance of a new ice-sheet in Scotland and with the limit of winter sea-ice in the North Atlantic moving as far south as the latitude of Spain (Lowe and Walker, 1984). The explanation for this apparent contradiction appears to lie in patterns of continental ice melt and melt-water flow in North America, probably responding to a major overall increase in global temperatures at about 11,000 BP. Ice retreat in North America resulted in a diversion of melt-water from the southward-flowing Mississippi eastwards into the Gulf of St Lawrence (Street-Perrott and Perrott, 1990). The resulting massive injection of cold freshwater into the North Atlantic led to major reductions in surface sea-temperature, with consequent adverse climatic conditions for north-west Europe.

Figure 2.8 The distribution of gorillas in Africa

(*Source:* Harcourt *et al.*, 1989)

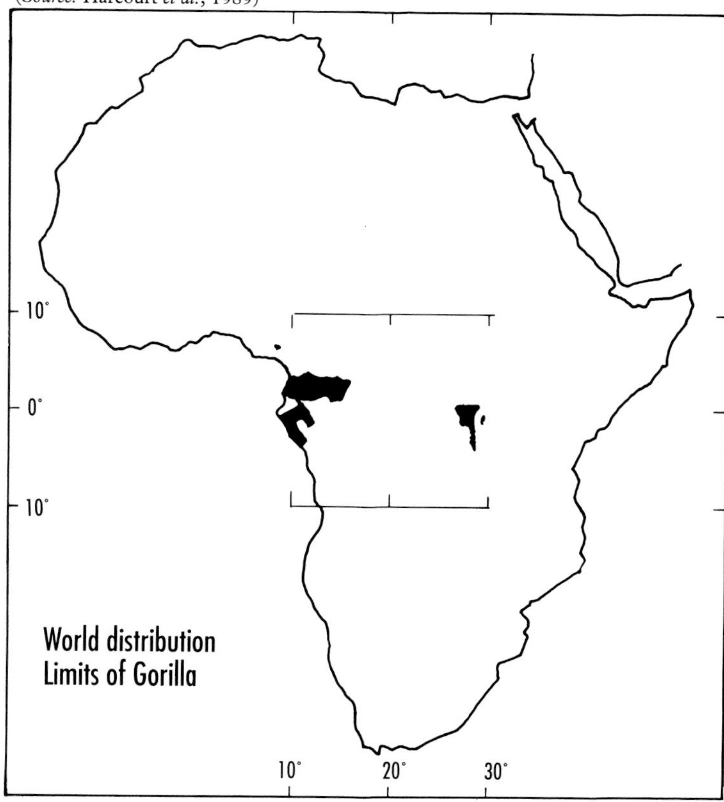

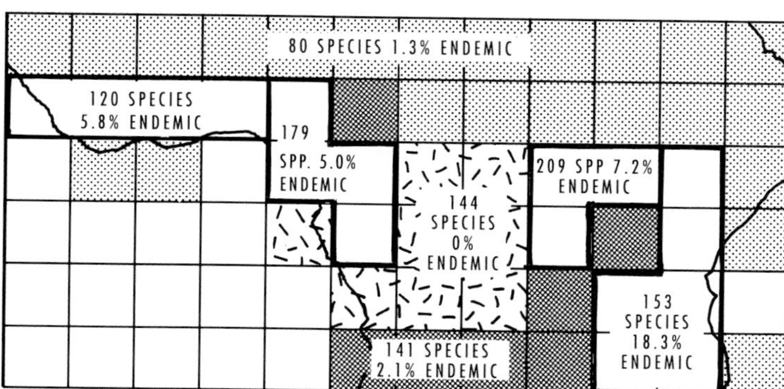

Figure 2.9 Avifaunal divisions of African forests, based on the distribution of passerine birds (*Source:* Diamond and Hamilton, 1980)

Moist lower montane forest was present on the hillslopes around Muchoya and Ahakagyezi Swamps from 11,000 BP until it was cleared by humans during comparatively recent years. The presence of this forest type is indicated by the relative abundance of such pollen types as *Alchornea*, *Cyathea comp.*, *Ilex* and *Macaranga capensis*. Elsewhere in Africa, the forest expansion which began at 12,500 BP continued, with, as already mentioned, a brief retreat in some areas at around 11,000–10,000 BP. Forest pollen reaches its maximum values in Lake Victoria sediments after 9500 BP (Kendall, 1969). Montane forest was extensive on Mt Elgon by 11,000 BP, contrasting with the largely treeless state of the mountain before 14,000 BP (Hamilton, 1987). Lowland forest was established around Lake Bosumtwi in Ghana, by 9000 BP, replacing earlier montane grassland containing small patches of montane-type forest (Maley and Livingstone, 1983).

It appears that the maximum postglacial spread of forest in tropical Africa was attained during the millennia after 10,000 BP. Subsequently, there has been some retreat; a number of sites in tropical Africa experienced a generally drier climate after about 4000 BP. Pollen diagrams from marine sediments off Senegal and Zaïre show replacement of more mesic by less mesic vegetation in neighbouring continental areas during the last few thousand years (Rossignol-Strick and Duzer, 1979; Giresse and Lanfranchi, 1984). At both Muchoya and Ahakagyezi, swamp forest replaced wetter vegetation types on the swamp surfaces at about this time and moisture-loving forest trees declined in relative abundance on surrounding hillslopes. It can be seen from Figure 2.3 that *Podocarpus* pollen increased dramatically in abundance at about 4000–3000 BP at Muchoya. The same phenomenon of increased *Podocarpus* after about 4000 BP is known from a number of other scattered sites in Africa, including Ethiopia, Mt Kenya, Rwenzori and Mt Elgon (Hamilton, 1982). An analysis of pollen diagrams from Mt Elgon shows that many other forest taxa either increased or decreased markedly in abundance on that mountain at the same time (the *Podocarpus* rises are especially noticeable because of the over-representation of this abundantly produced pollen type); furthermore these analyses support the conclusion that the *Podocarpus* increase was a response to drier climate (Hamilton, 1987).

There is abundant geomorphological evidence of a shift to a generally drier climate in tropical Africa after about 4000 BP. Lower lake levels are widely reported (e.g. Butzer *et al.*, 1972; Hecky and Degens, 1973; Fontes *et al.*, 1985; Ritchie *et al.*, 1985), Nile levels were reduced (Adamson *et al.*, 1980) and there was an expansion of the Sahara (Geyh and Jakel, 1974). As with many other climatic events in Africa, there seem to be associated climatic changes in Europe. For example, a notable feature in Ireland after 4000 BP is major expansion of blanket bog; the temporal correlation with significant climatic events further south suggests that climatic change was a major cause.

Some changes in the uppermost part of the Muchoya pollen diagram are attributable to forest clearance by man. These include rises in abundance of pollen from shrubs and herbs such as *Dodonaea, Plantago, Pteridium comp., Rumex* and *Vernonia comp.* pollen. This clearance began at about 2200 BP and was most likely due to the activities of iron-working agriculturalists, who are known from archaeological evidence to have been in the region at that time. The Ahakagyezi pollen diagram shown on Figure 2.4 does not include samples from the upper 7 m of sediment, but another pollen diagram is available from the same site covering this depth range (Hamilton *et al.*, 1986). These uppermost samples at Ahakagyezi show signs of very early forest clearance, dating back to before 5000 BP, but this is disputed by Taylor (1988), who produces good evidence that forest clearance started much later, at about the same time as at Muchoya (around 2200 BP).

To summarise, it seems likely that agriculture in the forest zone of Africa dates back to, at most, only about 4000 BP and that major forest clearance in East Africa, the part of the continent where the story is best known, occurred during about the last 2000 years only. In the light of this, it may seem surprising that the influence of humans on African forest seems to be so widespread, but the situation can be compared to north-western Europe, where agriculture is only slightly older and where all remaining forests are man-modified, often to a very great extent. An unknown factor is the degree of penetration of pre-agricultural man into the forests and the extent of his influence on its structure and composition.

References

Adamson, D. A., Gasse, F., Street, F. A. and Williams, M. A. J. (1980) Later Quaternary history of the Nile. *Nature, London* **288**: 50–5.

Agwu, C. D. C. and Beug, H. J. (1984) Palynologische Untersuchungen and marinen Sedimenten vor der Westafrikanischen Kuste. *Palaeoecology of Africa* **16**: 37–52.

Ashton, P. S. (1982) Dipterocarpaceae. *Flora Malesiana* **9**: 237–552.

Assemien, P., Filleron, J. C., Martin, L. and Tastet, J. P. (1970) Le Quaternaire de la zone littorale de Côte d'Ivoire. *Bulletin Asequa* **25**: 65–78.

Berger, W. H. (1985) On the time-scale of deglaciation: Atlantic deep-sea sediments and Gulf of Mexico. *Palaeogeography, Palaeoclimatology, Palaeoecology* **5**: 167–84.

Bonnefille, R. (1983) Evidence for a cooler and drier climate in the Ethiopian uplands towards 2.5 Myr ago. *Nature, London* **303**: 487–91.

Bonnefille, R. (1984) The evolution of the East African environment. In: *The Evolution of the East Asian Environment* Vol. II. Whyte, R. O. (ed.), pp. 579–612. Centre for Asian Studies, University of Hong Kong.

Bonnefille, R. (1987) Evolution des milieux tropicaux africains depuis le début du Cénozoique. *Memoire et Travaux de l'Institut Montpellier* **17**: 101–10.

Bonnefille, R. and Letouzey, R. (1976) Fruits fossiles d'*Antrocaryon* dans la vallée de l'Omo (Ethiopie). *Adansonia* **16**: 65–82.

Butzer, K. W., Issac, G. L., Richardson, J. L. and Washbourn-Kamau, C. K. (1972) Radiocarbon dating of East African lake levels. *Science, New York* **175**: 1069–76.

Chesters, K. I. M. (1957) The Miocene flora of Rusinga Island, Lake Victoria. *Palaeontographica* **101**: 30–71.

Collins, N. M., Sayer, J. A. and Whitmore, T. C. (1991) *Conservation Atlas of Tropical Forests. Asia and the Pacific.* Macmillan, London in association with IUCN, Gland, Switzerland.

Coetzee, J. A. (1967) Pollen analytical studies in East and Southern Africa. *Palaeoecology of Africa* **3**: 1–146.

Dechamps, R. and Maes, F. (1985) Essai de réconstitution des climats et des végétations de la basse vallée de l'Omo au Plio-Pléistocène à l'aide de bois fossiles. In: *L'Environnement des Hominidés au Plio-Pléistocène.* Beden, M. M. *et al.* (eds), pp. 175–222. Masson, Paris, France.

de Ploey, J. (1968) Quaternary phenomena in the Western Congo. In: *Means of Correlation of Quaternary Successions. Vol. 8 Proceedings of the VIIth International Congress of the Association for Quaternary Research.* Pp. 501–17. University of Utah Press, USA.

Diamond, A. W. and Hamilton, A. C. (1980) The distribution of forest passerine birds and Quaternary climatic change in tropical Africa. *Journal of Zoology, London* **191**: 379–402.

Fontes, J.-C., Gasse, F., Callot, T., Plazia, J.-C., Carbonell, P., Dupeuble, P. A. and Kaczmarska, I. (1985) Freshwater to marine-like environments from Holocene lakes in northern Sahara. *Nature, London* **317**: 608–10.

Gautier-Hion, A., Bourlière, F., Gautier, J.-P. and Kingdon, J. (eds) (1988) *A Primate Radiation: Evolutionary Biology of the African Guenons.* Cambridge University Press, Cambridge, UK. 567 pp.

Geyh, M. A. and Jakel, D. (1974) Late glacial and Holocene climatic history of the Sahara desert derived from a statistical assay of 14C dates. *Palaeogeography, Palaeoclimatology, Palaeoecology* **15**: 205–8.

Giresse, P. and Lanfranchi, R. (1984) Les climats et les océans de la région congolaise pendant l'Holocène. Bilans selon les échelles et les méthodes de l'observation. *Palaeoecology of Africa* **16**: 77–88.

Haffer, J. (1974) Avian speciation in tropical South America. *Publications of the Nuttall Ornithological Club* No. 14. Cambridge, Massachusetts, USA.

Hall, J. B. and Swaine, M. D. (1981) *Distribution and Ecology of Vascular Plants in a Tropical Rain Forest: forest vegetation in Ghana.* Junk, The Hague, The Netherlands. 383 pp.

Hamilton, A. C. (1972) The interpretation of pollen diagrams from highland Uganda. *Palaeoecology of Africa* **7**: 45–149.

Hamilton, A. C. (1981) The Quaternary history of African forests: its relevance to conservation. *African Journal of Ecology* **19**: 1–6.

Hamilton, A. C. (1982) *Environmental History of East Africa: A Study of the Quaternary.* Academic Press, London, UK. 328 pp.

Hamilton, A. C. (1987) Vegetation and climate of Mt Elgon during the late Pleistocene and Holocene. *Palaeoecology of Africa* **18**: 283–304.

Hamilton, A. C. (1988) Guenon evolution and forest history. In: *A Primate Radiation: Evolutionary Biology of the African Guenons.* Gautier-Hion, A., Bourlière, F., Gautier, J.-P. and Kingdon, J. (eds), pp. 13–34. Cambridge University Press, Cambridge, UK.

Hamilton, A. C. and Bensted-Smith, R. (eds) (1989) *Forest Conservation in the East Usambara Mountains, Tanzania.* IUCN, Gland, Switzerland and Cambridge, UK. 392 pp.

Hamilton, A. C. and Perrott, R. A. (1978) Date of deglacierisation of Mount Elgon. *Nature, London* **273**: 49.

Hamilton, A. C., Taylor, D. and Vogel, J. C. (1986) Early forest clearance and environmental degradation in south-west Uganda. *Nature, London* **320**: 164–7.

Harcourt, A. H., Stewart, K. J. and Inahoro, J. M. (1989) Gorilla quest in Nigeria. *Oryx* **23**(1): 7–13.

Hecky, R. E. and Degens, E. T. (1973) Late Pleistocene-Holocene chemical stratigraphy and palaeolimnology of the Rift Valley lakes of Central Africa. *Technical Report Woods Hole Oceanographic Institute, Massachusetts*. 93 pp.

Hurni, H. (1981) Simien Mountains – Ethiopia: palaeoclimate of the last cold period (Late Würm). *Palaeoecology of Africa* **13**: 127–37.

Kadomura, H. (1982) Summary and conclusions. In: *Geomorphology and Environmental Changes in the Forest and Savanna of Cameroon*. Kadomura, H. (ed.), pp. 90–100. Hokkaido University, Japan.

Keay, R. W. J. (1959) Derived savanna – derived from what? *Bulletin de l'Institut Français d'Afrique Noire* **21**: 427–38.

Kendall, R. L. (1969) An ecological history of the Lake Victoria basin. *Ecological Monographs* **39**: 121–76.

Kingdon, J. S. (1980) The role of visual signals and face patterns in African forest monkeys (guenons) of the genus *Cercopithecus*. *Transactions of the Zoological Society, London* **35**: 425–75.

Kingdon, J. S. (1990) *Island Africa*. Collins, London, UK. 287 pp.

Leakey, R. E. (1981) *The Making of Mankind*. Book Club Associates, London, UK. 256 pp.

Livingstone, D. A. (1962) Age of deglaciation in the Ruwenzori Range, Uganda. *Nature, London* **194**: 859–60.

Livingstone, D. A. (1967) Postglacial vegetation of the Ruwenzori Mountains in Equatorial Africa. *Ecological Monographs* **37**: 25–52.

Livingstone, D. A. (1971) A 22,000-year pollen record from the plateau of Zambia. *Limnology, Oceanography* **16**: 349–56.

Livingstone, D. A. (1980) Environmental changes in the Nile headwaters. In: *The Sahara and the Nile*. William, M. A. J. and Faure, H. (eds), pp. 339–59. Balkema, Rotterdam, The Netherlands.

Lowe, J. J. and Walker, M. J. C. (1984) *Reconstructing Quaternary Environments*. Longman, London, UK. 389 pp.

Mahaney, W. C. (1982) Chronology of glacial and periglacial deposits, Mt Kenya, East Africa: descriptions of type sections. *Palaeoecology of Africa* **14**: 25–43.

Maley, J. and Livingstone, D. A. (1983) Extension d'un élément montagnard dans le sud du Ghana (Afrique de l'Ouest) au Pléistocène Supérieur et a l'Holocène inférieur: premières données polliniques. *Comptes Rendus des Séances de l'Academie des Science, Paris* Serie **II**: 1287–92.

Pokras, E. M. and Mix, A. C. (1985) Eolian evidence for spatial variability of late Quaternary climates in tropical Africa. *Quaternary Research* **24**: 137–49.

Ritchie, J. C., Eyles, C. H. and Haynes, C. V. (1985) Sediment and pollen evidence for an early to mid-Holocene humid period in the eastern Sahara. *Nature, London* **314**: 352–5.

Rognon, P. and William, M. A. J. (1977) Late Quaternary climatic changes in Australia and North Africa: a preliminary interpretation. *Palaeogeography, Palaeoclimatology, Palaeoecology* **21**: 285–327.

Rossignol-Strick, M. (1983) African monsoons, an immediate climate response to orbital insolation. *Nature, London* **304:** 46–9.

Rossignol-Strick, M. and Duzer, D. (1979) West African vegetation and climate since 22,500 BP from deep-sea cores. *Pollen Spores* **21:** 105–34.

Rossignol-Strick, M., Nesteroff, W., Olive, P. and Vergnaud-Grazzini, C. (1982) After the deluge: Mediterranean stagnation and sapropel formation. *Nature, London* **295**: 105–10.

Shackleton, N. J. *et al.* (1984) Oxygen isotope calibration of the onset of ice-rafting and history of glaciation in the North Atlantic region. *Nature, London* **307**: 620–3.

Sowunmi, M. A. (1981) Nigerian vegetational history from the Late Quaternary to the present day. *Palaeoecology of Africa* **13**: 217–34.

Stein, R. and Sarnthein, M. (1984) Late Neogene events of atmospheric and oceanic circulation offshore Northwest Africa: high resolution record from deep-sea sediments. *Palaeoecology of Africa* **16**: 9–36.

Street-Perrott, F. A. and Perrott, R. A. (1990) Abrupt climate fluctuation in the tropics: the influence of Atlantic Ocean circulation. *Nature, London* **343**: 607–12.

Talbot, M. R. (1983) Lake Bosumtwi, Ghana. *Nyame Akuma* **23**: 11.

Talbot, M. R., Livingstone, D. A., Palmer, P. G., Maley, J., Melack, J. M., Delibrias, G. and Gulliksen, S. (1984) Preliminary results from sediment cores from Lake Bosumtwi, Ghana. *Palaeoecology of Africa* **16**: 173–92.

Taylor, D. (1988) *The Environmental History of the Rukiga Highlands, South-west Uganda, During the Last 40,000–50,000 Years*. Unpublished D. Phil University of Ulster, Northern Ireland.

Van Campo, E., Duplessy, J. C. and Rossignol-Strick, M. (1982) Climatic conditions deduced from 150-Kyr oxygen isotope-pollen record from the Arabian Sea. *Nature, London* **296**: 56–9.

Van Zinderen Bakker, E. M. and Coetzee, J. A. (1972) A re-appraisal of late-Quaternary climatic evidence from tropical Africa. *Palaeoecology of Africa* **7**: 77–99.

Williamson, P. G. (1985) Evidence for an early Plio-Pleistocene rainforest expansion in East Africa. *Nature, London* **315**: 487–9.

Yemane, K., Bonnefille, R. and Faure, H. (1985) Palaeoclimatic and tectonic implications of Neogene microflora from the Northwestern Ethiopian highlands. *Nature, London* **318**: 653–6.

Authorship

Alan C. Hamilton, WWF-UK.

3 Biological Diversity

Introduction

The tropical moist forests of Africa, like those in Asia and Latin America, are the richest ecosystems in the region. They are estimated to house more than half of Africa's biota. The fauna of the region is by far the richest of the African continent, with the major block, the Guineo-Congolean region, holding some 84 per cent of African primate species (see chapter 4), 68 per cent of passerine birds (Crowe and Crowe, 1982) and 66 per cent of butterfly species (Carcasson, 1964). The forests are estimated to contain over 8000 plant species, a floristic diversity rivalled in Africa only by the Mediterranean-climate Cape floristic region, which may itself be botanically the richest region on earth (White, 1983). This richness is largely made up of species confined to tropical moist forest, and indeed to species endemic to these regions, although a notable part of the fauna consists of species that are also widespread outside the forests. The mammals include the elephant *Loxodonta africana*, buffalo *Syncerus caffer* and leopard *Panthera pardus*.

Although the forests are rich in species compared with other regions and biotopes in Africa, they are regarded as biotically rather impoverished compared with equivalent areas in Asia and Latin America. The precise reasons for this are subject to debate. The past history of the forests has undoubtedly played a major part, in particular the effects of the interaction of climatic change and topography on the extent and distribution of the forests through time (Hamilton, 1981 and see chapter 2). In addition, present environmental conditions, particularly rainfall, have influenced the often complex patterns of distribution of different taxonomic groups within the forests.

It is generally argued that one of the most significant factors affecting the present day distribution of the fauna and flora of these forests was a major arid phase during the Quaternary, which finished some 12,000 years ago. During this period, moist forests were confined to a number of relatively small refugia isolated from each other by areas where the climate was too dry for such forest to survive (see Figures 2.6 and 2.7). Contraction of rain forest to small areas would probably have led to the extinction of a significant number of taxa, providing a possible explanation of the lower diversity of these forests compared with those in Latin America and Asia where, it is argued, such contraction was not nearly so severe.

There is strong evidence that moist forests were highly fragmented and covered a far smaller area then than they do now or did in the recent past, and evidence from East Africa suggests that forest cover in total has declined over the past 4 million years. What is less demonstrable, through lack of fossil evidence, is that this has led to appreciable extinction of taxa. It is interesting to note that, in plants at least, the endemic genera of the Afrotropical moist forests generally contain few species, in marked contrast to endemic genera in other tropical moist forest regions, which often have a large number of species. It is tempting to speculate that conditions in African forests have historically militated against extensive speciation within genera, rather than that conditions have led to large-scale extinctions.

The precise locations and extent of the so-called Pleistocene forest refugia are also the subject of discussion. They have tended to be identified on the basis of the present distributions of animal and plant species but are also cited as explanations of these distributions. There is thus a distinct danger of circularity in the argument. Nevertheless, it is worth stressing that the fossil evidence for changes in the tropical forest distribution in Africa is considerably better than in the case of South America and Asia (Hamilton, *in litt.*).

Whatever their history, the present day tropical moist forests of the Afrotropical region can be divided into three major blocks: the Central and West African area, the East African coastal region and Madagascar. Each of these has its own distinctive floral and faunal characteristics, particularly Madagascar, which is believed to have been separated from the African mainland for around 160 million years (Rabinowitz et al., 1983) and which possesses a unique flora and fauna with very high levels of endemism.

The Central and West African Rain Forests

The Central and West African forests extend discontinuously from Senegal in West Africa to extreme western Kenya and northern Angola. The forests in this vast area have a large proportion of their animal and plant species in common and the lowland forests are botanically classified as one region, the Guineo-Congolean regional centre of endemism (White, 1983 and see Figure 3.1). Overall this area is estimated to hold around 8000 plant species, some 80 per cent of which are endemic. Endemism at the generic level is also high at around 45 per cent.

These forests are, however, far from homogeneous and have many localised species occurring in them; species in different taxonomic groups often share similar patterns of distribution and on the basis of this the area can be divided into several sub-units which have more or less distinctive floral and faunal characteristics. The principal divisions are between the West and Central African forests, and also between lowland and montane forest.

The West African Rain Forests

The West African or Upper Guinean rain forests extend along the coastal region of West Africa, from Senegal in the west to Togo in the east. They are separated from the Central African rain forests by the Dahomey Gap in Benin, where savanna extends to the coast. They are of much lesser extent than the Central African forests and support fewer species, with, for example, around 750 butterfly species compared with more than 1100 in Central Africa

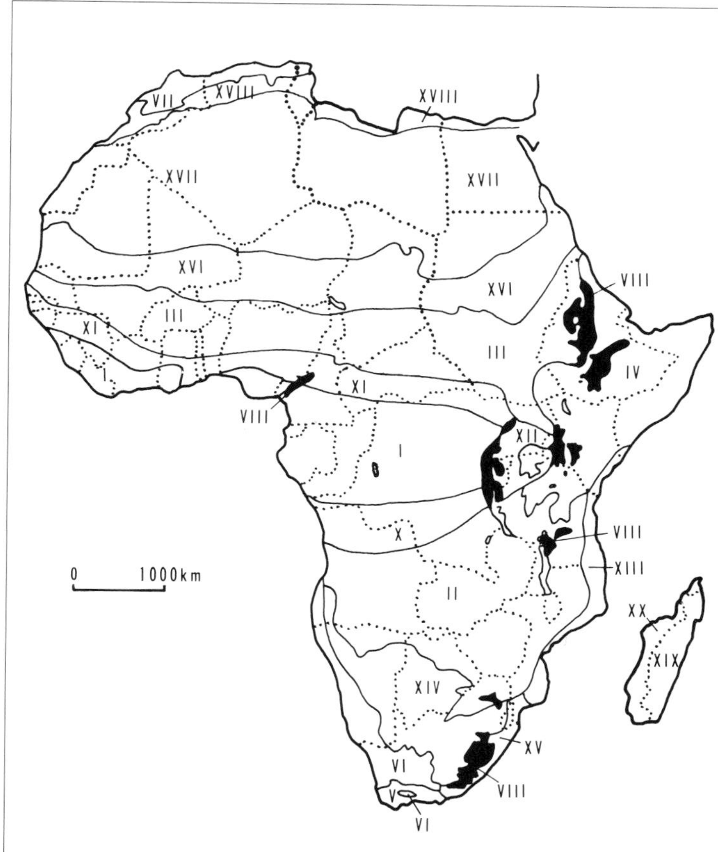

Figure 3.1 Main phytochoria of Africa and Madagascar

I. Guineo-Congolian regional centre of endemism. II. Zambezian regional centre of endemism. III. Sudanian regional centre of endemism. IV. Somalia-Masai regional centre of endemism. V. Cape regional centre of endemism. VI. Karoo-Namib regional centre of endemism. VII. Mediterranean regional centre of endemism. VIII. Afromontane archipelago-like regional centre of endemism, including IX. Afroalpine archipelago-like region of extreme floristic impoverishment (not shown separately). X. Guinea-Congolia/Zambezi regional transition zone. XI. Guinea-Congolia/Sudania regional transition zone. XII. Lake Victoria regional mosaic. XIII. Zanzibar-Inhambane regional mosaic. XIV. Kalahari-Highveld regional transition zone. XV. Tongaland-Pondoland regional mosaic. XVI. Sahel regional transition zone. XVII. Sahara regional transition zone. XVIII. Mediterranean/Sahara regional transition zone. XIX. East Malagasy regional centre of endemism. XX. West Malagasy regional centre of endemism (*Source:* White, 1983)

(Carcasson, 1964), around 200 passerine birds compared with up to 400 (Crowe and Crowe, 1982), and some 14 primate species compared with nearly 40. This may be partly because of the relative lack of land at high altitudes in West Africa. Although there are mountainous forested areas here, they are relatively small and low (maximum altitude of 1752 m on Mt Nimba) and their fauna and flora bear close resemblance to those in the surrounding lowlands, with relatively few endemics. In contrast there are major upland areas in the Central African forest block with their own distinctive faunas and floras, including many endemics. These make a significant contribution to the biological diversity of the region as a whole.

It is also argued that Pleistocene forest refugia in West Africa were considerably smaller than those in Central Africa; this would almost certainly have led to the loss of a considerable number of taxa during periods of forest contraction. Colonisation by the same or related taxa from the large Central African refugia would have been hampered by the distance between the two areas and the various physical barriers to dispersal discussed below. Nevertheless these forests are still diverse, with many notable endemic species.

Floristically there is one endemic family in West Africa, the Dioncophyllaceae, a small family of unusual lianas comprising three monotypic genera. There is a significant number of endemic species, with, for example, an estimated 700 of the 1300 species recorded in the Taï Forest in Côte d'Ivoire being endemic to the Guinean rain forests. Of these around 200 were believed confined to the region west of the Sassandra River (Guillaumet, 1967). The majority of Guinean endemic species belong to genera also found in the Central African rain forests. Notable exceptions include *Pitcairnia feliciana*, endemic to Guinea and the only member of the pineapple family Bromeliaceae occurring outside the Americas, and *Dinklageodoxa*, the only climber in the family Bignoniaceae indigenous to Africa (Brenan, 1978).

Endemic mammal species include two carnivores, Johnston's genet *Genetta johnstoni* and the rare Liberian mongoose *Liberiictis kuhnii*, in a monotypic genus, and four antelopes: the pygmy antelope *Neotragus pygmaeus* and three species of duiker, *Cephalophus zebra*, *C. niger* and the highly threatened Jentink's duiker *C. jentinki*. The latter is the largest of this widespread genus of forest antelopes. Noteworthy endemic rodents are the primitive gliding Pel's anomalure *Anomalurus peli*, the slender-tailed giant squirrel *Protoxerus aubinii* and the splendid squirrel *Epixerus ebii*. An important relict species confined to the Nimba massif on the borders of Guinea, Liberia and Côte d'Ivoire is the Mount Nimba lesser otter shrew *Micropotamogale lamottei*, a primitive insectivore whose closest relative, the congeneric *Micropotamogale ruwenzorii*, is found in the Rwenzori mountains 4500 km to the east. Several species of primate are also confined to these forests, with the precise number depending on the classification adopted — primate taxonomy is notoriously unstable and there is little general agreement on the taxonomic status of many primate populations, particularly in the genera *Cercocebus*, *Cercopithecus* and *Procolobus*. Generally ac-cepted species endemic to the Guinean forests are the diana monkey *Cercopithecus diana*, the spot-nosed guenon *C. petaurista*, and the western red colobus *Procolobus badius badius*, with the sooty mangabey *Cercocebus torquatus atys* and Geoffroy's black-and-white colobus *Colobus polykomos vellerosus* sometimes also considered distinct species.

Twenty-eight bird species are endemic to the Guinean forests. Notable among these are the rufous fishing owl *Scotopelia ussheri*, the white-breasted guineafowl *Agelastes meleagrides*, the western wattled cuckoo-shrike *Campephaga lobata*, the yellow-throated olive greenbul *Criniger olivaceus*, the spot-winged greenbul *Phyllastrephus leucolepis*, the white-necked picathartes *Picathartes gymnocephalus*, the Nimba flycatcher *Melaenornis annamarulae* and the Gola malimbe *Malimbus ballmanni*. Most of these species have closely related equivalents, with which they form so-called superspecies, in the Central African forests, emphasising the similarities between the two areas.

The amphibian fauna of West Africa is also distinctive and clearly illustrates the different faunal blocks. Schiotz (1967) has analysed the treefrog fauna (family Hyperoliidae) in detail. Sixteen species considered true tropical rain forest forms occur in the Guinean forests. Of these only four have also been recorded east of the Dahomey Gap. The fauna of the Guinean block is itself not uniform, as there is a notable division between the eastern and western parts of the forest. This division is marked by the so-called Baoulé-V centred on the Bamanda River in central Côte d'Ivoire. This is an area where the savanna lying north of the rain forest makes a marked V-shaped southward incursion into the forest. It is argued that this probably represented a complete break in the forest in the relatively recent past, allowing forest forms on either side to evolve in isolation from each other. Four species are confined to the western 'Liberian' forest block, and three are known only from the eastern 'Ghanaian' block.

The Hyperoliidae show this distinctive pattern with a notable number of localised species, because they are a rapidly evolving group. This is particularly true for the genus *Hyperolius*, which comprises half of the Guinean forest species and is regarded as the most progressive genus in the family. West African amphibians belonging to other families are generally considered to be more conservative and show a rather different pattern, with species common to the Guinean and Congolean forest blocks outnumbering those confined to the area west of the Dahomey Gap. Nevertheless, the Guinean forests have several endemic forms, notably five species in the genus *Phrynobatrachus*.

The faunal divide of the Baoulé-V is also manifest in the primates, another apparently rapidly evolving group, although here the populations on either side of the V are only differentiated at subspecific level.

The effectiveness of these faunal divides varies from group to group. Carcasson (1964) notes that the butterfly fauna of West Africa is essentially the same as, though poorer than, the fauna of the Congolean forest, with very few endemic forms. In addition, several West African forms, such as the olive colobus *Procolobus verus*, extend into Nigerian forests east of the Dahomey Gap, some extending as far as the Cross River in eastern Nigeria, which also serves as the western limit for several Congolean forest animals. It is argued that for many mammals, which are poor swimmers, rivers such as the Cross may serve as more effective barriers to dispersal than woodland or savanna areas such as the Dahomey Gap, while the reverse will be true for other groups such as amphibians.

The area between the Cross River and the Dahomey Gap thus serves as something of a transition zone between the Guinean and Congolean forest blocks, although it is predominantly Congolean in character. At least 38 plant species, including one monotypic genus, *Psammetes*, and two animal species, one bird – the Ibadan malimbe *Malimbus ibadanensis* – and one mammal – the white-throated guenon *Cercopithecus erythrogaster* – are apparently endemic to this region; both the animals are regarded as endangered (Collar and Stuart, 1985; Lee *et al.*, 1988).

The Central African Rain Forests

The Central African, or Congolean, rain forests are far more extensive than those of West Africa, and house in total a considerably greater number of species. These forests may be divided on the basis of their fauna and flora into several distinct regions. There are three distinct highland areas consisting of the Cameroon highlands in Cameroon and eastern Nigeria, the Albertine Rift highlands in eastern Zaïre, Rwanda, Burundi and Uganda, and the little-known forests of the Angolan escarpment. The lowland forests can also be divided into three major areas comprising: the western coastal region; the lowland forests of eastern Zaïre east of the Zaïre River; and the so-called Cuvette Centrale, here taken to mean the area lying within the great curve of the Zaïre River.

The Highland Congolean Forests

Montane forest regions in general have a lower overall diversity than equivalent lowland areas, but often have a higher number of endemic species. The number of endemics appears to be related to the age of the montane region, its size, climatic history and degree of isolation from other such areas.

The Cameroon Highlands This is an extensive region of volcanic uplands in western Cameroon and eastern Nigeria, formed over the past 100 million years. The offshore island of Bioko (Fernando Poo) forms part of the same montane region.

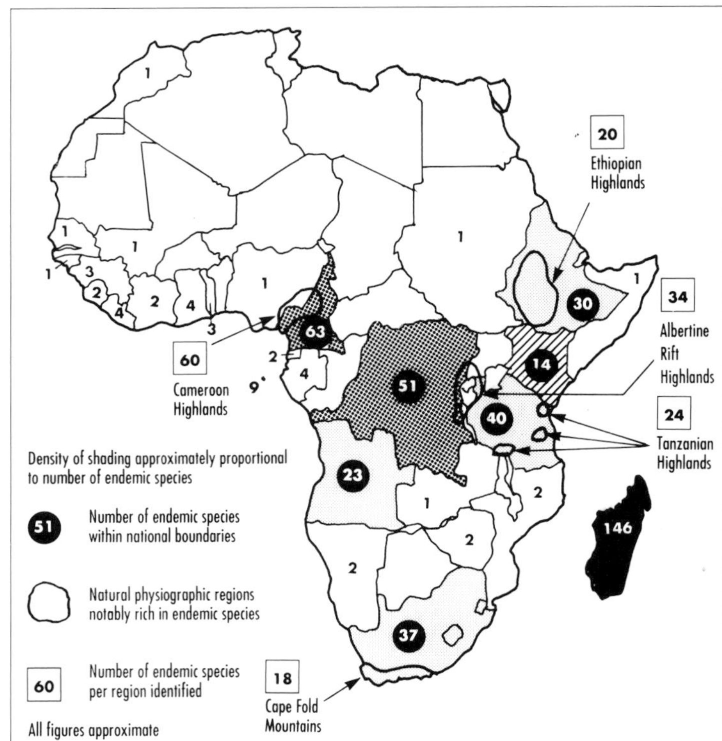

Figure 3.2 Number of endemic amphibian species in Africa

(*Source:* Data from Frost, 1985; figure by WCMC Biodiversity Project)

Floristically this area is of considerable importance. Some 130 species are believed endemic to eastern Nigeria and over 150 to north-western Cameroon (Brenan, 1978). A large percentage of these are montane forest species, with, for example, 45 species believed endemic to Mt Cameroon alone. Bioko has around 50 endemic species, including four members of the genus *Leptonychia* (family Sterculiaceae – the cocoa and cola-nut family). The area is also particularly important for endemic birds and amphibians. Of the 53 montane bird species found in the region, 20 are endemic. Of these, eight are independent species with no close relatives in their genera (Jensen and Stuart, 1986). Three of these, the Mount Kupe bush-shrike *Malaconotus kupeensis*, the white-throated mountain babbler *Lioptilus gilberti* and Bannerman's weaver *Ploceus bannermani*, are considered threatened. Of the remaining twelve endemic species, six are very closely related to (that is, form superspecies with) forms largely confined to other montane regions in Africa while the remainder are highland representatives of essentially lowland forms.

Around 60 species of amphibian are endemic to the Cameroon Highlands, this being the richest assemblage of locally endemic species in continental Africa (Figure 3.2). Although some of these are grassland species, the majority are found in montane or submontane forest. Three genera are confined to the region, all in the family Bufonidae. These are the highly distinctive monotypic *Didynamipus*, whose sole species, *D. sjostedti* is found in extreme south-west Cameroon and Bioko; *Werneria* with four species, and *Wolterstorffina* with two species, all confined to Cameroon. These genera are thought likely to be derived from ancient stock of South American origin before the two continents divided. The remaining forms are generally related to lowland forms in the adjacent Congolean forest block rather than to species in montane regions further east. In contrast, the mammal and reptile fauna of the region is relatively impoverished, with fewer species than in the adjacent lowland forests and a relatively low number of endemics.

Among the latter are Preuss's guenon *Cercopithecus preussi*, two rodents (*Praomys hartwigi* and *Paraxerus cooperi*), three chameleon species, all in the genus *Chamaeleo*, and five skinks in the genus *Panaspis*. The last named has been relatively little studied and it is thought likely that more species remain to be discovered (Gartshore, 1986).

The Albertine Rift Highlands The montane and sub-montane forests of the Albertine Rift in eastern Zaïre, Rwanda, Burundi and western Uganda are also relatively rich in endemic species. For plants, this is the richest montane area for forest species in Africa (Hamilton, *in litt.*). The area around Lakes Edward and Kivu in Zaïre has at least 62 endemic plants, with a further 26 species apparently endemic to Rwanda and Burundi.

As with the Cameroon Highlands, the area has a significant number of endemic birds and amphibians. Some 36 bird species are confined to the region, as are at least 34 amphibians, notably in the genera *Hyperolius* (eight species), *Phrynobatrachus* (seven species) and *Schoutedenella* (six species). Again the area is less important for endemic mammals, with notable exceptions being the Rwenzori otter-shrew *Micropotamogale ruwenzorii*, L'Hoest's monkey *Cercopithecus lhoesti* and the rodents *Dendromus kahuziensis*, *Funisciurus carruthersi* and *Heliosciurus ruwenzorii*. Several other species largely confined to this region also range into the eastern Zaïrean lowland forests. Examples include Thomas's bushbaby *Galagoides thomasi* and the owl-faced guenon *Cercopithecus hamlyni*. There is little information on reptiles or invertebrates, although at least three chameleon species and the vulnerable cream-banded swallowtail *Papilio leucotaenia* are confined to the region (Collins and Morris, 1985).

Fruiting bodies of the rain forest fungus Dictyophora phalloidea *found in the Korup National Park, Cameroon.* M. Rauktari

The Angolan Escarpment Forests The forests of the Angolan scarp in north-western Angola are far less extensive than those of the Cameroon highlands or Albertine Rift; they are also generally lower in altitude and show less typically montane character. They are poorly known biologically but the indications are that they support a significant number of endemic species, with at least seven birds confined to the region. These include the Gabela helmet-shrike *Prionops gabela*, Monteiro's bush-shrike *Malaconotus monteiri*, the Gabela akalat *Sheppardia gabela* and Pulitzer's longbill *Macrosphenus pulitzeri*. A number of amphibians are apparently confined to Angola, with several of these recorded from the scarp forests. However, most of these records are from early collections and it is thought likely that a proportion are synonyms of more widespread species. These forests are considered a high priority for further research (Collar and Stuart, 1988).

The Lowland Congolean Forests
The lowland forests of the Congo Forest Block are the most extensive and undoubtedly the most species-rich in Africa. Two areas stand out as being particularly diverse. These are the forests of the western lowlands, bounded in the west by the Atlantic Ocean and in the east by the Zaïre River and its affluent the Sangha, and the forests of eastern Zaïre between the Zaïre River and the Albertine Rift highlands. Compared with these, the forests of the Cuvette Centrale, lying within the sweep of the Zaïre River, are apparently somewhat less species-rich.

The Western Equatorial Forests These forests may be the richest in Africa. In Cameroon (chapter 13) a study by Gentry reported over 200 woody plant species in 0.1 ha, a diversity matched only in exceptionally rich sites in South America. The area contains many species widespread in the African lowland tropical moist forests, but also has a significant number of endemics. Among plants, Brenan (1978) cites an estimated 26 genera and over 600 species endemic to Cameroon and Bioko, with a further 28 genera and over 1000 species endemic to Gabon and Equatorial Guinea; many of these are tree species. Of particular note are 28 species of *Beilschmiedia* apparently endemic to lowland Cameroon.

Primates endemic to this region include the angwantibo *Arctocebus calabarensis*, Allen's bushbaby *Galagoides alleni*, the red-capped mangabey *Cercocebus torquatus*, three *Cercopithecus* species (the sun-tailed guenon *C. solatus*, moustached guenon *C. cephus* and red-eared guenon *C. erythrotis*), the black colobus *Colobus satanas*, the mandrill *Mandrillus sphinx* and drill *M. leucophaeus*. Other notable endemic mammals include the primitive rodent *Zenkerella insignis*, the pygmy squirrel *Myosciurus pumilio* and the bat *Kerivoula muscilla*. The area is a centre of diversity for amphibians although less important than the adjacent highlands. Several genera have speciated in both upland and lowland areas, including *Cardioglossa, Astylosternus*, *Hyperolius* and *Phrynobatrachus*. Of particular interest is the genus *Conraua*, with the lowland *Conraua goliath*, the world's largest anuran (frogs and toads), replaced in the uplands by *C. robusta*.

The butterfly fauna in this region is the most diverse in Africa, with over 1100 species recorded. The great majority of these are widespread forms, also being found in either the Upper Guinean forests or the Zaïrean forests further east, or both (Carcasson, 1964). The spectacular African giant swallowtail, *Papilio antimachus*, the largest butterfly on the continent, is an example of a widespread but rarely seen species (Collins and Morris, 1985). *Graphium aurivilliusi* is a swordtail butterfly known only from the type series – a common occurrence among the relatively little-studied invertebrates.

The Eastern Zaïrean Lowland Rain Forests These forests rival those of the western lowlands in diversity. However, detailed floristic analysis is difficult, because studies do not generally differentiate between these forests and those of the Cuvette Centrale. Nevertheless, Zaïre as a whole is believed to have the largest flora of any continental African country, with a high level of endemism. The two most important areas for endemics are the area identified in Brenan (1978) as the 'forestier central' which includes these lowland rain forests as well as part of the Cuvette Centrale, and the Haut Katanga, which is in the Zambezian domain and is not rain forest.

The area is better known for its fauna. As well as housing most of the widespread Guineo-Congolean species, the forests also possess a large number of more localised forms, sharing many species with the adjacent Albertine Rift highlands (*Cercopithecus hamlyni*, *Galagoides thomasi*, and the giant genet *Genetta victoriae*), others with the Cuvette Centrale (most notably the Congo peacock *Afropavo congensis*) and yet others with the western lowland forests (the grey-cheeked mangabey *Cercocebus albigena*, and the royal antelope *Neotragus batesi*). These forests also have a number of endemic species, of which the most notable is the okapi *Okapia johnstoni* (see case study in chapter 32). Other endemics include the aquatic genet *Osbornictis piscivora* and two weaverbirds, *Ploceus aureonucha* and *P. flavipes*, both apparently confined to the Ituri forest. This forest has populations of 13 diurnal primate species, the richest known assemblage in mainland Africa.

The Cuvette Centrale This region, constituting the low-lying area of the Zaïre Basin within the curve of the river, is generally regarded as impoverished compared with forests on the rim of the basin. Several species are found in both western and eastern forests, but seem to be absent from this central area. These include the gorilla *Gorilla gorilla*, the chimpanzee *Pan troglodytes* (also found in the Guinean forests) and the western needle-clawed galago *Galago elegantulus*, as well as the royal antelope and grey-cheeked mangabey. Two of these species – the chimpanzee and grey-cheeked mangabey – are, however, replaced in the Cuvette by closely related species, the pygmy chimpanzee or bonobo *Pan paniscus* and black mangabey *Cercocebus aterrimus*, respectively. Several authorities consider these to be only separable from their sibling taxa at subspecific level.

It is argued that the principal reason for this impoverishment is that the present-day forests of the Cuvette Centrale are mostly very recent in origin, the area having experienced a dry phase until perhaps only 12,000 years ago. Thus their fauna will consist largely of species which have since immigrated from surrounding forest regions. The Zaïre River, which separates the Cuvette Centrale both from the western forests and the eastern Zaïrean forests, will have acted as an effective barrier to dispersal for many species. Minor forest refugia may have existed in the Cuvette Centrale during the dry phase, probably sustained by groundwater. In general the number of species endemic to the Cuvette Centrale appears low; for example, there are fewer than ten amphibians. This may corroborate the view that the forests are recent in origin, although it may also be at least in part a reflection of the fact that they are still relatively little known zoologically.

THE EAST AFRICAN COASTAL RAIN FORESTS

The moist forests of eastern Africa are of far lesser extent than those of the Guineo-Congolean region. They occur principally in two areas: lowland coastal regions, mainly in the southern half of Kenya and in Tanzania, and upland areas, chiefly the Usambara and Uluguru mountains of Tanzania. Their fauna and flora differ greatly from those of the Guineo-Congolean block. They are certainly of lower overall diversity than the latter, but have a significant number of endemic species, particularly of birds, amphibians and invertebrates. The flora of this region shows affinities with the moist forests further west at generic rather than specific level and has a very large number of endemic species. Moreover, many of the widespread genera of the Congolean forest block are absent. This indicates that the two forest blocks have been isolated for a long period of time – probably at least 500,000 years. Brenan (1978) has suggested that the eastern forests may represent the fragmented relics of a primitive and formerly more widespread forest flora not clearly recognisable elsewhere in Africa today.

The East Coast Lowland Forests

The lowland coastal vegetation of Kenya and Tanzania (including Zanzibar) has around 100 endemic plant species and five endemic genera: *Angylocalyx*, *Asteranthe*, *Lettowianthus*, *Mkilua* and *Ophrypetalum* (Brenan, 1978). Two of the most important forests in this region are Sokoke Forest and the Shimba Hills in Kenya. Two bird species, the Sokoke scops owl *Otus ireneae* and Clarke's weaver *Ploceus golandi*, are wholly endemic to Sokoke, with two mammals, the golden-rumped elephant-shrew *Rhynchocyon chrysopygus* (sometimes considered a distinctive subspecies of *R. cirnei*) and Aders's duiker *Cephalophus adersi*, being nearly so. Several other localised East African forest bird species have important populations in Sokoke. The only two amphibians believed endemic to Kenya, *Afrixalus sylvaticus* and *Hyperolius rubrovermiculatus*, are confined to the Shimba Hills, which also house several restricted range species such as the black-and-rufous elephant-shrew *Rhynchocyon petersi*, east coast akalat *Sheppardia gunningi* and plain-backed and Uluguru violet-backed sunbirds *Anthreptes reichenowi* and *A. neglectus*.

The East African Upland Forests

The upland forests of East Africa are in the main situated on scattered ancient crystalline massifs. Of these, the Usambara mountains are of quite exceptional importance for endemic plants. Polhill (1968) lists 112 tree and shrub species endemic to the mountains (although around 20 of these were considered 'imperfectly known species') with another 30 or so nearly endemic, although Hamilton (*in litt.*) notes that there are relatively few strict endemics as most species range beyond the Usambaras proper. Polhill's totals for endemics and near endemics include 50 tree species over 10 m tall, of which three are in monotypic genera: *Cephalosphaera usambarensis*, *Englerodendron usambarense* and *Platypterocarpus tanganyikensis*. In addition, nine African violets *Saintpaulia* spp. are endemic to the mountains.

Two bird species are endemic to the Usambaras, the Usambara eagle owl *Bubo vosseleri* and the Usambara ground robin *Dryocichloides montanus*, with several other near endemic species found. Two mammal species are believed endemic: Swynnerton's squirrel *Paraxerus vexillarius* and a white-toothed shrew *Crocidura tanzaniana*. Around 14 endemic lizards occur along with six amphibians, with an additional 13 of the latter found more widely in the East African highlands. Levels of endemism in those invertebrate groups which have been studied are similarly high, with, for example, 27 recorded endemic wasps in the family Sphecidae and 26 carnivorous snails in the family Streptaxidae, including 18 species in the genus *Gulella* (Wells *et al.*, 1983).

Second in importance are the Uluguru Mountains south of the Usambaras. These mountains have around 80 endemic trees and shrubs (although 24 of these are listed by Polhill (1968) as imperfectly known) and nine near endemics. Three species in monotypic genera are apparently endemic: *Dionychastrum schliebenii*, *Pseudonesohedyotis bremekampii* and *Rhipidantha chorantha*. The area also has three endemic African violets. There are two endemic bird species, the Uluguru bush shrike *Malaconotus alius* and Loveridge's sunbird *Nectarinia loveridgei* along with two mammals, the Uluguru golden mole *Chlorotalpa tropicalis* and the shrew *Crocidura telfordi*. Five reptiles are endemic to the range, as are seven amphibians, with an additional 14 or so amphibians restricted to the Eastern Highlands but not confined to the Ulugurus. Levels of endemicity among invertebrates are similarly high with, for example, 40 known endemic carabid beetles and 41 out of the 43 members of the family Pselaphidae in the mountains being apparently endemic.

Smaller patches of highland forest occur in isolated outcrops from Mt Mulanje in Malawi to Mt Kulal in northern Kenya. Endemics, often in danger of extinction, occur in every case. The Taita Hills in southern Kenya, for example, are home to at least three plants, one butterfly, an amphibian, a reptile and three bird species known nowhere else (Collins and Clifton, 1984). They are now confined to less than 5 sq. km of relict forest.

MADAGASCAR

Madagascar has been separated from the African mainland for about 160 million years and has evolved its own highly distinctive flora and fauna. The island as a whole has one of the richest floras in the world (Jenkins, 1987). Phytogeographically, the island is divided into two regional centres of endemism, the western and the eastern. Tropical moist forest is virtually confined to the eastern centre, which has a westward extension in the northern part of the island known as the Sambirano. The flora of the whole island is thought to number 10,000–12,000 species of which between 55 and 80 per cent have variously been estimated to be endemic. Of the total, an estimated 6000 species in 500–1000 genera are found in the moist forests of the eastern domain. Around 90 per cent of the species and a third of the genera are believed endemic to this region. The flora has much stronger affinities with the Sudano-Zambezian element of the African mainland flora than with Guineo-Congolian forest flora.

The island's fauna is very distinctive, with, as might be expected, the eastern and western domains showing strong affinities with each other. Diversity in mammals and birds is relatively low compared with continental Africa, while in reptiles and amphibians it is at least comparable. Endemism in most groups is very high for Madagascar as a whole. Notable are the primates with 30 species in five families, all of which are in the suborder Lemuroidea, otherwise found only on the Comoros where two species of the genus *Lemur* are present, probably having been introduced there from Madagascar by man. Fifteen lemur species are confined to the moist forests of the eastern domain, with a further four shared with the drier deciduous forests of the west. Notable species endemic to the eastern region include four monotypic genera: the indri *Indri indri*, hairy-eared dwarf lemur *Allocebus trichotis*, ruffed lemur *Varecia variegata* and aye-aye *Daubentonia madagascariensis*, this last in a monotypic family, the Daubentoniidae (Jenkins, 1987; Harcourt and Thornback, 1990). Four of the eight native carnivores are apparently endemic to the eastern forests, each in a monotypic genus: *Fossa fossana*, *Eupleres goudotii*, *Galidictis striata* and *Salanoia concolor*. Other noteworthy species include around 15 species of tenrec, most in the genera *Microgale*, and the sucker-footed bat *Myzopoda aurita* in a monotypic family.

Of the 105 birds endemic to Madagascar, 83 are found in the eastern region and 30 are apparently confined to this area. These include the extremely rare Madagascar serpent eagle *Eutriorchis astur*, and four of the five members of the endemic Madagascan family, the Brachypteraciidae or ground-rollers: *Atelornis pittoides*, *Brachypteracias squamiger*, *Brachypteracias leptosomus* and *Atelornis crossleyi*. Madagascar has a high diversity of reptiles and amphibians, with at least 260 species of the former and 150 of the latter. All but two of the amphibians are endemic to Madagascar and approximately a third are in an endemic subfamily, the Mantellinae. Around 60 per cent of the amphibians are confined to low or medium altitude moist forest, with a further 30 per cent occurring in montane forest regions. The genera *Mantidactylus* and *Boophis* are particularly diverse. The reptiles are more evenly distributed on the island, with a large number of species in the arid southern and seasonal western regions. Nevertheless the eastern forests still hold a large number of species, particularly geckos (notably *Phelsuma*, *Lygodactylus*, *Uroplatus* and *Phyllodactylus*) and chameleons (*Chamaeleo* and *Brookesia*).

Among invertebrates, the island has a particularly important terrestrial molluscan fauna with around 360 endemic species out of a total of about 380. There are 11 endemic genera but no endemic families. The molluscan fauna is particularly rich in areas with calcareous soils, most of which are in the drier western domain, although some areas of moist forest are very important, notably the Tsaratanana massif which has a large number of species in the endemic Madagascan genera *Ampelita* and *Acroptychia*.

Mangroves

Africa and Madagascar possess extensive tracts of mangrove, particularly on their western coasts. In common with mangroves elsewhere, these forests are of far lower diversity than terrestrial tropical moist forests. There are five mangrove species in West Africa, all different from those in the east and on Madagascar. They are *Rhizophora mangle*, *R. harrisonii*, *R. racemosa*, *Avicennia germinans* (= *A. africana*, *A. nitida*) and *Laguncularia racemosa*. All five are widely distributed on the eastern coast of South America and on neighbouring islands. Madagascar and East Africa share a rather more diverse mangrove flora with nine species, all widespread in the Indian Ocean and most extending into the Pacific. These are: *Rhizophora mucronata*, *Avicennia marina*, *Sonneratia alba*, *Ceriops tagal*, *Bruguiera gymnorrhiza*, *Xylocarpus granatum*, *X. moluccensis*, *Lumnitzera racemosa* and *Heritiera littoralis* (White, 1983). The *Ceriops* on Madagascar is sometimes considered an endemic species, *C. boiviniana*.

Biodiversity and Conservation

In view of their richness, the conservation of the moist forests of the Afrotropical region is of global significance for the maintenance of biological diversity. As has been discussed in this chapter, many of the species occurring in these forests occupy localised ranges, often in areas where forest is inadequately protected and disappearing at a rapid rate. Areas with a particularly high number of threatened forest species include the Cameroon Highlands, Madagascar, the Upper Guinea forests and the East African coastal forests. The survival of such species is self-evidently dependent on the survival of the forests they inhabit and these areas must thus be considered among the highest priorities for immediate conservation action in the Afrotropical region. However, it must also be remembered that in the medium to long term, adequate protection of the lowland forests in the Congo forest block is perhaps of greater importance in the maintenance of biological diversity as these forests are the richest in species in Africa. At present, because the species occurring in them are still comparatively widespread, as are the forests themselves, they hold a relatively low proportion of species identified as threatened and are therefore generally not considered as of the highest priority for species conservation.

Eliurus myoxinus, *an endemic but widespread rodent in Madagascar. This is one of only ten indigenous rodent species found on the island. All are endemic.* C. Harcourt

References

Brenan, J. P. M. (1978) Some aspects of the phytogeography of tropical Africa. *Annals of the Missouri Botanical Garden* **65**: 437–78.

Carcasson, R. H. (1964) A preliminary survey of the zoogeography of African butterflies. *East African Wildlife Journal* **2**: 122–57.

Collins, N. M. and Clifton, M. P. (1984) Threatened wildlife in the Taita Hills. *Swara* 7(5): 10–14.

Collins, N. M. and Morris, M. G. (1985) *Threatened Swallowtail Butterflies of the World. The IUCN Red Data Book.* IUCN, Cambridge, UK and Gland, Switzerland. vii + 401 pp. + 8 pls.

Collar, N. J. and Stuart, S. N. (1985) *Threatened Birds of Africa and Related Islands. The ICBP/IUCN Red Data Book.* ICBP, Cambridge, UK and IUCN, Cambridge and Gland, Switzerland.

Collar, N. J. and Stuart, S. N. (1988) *Key Forests for Threatened Birds in Africa.* ICBP Monograph No. 3. ICBP, Cambridge, UK. 102 pp.

Crowe, T. M. and Crowe, A. A. (1982) Patterns of distribution, diversity and endemism in Afrotropical birds. *Journal of Zoology, London* **198**: 417–42.

Frost, D. R. (ed.) (1985) *Amphibian Species of the World: a taxonomic and geographical reference.* Allen Press Inc. and The Association of Systematics Collections, Lawrence, Kansas, USA.

Gartshore, M. E. (1986) Status of the montane herpetofauna of the Cameroon highlands. In: Stuart, S. N. (ed.) *Conservation of Cameroon Montane Forests.* ICBP, Cambridge, UK. 204–240.

Guillaumet, J. L. (1967) *Recherches sur la Végétation et la Flore de la Région du Bas-Cavally (Côte d'Ivoire).* Mémoires ORSTOM No. 20, Paris, France.

Hamilton, A. C. (1981) The Quaternary history of African forests: its relevance to conservation. *African Journal of Ecology* **19**: 1–6.

Harcourt, C. and Thornback, J. (1990) *Lemurs of Madagascar and the Comoros. The IUCN Red Data Book.* IUCN, Gland, Switzerland and Cambridge, UK.

Jenkins, M. D. (1987) *Madagascar: An Environmental Profile.* IUCN/UNEP/WWF, Cambridge, UK.

Jensen, F. P. and Stuart, S. N. (1986) The origin and evolution of the Cameroon montane forest avifauna. In: Stuart, S. N. (ed.) *Conservation of Cameroon Montane Forests.* ICBP, Cambridge, UK. 28–37.

Lee, P. C., Thornback, L. J. and Bennett, E. L. (1988) *Threatened Primates of Africa. The IUCN Red Data Book.* IUCN, Gland, Switzerland and Cambridge, UK.

Polhill, R. M. (1968) Tanzania. In: Hedberg, I. and Hedberg, O. (eds) *Conservation of Vegetation in Africa South of the Sahara.* Acta Phytogeographica Suecica **54**.

Rabinowitz, P. D. , Cotlin, M. F. and Falvey, D. (1983) The separation of Madagascar and Africa. Science 220: 67–69.

Schiotz, A. (1967) The treefrogs (Rhacophoridae) of West Africa. *Spolia Zoologica Musei Hauniensis* **25**: 1–346.

Wells, S. M., Pyle, R. M. and Collins, N. M. (1983) *The IUCN Invertebrate Red Data Book.* IUCN, Cambridge, UK and Gland, Switzerland.

White, F. (1983) *The Vegetation of Africa: a descriptive memoir to accompany the Unesco/AETFAT/UNSO vegetation map of Africa.* Unesco, Paris, France. 356 pp. + 4 maps.

Authorship

Martin Jenkins, Cambridge with contributions from Alan Hamilton WWF-UK.

The okapi Okapia johnstoni *is one of the most notable endemic species of the eastern Zaïrean lowland rain forests.* J. and T. Hart

4 Case Studies in Conserving Large Mammals

Introduction

From an aircraft flying overhead, much of the African rain forest canopy may appear intact, verdant and, therefore, rich in wildlife. However, the reality on the ground is often quite different. Large mammals are hunted extensively for their meat, hides and tusks, sometimes to the point of being wiped out over large areas. In addition, the rate at which the tropical forest is disappearing is having a devastating effect on wildlife. The combined effects of hunting and the destruction of forest on elephants and primates are considered in detail in this chapter.

In much of Africa, ivory poaching is a major problem involving networks of criminals at all levels of society both in and outside the continent. Although it is easier to kill elephants in the savanna areas, heavily armed poachers are eradicating elephants from even the remotest forests in Central Africa. In West Africa too, poaching is a problem, but the elephants here are probably more affected by the disappearance of their forest habitat.

Primates are also under severe threat, with 55 per cent of continental Africa's forest species listed as vulnerable to, or in danger of, extinction. Hunting for meat is a major danger: the larger monkeys and apes are a favourite target for well-armed hunters. Total destruction of the forest obviously has a devastating effect on the primates, but even the disturbance caused by logging can cause their numbers to decline. It is not just single species at risk: whole communities of primates are under pressure.

Elephants and primates can tolerate some degree of disturbance to the forest by shifting agriculturalists. For example, elephants prefer to feed in the secondary vegetation that springs up on abandoned fields and villages, while moderate forest disturbance may actually increase local diversity and abundance of primates. These conspicuous species, which also play vital ecological roles, can provide the flagships for broader conservation programmes. The preservation of elephants and primates requires secure protected areas but, possibly more importantly, the local people themselves must be interested in and involved with the conservation of the forests and their wildlife.

Forest Elephants

Little is known about forest-dwelling elephants. Our ignorance is epitomised by the confusion which surrounds the subspecies (or races) of elephant which live in the forest. Textbooks propagate the myth that the savanna elephant *Loxodonta africana africana* lives in the African savannas, and the forest elephant *Loxodonta africana cyclotis* lives in the forests. Within the past few years it has become clear that both subspecies are found within the forest zone (Western, 1986; Carroll, 1988). The relative distribution of the two subspecies is not yet understood: do they occupy the same areas, or do they inhabit separate patches of the forest? Recent evidence suggests that *L. a. africana* may be more common in the equatorial forests than *L. a. cyclotis*, which may be a creature of the forest edge rather than the true forest. *L. a. cyclotis* is smaller than *L. a. africana* and it has slender tusks which are generally straight or only slightly curved. In addition, its ears are smaller and more rounded in comparison to those of *L. a. africana*, which has almost triangular shaped ears.

Many people believe that a third type of forest elephant exists: the pygmy elephant. It is often called *L. africana pumilio* or *L. pumilio*, but its existence has yet to be proved to the satisfaction of taxonomists (Haltenorth and Diller, 1977). While it is certain that small elephants are found in forests, they may simply be juvenile elephants living apart from their natal groups (Western, 1986).

The Role of Elephants in the Forest Ecosystem

The forest gives the impression of a paradise of super-abundant food. But this is not the case for plant-eating animals. Most plants protect themselves from herbivores – whether vertebrate or invertebrate – with poisonous chemicals, indigestible compounds, or physical defences such as thorns. These either poison the animal which eats them or interfere with its digestive process. Elephants have a digestive system which makes them particularly susceptible to toxins and tannins (Olivier, 1978). They must search for plants and plant parts which contain only small amounts of such chemicals, or for those which are not protected at all. For example, the fast-growing plant species which spring up in abandoned villages and fields usually lack toxins and tannins. Therefore elephants prefer to feed in secondary forests, those which have been disturbed by former human occupation.

Elephants play a key role in the forest ecosystem (Carroll, 1988; Western, 1989). They make paths and mud wallows; on some hydromorphic soils they create large, open grassy areas. Their browsing retards the closure of canopy gaps caused by fallen trees. These gaps are usually occupied by light-loving plants which cannot grow in the gloom of the forest, so elephant browsing helps to increase the diversity of plants growing at any one time in the forest. Their browsing also retards the development of secondary forest on abandoned villages and fields.

Perhaps their most important role is as seed dispersers (Alexandre, 1978; Lieberman *et al.*, 1987). Elephants consume fruits in large quantities. By the time a fruit has been digested and the seeds have passed out in the faeces, the animal may be many kilometres from the spot where the fruit was eaten. Furthermore, passage through the elephant's digestive system seems to improve the chances of successful germination of some plants (Lieberman *et al.*, 1987). Thus the elephant benefits by eating the fruit and the tree profits by having its seeds carried away to a new place where they may germinate. Some trees, such as *Tieghemella heckelii*, *Panda oleosa*, *Klainedoxa gabonensis* and *Balanites wilsoniana*, have very large fruits or seeds which can be swallowed only by elephants (Lieberman *et al.*, 1987; Struhsaker, 1987). It is said that 30 per cent of the tree species in the Taï Forest in Côte d'Ivoire are dispersed by elephants (Alexandre, 1978).

Twenty thousand years ago the global climate was quite different from today's. Ice caps had extended over Europe and North America, while Africa was cooler and drier than at present. The African forests had almost disappeared, being found only in isolated refuge patches (Hamilton, 1982 and chapter 2). The Zaïre Basin was covered by dry forests. Later, as the climate became warmer and wetter, the forest spread out from the refuges to recolonise the lands it had formerly covered. It is certain that elephants played an important role in modifying the structure and composition of the expanding forests. For example, by dispersing tree seeds they would have accelerated the spread of forest. Thus elephants played an important part in the evolution of today's forests and their disappearance could have profound implications for the future of those forests. Those trees which need elephants to disperse their seeds would disappear. Indeed, the poor regeneration of some forest trees in Côte d'Ivoire, Ghana and Uganda may be due to the decline of elephants (Lieberman *et al.*, 1987; Struhsaker, 1987).

The Numbers and Distribution of Forest Elephants

The forests account for about one-third of the total area occupied by African elephants. In West Africa the forest elephant populations are small (Table 4.1), fragmented and vulnerable to habitat loss and poaching. West Africa accounts for only about 3 per cent of the continent's elephants.

Until quite recently, elephants were found throughout the unbroken expanse of forest which covers Central Africa, from the Atlantic coast of Gabon and Cameroon to the mountains of eastern Zaïre. Today they are still found in many parts of the forest zone. In 1989 it was estimated that 214,000 elephants lived in the Central African forests. This accounts for 35 per cent of the continental total. Most are found in Zaïre, Gabon and Congo (Table 4.1).

Table 4.1 Summary of 1989 estimates of numbers of elephants in the West and Central African countries considered in this Atlas

West Africa		*Central Africa*	*Total*	*In forests*
Benin	2,100	Cameroon	22,000	15,500
Ghana	2,800	CAR	23,000	2,200
Guinea	560	Congo	42,000	40,500
Guinea-Bissau	40	Equatorial Guinea	500	500
Côte d'Ivoire	3,600	Gabon	74,000	68,700
Liberia	1,300	Zaïre	112,000	86,900
Nigeria	1,300	TOTAL	273,500	214,300
Senegal	140			
Sierra Leone	380			
Togo	380			
TOTAL	12,600			

(*Source*: West African figures were calculated using a computer model (Burrill and Douglas-Hamilton, 1987; Douglas-Hamilton, 1989); savannas of Central Africa: same computer model; Central African forests: based on a modification of the model (Michelmore *et al.*, 1989))

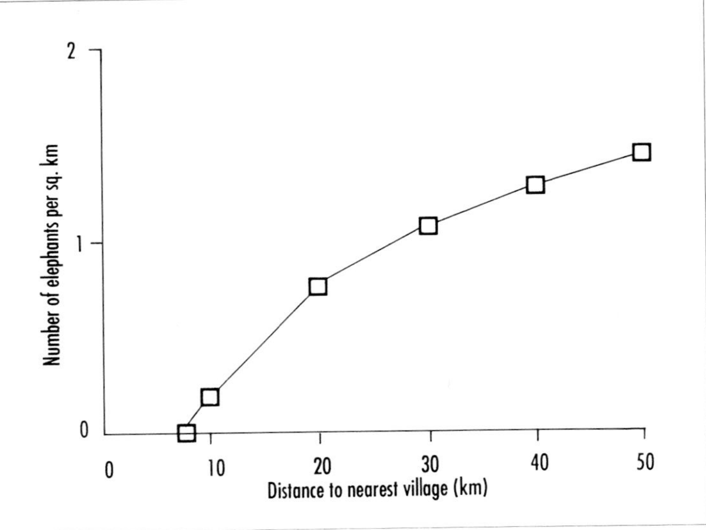

Figure 4.1 Elephant density (numbers of elephants per sq. km of forest) increasing with distance from the nearest village

(*Source:* Adapted from Barnes *et al.*, 1991)

Today ivory poaching is the most important factor determining the abundance and distribution of elephants. Heavy poaching has eradicated elephants from many forests. In places where there is little or no ivory poaching, the distribution of elephants is governed by both the past and present distribution of human activities (Barnes *et al.*, 1991). Elephants are attracted to lush secondary vegetation which occurs in areas of past human habitation. On the other hand, they avoid roads, villages and towns. On a walk from a village into the forest there are, at first, no signs of elephants. After about 10 or 15 km – depending upon the degree of human disturbance – there will begin to be tracks and droppings visible, and deeper into the uninhabited forest signs of elephants become progressively more abundant (Figure 4.1).

Over most of the equatorial forests the human population is concentrated along the main roads, and also along some of the larger rivers. Huge areas in between remain largely uninhabited. Therefore one can imagine people living and cultivating their crops in the roadside band, leaving the deep forests free for elephants. This habitat partition is not complete because villagers like to walk far away from their villages in search of fish or game, while elephants sometimes approach villages at night to feed on their crops. Crop-raiding elephants can be devastating: in former times elephants' depredations kept small, isolated communities on the brink of starvation. If elephants are to be conserved, the interests of people who lose crops to elephants must be protected.

Economic development can have mixed consequences for elephants. Clear-felling for large commercial plantations destroys the forest habitat. Roads open up new areas of forest for disturbance and poaching. Road and railway construction and mining often result in the decimation of elephants because the employees turn to poaching in their spare time. However, if the companies take the responsibility for controlling the leisure activities of their employees and insist that they observe the law, these forms of economic development need have little deleterious effect upon elephants.

Logging can have a positive or negative impact on elephant numbers. Timber companies have to construct roads through the forest so that they can extract the logs. These roads provide poachers with access to places which were previously out of reach. Nevertheless, if the timber companies control access to their road networks and destroy the roads and bridges when they finish logging, poaching can be controlled. Logging can improve the habitat for elephants.

Selective logging, in which between one and five trees per hectare are felled, simulates the effect of natural tree-fall, creating gaps in the canopy. Tangled masses of secondary plants spring up and attract elephants. In logging concessions where the company prevents poaching one can often find surprisingly large numbers of elephants.

Ivory Poaching

In the 1970s and 1980s a crime wave swept across Africa. The crime was ivory poaching, and the criminals were the poachers and ivory traffickers. The surge in poaching was stimulated by the rising international price of ivory. There has always been hunting for ivory in Africa, but the surge in the 1970s was unprecedented because it involved a huge network of national and international criminals. Local and government officials (some very highly placed), the army, police, gendarmes, customs officials, wildlife and forestry officers, peasants and merchants were all involved. Ivory was given fraudulent certificates or smuggled out of the country of origin. Much of the ivory passed at some stage through Hong Kong. During the murky journey from Africa to the Far East illegal consignments appear to have acquired legal certificates, so that by the time the ivory reached its destination it was impossible to distinguish between ivory of legal and illegal origin.

Many conservationists had assumed that the vast equatorial forests provided a safe refuge for elephants. They thought that the wave of poaching had swept across the open savannas leaving the forest elephants in peace. This assumption is now known to be false (WCI, 1989). For years bands of well-armed and well-organised poachers have been going deep into the remotest forests of Zaïre, and have eradicated elephants from large areas. In Zaïre and northern Congo whole villages gave up their normal pursuits and turned to ivory hunting. In northern Gabon some cocoa-farmers have abandoned their fields to hunt for elephants. In Cameroon pygmies were given rifles and commissioned to hunt for ivory. Throughout Central Africa, except for Gabon, automatic weapons are commonly used to hunt elephants.

Between 1979 and 1988 about 11 tonnes of ivory were exported from the forest countries of West Africa (less than 0.2 per cent of the continental total). During the same period 2825 tonnes left the forest countries of Central Africa, which was about 39 per cent of the continental total (Luxmoore *et al.*, 1989). These figures represent the deaths of at least 800 elephants in West Africa and more than 120,000 elephants – possibly as many as a quarter of a million – in Central Africa during that decade. These figures do not distinguish between ivory from forest or savanna. Nevertheless, they illustrate the scale of the killing, most of it illegal. The most recent estimates (Michelmore *et al.*, 1989) suggest that at least half the forest elephants of Cameroon, Central African Republic, Equatorial Guinea and Zaïre have been killed since the poaching surge began in the mid-1970s. Congo has lost about a third of its elephants, while Gabon has probably lost only a small proportion. Nearly all have been killed for their ivory; relatively few have been killed legally, after crop-raiding, or by sport-hunters.

Threats to Elephants

The immediate threat to elephants in the African forests is ivory poaching. Poverty, corruption, the breakdown of law and order, the availability of modern weapons, the lack of support for government wildlife agencies, and the overseas demand for ivory all play a part. If ivory poaching continues on a large scale, elephants will disappear, first from the West African forests and then from the equatorial forest block.

In the long term, human population growth and the expansion of settlement and agriculture will result in loss of habitat and increasing conflict between people and elephants (e.g. crop-raiding). There will be increased disturbance caused by logging and mining. The combination of accelerating deforestation and growing human populations will mean that per capita forest resources will shrink rapidly. This has already reached an acute stage in West Africa. The same process is well under way in Central Africa, despite the widespread illusion that the equatorial forests will persist for ever (Barnes, 1990). There will be less and less room for elephants.

At the same time, the large blocks of forest we see today will be fragmented by roads, railways, villages and towns. The effect of fragmentation is illustrated in Figure 4.2; large populations of animals become progressively isolated into smaller groups. As their refuges diminish (the ratio of edge to area of their habitat increases) they become increasingly vulnerable to human disturbance. This process is already in its final stages in the West African countries, and in south-west and south-central Cameroon (WCI, 1989). It is far advanced in Zaïre where the fragmentation has been caused by poachers rather than by settlement. Economic development in the equatorial forests with the spread of logging, mining, roads, railways and towns into them, will accelerate the fragmentation of elephant populations.

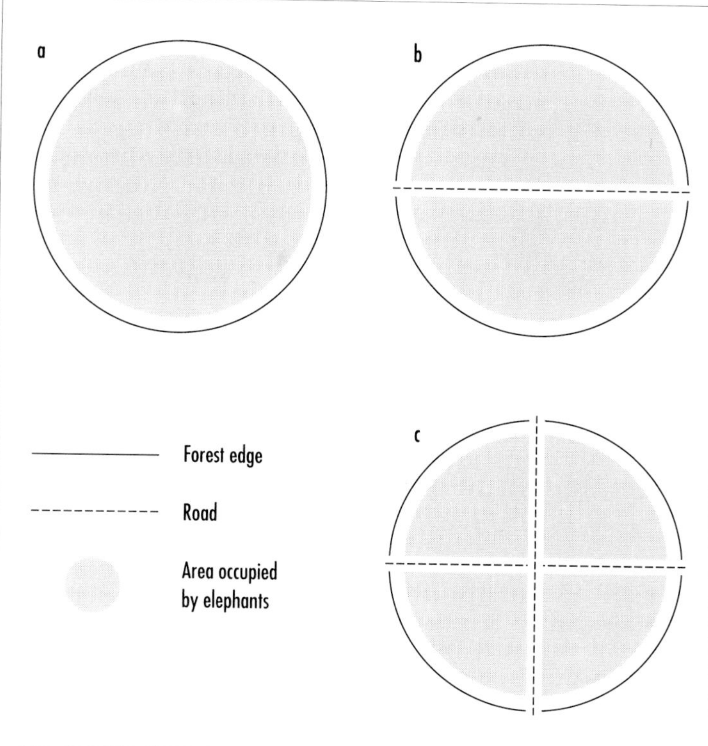

Figure 4.2 Schematic representation of the process of fragmentation and its effects on elephants

Figure (a) shows a forest in which elephants dwell. Since the forest is surrounded by human habitation the elephants avoid its edges. Figure (b) shows what happens when a road is built. Not only is the forest split in two parts, but the total area used by elephants has decreased because they avoid the band on either side of the road. Figure (c) shows the results of constructing another road. Now the forest has been split into four parts and the total area used by elephants has shrunk further. At first elephants might slip backwards and forwards across the roads, but as traffic increases and more villages are built along the roads, the fragments become completely isolated from each other. In time, if the number of elephants in the fragment is small, genetic problems due to inbreeding will occur.

Management and Conservation

Control poaching Clearly the immediate need is to control ivory poaching. This requires determined action by governments to deter poachers in the field, prevent the illicit transport of ivory within each country and stop the smuggling of ivory within and out of Africa. It also requires international action, either to control the ivory trade rigorously, or to close it down altogether. The international community must understand that in the rain forest countries most of the raw material for the ivory trade is obtained through violent crime.

National parks The anticipated deforestation and fragmentation described above suggests that the long-term future of elephants must lie in a network of large and well-protected national parks or game reserves. Existing protected areas must be strengthened and new ones created. They should be designed taking into account the needs and aspirations of the rural communities. For example, to minimise conflict between people and elephants, each protected area should be surrounded by a large buffer zone where there is no settlement or agriculture, but where hunting, fishing and gathering are conducted by the local populace. In those countries which allow sport-hunting of elephants, buffer zones can serve as controlled elephant hunting areas, with the profits going to the nearby villages rather than to central government accounts. The forests in West and Central Africa which contain priority populations of elephants are shown in Table 4.2. The conservation of these populations is necessary to ensure the preservation of the genetic and behavioural diversity of the species.

Forest elephants (Loxodonta africana cyclotis) at a salt lick in the Dzanga-Sangha National Park in the Central African Republic. G. Renson

Table 4.2 Priority populations of elephants in West and Central African forests

West Africa	
Country	*Population*
Côte d'Ivoire	Taï
Liberia	Mano and Lofa River forests
Ghana	Bia
Central Africa	
Country	*Population*
Cameroon	Korup
	Dja
	Mbam and Djerem
	Forests in the extreme south-east
CAR	Bayanga/Dzanga-Sangha
Congo	Nouabalé
	Odzala
	Lac Télé
	Conkouati
Gabon	Minkebe
	Petit Loango
	Lopé
Zaïre	Kahuzi-Biéga
	Salonga
	Maïko
	Lomami

(*Source*: AECCG, 1990)

Minimise damage to forests As the rain forest countries develop their economies, biologists and government wildlife departments will have to work with the managers of development agencies and the companies involved in logging, mining, oil-drilling and road and railway construction to minimise the deleterious impacts upon the forest ecosystem. The development agencies and industrial concerns must assume responsibility for ensuring that the forest ecosystem continues to function alongside their installations.

Wildlife departments The conservation of elephants, both inside and outside protected areas, requires strong and effective wildlife departments with the firm political support of their respective governments. Until now the wildlife departments, especially in Central Africa, have been starved of funds, equipment and personnel and lack the professional training and experience to cope with today's problems, let alone the challenge of the future.

Forest Primates

The African tropical moist forest zone is home to a very diverse assemblage of primates. In a conservative classification (Oates, 1986), there are 63 non-human primate species in mainland Africa. If the moist forest zone considered here (Madagascar is excluded) is broadly defined as including the Guineo-Congolian lowland rain forest zone of White (1983), together with tropical montane forests and the outlying patches of moist forest in coastal East Africa, then 53 African primate species (84 per cent) have all or most of their geographical distribution within tropical moist forest.

The African forest primates range in size from the nocturnal dwarf galago *Galagoides demidoff* (called *Galago demidovii* in earlier publications), which has an average adult body weight of only 70 g (Nash *et al.*, 1989) and is therefore one of the world's smallest primates, to the gorilla *Gorilla gorilla*, which has an adult male weight of 160 kg (Harvey *et al.*, 1987) and is the largest living primate species. Many classifications place the African forest primates into three families, five subfamilies and about a dozen genera (Table 4.3). Where there has not been heavy hunting by humans, between 10 and 15 primate species commonly live together in areas of African moist lowland and medium-altitude forests within 8° of the Equator. The most species-rich communities occur in the central forests (that is, from Cameroon to the Great Rift Valley), in areas where topography, river courses or moderate disturbance by humans or large mammals have produced a mosaic forest habitat. Montane forests above about 1500 m are generally less species-rich.

Threats to Species Survival

The 1986–90 IUCN/SSC Primate Specialist Group Action Plan for African Primate Conservation (referred to here as the PSG Action Plan), rated 29 of the 53 African forest primates (55 per cent) as vulnerable to, or in danger of, extinction (Oates, 1986). The IUCN Red Data Book on Threatened Primates of Africa (Lee *et al.*, 1988, referred to here as the Red Data Book) lists 47 predominantly forest-living primates, of which 21 (45 per cent) are

The eastern lowland gorilla Gorilla gorilla graueri, *is found only in Zaïre. In 1980, it was estimated that 3000–5000 individuals survived. The main threat to them is forest clearance.* N. Ellerton

Table 4.3 Primate species occurring in tropical lowland rain forest, montane forest and swamp forest in continental Africa. Classification as in Oates (1986)

	DEGREE OF THREAT	
	Oates (1986)*	Lee *et al.* (1988)**
Family Lorisidae		
Subfamily Lorisinae		
Arctocebus calabarensis	2	K
Perodicticus potto	1	nt
Subfamily Galaginae†		
Galago alleni	1	nt
Galago demidovii	1	nt
Galago inustus	2	nt
Galago thomasi	2	K
Galago zanzibaricus	3	V
Galago elegantulus	1	nt
Family Cercopithecidae		
Subfamily Cercopithecinae		
Tribe Papionini		
Cercocebus atys	2	nt
Cercocebus torquatus	3	V
Cercocebus galeritus	3	nt
Cercocebus albigena	1	nt
Cercocebus aterrimus	3	K
Mandrillus sphinx	3	V
Mandrillus leucophaeus	5	E
Tribe Cercopithecini		
Cercopithecus diana	4	V
Cercopithecus salongo	4	V
Cercopithecus neglectus	2	nt
Cercopithecus hamlyni	4	V
Cercopithecus lhoesti	3	V
Cercopithecus preussi	5	E
Cercopithecus solatus	4	V
Cercopithecus albogularis	1	nt
Cercopithecus mitis	1	nt
Cercopithecus nictitans	2	nt
Cercopithecus petaurista	1	nt
Cercopithecus sclateri	6	–
Cercopithecus erythrogaster	5	E
Cercopithecus erythrotis	4	E
Cercopithecus cephus	1	nt
Cercopithecus ascanius	1	nt
Cercopithecus campbelli	1	–[a]
Cercopithecus mona	1	nt
Cercopithecus pogonias	1	–[a]
Miopithecus talapoin	3	–
Miopithecus sp.	1	nt[b]
Allenopithecus nigroviridis	3	K
Subfamily Colobinae		
Procolobus badius	3	V
Procolobus pennanti	5	E
Procolobus rufomitratus	3	V
Procolobus kirkii	6	E
Procolobus gordonorum	6	E
Procolobus verus	3	R
Colobus polykomos	3	nt
Colobus vellerosus	3	nt
Colobus guereza	1	nt
Colobus satanas	4	E
Colobus angolensis	1	nt
Family Pongidae		
Pan troglodytes	3	V
Pan paniscus	4	V
Gorilla gorilla	4	V

(*Source:* Oates, 1986; Lee *et al.*, 1988)

* 1 - not known to be threatened; 2 - rare; 3 - vulnerable; 4 - highly vulnerable; 5 - endangered; 6 - highly endangered.

** nt - not threatened; K - insufficiently known (but suspected to be rare or under threat); R - rare; V - vulnerable; E - endangered.

† The taxonomy of the galagos is in the process of revision. Four of the listed species (excluding *elegantulus* and *inustus*) are now in the genus Galagoides (see Nash *et al.*, 1989). *Galago inustus* has now been renamed *G. matschiei.*

[a] The Red Data Book does not list *campbelli* and *pogonias* as distinct species. Since *C. mona* is sympatric in Ghana with *campbelli* and in Nigeria-Cameroon with *pogonias*, they are considered here to be distinct species.

[b] This refers to the northern *Miopithecus* which Oates (1986) regards as a questionably distinct species.

regarded as vulnerable or endangered. The two lists are compared in Table 4.3, and some of their discrepancies are discussed below. The Red Data Book entries indicate that forest destruction and hunting for meat, often acting in concert, are the chief problems facing the great majority of the threatened primate species. In this regard, primates are little different from other forest mammals of similar size.

Moderate forest disturbance resulting, for instance, from shifting cultivation in areas of low-density human population does not appear to pose a major threat to most primates and may actually increase local diversity and abundance (Oates *et al.*, 1990b). However, heavy logging produces large declines in the abundance of most species (Skorupa, 1988; see case study in chapter 31), while intensive agriculture resulting in the elimination of most natural forest is obviously incompatible with the survival of forest primates. For instance, at least 12 primate species occur in Rwanda's Nyungwe Forest, but none of these species occurs in the adjacent farmland which supports one of the densest human populations in Africa.

Hunting by man has been cited as a major cause of decline in the populations of many African forest-living monkeys and apes. There are few parts of the African forest zone where humans do not live or hunt, and most forest-living people in Africa, unlike many grassland people, readily eat monkeys and will often eat ape-flesh also (Harcourt and Stewart, 1980; Carpaneto and Germi, 1989). Monkeys and apes are active during the day, are relatively conspicuous animals and are therefore a frequent target of hunting; they suffer population declines where the density of hunters is high and modern weapons are available. In unlogged forests of Cross River State, Nigeria, for example, hunting is intense and few primates are seen (Harcourt *et al.*, 1989; Oates *et al.*, 1990a); in contrast, monkeys are relatively abundant in forests on the neighbouring island of Bioko. In 1969–79, Bioko suffered from harsh dictatorial rule during which it lost an estimated 30–50 per cent of its human population; in addition in 1974, nearly all shotguns were removed from the civilian population (Butynski and Koster, 1989). In general, species with large body sizes are more vulnerable to hunting than small species. Large primates provide more meat and are therefore a more valuable quarry for rural hunters to whom guns and ammunition are expensive items; also large primates often occur at relatively low population densities and have slow breeding rates, so that a small amount of hunting can have a major impact on population viability. Least affected by hunting are the small, nocturnal prosimians and the smallest *Cercopithecus* species.

Interacting with habitat destruction and hunting as threats to the survival of forest primates are factors of distribution and ecology. Species with restricted geographical distributions inevitably tend to be more vulnerable than those that are widespread, and those with narrow niches tend to be more vulnerable than generalists with broad habitat tolerances. The interaction of these factors has been shown in graphical form by Wolfheim (1983). Threatened species are frequently those with a geographical range of less than 100,000 sq. km (Happel *et al.*, 1987).

The degree to which primate populations are threatened with extinction can therefore be predicted to some extent from a limited set of intrinsic and extrinsic circumstances, or vulnerability factors: habitat destruction and hunting are usually the major threats, and vulnerability to these threats is increased by large body size, ecological specialisation, and small geographical range. Each of six African species rated as endangered in the PSG Action Plan has a small range and suffers from destruction of its habitat or hunting or both (Table 4.4). The Zanzibar (or Kirk's) red colobus

Table 4.4 Vulnerability factors and threats affecting six endangered African forest primate species

Species	*Location*	*Large body size*[1]	*Specialised niche*	*Restricted range*[2]	*Intensive hunting*	*Serious habitat loss*
Mandrillus leucophaeus	SE Nigeria W Cameroon Bioko	+	+	+[3]	+	
Cercopithecus preussi	SE Nigeria W Cameroon Bioko		+	+	+	
Cercopithecus sclateri†	SE Nigeria		?	+	+	+
Cercopithecus erythrogaster	SW Nigeria Benin Togo?		+	+	+	+
Procolobus pennanti	W Cameroon Bioko Congo SE Nigeria?	+	+	+	+	
Procolobus kirkii	Zanzibar	+	+	+		+
Procolobus gordonorum	Uzungwa Mts	+	+	+	(+)	+

(*Source:* Oates, 1986)

[1] Adult female body mass greater than 5 kg.

[2] Geographical range covers less than 100,000 sq. km.

[3] Wolfheim (1983) gives the range of *M. leucophaeus* as 150,000 sq. km; by our calculations its range is not more than half that size.

† Note that PSG Action Plan treats this as a full species, following Kingdon (1980), while the Red Data Book treats it as a subspecies of *Cercopithecus erythrotis*.

Table 4.5 Endangered subspecies of African forest primates

Subspecies	*Location*
Cercocebus galeritus galeritus	Lower Tana River, Kenya
Cercocebus galeritus sanjei	Uzungwa Mountains, Tanzania
*Procolobus badius waldroni**	Eastern Côte d'Ivoire, Western Ghana
*Procolobus pennanti pennanti**	Bioko (Fernando Poo)
*Procolobus pennanti preussi**	Western Cameroon, Nigeria?
*Procolobus pennanti bouvieri**	Lefini Reserve, Congo
*Procolobus rufomitratus rufomitratus**	Lower Tana River, Kenya
Pan troglodytes verus	West Africa (Senegal to Nigeria)
Gorilla gorilla berengei	Virunga Volcanoes
Gorilla gorilla graueri	Eastern Zaïre

(*Source:* Lee *et al.*, 1988)
* In the superspecies *[badius]*

Procolobus [badius] kirkii probably has the smallest total remaining population of any African primate species (the Red Data Book records 'almost 1500' individuals in 1981) and it is subject to all three vulnerability factors. However, a species can be endangered even if it is not intrinsically especially vulnerable. For instance, the natural range of Sclater's guenon *Cercopithecus sclateri* appears to have been a relatively small area between the Niger and Cross Rivers in southern Nigeria, but this species is small-bodied and (as far as is known) not particularly specialised in its habitat requirements. But it inhabits an area with one of the densest human populations in Africa where little forest remains and where hunting is intense in almost all the remaining forest fragments.

Additional special threat factors affect some species. The great apes, for instance, are threatened not only by habitat destruction and hunting, but also because of their close relationship to humans. The close genetic similarity of the chimpanzee *Pan troglodytes* to humans makes it a useful 'model' in biomedical research. Demand for chimpanzees in biomedical laboratories has put extra pressures on wild populations, where mothers are shot so that their young may be captured (Luoma, 1989). Despite the chimpanzee's inclusion in Appendix I of CITES and the recent upgrading of wild populations to endangered status on the US List of Endangered and Threatened Wildlife, there are fears that continuing biomedical demand will result in efforts to circumvent legislation by, for instance, the establishment of laboratories in the African countries where the species occurs.

Human fascination with our closest relatives has maintained a demand for both chimpanzees and gorillas by pet-keepers, circuses and zoos, and this demand inevitably increases threats to wild populations. Young chimpanzees are still being offered for sale on the streets of many African cities, while unscrupulous dealers and zoos continue to circumvent regulations designed to protect gorillas (Anon., 1989).

Threatened Subspecies

One of the dangers in ignoring subspecies in conservation planning was recently highlighted by a new taxonomic study of the tuatara *Sphenodon punctatus* in New Zealand. It was found that one neglected subspecies of this reptile may be extinct, and that another neglected population, now reduced to 300 individuals, should probably be regarded as a distinct species (Dougherty *et al.*, 1990).

The PSG Action Plan and the Red Data Book recognise a number of distinct subspecies of African primate that are threatened. Forest-living subspecies listed as endangered in the Red Data Book are shown in Table 4.5; several other subspecies are listed as vulnerable. Presently unlisted, poorly known, but possibly in danger is the eastern subspecies of the diana monkey *Cercopithecus diana roloway* which occurs in eastern Côte d'Ivoire and western Ghana (along with Mrs Waldron's red colobus *Procolobus [badius] badius waldroni*, which is listed).

Subspecies characteristically inhabit small geographical areas and because of this they may be threatened by special factors uniquely affecting that area. For instance, the Tana River red colobus *Procolobus [badius] rufomitratus rufomitratus* and the Tana River mangabey *Cercocebus galeritus galeritus* are dependent on forest that has regenerated on low-lying riverside levees produced by changes in river course. The dynamic relationship between the river, the forests and the primate populations is being disturbed by the increasing farming and burning of low-lying areas where forest regeneration might otherwise occur, and by changes in the river's flood regime caused by upstream dam construction (Marsh, 1986).

Threatened Communities

Species and subspecies of primate, like those of other animals and plants, typically occur in geographical association with particular other species or subspecies. Several such distinct regional communities of forest primates can be recognised in Africa, each containing a number of unique species or subspecies which have broadly overlapping distributions. These communities appear to have evolved at least in part through historical processes of large-scale climatic change which have caused related changes in both the distribution of forest types and in sea level (see chapter 2).

The PSG Action Plan described five major regional communities of primate species in the lowland rain forest zone: Upper Guinea, Cameroon, Western Equatorial Africa, Congo Basin, and Eastern Zaïre (see Figure 4.3). Each region supports 12–20 primate species, and between them the five regions support more than 80 per cent of all African forest primates. For this reason, and

Figure 4.3 The five distinct regional communities of African primates mentioned in this chapter

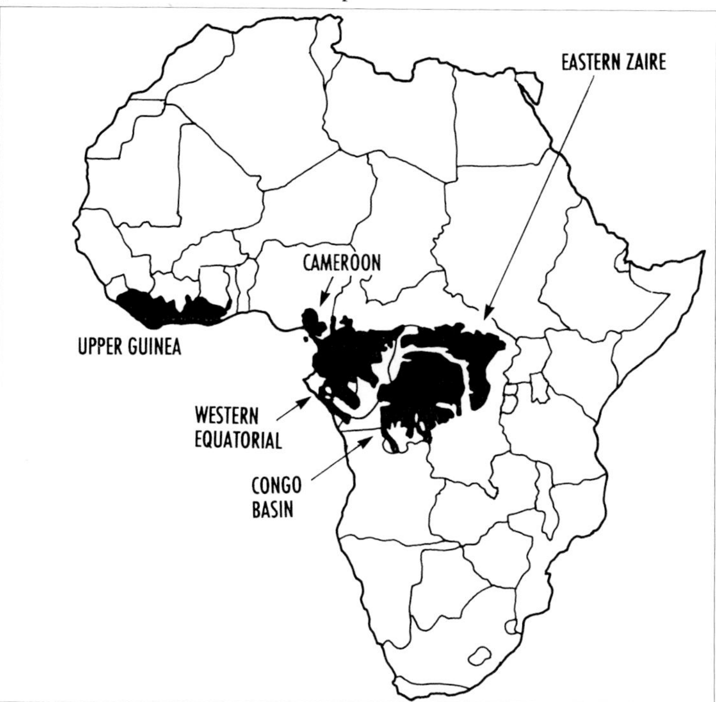

Table 4.6 Existing and planned protected areas in continental Africa of particular significance for forest primate conservation

Burundi
- Kibira National Park

Cameroon
- Korup National Park
- Dja Faunal Reserve

Central African Republic
- Dzangha-Sangha Dense Forest Faunal Reserve
- Dzangha-Ndoki National Park

Congo
- Northern forest region

Côte d'Ivoire
- Taï National Park

Equatorial Guinea
- Sur de la Isla de Bioko (Gran Caldera)

Gabon
- Lopé Reserve, including Forêt des Abeilles

Ghana
- Bia National Park and Bia Game Production Reserve
- Nini-Suhien National Park
- Ankasa Game Production Reserve

Kenya
- Tana River National Primate Reserve

Liberia
- Sapo National Park
- Gola/Kpelle National Forests

Nigeria
- Okomu Game Reserve
- Cross River National Park (Oban + Okwangwo Divisions)*
- Stubbs Creek Game Reserve*

Rwanda
- Volcanoes National Park
- Nyungwe Forest Reserve

Sierra Leone
- Gola Strict Nature Reserves*
- Tiwai Island Game Sanctuary

Tanzania
- Uzungwa National Park
- Jozani Forest Reserve
- Mahale Mountain National Park

Uganda
- Bwindi (Impenetrable) National Park*
- Kibale Forest Reserve
- Gorilla Game Reserve

Zaïre
- Ituri Forest*
- Kahuzi-Biega National Park
- Lomako National Park*
- Maïko National Park
- Salonga National Park
- Virunga National Park

(*Source:* MacKinnon and MacKinnon, 1986; Oates, 1986)
* proposed

because many species are endemic to a single region, effective conservation of the African forest primate fauna as a whole requires effective conservation measures in each different forest region. Under most active threat at the present time are the Upper Guinea and Cameroon forest regions.

The Upper Guinea forests coincide over much of their area with a relatively dense human population, and they are becoming increasingly fragmented. This is threatening two subregional primate communities, because the primates of the eastern part of the Upper Guinea region (the forests of eastern Côte d'Ivoire and western Ghana) are distinct at the subspecies – and in a few cases at the species level – from those in the western part of the region (between Sierra Leone and western Côte d'Ivoire). The eastern subregion, home of the roloway monkey and Mrs Waldron's red colobus, is probably most acutely threatened.

The Cameroon region, comprising the area between the Cross River in Nigeria and the Sanaga River in Cameroon, plus the island of Bioko, is vulnerable because it is small in size and has high-density human populations both scattered through it and threatening its periphery. Endemic to this region are the drill *Mandrillus leucophaeus*, Preuss's guenon *Cercopithecus preussi* and two forms of Pennant's red colobus *Procolobus [badius] pennanti*.

More vulnerable than the primates in any of the major lowland forest regions are those inhabiting peripheral forest areas. In the mountains and coastal regions of eastern Africa there are many small forest areas that contain unique primates and that are vulnerable because of their size and their isolation. Among the most important are the Virunga Volcanoes, the Uzungwa Mountains, the Tana River and Zanzibar.

In West Africa, an especially threatened peripheral area is that in southern Nigeria and Benin, between the main Upper Guinea and Cameroon forests. The white-throated guenon *Cercopithecus erythrogaster* occurs only in this area, in the increasingly fragmented forests from the Niger Delta westwards, while Sclater's guenon *C. sclateri* is found only in the very densely populated area north and east of the delta.

Protected Areas and Other Conservation Strategies

The PSG Action Plan and the IUCN Review of the Protected Areas System in the Afrotropical Realm (MacKinnon and MacKinnon, 1986) list a variety of existing and proposed national parks and other protected areas which between them could protect the great majority of the African forest primate fauna. Some of these sites have particular significance for the conservation of primates (see Table 4.6), and may not be stressed in other contexts. For instance, Okomu in Nigeria is an important site for the protection of *Cercopithecus erythrogaster*, while Stubbs Creek contains a population of *C. sclateri*. The Lopé Reserve in Gabon contains part of the Forêt des Abeilles, the only known locality for *C. solatus*. Zanzibar's Jozani Forest is a key site for the protection of Kirk's red colobus and the Tana Reserve in Kenya is critical for the protection of the endangered Tana River red colobus and crested mangabey. The Gran Caldera on the island of Bioko, Equatorial Guinea, is home to at least six monkey species including the endangered Pennant's red colobus (which may now survive only in and near this locality: Butynski and Koster, 1989). The Ituri Forest in Zaïre has 12 sympatric monkey species including the rare owl-faced guenon *Cercopithecus hamlyni*; this may be the richest monkey-species assemblage in Africa (Hart *et al.*, 1986 and case study in chapter 32).

It may, however, be dangerous to stress the importance of certain protected areas, while ignoring others. No protected area can be regarded as completely secure in the long term from destructive intrusion, and even the most secure reserve is likely to change through time as a result of the immigration and extinction of some species and changes in climate. Any long-term plan for African forest primate conservation must therefore aim to establish, where possible, several protected areas within the range of each species (or each set of co-occurring species: the regional communities described above).

On the other hand, there is clearly a need in the short term to maintain or establish *effective* protected areas for primates in areas where current threats are more acute such as in southern Nigeria, and many of the forest islands in eastern Africa. The existence in law of large numbers of protected areas in the African forest zone is not in itself sufficient to conserve primates and their forest habitats. The protected areas must actually be protected, which requires trained personnel, the money to pay for both the personnel and the facilities and equipment they need. The support of local communities for conservation efforts is also required. In tropical Africa some or all of these commodities are frequently lacking and, as Watts (1989) has pointed out, economic, social and political factors often make these deficiencies difficult to remedy.

There are exceptions. Examples of successful conservation projects are provided by the very successful Mountain Gorilla Project (see case study in chapter 12), together with the Zaïre Gorilla Conservation Project. These schemes have helped stabilise the threatened gorilla population in the Virunga Volcanoes through a combination of increased protection, tourism development and education (Aveling and Aveling, 1989; Vedder, 1989). Unfortunately, rather few other forest primate populations have the popular appeal, tourism potential and, therefore, revenue-generating potential of the Virunga gorillas and their spectacular habitat.

We must acknowledge that there are no easy remedies to the threats facing many African primates. In the long run the most effective measures will be those which assist African nations and people to develop their own solutions to conservation problems.

References

AECCG (1990) *Elephant Action Plan*. Fourth Edition, March 1990. AECCG/CITES.

Alexandre, D. Y. (1978) Le rôle disseminateur des éléphants en forêt de Taï, Côte d'Ivoire. *La Terre et la Vie* **32**: 47–72.

Anon. (1989) Two gorillas arrive at Mexican Zoo. *International Primate Protection League Newsletter* **16**(2): 3–4.

Aveling, C. and Aveling, R. (1989) Gorilla conservation in Zaïre. *Oryx* **23**: 64–70.

Barnes, R. F. W. (1990) Deforestation trends in tropical Africa. *African Journal of Ecology* **28**: 161–73.

Barnes, R. F. W., Alers, M. P. T. and Blom, A. (1991) Man determines the distribution of elephants in the relatively undisturbed forests of N. E. Gabon. *African Journal of Ecology* **29**: 54–65.

Burrill, A. and Douglas-Hamilton, I. (1987) *African Elephant Database Project: Final Report*. UNEP/GRID, Nairobi, Kenya.

Butynski, T. M. and Koster, S. H. (1989) *The Status and Conservation of Forests and Primates on Bioko Island (Fernando Poo), Equatorial Guinea*. WWF Unpublished Report, Washington, DC, USA. 64 pp.

Carpaneto, G. M. and Germi, F. P. (1989) The mammals in the zoological culture of the Mbuti pygmies in north-eastern Zaïre. *Hystrix* **1**: 1–83.

Carroll, R. W. (1988) Elephants of the Dzangha-Sangha dense forests of south-western Central African Republic. *Pachyderm* **10**: 12–15.

Dougherty, C. H., Cree, A., Hay, J. M. and Thompson, M. B. (1990) Neglected taxonomy and continuing extinctions of tuatara (*Sphenodon*). *Nature* **347**: 177–9.

Douglas-Hamilton, I. (1989) Overview of status and trends of the African elephants. In: *The Ivory Trade and the Future of the African Elephant*. Cobb, S. (ed.). Ivory Trade Review Group, Oxford, UK.

Haltenorth, T. and Diller, H. (1977) *A Field Guide to the Mammals of Africa including Madagascar*. Collins, London, UK.

Hamilton, A. C. (1982) *Environmental History of East Africa*. Academic Press, London, UK. 328pp.

Happel, R. E., Noss, J. F. and Marsh, C. W. (1987) Distribution, abundance and endangerment of primates. In: *Primate Conservation in the Tropical Rain Forest*. Marsh, C. W. and Mittermeier, R. A. (eds), pp. 63–82. Alan R. Liss, New York, USA.

Harcourt, A. H. and Stewart, K. J. (1980) Gorilla-eaters of Gabon. *Oryx* **15**: 248–51.

Harcourt, A. H., Stewart, K. J. and Inaharo, I. M. (1989) Gorilla quest in Nigeria. *Oryx* **23**: 7–13.

Hart, J. A., Hart, T. B. and Thomas, S. (1986) The Ituri Forest of Zaïre: primate diversity and prospects for conservation. *Primate Conservation* **7**: 42–4.

Harvey, P. H., Martin, R. D. and Clutton-Brock, T. H. (1987) Life histories in comparative perspective. In: *Primate Societies*. Smuts, B. B., Cheney, D. L., Seyfarth, R. M., Wrangham, R. W. and Struhsaker, T. T. (eds), pp. 181–96. University of Chicago Press, Chicago, USA.

Kingdon, J. S. (1980) The role of visual signals and face patterns in African forest monkeys (guenons) of the genus *Cercopithecus*. *Transactions of the Zoological Society, London* **35**: 425–75.

Lee, P. C., Thornback, J. and Bennett, E. L. (1988) *Threatened Primates of Africa: The IUCN Red Data Book*. IUCN, Gland, Switzerland and Cambridge, UK.

Lieberman, D., Lieberman, M. and Martin, C. (1987) Notes on seeds in elephant dung from Bia National Park, Ghana. *Biotropica* **19**: 365–9.

Luoma, J. R. (1989) The chimp connection. *Animal Kingdom* **92**(1): 38–51.

Luxmoore, R., Caldwell, J. and Hithersay, L. (1989) The volume of raw ivory entering international trade from African producing countries from 1979–1988. In: *The Ivory Trade and the Future of the African Elephant*. Cobb, S. (ed.). Ivory Trade Review Group, Oxford, UK.

Marsh, C. W. (1986) A resurvey of Tana River primates and their habitat. *Primate Conservation* **7**: 72–82.

MacKinnon, J. and MacKinnon, K. (1986) *Review of the Protected Areas System in the Afrotropical Realm*. IUCN, Gland, Switzerland and Cambridge, UK.

Michelmore, F., Beardsley, K., Barnes, R. and Douglas-Hamilton, I. (1989) Elephant population estimates for the central African forests. In: *The Ivory Trade and the Future of the African Elephant*. Cobb, S. (ed.). Ivory Trade Review Group, Oxford, UK.

Nash, L. T., Bearder, S. K. and Olson, T. R. (1989) Synopsis of *Galago* species characteristics. *International Journal of Primatology* **10**: 57–80.

Oates, J. F. (1986) *Action Plan for African Primate Conservation 1986–90*. IUCN/SSC Primate Specialist Group, Stony Brook, New York, USA.

Oates, J. F., White, D., Gadsby, E. L. and Bisong, P. O. (1990a) *Conservation of Gorillas and Other Species. Appendix to Feasibility Study, Cross River National Park (Okwangwo Division)*. Unpublished report to WWF-UK.

Oates, J. F., Whitesides, G. H., Davies, A. G., Waterman, P. G., Green, S. M., Dasilva, G. L. and Mole, S. (1990b) Determinants of variation in tropical forest primate biomass: new evidence from West Africa. *Ecology* **71**: 328–43.

Olivier, R. C. D. (1978) *On the Ecology of the Asian Elephant,* Elephas maximus *Linnaeus, with Particular Reference to Malaya and Sri Lanka.* Unpublished PhD thesis, University of Cambridge, UK.

Skorupa, J. P. (1988) *The Effects of Selective Timber Harvesting on Rain Forest Primates in Kibale Forest, Uganda.* PhD dissertation, University of California, Davis, USA.

Struhsaker, T. T. (1987) Forestry issues and conservation in Uganda. *Biological Conservation* **39**: 209–34.

Vedder, A. (1989) In the hall of the mountain king. *Animal Kingdom* **92**(3): 31–43.

Watts, D. P. (1989) Review of Threatened Primates of Africa: The IUCN Red Data Book. *International Journal of Primatology* **10**: 383–5.

WCI (1989) The status of elephants in the forests of central Africa: results of a reconnaissance survey. In: *The Ivory Trade and the Future of the African Elephant.* Cobb, S. (ed.). Ivory Trade Review Group, Oxford, UK.

Western, D. (1986) The pygmy elephant: a myth and a mystery. *Pachyderm* 7: 4–5.

Western, D. (1989) The ecological value of elephants: a keystone role in Africa's ecosystems. In: *The Ivory Trade and the Future of the African Elephant.* Cobb, S. (ed.). Ivory Trade Review Group, Oxford, UK.

White, F. (1983) *The Vegetation of Africa: a descriptive memoir to accompany the Unesco/AETFAT/UNSO vegetation map of Africa.* Unesco, Paris, France. 356 pp.

Wolfheim, J. H. (1983) *Primates of the World: Distribution, Abundance and Conservation.* University of Washington Press, Seattle, USA.

Authorship

Richard Barnes, Wildlife Conservation International, University of California, San Diego, USA for the section on forest elephants and John Oates, Hunter College, City University of New York for primate conservation.

5 Forest People

Introduction

The majority of people living in Central Africa rely on the resources of the forest for a significant proportion of their livelihood. The major groups of forest-dwelling peoples are the traditional shifting cultivators of the Bantu and Sudanic speaking groups and the many loosely linked tribes collectively known as pygmies, who are mainly hunter-gatherers. The shifting cultivators are predominant, clearing and planting forest land which they cultivate for one to three years before moving on. Most households clear some new area of forest each year. In addition to cultivation, nearly all forest-living farmers supplement their diet and income by fishing, hunting and gathering forest products. Bantu people living along rivers often specialise in fishing and nearly all farmers fish seasonally. Hunting forest animals, especially duikers and monkeys, provides the major source of protein for many families, while commercial hunting of forest animals to supply protein to large towns and cities is also common among African forest-living people. Moreover, gathered forest products, including honey, nuts, fruits, leaves and insects (mainly caterpillars and termites), provide important supplements to diets that would otherwise lack diversity and essential nutrients. Additionally, these people rely on the forest for their building materials and their firewood.

Before the colonial period, the Bantu and Sudanic people were primarily subsistence farmers, cultivating crops and extracting from the forest only those resources required for their own maintenance. After the First World War, when colonists introduced new crops, local people were induced to produce surplus crops for cash. Thus today, in addition to their traditional swidden gardens, which normally contain cassava, plantains, corn, taro and yams, most households cultivate cash crops such as rice, peanuts, coffee, cacao and oil palms.

Although in many areas of Central Africa road systems and river transport are poor, almost all forest farmers live along a river or road offering some access to outside markets. Many also live near a commercial operation which can offer opportunities for casual employment. Consequently, a substantial proportion of the people eke out their living by working at least seasonally for commercial coffee, oil palm, cacao or rubber plantations or for mining or logging operations.

Most Bantu and Sudanic farmers of Central Africa reside in small villages with between 10 and 250 inhabitants. Village residence is usually determined by clan affiliation. Chieftainships were frequently created by colonial powers for administrative purposes, often in ignorance of cultural affinities and, therefore, not always representing traditional tribal affiliations. Moreover, because the majority of Central African tribal peoples were divided into lineages without chiefs, kings or any form of centralised authority, modern day chiefs and other government officials are not always effective leaders nor trusted representatives of tribal opinion. Cultural identity is based on language, kinship, oral history and cultural practices, such as initiation ceremonies, body markings, marriage and kinship rules and often centred on a specific area of forest. The forest nearly always figures prominently in the history as well as magico-religious myths and ceremonies of its peoples and is thus important to their sense of identity and psychological well-being.

African Pygmies

Pygmies are distributed across the forested regions of Central Africa. They are short in stature and traditionally live by specialising in hunting and gathering wild forest resources which they consume themselves or trade to neighbouring Bantu and Sudanic-speaking farmers in exchange for cultivated foods. There has been a long history of contact and extensive economic and political relations between pygmies and these farmers in the African rain forests for at least 2000 and possibly as long as 4000 years (Ehret and Posnansky, 1982).

There are around 200,000 pygmies distributed discontinuously across the nine African countries of Rwanda, Burundi, Uganda, Zaïre, Central African Republic, Cameroon, Equatorial Guinea, Gabon and Congo (Waehle, 1991). They live in innumerable distinct ethnic groups that are separated by geography, language, custom and technology. Pygmies in most areas are unaware of the existence of pygmies in other areas and there is no sense of solidarity between different populations. The one characteristic that is common to them all, no matter their location or level of acculturation, is their disdain for the term pygmy. Without exception they prefer to be called by their appropriate ethnic name (such as Mbuti, Efe, Aka, Asua) and consider the term pygmy as pejorative. That there is no one generic term other than this European word – derived from the Greek *pyme*, meaning a unit of measure whose length was from the elbow to the knuckle – bears testimony to the absence of any pan-pygmy awareness. Unfortunately, until the people themselves generate a different term, we are forced to use the word pygmy.

Pygmies have long been considered the original inhabitants of African tropical rain forest. However, there is no suitable definition or precise description of the African people referred to by most of the world as pygmy, since there is no physical or cultural feature that distinguishes them absolutely from other Africans. While they are known for their short stature, the average height of many pygmy populations overlaps with that of other populations in Africa and

A Bambuti from Semliki Forest in Uganda. S. Bearder

in other tropical forest areas of the world (Bailey, 1991). Genetically, there is no evidence that pygmies are distinct from other Africans; there is no 'pygmy marker' that is common to all pygmies and exclusive of all other Africans (Cavalli-Sforza, 1986). Similarly, linguistically and culturally pygmies cannot be considered distinct from other Africans; there is no distinctive 'pygmy language family' and pygmies across Central Africa exhibit a broad range of cultural adaptations, many similar to those of Bantu and other African farmers.

Contrary to many romanticised accounts of pygmy life, there are no people living today in Central Africa as pure hunter-gatherers, independently of agriculture. All evidence suggests that this has been true for many hundreds of years (Bahuchet and Guillaume, 1982) if indeed pygmies ever lived in the forest without access to agricultural foods (Bailey and Peacock, 1988; Bailey *et al.*, 1989). Today, most pygmies are hunter-gatherers. They specialise in extracting resources from the forest. They consume some of those products themselves and some they trade with others to acquire cultivated foods, iron implements and other merchandise. Everywhere that pygmies have been carefully studied, including the most remote corners of their geographic distribution, researchers have found them relying on cultivated foods for at least 50 per cent of their diet (Bahuchet, 1985; Bailey and Peacock, 1988). Moreover, pygmies everywhere have extensive relations with neighbouring Bantu and Sudanic-speaking farmers, relations that extend beyond economic trade to include all aspects of political, religious and social life. Indeed, it is not possible to consider pygmy culture and subsistence in isolation from the African farmers with whom they trade and live.

In most areas of Central Africa, specific clans of pygmies have traditional relations with specific clans of farmers and these relationships are passed from one generation to the next creating a complex web of economic and social exchange that leads to high levels of cooperation and support. Pygmies provide forest products – protein-rich meat in particular – to farmers while the farmers provide much needed starch to pygmy foragers. The meat, honey and medicinal products from the forest are significant contributions to farmers' survival, while these days pygmies would be hard pressed without the iron implements and the political representation provided by farmers. In most areas, pygmies are viewed by farmers as essential to successful ceremonies, while farmers can have considerable control over many crucial pygmy events, including marriage, circumcision and burial. Relations between pygmies and farmers are so extensive that elaborate systems tie the two groups together in a web of kinship that ensures social and economic interdependence. In some areas there is intermarriage – pygmy women marry farmer men but farmer women never marry pygmy men (Bailey, 1988).

Close relations between pygmies and farmers extend to their perceptions of rights to land. Each farmer clan has rights recognised by all neighbouring farmer clans to a specific area of forest which they may clear to cultivate their crops, or where they may hunt, fish, gather and extract required materials. Each clan of pygmies also has recognised rights to exploit the same area of forest as the farmer clan with which it is traditionally associated. The farmers assist their pygmy partners in maintaining exclusive rights to this area and violations by either pygmies or other farmers are contested through negotiation or sometimes violence. In this way, most, if not all areas of forest in Central Africa are claimed by indigenous people and elaborate informal mechanisms exist to guarantee specific land rights.

It should be clear that, for the purposes of designing programmes for development or conservation, pygmies cannot be considered in isolation from indigenous forest farmers. Central African farmers and pygmies exist together, are interdependent and should be considered as an integrated economic and social system. This is a system that is generally not recognised by African governments and is minimally integrated into the formal politics and economy of the national societies. Yet, for the people themselves, the system facilitates the spreading of risk in an uncertain environment and offers mutual support to indigenous people vulnerable to unpredictable changes brought by outside agents.

While most pygmies in Central Africa still live within the traditional farmer-pygmy relationship, most also engage in activities outside that relationship and, like their farming partners, have managed to adapt in myriad ways to changes caused by development and commercialisation. This is true not just in individual localities where development has been more extensive, but in every area of Central Africa. Any one population of pygmies spans the full range of levels of acculturation and adaptation to changing conditions.

Commercial Hunting and Employment

The growing population around the edges of the Zaïre Basin means a rising demand for meat from the forest. Increasingly, pygmies are becoming commercial hunters, spending greater proportions of their time hunting forest game and selling larger quantities of meat to traders who come great distances from towns and cities located at the edges of the forest. The effect is to break down the traditional farmer-pygmy relationship, to bring pygmies into the money economy and inevitably to cause the depletion of wild game, thus endangering not only the forest fauna but also the subsistence base and basic way of life of pygmies and their farmer partners (Hart, 1979; Bailey, 1982; Bahuchet, n.d.).

Many pygmies also work on a casual, sporadic basis for commercial coffee, rubber or palm plantations or for logging companies. None is in a position of authority or high salary. They generally work seasonally, planting, weeding or harvesting on plantations or identifying trees and supplying other workers with meat on logging operations.

Farming and Settlement

In recent years, some pygmies have become sedentary, living a settled life as farmers in villages. In some regions, insufficient areas of forest remain to support the pygmies' specialised hunting and gathering. In others, overhunting has depleted forest game. Moreover, in every region there have been periodic formal campaigns by national governments to force pygmies, or induce them with gifts, to settle in villages and become sedentary farmers. Missionaries in almost every region have also been active in this regard. Many reasons are given for the need for these programmes, but the most often cited are three: first, pygmies are at a primitive stage of evolution and intervention is needed to bring them into the modern economy; second, pygmies must be brought into the mainstream of the national culture and economy to become productive members of the society; and third, pygmies must become independent of their farmer 'patrons', who exploit them unfairly. Those who design and implement these settlement programmes do not recognise the economic or social value of the traditional farmer-pygmy relationship, nor do they appreciate the contribution that forest nomads make to the national economy by efficiently exploiting forest resources on a sustainable basis. The pygmies themselves are seldom, if ever, consulted or given a decision making role in the design and implementation of these programmes.

Most settlement programmes have failed. The pygmies return to the forest when the gifts run out or they abandon their gardens when the first good honey season begins. Nevertheless, increasingly there are pygmies who have voluntarily turned to farming and who live in villages along the roads. Like traditional African farmers, they spend at least some time in the forest and depend upon it for a significant supplement to their mixed farming subsistence. A few such sedentary farming pygmies, again like their farmer neighbours, grow some cash crops in addition to their subsistence crops, but this is far from common in any region.

A very small number of pygmies have moved into towns and can even be seen in major cities. Some are hired as guards, armed with bow and arrow or crossbow, to protect stores in urban settings. Others become homeless beggars, curiosities for foreign tourists and African urban dwellers. The great majority return to the forest after a short time.

Education and Health

Very few pygmies are literate. Because of their mobile existence, they seldom attend school for more than a few weeks. In many areas there is overt discrimination against pygmies in schools by both teachers and farmers, who value pygmies' skills in the forest but belittle their capabilities to learn in school. In almost every region of Central Africa there are a few literacy programmes exclusively for pygmies. These are often associated with settlement schemes initiated and administered by missionaries. Thus far, they have had limited success as pygmies strive to maintain their mobility.

Health facilities are poor throughout rural Africa, but especially for people living in remote areas. Dispensaries are usually available but are rarely supplied with medicines. In many areas local tradesmen with no medical knowledge are the main suppliers of antibiotics, antimalarials and other drugs. Virtually all rural people have indigenous health care systems with traditional healers using herbs and divination techniques to cure natural and supernatural (for example, witchcraft and sorcery) illnesses. In forest regions, the local pygmies are either the principal traditional healers or play important roles.

Pygmies tend to use dispensaries and other sources of western medicine less than their farmer partners. This is no doubt due in part to pygmies' high mobility and tendency to be further from the source, but other factors contribute to their lower reliance on non-traditional health care. In some areas they are discriminated against by health care workers. They are less integrated into the cash

A Mbuti family in the south-western forests of the Central African Republic.

G. Renson

THE OKAPI PROJECT

When Stanley's Emin Pasha Relief Expedition completed its crossing of eastern Zaïre in 1885, the surviving expedition members turned on the hated rain forest – 'this Green Hell' – and accused it of the murder of hundreds of their company. They called it 'a wilderness of fungi and wood-beans, infinitely sullen, remorseless and implacable' (Stanley, 1887). Yet for many thousands of years, these same forests have been the secure home of Africa's nomadic pygmies, hunter-gatherer people whose existence is finely tuned to life beneath the forest canopy.

The first Central African conservation project which aims to 'meet the needs of sustaining the traditional lifestyle of these people' is the Okapi Conservation Project (Stone, 1987), in the Ituri forest of north-east Zaïre. The Ituri is home to four major pygmy hunter-gatherer groups – the Efe, Mbuti, Tswa and Aka – numbering many hundreds of people (Wilkie, 1990). Due to the rugged and inaccessible nature of the forest, these pygmies are some of the least acculturated in Zaïre. However, in recent years, gold prospecting, immigration, commercialisation of the wild bush-meat trade, timber exploitation and charcoal production have all started to threaten the Ituri forest (Hall, 1990). The Bantu population is expanding fast, especially in the west, around the town of Wamba, and in the south as people spill over from the highly populated Kivu region. While timber and charcoal exploitation is concentrated around the three roads that run through the Ituri, gold prospectors have penetrated the heart of the Ituri forest itself (Wilkie, 1990).

All these factors, though contributing little as yet to the destruction of the Ituri forest, may lead to the rapid breakdown of the traditional existence of the pygmies. For instance, gold-mining introduces large Bantu populations into the deepest reaches of the forest. The gold-miners pay porters to bring in food from the nearest road-side villages and finance the hunting of forest animals around the camps (Wilkie, 1990). Those best fitted for both tasks are the pygmies of the Ituri, who have an intimate knowledge of the topography of the forest and are expert hunters of its wildlife with their nets and spears or with bow and arrow. Thus the pygmies have been drawn into the thriving market economy for bush-meat both within the forest and in nearby towns.

As a result, the normal hunting-gathering cycle of many pygmy groups – which follows the yearly abundance of honey, good hunting or the cropping of their neighbours' farms – has been disrupted and is now ruled by the occurrence of commercial opportunities in the gold-camps or along the roads. This has lead to the breakdown of the traditional pygmy relationship based on game and labour exchange with neighbouring Bantu villages. Though the Efe pygmies are often cited as an example of traditional hunters-gatherers, they obtain at least 50 per cent of their food from the fields of their cultivating neighbours. All the Ituri pygmy groups have a similar, mutually beneficial relationship with the Bantu. The breakdown of this traditional system strikes at the very heart of the culture of the pygmies and their farming partners.

The Okapi Project aims to use the very skills that can make the pygmies a threat to the wildlife of the Ituri, while at the same time 'protecting the land-tenure and subsistence rights of the indigenous human populations' (Wilkie, 1990). Because the Ituri harbours such an abundance of forest mammals, the creation of an Okapi Rainforest Reserve is the first concern of the project. Yet for pygmy and farmer alike, the Ituri is a vital, renewable resource and, as forest animals are the prime source of protein in their diets, they will need to continue exploiting the forest after the establishment of the reserve.

Recognising this fact, WWF and the Institut Zairois pour la Conservation de la Nature (IZCN) have designed a multiple-use rain forest reserve, with a smaller core national park wilderness area and surrounding buffer zone. Human activity will be zoned – minimal within the core park area, increasing towards the surrounding buffer zones. Under an IZCN programme to supply the rare okapi *Okapia johnstoni* to zoos, capture zones for this species have been set (Hall, 1990). Pygmies have long been employed by the Okapi Research Station to track and trap the elusive okapi (see case study on the Ituri forest in chapter 32). As a result, both the pygmies and the Okapi Project benefit; the pygmies receive regular wages for conservation-oriented work which uses their forest skills and alerts them to the benefit of conserving the wildlife of the Ituri, while the Okapi Project is assured of the support of the pygmies in all its conservation activities.

At present, though the Ituri is not yet an official park/reserve, it is treated like one by all concerned. The process of gazettement has been long and delicate, but an official park proposal has recently been completed. With the inauguration of the reserve, the opportunities for involvement of the local pygmies should increase. With their intimate knowledge of and respect for the forest, the pygmies should prove ideal park guards, scientific research counterparts and tourist guides for the increased volume of 'ecotourists' (both national and foreign) who will visit the park. In this way, the Okapi Project should reinforce the sanctity of the pygmies' self-sufficient existence by involving them as partners in conserving the Ituri forest. It seems to be working. Even the most acculturated Mbuti pygmies, who work for the Okapi Research Station, wear 'European' clothing and are closely involved in the intruding cash-economy, still retain much of their traditional way of life and intimacy with the forest. During celebrations at the start of the honey season – a ritual at which pygmy camps gather to dance, socialise and reinforce their traditions – all the pygmies, regardless of their 'profession' become deeply involved in the day's events. Working for the Research Station, in the outside world, does not appear to detract from the forest world of the individual Mbuti and the reverence and respect they retain for it.

To succeed in protecting the rain forest and the way of life of Africa's pygmies, projects like the Okapi will need to involve the pygmies at all levels. In this respect, the Okapi project has one major shortcoming: it aims to ensure that 'the majority of the indigenous population inhabit areas outside the reserve' to minimise the impact of 'their subsistence practices . . . upon the reserve's protected biota' (Wilkie, 1990). Bearing in mind the low density of the pygmy population and the fact that they adhere to well-respected traditional territories (Turnbull, 1961), it is to be hoped that it will be possible for them to remain within the reserve. The major conservation challenge is not only to protect the biodiversity of the Central African rain forest, but to preserve the cultural variety or ethnodiversity of its people.

Source: Damien Lewis

economy and so have fewer means of paying for medicines. Being the primary traditional healers in many areas, they are more likely to rely exclusively upon this traditional health care.

Both African farmers and pygmies are less well nourished – judged by weight to height ratio and skinfold thickness – than western populations. While farmers tend to be better nourished than their pygmy partners, there is evidence that pygmies experience less dramatic fluctuations in body weight than do farmers. There is a high prevalence of parasitic and infectious diseases among both farmers and pygmies. Malaria, tuberculosis, amoebiasis and filariasis are all very prevalent. In some areas river blindness caused by filaria afflicts up to 20 per cent of adults. Manioc is the staple food for many forest-dwelling Africans; as a result goitre is highly prevalent. Hypertensive and coronary heart disease and chronic diseases more typical of industrialised countries are rare. Pain and secondary infection caused by trauma are common occurrences in the lives of forest people. Hernias among both sexes are a common ailment. Accidents with machetes and other tools are not infrequent among farmers, while pygmies experience trauma to their feet and other body parts almost routinely as part of their forest foraging existence.

Fertility and Mortality

Africans are known to experience the highest average fertility rates in the world, with each woman in many countries bearing an average of 6.5–8.0 children. In contrast, in forested parts of Central Africa many populations have astonishingly low average fertility, due primarily to high rates of sterility. In many areas, 25–45 per cent of postmenopausal women have had no or just one live birth (Romaniuk, 1967; Voas, 1981; Caldwell and Caldwell, 1983). While there are many possible causes of the high rates of infertility, the most likely is infection with gonorrhoea causing tubal occlusions and blockage (Belsey, 1976).

Many farmer and pygmy populations that are the traditional inhabitants of the Central African rain forest have high rates of infertility and each woman has an average of only 2.5–3.5 children. The more recent immigrants to the forest and the populations on the edge of the forest, on the other hand, tend to have more offspring. Consequently, populations expanding into the forest are growing populations, while the indigenous populations may be declining or, at best, stable. Each pygmy population tends to have a fertility rate similar to the farmer population with which it associates. In many areas, Africans consider pygmies as highly fertile but this is not supported by the evidence at hand.

Infant and child mortality rates of forest-dwelling farmers and pygmies are poorly known. It appears that in areas where fertility rates are low, infant and child mortality rates are surprisingly low (Bailey, 1989), whereas in areas with higher fertility, mortality rates climb. The principal causes of infant and childhood death are infectious and parasitic diseases, including malaria, tuberculosis and amoebiasis. There is no evidence that pygmies differ from farmers in their mortality rates. This suggests that the two groups are equally exposed and susceptible to the same diseases, although this has not been studied systematically (cf. Mann *et al.*, 1962; Hewlett *et al.*, 1986; Dietz *et al.*, 1989).

Protection of the Forests

The pygmies' nomadic way of life is an effective strategy for exploiting the tropical rain forest in a sustainable way and the forest itself is vital to the economic, social and psychological well-being of the indigenous peoples. Protection of forest areas as reserves and parks is not incompatible with the continued presence of forest people (see the Okapi Project case study). Indeed they can enhance efforts to protect forest flora and fauna. Living at low densities as they do, the pygmies are unlikely to over-exploit the resources on which they depend. It should, therefore, not be necessary to remove them from protected areas, nor should there be any need to place severe restrictions on their rights to forest resources.

In the forest areas of Central Africa, tourism is only nascent at present, but it does exist (see case study on tourism in Rwanda in chapter 9) and is sure to increase with the growing popularity of 'ecotourism' and 'ethnotourism' in the developed countries. Often indigenous groups have been permitted to remain in protected areas as long as they remain 'traditional' – a term usually defined by policy makers without consultation with, or extensive historical knowledge of, the peoples themselves. Such restrictions lead to 'enforced primitivism' (Goodland, 1982), whereby tribal people are expected to remain traditional, enhancing their value as a tourist attraction while the rest of the world passes them by. If forest people are made part of tourist strategies rather than being manipulated by those seeking unfair profits, tourism can enhance cultural awareness and the knowledge of ethnic history while avoiding the 'people in a zoo' phenomenon. The durable success of a tourist industry in any Central African country depends upon the enthusiastic participation of indigenous peoples who will be crucial for maintaining the cultural and environmental integrity of the region.

References

Bahuchet, S. (1985) *Les Pygmées Aka et la Forêt Centrafricaine: Ethnologie Ecologique*. SELAF, Paris, France.

Bahuchet, S. (n.d.) *Les Pygmées d'Aujourd'hui en Afrique Centrale*. Unpublished manuscript.

Bahuchet, S. and Guillaume, H. (1982) Aka-farmer relations in the northwest Congo basin. In: *Politics and History in Band Societies*. Leacock, E. P. and Lee, R. B. (eds), pp. 189–211. Cambridge University Press, Cambridge, UK.

Bailey, R. C. (1982) Development in the Ituri Forest of Zaïre. *Cultural Survival Quarterly* **6**(2): 23–5.

Bailey, R. C. (1988) The significance of hypergyny for understanding subsistence behaviour among contemporary hunters and gatherers. In: *Diet and Subsistence: Archaeological Perspectives*. Kennedy, B. V. and LeMoine (eds), pp. 57–65. Calgary University Press, Calgary, Canada.

Bailey, R. C. (1989) *The Demography of Foragers and Farmers in the Ituri Forest, Zaïre*. Paper presented at the 88th Annual Meeting of the American Anthropological Association, Washington, DC, USA.

Bailey, R. C. (1991) The comparative growth of Efe pygmies and African farmers from birth to age five years. *Annals of Human Biology* **18**(2): 113–20.

Bailey, R. C. and Peacock, N. (1988) Efe pygmies of northeast Zaïre: subsistence strategies in the Ituri Forest. In: *Coping with Uncertainty in Food Supply*. de Garine, I. and Harrison, G. A. (eds), pp. 88–117. Oxford University Press, Oxford, UK.

Bailey, R. C., Head, G., Jenike, M., Owen, B., Rechtman, R. and Zechenter, E. (1989) Hunting and gathering in tropical rain forest: is it possible? *American Anthropologist* **91**(1): 59–82.

Belsey, M. A. (1976) The epidemiology of infertility: a review with particular reference to sub-Saharan Africa. *Bulletin of the World Health Organisation* **54**: 319–41.

Caldwell, J. C. and Caldwell, P. (1983) The demographic evidence for the incidence and cause of abnormally low fertility in tropical Africa. *World Health Statistics Quarterly* **36**(1): 2–34.

Cavalli-Sforza, L. L. (ed.) (1986) *African Pygmies*. Academic Press, New York, USA.

Dietz, W. H., Marino, B., Peacock, N. R. and Bailey, R. C. (1989) Nutritional status of Efe pygmies and Lese horticulturalists. *American Journal of Physical Anthropology* **78**: 509–18.

Ehret, C. and Posnansky, M. (eds) (1982) *The Archaeological and Linguistic Reconstruction of African History*. University of California Press, Berkeley, USA.

Goodland, R. (1982) *Tribal Peoples and Economic Development*. International Bank for Reconstruction and Development/World Bank, Washington, DC, USA.

Hall, J. S. (1990) *Conservation Politics in Zaïre – the Protection of Ituri Forest*. Unpublished report to WWF, Gland, Switzerland.

Hart, J. A. (1979) *Nomadic Hunters and Village Cultivators: a Study of Subsistence Interdependence in the Ituri Forest, Zaïre*. University Microfilms, Ann Arbor, USA.

Hewlett, B. S., van de Koppel, J. M. H. and van de Koppel, M. (1986) Causes of death among Aka pygmies of the Central African Republic. In: Cavalli-Sforza, L. L. (ed.), *loc. cit.*, pp. 45–63.

Mann, G. V., Roels, A. O., Price, O. L. and Merrill, J. M. (1962) Cardiovascular disease in African pygmies; a survey of health status, serum lipids and diet of pygmies in the Congo. *Journal of Chronic Diseases* **15**: 341–71.

Romaniuk, A. (1967) *La Fécondité des Populations Congolaises*. Mouton, Paris, France.

Stanley, H. M. (1887) *In Darkest Africa*. London, UK.

Stone, D. (1987) *WWF List of Approved Projects, Volume 4 Africa and Madagascar*. WWF, Gland, Switzerland.

Turnbull, C. M. (1961) *The Forest People*. Triad/Paladin, Grafton Books, London, UK.

Voas, D. (1981) Subfertility and disruption in the Congo Basin. In: *African Historical Demography*. Centre for African Studies, University of Edinburgh, UK, pp. 777–802.

Waehle, E. (1991) The Central Africa rainforest and its inhabitants are under siege. In: IWGIA Yearbook 1990.

Wilkie, D. S. (1990) *Human Settlement and Forest Composition within the Proposed Okapi Rainforest Reserve in Northeast Zaïre*. Project Summary, WWF Project 3249.

Authorship

Robert Bailey of the Department of Anthropology, University of California, Los Angeles; Serge Bahuchet, of the Centre Nationale de Recherche Scientifique, Paris; Barry Hewlett, of the Department of Anthropology, Tulane University, New Orleans, and Damien Lewis, London.

6 Population, Environment and Agriculture

Introduction

A World Bank study of the linkages between rapid population growth, agricultural stagnation and environmental degradation in sub-Saharan Africa shows that these phenomena are mutually reinforcing. Rapid population growth is the principal exogenous factor which has stimulated the increase in environmental degradation, contributing to agricultural stagnation relative to population size. This is because population growth has been such that Africans have been unable to adapt their traditional agricultural land-use and wood-use practices fast enough to respond to the pressure of more people.

For several years there has been concern that Africa's high rate of population growth – now 2.9 per cent (PRB, 1990) – cannot be fed or employed by African economies, which are unlikely to grow by more than 3 or 4 per cent a year. In addition, more people require more land, water and fuelwood, while such resources are finite. On the other hand, agriculturalists and agricultural economists have observed that the intensification of farming occurs as populations grow denser. This has been found to be the case in many developing countries (Boserup, 1965; Binswanger and Pingali, 1989; Lele and Stone, 1989). Only if there is a constraint on land will farmers have an incentive to intensify their agricultural production. If land is free or very cheap, it makes more sense for the farmer to extend his holdings and to minimise the use of other inputs such as capital and labour. Under these conditions, shifting cultivation and nomadic or transhumant (that is, the seasonal movement of animals) livestock raising are sensible methods from the perspective of the farmer or herdsman and these have, indeed, predominated in most of sub-Saharan Africa.

However, these traditional practices of shifting cultivation and livestock husbandry change when populations become more dense. This can be seen in the highlands of Kenya, the Kivu Plateau in eastern Zaïre, in parts of Nigeria and, most particularly, in Rwanda. In the latter country, there is now a relatively intense traditional agricultural system, resulting largely from the scarcity of land. In most of sub-Saharan Africa land has been abundant until recently and in many countries this is still the case. Given this situation, it has been, and continues to be, very difficult to stimulate interest in reducing population growth. African leaders and agriculturalists usually consider that the priority must be to get economies growing faster and agriculture expanding more quickly. If this can be done, then rapidly expanding populations can be fed and incomes can be increased. In most cases, African leaders have resisted recommendations to reduce population growth in their country.

In summary, the traditional structure of the rural economy – its farming and livestock husbandry methods, the dependency on wood for energy and building material, the land tenure arrangements and the burdens of rural women – worked well when population densities were low and populations were growing slowly. However, under stress from high population growth, these traditional practices have led to the degradation of natural resources. This in turn has contributed to agricultural stagnation.

Shifting Agriculture and Pastoralism

For centuries, shifting cultivation and transhumant pastoralism have been appropriate systems for people throughout most of sub-Saharan Africa, enabling them to derive a sustainable livelihood from nature. The ecological and economic systems were in equilibrium. The principle was that people moved on to new land when soil fertility declined or pasture vegetation disappeared. Land left fallow had its fertility reconstituted over many years through vegetative growth and decay. Typically, this involved two to four year cultivation periods and even shorter grazing periods. Land was left fallow for between 10 and 20 years. Adjustments took place as and when they became necessary but the pace of adjustment required was slow. The inter-cropping practised by Rwanda's farmers was an early traditional adaptation which occurred when shifting was constrained by the high population density.

Although new land has been opened up in response to population growth, land for cultivation has become increasingly scarce in most regions of Africa. On average, arable land has declined from 0.5 ha per capita in 1965 to 0.3 ha in 1987, although there are considerable differences between countries. Everywhere, however, fallow periods are gradually being reduced as populations expand and new land becomes scarce. In many countries, such as Kenya, Rwanda and Liberia, fallow periods are no longer sufficient to allow fertility to be restored. Many people are forced to remain on the same parcel of land where they maintain their traditional farming methods. Shifting cultivation usually involves the annual burning of vegetation on newly opened-up land. When farmers are unable to open new land, but continue the annual burning before cultivation, soil fertility quickly declines. In this situation fertility is not restored and crop yields decline. When this becomes very serious, people migrate to marginal farming land in semi-arid areas or into tropical forests and try to establish farms there. Crop yields in these areas are low because soils are less fertile or rainfall is less abundant, or both. These problems are gravest in parts of the Sahel, parts of mountainous East Africa and in the dry belt stretching from the coast of Angola through Botswana, Lesotho and southern Mozambique.

Table 6.1 Food consumption, agriculture, population and the environment. A comparison of present levels and preliminary

Country	*Agricultural production growth rates (per cent per annum)*			*Population growth rate (per cent per annum)*			*Per capita calorie consumption per day*			*Percent of population food insecure*[1]		*Reforestation rates per annum*[2] *(per cent)*	
	1965–73	1980–8	Target 1990–2020	1965–73	1980–8	Target 2020	1965	1988	Target 2010	1980/81	Target 2020	1980s	Target 1990–2020
Total													
Sub-Saharan Africa	2.2	1.8	4.0	2.6	3.2	2.2	2092	2095	2400	25	10	(0.5)	1.0
Sahelian Countries													
Burkina Faso	–	6.4	4.0	1.9	2.6	2.3	2009	2139	2400	32	5	(1.7)	1.5
Chad	–	2.6	3.0	1.9	2.4	2.2	2399	1717	2200	54	20	(0.6)	1.5
Mali	0.9	0.3	2.0	2.1	2.4	2.7	1858	2073	2300	35	20	(0.5)	1.5
Mauritania	(2.1)	1.5	2.0	2.2	2.6	2.6	2064	2322	2400	25	10	(2.4)	1.5
Niger	(2.9)	2.8	3.0	2.3	3.5	3.1	1994	2432	2450	28	5	(2.6)	1.5
Coastal West Africa													
Benin	n.a.	4.2	4.0	2.7	3.2	1.9	2009	2184	2400	18	0	(1.7)	1.5
Cape Verde	n.a.	n.a.	3.0	2.0	2.2	1.6	1766	2717	2800	n.a.	0	n.a.	-
Côte d'Ivoire	4.9	1.6	4.0	4.1	4.0	2.7	2359	2562	2700	8	0	(5.2)	1.5
Gambia	4.5	7.1	4.5	2.8	3.3	2.4	2194	2517	2700	19	5	(2.4)	1.5
Ghana	4.5	0.5	4.5	2.3	3.4	1.9	1950	1759	2400	36	10	(0.8)	1.5
Guinea	n.a.	n.a.	5.0	1.8	2.4	2.4	1923	1776	2400	n.a.	0	(0.8)	1.5
Guinea-Bissau	n.a.	5.7	5.0	1.1	1.7	1.9	1910	2186	2400	n.a.	0	(2.7)	1.5
Liberia	6.5	1.2	4.0	2.9	3.2	1.8	2154	2381	2500	30	5	(2.3)	1.5
Nigeria	2.8	1.0	4.0	2.5	3.3	2.1	2185	2149	2400	17	5	(2.7)	1.5
Senegal	0.2	3.2	4.0	2.3	3.0	2.5	2479	2350	2500	21	0	(0.5)	1.5
Sierra Leone	1.5	1.6	4.0	1.9	2.4	2.4	1837	1854	2400	23	5	(0.3)	1.5
Togo	2.6	4.2	3.0	3.8	3.5	2.1	2378	2207	2400	29	10	(0.7)	1.5
Central Africa Forest Zone													
Angola	0.2	n.a.	4.0	2.0	2.5	2.5	1897	1880	2400	n.a.	5	(0.2)	1.0
Cameroon	4.6	2.4	4.5	2.4	3.2	2.4	2079	2028	2400	9	0	(0.4)	1.0
Central African Rep.	2.1	2.6	4.5	1.5	2.7	1.8	2135	1949	2400	39	5	(0.2)	1.0
Congo	4.1	2.0	4.5	2.5	3.5	2.7	2259	2619	2700	27	0	(0.1)	0.0
Equatorial Guinea	n.a.	n.a.	4.0	1.7	1.9	1.6	n.a.	n.a.	2400	n.a.	0	(0.2)	0.0
Gabon	n.a.	n.a.	4.0	1.9	3.9	2.5	1881	2521	2600	0	0	(0.1)	0.0
Zaïre	n.a.	3.2	5.0	2.3	3.1	2.0	2187	2163	2400	42	10	(0.2)	0.0
Northern Sudanian													
Djibouti	n.a.	n.a.	3.0	8.1	3.0	2.1	n.a.	n.a.	2400	0	0	n.a.	n.a.
Ethiopia	n.a.	(1.1)	3.0	2.6	2.9	3.0	1824	1749	2200	46	20	(0.3)	1.5
Somalia	n.a.	3.9	3.0	2.6	3.0	2.5	2167	2138	2400	50	15	(0.1)	1.5
Sudan	n.a.	2.7	4.0	2.5	3.1	1.8	1938	2208	2400	18	0	(0.2)	1.5
East Africa Mountain and Temperate Zones													
Burundi	4.7	3.1	3.0	1.7	2.8	2.4	2391	2343	2400	26	5	(2.7)	1.0
Kenya	6.2	3.3	4.0	3.4	3.8	1.9	2289	2060	2400	37	10	(1.7)	1.5
Lesotho	n.a.	1.8	3.0	2.1	2.7	1.5	2065	2303	2500	n.a.	0	n.a.	n.a.
Madagascar	n.a.	2.2	4.0	2.3	3.3	1.5	2462	2440	2500	13	0	(1.2)	1.5
Malawi	n.a.	2.7	4.0	2.8	2.8	2.9	2244	2310	2400	24	5	(3.5)	1.5
Rwanda	n.a.	0.3	3.0	3.1	3.8	3.0	1665	1830	2300	24	10	(2.3)	1.5
Swaziland	8.0	3.9	4.0	2.6	3.4	2.0	2100	2578	2600	0	0	0.0	1.5
Tanzania	3.1	4.0	4.0	3.2	3.5	2.3	1832	2192	2400	35	10	(0.3)	1.5
Uganda	3.6	(0.3)	4.5	3.4	3.2	2.7	2360	2344	2500	46	10	(0.8)	1.5
Zambia	2.0	4.1	5.0	3.0	3.7	2.6	n.a.	n.a.	2400	48	10	(0.3)	1.5
Zimbabwe	n.a.	2.5	4.5	3.5	3.7	1.4	2105	2132	2400	n.a.	0	(0.4)	1.5
Other South East Africa													
Botswana	12.4	(5.9)	2.0	3.1	3.4	1.4	2019	2201	2400	n.a.	10	(0.1)	1.5
Comoros	n.a.	n.a.	3.0	2.3	3.6	2.3	2296	2109	2300	n.a.	10	n.a.	n.a.
Mozambique	n.a.	(0.8)	4.0	2.3	2.7	2.3	1979	1595	2200	49	20	(0.8)	1.5
Mauritius	n.a.	4.0	5.0	1.6	1.0	0.5	2271	2748	2900	9	0	0.0	1.5
Comparison													
India	n.a.	2.3	–	2.3	2.2	–	–	2238	–	–	–	–	–
China	n.a.	6.6	–	2.7	1.3	–	–	2630	–	–	–	–	–

1 Defined as percentage who do not have access to enough food for an active and healthy life.
2 A number in brackets means deforestation. The rate is measured as the percentage of total forested area which is reforested (deforested) per annum.
n.a . data not available
Figures in brackets are negative rates

indicative targets for the year 2020

Per capita arable land area (ha)		*Percentage of total land under crops*		*Percentage wilderness area to total area*	
1965	1987	1987	Minimum Target 2020	Present %	Minimum Target %
0.5	0.3	6.6	8.9	27	24.5
0.5	0.4	11	12	3	3
0.9	0.6	3	5	52	45
0.4	0.3	2	5	49	45
0.2	0.1	0	1	74	69
0.6	0.5	3	6	53	45
0.6	0.4	17	19	15	14
0.2	0.1	10	10	0	0
0.6	0.3	11	13	10	10
0.3	0.2	17	23	0	0
0.3	0.2	12	15	0	0
0.4	0.2	6	8	0	0
0.5	0.4	12	14	0	0
0.3	0.2	4	9	17	17
0.5	0.3	34	34	2	2
0.1	0.8	27	29	11	11
0.6	0.5	25	25	0	0
0.7	0.4	26	27	0	0
0.6	0.4	3	5	26	24
1.0	0.6	15	17	3	3
1.0	0.7	3	7	39	32
0.6	0.3	2	5	42	37
0.8	0.6	8	10	0	0
0.4	0.4	2	5	35	30
0.3	0.2	3	5	6	6
n.a.	n.a.	n.a.	n.a.	0	0
0.5	0.3	13	17	22	18
0.3	0.2	1	2	24	22
0.9	0.5	5	9	40	35
0.3	0.3	52	52	0	0
0.2	0.1	4	4	25	25
0.4	0.2	11	11	80	75
0.4	0.3	5	10	2	2
0.5	0.3	25	28	10	8
0.2	0.2	45	45	0	0
0.4	0.2	10	10	0	0
0.3	0.2	6	10	10	10
0.6	0.4	34	34	4	4
1.3	0.7	7	11	24	20
0.5	0.3	7	10	0	0
1.9	1.2	2	4	63	58
0.4	0.2	44	44	n.a.	n.a.
0.3	0.2	4	7	9	9
0.1	0.1	58	58	n.a.	n.a.
0.3	0.2	57	–	–	–
0.6	0.4	11	–	–	–

(*Source:* World Bank statistics)

The crisis is most acute in countries such as Burundi, Ethiopia, Ghana, Kenya, Nigeria, Rwanda and Togo, which already have low per capita arable land and high population growth (see Table 6.1). In these countries, populations are already unsustainably large, soil degradation or desertification are advanced and agricultural productivity is stagnating or declining. Population pressure is causing traditional farmers to intensify production, but, for the most part, this is occurring much too slowly.

There are other parts of the continent in which land appears to be more abundant relative to population. These include Central Africa, humid West Africa and Southern Africa. However, much of the potential arable land in Central Africa and humid West Africa is under tropical forest. A large proportion of West Africa's forests are secondary, having been logged or cut by shifting cultivators in the past. Although less diverse than primary forests, these secondary forests still have considerable biological and economic value. In Central Africa, in contrast, there remain vast areas of primary forest. To preserve biodiversity, maintain rainfall and the humid climate on which its agriculture is based, much of this area should not be cultivated. Instead, the humid forests have to be preserved. Historically, this land provided poor land for cropping. Therefore, even in those countries where land is apparently more abundant, the problem of an expanding population moving to unsuitable land can already be observed. In many cases, it is the tropical forests that suffer from this expansion.

More contentious is the observation that people may be having as many children as possible to provide more labour for farming and for water and fuelwood gathering tasks. In this way, the traditional role of women indirectly may be helping to maintain the extraordinary fertility rate in sub-Saharan Africa. This rate – the total number of children the average woman has in a lifetime – is now about 6.6, compared to four in other developing countries. Of course there are many other factors which contribute to such a high fertility rate. Traditional male attitudes favour large families. Poor health services and high infant mortality also induce people to want large families, perhaps to ensure that some offspring remain to care for them in their old age. This is Africa's social security system. The relative importance of these various factors has not been established and may never be. Nevertheless, the traditional role of women in farming and the constraints under which they exist may be reducing their demand for small families.

Traditional Land Tenure Systems

Traditional land tenure systems provide security of tenure through customary rules of community land ownership and distribution of land to individuals within the community. As population pressures slowly increased, these systems appear in some cases to have evolved to individually managed semi-permanent holdings. Traditional tenure arrangements provided sufficient protection to farmers settling on this semi-permanent basis to persuade them to invest in the land. However, most African governments and most aid donors have mistakenly believed that community ownership of farm land did not provide adequate security of land tenure and that it discouraged farmers from investing in the land. Land markets were difficult to develop under traditional tenure arrangements and ownership was not established as collateral for credit.

The response of many governments has been to nationalise ownership of the land, with the government owning it all. They then allow *de facto* customary law to guide use of some land while arbitrarily allocating other land to private investors, political elites and public projects. This has reduced land tenure security rather than increasing it. Investment by farmers in the land then becomes risky since governments can, and do, reallocate land to serve larger national purposes. Combined with this government intervention has been a slow breakdown in customary law, including that affecting land rights. In many cases, these two factors have caused customary land tenure management to break down entirely and an 'open access' form of tenure occurs in which settlement by anyone is permitted. Open access to forest and pasture areas results in rapid environmental destruction that is akin to the 'tragedy of the commons' (Noronha, 1985; Magrath, 1989). Each individual has an incentive to exploit the land and the resources on it as quickly as possible, moving on when the resources are mined and the land more or less unusable.

A second type of response, shown by governments such as those of Kenya, Zimbabwe and Côte d'Ivoire, has been to distribute individual land titles to address the problem of open access and the perceived deficiencies of traditional land tenure systems. However, land titling has permitted the political and economic elite, who maintain control of the title distribution mechanism, to grab land from traditional owners. Land distribution becomes more skewed as a result. The wealthy few own much land, while the millions of poor have little or none.

Where traditional land tenure systems are allowed to function unmolested, they appear to evolve with population pressure. In particular, they appear to give more individual proprietorship over time. However, this evolution has been slow. Partly because it has been slow, governments have felt compelled to intervene as indicated above. With population pressure, migration and breakdown in respect for customary law, the law regulating land rights is increasingly unable to ensure tenure security. Open access and its destructive consequences occur when traditional land tenure systems break down as well as when governments claim ownership.

Traditional Fuelwood Gathering

The heavy dependency on wood for fuel and building material has combined with population growth to contribute to the increasing rate of forest and woodland destruction. This is exacerbated by destructive commercial logging practices and the inadequacy of forest management. Wood has been treated as a free resource, taken largely from land to which everyone has the right of access. In many countries, because fuelwood can generally be collected free, a market has not developed for it despite its increasing scarcity. Even when extreme shortage of fuelwood has resulted in a market, the price has been lower than social value requires because most supplies come from open access forests. Alternative energy supplies, such as oil, are costly as they are not available in open access areas and are not free for the taking by individuals. Despite dwindling supplies of wood, other fuels are not substituted in significant quantities. Increasing destruction of the forests and woodlands accelerates soil degradation and this negatively affects agriculture. The environmental destruction also results in the loss of plant and animal diversity.

Other Factors

Some of the most obvious constraints on agricultural development include civil wars, adverse policies such as poor price/exchange rate and tax policy, lack of rural infrastructure, falling international prices (true of most of Africa's agricultural exports), lack of private investment in agricultural marketing and processing and poor agricultural research and extension (see World Bank, 1989). By preventing significant gains in agricultural productivity, these factors compel growing populations, as a survival strategy, to exploit ever more extensively the natural resources available to them. The predominance of shifting cultivation, of traditional farming methods, the separation of farming into male and female occupations and fuelwood dependency will continue in these circumstances.

Population Growth

Agricultural stagnation and environmental degradation are probably inhibiting the demographic transition to lower fertility rates because they inhibit economic development which is the motor for this transition. The extraordinarily high fertility rate which characterises Africa is the result of many factors (Cain, 1984; Cochrane and Farid, 1989; Caldwell and Caldwell, 1990). The basic problem is low demand for smaller families. Childbearing enhances the status of both men and women in African societies. Many Africans define themselves spiritually through their ancestors. Having children who will, in turn, revere them, is part of this process. In addition, there is some evidence that the desire for large families is economically inspired. Often, women farmers can only add more children as a way of assisting with farming, wood gathering and water fetching. In some traditional land tenure systems, the amount of land provided for a family by the clan chief is a function of family size or family labour. This also stimulates demand for larger families. High infant mortality rates in this situation encourage people to have more children, to assure that some remain alive. Finally, poor education systems, especially for young women, do not prepare young Africans to make knowledgeable decisions about family planning.

Synthesis

Traditional practices and systems evolve over time. However, in Africa, farming and land management techniques and fuelwood gathering have not adapted to the incredibly rapid rate of population growth. The result is agricultural stagnation and environmental degradation. Table 6.1 provides a quantitative assessment of what has transpired. Since 1965, agriculture in sub-Saharan Africa has grown at about 2 per cent per annum. Population has grown at about 2.8 per cent per annum. More recently, in the 1980s, agricultural growth declined slightly from the long term average, while annual population growth increased to 3.2 per cent. Scrutiny of the table reveals that per capita calorie consumption has stagnated at very low levels and that a large number of Africans (100 million) are food insecure. The food gap (consumption minus production) which is filled by

As wood becomes scarce people, usually women and children, have to travel further to collect it. Here a young girl is carrying firewood on Anjouan, Comoros. I. Thorpe

In most of Africa, as here in Madagascar, a vast proportion of the population are under 15 years old. The rapid population growth is placing increasing demands on resources in the region. C. Harcourt

food aid and imports, or by some people going without, is increasing rapidly. These are the outcomes of the phenomena described above.

Sub-Saharan Africa's forest cover of about 6,600,000 sq. km is disappearing at the rate of 32,000 sq. km per annum. The rate of destruction is increasing with time. Fuelwood consumption is increasing at about the rate of population growth. There are vast areas of soil erosion and up to 80 per cent of Africa's pastures show signs of degradation. Significant declines in rainfall are being recorded in many countries and, although unproven, this is believed to be related to forest and vegetation destruction. This is having a serious impact on crops and on water availability.

Solutions

The outside world's response to these problems is not obvious. Land titling to resolve the land tenure issue has not worked well where it has been tried in Africa. The introduction of modern agricultural technology in the form of high yield seeds, fertilisers and farm mechanisation has met stiff resistance from farmers. Population control programmes, based on the supply of family planning services and the distribution of contraceptives, have not been very successful except in three or four African countries. Soil conservation and forestry protection projects have not had much success either. New approaches are needed.

For each African country, Table 6.1 provides targets for desirable and achievable population growth rates, food consumption, agricultural growth and environmental protection (i.e. protected areas). For Africa as a whole the objective should be to reduce annual population growth to about 2.2 per cent by the year 2020. (Note that the expected impact of AIDS on population growth has been incorporated in the projections. See case study.) Agricultural production must grow at about 4 per cent per annum during the period 1990 to 2020 (World Bank, 1989). Per capita daily calorie consumption should increase from its present level of 2100 to 2400 calories by the year 2010. The percentage of the population who are food insecure should drop from 25 per cent to 10 per cent by the year 2020. The annual reafforestation rate should increase from its present negative 0.5 per cent to a positive 1 per cent. Cropped land should increase from only 6.6 per cent of Africa's total land area to 8.9 per cent. This would permit the maintenance of approximately 25 per cent of sub-Saharan Africa's total land area under extensively used natural vegetation, compared to about 27 per cent today.

There are of course enormous country variations in what is attainable. Targets must be adjusted to each country's circumstances and potential. Nevertheless, the targets are extremely ambitious. However, the solutions suggested below build on the advantages obtainable from reduced population growth, environmental protection and expanded agricultural production. In particular, with expanding agricultural output and higher incomes, population growth rates are likely to fall more quickly. With agricultural intensification and falling population growth rates, environmental protection becomes more feasible.

Reduce population growth The key change is to focus on increasing demand for smaller families rather than emphasising only the supply of family planning services and contraceptives, though that supply must follow demand. To alter the demanded family size, education is required and this must be directed at both men and women. However, a focus on young women may have the greatest payoff since they are relatively neglected in education programmes while it is their decisions regarding childbearing which are the most important. Improving women's education, providing health services and providing fiscal incentives for small families are all means to increase the demand for them. Community groups, stores and pharmacies should all be mobilised in this effort. The present investment in population programmes needs to be multiplied by at least a factor of five. Governments alone will not be adequate. Non-governmental organisations (NGOs) and the private sector must also be brought in. Successful efforts to intensify agriculture, reduce women's work burdens and stabilise land tenure will also, in the long run, stimulate demand for smaller families. Increased economic growth will have the same effect.

Promote sustainable agriculture Farm productivity per unit area must be increased to permit greater output with little increased farming area. However, this increased productivity must be obtained with minimum destruction of the environment. There are numerous environmentally benign agricultural technologies which have been

HIV AND AIDS

By 1 May 1991, the number of acquired immune deficiency syndrome (AIDS) cases reported to the World Health Organisation (WHO) from Africa amounted to 90,646 (WHO, 1991a). WHO estimates that there are, in early 1991, almost six million people in Africa who are infected by human immunodeficiency virus (HIV), most of whom will die over the next five to ten years. So far, the pandemic of HIV has been confined to cities in most of Africa. Indeed, in some major urban centres, between one-quarter and one-third of all men and women aged between 15 and 49 have been infected and AIDS deaths may reduce expected population growth by more than 30 per cent (WHO, 1991b). In addition, recent reports have shown an alarming increase in HIV infection in rural areas. The pandemic of AIDS affects mainly the people in their prime productive and reproductive ages (20–49). In Africa, the pattern of HIV infection covers both men and women and, increasingly, children. It is estimated that one in forty adult men and women in sub-Saharan Africa are infected with HIV (WHO, 1990). Since 1987 the main focus of infection has been in East and Central Africa, but there has recently been a marked increase in cases in West Africa as well (WHO, 1991b).

With the considerable uncertainty about the behavioural and biological determinants of AIDS, future prospects of this disease and its impact on mortality in Africa can be examined only through simulation models. Bongaarts and Way (1989) simulated the mortality implications of AIDS in Africa. By approximating the trends in Uganda, they provide 'high' and 'low' prevalence scenarios for Africa by the year 2000. They estimate that the prevalence of this disease is likely to double in 25 years.

Their projections are as follows:

Projections (Year 2000)	HIV Seroprevalence (per cent of adults)	Aids death rate (per 1000)
High	21.0	12.0
Low	2.7	1.5

In the case of 'high' levels, the mortality due to AIDS would exceed mortality due to 'all other causes' of death in Africa. However, even then the growth rate of the population would be positive, albeit greatly reduced. It is extremely difficult to assess the possible role of this factor for the future population of Africa or of any specific country. Much depends on behavioural changes that might curtail the spread of HIV, and on breakthroughs in the development of a vaccine or any other form of treatment.

What impact would the spread of AIDS have on forest resources? Under the existing conditions of greater spread of AIDS in cities, the trend of urbanisation may decline with more people remaining in rural areas. In such a case, the pressure to find new areas for cultivation would persist. However, if rural areas also begin to be affected substantially by this pandemic, the pressure on the use of forest resources is expected to decline. This is perhaps more likely, for the young adults who might venture into forests in search of new areas for cultivation are also the prime victims of the disease.

developed experimentally on a small scale in sub-Saharan Africa. Examples include contour farming to prevent soil erosion, mulching, minimum tillage, intensive fallowing, crop rotations which assure constant vegetation cover, terracing and embankments, integration of livestock and cropping to maintain soil fertility, agroforestry, integrated pest management and water harvesting. Small scale irrigation has enormous potential in some countries. Behind each of these projects lies a considerable body of agricultural knowledge which as yet finds little application in Africa outside a few NGO projects. These technologies need to be mastered by national agricultural research and extension systems which should adapt them and introduce them more widely to African farmers. The various African national and international (but working within Africa) agricultural research establishments must develop and test such technologies. The knowledge must then be developed and communicated to farmers on a wide scale. This will require greatly improved research and extension services.

Farmers, however, have not demanded the new methods. The reason is that there has been little incentive for individual farmers to introduce such technology in place of traditional agriculture. As Boserup (1965) suggested, as long as there is free land to open up, labour and capital investment in more intensive agriculture is likely to make little sense from the farmers' perspective. Land tenure reform may be the most important factor.

Agricultural intensification on a wide scale, therefore, requires not only better research and extension, but also policies which induce farmers to remain in one place and intensify production. The first such policy must be the protection of forest and pasture areas from cultivation, which requires the creation of parks, reserves and community owned pasture land, as well as avoiding infrastructure investment in forest and pasture areas as this encourages settlement. Intensive farming must also be profitable. In the short term, subsidies for farm inputs needed for the introduction of intensive sustainable agricultural techniques may be necessary while shifting cultivation is taxed (World Bank, 1989). For example, land tenure reform might be financed by governments. Soil conservation efforts, research and extension efforts to introduce the sustainable agriculture technologies might be fully financed by governments. What these measures do is to create an artificial scarcity of cropped land while increasing the profitability of intensification, hence creating a demand by the farmers for intensification.

However, these sustainable technologies are unlikely to be sufficient to permit most African countries to achieve agricultural growth rates of 4 per cent. Improved crop varieties, fertiliser and farm mechanisation technologies will still be necessary. These are perfectly consistent with agricultural intensification. In some cases they may be environmentally damaging, so some trade-off with environmental protection will be necessary. In addition, new crop land will inevitably be opened up by an expanding agricultural population. It would be unrealistic and unnecessary to eliminate this completely. Some opening of new crop land will, in fact, be necessary to achieve the ambitious 4 per cent per annum agricultural growth target.

Land tenure security Given the failures of the past, providing improved land tenure security to encourage farmers to stay in one place and invest in that land will be difficult. Where traditional land tenure systems are evolving to provide greater individual ownership, these systems should be protected by law. Governments should hand over state-owned farm, pasture and forest land to traditional owners, who should be given clear responsibility for conservation and protection. Land tenure reform on a massive scale is urgently required to eliminate 'open access' systems.

Women's time constraint Research, extension, infrastructure and education initiatives are required to reduce women's time constraints. This can be accomplished through the widespread introduction of fuel efficient stoves, energy substitutes for domestic fuelwood, easier access to water by investment in domestic water supplies, the introduction of improved crop husbandry and better agricultural hand tools to women farmers, and the availability of credit and land to women as an incentive for production. These interventions will make women's farming less environmentally destructive, discourage shifting cultivation and increase food production, while reducing the incentive for women to have large families as help from children will no longer be essential.

Fuelwood Large scale investments in fuelwood plantations and tree farming will be needed. These investments will have to be undertaken by individuals, especially farmers, as well as schools, community groups and private enterprises. It must be financially worthwhile to grow and sell fuelwood and poles for building. As populations grow, trees are felled and fuelwood becomes scarcer, thereby creating a market value for the wood. However, this market is developing too slowly. Development will accelerate if cutting in protected forest areas is restricted. Marketing of farm fuelwood should not be restricted, licensed or taxed; nurseries should be established to grow and distribute appropriate species; government research and extension services should incorporate tree farming as a major theme and land tenure reform should provide ownership of forests to farmers who are then more likely to invest in the forest rather than simply mine it for fuelwood. There is need for more efficient wood and charcoal stoves (to reduce demand for wood per person) which can be made and sold by local artisans.

Natural Resource Management and Environmental Protection

It is clear that environmental protection and management is now necessary in every country throughout sub-Saharan Africa. The core of the actions needed, as identified above, is to develop sustainable agriculture, stimulate demand for smaller families, provide land tenure reform, address women's constraints and invest in fuelwood. However, these actions are not sufficient to arrest environmental destruction. Institutional and human capacity must be developed to manage protected areas, and land-use evaluation will be necessary to identify new areas to be protected. Local communities should be allowed to continue traditional use of resources inside protected areas and to collect revenue from their use. Policies regulating and taxing logging should be developed and governments must be capable of implementing these policies, but revenue from logging should be partly redirected to the forest communities. Agriculture and social developments can be provided outside protected areas to attract settlement of people there. Legal frameworks must be established to permit this. Community organisations which manage natural resources must be allowed to evolve autonomously without government management. Other elements of environmental action plans will include watershed protection, establishment of industrial wood plantations to take the pressure off natural forests, management of off-shore and lake fisheries to avoid over-exploitation, prevention of coastal erosion, control of water pollution and environmental monitoring.

Conclusions

There are important linkages between agricultural production, population growth and environmental protection. Protection of the environment is needed for the long-term growth of agriculture and the economy but will be very hard to achieve if present rates of population increase continue. Population growth is unlikely to decline unless agriculture, and the economies dependent on agriculture, grow. Agriculture, in turn, is increasingly constrained by rapid population growth.

Aid donors and governments need to undertake planning and analysis in a more multi-sectoral context. Agricultural sector work and sector planning, in isolation from population planning and environmental policy, do not make sense. However, greater focus on the issues raised here should not suggest that the more traditional issues such as agricultural policy, better governance and the creation and dissemination of improved agricultural technology are not important. The issues interact in a very complex manner and it is important to understand and act upon this.

References

Binswanger, H. and Pingali, D. (1989) *The Evolution of Farming Systems and Agricultural Technology in Sub-Saharan Africa.* Discussion Paper 23, Agriculture and Rural Development Department, World Bank, Washington, DC, USA.

Bongaarts, J. and Way, P. (1989) *Geographic Variation in the HIV Epidemic and the Mortality Impact of AIDS in Africa.* Proceedings of the International Population Conference. New Delhi, India.

Boserup, E. (1965) *The Conditions of Agricultural Growth: The Economics of Agrarian Change under Population Pressure.* George Allen and Unwin Ltd, London, UK.

Cain, M. (1984) *Women's Status and Fertility in Developing Countries.* World Bank Staff Working Papers No. 682. World Bank, Washington, DC, USA.

Caldwell, J. and Caldwell, P. (1990) High Fertility in Sub-Saharan Africa. *Scientific American* **262**(5): 82–9.

Cochrane, S. and Farid, S. (1989) *Fertility in Sub-Saharan Africa: Analysis and Explanation.* World Bank Discussion Papers No. 43. World Bank, Washington, DC, USA.

Lele, U. and Stone, S. (1989) *Population Pressure, the Environment and Agricultural Intensification: Variations on the Boserup Hypothesis.* Discussion Paper, World Bank, Washington, DC, USA.

Magrath, W. (1989) *The Challenge of the Commons: The Allocation of Nonexclusive Resources.* Environment Department Working Paper No. 14, World Bank, Washington, DC, USA.

Noronha, R. (1985) *A Review of the Literature on Land Tenure Systems in Sub-Saharan Africa.* Discussion Paper 43, Agriculture and Rural Development Department, World Bank, Washington, DC, USA.

PRB (1990) *1990 World Population Data Sheet.* Population Reference Bureau, Inc. Washington, DC, USA.

WHO (1990) *Update: AIDS Cases Reported to Surveillance, Forecasting and Impact Assessment Unit (SFI) of Global Programme on AIDS.* 1 August 1990. World Health Organisation, Geneva, Switzerland.

WHO (1991a) *Update: AIDS Cases Reported to Surveillance, Forecasting and Impact Assessment Unit (SFI) of Global Programme on AIDS.* 1 May 1991. World Health Organisation, Geneva, Switzerland.

WHO (1991b) The Global HIV/AIDS Situation. *In Point of Fact* No. 74. World Health Organisation, Geneva, Switzerland.

World Bank (1989) *Sub-Saharan Africa: From Crisis to Sustainable Growth.* World Bank, Washington, DC, USA.

Authorship

Kevin Cleaver, World Bank, Washington, DC with contributions from Gotz Schreiber, World Bank, Washington, DC and Per Ryden, IUCN, Gland, Switzerland.

7 Timber Trade

Introduction

Southeast Asia currently dominates the world's international tropical timber trade. However, tropical timber harvesting has a much longer history in Africa. As early as 1672, having obtained its first charter from the English monarch Charles II, the Royal African Company was trading in West African mahogany (*Khaya* and *Entandrophragma* spp.). By the early 18th century, imports which had previously emanated from the Caribbean were being replaced by those from the coastal forests of French West Africa and Gambia. In 1823, for example, shipbuilders in Liverpool in the UK were importing African oak, known in the trade as iroko (*Milicia excelsa*). By the 1880s, the British were exporting African mahoganies from the Gold Coast (present day Ghana) and Nigeria to Europe.

With the advent of European colonialism at the end of the 19th century, the trade received an important boost. Latham (1957) describes Miller Bros – a British firm, which originally dealt in West Indian exports – establishing itself in the 1890s in Benin City in Nigeria to ship mahogany to Liverpool. By the end of the 19th century British export houses in the Gold Coast and the Niger delta operated forest concessions instead of simply buying up logs from native producers, and both British and French interests harvested ebony (*Diospyros* spp.) in French West Africa. Many hitherto unknown species came on stream at the turn of the century, for instance, okoumé (*Aucoumea klaineana*), makoré (*Tieghemella* spp.) and, to a lesser extent, utile (*Entandrophragma utile*). European tropical timber imports rose significantly during this period. In France, for instance, they increased from 9000 tonnes in 1890 to 110,000 tonnes in 1913.

After 1945 the tropical timber trade evolved rapidly, partly facilitated by increased mechanisation of logging operations. Once a relatively minor import for specialist purposes, tropical timber became a major source of utility grade lumber, controlled principally by European interests (Nectoux and Dudley, 1987). Although a larger range of timbers began to be imported into Europe, the trade still concentrated on a few well-known species, including African mahoganies, obéché (*Triplochiton scleroxylon*) and okoumé. Between 1950 and 1960, African tropical hardwoods became increasingly important to the European timber trade. The proximity of European markets was a key factor, but also the timbers had superior technical properties and were cheap and fashionable. In addition, African timber became available at a time when hardwoods were in short supply in some European countries.

By 1962, nearly 30 per cent of European tropical timber imports came from West Africa. But by 1984 this percentage had shrunk by half due to the opening up in 1965 of forests in Indonesia and East Malaysia where more homogeneous stands of desirable tree species yield four to five times more timber per unit area than those in West Africa. A lack of investment in Africa, due to the absence of a strong domestic private sector, was also a factor. In addition, with little domestic processing infrastructure, Africa was less able to take advantage of new techniques for producing plywood, fibreboard and other manufactured products.

Today, the export trade in African timbers still involves mainly West and Central African countries. A loose distinction can be drawn between countries whose forests have already been exploited for timber (or over-exploited, as in the case of Nigeria which became a net timber importer in the 1970s), and those with large areas of undisturbed forest, such as central Zaïre. Gabon is a major exporter and, although its coastal forests have been heavily logged, it is still 85 per cent forested (see chapter 19).

Whether or not a country's forests are exploited depends more upon forest accessibility, proximity to ports and timber quality, than on specific government forestry policies. In Zaïre, the Zaïre River and its tributaries provide a relatively easy means of moving timber to Kinshasa, but thereafter it must be transported overland to the ports at Matadi and Boma if it is to be shipped overseas. Timber harvesting to date has thus been quite limited. In Gabon, which until 1980 possessed only 100 km of paved roads, exploitation was concentrated in the coastal zone, but the opening of the trans-Gabonese railway has allowed logging to take place over extensive areas of the interior forests in recent years.

East African trade in wood products is of little significance globally. However, special purpose timbers such as African blackwoods (for example, *Dalbergia melanoxylon*, which is used for making musical instruments) are exploited from drier forest types. Exports are of only marginal significance to both the countries of origin and the recipients.

Species in Trade

Export shipments of African logs were initially restricted to cabinet woods such as sapele *Entandrophragma cylindricum*, sipo or utile, the mahoganies (for example, *Khaya ivorensis*) and iroko. Qualities sought by early traders included dimensional stability, decorative figure and ease of working. European markets were looking for alternatives to depleted supplies of West Indian mahogany (*Swietenia* spp.), oak or teak *Tectonia grandis*. After the Second World War the list expanded to include white woods of relatively low density such as obéché, the terminalias and opepe *Nauclea diderrichii*.

Regional variation in species extraction is not uncommon. For instance, *Ceiba pentandra*, a low quality veneer timber, is exploited in the coastal zone of Côte d'Ivoire, but not inland since transport costs would render its cutting unprofitable (Poore, 1989). Generally, remote forests are exploited more selectively. Although more diverse in terms of timber characteristics than the timbers coming from the Far East, the number of African species traded has been, and continues to be, small considering the number and variety of species found in African rain forests. Only 15 or so African timber species are well established in European markets.

For most countries, one to three well-established species account for 50–80 per cent of the total log volume harvested and have done so since commercial logging began. In French Equatorial Africa okoumé was the principal species shipped from present day Gabon and limba *Terminalia superba* was exported from present day Congo. Limba was the only species exported from the Belgian Congo (now Zaïre) until the 1950s. In Zaïre, gedu nohor *Entandrophragma angolense* is now one of the principal species exploited for timber (UICN, 1990a). Today in Cameroon, 70 per cent of timber production is accounted for by about 15 species, out of a possible 300 usable species. In 1989 in Ghana, obéché accounted for 56.6 per cent of log exports and 41.8 per cent of lumber exports. In Liberia, although over 40 species are exported, five make up 63 per cent of production. Between 1986 and 1987, one species alone, niangon *Heritiera utilis* accounted for 45 per cent of total production. In Congo about 50 species are used in trade when more than 300 are potentially marketable.

International statistics on the volumes of individual species shipped by destination are notoriously unreliable. Of the five major species imported into Europe, okoumé has a clear market advantage in France (imported in large quantities from Gabon), Greece and the Netherlands. Obéché is imported in large quantities by Germany (mainly from Ghana). Both species are imported for the manufacture of plywood. Okoumé is favoured for face veneer and obéché for the core. The other species are used for both sawnwood and veneer. Niangon, principally from Liberia, is preferred for internal joinery work in France. Liberia is also the sole producer of gola *Tetraberlinia tubmaniana* exported particularly to Turkey for use in plywood. The Netherlands is responsible for significant trade in azobé *Lophira alata*. This species accounts for almost 80 per cent of the total tropical log imports of the country. A heavy, durable wood, imported mainly from Cameroon, it is used in sea defence works, an example of a species being imported for a specialist use.

The Timber Committee of UNECE/FAO has compiled a table (Table 7.1) showing percentages of total imports by several European countries: France, Italy, Federal Republic of Germany, Spain, Greece and Netherlands.

Table 7.1 Sawnwood imports to Europe declared from tropical Africa by trade name

Per cent of total imports	*Trade name*
15–20	Obéché
10–15	Sapele and Utile
5–10	Iroko, Makoré and Mahogany
2–5	Afrormosia, Agba, Azobé and Framiré
1–2	Afzelia, Afara, Niangon, Tiama, Guarea and Tchitola
<1	Okoumé, Bété, Ilomba, Abura, Antiaris, Fromager, Kosipo, African walnut and Ozigo

(*Source:* UNECE/FAO, 1990)

This tree is being felled within a logging concession in Tano-Ehuro Forest Reserve in Ghana. The company involved, African Timber and Plywood (Ghana) Ltd, is developing a programme for sustainable logging in this and other areas of Ghana. P. Chambers

To protect their timber industries, some tropical governments have introduced log export bans for selected species. Ghana announced recently that in addition to imposing export levies, from January 1994 it will curtail and eventually ban exports of unprocessed timber. Levies applied to, among others, afrormosia, utile and makoré are designed to ensure sustainability and encourage value-added production of timber products in Ghana, although the action may be too late for afrormosia which has an estimated commercial resource life of two to three years.

Lesser-known species

Log export bans combined with tax incentives have been used to try and stimulate the interest of loggers in what are termed lesser-known species. So far, Côte d'Ivoire, assisted by the French Technical Centre for Tropical Forestry (CTFT) has tried the hardest to increase exports of these species. Efforts in this direction are also now being made by Liberia and Cameroon.

In 1979, 53 per cent of Côte d'Ivoire's timber exports (1.47 million cu. m) comprised 25 lesser-known species not exported before 1973. The harvesting of a greater range of species enables more efficient use to be made of logging infrastructures but it results in more severe disturbance of the forest. It can also lead to pressure to re-enter logged coupes, resulting in poor regeneration of more valuable species and non-recovery of the forest after the first cut. Improved operational management is therefore crucial if lesser-known species are to be promoted on a significant scale. In general, there has been little development in the use of lesser-known species of African tropical hardwoods. Consumers tend to buy what is stocked and individually have little influence upon source, species or grade.

Forest Policies and Economics

African governments tend to treat their countries' forests simply as a source of revenue and foreign currency. They may have little alternative given their high external debts. Declining prices of other export commodities, particularly oil, cacao and coffee have recently exacerbated the situation. In 1988, for example, Ghana needed almost 60 per cent of its export earnings to service its debts. Yet African governments have often failed to obtain a reasonable percentage of the financial benefits accruing from timber harvests; certainly not enough to offset the ecological, economic and social costs of logging. Most profits have gone instead to loggers and timber traders.

Under well regulated management the forests of the Central African countries could produce high sustainable yields of timber. However, this will be possible only if forest services are greatly strengthened and the infrastructure of the forested regions, especially roads, railways and ports, is improved. Expanding timber production without meeting these prerequisites will be a recipe for an anarchic timber industry with large windfall profits for a few individuals and major resource depletion for the country. Cameroon has a production target of 4 million cu. m by the year 2000 and 5 million cu. m by 2010 (Gartlan, 1989), Congo one of 2 million cu. m by 2000 (UICN, 1990). Zaïre at one stage discussed a target of 6 million cu. m by the year 2000.

It is difficult to judge who really profits from the timber industry. The term 'rent' is used by economists for the difference between the price obtained for a product and its total production costs. Repetto (1988) calculated the rents for different categories of timber in a number of African countries at various times in their recent history (Table 7.2). A rational allocation of these rents would allow a reasonable profit to the loggers and traders with the balance being retained by the government to invest in development or to compensate for the external costs resulting from logging. In reality most of the rents are kept by the loggers. The latter argue that poor infrastructures, corruption and politically inspired disruption of their operations cause them to incur costs which do not appear in the abstract calculations of economists.

Table 7.2 gives some indication of the high rents available to loggers in Africa during the 1970s. The figures for Côte d'Ivoire are especially interesting since the high rents were in spite of export taxes which ranged from 25 per cent to 45 per cent of the value of the timber. Now, with the more valuable forest resources depleted, rents and thus profits per cubic metre in Côte d'Ivoire are much lower than those of countries which still hold large forest reserves.

The contributions made by the forest-based sector to different countries' economies vary considerably, as a brief look at Côte d'Ivoire, Gabon and Ghana shows. (For a full account, see Repetto, 1988; Repetto and Gillis, 1989.) In Côte d'Ivoire, up until 1981, value added in the forest sector was steady at about 6 per cent of GDP. In 1973, logs and wood product exports were worth 35 per cent of total export earnings; but this figure fell to only 11 per cent in 1980 due to a rapidly declining timber resource. Until 1984, forest products were the third main source of export revenue, a place now occupied by petroleum. In 1980, more than 5 million cu. m of industrial roundwood were extracted, but this had dropped to about 3 million cu. m by 1988 (FAO, 1990 and see Figure 7.1). Meanwhile, tax incentives favour continued logging and further reduce government income from the forestry sector. French logging companies, which are not taxed by the French government for income earned abroad, have been favoured by these incentives which appear to run counter to the long-term interests of Côte d'Ivoire itself.

In Gabon, as late as 1963, the forest sector accounted for 80 per cent of exports, thereafter ceding its place to oil and uranium. It remains a major source of employment – 28 per cent of the labour force works in logging and wood processing activities – although it brings in only 1–2 per cent of government revenues. In the mid-1970s, Gabon's rent capture policies were among the weakest in Africa. Even now, timber is only lightly taxed; export taxes are 20 per cent of export prices for logs and 12 per cent for sawn timber and other processed products. These are among the lowest timber taxes in the tropical world. Even lower rates apply for some species and concession fees are also low. As in Côte d'Ivoire, tax exemptions and tax holidays further reduce forest income potential for the government.

For Ghana, between 1974 and 1984, the forest sector contributed between 4.9 per cent and 6.2 per cent to the country's GDP. Although GDP overall declined, the forest sector's contribution to it remained constant. Estimates for the years preceding 1971 suggest that rents ranged from 26 per cent to 80 per cent of the value of log output. However, the Ghanaian government captured only 38 per cent of the timber rent available. Before 1965, timber exports provided up to 20 per cent of total export earnings, but this declined drastically thereafter to 13 per cent in 1972 and 4 per cent in 1980. Recent moves to stamp out corrupt trade practices aim to increase government revenues from timber harvesting.

One of the problems with forest fees in Africa is that they are highly complex. A 1988 IIED study of a Central African country, revealed that 53 procedures must be followed before a log can be exported (IIED, 1988). This sort of bureaucracy results in excessive paperwork and delays and encourages evasion. As a result, fees are forfeited by governments. In addition, because procedures for raising forest fees are cumbersome, often requiring new legislation,

Table 7.2 Theoretically available profits in the 1970s from log harvesting (US$ per cu. m)

	Highest	*Middle Valued Species*	*Lowest*
1979			
Liberia	98	41	
1973			
Liberia	89	58	25
Côte d'Ivoire	47	31	17
Gabon	89	54	22
Cameroon			
(a) Douala	61	32	14
(b) Pointe Noire*	52	23	7
Congo			
(a) South	81	52	23
(b) North	69	42	13
1971–2			
Ghana	79	28	

* Exports via Pointe Noire, Congo.
(*Source:* Repetto, 1988)

the temptation for governments is to leave them as they are, with the result that inflation reduces their value in real terms. Moreover, forest fees are often collected only partially or not at all. The task generally falls within the remit of the Ministry of Finance, but lack of incentives means that under-collection is common. The World Bank has estimated that in Congo only about one-fifth of volume-based forest revenue was collected and that in Ghana, the government was receiving a mere sixth of what was due to it in forest fees. It suggests that until the forestry departments of West and Central African countries can be strengthened, a simpler system of forest fees which places greater emphasis on concession fees, would be more efficient. For example, an annual concession fee, set by competitive bidding, could become the major source of revenue.

The weakness of forest departments is a major problem. Salaries are so low that employees are easily tempted to accept bribes for approving logging plans they have not even seen, or for accepting volume return forms filled out by company scalers without verifying them. Logs may be underscaled, underreported or misclassified. Transfer pricing is widely practised: logs or finished products are sold supposedly at one price, while the buyer in the importing country actually pays a higher price, the difference frequently being paid into the exporter's foreign bank account. The African government therefore loses out on export taxes. In Gabon, although standing timber sales are provided for in the legislation, the forestry department is unable to mark the trees or control cutting, and so the system has not been implemented.

Corruption plays a part at government level too. In Zaïre the government introduced new regulations in 1985 in a bid to increase value-added income. Exporters must now obtain licences to operate concessions and these are supposed to be given only to applicants with established or intended processing capacity and are thereafter subject to quotas. There has, however, been only limited success in applying the regulations and political interference with applications for export licences is said to be rife.

It is essential that tropical governments secure maximum revenue from the harvesting of their timber resources. Proper pricing policies which reflect the full value of forest products can encourage efficient forest management and conservation. Without effective pricing, the revenue to finance the management and protection of forests is not generated and so they are exploited rather than managed. Countries such as Congo and Zaïre which still possess large untouched forest areas could benefit enormously from a substantial overhaul of the management and economics of timber harvesting.

Figure 7.1 Production of industrial roundwood in Côte d'Ivoire 1977–88 *(Source: FAO, 1990)*

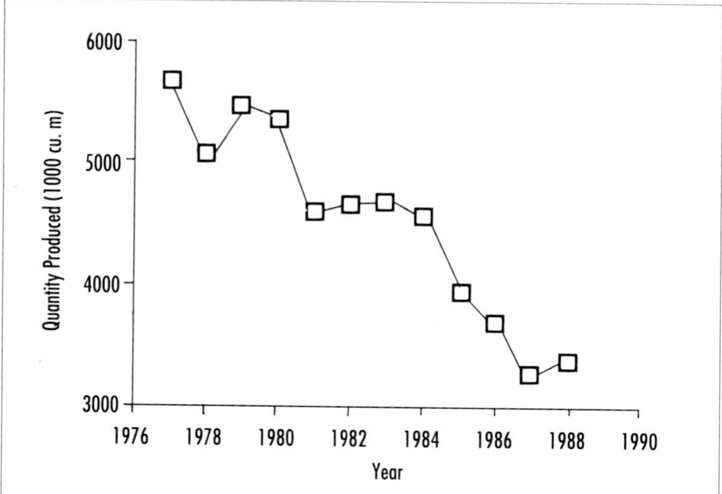

In recent years the volume of timber exported from much of Africa, even from regions with abundant timber resources, has declined. Many timber companies are closing down and some have gone bankrupt. This does not accord well with the view that huge profits are to be made from logging. The situation is highly complex. It may be true that large gains have been made at certain periods in the past and this may have encouraged some inefficiency in the industry. But in the depressed economic conditions of the late 1980s and early 1990s the timber industry is finding it difficult to survive, it feels its future is threatened and is disinclined to invest. A vicious circle ensues with the result that African countries are deprived of the benefit of a resource which could do much to fuel their economies. A consequence of this present insecurity is that many of the old established companies are leaving Africa. These companies have considerable expertise in forestry, provide employment to many Africans and in many cases have made a valuable contribution to economic and social development of the societies in which they have worked for many decades. The responsible companies are being replaced by speculators whose only interest is in reaping maximum profits for as long as the resources last, but who make no long-term commitment to the countries where they operate. Africa still has immensely valuable forest resources but in very few countries is there a solid base for the rational and sustainable development of these resources.

Domestic Processing and Value Added

Many industrialised countries, notably those of the EC, Japan, Korea, Taiwan and the USA, have erected trade barriers against processed wood products in order to protect their own wood processing industries (Nectoux and Kuroda, 1988). To counteract these, producer countries have sometimes banned log exports, reduced or eliminated taxes on processed wood exports and offered incentives to domestic forest product industries (Repetto, 1988).

Ghana has done all three of these and by 1982 had 95 sawmills, ten veneer and plywood plants and 30 wood-processing plants. However, log export bans were not enforced and removing export taxes for sawn logs meant little in the light of the country's currency overvaluation. In 1987, wood exports still consisted of 60 per cent logs and only 30 per cent sawn timber. Similarly, although subsidised credit for local investment in sawmills and plymills and income tax incentives for investment in processing plants had the desired impact – by the late 1960s there was substantial investment – wood-processing industries operated inefficiently.

In Liberia value added from timber processing has been a rising percentage of GDP since 1973. In 1977 two policies were introduced to promote forest-based industrialisation: a big increase in the Industrialisation Incentive Fee (IIF) for log exports and enactment of a Forest Products Fee (FPF) for sawnwood exports. The IIF is applied to all exported logs, ranging from US$2–4 for low-value species, to US$75 for high-value species. The FPF similarly differentiates between species and favours domestic processing.

These measures can be counterproductive if the domestic industry is inefficient. For instance, in Cameroon 3 cu. m of raw logs are used to produce 1 cu. m of sawnwood, equivalent in export value to only 2 cu. m of logs. In Côte d'Ivoire timber concessionaires have erected plymills in order to qualify for log export quotas. They also trade the quotas among themselves. However, their conversion ratios are often as low as 40 per cent, and because *ad valorem* plywood export taxes are only 1–2 per cent, as opposed to the 25–45 per cent imposed on logs, the government forfeits a considerable amount of revenue. Exports of processed wood from Cameroon remain low in comparison with exports of unprocessed logs (Figure 7.2).

TIMBER SPECIES

The following is a list of the most common tree species in the African timber trade, with their trade names, Latin names and distribution.

Abura/*Mitragyna ciliata*— found mainly in West Africa from Sierra Leone to the Congo region and Angola.

Afara or limba/*Terminalia superba*— widely distributed from Sierra Leone to Angola and Zaïre.

African blackwood/*Dalbergia melanoxylon*— an extensive range in savanna regions from Sudan southward to Mozambique, westward to Angola, and then northward to Nigeria and Senegal.

African ebony/*Diospyros* spp.— found in Equatorial West Africa.

African mahogany/*Khaya grandifoliola*, *K. senegalensis*, *K. ivorensis* and *K. anthotheca*— found in tropical West Africa from the Guinea Coast to Cameroon and extending eastward through the Congo basin to Uganda and parts of Sudan. The name mahogany is commonly used for a range of dark red timber from various species of Meliaceae in the genera *Khaya* and *Entandrophragma*.

African walnut or lovoa or tigerwood/*Lovoa trichilioides*— found in tropical West Africa, from Sierra Leone to Gabon.

Afrormosia/*Pericopsis elata* syn. *Afrormosia elata*— found in West Africa, but mainly in Ghana and Côte d'Ivoire.

Afzelia/*Afzelia* spp.— found in West, Central and East Africa.

Agba/*Gossweilerodendron balsamiferum*— found in tropical West Africa, from Nigeria southwards to the Congo Basin.

Antiaris/*Antiaris* spp.— found in West, Central and East Africa.

Azobé or ekki/*Lophira alata*— found in West Africa and extending into the Congo Basin.

Bété/*Mansonia altissima*— found from Côte d'Ivoire to Cameroon.

Ceiba or silk-cotton-tree/*Ceiba pentandra*— also known as fromager, enia, odouma, etc. Widely distributed in West Africa.

Gedu nohor/*Entandrophragma angolense*— also known as tiama, kalungi, etc. From West, Central and East Africa.

Gola/*Tetraberlinia tubmaniana*— found in Liberia.

Guarea/*Guarea cedrata* and *G. thompsonii*— the range of both species overlaps in Côte d'Ivoire, Ghana and southern Nigeria. The former reaches into Cameroon, the latter into Liberia.

Idigbo/*Terminalia ivorensis*— also known as, for example, framiré in Côte d'Ivoire, and emeri in Ghana. Found in tropical West Africa from Guinea to Cameroon.

Ilomba/*Pycnanthus angolensis*— found in West Africa.

Iroko/*Milicia excelsa* and *Chlorophora regia*— also known as odoum, oroko, kambala, etc. The two species between them extend across the entire width of tropical Africa.

Kosipo or sipo/*Entandrophragma candollei*— found in West Africa to Angola and the Congo region.

Limba or afara/*Terminalia superba*— widely distributed from Sierra Leone to Angola and Zaïre.

Makoré/*Tieghemella heckelii* and *T. africana*— both species are found from Sierra Leone to Cameroon, Gabon and south to Cabinda.

Niangon/*Heritiera utilis* and *H. densiflora*— found in West Africa from Sierra Leone to Ghana (*H. utilis*), Cameroon and Gabon (*H. densiflora*).

Obéché/*Triplochiton scleroxylon*— widely distributed in tropical West Africa from Guinea to Cameroon.

Okoumé/*Aucoumea klaineana*— distribution confined to Gabon, Rio Muni and Congo-Brazzaville.

Opepe/*Nauclea diderrichii* syn. *Sarcocephalus diderrichii*— widely distributed from Sierra Leone to the Congo region and eastward to Uganda.

Ozigo or adjouaba/*Dacryodes* spp.— found in West Africa.

Sapele/*Entandrophragma cylindricum*— distribution ranges from Côte d'Ivoire to Cameroon and eastward through Zaïre to Uganda.

Tchitola/*Oxystigma oxyphyllum*— occurs in tropical West Africa from Nigeria to Gabon and the Congo region.

Utile/*Entandrophragma utile*— known as sipo in Côte d'Ivoire; occurs principally in West and Central Africa.

Source: Chudnoff, 1984

Figure 7.2 Exports of processed and unprocessed wood from Cameroon 1977–88 (*Source:* FAO, 1990)

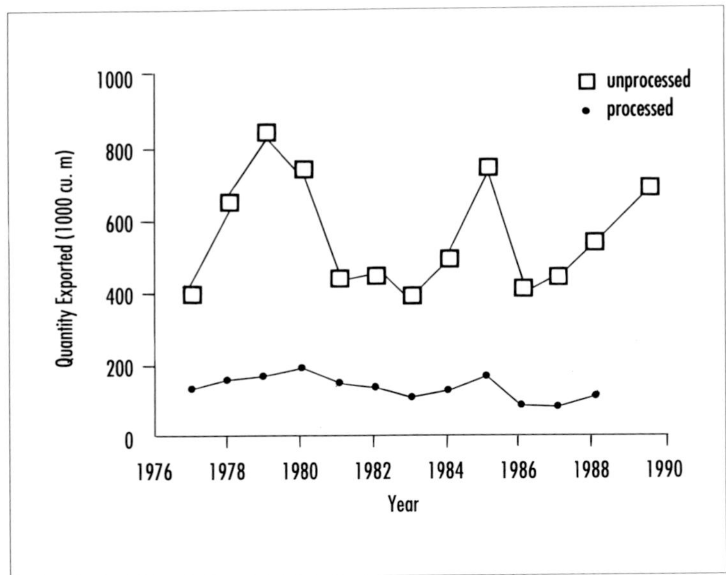

Figure 7.3 Exports of processed and unprocessed wood from Gabon 1977–88 (*Source:* FAO, 1990)

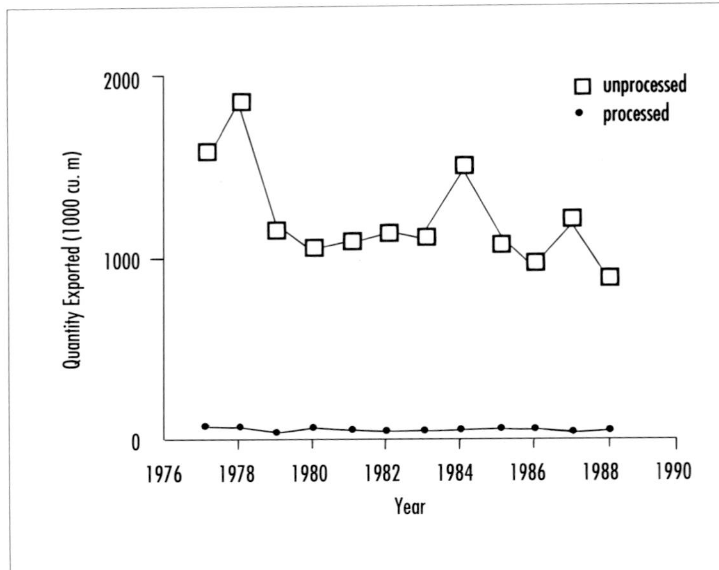

Sawmills like this one at Akim Oda in the eastern region of Ghana are found throughout the forest zone of the country. Wood is cut for export, for furniture, plywood and veneers.
I. and D. Gordon

Gabon has so far instituted neither log export quotas nor bans. The government stipulates that concessionaires with contract areas exceeding 150 sq. km deliver 55 per cent of their harvest to the local timber-processing industry. However, the high percentage of unprocessed logs that are exported suggests that this policy is not rigorously enforced. Gabonese mills are not efficient enough to compete in world markets. As recently as 1984, 89 per cent of Gabonese timber exports were in log form (see Figure 7.3).

Potentially, African countries could increase revenue from their timber resources if they were to make a successful transition to producing more value-added products such as furniture or building parts. Many are interested in doing so. Problems will lie in the region's lack of kiln drying and preservation capacity, and the absence of the technical and management skills needed to produce more complex products. Additionally, individual countries will have to conduct their own promotion aimed at architects, builders, furniture makers and DIY outlets in order to raise the profile of and demand for their products. They will also need to assure a steady supply of timber with which to fuel these value-added industries.

It is possible to improve trade between the exporters and importers. For instance, in 1988 Ghana held a Timber Products Show in London, with the aim of stimulating business between Ghanaian and British wood manufacturing companies. A year later the UK Upton Grey Agency reported deliveries of mahogany furniture blanks to a London furniture agency and plans further shipments worth £20 million annually. Upton Grey attributes the breakthrough to having its own staff on the factory floor in Ghana to advise on quality control and tooling (*Tropical Timber Trades Journal*, various issues). Industrialised countries could assist such efforts by reducing import duties for value-added timber products exported from developing countries.

References

Chudnoff, M. (1984) *Tropical Timbers of the World.* United States Department of Agriculture, Washington, DC, USA.

FAO (1990) *FAO Yearbook of Forest Products 1977–1988.* FAO Forestry Series No. 23 and FAO Statistics Series No. 90. FAO, Rome, Italy.

Gartlan, S. (1989) *La Conservation des Ecosystèmes forestiers du Cameroun.* IUCN, Gland, Switzerland and Cambridge, UK.

IIED (1988) *Zaïre Forest Policy Review: Summary Report.* IIED, Washington, DC, USA.

Latham, B. (1957) *Timber: Its Development and Distribution – A Historical Survey.* Harrap and Co, London, UK.

Nectoux, F. and Dudley, N. (1987) *A Hardwood Story, Europe's Involvement in the Tropical Timber Trade.* Friends of the Earth, London, UK.

Nectoux, F. and Kuroda, Y. (1988) *Timber from the South Seas: An Analysis of Japan's Tropical Timber Trade and Its Environmental Impact.* WWF International, Gland, Switzerland.

Poore, D. (ed.) (1989) *No Timber Without Trees.* IIED/Earthscan, London, UK.

Repetto, R. (1988) *The Forest for the Trees: Government Policies and the Misuse of Forest Resources.* World Resources Institute, Washington, DC, USA.

Repetto, R. and Gillis, M. (eds) (1989) *Public Policies and the Misuse of Forest Resources.* Cambridge University Press, New York and Cambridge, UK.

Tropical Timber Trades Journal, Volumes 1-5.

UICN (1990a) *La Conservation des Ecosystèmes forestiers du Zaïre.* Basé sur le travail de Charles Doumemnge IUCN, Gland, Suisse et Cambridge, Royaume-Uni 242pp

UICN (1990b) *La Conservation des Ecosystèmes forestiers du Congo.* Basé sur le travail de Philippe Hecketsweiler IUCN, Gland, Suisse et Cambridge, Royaume-Uni 187pp.

UNECE/FAO (1990) *Study of the Trade and Markets for Tropical Hardwoods in Europe.* Report by UNECE/FAO Agriculture and Timber for ITTO.

Authorship

Jacqueline Sawyer, IUCN, Gland, Switzerland with contributions from H. Stoll, Bremen, Germany, Chris Elliot, WWF-International, Gland, Switzerland and Peter Burgess, Suffolk, UK.

8 Forest Management

Introduction

Why is natural forest management for timber considered so important? Surprisingly, the most significant contributions of moist forests to the African national economies, are the non-industrial goods and benefits derived from trees and forests, rather than the timber. These include the best upland soils, a variety of foods including bushmeat (often accounting for a major share of the animal protein intake of the rural population), fuelwood and charcoal, framing, panelling and thatching materials for rural houses, agricultural and household implements and a host of environmental and other benefits (FAO, 1989). Hardly any tropical timber is in reality supplied by managed natural forest in Africa at the moment. Rather, the timber derives from forests which are being logged without a management plan, or are being converted to another use in either a planned or an unplanned manner.

However, the latter sources of timber are expected to decline dramatically. In West African countries nearly all moist forest has already been logged over at least once; in Central African countries there are still unlogged forests in remote areas, but accessible forests have been cut over several times. As timber exploitation in many of the remote forest areas is economically unattractive because of transport costs, tropical timber will in the future be derived from managed natural forest, or from managed secondary regrowth, agroforestry and plantations (Grainger, 1987), or most probably, from a combination of both.

Natural forest management for sustained timber production can be wholly or largely compatible with soil and water conservation, with production of most non-timber forest products and, to a certain extent, with nature conservation, depending on the intensity of management practised. This is especially important in the light of serious concerns about the environmental consequences of deforestation and forest degradation for the agricultural viability of adjacent lands. In Ghana, Côte d'Ivoire and Liberia, fears exist that the Sahel will spread south and the Harmattan winds penetrate further, leading to a decrease in dry-season humidity that would devastate crops such as cacao. As much land as possible must be kept under forest to avoid this. Natural forests managed for timber could be major components of this forest estate.

History

The first forest management proposals for Africa were drawn up by Büsgen and Jentsch in 1908–9 for the then German colonies of Cameroon and Togo. They established a number of forest reserves and conducted many scientific investigations in the primary forests. In the rest of tropical Africa, the origins of forestry activity date back to 1920–30 in British West Africa, 1930 in the Côte d'Ivoire and around 1953 in Liberia (Lamprecht, 1989). Early attempts to involve loggers in silviculture failed both in Nigeria (1906) and in francophone Africa (Philip, 1986a, cited in FAO, 1989).

The traditionally very selective timber exploitation in African forests does not favour regeneration and growth of the commercial species, most of which are reasonably fast-growing, so-called 'gap opportunist' or 'non-pioneer light-demanding' species. Although able to germinate and survive in the shade to varying extents, they need substantial light in order to grow up to the main canopy (e.g. Hawthorne, 1989). Most silvicultural systems have concentrated on providing the right light conditions both for recruitment and regeneration of these species.

Thinning operations are widely practised today but natural regeneration itself has not yet been mastered. Uncertainty as to the effectiveness of silvicultural treatments remains because even management systems with a long research history, such as the tropical shelterwood system, have not yet gone through a full rotation (60–90 years).

Silviculturalists in Nigeria and Côte d'Ivoire experimented with natural regeneration and line planting for much of the first half of the 20th century. Many other African forestry departments tried to take up the challenge of silviculture in moist forest beginning in the 1950s (FAO, 1989). Natural forest management was practised in Nigeria (Tropical Shelterwood System (TSS), 2000 sq. km); Ghana (TSS, hundreds of sq. km and various selection systems, 310 sq. km); Uganda (selection systems later converted to uniform systems on several hundred sq. km); Gabon (stand improvement, 1300 sq. km), and Côte d'Ivoire ('Amélioration des Peuplements Naturels', 500 sq. km). (See the case study on African silvicultural systems.) Such systems have, however, been progressively abandoned in favour of plantations (see the case study, Natural forest management or plantations?) except in Ghana where at least some of the prescribed management activities are still being carried out in some forest reserves.

There have been some promising recent initiatives in forest management. In 1976, an important programme for the study of moist forest development relative to different silvicultural interventions was set up in Côte d'Ivoire by the Société Ivoirienne de Développement des Plantations Forestières (SODEFOR) with the technical support of the French Technical Centre for Tropical Forestry (CTFT). The significance of this project (see case study overleaf), compared to previous experiments, was that it allowed accurate measurement of the impact of silvicultural operations.

Present Status of Forest Management

In Africa most of the forested land is nominally under government control. Good forest management depends, therefore, on the effective implementation of appropriate government policies. But, although many countries have committed themselves to sustained-yield policies in their forest legislation, little of this commitment can be traced in forest department programmes. African govern-

AFRICAN SILVICULTURAL SYSTEMS

Tropical Shelterwood System (TSS) This system was designed in Nigeria on the basis of tests which had been carried out for 20 years. The objective was to enhance the natural regeneration of valuable species before exploitation by gradually opening up the canopy (poisoning undesirable trees, cutting climbers) to obtain at least 100 one metre high seedlings per hectare over five years. The forest thus worked was logged in the sixth year; then cleaning and thinning operations were carried out over 15 years.

The Nigerian Forestry Department treated 2000 sq. km of forest in this way between 1944 and 1966, when the method was given up. The main problems encountered were the exuberant spreading of climbers following the opening of the canopy and the failure of the seedlings of valuable species to grow adequately.

Some Ghanaian foresters claim that TSS needs to be re-evaluated in the light of increased marketability of timber species, as too much emphasis was placed on African mahoganies in previous evaluations. Indeed, stands treated under the shelterwood system in Bobiri Forest Reserve in Ghana, show impressive stocking and growth rates of timbers such as *Piptadeniastrum africanum*, *Triplochiton scleroxylon* and *Terminalia ivorensis*.

In 1950, the Forestry Department of Côte d'Ivoire gave up line planting for the Amélioration des Peuplements Naturels (APN), a TSS-related technique. This was done both because initial results of the TSS in Nigeria seemed appealing and because increasing domestic timber consumption required a geographic dispersal of activities and widening of the range of species regenerated. The APN method was applied from 1950 to 1960 on large areas of forests which had been logged-over and were well-stocked with valuable trees. The aim was to favour the growth of these stems and also to ensure regeneration through natural seeding of the valuable species by removing climbers and opening up the canopy. It was abandoned in 1960 when results were judged to be unsatisfactory.

Selection systems Various systems have been applied in Ghana since 1960. Their objective is to assure regeneration of forests well-stocked with valuable species. Harvesting was meant to occur about every 25 years, after the Forestry Department had marked the stand to retain some well-distributed seed trees. However, in 1970 the felling cycle was reduced temporarily to 15 years in response to public allegations that overmature timber was going to waste in Ghana's forest. This relatively short rotation has been found to cause considerable felling damage. Regeneration has been poor and less valuable shade-tolerant species dominate because of insufficient opening up of the canopy. Plans now exist to re-establish a longer harvesting cycle of around 30 years.

'Improvement of stand dynamics' This system was used in Gabon in forests rich in okoumé, a vigorous 'pioneer' species able to recolonise savanna areas in the absence of fire and accounting for most of the country's timber exports. Its objective is to accelerate the growth of all-sized stems of valuable species in naturally well-stocked stands, without trying to provoke regeneration through natural seeding. The species grow in patches or clumps presumably due to natural seeding of forest trees in clearings or gaps. The objective was to let these stands attain commercial diameters as quickly as possible through thinning operations, but the production gain was never measured. After treating about 1200 sq. km of forest, including about 120 sq. km of pure okoumé stands, the Forestry Department gave up the technique in 1962 and switched to plantations.

ments consider agriculture to be more important than forestry. Forests are treated as convertible rather than renewable resources. As a consequence, forest legislation is not applied and forest protection not enforced (see chapter 7). Moist forest cover in the six African ITTO member countries (excluding Zaïre which joined ITTO during the preparation of this Atlas) that account for more than 90 per cent of African timber exports, is estimated at 674,000 sq. km. However, only 69,000 sq. km have been reserved as permanent timber production and watershed protection forests, and of this only 40,000 sq. km are actually forested.

The control of logging is equally problematic. In moist forests in Africa, heterogeneity and low commercial timber volume effectively preclude the use of systems which depend upon regeneration from seedlings; future timber harvests depend, therefore, on advance growth of commercial species left undamaged after logging. However, measures to limit felling damage, such as directional felling and marking of residuals, are not required in most of the present concession agreements. Even when they are required the lack of enforcement does not encourage compliance by the loggers. One consequence of exhausting a forest concession and moving on to a new concession, has been that the infrastructure – including such items of primary importance as schools and medical dispensaries – established during logging operations are left to deteriorate and the people lose their jobs.

In spite of these problems, many forests exploited for timber are not being degraded. This is especially true in areas where there is little settlement, where the topography is not too incised so that there are few erosion problems with roads and log landings and where the market is so distant as to allow exploitation of only the highest value species.

NATURAL FOREST MANAGEMENT OR PLANTATIONS?

Various forest management systems practised up to 1960 have since been largely abandoned in favour of plantations. Plantations have performed quite well as a source of industrial timber in a number of cases. With the best African timber species on good sites, a mean annual increment of 8 cu. m per ha may be expected at the age of 30–45 years; but the high establishment cost of US$1000–3000 per ha renders the profitability of such undertakings arguable, especially if the value of the natural forest that has been cleared to make room for plantations is counted as a cost. Profits are further eroded because thinnings are unsaleable in most countries, as secondary bush provides low-cost alternatives for the supply of poles and fuelwood.

Plantations are high-risk investments, not only because of pest and disease problems such as the *Hypsipyla* spp. borers which attack most Meliaceae plantations, but also because of discontinuity of maintenance caused by a lack of funds. Thus, valuable and expensive plantations in Gabon, Liberia and Cameroon have largely gone untended due to cuts in the budgets of forestry departments.

Managed natural forest does not require such continuous or high investment. Indeed, investments in the natural forest are usually compounded over longer periods (Schmidt, 1987). The economics of management of natural forest would improve even further if a wider range of natural-forest species could be marketed, and if revenues could be derived from non-timber products and services (Pearce, 1990).

A PROMISING PROJECT IN CÔTE D'IVOIRE

The Société de Développement des Plantations Forestières (SODEFOR) and French Centre Technique Forestier Tropical (CTFT) pilot project in Côte d'Ivoire covers 120 sq. km at three field stations characteristic of three ecological areas of moist forests. Silvicultural practices have consisted of traditional exploitation of economic species and thinning of residual stands by poison girdling. Two thinning regimes modelled on experience gained in Peninsular Malaysia were tested. These involved the removal of either 30 or 45 per cent of the total basal area, beginning with the felling of the tallest trees in the residual forest until the desired percentage of basal area was reached. The objective of the thinning was to favour valuable trees of greater than 10 cm dbh (diameter at breast height). No measures to favour regeneration through natural seeding are envisaged. Preliminary investigations seem to indicate that the induced natural regeneration is adequate.

After four years, for the 73 species measured there was a volume increase for the stems over 10 cm dbh of 3.0 to 3.5 cu. m per ha per year against 2 cu. m in the control stands, a gain in growth of 50 to 75 per cent. Measurements taken every year showed that the annual volume gain increases with time and that the impact of the thinning operations will probably be felt for at least ten years. The largest yearly diameter increases, averaging 1.0 cm per year, were found in species such as *Triplochiton scleroxylon*, *Terminalia superba* and *Heritiera utilis*.

The economics of the operation are promising, with 1 cu. m produced for every US$5.6 invested, as compared with US$7.4 per cu. m for plantation-grown timber. If a wider range of species from the natural forest can be introduced in local markets, the economics of the operation would improve. The 100 sq. km Yapo forest is being managed experimentally on the basis of the results of this research project (Schmidt, 1987). Plans are now advanced to extend this management to a further 5000 sq. km of forest through a series of joint ventures with the private industrial sector. In addition, another 3000 sq. km will be managed directly by SODEFOR with assistance from aid agencies. Advocates of the scheme acknowledge that the forests will not be completely natural but their future may be more secure than that of forested national parks. The latter are often not seen by local people as yielding any tangible benefits, except to foreign tourists, and are consequently threatened by encroachment and poaching.

Indirect Conservation Impact of Forest Harvesting

One of the most economically valuable substances produced by rain forests is the relatively fertile soil which they are instrumental in regenerating. These soils are exploited by shifting cultivation which is by far the most widespread type of land use in the forest. In large parts of the region, it is also the agricultural practice most adapted to local circumstances of low population density and therefore relative abundance of forested lands. The virtual absence of agricultural inputs and the prevalence of soils which are unsuitable for permanent cropping make shifting agriculture a rational option for African farmers.

However there are important differences in farming practices between native shifting cultivators and migrant slash-and-burn cultivators. The former use methods that involve selective felling, light burning and no tillage and they retain a variety of fallow and high forest trees (Kahn, 1982; de Rouw, 1987). For instance in the south-west of Côte d'Ivoire, smallholders successfully cultivate food crops in swiddens in association with more permanent tree crops of coffee and cacao in a rain forest environment. In contrast, the slash-and-burn cultivation practised by migrants from savanna regions is much more destructive. It involves clear felling, excessive burning and soil tillage.

Wilkie (1988) provides a similar picture of indigenous shifting cultivators in the Congo basin, who prefer to clear 15–20 year old secondary forest rather than primary forest. The forest growing in these patches is composed of softwoods such as the parasol tree *Musanga cecropioides*, which can be fairly easily felled and cleared with simple tools. Cassava cuttings and plantain sprouts are planted directly after slashing; two months later, a controlled burn during which the soil stays remarkably moist and cool, is carried out. This allows the cassava and plantain to survive and makes optimum use of the wood ash fertiliser. After the remaining debris is cut, piled into mounds and burnt, other crops such as peanuts, squash, corn and sugarcane are planted. This mimics natural vegetation succession and quickly covers the soil. Such procedures, which used to be dismissed as primitive and wasteful, have now been recognised as local adaptations to rainfall intensity, soil erosion risks and soil nutrient deficiencies which cannot be solved by applications of organic fertiliser. Indeed, scientists now legitimise these practices by giving them labels such as 'minimum tillage' and 'intercropping' (Richards, 1985).

The slash-and-burn practices of migrant farmers, on the contrary, often provoke severe forest degradation in relatively short periods of time. For example, in the less populated western part of Kivu province in Zaïre the construction of a new road, in combination with the establishment of a coffee plantation, led to the loss of about eight additional hectares of forest to extensive slash-and-burn foodcropping for every hectare of coffee planted (GTZ/IZCN, 1986). This pattern of deforestation, involving labour-intensive cash crop development accompanied by land-extensive slash-and-burn foodcropping seems to have played an important role in other forest areas in Africa: for instance, in the cacao regions of Cameroon, Ghana and Côte d'Ivoire.

The single greatest threat to the rain forests of Africa is the proliferation of destructive slash-and-burn agriculture by migrant populations. This is frequently associated with the opening-up of new forest areas by logging.

Hunting is also closely associated with logging. The proliferation of modern hunting equipment, such as guns and powerful lights, and the emergence of important urban and sometimes even export markets for bushmeat, has exacerbated this problem. The flourishing bushmeat trade is an important economic factor in attracting people into the forest. A timber worker can make more money by poaching a chimpanzee *Pan troglodytes* in Zaïre than he can from two months' hard work with a logging company. As soon as a new road is opened up, hunters use it in search of new game unaccustomed to man and thereby easier to kill with inaccurate home-made guns.

The question whether or not selective felling is compatible with wildlife conservation is often rendered academic by the uncontrolled hunting that logging roads almost always facilitate (Shelton, 1985, cited in Skorupa, 1988). Even in areas where logging has not been widespread, hunters have eliminated larger primates from tropical forests (Terborgh, 1986b, cited in Skorupa, 1988). Logging roads greatly increase the threat of hunting to animal populations. The hunting taboos that were traditionally respected by the pygmies and other forest peoples, especially in the case of rare animals thought to be endowed with supernatural powers, such as okapi *Okapia johnstoni* (but see case studies in chapters 5 and 32), chimpanzee and some duikers *Cephalophus* spp., are now gradually being abandoned in the face of increasing commercialisation and the need for money to purchase consumer goods.

As a consequence of the low intensity of logging in Africa, conflicts with local people are limited. The widespread animosity between loggers and forest dwellers in Southeast Asia, where people see their livelihoods threatened by the environmental consequences of intensive logging operations, does not exist in Africa. On the contrary, interactions may be quite positive, with loggers providing employment, farm-to-market roads, schools and dispensaries. The problem is that, in the absence of any forest management, the logger is bound to move on some day, leaving the infrastructure he created to deteriorate.

Localised problems do occur; for instance, in the south of Cameroon concessionaires have been prevented from logging moabi *Baillonella toxisperma*, by a violent reaction from the local population who use the seeds of this species for cooking oil. Another case is the refusal of loggers involved in salvage fellings in cacao areas in Ghana to pay proper compensation to farmers. Little is known about the impact of logging on forest-dwellers such as the pygmies and other tribes living in Cameroon, Gabon, Congo, Central African Republic and Zaïre, but colonisation of the forest by Bantu communities attracted by logging infrastructure disrupts locally adapted lifestyles and cultures.

In Africa, many tree species, including most of the commercially important timbers, are much more wide-ranging than in Southeast Asia and Latin America. Furthermore, many stems of commercial species are left standing during selective logging operations, because of apparent defects, low diameter (minimum exploitable diameters in African concession agreements are higher than in Southeast Asia and Latin America) and difficult topography. In consequence, no commercial timber species are thought to be in danger of extinction through logging. However, some well-known species are threatened by a combination of logging and forest conversion in parts of their geographic range. These include *Entandrophragma angolense*, *Gossweilerodendron balsamiferum*, *Irvingia gabonensis*, *Milicia excelsa*, *Nesogordonia papaverifera* and *Pericopsis elata* (FAO, 1986).

The direct impact of logging on animal populations gives more reason for concern in countries where few totally protected areas have been established in the rain forest, for example, Congo, Côte d'Ivoire, Gabon and Liberia. Results of the fairly extensive research on the impact of logging on wildlife conservation done in Southeast Asia, suggest that, provided pockets of untouched forest are left in the exploited area, logged forest will be colonised by most species of birds and mammals within a few years of logging (Wilson and Wilson, 1975; Johns, 1985). This is also likely to apply to the African situation where logging intensity is lower (but see the case study on the Kibale Forest in chapter 31). The conservation value of logged forests will be much greater in situations where systems of totally protected areas are located within the production forest estate.

There seems to be considerable potential for conserving animals such as primates in logged African forests providing hunting and agricultural encroachment can be controlled. However, current standards of practice in mechanised logging operations are rarely high enough to guarantee compatibility of timber exploitation and wildlife conservation (Skorupa, 1988). Although these standards could sometimes be improved by using different machinery, such as wheeled instead of tracked skidders, operator performance is generally the more important variable, and both training and other incentives are necessary to improve the latter (Jonsson and Lindgren, 1990).

For primate conservation, overall logging damage seems to be more important than specific variables such as the density of fig trees remaining. Damage should therefore be minimised by practices such as directional felling, and, perhaps, climber cutting (Sayer *et al.*, 1990). However, lianas are often important sources of food for arboreal vertebrates, and many of them may be key resources for wildlife, due to their aseasonal phenology and relatively high abundance in regions with distinct dry seasons. As a consequence, the indiscriminate application of pre-harvest climber cutting, a silvicultural practice intended to reduce felling damage and post-logging competition with residuals of commercial species, might be extremely harmful to wildlife (Skorupa, 1988).

Selective logging poses a threat to Africa's lowland rain forest birds (Diamond, 1985). Conservation of the most specialised species has been found to be 'incompatible with modern logging methods'. Frugivores are considered to be especially threatened by the changed forest structure induced by selective logging and by the disappearance of the seed dispersers such as large mammals and birds (Thiollay, 1985), although the latter is not due to the logging itself, but to the hunting often associated with it.

The Future of Forest Management

Natural forest management for timber production can be practised at various levels of intensity. As a minimum, it requires demarcation and protection, inventory and the regulation and control of exploitation of the forest. More intensive management involves silvicultural interventions such as the release of regenerating timber trees by clearing unwanted competitors and cutting climbers. The high costs involved in silvicultural operations can only be justified in the case of forests that are rich in exploitable timber and where the cost of transport to markets is low. In Africa, few forests fulfil these requirements, and extensive management is often the only feasible option. Therefore, efforts to enhance forest management should concentrate on improved control of forest access and harvesting, and a satisfactory economic and policy environment for the logging and timber processing industry. These proposed improvements are amplified in the five measures discussed overleaf.

A tree nursery in Korup, Cameroon which will provide indigenous tree species for reforestation projects. M. Rauktari

Forest Management

An important advantage of forests extensively managed for timber production is that they can also be managed for other uses. They can provide a range of non-timber benefits, such as watershed protection, nature conservation and non-timber forest products. If these non-timber benefits are more important than timber products, for instance in areas where forest cover has substantially decreased in recent times, conflicts between timber production and other management objectives may occur. The trade-offs must then be assessed, and management plans adapted accordingly.

Recently, many people have suggested that tropical moist forests should be managed for non-timber forest products, to the exclusion of timber. Two arguments should suffice to counter this suggestion. First, although the forest legislation guaranteeing local people's rights to continue to harvest non-timber products is not always respected, there is no intrinsic reason why logging and harvesting of non-timber products should not be compatible. Second, the management history of non-timber products has been depressingly similar to that of timber: overhunting and overharvesting have been the rule rather than the exception (see case studybelow).

The popular image of the big bad logger or colonist farmer clear-felling the forest at the expense of the poor defenceless hunter-gatherer tribesman is an oversimplification. This image is false in Africa. No logger spends money felling trees he has to leave in the forest because no market will accept them. Furthermore, most African forest dwellers are farmers who also gather forest products, and the few true African forest dwellers (mainly the pygmies) have lived in symbiosis with these farmers since time immemorial, bartering bushmeat and labour against farm produce to increase the quality and security of their livelihoods (Ichikawa, 1983; see chapter 5). In fact, most forest-dwelling populations are not as dependent on primary forest as is often suggested. In the Mayombe region of Congo, for instance, local populations depend to a far greater extent on secondary forest for their livelihood (for example, hunting, collecting various other foodstuffs and gathering thatching material), even though the primary forest is still nearby (Michon-de Forestal, 1987).

Five Measures Required to Improve Forest Management

1 Control of access and support measures

Since the development of intensive agricultural practices is not yet – and in some cases might never be – feasible in most African rain forest areas, the only way to safeguard forests against encroachment is to plan infrastructure and other development so as to minimise access to these forests.

The decision to open up forests with large-scale infrastructure investment in road building, railway construction, river access and port improvement is a policy to speed up deforestation in that area, whether or not this is intentional (Schmithüsen, 1989). This situation is caused by the fact that benefits from 'creaming', existing or artificially induced pressure for land clearing for agriculture or pasture, and the possibility of transferring public forest land into some form of private tenure through deforestation, are all powerful incentives against preservation of accessible forest ecosystems.

Law enforcement alone is neither effective nor desirable. Measures restricting access should be complemented by positive interventions supporting local livelihoods, an approach that has been pioneered in a nature conservation context by projects such as Korup National Park in Cameroon (see case study in chapter 13). Thus, where developments such as road construction and the establishment of large cash crop estates are likely to result in large concentrations of people, additional measures to reduce the possible negative impact of encroachment by forest farmers should be taken (Doumenge *et al.*, 1989).

Clearing of primary forest for agriculture will in some circumstances be necessary. In countries such as Zaïre, where extensive areas with fertile alluvial soils are still covered in forest, clearing of the forest to grow food for an increasing population will undoubtedly have a role to play in the country's future development. A careful choice has to be made as to which areas to preserve intact, which to reserve for timber production, and which to convert to other uses (Poore and Sayer, 1987).

Non-Timber Forest Products as an Alternative?

There is a strong economic case for basing forest management on non-timber products (Peters *et al.*, 1989). But the potential of non-timber forest products to generate revenue, particularly from exports, has sometimes been exaggerated and the problems involved in the management of forests for these products have been underestimated. Peters *et al.* analysed the economic returns of a 1 ha patch of rain forest near a village 30 km from the town of Iquitos in Amazonian Peru, using hypothetical market prices for fruits and forest-gate prices for timber. This showed that non-timber forest products could provide a sustainable yield ten times as high as that of timber. But what holds true for an intensively managed forest near a town, cannot necessarily be applied to the whole of the tropical rain forest.

Southeast Asia has a long history of successful export of non-timber forest products such as rattans, resins, and gums to Europe. However, African non-timber forest products have generally been important only for subsistence and local cash economies. Exceptions were a short-lived rubber boom (from *Landolphia heudelotii*) in the 1890s, and the relatively small but steady export of medicinal barks from Cameroon and medicinal plants from Zaïre. The latter country exports seeds of *Strophanthus* spp., containing cardiac glycosides such as ouabain and cymarin, and *Rauvolfia vomitoria* and other species containing a latex rich in the alkaloids reserpine and rescinnamine.

Non-timber forest products are frequently harvested unsustainably. This is particularly the case with tree barks and gums. Thus, *Prunus africana*, a well-known medicinal tree species, has been severely overexploited in the mountain forests of Western Cameroon. Sustainable harvesting is possible in principle, by stripping only part of the bark and leaving the tree to recover until the next harvesting cycle. Similarly, rubber from *Landolphia heudelotii* was obtained mainly by so-called slaughter tapping (Richards, 1985).

Many species yielding non-timber forest products have been brought close to extinction through over-exploitation in the wild. The best option for their continued production is through domestication. Cinnamon, trees yielding resins such as damar and, increasingly, rattans are the most well-known Southeast Asian examples of this pattern: initially collected in the forest, they are now grown in plantations. Light-demanding species that thrive in secondary environments, such as the *Strophanthus* and *Rauvolfia* mentioned above, are particularly easy to domesticate.

Recently, efforts have been made to include agricultural expansion zones in forest management plans. Examples are in the Forest Management Unit 6 in the Chaillu, Congo; in Akonolingga Forest Reserve in Cameroon; and in an IUCN project in the Usambara mountains in Tanzania (see case study in chapter 17). Such zones should be agreed through consultation with local populations and clearly demarcated on the ground, if necessary by buffers of fast-growing plantations. This model was previously successful in Ghana and Côte d'Ivoire, where the forest reserves established before independence in consultation with the local chiefs are still present, the straight lines of their boundaries clearly visible on satellite imagery.

2 Control and improvement of logging

Extensive management is the only feasible system for most of the African rain forests devoted to permanent timber production. Therefore, the integration of harvesting and silviculture is one of the most important and, at the same time, most difficult elements of forest management (Catinot, 1986, cited in FAO, 1989).

Protection of the forest, realistic assessment of the annual cut, arrangement and demarcation of annual coupes, pre-felling inventory and choice of silvicultural system, marking of trees for retention or felling, exploitation of coupe to acceptable damage limits, post-felling inventory, check of coupe harvest by species, silvicultural treatment of relic stand, continuous forest inventory, maintenance of main roads and erosion control on subsidiary roads should all be the subject of intensive on-the-ground control (Burgess, 1991). Modern technology such as satellite imagery might make this cheaper, but logging control will remain costly and, therefore, a firm financial commitment on the part of the government is indispensable.

Jonsson and Lindgren (1990) identified a combination of measures, including improved concession agreements, information dissemination and better planning and control of harvesting operations, which would bring about more efficient and ecologically sound logging practices. This is a matter of long-term commitment to improving human resources rather than extra investment in machinery. Present logging technology is acceptable and appropriate in most cases if properly applied. The increased cost of planning and control is easily offset by the financial benefits of more efficient operations.

3 Control of hunting in logging concessions

Loggers cannot be held solely responsible for the wildlife in their concessions, but it is imperative that they are aware of prevailing hunting patterns and that they respect government schemes such as gun control, bushmeat farming, and policies that invest traditional hunters with the means to exclude competing commercial hunters (Skorupa, 1988). In Gabon, where logging is allowed in wildlife reserves, the loggers in Lopé Reserve are to be asked to accept responsibility for the control of hunting in their concession areas.

4 Economically rational decision making

Valuation of forests for timber production should be based on a fully socio-economic (rather than purely financial) basis. Not only timber but non-timber products and environmental services should be assessed as far as possible. Where forests are managed for multiple uses such as timber production, watershed protection and wildlife conservation, the trade-offs between the various management objectives should be assessed (Pearce 1990). Forests of marginal timber value should be managed for other purposes of greater value, such as biological diversity, or soil and water conservation (Repetto and Gillis, 1988).

Risk analysis should play an important role in economic assessment, as the risks of both fire and encroachment are known to be higher in logged-over forest. A clear lesson was provided by the fires which destroyed millions of hectares of moist forest in Borneo in 1983, and where logged-over forest was particularly severely affected.

5 Equitable distribution of benefits

In many countries, profit levels in tropical moist forest logging have been extremely high (Jonsson and Lindgren, 1990). Excessive profit levels are known to lead people to engage in rent-seeking behaviour – that is, getting as much profit as possible as soon as possible (see chapter 7).

A recent study on incentives for better forest management commissioned by the International Tropical Timber Organisation has found that present consumer prices are high enough to provide just compensation for people involved in or affected by logging. However, available benefits get distributed very unevenly. As Schmithüsen (1989) put it: 'Because of institutional weakness, local people cannot articulate their interests if they coincide with forest conservation, and if their interests do not, this same weakness does not allow for social and economic compensation to accrue to them.' The need is therefore to strengthen local people's power rather than to increase timber prices.

Conclusion

At present, there is little sustained-yield forest management in African moist forests. Forest is being destroyed largely by unsustainable farming practices, with access frequently being provided by logging roads. The extraction of timber and other products is not accompanied by reinvestment of benefits to ensure the capacity of the forest for future production. It is only in remote areas with low population pressure that forests are allowed to regenerate after logging. If the measures described above are not all taken in a coordinated way, sustained-yield management will not come about, at least not on the scale required.

From this review of complex and interrelated issues one conclusion is obvious: there are no simple solutions. What is needed to guarantee sustainable management of forests is a carefully designed system of checks and balances, aiming at an equitable distribution of benefits among all parties involved, while guaranteeing the future of the resource. At present, such systems are absent in most parts of Africa.

References

Burgess, P. (1991) Natural rain forest management. In: *The Conservation Atlas of Tropical Forests. Asia and the Pacific.* Collins, N. M., Sayer, J. A. and Whitmore, T. C. (eds), pp. 43–50. Macmillan, London, in association with IUCN, Gland, Switzerland.

de Rouw, A. (1987) Tree management as part of two farming systems in the wet forest zone (Ivory Coast). *Acta Oecologica, Oecologia Applicata* **8**(1): 39–51.

Diamond, A. W. (1985) Threats to tropical forest birds and critical sites for their conservation. In: *Conservation of Tropical Forest Birds.* Diamond, A. W. and Lovejoy, T. E. (eds). International Council for Bird Preservation Technical Publication No. 4. ICBP, Cambridge, UK.

Doumenge, C., Kitalema, N. and Rietbergen, S. (1989) *Plan d'Action Forestier Tropical, Zaïre. Conservation des Ecosystèmes Forestiers, Volet Parcs et Réserves.* Département des Affaires Foncières, de l'Environnement et de la Conservation de la Nature, Kinshasa, Zaïre.

FAO (1986) *Databook on Endangered Tree and Shrub Species and Provenances.* FAO Forestry Paper No. 77, Rome, Italy.

FAO (1989) *Management of Tropical Moist Forests in Africa.* FAO Forestry Paper No. 88, Rome, Italy.

Grainger, A. (1987) Tropform: a model of future tropical timber

A road being built through Zaïre's rain forest. This opens the way to settlers and increased exploitation of the forest. C. Doumenge

hardwood supplies. *Proceedings of the Symposium in Forest Sector and Trade Models*. University of Washington, Seattle, USA.

GTZ/IZCN (1986) *Rapport Final du Séminaire Préliminaire sur la Conservation de la Nature Intégrée au Développement*. Bakau, 2–7 December 1985. GTZ/IZCN, Kinshasa, Zaïre.

Hawthorne, W. (1989) *The Regeneration of Ghana's Forests*. Interim report March 1989. ODA Forest Inventory Project Report, Kumasi, Ghana.

Ichikawa (1983) An examination of the hunting-dependent life of the Mbuti pygmies, Eastern Zaïre. *African Study Monographs* **4**: 55–76.

Johns, A. D. (1985) Selective logging and wildlife conservation in tropical rainforest: problems and recommendations. *Conservation Biology* **31**: 355–75.

Jonsson, T. and Lindgren, P. (1990) *Logging Technology for Tropical Forests – For or Against?* Report from the ITTO pre-project 'Improvement of Harvesting Systems for the Sustainable Management of Tropical Rain Forests'. Forest Operations Institute, Kista, Sweden.

Kahn, F. (1982) *La Reconstitution de la Forêt Tropicale Humide, Sud-Ouest de la Côte d'Ivoire*. Collection Mémoires No 97. Editions de l'Office de la Recherche Scientifique et Technique Outre-Mer, Paris, France.

Lamprecht, H. (1989) *Silviculture in the Tropics*. Deutsche Gesellschaft für Technische Zusammenarbeit (GTZ) GmbH, Eschborn.

Michon-de Forestal, G. (1987) *Utilisation et Role de l'Arbre et des Végétations Naturelles dans les Systèmes Agraires du Mayombe (Sud-Congo): Perspectives pour le Développement d'Agroforesterie Paysannes Integrées*. Report to Unesco, Paris, France.

Pearce, D. (1990) *An Economic Approach to Saving the Tropical Forests*. Paper prepared for University of Oxford and Oxford Economic Research Associates, UK.

Peters, C. M., Gentry, A. H. and Mendelsohn, R. O. (1989) Valuation of an Amazonian rainforest. *Nature* **339**: 29.

Poore, D. and Sayer, J. A. (1987) *The Management of Tropical Moist Forest Lands: Ecological Guidelines*. IUCN, Gland, Switzerland and Cambridge, UK.

Repetto, R. and Gillis, M. (eds) (1988) *Public Policy and the Misuse of Forest Resources*. Cambridge University Press, Cambridge, UK.

Richards, P. (1985) *Indigenous Agricultural Revolution: Ecology and Food Production in West Africa*. Unwin Hyman, London, UK.

Sayer, J. A., McNeely, J. A. and Stuart, S. N. (1990) The Conservation of Tropical Forest Vertebrates. In: *Vertebrates in the Tropics*. Peters, G. and Hatterer, R. (eds). Museum Alexander Koenig, Bonn, Germany.

Schmidt, R. S. (1987) *Tropical Rainforest Management: A Status Report*. FAO, Rome, Italy.

Schmithüsen, F. (1989) Tropical forest conservation and protection: political issues and policy considerations. In: *Deforestation or Development in the Third World?* Palo, M. and Salmi, J. (eds). Vol. III, Research Notes of the Finnish Forest Research Institute, Helsinki.

Skorupa, J. P. (1988) *The Effects of Selective Timber Harvesting on Rainforest Primates in Kibale Forest, Uganda*. Unpublished PhD thesis, University of California, Davis, USA.

Thiollay, J.-M. (1985) The West African forest avifauna: a review. In: *Conservation of Tropical Forest Birds*. Diamond, A. W. and Lovejoy, T. E. (eds), International Council for Bird Preservation Technical Publication No. 4. ICBP, Cambridge, UK.

Wilkie, D. S. (1988) Hunters and farmers of the African forest. In: *People of the Tropical Rainforest*. Denslow, J. S. and Padoch, C. (eds). University of California Press, Berkeley and Smithsonian Institution, Washington, DC, USA.

Wilson, C. C. and Wilson, W. D. (1975) The influences of selective logging on primates and some other animals in East Kalimantan. *Folia Primatologica* **23**: 245–75.

Authorship

Simon Rietbergen, IIED, London.

9 The Protected Areas System

Introduction

The governments of African countries are increasingly aware of the urgent need to conserve examples of their remaining tropical forests. Protected areas constitute the most widespread mechanism for achieving this, yet several problems face these areas in many African countries. This chapter outlines the role of protected areas and describes the history of forest protection in Africa. The extent and effectiveness of the current network of protected forests is assessed and criteria for expanding the protected area system and reconciling it with local development needs, are discussed. Three case studies illustrating the benefits and problems of reconciling local interests with protected areas, are included. Finally, the prospects for forest conservation across Africa are assessed. It is concluded that protected areas will be successful in conserving tropical forests only if they are able to meet the legitimate development aspirations of the people that live in and around them.

Concept and Function of Protected Areas

Protected areas can be defined as predominantly natural areas, safeguarded by law or custom, where species and ecosystems are conserved for current and future generations. Tropical rain forests contain at least half of the world's species (Sutton *et al.*, 1984), so to establish and maintain tropical forest conservation areas is particularly important for protecting biological diversity. Large numbers of tropical forest species are already included within Africa's protected areas. For example, there are 1300 species of plants in Côte d'Ivoire's Taï Forest, of which 54 per cent are endemic to the Upper Guinea sub-region (Guillaumet, 1967). Similarly, Cameroon's Korup National Park contains approximately 500 tree species (Gartlan and Agland, 1981) and is one of the more important sites for primate conservation on the continent (Oates, 1986). It has been shown that Africa's protected area system protects the majority of the continent's bird species (Sayer and Stuart, 1989).

The primary function of protected areas is seen by some as the conservation of species and habitats (e.g. Bell, 1983; 1987), but protected areas also provide an array of economically important goods and services (Prescott-Allen and Prescott-Allen, 1982; MacKinnon *et al.*, 1986; McNeely, 1988; summarised in Tables 9.1 and 9.2). Direct benefits of protected areas – those that can be readily quantified in economic terms – include protecting renewable resources harvested within or beyond the boundaries of these areas, supporting nature-related recreation and tourism and conserving genetic resources. Indirect benefits, such as stabilising local climate, protecting watersheds and preventing soil erosion, are more difficult to evaluate and are therefore often overlooked in cost-benefit analyses. Nevertheless, when they are taken into account, such indirect benefits often outweigh the direct benefits of conserving wildlife (McNeely, 1988).

There are also a number of costs associated with protecting tropical forests. Direct costs include the loss of crops and livestock in adjacent agricultural areas to animals living in reserves. Important indirect costs include the lost opportunities for gaining immediate benefits by, for example, logging an area or converting it to an alternative form of land use. The burden of such costs falls largely on those communities living in the vicinity of protected areas, while the benefits of conservation are sometimes enjoyed mainly at national or international level (Bell, 1987). This means that local communities often pay a considerable price for protected areas and

Table 9.1 Direct benefits of protected areas

Direct benefits are immediate and can be easily observed and measured. They can include:

Protecting renewable resources Protected area management often involves the sustainable exploitation of resources such as timber and so-called 'secondary' forest products. For example, honey-gathering is a culturally and economically important activity in many protected African forests, including Mt Kilum in Cameroon (Stone, 1990), Mau Forest in Kenya and Nyungwe Forest in Rwanda.

Supporting tourism and recreation Wildlife-related tourism is a large, rapidly growing industry that is playing an increasingly important role in earning foreign exchange for developing countries (see case study on Nyungwe Forest). For instance, revenue from an estimated 6500 visitors to Rwanda's mountain gorillas earned the Volcanoes National Park more than US$800,000 in 1989 (compared with barely US$10,000 in 1979; Wells *et al.*, 1990). Moreover, these foreign tourists also spend money elsewhere in the country and the total income earned from tourism was as high as US$17 million in 1989 (Monfort, 1990), which makes tourism the third largest earner of foreign revenue in Rwanda.

Conserving genetic resources People currently use some 15,000 species of wild plants and animals and it is likely that many more will be of use in the future (McNeely, 1988). In addition, genetic turnover in domestic species is extremely rapid and modern agriculture is highly dependent on using wild genetic resources for improving crop yields and resistance to pests (Prescott-Allen and Prescott-Allen, 1982). As remaining natural habitats dwindle, protected forest areas are set to become extremely important as *in situ* gene banks.

Table 9.2 Indirect benefits of protected areas

Protected forests provide a number of indirect benefits which, although difficult to measure in economic terms, are often essential in maintaining other forms of land use.

Stabilising water catchments Natural forests play a vital role in regulating water flow to surrounding agricultural areas. Rivers from the Taï Forest, for instance, have twice the flow of water midway through the dry season, and three to five times the flow at the end of the dry season, as do rivers from a nearby coffee plantation (Dosso *et al.*, 1981). Forests thus mitigate the effects of seasonal droughts and, by their ability to soak up water during rainy seasons, also serve to minimise soil erosion on surrounding hillsides.

Maintaining climatic stability There is considerable evidence that forests regulate rainfall and humidity levels in their vicinity by returning water vapour to the atmosphere at a steady rate (Dickenson, 1981). Thus it is likely that the recent reduction in mistiness at high altitudes in the East Usambaras, Tanzania, for example, is the result of extensive forest clearance (Hamilton and Bensted-Smith, 1989). Retaining forest cover also helps keep ambient temperatures relatively low, benefiting agriculture.

Protecting soils Exposure of tropical soils leads to their rapid degradation through leaching of minerals and accelerated erosion, particularly of nutrient-rich topsoil. A study of erosion in Kenya revealed that soil loss from agricultural land is more than 100 times greater than that from intact forest (Dunne, 1979).

Additional biological services Tropical forests often act as important reservoirs for birds and other animals which control pests in nearby agricultural areas. Similarly, several economically important crops rely on forest species for pollination or germination. Forest protected areas play a key role in maintaining these important agents and the species on which they, in turn, rely.

Providing facilities for scientific research and education Little is understood about maximising agricultural productivity on marginal tropical soils, but studying ecological processes in intact forest areas may result in valuable contributions to such applied research. In addition, tropical forests provide unparalleled opportunities for education and training of local and foreign students, as well as for fundamental research in fields such as evolutionary biology. For example, scientists at the Makerere University Biological Field Station in Kibale Forest, Uganda, have carried out important applied research on the relationships between forest regeneration and clearing size (Skorupa and Kasenene, 1983). Kibale is also a centre for the study of primate ecology and behaviour and the station provides basic field experience for large numbers of undergraduates and secondary school students.

are therefore understandably resentful towards them. Conflicts between protected areas and nearby rural communities can only be reconciled by more effective integration of conservation programmes with the legitimate development aspirations of local people.

In recognition of the importance to conservation of sustained resource use as well as resource protection, IUCN's Commission on National Parks and Protected Areas (CNPPA) distinguishes eight different types of conservation areas (Table 9.3; IUCN, 1984). Their management objectives vary from strict protection to sustainable exploitation. The range of more commonly used categories runs from Strict Nature Reserves (Category I) and National Parks (Category II), which are protected from exploitation, to Multiple-Use Management Areas (Category VIII), where conservation is combined with sustainable resource use.

History of Protected Areas in Africa

Although nature reserves were not formally established in Africa until the end of the 19th century, several African cultures have a long tradition of protecting forests. For instance, 'Kaya' (homestead) forests situated along the Kenyan coast have been safeguarded for more than 400 years by the Mijikenda tribe (see case study in chapter 17). These forests were originally the locations of fortified villages where the people hid from their enemies and, until recently, the abandoned groves were rigorously protected by tribal elders as sites for burials and traditional ceremonies. Indeed, religious and cultural factors have been important in preserving many other forests in Africa. For instance, a relict grove at Zagné in Côte d'Ivoire has long been protected by local villagers because they revere the mona monkey *Cercopithecus mona* and spot-nosed guenon *C. petaurista* that live there (Kingdon, 1990). Similarly, the forested slopes of Mt Kilum in Cameroon have traditionally been protected from cultivation by a figure known as Mabu, a fierce deity in the religion of the local Oku tribe (Stone, 1989). Ghana has similar sacred sites (see case study on Boabeng-Fiema Monkey Sanctuary in chapter 21).

Table 9.3 Categories and management objectives of protected areas

I *Scientific Reserve/Strict Nature Reserve:* to protect nature and maintain natural processes in an undisturbed state in order to have ecologically representative examples of the natural environment available for scientific study, environmental monitoring, education and for the maintenance of genetic resources in a dynamic and evolutionary state.
II *National Park:* to protect natural and scenic areas of national or international significance for scientific, educational and recreational use.
III *Natural Monument/Natural Landmark:* to protect and preserve nationally significant natural features because of their special interest or unique characteristics.
IV *Managed Nature Reserve/Wildlife Sanctuary:* to assure the natural conditions necessary to protect nationally significant species, groups of species, biotic communities, or physical features of the environment, where these require specific human manipulation for their perpetuation.
V *Protected Landscape:* to maintain nationally significant natural landscapes which are characteristic of the harmonious interaction of man and land while providing opportunities for public enjoyment through recreation and tourism within the normal lifestyle and economic activity of these areas.
VI *Resource Reserve:* to protect the natural resources of the area for future use and prevent or contain development activities that could affect the resource pending the establishment of objectives which are based upon appropriate knowledge and planning.
VII *Natural Biotic/Anthropological Reserve:* to allow the way of life of societies living in harmony with the environment to continue undisturbed by modern technology.
VIII *Multiple-Use Management Area/Managed Resource Area:* to provide for the sustained production of water, timber, wildlife, pasture and outdoor recreation, with the conservation of nature primarily oriented to the support of economic activities (although some zones may also be designed within these areas to achieve specific conservation objectives).

(*Source:* adapted from IUCN, 1984)

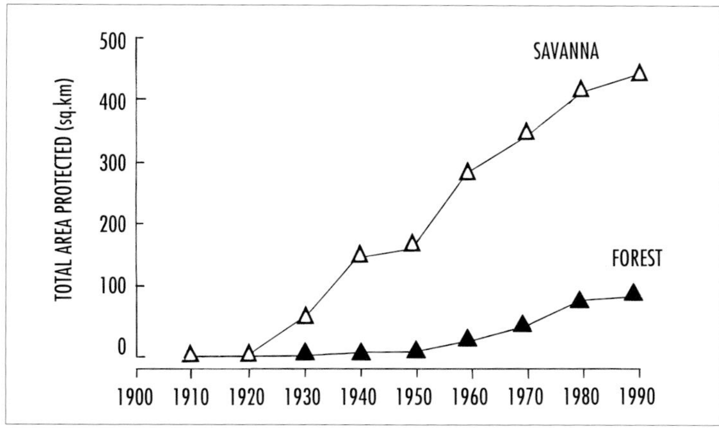

Figure 9.1(a) Total areas of rain forest and savanna encompassed within Africa's protected areas network, 1900–89

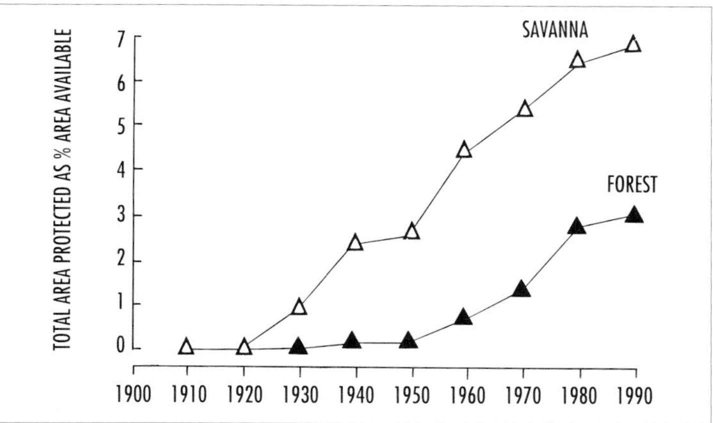

Figure 9.1(b) Total areas protected over the same period, expressed as a percentage of the total covered by rain forest and open woodland/savanna phytochoria (from White, 1983)

The colonial era saw the establishment of many protected areas in Africa. This process has gathered momentum following the independence of African countries, but until recently emphasis has consistently been placed on protecting large mammals in relatively open habitats, first for hunting and later for game viewing. Thus although Africa's first national park (Albert National Park, established in 1925 in what was then Belgian Congo) included important tracts of rain forest, the protection of forests in Africa has consistently lagged behind that of savanna (see Figure 9.1). Forest species tend to be less widely represented in protected areas than their taxonomic counterparts in open savanna habitats (MacKinnon and MacKinnon, 1986).

Moreover, the siting of forest protected areas has often been influenced primarily by socio-economic rather than biological considerations (Clarke and Bell, 1986; Leader-Williams *et al.*, 1990). Protected areas are frequently established in areas not in demand for other forms of land use. The establishment of protected area networks by default often leaves important gaps in the coverage of species and habitats. For example, the vast Salonga National Park in Zaïre was gazetted primarily in order to protect the bonobo or pygmy chimpanzee *Pan paniscus* (found in no other conservation area), but subsequent investigations have revealed that the centre of the species range lies outside the park. Similarly, there are considerable shortfalls in forest protection in the protected area networks of Malawi (Clarke and Bell, 1986) and a number of Central African states (UICN, 1990c). However, recognition of these gaps has been accompanied by plans for several countries, such as Gabon, Congo and Cameroon (Gartlan, 1989; UICN, 1990a; UICN, 1990b), aimed at ensuring better coverage of forest types within protected area networks.

Almost all countries in the Afrotropical realm now participate in international and regional conventions concerned with protecting natural areas, the exceptions being Angola, Equatorial Guinea and São Tomé and Príncipe (see Tables 9.4 and 9.5). These treaties and programmes provide powerful forces for conserving some of the region's most important sites by strengthening the position of the responsible national authorities and attracting financial support from international sources. Nevertheless, opportunities to augment the national resources available for managing these sites are seldom fully exploited.

The World Heritage Convention boasts the greatest number of member countries (22) from the region considered in this Atlas, but half of these have not yet had any natural properties inscribed on the World Heritage List. Of the 21 natural properties in the region listed under the convention, nine contain tropical moist forest (Table 9.4). Notable is Zaïre's 36,000 sq. km Salonga National Park, the largest rain forest national park in the world. In addition a network of *Biosphere reserves (Unesco MAB Programme)* has been set up in 18 countries of the region, with 20 sites in 15 nations containing tropical moist forest. Although based on sound principles, the MAB Programme is not primarily a conservation instrument; implementation of the programme is the responsibility of national MAB committees, which often lack management authority. Six countries within the region have signed the *Convention on Wetlands of International Importance especially as Waterfowl Habitat (the Ramsar Convention)*, but only three Ramsar sites contain extensive stands of rain forest, all in Gabon (Wonga-Wongué and Setté-Cama Nature Reserves and Petit Loango Faunal Reserve). Two other countries – Senegal, with two sites, and Guinea-Bissau – have Ramsar sites that contain mangroves. Figure 9.2 (page 79) shows the position of the World Heritage, Biosphere and Ramsar sites in Africa. Most countries (30) are party to the *African Convention on the Conservation of Nature and Natural Resources* (1968) and the wording of this convention has inspired much wildlife protection and protected area legislation in the region. Lastly, a new *Convention on the Conservation of Biological Diversity* is being negotiated as part of the process leading up to the United Nations Conference on Environment and Development in Rio de Janeiro in June 1992.

The Coverage of Africa's Tropical Forests by Protected Areas

No comprehensive survey exists of the extent of tropical moist forest within Africa's protected areas. Indeed, this would be difficult to achieve with any degree of accuracy as most protected areas include a variety of vegetation types. Although information is available on the total area protected, it is hard to obtain information on the extent of rain forest within protected areas. This is especially the case in Eastern and Southern Africa and for the Indian Ocean Islands where forest is often restricted to tiny remnant patches within very extensive savanna protected areas. Table 9.6 gives information on the extent of these protected areas which lie within the forest zone as defined in this Atlas (see Table 1.2, p. 14) and which are known to contain at least some closed broadleaved forest.

For the forested countries of West and Central Africa it is possible to get a reasonable approximation of the area of rain forest receiving protection. In these regions the protected areas in the rain forest tend to be largely forested. Table 9.6 shows that about 20 per cent of the remaining forest in West Africa is now protected and 7 per cent of that remaining in Central Africa. The apparently high figure for West Africa reflects the fact that all forest outside protected areas is rapidly being cleared. The WCMC Protected Areas Data Unit is aware of proposals to give total protection to a

Table 9.4 State parties to international and regional conventions or programmes concerned with the conservation of natural areas. Where applicable, the number of tropical moist forest sites recognised under respective conventions is given in brackets

	World Heritage Convention	*International Biosphere Reserves*	*Ramsar Convention*	*Regional African Convention*
Angola				
Benin	14 Jun 82			S
Burundi	19 May 82			S
Cameroon	7 Dec 82 (1)	(1)		D
Central African Rep.	22 Dec 80	(1)		D
Comoros				S
Congo	10 Dec 87	(2)		D
Côte d'Ivoire	9 Jan 81 (2[a])	(1)		D
Equatorial Guinea				
Ethiopia	6 Jul 77			S
Gabon	30 Dec 86	(1)	30 Dec 86 (3)	S
Gambia	1 Jul 87			D
Ghana	4 Jul 75	(1)	22 Feb 88	D
Guinea	18 Mar 79 (1[a])	(2)		S
Guinea-Bissau			14 May 90 (1)	
Kenya	5 Jun 91	(2)	5 Jun 91	D
Liberia				D
Madagascar	19 Jul 83	(1)		D
Malawi	5 Jan 82			D
Mauritius		(1)		S
Mozambique	27 Nov 82			D
Nigeria	23 Oct 74	(1)		D
Rwanda		(1)		D
São Tomé and Príncipe				
Senegal	13 Feb 76	(2)	11 Jul 77 (2)	D
Seychelles	9 Apr 80 (1)			D
Sierra Leone				S
Somalia				S
Sudan	6 Jun 74			D
Tanzania	2 Aug 77 (1)			D
Togo				D
Uganda	20 Nov 87	(1)	4 Mar 88 (1)	D
Zaïre	23 Sep 74 (3)	(2)		D
Zimbabwe	16 Aug 82			

[a] Mount Nimba World Heritage Site straddles Côte d'Ivoire and Guinea
D = deposit of instrument; S = signatory

(*Sources:* IUCN, 1989; 1990)

further 4 per cent of West Africa's rain forests but to only an additional 0.5 per cent of those of Central Africa.

The coverage of tropical forests by Africa's protected area network has been reviewed by a detailed analysis of the extent of protection of the different biogeographical units (termed phytochoria) within the Afrotropical realm (MacKinnon and MacKinnon, 1986). This review assessed needs for further conservation in each phytochorion, on the basis of both current coverage by protected areas and species richness and endemism within the different units. Of all the 17 phytochoria considered, the three predominantly forested units – that is, Guineo-Congolian, East Malagasy and Afromontane areas – were found to be those in most urgent need of additional conservation action (MacKinnon and MacKinnon, 1986). Considerable extension of the protected area network is required if Africa's tropical forests are to be effectively conserved.

A second important shortcoming of the current protected area system is that management authorities are often weak and are unable to limit destructive activities within the boundaries of reserves. For instance, São Tomé and Príncipe has no enabling legislation for the establishment of protected areas, while in Somalia the protected area legislation is largely obsolete. In several countries protected area legislation emphasises the protection of animals with little or no provision for habitat protection. Thus in Gabon, timber may be exploited within faunal reserves.

Even where there is legal protection this does not necessarily guarantee the survival of tropical forests. For example, Taï National Park has been described as 'probably the most important protected area in the whole of West Africa' (MacKinnon and MacKinnon, 1986), but it continues to be damaged by poaching, gold prospecting and agricultural encroachment. IUCN's Commission on National Parks and Protected Areas maintains a register of Threatened Protected Areas of the World; this includes Mt Nimba Strict Nature Reserve in Côte d'Ivoire and Guinea, which is threatened by iron ore mining, and Kahuzi-Biega National Park in Zäire, which is at risk from road construction. Logging, land clearance for cultivation and undercropping of the canopy with cardamom have already destroyed considerable parts of the forest reserves of Tanzania's Usambara Mountains (Hamilton and Bensted-Smith, 1989; Wells *et al.*, 1990), while all primary forest remaining in Gabon's 5000 sq. km Lopé Reserve will have been logged in less than ten years (UICN, 1990b).

Criteria for Extending the Protected Area Network

While biogeographical studies at a continental level indicate overall gaps within protected area networks (see above and Harrison *et al.*, 1982; MacKinnon and MacKinnon, 1986), comparisons of species lists for individual forests enable particular sites to be identified as priorities for further protection (e.g. see Rylands, 1990). For Africa's forests such fine-scale analyses have been carried out for selected groups such as primates, plants and birds (Oates, 1986; IUCN, 1987; Collar and Stuart, 1988). These studies have identified a number of forests as being particularly urgent priorities for conservation action. For example, many Afromontane forests are extremely rich in endemic species but currently receive little effective protection. These include Mt Cameroon, the Itombwe mountains in Zaïre, the Usambara and Uzungwa mountains in Tanzania and the forests of the Angola Scarp (MacKinnon and MacKinnon, 1986; Collar and Stuart, 1988). Special importance has also been placed on remaining lowland rain forests west of the Dahomey gap, such as Gola in Sierra Leone (Davies, 1987) and Lofa-Mano in Liberia (MacKinnon and MacKinnon, 1986; Collar and Stuart, 1988). Key unprotected forests in Madagascar include Masoala, 'Sihanaka' and those around Maroantsetra (Jenkins, 1987; Collar and Stuart, 1988). Other critically important but as yet largely unprotected forests in the Afrotropical realm include those of south-west São Tomé, the Ituri Forest in Zaïre and Arabuko-Sokoke Forest in Kenya (see Figure 10.2 and Table 10.5 on pp. 90–2; MacKinnon and MacKinnon, 1986; Collar and Stuart, 1988).

Once critical sites for the establishment of protected areas have been identified, several theoretical considerations bear upon the design of those reserves (see Leader-Williams *et al.*, 1990 for a detailed review). For instance, the theory of island biogeography suggests that species richness in reserves might be maximised by making protected areas as large as possible (Diamond, 1975). However, protected areas are only successful insofar as their contents survive, and attention has now switched to using Population Vulnerability Analysis (PVA) to determine what size conservation areas need to be to support Minimum Viable Populations (MVPs) of species (Soulé, 1986a, b). Extinction is considered to be more likely for relatively small populations, so it is important that populations contained in protected areas are sufficiently large to minimise the chance of extinction through processes such as genetic deterioration (for example, through inbreeding depression) and extrinsic catastrophic events. PVA suggests that the exact size of MVPs will vary between species and habitats, but in general populations of around 1000 (including non-breeding individuals) are thought to be necessary to sustain most species (Soulé, 1986a).

It is becoming clear that protected areas are often too small to protect MVPs of larger species that exist at low densities (Leader-Williams *et al.*, 1990), but in most African countries opportunities for developing vast new protected forest areas are extremely limited. Even where protected areas are sufficient to hold MVPs of large species, the effective protection of such enormous expanses has often proved impossible (see examples above and Leader-Williams and Albon, 1988). One important short-term way of improving protection within protected areas is to increase the funding available for law enforcement patrols (Bell and Clarke, 1986; Harcourt, 1986; Leader-Williams and Albon, 1988; Parker and Graham, 1989). But the long-term survival of Africa's protected forests ultimately depends on the positive support of the people living in and around them. This realisation has brought about a dramatic revision of traditional preservationist approaches towards wildlife conservation (Bell, 1987; Western and Pearl, 1989; Wells *et al.*, 1990). Integration of conservation with local development is now seen as essential for maintaining existing protected areas. The extension of protected areas to incorporate surrounding multiple-use zones is widely believed to be the key to success.

Protected areas in Africa will become increasingly isolated as surrounding land is cultivated. This is what happened here around Bwindi Forest Reserve in Uganda. Coniferous trees mark the boundary of the reserve. C. Harcourt

Table 9.5 International and regional conventions and programmes relevant to the protection of Africa's tropical forests

The World Heritage Convention (1972) provides for the designation of natural and cultural areas of 'outstanding universal value' as World Heritage sites, in order to promote their significance at local, national and international levels. It imposes a legal duty on contracting parties to do their utmost to protect their natural and cultural heritage; this obligation extends beyond sites inscribed on the World Heritage List. The Convention also has provision for aid and technical cooperation to be offered to contracting parties for the protection of their World Heritage sites.

The Unesco Man and the Biosphere Programme provides for the establishment of a worldwide system of 'biosphere reserves' representative of natural ecosystems, to conserve genetic diversity and to promote monitoring, research and training. Particular emphasis is placed on the restoration of degraded ecosystems to more natural conditions, harmoniously integrating traditional patterns of land use within a conservation framework and involving local people in decision-making processes.

The Convention on Wetlands of International Importance especially as Waterfowl Habitat (Ramsar Convention, 1971) provides the framework for international cooperation to conserve wetlands. Contracting parties accept an undertaking to promote the wise use of all wetlands and to designate one or more wetlands for inclusion in a List of Wetlands of International Importance.

The African Convention on the Conservation of Nature and Natural Resources (1968) emphasises the need for special conservation measures for particular species. Three categories of protected area are defined under the Convention: strict nature reserve, in which natural resources are totally protected; national park, in which wildlife is protected for the benefit of the public; and special reserve, in which wildlife is protected but natural resources may be harvested. Countries have tended to integrate these categories into their respective legislation, using a variety of terms.

Table 9.6 Protected area coverage of tropical moist forest (for definition see Table 1.2, chapter 1)

	Land area ('000 sq. km)	Approximate original extent of closed canopy tropical moist forests (sq. km)[1]	Remaining area of tropical moist forest (sq. km)[2]	Total area of protected[3] areas with tropical moist forest			Existing tropical moist forest protected areas as a percentage of:			Existing and proposed tropical moist forest protected areas as a percentage of:		
				Existing (sq. km)	Proposed (sq. km)	Totals (sq. km)	Land area	Original moist forest	Remaining moist forest	Land area	Original moist forest	Remaining moist forest
West Africa												
Benin	111	16,800	424	none	none	–	–	–	–	–	–	–
Côte d'Ivoire	318	229,400	27,464	7,095	none	7,095	2.2	3.1	25.8	2.2	3.1	25.8
Gambia	10	4,100	497	100	none	none	1	2.4	20	1	2.4	20
Ghana	230	145,000	15,842	946	379	1,325	0.4	0.7	6	0.6	0.9	8.4
Guinea	246	185,800	7,655	140	none	140	0.06	0.08	1.8	0.06	0.08	1.8
Guinea-Bissau	28	36,100	6,660[4]	none	none	–	–	–	–	–	–	–
Liberia	96	96,000	41,238	1,308	687	1,995	1.4	1.4	3.2	2.1	2.1	4.8
Nigeria	911	421,000	38,620	2,158	4,060	6,218	0.24	0.5	5.6	0.68	1.58	16.1
Senegal	193	27,700	2,045	846	1	847	0.4	3.1	41.4	0.4	3.1	41.4
Sierra Leone	72	71,700	5,064	12	992	1,004	0.02	0.02	0.24	1.4	1.4	19.8
Togo	54	18,000	1,360	none	none	–	–	–	–	–	–	–
Central Africa												
Burundi	26	10,600	413	379	none	379	1.5	3.6	91.8	1.5	3.6	91.8
Cameroon	465	376,900	155,330	11,266	none	11,266	2.4	3	7.3	2.4	3	7.3
Central African Republic	623	324,500	52,236	4,359	10,500	14,859	0.7	1.3	8.3	2.4	4.6	28.4
Congo	342	342,000	213,400[4]	12,148	none	12,148	3.6	3.6	5.7	3.6	3.6	5.7
Equatorial Guinea	28	26,000	17,004[5]	3,150	none	3,150	11.3	12.1	18.5	11.3	12.1	18.5
Gabon	258	258,000	227,500[6]	17,900	none	17,900	6.9	6.9	7.9	6.9	6.9	7.9
São Tomé and Príncipe	1	960[*]	299	none	none	–	–	–	–	–	–	–
Rwanda	25	9,400	1,554	150	none	150	0.6	1.6	9.7	0.6	1.6	9.7
Zaïre	2,267	1,784,000	1,190,737[5]	63,130	none	63,130	2.8	3.5	5.3	2.8	3.5	5.3
Southern Africa												
Angola	1,247	218,200	29,000[4]	nd	–	–	–	–	–	–	–	–
Malawi	94	10,700	320[7]	nd	–	–	–	–	–	–	–	–
Mozambique	782	246,900	9,350[4]	nd	–	–	–	–	–	–	–	–
Zimbabwe	387	7,700	80[8]	nd	–	–	–	–	–	–	–	–
Eastern Africa												
Djibouti	23	300	10[4]	nd	–	–	–	–	–	–	–	–
Ethiopia	1,101	249,300	27,500[4]	11,574[11]	–	–	–	–	–	–	–	–
Kenya	570	81,200	6,900[4]	13,148[11]	–	–	–	–	–	–	–	–
Somalia	627	21,200	14,800[4]	nd	–	–	–	–	–	–	–	–
Sudan	2,376	27,000	6,400[4]	nd	–	–	–	–	–	–	–	–
Tanzania	886	176,200	14,400[4]	77,008[11]	–	–	–	–	–	–	–	–
Uganda	200	103,400	7,400[9]	4,483	310	4,793	2.2	4.3	60.6	2.4	4.6	64.8
Indian Ocean Islands												
Comoros	2	2,230[*]	160[4]	nd	–	–	–	–	–	–	–	
Mauritius	2	1,850[*]	30[4]	nd	–	–	–	–	–	–	–	
Réunion	2.5	2,500[*]	820[4]	nd	–	–	–	–	–	–	–	
Seychelles	0.3	270[*]	30[4]	nd	–	–	–	–	–	–	–	
Madagascar	582	275,086	41,715[10]	5,788	none	5,788	1	2.1	13.9	1	2.1	13.9

1 Taken from White (1983) as in MacKinnon and MacKinnon (1986) except for Gabon and Liberia where these authors indicate that both countries were originally completely forested but the figures they give are country areas rather than land areas. 'Original' cover includes mosaics.
2 Figures given here are derived from maps in chapters 11–32 unless stated otherwise.
3 Not including forest reserves. Please note that these totals are for protected areas which contain at least some tropical moist forest as determined on the maps in this Atlas (except in Kenya, Ethiopia and Tanzania; see 11); it is not possible to take account of fragmentation of forest within each protected area. In many cases, therefore, the forest coverage will be over-optimistic.
4 Figures from FAO (1988).
5 Including 7,945 sq. km of degraded lowland rain forest in Equatorial Guinea and 86,547 sq. km in Zaïre.
6 Figure from UICN (1990b).
7 Figure from Dowsett-Lemaire (1989, 1990).
8 Figure supplied by T. Muller (*in litt.*).
9 From Howard (1991).
10 This figure has been calculated by adding Green and Sussman's (1990) figure for eastern rain forest to that calculated for mangroves from Map 26.1 plus an estimated 400 sq. km for forest remaining in the Sambirano region.
nd No data – protected area maps within these country chapters are sketch maps only.
11 Percentage of forest cannot be realistically calculated for Kenya, Ethiopia and Tanzania. This is because although there are protected areas with forest within their boundaries, the forest are often fragmented and small in size and only cover a fraction of the size of the actual protected area.
* It has been assumed these islands were originally completely forested. MacKinnon and MacKinnon (1986) give no data for them.

Reconciling Development and Tropical Forest Conservation

The mutual interdependence of conservation and development forms the central theme of the World Conservation Strategy, launched by IUCN, WWF and UNEP in 1980. This groundbreaking initiative defined conservation as 'the management of human use of the biosphere so that it may yield the greatest sustainable benefit to present generations while maintaining its potential to meet the needs and aspirations of future generations' (IUCN, 1980). Such a pragmatic approach to conservation was reiterated by the final report of the World Commission on Environment and Development (WCED, 1987) and forms the basis for the National Conservation Strategies developed for Madagascar, Nigeria and Zimbabwe and those in preparation for Guinea-Bissau and Ethiopia.

The importance of forest conservation to the development process is stressed by the Tropical Forestry Action Plan (FAO, 1985). This proposes a number of ways to conserve tropical forests, including minimising the damage caused by selective timber extraction and reducing demand on remaining forests by improving agroforestry and reforestation. These and other measures have also been proposed at national level to promote the diversification and sustainability of forest use in several Central African countries (UICN, 1990c) and in Uganda (Tabor *et al.*, 1990).

On a local scale, a key tool for promoting a balance between conservation and resource use is the concept of the Biosphere reserve, first launched in 1971 (Table 9.5; Batisse, 1986). Biosphere reserves aim to integrate rural development with conservation by establishing multiple-use buffer zones around undisturbed core areas of high biological diversity. In Ghana, for instance, the proposed conversion of a forest reserve to a national park is being accompanied by the development of a neighbouring reserve as a multiple-use area where local people can harvest game meat and medicinal plants (see case study on Kakum and Assin-Attandanso forest reserves below, and case studies on projects in Cameroon and Nigeria in chapters 13 and 27 respectively). Similarly, an EEC-funded buffer zone project will promote agroforestry and sustained yield charcoal production around the Dja Biosphere Reserve in Cameroon. An equivalent approach has already proved successful in Burundi, where two community-based projects have reduced threats to important forest reserves by combining improved law enforcement with agroforestry extension programmes around the reserves (Wells *et al.*, 1990).

Other local initiatives seek to provide direct economic incentives for the conservation of African forests. For example, in Liberia local perceptions of Sapo National Park have greatly improved as a result of a WWF-sponsored agriculture project and a development fund financed by tourism in the park (see case study in chapter 25). Likewise in Madagascar a rural community supported the creation

COMMUNITY INVOLVEMENT IN KAKUM AND ASSIN-ATTANDANSO FOREST RESERVES, GHANA

A community-oriented conservation programme provides the focus for converting former timber concession lands in southern Ghana into a national park and wildlife reserve. It is proposed to transfer the Kakum Forest Reserve (213 sq. km) and the adjoining Assin-Attandanso Forest Reserve (154 sq. km) from the Department of Forestry to the Department of Game and Wildlife. Local farmers are giving strong and unprecedented community support for the proposed national park because few tangible benefits have resulted from earlier commercial logging in the reserves. Under the circumstances local people feel that there is everything to gain and nothing to lose from the national park development project.

Despite extensive commercial logging in the reserves, the forest landscape remains largely intact. The diversity of plants appears to be quite substantial, and the University of Ghana-Legon has begun a floral inventory of the reserves. The fauna is rich and varied and includes forest elephants *Loxodonta africana cyclotis*, perhaps nine species of primates, numerous forest antelope (including bongo *Tragelaphus euryceros*, bushbuck *Tragelaphus scriptus* and several duikers), forest buffalo *Syncerus caffer nanus*, bushpig *Potamochoerus porcus*, giant forest hog *Hylochoerus meinertzhageni*, giant flying squirrels and many small mammals. Prominent among the many bird species are grey parrots *Psittacus erithacus* and three species of hornbills, including the vulture-sized black-casqued hornbill *Ceratogymna atrata*. The Nile monitor lizard *Varanus niloticus* is relatively common in the reserves, while the hinged tortoise *Kinixys* sp. and many species of frogs and snakes are also present.

The development programme envisaged for the proposed Kakum National Park and Assin-Attandanso Game Production Reserve will follow that prescribed for protected and multiple-use areas under the Unesco Man and the Biosphere Programme model. Management strategies will target watershed protection, wildlife resource conservation and rural development, within the framework of a UNDP-funded regional development project. Nature tourism will be integrated with community education and wildlife research to enhance local support. While the Kakum Forest Reserve will be developed as a national park, it is envisaged that the Assin-Attandanso Reserve will become a multiple-use forest. Local community access to important forest resources is proposed through the development of harvest schedules for the sustainable use of wildlife, especially locally abundant game animals and medicinal plants. An on-site plant medicine research scheme is planned.

A potential source of conflict between local inhabitants and the reserves is wildlife damage to surrounding farmlands. Destruction of cash and subsistence crops by elephants, antelopes, primates and other forest-dwelling wildlife is already substantial. Any increase in crop depredation could exacerbate the legitimate concerns of local farmers and alter the current climate of community support. An experimental buffer zone around the reserves will be used to try to combat this problem. It is believed that tree plantations adjacent to the reserve boundaries may act as passive barriers to elephant movement. The elephants rarely travel more than 200 m beyond the forest and, unlike other Ghanaian elephants, do not eat or otherwise damage cacao trees. Thus the establishment of cacao and fuelwood plantations along the reserve boundaries should prevent elephants from entering the surrounding agricultural landscape. A buffer zone of trees should also help maintain forest microclimates within the reserves and mitigate the adverse edge effects associated with increased wind and light penetration along the forest boundaries.

Long-term protection of the rich biological diversity of these forests will ultimately depend upon the active support of local people. The goal of this project, therefore, is the protection of native forest landscapes and biodiversity for the mutual benefit of resident human and wildlife communities.

Source: Joseph P. Dudley

HUNTING IN KORUP NATIONAL PARK, CAMEROON

Hunting and trapping are important economic and cultural activities which, if correctly managed, could be legitimate sustainable uses of Africa's tropical forests. However, little is known about the effect on wildlife in forest areas. Research undertaken in and around Cameroon's Korup National Park during 1988 sheds some light on the problems that will have to be overcome if sustainable hunting is to play a role in forest conservation.

Fears were expressed that the gazetting of the park in 1986 might cause severe hardship to local communities. Initially it had been assumed that local villagers hunted primarily for subsistence. However, bushmeat commands high prices in Cameroonian towns and cities and it was soon realised that people living around Korup hunted more for money than for meat. Although villagers earned cash through a wide array of other activities, approximately 70 per cent of all households were involved in hunting and trapping, and revenue earned through bushmeat accounted for 56 per cent of the villagers' total income. A complete hunting ban would therefore have had a serious impact on these communities.

All of the forest's larger mammals were hunted, except elephant and buffalo. The species taken most commonly were blue and bay duikers (*Cephalophus monticola* and *C. dorsalis*), which together made up 49.3 per cent of all the animals killed by hunters. The brush-tailed porcupine *Atherurus africanus* made up 13 per cent of the take, followed by two species of primate, Preuss's red colobus *Procolobus [badius] pennanti preussi* (7.2 per cent) and drill *Mandrillus leucophaeus* (5.7 per cent). The remaining 24.8 per cent of the total animal offtake was divided between 18 species of mammal and four reptiles.

It is difficult to draw firm conclusions about the sustainability of hunting in Korup since there is no information on live animal densities. Moreover, estimates for other African forests vary widely. For example, blue duiker densities of 70 per sq. km have been reported for forest in north-east Gabon (Dubost, 1980). If densities in Korup were similar, annual offtake would be only about 8 per cent, which would be sustainable. However, in north-east Zaïre, blue duiker densities were estimated at only 15 per sq. km (Hart and Petrides, 1987), while the estimated total density of all antelope in forest in south-west Gabon was only 0.85 animals per sq. km (Prins and Reitsma, 1989). The total weight of animals hunted (the biomass offtake) in Korup was estimated at 217 kg per sq. km per year (see Table 9.7). Based on a figure of available biomass of approximately 1050 kg per sq. km in south-west Gabon (Prins and Reitsma, 1989), this suggests a 20 per cent harvest in Korup. However, over half of the total of animals hunted in the Gabon study area was made up of elephants, indicating that the Korup offtake may exceed 40 per cent. Without estimates of animal densities in Korup, it can only be inferred that present hunting levels are likely to be unsustainable. This impression is reinforced in more densely inhabited forest areas around the national park, where hunting and trapping are still carried out, but where prey densities have been so reduced that bushmeat sales are no longer economically important to local communities.

The bay duiker Cephalophus dorsalis *is the most commonly hunted species in Korup National Park.* A. Dunn

Table 9.7 Numbers and Biomass of species hunted in Korup National Park

Species	*Estimated number of animals taken per sq. km per annum*	*Estimated Biomass removed per sq. km per annum*
Cephalophus monticola	5.76	26.5
Cephalophus dorsalis	5.88	91.7
Cephalophus ogilbyi	0.69	11.0
Cephalophus silvicultor	0.13	8.1
Haemoschus aquaticus	0.50	6.3
Potamochoerus porcus	0.29	17.5
Procolobus [badius] pennanti	1.71	16.2
Mandrillus leucophaeus	1.34	16.7
Cercopithecus nictitans	0.89	6.2
Cercopithecus mona	0.67	2.2
Cercopithecus erythrotis	0.22	0.6
Cercocebus torquatus	0.07	0.6
Cercopithecus pogonias	0.04	0.2
Atherurus africanus	3.03	8.1
Others:		
mammals and reptiles	2.40	5.1
Total	23.62	217.0

(*Source:* Mark Infield)

Overhunting in Korup National Park will have a drastic effect on the local economy. Reduced stocks of wildlife could mean that hunting in the park will rapidly cease to be economically viable – as is already the case around its periphery. In the short term, many animal species will probably persist despite depletion, provided that their habitat remains intact. In the long term, however, the collapse of hunting could have severe negative consequences for forest biodiversity, since it might increase existing economic pressures to convert primary forest to intensive cash crop production (but see case study in chapter 13).

A good way to resist such pressure is to establish a more sustainable basis for the local hunting industry in the forests around the national park. This would require the development of techniques for monitoring wildlife populations in forests and for determining what level of hunting is indeed sustainable. In addition, mechanisms should be developed to enable local communities to regulate their own hunting activities. Most importantly, a dramatic rethinking of long-held attitudes is needed if hunting is to be turned from a marginal and often illegal activity into an attractive rural enterprise that helps conserve forest biodiversity.

Source: Mark Infield

TOURISM AS A CONSERVATION STRATEGY IN THE NYUNGWE FOREST, RWANDA

The Nyungwe Forest Reserve is the largest montane forest block in East Africa, covering about 900 sq. km of south-west Rwanda. Mountain gorillas *Gorilla gorilla berengei* have long attracted visitors to Rwanda's northern Volcanoes National Park (see case study in chapter 12). Now a tourism initiative in Nyungwe, started in September 1988 by the Nyungwe Forest Conservation Project (PCFN), financed by USAID and sponsored by Wildlife Conservation International, is already drawing 4000 visitors a year.

The tourism programme, devised by Dr Amy Vedder, provides low impact exploitation of Nyungwe while giving the Rwandan government an economic incentive to protect the forest. Visitors are attracted by the beauty and diversity of the forest and its extraordinary variety of animals and plants, with orchids, butterflies, birds and primates being particularly abundant. But Nyungwe is also crucial to the well-being of the local population. It protects the watershed, regulating stream flow and limiting soil erosion so that rivers remain clear downstream. The forest is also a source of many natural products, including wood for fuel and construction, honey, bamboo, natural ropes, medicinal plants and thatching material.

However, there are enormous pressures on Nyungwe. At the beginning of this century forests ran continuously along the north–south backbone of western Rwanda, covering nearly one-third of the country. By 1990 this had been reduced by nearly 80 per cent, with only four isolated forest blocks remaining. With population densities of up to 800 people per sq. km and over 90 per cent of the nation depending on farming for a living, the demand for land is intense. Hunting pressure in the forest is very high:buffalo are now extinct and elephant, giant forest hog, leopard *Panthera pardus* and three species of duiker are extremely rare. Valuable tree species, most of which regenerate very slowly, risk being overexploited. Thousands of gold panners destroy stream-beds in search of the small quantity of gold found in the forest.

The PCFN tourism programme is located in two areas of Nyungwe. The central site is the base for a network of over 25 km of trails, primitive campsites and guided visits to see monkeys. The western site is a picnic and camping area with the potential for the development of monkey visits and trail hiking as well. Initially, most visitors were attracted to the forest by the spectacular and acrobatic Angola black-and-white colobus monkeys *Colobus angolensis*, which live in extremely large groups of up to 400 animals. But now about half of the visitors choose to walk one of the scenic trails where they may see monkeys and other forest attractions on their own. In addition, people may be guided to grey-cheeked mangabeys *Cercocebus albigena* and blue monkeys *Cercopithecus mitis*.

The benefits of tourism are considerable. The presence of visitors has already reduced poaching pressure and the calls of leopards and francolins *Francolinus* sp. are now heard frequently in tourist areas. Moreover, in addition to foreign and local visitors, national and regional government officials (including the President of Rwanda, government ministers and the local governor) visit the forest.

Currently the programme earns about US$15,000 annually, which is more than sufficient to pay Rwandan staff. Furthermore, the presence of the project has influenced local officials to reflect on the potential for additional tourism in the region. Discussions have been held about the possibilities of hotels in or near the forest to accommodate visitors who prefer not to camp. Other proposed attractions include tea factory tours, beaches around Lake Kivu and the possibility of visits to see chimpanzees *Pan troglodytes* in a nearby gallery forest.

A mountain gorilla Gorilla gorilla berengei *female and infant in a nest which overlooks the cleared, cultivated land which surrounds the Volcanoes National Park in Rwanda.* C. Harcourt

Clearly tourism alone cannot protect the forest. Its scope is too small and there is too little local involvement. That is why PCFN emphasises conservation education as well. Local people are dependent on forest resources. Since more than half the forest will be managed as a multiple-use zone, that use must be wise and sustainable. Forest products such as honey, baskets and carvings are being marketed, but care must be taken to ensure that they are not being overharvested. PCFN aims to establish a dialogue with the local population, in which the values of the forest and the problems of its conservation and use are openly discussed. Secondary schools will be visited to teach the students more technical aspects of forest ecology and value.

A guard force patrols the forest and enforces strict laws. The forest has been clearly delineated from farmers' fields by a buffer zone of trees planted around its perimeter. As they mature, these trees should also supply local people with wood and so reduce demands upon the forest. Yet despite its success, Nyungwe's conservation programme still lacks some essential components. There is no demarcation between the multiple-use area and the central protected zone. Local people are not being provided with enough alternatives to using forest resources. And all money from tourism, except that needed for local salaries, goes to central government rather than to local development. It is imperative that some of the earnings from tourism directly benefit the local people, since the responsibility for the long-term protection of the forest ultimately rests with them.

Nevertheless, the strength of tourism as a conservation strategy is that it draws government attention to the forest as a source of foreign exchange. The areas that it touches are immediately protected because of the continuous presence of project personnel and tourists. But tourism is a fickle business and the conflict that rocked Rwanda may have had lasting effects on tourist confidence there. Rwanda could suffer greatly as a result, since tourism has become the third highest source of foreign income, after coffee and tea. A hard lesson must be learnt: tourism in good times can lend credence to forest conservation, but ultimately, such conservation depends on local people and their perceived value of the forest.

Source: Katherine Offutt

of the Beza-Mahafaly Special Reserve in exchange for direct benefits such as repairs to an important access route, improved irrigation and the building of a school (Wells *et al.*, 1990).

Several general lessons can be gleaned from current attempts to integrate conservation with rural development (Sayer, 1991). First, the results of such efforts depend greatly on local social and economic factors. Initiatives to encourage sustainable resource use are frequently most successful where people have previously gained little from a forest (see case study on Kakum and Assin-Attandanso forest reserves) and where human population density is relatively low. For example, hunting and trapping are important activities in Africa (Marks, 1989), yet high population densities may mean that existing harvests from West African forests such as Korup are unsustainable (see case study on hunting in Korup National Park). In contrast, hunting levels in less densely populated areas of Central Africa, such as Lopé in Gabon or Salonga in Zaïre, may be sustainable.

Second, it is possible to encourage an array of income-generating activities within any one forest. For instance, wildlife-related tourism is widely regarded as an important and sustainable source of foreign exchange for developing countries (see Table 9.1, the case study on tourism in the Nyungwe Forest and the case study in chapter 12). Yet tourism is extremely vulnerable to the political instability which besets many African countries (see case study on Nyungwe Forest and chapter 25). To ensure that such external factors do not eliminate all direct economic incentives for conservation it is important to diversify the use of resources within forests.

Third, active local involvement in planning lies at the core of successful integration of development and conservation (Bell, 1987; Sayer, 1991). Thus the direct management of wildlife resources by and for local people is seen as pivotal to the success of such initiatives as the Communal Area Management Plan for Indigenous Resources (CAMPFIRE) in Zimbabwe (Martin, 1986; Cole, 1990). Similar attempts to involve local communities in the management of protected forests include the involvement of villagers in marking the boundary of the Kilum Mountain Forest Reserve in Cameroon (Stone, 1990) and the decision of local people to create the Beza-Mahafaly Special Reserve in south-western Madagascar (Wells *et al.*, 1990).

A final problem faces many projects which stress utilitarian reasons for protecting forests: the direct economic benefits of conservation may often prove to be less than those of converting forests to other forms of land use, particularly where governments are of necessity concerned with immediate rather than long-term financial returns (Bell, 1987; Bodmer *et al.*, 1990; cf. Peters *et al.*, 1990). The fate of Africa's forests must not be dictated simply by the vagaries of economics; therefore, two important factors must be acknowledged and integrated into conservation planning. First, local people – as well as foreigners – often support conservation for aesthetic rather than economic reasons (Collar, 1986; Bell, 1987). This non-utilitarian motivation may help counterbalance the economic incentives for destroying forests, in which case it is essential that conservation schemes ensure local communities have continued access to protected areas (Bell, 1987). Second, many of the more indirect benefits of protected forests, such as climate regulation and species conservation, are enjoyed at international level, while most of the costs are incurred by local people. Thus it is perhaps appropriate that any net costs of protecting forests should be met by the international community, through mechanisms such as debt relief (Ayres, 1989; Cartwright, 1989).

Conclusion

The present coverage of tropical forests within Africa's protected area network is generally inadequate. Many critical habitats are poorly represented in protected areas and there is a clear need to extend the system to include additional sites. Moreover, existing protection within nominal protected areas is often ineffective. This is partly because the costs of forest conservation currently fall disproportionately on nearby communities, leading to antagonistic relations between protected areas and local people. Reversing this pattern, by means of integrating conservation with small-scale rural development, is therefore essential if local support for forest protection is to be increased and the existing protected area network expanded. Given the rapidly increasing rate of forest loss across the continent, conservation policies developed in the next few years will in large part determine the fate of Africa's tropical forests.

References

Ayres, J. M. (1989) Debt-for-equity swaps and the conservation of tropical rain forests. *Trends in Ecology and Evolution* **4**: 331–2.

Barnes, R. F. W. (1990) Deforestation trends in tropical Africa. *African Journal of Ecoloy* **28**: 161–73.

Batisse, M. (1986) Developing and focusing the Biosphere reserve concept. *Nature and Resources* **12**: 2–11.

Bell, R. H. V. (1983) Decision-making in wildlife management with reference to the problem of overpopulation. In: *Management of Large Mammals in African Conservation Areas*, pp. 145–72. Owen-Smith, R. N. (ed.). Haum Educational Publishers, Pretoria, South Africa.

Bell, R. H. V. (1987) Conservation with a human face: conflict and reconciliation in African land use planning. In: *Conservation in Africa: People, Policies and Practice*, pp. 79–101. Anderson, D. and Grove, R. (eds). Cambridge University Press, Cambridge, UK.

Bell, R. H. V. and Clarke, J. E. (1986) Funding and financial control. In: *Conservation and Wildlife Management in Africa*, pp. 543–55. Bell, R. H. V. and McShane-Caluzi, E. (eds). US Peace Corps.

Bodmer, R. E., Fang, T. G. and Moya, L. (1990) Fruits of the forest. *Nature* **343**: 109.

Cartwright, J. (1989) Conserving nature, decreasing debt. *Third World Quarterly* **11**: 114–26.

Clarke, J. E. and Bell, R. H. V. (1986) Representation of biotic communities in Protected Areas: a Malawian case study. *Biological Conservation* **35**: 293–311.

Cole, M.-L. (1990) A farm on the wild side. *New Scientist* **1733**: 62–5.

Collar, N. J. (1986) Species are a measure of man's freedom: reflections after writing a Red Data Book on African birds. *Oryx* **20**: 15–19.

Collar, N. J. and Stuart, S. N. (1988) *Key Forests for Threatened Birds in Africa*. ICBP, Cambridge, UK. 102 pp.

Davies, A. G. (1987) *The Gola Forest Reserves, Sierra Leone. Wildlife Conservation and Forest Management*. IUCN, Gland, Switzerland and Cambridge, UK. 126 pp.

Diamond, J. M. (1975) The island dilemma: lessons of modern biogeographic studies for the design of nature reserves. *Biological Conservation* **7**: 3–15.

Dickenson, R. E. (1981) Effects of tropical deforestation on climate. In: *Blowing in the Wind: Deforestation and Long-range Implications*. Studies in Third World Societies, **14**. College of William and Mary, Williamsburg, USA.

Dosso, H., Guillaumet, J. L. and Hadley, M. (1981) The Taï project: land use problems in a tropical rain forest. *Ambio* **10**: 120–5.

Dowsett-Lemaire, F. (1989) The flora and phytogeography of the evergreen forests of Malawi. I: afromontane and mid-altitude forests. *Bulletin du Jardins Botanique National de Belgique* **59**: 3–131.

Dowsett-Lemaire, F. (1990) The flora and phytogeography of the evergreen forests of Malawi. II: lowland forests. *Bulletin du Jardins Botanique National de Belgique* **60**: 9–71.

Dubost, G. (1980) L'écologie et la vie sociale du céphalophe bleu (*Cephalophus monticola* Thunberg), petit ruminant forestier africain. *Zeitschrift für Tierpsychologie* **54**: 205–66.

Dunne, T. (1979) Sediment yield and land use in tropical catchments. *Journal of Hydrology* **42**: 281–300.

FAO (1985) *Tropical Forestry Action Plan*. FAO, Rome, Italy. 159 pp.

FAO (1988) *An Interim Report on the State of Forest Resources in the Developing Countries*. FAO, Rome, Italy. 18 pp.

Gartlan, S. (1989) *La Conservation des Ecosystèmes forestiers du Cameroun*. IUCN, Gland, Switzerland and Cambridge, UK.

Gartlan, J. S. and Agland, P. C. (1981) *A Proposal for a Program of Rain Forest Conservation and National Park Development in Cameroon, West-Central Africa*. Presented to Gulf Oil Corporation and Société Nationale Elf Acquitaine. WWF (unpublished).

Green, M. J. B., Paine, J. and McNeely, J. (1991) The protected areas system. In: *The Conservation Atlas of Tropical Forests, Asia and the Pacific*. pp. 60–7. Collins, N. M., Sayer, J. A. and Whitmore, T. C. (eds). Macmillan, London, UK. in association with IUCN, Gland, Switzerland. 256pp.

Green, G. M. and Sussman, R. W. (1990) Deforestation history of the eastern rain forests of Madagascar from satellite images. *Science* **248**: 212–15.

Guillaumet, J. L. (1967) Recherches sur la végétation et la flore de la région du Bas-Cavally (Côte d'Ivoire). *Mémoire ORSTOM* **20**: 1–247.

Figure 9.2 Distribution of Conservation Areas designated under International Conventions and Programmes.

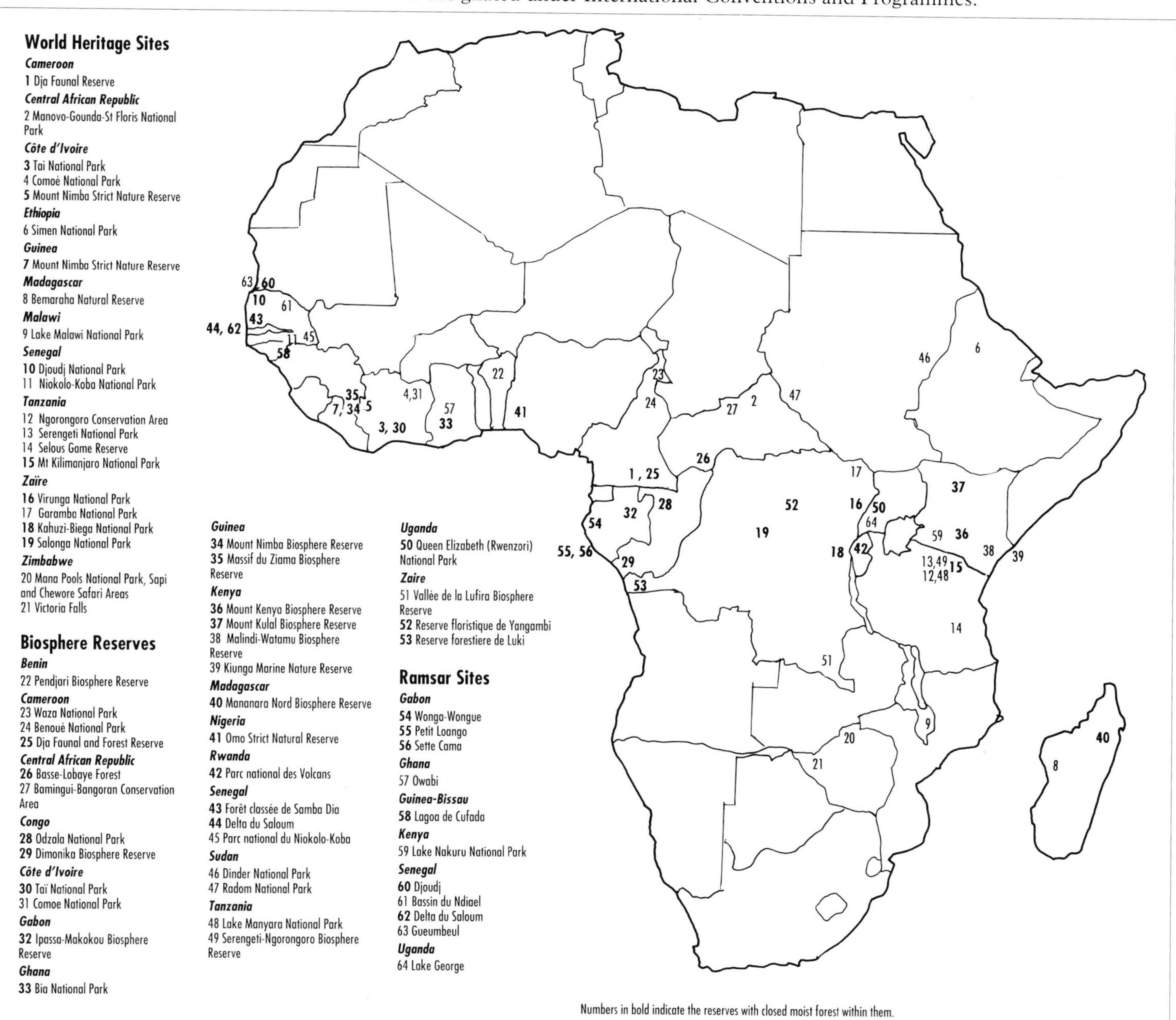

Hamilton, A. C. and Bensted-Smith, R. (eds) (1989) *Forest Conservation in the East Usambara Mountains, Tanzania*. IUCN, Gland, Switzerland and Cambridge, UK. 392 pp.

Harcourt, A. H. (1986) Gorilla conservation: anatomy of a campaign. In: *Primates: the Road to Self-sustaining Populations*, pp. 31–46. Benirschke, K. (ed.). Springer-Verlag, New York, USA.

Harrison, J., Miller, K. and McNeely, J. (1982) The world coverage of protected areas: development goals and environmental needs. *Ambio* **11**: 238–45.

Hart, J. and Petrides, G. (1987) In: *People and the Tropical Forest*. Department of State, Washington, DC, USA.

Howard, P. C. (1991) *Nature Conservation in Uganda's Tropical Forest Reserves*. IUCN, Gland, Switzerland and Cambridge, UK. 330 pp.

IUCN (1980) *World Conservation Strategy: Living Resource Conservation for Sustainable Development*. IUCN/UNEP/WWF, Gland, Switzerland.

IUCN (1984) Categories, objectives and criteria for protected areas. In: *National Parks, Conservation and Development: the Role of Protected Areas in Sustaining Society*, pp. 47–53. McNeely, J. A. and Miller, K. R. (eds). Smithsonian Institution Press, Washington, DC, USA.

IUCN (1987) *Centres of Plant Diversity. A Guide and Strategy for their Conservation*. IUCN Threatened Plants Unit, Kew, UK.

IUCN (1989) *Status of Multilateral Treaties in the Field of Environment and Conservation*. IUCN Environmental Policy and Law Occasional Paper No. 1, 2nd edition.

IUCN (1990) *1990 United Nations List of National Parks and Protected Areas*. IUCN, Gland, Switzerland and Cambridge, UK. 275 pp.

Jenkins, M. D. (1987) *Madagascar: an Environmental Profile*. IUCN/UNEP/WWF. IUCN, Gland, Switzerland and Cambridge, UK. 374 pp.

Kingdon, J. (1990) *Island Africa*. Collins, London, UK. 287 pp.

Leader-Williams, N. and Albon, S. D. (1988) Allocation of resources for conservation. *Nature* **336**: 533–5.

Leader-Williams, N., Harrison, J. and Green, M. J. B. (1990) Designing protected areas to conserve natural resources. *Science Progress, Oxford*. **74**: 189–204.

MacKinnon, J. and MacKinnon, K. (1986) *Review of the Protected Areas System in the Afrotropical Realm*. IUCN/UNEP, Gland, Switzerland. 259 pp.

MacKinnon, J. R., MacKinnon, K., Child, G. and Thorsell, J. (1986) *Managing Protected Areas in the Tropics*. IUCN, Gland, Switzerland. 284 pp.

Marks, S. A. (1989) Small-scale hunting economies in the tropics. In: *Wildlife Production Systems*, pp. 75–95. Hudson, R. J., Drew, K. R. and Baskin, L. M. (eds). Cambridge University Press, Cambridge, UK.

Martin, R. B. (1986) Communal area management plan for indigenous resources (Project CAMPFIRE). In: *Conservation and Wildlife Management in Africa*, pp. 279–95. Bell, R. H. V. and McShane-Caluzi, E. (eds). US Peace Corps.

McNeely, J. A. (1988) *Economics and Biological Diversity. Developing and Using Economic Incentives to Conserve Biological Resources*. IUCN, Gland, Switzerland. 236 pp.

McNeely, J. A. and Miller, K. R. (eds) (1984) *National Parks, Conservation and Development: the Role of Protected Areas in Sustaining Society*. Smithsonian Institution Press, Washington, DC, USA. 838 pp.

Monfort, A. (1990) *Rapport Annuel du Projet Tourisme et Parcs Nationaux*. ORTPN, Kigali, Rwanda.

Oates, J. F. (1986) *Action Plan for African Primate Conservation: 1986–1990*. IUCN/SSC Primate Specialist Group. Stony Brook, New York, USA.

Parker, I. S. C. and Graham, A. D. (1989) Men, elephants and competition. *Symposium Zoological Society of London* **61**: 241–52.

Peters, C. M., Gentry, A. H. and Mendelsohn, R. O. (1989) Valuation of an Amazonian rainforest. *Nature* **339**: 655–6.

Prescott-Allen, R. and Prescott-Allen, C. (1982) *What's Wildlife Worth? Economic Contributions of Wild Plants and Animals to Developing Countries*. Earthscan, London, UK. 92 pp.

Prins, H. H. T. and Reitsma, J. M. (1989) Mammalian biomass in an African equatorial rain forest. *Journal of Animal Ecology* **58**: 851–61.

Rylands, A. B. (1990) Priority areas for conservation in the Amazon. *Trends in Ecology and Evolution* **5**: 240–1.

Sayer, J. A. (1991) *Rainforest Buffer Zones; Guidelines for Protected Area Managers*. IUCN, Gland, Switzerland.

Sayer, J. A. and Stuart, S. N. (1989) Biological diversity and tropical forests. *Environmental Conservation* **15**: 193–4.

Skorupa, J. P. and Kasenene, J. M. (1983) Tropical forest management: can rates of natural treefalls help guide us? *Oryx* **18**: 96–101.

Soulé, M. A. (ed.) (1986a) *Viable Populations for Conservation*. Cambridge University Press, Cambridge, UK. 189 pp.

Soulé, M. A. (ed.) (1986b) *Conservation Biology. The Science of Scarcity and Diversity*. Sinauer, Sunderland, USA. 584 pp.

Stone, R. D. (1990) The view from Kilum mountain. *WWF News* **84**: March/April 1990. WWF, Gland, Switzerland.

Sutton, S. L., Whitmore, T. C. and Chadwick, A. C. (eds) (1984) *Tropical Rain Forests: Ecology and Management*. Blackwell, Oxford, UK.

Tabor, G. M., Johns, A. D. and Kasenene, J. M. (1990) Deciding the future of Uganda's tropical forests. *Oryx* **24**: 208–14.

UICN (1990a) *La Conservation des Ecosystèmes forestiers du Congo*. Basé sur le travail de P. Hecketsweiler. UICN, Gland, Switzerland and Cambridge, UK. 187 pp.

UICN (1990b) *La Conservation des Ecosystèmes forestiers du Gabon*. Basé sur le travail de C. Wilks. UICN, Gland, Switzerland and Cambridge, UK. 215 pp.

UICN (1990c) *La Conservation des Ecosystèmes forestiers d'Afrique Centrale*. UICN, Gland, Switzerland and Cambridge, UK. 124 pp.

WCED (1987) *Our Common Future*. Oxford University Press, Oxford, UK. 383 pp.

Wells, M., Brandon, K. and Hannah, L. (1990) *People and Parks. Linking Protected Area Management with Local Communities*. Report presented to World Bank conference, Abidjan, Côte d'Ivoire. November 1990.

Western, D. and Pearl, M. (eds) (1989) *Conservation for the Twenty-first Century*. Oxford University Press, New York, USA. 365 pp.

White, F. (1983) *The Vegetation of Africa: a descriptive memoir to accompany the Unesco/AETFAT/UNSO vegetation map of Africa*. Unesco, Paris, France. 356 pp.

World Bank (1990) *World Development Report 1990*. Oxford University Press, Oxford, UK. 260 pp.

Authorship

Andrew Balmford and Nigel Leader-Williams of the Large Animal Research Group, Department of Zoology, Cambridge University, and Michael Green, WCMC, Cambridge, with contributions from J. P. Dudley, Ghana, Katie Offert, Uganda, Mark Infield, WWF-International, Gland, Gil Isabirye-Basuta, Kibale, Uganda and Alison Rosser, Cambridge.

10 A Future for Africa's Tropical Forests

Introduction

Africa's rain forests are the product of a history of extreme climatic variation and of human influences dating back thousands of years. The conservation movement today tends to focus on the dramatic impact of industrial logging or the pervasive degradation caused by shifting cultivation, but these must be seen in the context of much more fundamental problems, both within Africa and in the world at large, which have far-reaching impacts upon the continent's forests. Nature conservation in Africa is not just be a question of restricting the activities of poor peasants or defending the boundaries of isolated national parks against incursions by loggers. Conservation must be part of a broader process of managing the whole landscape. A balance must be achieved between the production of the goods and services needed to improve people's material well-being and the protection of the forests and soils and their wealth of biological diversity so that the welfare of future generations is assured (Poore and Sayer, 1991).

Prehistory of Africa's Forests

Most of Africa's forests are young on the geological time scale. During the periods of glacial advance in the northern hemisphere in the Pleistocene the climate of Africa was drier and cooler (see chapter 2). Deserts were more extensive and most of the area that today enjoys a climate suitable for rain forests was dominated by drier deciduous forests and woodlands.

At the height of the last glacial advance, around 18,000 BP, rain forests in Africa were restricted to a number of 'refugia' isolated from one another by areas that provided an unsuitable habitat for most rain forest species. As the ice retreated in the north, the forests spread out from these refugia and the present composition of the forests is the result of colonisation and *in situ* evolution that has occurred during the period of relative climatic stability of the last 10,000 years. The Pleistocene climatic changes are believed to have caused more drastic reductions of the forests in Africa than of those in Southeast Asia and South America and this is one explanation for the relative poverty of plant and animal species in African forests.

Throughout the past 10,000 years, humans have had a major influence on the African scene. Hunter-gatherer communities were ubiquitous. Through their use of fire they began to convert the dry forests in areas of seasonal rainfall into the savannas and grassy plains that are characteristic of present day Africa. The development of agriculture and the introduction, 3000–4000 years ago, of cattle, sheep and goats had major impacts upon the drier parts of the continent (Deshler, 1963).

One thousand years ago the drier areas with more seasonal rainfall were already occupied by humans with advanced societies, complex social structures and trade relations that covered long distances (Fage, 1969). In contrast, the humid forest regions provided a less attractive environment for such societies. Some indigenous crops of rather low productivity, such as yams, have been cultivated there for 4000–5000 years (Shaw, 1978), but in general the climate was unhealthy for both introduced cereal crops and livestock, and for dense, sedentary human populations. At the time of the first European ventures to the coasts of sub-Saharan Africa in the 16th century, the forests of the high rainfall areas had been little modified and their sparse human populations still subsisted largely by hunting and gathering (Aubréville, 1949). This changed rapidly when the Portuguese introduced South American maize, manioc, plantains and yams. These soon became the staple diet in the forest zones of Africa (Fage, 1969) and provided the stimulus for colonisation and agricultural expansion in the forests, which began in the 16th century and continues today.

European Influence on Forest Depletion

Subsequently, the development of trade with the outside world has had a profound and often devastating impact both on the people of Africa and their forest resources. The slave trade was the stimulus for the evolution of militaristic states in coastal Africa and for the radical redistribution of population throughout the continent. This promoted a mixing of ethnic groups and the almost total depopulation of large areas.

In the 19th century the slave trade was superseded by trade in commodities, mainly those produced in the forest belt. Ivory was one of the most important and its collection provided a major incentive for humans to penetrate the forests of West Africa. Gold and other commodities were also obtained from the forest zone and in the middle of the 19th century the first timber from West Africa was exported to Europe (Martin, 1991). Cacao and palm oil were among the earlier agricultural commodities to be traded from the forest belt and the prospects for great expansion of trade

Podding cacao beans in the Ashanti region of Ghana. D. and I. Gordon

in these and other products provided the stimulus for the establishment of European colonies in the 1890s. The perennial border conflicts, secessionist movements and racial tensions which undermine the stability of modern Africa are a legacy of the arbitrary decisions taken at a conference in Berlin in 1894/5, at which the European powers divided up the continent.

During the early years of the 20th century, colonial infrastructures were put in place throughout most of Africa. Where rivers and footpaths had earlier provided the only trade routes, roads and railways were rapidly established to give access to even the most remote regions. At this time much of the forest had already suffered some disturbance from farmers and hunters. Indeed virtually all the forests of the continent had probably been cleared at some time in the preceding centuries for temporary cultivation. But the vegetation had recovered rapidly from these small, localised disturbances and the total extent of the closed broadleaved forest was probably little reduced from its climatic maximum. However, the stage had been set for much more rapid change.

The change came from two sources. Foreign based enterprises, operating under the umbrella of colonial regimes, developed plantations of palm oil, rubber, cacao, coffee and a variety of lesser crops. These occupied a tiny area of forest land, employed fairly large numbers of people and were environmentally quite benign. Foreign enterprises also began to harvest the prime quality hardwoods from the forests (see chapter 7). Demand in Europe was for the decorative, deep red African mahoganies (*Khaya* spp. and *Entandrophragma* spp.) used for furniture manufacture. At first only a few species were selectively felled and only from the most accessible forests near the coast. In parallel to this, improvements in health services and agriculture fuelled the great increase in population which continues to accelerate today. Most of these people depended on extensive agriculture and they began to have an increasing impact on the forest (see chapter 6). However, forest and soil degradation was still much more pronounced in the seasonal rainfall savannas around the periphery of the rain forests, while the rate of clearance of the forests themselves was relatively low.

Development and the Environment

By the Second World War much of the forest of western and eastern Africa and Madagascar had already been penetrated and fragmented but the Central African forests were still mostly intact and large areas of forest persisted in the region of Liberia, Côte d'Ivoire and Ghana. This was still the situation as the countries of Africa attained their independence in the late 1950s and 1960s. Independence coincided with the wide availability of improved medical services and the introduction of drugs to treat the major endemic diseases of the forest zone, most notably malaria. This led to an acceleration in the rate of population increase. At the same time, mobile chain saws and heavy vehicles made the logging and clearing of the forests much easier and economic growth in Europe provided a rapidly expanding market for utility grade timber and commodity crops from the forests. As a result of these pressures, the rain forests of Africa have suffered more radical change in the past 30 years than they had throughout their 10,000 year post-glacial history.

Massive aid projects in the post-colonial period concentrated on maximising growth in the economies of the newly independent countries. Large investments were made in improving the infrastructure, especially roads, modernising agriculture and creating industrial employment. In many countries, this resulted in the opening up of forest areas to logging and agricultural colonisation. The proliferation of livestock in the Sahel caused desertification, which was exacerbated by drought. The latter probably related to deforestation in the equatorial zones from which the Sahel receives its rainfall. The result was a large scale movement of people from the Sahel to more rapidly developing regions in the forest zones. For instance, the population of Côte d'Ivoire increased from three million at independence in the 1960s to over 12 million in 1990, largely as a result of immigration from Mali and Burkina Faso. The immigrants found ready employment in the industrial plantations in the forest zone of Côte d'Ivoire and their friends and families supplemented their incomes by practising low grade agriculture, all at the expense of the forest. In consequence, Côte d'Ivoire has suffered the highest deforestation rate of any African country (see chapter 16).

The population of Africa continues to grow more rapidly than that of any other major region. Any strategy for the future of the forests has to take account of the fact that the population will double in the next two to three decades. Pressures on resources will inevitably increase, not only from this growth in numbers of people but from the increasing per capita consumption of goods that will occur with development. Understanding the demographic processes underlying the growth can help us to plan for its impact on forests. Figure 10.1 shows the relation of predicted urban and rural growth for a selection of forest and non-forest countries. It shows that in the medium term most growth will occur in urban areas and that in the most important forest countries relatively little growth will occur in rural areas. However, even urban populations place demands upon forest resources which, as chapter 6 points out, will have to be met by the intensive exploitation of small areas of land. The problems of shifting cultivation could be solved if more people obtained their food from intensive agriculture outside the forest zone. However, this will happen only if the general political and economic state of the continent becomes more robust.

In the early post-colonial days the environment was seen as an inexhaustible resource while governments saw the alleviation of poverty as the primary development objective. Investments were judged on their contribution to gross national product and projects on their internal rate of return. The prevailing attitude was that nature conservation was primarily an aesthetic concern and could be dealt with by establishing national parks to protect the big game animals which were a valuable resource for tourism (Newby and Sayer, 1976 and see chapter 9).

When forest services were set up in the early colonial days the colonial powers modelled them on those of India and Europe. Their principal role was to conserve and regulate the use of the forest resource. Their senior staff, in both anglophone and francophone Africa, were known as 'conservators'. But after independence the tendency was to see forest departments as agents for development, not for conservation. Conservation projects did not produce an immediate measurable economic return and were not attractive to aid agencies. Aid focused on the more 'commercial' components of forestry. In order to support commercial forestry efficiently, and make neat, self-contained development-oriented projects, many aid agencies promoted the establishment of autonomous forest enterprises to manage logging, plantations, forest inventories, planning and the like. This made sense to the aid donors; their projects were easy to manage and the results were visible and measurable. But for the forest departments of Africa it was a disaster. At a time when pressures on the forests were unprecedented they found their best staff deserting them, lured away by the attractions of air-conditioned offices, vehicles, liberal travel expenses and trips to overseas conferences, which were available to those who worked with the project-supported autonomous forest enterprises. In other words, just as government institutions were weakened by the difficult transition from colonial administration to independence, so forest departments were emasculated by aid agencies in their search for operational efficiency.

Figure 10.1 Predicted growth in total population and in urban and rural populations in selected African countries, 1990–2025

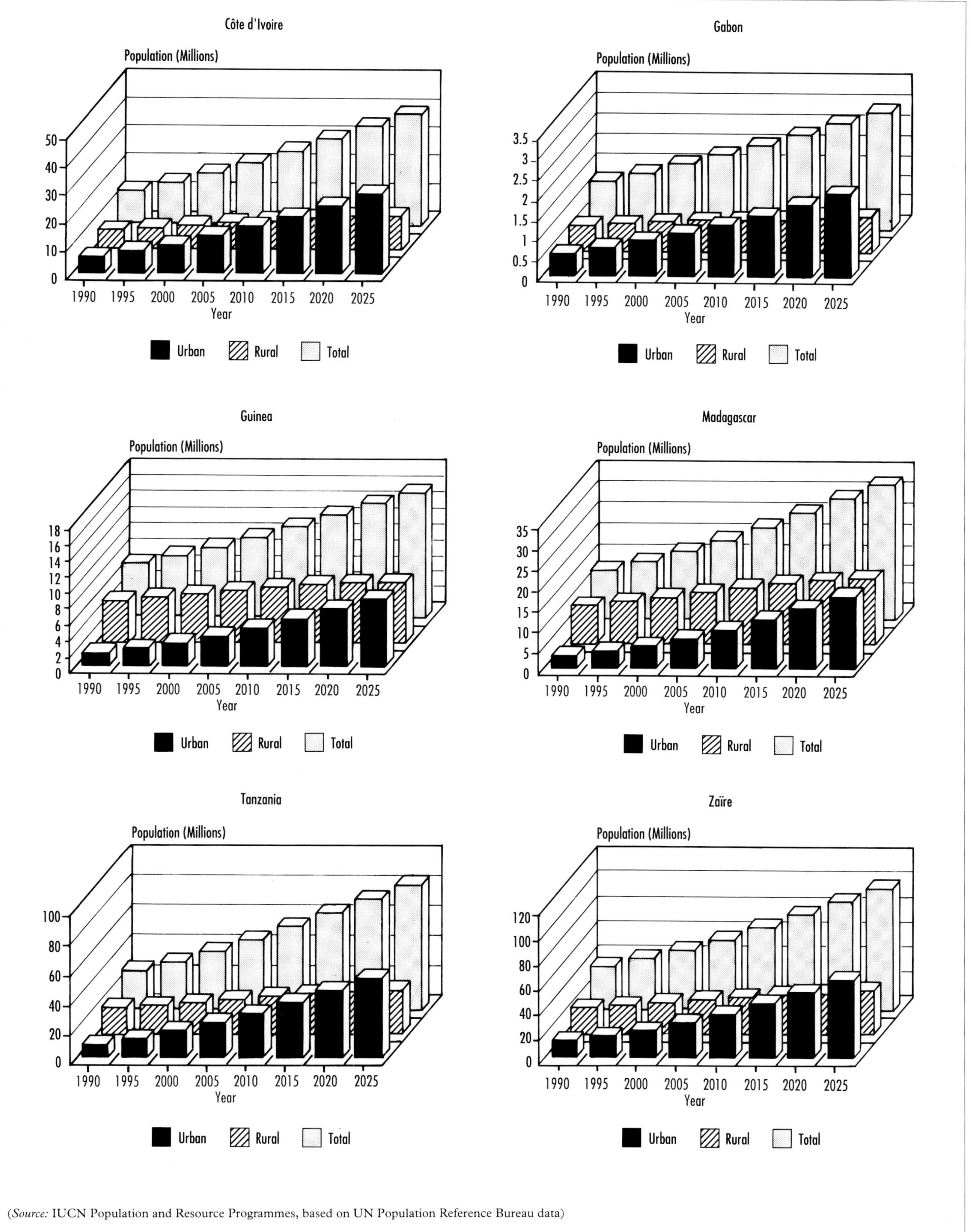

(*Source:* IUCN Population and Resource Programmes, based on UN Population Reference Bureau data)

Charcoal being made near Analamazaotra (Périnet) Special Reserve in Madagascar for urban markets. This is a serious threat to forests, particularly when large cities are located nearby. J. Sayer

While government forestry institutions were seriously weakened in the Africa of the 1960s and 1970s, the nature conservation programmes of these same countries generally fared somewhat better. Nature conservation, and especially national parks, were given strong political support in post-colonial Africa. Africa's wildlife was appreciated as a rich part of the heritage of the newly independent countries. President Nyerere of Tanzania captured the spirit of this new African commitment to conservation in his 1961 Arusha declaration (IUCN, 1963), which committed African states to conserve nature and natural resources. The 1968 Algiers Convention on Nature Conservation was one of the first pan-African agreements in the post-colonial era and has done much to shape conservation legislation and programmes in many African countries. The cultural significance of wildlife and wild nature for many Africans was captured by President Mobutu Sésé Séko of Zaïre, who, while addressing the 1975 IUCN General Assembly at Kinshasa, referred to Zaïre's national parks as 'the cathedrals of my country'. At this same meeting, President Mobutu proposed a World Charter for Nature; this was adopted by the General Assembly of the United Nations in 1982. But the real successes of conservation in Africa occurred on the ground. Wildlife and national parks departments attracted dedicated, enthusiastic people, both Africans and foreigners, who struggled, often in difficult and dangerous conditions, to establish an African protected areas network unequalled elsewhere in the world. At first, parks and reserves were located mainly in the savanna zones where they protected the conspicuous plains game which became a major tourist attraction. However, there has been significant growth in forest protected area in post-colonial times. Some of the largest and most important rain forest national parks in Central and West Africa – the Salonga and Maiko in Zaïre, Korup in Cameroon – were established in the 1970s and 1980s.

Initially, outside support for conservation programmes came mostly from non-governmental organisations and foundations. It concentrated on maintaining the status quo and principally on conserving plains game. In the early 1960s, WWF launched a major campaign to save the rhino and North American and British foundations gave support to research facilities in the savanna habitats of the Serengeti National Park in Tanzania and the Queen Elizabeth National Park in Uganda. At the same time, primatologists, ornithologists and botanists were beginning to unravel the fascinating biology of some of Africa's forested areas. Landmark studies were undertaken on the gorillas of eastern Zaïre (Schaller, 1963) and chimpanzees in western Tanzania (Goodall, 1988). Important programmes of ecological research were launched in the Taï National Park, Côte d'Ivoire (Guillaumet *et al.*, 1984) and the Makokou Reserve in Gabon. The latter site is now one of the best documented tropical research sites and a Unesco (1987) review lists many hundreds of papers on the biology of its forests.

Many of the conservation programmes in the forests were the result of the initiatives taken by these researchers. Dedicated individuals struggled to establish protected areas often with only limited financial support from international conservation bodies. Africa entered the 1980s with a good network of national parks and equivalent reserves in the forest zones. Several of these areas were under threat from poaching and encroachment but on the whole their integrity was being maintained and, in many countries, the networks were being extended. At the same time the forests outside protected areas were under greater threat than at any time in their history and the ability of governments to counter these threats was weak and declining. At this time the development assistance agencies began to recognise that environmental degradation, especially in Africa, was depleting the soil, water and forest resources upon which all of their development programmes depended (WCED, 1987). It was soon accepted that environmental conservation should be a major component of any aid programme, and some of the more progressive agencies are beginning to acknowledge that support for conservation may be the single most important contribution that they can make to improving the quality of life of the people of developing countries.

Now conservation has become a priority for the official development agencies, the amounts of money available are vastly increased. People who have observed the variable record of big aid projects in achieving development are worried about the impact of these heavy-handed interventions on embryonic conservation programmes.

The Present State of Africa's Forests

Assessments of the extent of tropical forests in Africa remains surprisingly imprecise. Even nations such as Congo and Gabon in which the forests represent a major economic resource are poorly mapped. Nevertheless, more information has emerged over the past decade or so and the maps in the following chapters represent a significant advance on the estimates available until now. In the past, even when maps were available, they were often not accompanied by accurate measures of forest cover. Remote sensing and computerised Geographic Information Systems (GIS) have significantly improved accuracy and accountability. GIS has enabled us to overlay the boundaries of chosen forest types from maps of potential vegetation on to maps of forest cover, thus differentiating moist from dry forests and true forests from woodlands and thicket. Many earlier maps have depicted forests in areas which would, using White's (1983) categories, be classed as woodlands in this Atlas (see Table 1.2 on p. 14).

Table 10.1 is a compilation of tropical rain forest cover statistics derived from maps in this Atlas and from FAO (1988). Where country maps were based on especially weak sources we have refrained from inferring a figure from them but, in many instances it has proved possible to find more recent information than that for 1980 given in FAO (1988). While palaeoclimatological study has demonstrated that the forest boundary is ever moving (see chapter 2), it is interesting to compare present forest area with the extent of the forests in the relatively recent past. White's map of Africa's vegetation (1983) permits calculation of the 'original' extent of tropical rain forests (as defined in Table 1.2) and the data derived from an analysis of this map, already carried out by MacKinnon and MacKinnon (1986), are presented in Table 10.1. It should be noted that White (1983) described a huge area of mixed forest and savanna woodland encircling the main forest block. There remains considerable uncertainty concerning the true extent of forest within that region and the 'original' forest areas given here are maximised (i.e. they assume these mosaic areas were once all forested).

The aggregate figures of FAO (1988) suggest that only 36 per cent of the original closed canopy broadleaved forest remained in 1980. The maps reproduced in this Atlas indicate that even less than this may now remain. A definite figure cannot be given as no information is available for most of southern and eastern Africa and the Indian Ocean islands. Note that we have not included dry deciduous or riverine forests on our maps, whereas these are generally incorporated into FAO's figures (see chapter 1). The Atlas maps are based upon our interpretation of the most recent and best available forest cover maps (see chapter 1 for an explanation of how the maps were produced). However, the quality of the source maps was variable and often the definitions of what the maps represented were difficult to interpret. The most striking picture to emerge from Table 10.1 is that only nine countries have more than one-fifth of their original forest cover remaining. Two of these, Reunion and São Tomé and Príncipe are small islands, while in Somalia, much of what FAO classifies as forest is not what would be considered as such here (see chapter 17). Table 10.1 shows that as many as 17 countries have less than 10 per cent of their original forest cover remaining.

A regional examination of the data shows that the patterns of deforestation differ widely. The whole of West Africa, except Liberia, has suffered severe deforestation, with only about 11–12 per cent remaining overall. The statistics in this Atlas are generally sound for this area being drawn in the main (Sierra Leone to Nigeria) from a UNEP/GRID dataset derived from AVHRR satellite imagery, checked against related Landsat scenes and by verification on the ground (Päivinen and Witt, 1989). The three most westerly countries, Guinea-Bissau, Senegal and The Gambia, were not included in the UNEP/GRID dataset but it is only the first of these for which a good map could not be obtained.

An oil palm plantation in Cameroon. Industrial plantations can employ large numbers of people on small areas of land and thus relieve the pressure on natural forests. J. Sayer

Table 10.1 Original extent of closed canopy moist forest in Africa, compared with remaining extent as judged from the maps in this Atlas and FAO (1988) statistics for 1980

	Approximate original extent of closed tropical moist forests (sq. km)[1]	*Remaining extent of moist forests (sq. km)*			*Percentage of moist forest remaining*	
		From atlas maps (unless otherwise stated); moist forests	*Publication date of maps*	*FAO (1988) data for 1980, closed broadleaved forests*	*From map data*	*From FAO (1988) data*
West Africa						
Benin	16,800	424	1989–90 and 1979	470	2.5	2.8
Côte d'Ivoire	229,400	27,464	1989–90	44,580	12	19.4
Gambia	4,100	497	1985	650	12.1	15.6
Ghana	145,000	15,842	1989–90	17,180	10.9	11.8
Guinea	185,800	7,655	1989	20,500	4.1	11.0
Guinea-Bissau	36,100	nd	1990	6,660	nd	18.4
Liberia	96,000	41,238	1989–90	20,000	43	20.8
Nigeria	421,000	38,620	1989–90	59,500	9.2	14.1
Senegal	27,700	2,045	1985	2,200	7.4	7.9
Sierra Leone	71,700	5,064	1989–90	7,400	7.1	10.3
Togo	18,000	1,360	1989–90	3,040	7.6	16.9
Total	1,251,600	140,209		182,180	11.5†	14.6
Central Africa						
Burundi	10,600	413	1984	150	3.9	1.4
Cameroon	376,900	155,330	1985	179,200	41.2	47.5
CAR	324,500	52,236	1985	35,900	16.1	11.1
Congo	342,000	nd	–	213,400	–	62.4
Equatorial Guinea	26,000	17,004[2]	1960	12,950	65.4[2]	49.8
Gabon	258,000	227,500[3]	–	205,000	85.2	76.8
Rwanda	9,400	1,554	(nd)	1,010	16.5	10.7
São Tomé and Príncipe	960*	299	1985	560	31.1	58.3
Zäire	1,784,000	1,190,737[7]	1990[3]	1,056,500	66.7[7]	59.2
Total	3,132,360	1,645,073		1,704,670	59.0†	54.4
Southern Africa						
Angola	218,200	nd		29,000	–	13.3
Malawi	10,700	320[4]		1,860	3.0	17.4
Mozambique	246,900	nd		9,350	–	3.9
Zimbabwe	7,700	80[5]		2,000	1.0	26.0
Total	483,500			42,210	–	8.7
Eastern Africa						
Djibouti	300	nd		10	–	3.0
Ethiopia	249,300	nd		27,500	–	11.0
Kenya	81,200	nd		6,900	–	8.5
Somalia	21,200	nd		14,800	–	69.8
Sudan	27,000	nd		6,400	–	23.7
Tanzania	176,200	nd		14,400	–	8.2
Uganda	103,400	7,400[8]		7,500	7.2	7.3
Total	658,600			77,510	–	11.8
Indian Ocean Islands						
Comoros	2,230*	nd		160	–	7.1
Mauritius	1,850*	nd		30	–	1.6
Réunion	2,500*	nd		820	–	32.8
Seychelles	270*	nd		30	–	11.1
Madagascar	275,086	41,715[9]	1985	103,000	15.2	37.4
Total	281,936			104,040	–	36.9
Grand Total	5,832,396			2,110,610	–	36.2

* To calculate original extent, it has been assumed that the islands were once completely forested. MacKinnon and MacKinnon (1986) give no figures for these areas.

† In both Central and West Africa, no data are available for one country, therefore the extent of forest originally in that country has been subtracted from the total original extent in the region before these percentages were calculated.

1 Taken from White (1983) as in MacKinnon and MacKinnon (1986) except for Gabon and Liberia where the figures for original forest extent given by MacKinnon and MacKinnon are too high (see explanation on Table 9.6).

2 Including 7,945 sq. km of degraded lowland rain forest.

3 Figure from UICN (1990).

4 Figure from Dowsett-Lemaire (1989, 1990).

5 Figure supplied by T. Muller (*in litt.*).

6 The digital dataset was completed in 1990 but is based on 1988 data.

7 Including 86,547 sq. km of degraded forest.

8 Figure from Howard (1991).

9 This figure has been calculated by adding Green and Sussman's (1990) figure for eastern rain forest to that calculated for mangroves from Map 26.1 plus an estimated 400 sq. km for forest remaining in the Sambirano region.

nd No data available.

(nd) No date.

In the Central African region, data have been drawn from a range of sources. The Zaïre map is very recent, being derived from a NASA interpretation of 1988 AVHRR imagery (Justice and Kendall, in press). Neighbouring Equatorial Guinea, Gabon and Congo are less well served and maps presented here are drawn from a very old source, a generalised one and a personal communication respectively. Cameroon and Central African Republic have excellent published maps of forest cover which are reasonably up to date. The overall pattern in Central Africa is one of huge tropical rain forests especially in the many inaccessible areas which are poorly served by road, rail or river. About 59 per cent of the original cover remains in this region.

In eastern Africa, data are generally of a poorer quality. However, the total forest area in the eastern region is relatively small and errors are of little significance to the overall African picture. Much of the original forest area has been lost, less than 10 per cent remaining.

The southern borders of the rain forest block are also poorly served with accurate maps. For Zimbabwe and Malawi only sketch maps have been provided in the Atlas but an accurate figure for forest extent has been given in the text and incorporated in Table 10.1. It has not been possible to find recent maps for Angola and Mozambique and the political situation within those countries means that area of forest cover can only be guessed at. Table 10.1 suggests between 5 and 8 per cent of forest remains, mainly in Angola.

The only Indian Ocean island with significant remaining forest cover is Madagascar and for this country, information has been used from a recent report by Green and Sussman (1990) with some additional data supplied by other sources. Madagascar and the other Indian Ocean islands are very important for their endemic species, but although they used to be completely forested, very little now remains. FAO (1988) suggests that about 15 per cent of forest remains on the Mascarene islands and the map in this Atlas indicates that Madagascar's forests also cover only 15 per cent of their original extent.

In general terms the reduction in forest extent has been greatest in those countries which had the least forest to start with. This is partly because conditions for forest in these countries were marginal, but also because population growth has been highest in non-forest areas.

The Tropical Forestry Action Plan

This new concern for the forest environment was manifested in the Tropical Forestry Action Plan (WRI, 1985). This was prepared by NGOs, the World Bank and UNDP and adopted by the representatives of tropical forest countries at a meeting of the FAO Committee on Forest Development in the Tropics in late 1985 (FAO, 1985). It committed donors to double their support for forestry over the following five years and singled out ecosystem conservation and forestry in rural development as major targets for investment.

By early 1991, a total of 38 African countries had committed themselves to the TFAP process. Five African countries had completed the TFAPs and were beginning implementation; four had completed TFAP sector reviews but had not begun implementation; 21 were developing TFAPs and eight were in the early stages of negotiations with donors (Table 10.2). In addition the countries

Table 10.2 Status of TFAPs in Africa, at February 1991

Country	*Planning phase completed*	*Forest Sector review completed*	*Forest Sector review underway*	*Interest expressed*
Angola				X
Burkina Faso			X	
Burundi			X	
Cameroon	X			
Cape Verde			X	
CAR			X	
Chad				X
Congo			X	
Côte d'Ivoire			X	
Equatorial Guinea			X	
Ethiopia			X	
Gabon			X	
Gambia				X
Ghana		X		
Guinea		X		
Guinea-Bissau			X	
Kenya			X	
Lesotho			X	
Liberia				X
Madagascar			X	
Malawi				X
Mali			X	
Mauritania		X		
Mauritius				X
Mozambique			X	
Niger			X	
Nigeria			X	
Rwanda			X	
Senegal			X	
Sierra Leone	X			
Somalia		X		
Sudan	X			
Tanzania	X			
Togo			X	
Uganda				X
Zaïre	X			
Zambia			X	
Zimbabwe				X
TOTALS	5	4	21	8

(*Source:* IUCN data, 1991)

Forest regeneration can occur if cleared areas are subsequently undisturbed. This is the trace of the main Itombwe road, abandoned in the early 1960s, but still marked on maps of Zaïre as a major road. R. Wilson

of CILSS (Inter-Governmental Committee for Drought Relief in the Sahel), SADCC (Southern Africa Development Cooperation Conference) and IGADD (Inter-Governmental Authority on Drought and Development in Eastern Africa) had all embarked upon regional TFAPs.

The results of the national TFAPs have been variable. Critics claim that they have not recognised the need for change and that they are simply advocating more of the same forestry assistance policies that have not succeeded in the past. The failure to address the needs of forest dwelling peoples in TFAPs and the neglect of biological diversity issues have come in for strong criticism (Colchester and Lohmann, 1990; Winterbottom, 1990). Advocates of TFAPs point out that having a process in place to increase and harmonise aid to forestry in 38 countries in five years is a considerable achievement. They note that governmental forestry institutions are conservative and bringing about change in these organisations is inevitably a slow process. In reality some TFAPs have been more effective than others, reflecting the fact that some countries are more amenable to change than others. Cameroon chose the path of expanded industrial logging and received very little donor support, whereas Tanzania adopted a strong social forestry and conservation line and was well supported by the donors.

There can be no doubt that the TFAP has given forest conservation and sustainable forestry a much higher profile in the programmes of both governments and aid agencies. For all its possible shortcomings, the TFAP has certainly done far more good than harm. In addition, it now exists as a process and can provide a framework for greater policy change and greater conservation efforts in the future.

Other Initiatives

Several other international initiatives are now focusing attention on the forests of Africa. National Conservation Strategies (NCS) are in preparation or exist in ten African countries and are under consideration in a further eight (Table 10.3). The NCS aims mainly to help countries re-examine their own policies on the conservation and sustainable development of natural resources. Environmental Action Plans (EAPs) are being prepared by the World Bank for 19 countries (Table 10.4). They are intended to identify projects for grants or loans to address important conservation issues. The EAP for Madagascar has already resulted in significant investments in forest conservation.

The new Global Environment Facility (GEF) being administered by the World Bank, UNDP and UNEP has US$250 million available for grants and soft loans to support biological diversity; Cameroon is one of six African countries which has been identified as a recipient. US$25 million is available for conservation schemes in Cameroon's forests. Several bilateral agencies have pledged greatly increased support for forest conservation and forestry. The question that still remains largely unresolved is how all the new money can be spent effectively to ensure a future for Africa's forests.

The Future for Protected Areas

The first law of conservation must be to ensure the integrity of existing protected areas and bring other sites of known value for biological diversity under conservation management. This means more trained and equipped forest managers and guards, more political support for protected areas and more measures to reconcile conflicts between protected areas and the traditional users of the forests. Several projects exist in Africa which attempt to protect critical forest sites while using development assistance to help local communities meet their needs in a sustainable, non-destructive way. Notable examples are the WWF Korup Project in Cameroon (see case study in chapter 13), the IUCN East Usambaras project in Tanzania (see case study in chapter 17), and the Wildlife Conservation International (WCI) gorilla conservation project in the Volcanoes National Park in Rwanda (see case study in chapter 12).

Much work has also been done to identify important sites of biological diversity. Many national studies are cited in the country chapters of this Atlas. A regional study of the Central African forest block carried out by IUCN on behalf of the EEC identified 104 sites in the seven countries of that region (UICN, 1989). Less than half of these sites have any conservation management at present. Considerable information on other sites is given in several IUCN publications (MacKinnon and MacKinnon, 1986; Stuart *et al.*, 1990 etc). Figure 10.2 and Table 10.5 show forest sites in Africa which are known to have special importance for conserving biological diversity. Many others will eventually be identified. If all of these known sites, covering perhaps 10–15 per cent of Africa's forests, could be managed in a natural or near-natural state for biological diversity conservation, then the immediate future of most of Africa's forest flora and fauna would be safe (Sayer and Stuart, 1988; Sayer *et al.*, 1990). Supporting protection of these sites could be the best way for the aid agencies to help conservation.

Table 10.3 Status of National Conservation Strategies, at February 1991

(Source: IUCN data, 1991)

Country	*Completed*	*In Preparation*	*Under Discussion*	*Suspended*	*No Progress*
Angola					X
Benin					X
Botswana	X				
Burkina Faso					X
Burundi					X
Cameroon					X
Cape Verde					X
CAR			X		
Chad		X			
Comoros					X
Congo					X
Côte d'Ivoire				X	
Djibouti					X
Equatorial Guinea					X
Ethiopia		X			
Gabon					X
Gambia					X
Ghana			X		
Guinea					X
Guinea-Bissau		X			
Kenya			X		
Lesotho					X
Liberia					X
Madagascar	X				
Malawi					X
Mali			X		
Mauritania		X			
Mauritius					X
Mozambique					X
Namibia			X		
Niger			X		
Nigeria	X				
Réunion					X
Rwanda					X
São Tomé and Príncipe					X
Senegal				X	
Seychelles				X	
Sierra Leone				X	
Somalia				X	
South Africa	X				
Sudan					X
Swaziland					X
Tanzania			X		
Togo			X		
Uganda				X	
Zaïre					X
Zambia	X				
Zimbabwe	X				
TOTALS	5	4	8	6	24

Table 10.4 Status of Environmental Action Plans, at February 1991

(*Source:* IUCN data, 1991)

Country	*Completed*	*Under Preparation*	*Under Discussion*
Benin			X
Burkina Faso		X	
Burundi			X
Congo			X
Côte d'Ivoire			X
Gambia			X
Ghana		X	
Guinea		X	
Guinea-Bissau			X
Lesotho	X		
Madagascar	X		
Mali			X
Mauritius	X		
Nigeria	X		
Rwanda		X	
Seychelles	X		
Somalia			X
Togo			X
Uganda			X
TOTALS	5	4	10

However, if these protected areas are merely islands of conservation in a totally transformed or degraded landscape then their long-term future will not be secure (Sayer and Whitmore, 1991). Many existing protected areas are small and their plant and animal populations are isolated. These small, fragmented populations are prone to extinction from a variety of chance factors and from genetic deterioration resulting from inbreeding (Whitmore and Sayer, 1992). Conservation objectives can be met only if very extensive areas outside parks and reserves are retained under some sort of forest cover. The conservation value of these forests depends upon how closely they resemble the native forests of the region, how diverse they are and how many indigenous species they contain. The better these criteria are fulfilled, the better the forests will act as buffers, protecting the core conservation areas in the national parks and equivalent reserves.

This is where development aid, the TFAP, the EAPs, and the GEF have their greatest role to play. They must support uses of forest which are compatible with biological diversity conservation. This may involve sustainable management of forests for timber, but it can also be management for many other food, fibre or medicinal products. Many of these products are already more valuable to forest dwelling peoples than timber whose value accrues mainly to urban entrepreneurs or foreign companies.

Within Africa, some regions are in much greater need of conservation action than others. In the central forest block of Cameroon, Gabon, Congo and Zaïre there is still time for careful zoning of forest land. Zoning plans should establish areas for protection, for timber production and for intensive agricultural development. In coastal West Africa, eastern Africa and Madagascar the destruction of the forests has advanced so far that a major effort should be made to conserve every remaining patch of native forest. But population pressure is such that it will not be possible to give total protection to these forests. Their conservation will only be achievable if it involves their careful, sustained use to produce the various products and services needed by people. Aid agencies have a major role to play in supporting development of such uses.

There have been innumerable studies of the plants and animals of Africa's forests. Their biological diversity is far better documented than that of Asia or South America. Virtually all the countries of the region have made a political commitment to conserve their fauna and flora. The richer countries of the north want to help Africa conserve its forests. The knowledge and resources are now available and the time has come to translate the many plans and strategies into practical action. It is against this that our descendants will judge the success or failure of our conservation efforts.

Figure 10.2 Critical forest sites in the Afrotropical region, listed in Table 10.5

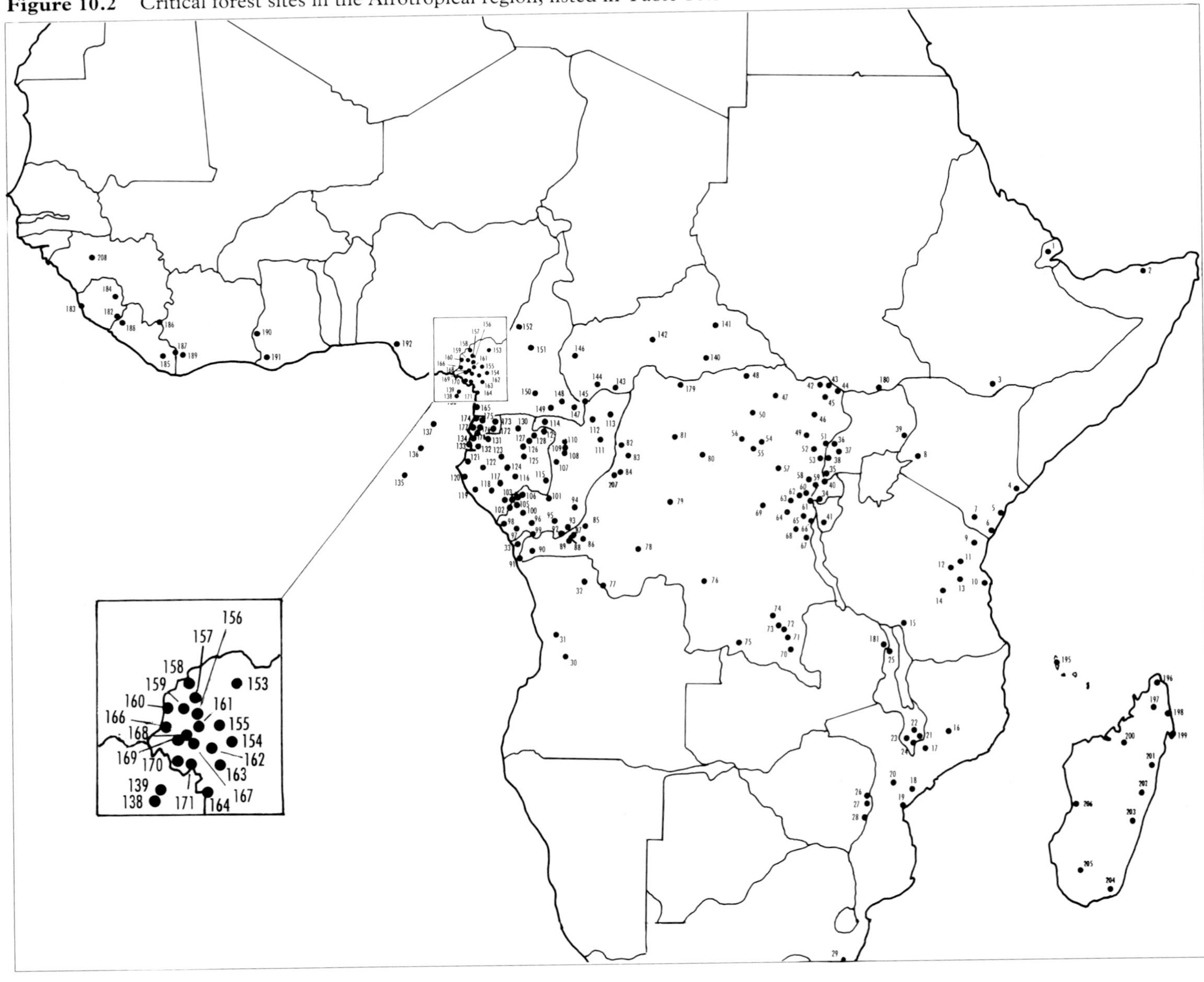

Table 10.5 Critical forest sites in the Afrotropical region

Sites		*Description*
DJIBOUTI		
1	Day	national park
SOMALIA		
2	Daloh	forest reserve
ETHIOPIA		
3	Neghelli region	unprotected
KENYA		
4	Lower Tana Riverine Forests	part in reserves
5	Sokoke	forest reserve, part nature reserve
6	Shimba hills	national reserve
7	Taita hills	national forest
8	Kakamega and Nandi	forest reserve, part nature reserve
TANZANIA		
9	Usambara mountains	national forest reserves
10	Pugu hills	national forest reserves
11	Nguru mountains	national forest reserves
12	Ukaguru mountains	national forest reserves
13	Uluguru mountains	national forest reserves
14	Uzungwa mountains	national forest reserves
15	Southern highlands	national forest reserves
MOZAMBIQUE		
16	Mount Namuli	?unprotected
17	Mount Chiperone	?unprotected
18	Inhamintanda	unprotected
19	Dondo	unprotected
20	Gorongosa mountains	unprotected
MALAWI		
21	Mt Mulanje	protection forest reserve
22	Mt Chiradzulu	protection forest reserve
23	Mt Soche	protection forest reserve
24	Mt Thyolo	part protection forest reserve
25	Nyika Plateau	national park
ZIMBABWE		
26	Vumba highlands	botanical reserves
27	Chimanimani hills	national park
28	Chirinda forest	botanical reserve
SOUTH AFRICA		
29	Ngoye forest	controlled by KwaZulu government
ANGOLA		
30	Bailundu highlands	unprotected
31	Amboin region	unprotected
32	Northern Angolan region	unprotected
33	Cabinda enclave	unprotected
RWANDA		
34	Nyungwe	reserve, part national park
40	Volcanoes	national park
UGANDA		
35	Impenetrable (Bwindi)	forest reserve, part natural reserve
36	Semliki	forest reserve, part animal sanctuary
37	Kibale	forest reserve, part natural reserve
38	Rwenzori	forest reserve
39	Mt Elgon	forest reserve
BURUNDI		
41	Bururi	natural forest reserve
SUDAN		
180	Imatong Hills	unprotected
ZAMBIA		
181	Nyika Plateau	national park
ZAÏRE		
42	Azandes	hunting reserve
43	Garamba	national park
44	Mondo	hunting reserve
45	Gangala-na Bodio	hunting reserve
46	Maika-Penge	hunting reserve
47	Epi	hunting reserve
48	Bili Uere	hunting reserve
49	Okapi (Ituri)	unprotected
50	Rubi-Tele	hunting reserve
179	Abumonbazi	unprotected
51	Semliki	unprotected
52	Mt Hoyo	reserve
53	Virunga	national park
54	Masako	reserve
55	Kongolo	reserve
56	Yangambi	biosphere reserve
57	Maiko	national park
58	Tongo	unprotected
59	Rutshuru	hunting reserve
60	Area west of L. Kivu	unprotected
61	Shushu	unprotected
62	Irangi	unprotected
63	Kahuzi-Bièga	national park
64	Manièma	unprotected
65	Itombwe	unprotected
66	Uvira	unprotected
67	Mt Kabobo	unprotected
68	Luama	hunting reserve
69	Lomami-Lualuba	unprotected
70	Kundelungu	national park
71	Lufira	biosphere reserve
72	Basse Kando - Bena Mulumbu	hunting reserve
73	Lubudi-Sampwé	hunting reserve
74	Upemba	national park
75	Kyamasumba-Kolwezi	unprotected
76	Bushimae	hunting reserve
77	Swa-Kibula	hunting reserve
78	Mangai	hunting reserve
79	Salonga	national park
80	Luo	unprotected
81	Lomako	unprotected
82	Ngiri	unprotected
83	Eala	botanical garden
84	Botende	reserve
207	Lake Tumba	unprotected
85	Maï-Mpili	unprotected
86	Bombo Lumène	hunting reserve
87	Ngaenke	unprotected
88	Nsélé	unprotected
89	Kisantu	botanical garden
90	Luki	biosphere reserve
91	Mangroves	unprotected
CONGO		
92	Patte d'Oie	unprotected
93	Tsiémé	unprotected
94	Léfini	faunal reserve
95	Bangou	unprotected
96	Loudima	faunal reserve
97	Dimonika/Londela-Kayes	biosphere reserve
98	Conkouati	faunal reserve

99	Boko-Songo	unprotected
100	Tsoulou	faunal reserve
101	Sces Ogooué-Zanaga	unprotected
102	Nyanga Nord	faunal reserve
103	Mt Fouari	faunal reserve
104	Nyanga-Sud	faunal reserve
105	Mt Mavoumbu	hunting reserve
106	Bowé de Kouyi	unprotected
107	Kéllé-Oboko II	unprotected
108	M'boko	hunting reserve
109	Lekoli-Pandaka	faunal reserve
110	Odzala	national park
111	Likouala and Lac Télé	unprotected
112	Nouabalé	unprotected
113	Ibenga-Motaba	unprotected
114	Mt Nabemba	unprotected

GABON

115	Leconi	unprotected
116	Soungou-Milongo	unprotected
117	Moukalaba-Dougoula	reserve
118	Mts Doudou	unprotected
119	Setté-Cama	reserve
120	Ozouri	unprotected
121	Wonga-Wongué	reserve
122	Ogooué-Onangué	national park
123	La Lopé	reserve
124	Forêt des Abeilles	unprotected
125	Mingouli	unprotected
126	Ipassa-Makokou	reserve
127	Mts de Belinga	unprotected
128	Grottes de Belinga	unprotected
129	Djoua	unprotected
130	Minkébé	unprotected
131	Tchimbélé	unprotected
132	Sibang	unprotected
133	Mondah	unprotected
134	Akanda	unprotected

SÃO TOMÉ AND PRÍNCIPE

136	São Tomé	unprotected
137	Príncipe	unprotected

CENTRAL AFRICAN REPUBLIC

140	Bangassou	unprotected
141	Kotto remnants	unprotected
142	Kaga-Bandoro remnants	unprotected
143	Basse-Lobaye	biosphere reserve
144	Ngoto and Mbaèré-Modingué	?unprotected
145	Dzanga-Sangha	?unprotected
146	Nana remnants	unprotected

CAMEROON

147	Lobéké	faunal reserve
148	Boumba Bek	unprotected
149	Nki	unprotected
150	Dja	faunal reserve
151	Mbam and Djerem	national park
152	Tchabal Mbabo	unprotected
153	Mt Oku	protected by prefectural orders
154	Bonepoupa	forest reserve
155	Mt Manengouba	unprotected
156	Bayang Mbo	forest reserve
157	Mawne	forest reserve
158	Takamanda	forest reserve
159	Nta Ali	forest reserve
160	Ejagham	forest reserve
161	Mts Bakossi	unprotected
162	Mt Nlonako	unprotected
163	Bonepoupa	forest reserve
164	Douala-Edea	faunal reserve
165	Campo	faunal reserve
166	Korup	national park
167	Mt Koupé	protected by local taboo
168	Barombi Mbo	forest reserve
169	Rumpi mountains	part forest reserves
170	Mokoko	forest reserve
171	Mt Cameroun	part forest reserve

EQUATORIAL GUINEA

172	Mongomo	unprotected
173	Acurenam-Ns	unprotected
174	Rio Ntem-Rio Uolo	protected
175	Mont Alen	protected
176	Monte Mitra	protected
177	Bata Rio-Uolo	unprotected
178	Rio Muni Estuary	protected
135	Annobon	protected
138	Caldera de Luba	protected
139	Pico Basile	protected

SIERRA LEONE

182	Gola	forest reserves
183	Freetown Peninsula	forest reserve
184	Loma Mountains	forest reserve

LIBERIA

185	Sapo	national park
186	Mt Nimba*	national forest and nature reserve
187	Grand Gedeh County/Grebo	national forest
188	Lofo-Mano	national forest

* shared with Guinea and Côte d'Ivoire

GUINEA

208	Fouta Djalon Plateau	part biosphere reserve

CÔTE D'IVOIRE

189	Tai-N'Zo	national park and faunal reserve

GHANA

190	Bia	national park
191	Nin-Suhien	national park

NIGERIA

192	Okumu	forest reserve
193	Obudu Plateau	forest reserve
194	Stubbs Creek	forest reserve

COMOROS

195	Mt Karthala	unprotected

MADAGASCAR

196	Montagne d'Ambre	national park and special reserve
197	Tsaratanana massif	natural reserve
198	Marojejy massif	natural reserve
199	Masoala peninsula	classified forest
200	Ankarafantsika	natural reserve
201	Zahamena	natural reserve
202	Périnet-Analamazaotra	special reserve
203	Ranomafana	classified forest
204	Andohahela	natural reserve
205	Zombitse	classified forest
206	Analabe	special reserve

(*Source:* IUCN data, 1991)

References

Aubréville, A. (1949) *Climats, Forêts et Désertification de l'Afrique Tropicale*. Société d'Editions Géographiques, Maritimes et Coloniale, Paris, France.

Colchester, M. and Lohmann, L. (1990) *The Tropical Forestry Action Plan: What Progress?* World Rainforest Movement, London, UK.

Deshler, W. (1963) Cattle in Africa, distribution, types and problems. *Geographical Review* **53**: 52–8.

Dowsett-Lemaire, F. (1989) The flora and phytogeography of the evergreen forests of Malawi. I: afromontane and mid-altitude forests. *Bulletin du Jardin Botanique National de Belgique* **59**: 3–131.

Dowsett-Lemaire, F. (1990) The flora and phytogeography of the evergreen forests of Malawi. II: lowland forests. *Bulletin du Jardin Botanique National de Belgique* **60**: 9–71.

Fage, J. D. (1969) *A History of West Africa*. Cambridge University Press, Cambridge, UK.

FAO (1985) *Tropical Forestry Action Plan, Committee on Forestry Development in the Tropics*. FAO, Rome, Italy.

FAO (1988) *An Interim Report on the State of Forest Resources in the Developing Countries*. FAO, Rome, Italy.

Goodall, J. (1988) *The Chimpanzees of Gombe*. Belknap/Harvard Press, Cambridge, Massachusetts, USA.

Green, G. M. and Sussman, R. W. (1990) Deforestation history of the eastern rain forests of Madagascar from satellite images. *Science* **248**: 212–15.

Guillaumet, J. L., Conturier, G. and Dosso, H. (1984) *Recherche et Aménagement en Milieu forestier Tropical Humide; le Projet Taï de Côte d'Ivoire*. Document Technique MAB No. 15. Unesco, Paris, France.

Howard, P. C. (1991) *Nature Conservation in Uganda's Tropical Forest Reserves*. IUCN, Gland, Switzerland and Cambridge, UK. 330 pp.

IUCN (1963) *Conservation of Nature and Natural Resources in Modern African States*. IUCN new series No. 1. IUCN, Morges, Switzerland.

MacKinnon, J. and MacKinnon, K. (1986) *Review of the Protected Area System in the Afrotropical Realm*. IUCN, Gland, Switzerland and Cambridge, UK.

Martin, C. (1991) *The Rainforests of West Africa, Ecology, Threats and Conservation*. Birkhauser, Basel, Boston, London.

Newby, J. E. and Sayer, J. A. (1976) *Wildlife, National Parks, Tourism and Recreation*. Paper presented at a consultation on the role of forestry in a rehabilitation programme for the Sahel, CILSS/UNSO/FAO, Dakar, Senegal.

Päivinen, R. and Witt, R. (1989) *The Methodology Development Project for Tropical Forest Cover Assessment in West Africa*. Unpublished report. UNEP/GRID, Geneva, Switzerland.

Poore, D. and and Sayer, J. A. (1991) *The Management of Tropical Moist Forest Lands: Ecological Guidelines*. 2nd edition. IUCN, Gland, Switzerland and Cambridge, UK.

Sayer, J. A. and Stuart, S. N. (1988) Biological diversity and tropical forests. *Environmental Conservation* **15**(3): 193–4.

Sayer, J. A., McNeely, J. A. and Stuart, S. N. (1990) The conservation of tropical forest vertebrates. In: *Vertebrates in the Tropics*. Peters, G. and Hutterer, R. (eds), pp. 407–19. Museum Alexander Koenig, Bonn, Germany.

Sayer, J. A. and Whitmore, T. C. (1991) Tropical moist forests: destruction and species extinction. *Biological Conservation* **55**: 199–213.

Schaller, G. B. (1963) *The Mountain Gorilla, Ecology and Behaviour*. University of Chicago Press, Chicago, USA.

Shaw, T. (1978) *Nigeria, its Archaeology and Early History*. Thames and Hudson, London, UK.

Stuart, S. N., Adams, R. J. and Jenkins, M. D. (1990) *Biodiversity in Sub-Saharan Africa and its Islands: Conservation, Management and Sustainable Use*. Occasional Papers of the IUCN Species Survival Commission No. 6. IUCN, Gland, Switzerland.

UICN (1989) *La Conservation des Ecosystèmes forestiers d'Afrique Centrale*. UICN, Gland, Switzerland and Cambridge, UK.

UICN (1990) *La Conservation des Ecosystèmes forestiers du Gabon*. Basé sur le travail de C. Wilks. UICN, Gland, Switzerland and Cambridge, UK. 215 pp.

Unesco (1987) *Makokou, Gabon*. Unesco, Paris, France.

WCED (1987) *Our Common Future*. World Commission on Environment and Development. Oxford University Press, Oxford, UK.

White, F. (1983) *The Vegetation of Africa: a descriptive memoir to accompany the Unesco/AETFAT/UNSO vegetation map of Africa*. Unesco, Paris, France. 356 pp.

Whitmore, T. C. and Sayer, J. A. (1992) *Tropical Deforestation and Species Extinction*. Chapman and Hall, London, UK.

Winterbottom, R. (1990) *Taking Stock, The Tropical Forestry Action Plan after Five Years*. The World Resources Institute, Washington, DC, USA.

WRI (1985) *Tropical Forests, A Call for Action. Parts I–III*. World Resources Institute, Washington, DC, USA.

Authorship

Jeff Sayer at IUCN, Gland, Switzerland with contributions from Claude Martin and Francis Kasisi of WWF-International, Gland, Jeff McNeely of IUCN and Martin Jenkins, Cambridge, UK.

PART II

11 Benin and Togo

BENIN
Land area 110,620 sq. km
Population (mid-1990) 4.7 million
Population growth rate in 1990 3.2 per cent
Population projected to 2020 11.7 million
Gross national product per capita (1988) US$340
Rain forest (see map) 424 sq. km
Closed broadleaved forest (end 1980)* 470 sq. km
Annual deforestation rate (1981–5)* 12 sq. km
Industrial roundwood production† 262,000 cu. m
Industrial roundwood exports† nd
Fuelwood and charcoal production† 4,738,000 cu. m
Processed wood production† 11,000 cu. m
Processed wood exports† nd

TOGO
Land area 54,390 sq. km
Population (mid-1990) 3.7 million
Population growth rate in 1990 3.6 per cent
Population projected to 2020 9.9 million
Gross national product per capita (1988) US$370
Rain forest (see map) 1360 sq. km
Closed broadleaved forest (end 1980)* 3040 sq. km
Annual deforestation rate (1981–5)* 21 sq. km
Industrial roundwood production† 183,000 cu. m
Industrial roundwood exports† nd
Fuelwood and charcoal† 683,000 cu. m
Processed wood production† 5000 cu. m
Processed wood exports† nd
* FAO (1988)
† 1989 data from FAO (1991)

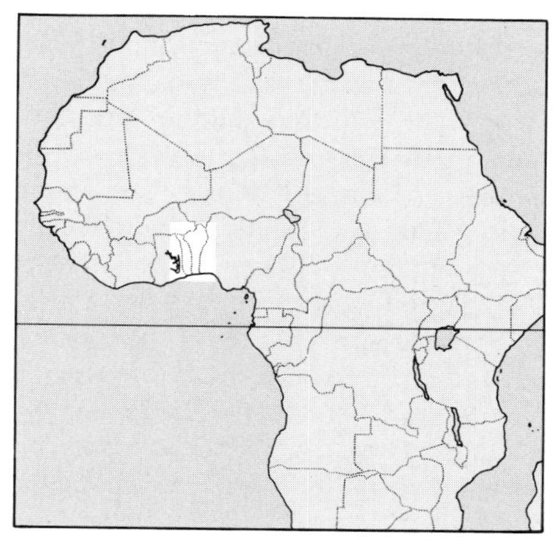

A long history of intense human activity, coupled with a relatively dry climate, meant that most of the closed forests of both Benin and Togo had already been lost when colonial administrations were imposed late in the 19th century. Now only tiny relict forest patches remain and the flora and fauna of both countries are seriously endangered.

Introduction

Benin and Togo are small, elongated countries that lie in the area where savannas have for a long time interrupted the forests which bordered the rest of the West African coast. This interruption in the forests, the so-called Dahomey Gap, may result directly from the dry climate (cold sea currents create an area of low rainfall along the 150 km coastline), or possibly from the concentration of human activity in an area where the drier conditions favour agriculture (Robbins, 1978). Benin, the larger of the two countries with an area of 112,620 sq. km, lies between 6°15' and 12°25'N and between 0°40' and 3°45'E. It is bordered by Nigeria to the east and by both Niger and Burkina Faso in the north. Its western neighbour, Togo, covering an area of only 56,790 sq. km, lies between latitudes 6°10' and 11°10'N and longitudes 0°4'W and 1°40'E. Ghana is on its western border.

In Togo, the sandy coastal plain rises gently until the 200 m contour is reached where a chain of mountains (Chaîne de Togo) enters the country from the west and crosses the interior obliquely. This same chain also crosses Benin (Chaîne de l'Atakora), reaching an altitude of around 650 m; a bit lower than in Togo where the highest peak is about 1000 m. In Benin the coastal plain is broken by the Lama Depression, a swampy clay plain between Abomey and Cotonou, as well as by a number of other river valleys. In the north of both countries, the land surface dips down again to the broad valley of the Pendjari River.

The coastal areas of the Dahomey Gap receive less than 1000 mm of rainfall, and their sandy soils may never have supported closed forest. Further inland, rainfall rises somewhat but never exceeds 1500 mm, and there is a marked dry season. The dry, dusty Harmattan wind blowing from the Sahara now seems to reach further south in the dry season, a phenomenon associated with drought and deforestation in the Sahel. Temperatures in the south are relatively constant throughout the year with daily maxima and minima close to 34°C and 22°C. In the north, daytime temperatures can reach as high as 43°C and fall to below 10°C at night.

In both countries, most people live in the coastal zone; indeed, in Togo, population density rises to 300 inhabitants per sq. km near the coast compared to five people per sq. km inland (FAO/UNEP, 1981). Around 60 per cent of the people in Benin and nearly 80 per cent of those in Togo live in rural areas. The population growth rate in both countries is more than 3 per cent, slightly higher in the more densely populated Togo (with a mean density of 68 people per sq. km) than in Benin (42 inhabitants per sq. km).

The Dahomey Gap was once occupied by powerful African kingdoms. Later, a succession of commercial and colonial settlements from Europe was installed on the coast. England, Holland, Portugal, France and Germany all established trading posts, with slaves and ivory the main commodities. When the European powers established the boundaries of their colonies in the 1890s, Germany acquired Togo, while the present-day state of Benin, then known as Dahomey, was included in French West Africa. After the First World War, Togo was placed under British and French administration. In 1956, the eastern part of Togo passed to the French and the western part joined the Gold Coast, now Ghana.

After independence in 1960 the countries took different paths. The intellectual Beninois enjoyed a remarkable succession of coups, finally settling for the Marxist-Leninist regime of Matthieu Kérékou in a 1972 military takeover, and adopting the name People's Republic of Benin in 1975. Togo took a more pragmatic, market-oriented, pro-western path and, aided by the export of rich phosphate deposits, enjoyed a degree of economic prosperity exceeding that of several of its neighbours whose territories are more richly endowed with mineral and agricultural resources.

Neither country has contained extensive closed evergreen forests in recent times but the riparian strips and isolated patches of more humid forest are the habitat of a variety of forest animals and plants and still contribute significantly to both countries' wood requirements.

The Forests

The predominant vegetation of both Benin and Togo was originally a dense semi-evergreen or deciduous forest. Small islands of more evergreen types occurrred on moist soils and in narrow strips of riparian forest along the rivers.

Many of the small patches of closed forest now remaining in otherwise intensively cultivated landscapes, are considered sacred and are protected by strong local traditions. None of these sacred forest patches covers more than 5 sq. km and most are less than 1 sq. km. The largest remaining natural forest area is the Lama Forest in south-central Benin. This covers an area of about 50 sq. km to the south of the city of Abomey. The forest grows on very heavy clay soils which are waterlogged in the rainy season and thus unattractive to farmers. Even so, it is much reduced from the original 163 sq. km forest reserve gazetted in 1946. Forest trees in the Lama Forest Reserve include *Triplochiton scleroxylon*, *Antiaris africana*, *Milicia excelsa*, *Afzelia africana*, *Ceiba pentandra*, and *Diospyros mespiliformis* (FAO/UNEP, 1981).

The other forest fragments in both countries contain these species together with the West African mahogany *Khaya grandifoliola*, and species such as *Cola grandifolia*, *Ceiba pentandra* and species of *Celtis*, *Holoptelea*, *Vitex* and others more commonly associated with the savanna zone. The original forests of Togo and Benin are described in more detail in Aubréville (1937).

Most of the forest patches occur between 7° and 9°N, roughly between Savalou and Bassila in Benin and between Kpalimé and Fazao in Togo. However, where topography and soils are suitable a few fragments persist up to 11°N and relict populations of forest species such as Geoffrey's black-and-white colobus monkeys *Colobus polykomos vellerosus* existed in such areas at least until recently.

The Precambrian mountain chain running north-east from Kpalimé in Togo and extending as the Atakora range in northern Benin has some local impact on climate and supports relict patches of forest with submontane characteristics. The finest examples of these lie at altitudes of 800–900 m on the Danyi plateau and on the Togo, Agou and Haïto mountains. Some of the best preserved of these forests are on steep, rocky hillsides, unsuitable for cultivation. Chimpanzees *Pan troglodytes* were said to persist in some of these areas until the 1970s.

Mangroves

Coastal currents have built up extensive sand bars along the shores of both countries. These protect brackish lagoons which are quite extensive in Benin. Small areas of mangrove exist in these lagoons and around the estuaries of the Mono and Ouémé rivers. Map 11.1 indicates that 69 sq. km remain in Benin, while none is shown in Togo. The mangroves are subject to considerable illegal hunting and fuelwood gathering and are under serious threat. Small populations of sitatunga *Tragelaphus spekei* occurred, at least until recently, and manatees *Trichechus senegalensis* may still survive in small numbers in remote parts of the lagoons (Sayer and Green, 1984). Although not important at the regional level, the mangroves are important in preventing coastal erosion and are significant nature conservation sites at the national level.

Forest Resources and Management

In 1980 FAO estimated that in Benin only 470 sq. km or 0.4 per cent of national territory remained under natural cover of closed broadleaved forest, while in Togo it was estimated that 3040 sq. km of closed broadleaved forest remained at that time. Only 140 sq. km of forest in Benin and 470 sq. km in Togo were considered to be undisturbed. However, these figures give an excessively favourable picture of the situation, particularly for Togo. In reality there are only tiny areas of forest in either country where there is even a remote chance of retaining the full range of natural flora and fauna. Map 11.1 shows that in 1979 there were 355 sq. km of dryland forest remaining in Benin but, as indicated on the source map (see Map Legend), all the forest patches are degraded. In contrast, the UNEP/GRID satellite imagery which has been used to plot the forest remaining in Togo, showed no areas of forest left in Benin, or none big enough to be depicted at the scale used. Map 11.1 indicates that 1360 sq. km of forest remain in Togo (see Table 11.1). The forests of both countries are evidently seriously endangered.

Table 11.1 Estimates of forest extent in Benin and Togo

	Area (sq. km)	*% of land area*
BENIN		
Rain forests		
Lowland	355	0.3
Mangrove	69	<0.1
Totals	424	0.4
TOGO		
Rain forests		
Lowland	1,360	2.5
Totals	1,360	2.5

(Based on analysis of Map 11.1. See Map Legend on p. 101 for details of sources.)

Benin has extensive classified forest reserves, some 21,440 sq. km in 1980 (FAO/UNEP, 1981). This includes 7750 sq. km of national parks and reserves (managed for sport-hunting). Forest reserves are open to specified uses by local people and to controlled logging. Local uses extend to clearance for temporary cultivation, on condition that forest is allowed to regenerate in the fallow period. Permits to fell timber are allocated on an individual tree basis by the Forest Department. In reality much of the forest reserve land lies in the savannas of the centre and north of the country. These areas were, until recently, infested with tsetse flies, the insect vector of sleeping sickness (trypanosomiasis), and simuliid flies, which transmit river blindness (onchocerciasis). This, coupled with the low fertility of the soils, made the areas unattractive to settlers and afforded protection to the forests. International campaigns to eliminate river blindness and the availability of drugs to treat trypanosomiasis in cattle, have removed this protection and the forests are now under more serious threat of agricultural conversion.

Togo has three national parks and 3360 sq. km of forest reserves where exploitation and cultivation are illegal. In addition an unspecified area is considered as 'protected forest' where some logging and agriculture are allowed under Forest Department control. In fact, the Forest Department tolerates a considerable level of activity in both categories of land, but does attempt to control the felling of certain more valuable timbers.

Much of the use of forest products occurs in the villages and is not recorded in national statistics. The volume of timber passing annually through the larger and better monitored sawmills in Benin was about 20,000 cu. m in the early 1980s (FAO/UNEP, 1981). This was mainly composed of West African mahogany, iroko *Milicia excelsa* and *Afzelia africana*. Most is used locally for construction and furniture manufacture. Some timber is exported illegally from Benin to Nigeria. Togo has been a net importer of timber for the past 20 years and Benin now relies largely on imports.

It is difficult to relate timber production figures to forest area. A relatively large proportion of timber in both countries comes from trees growing in farmland, along roadsides and as shade for cacao and coffee. A rather small proportion comes from forests that

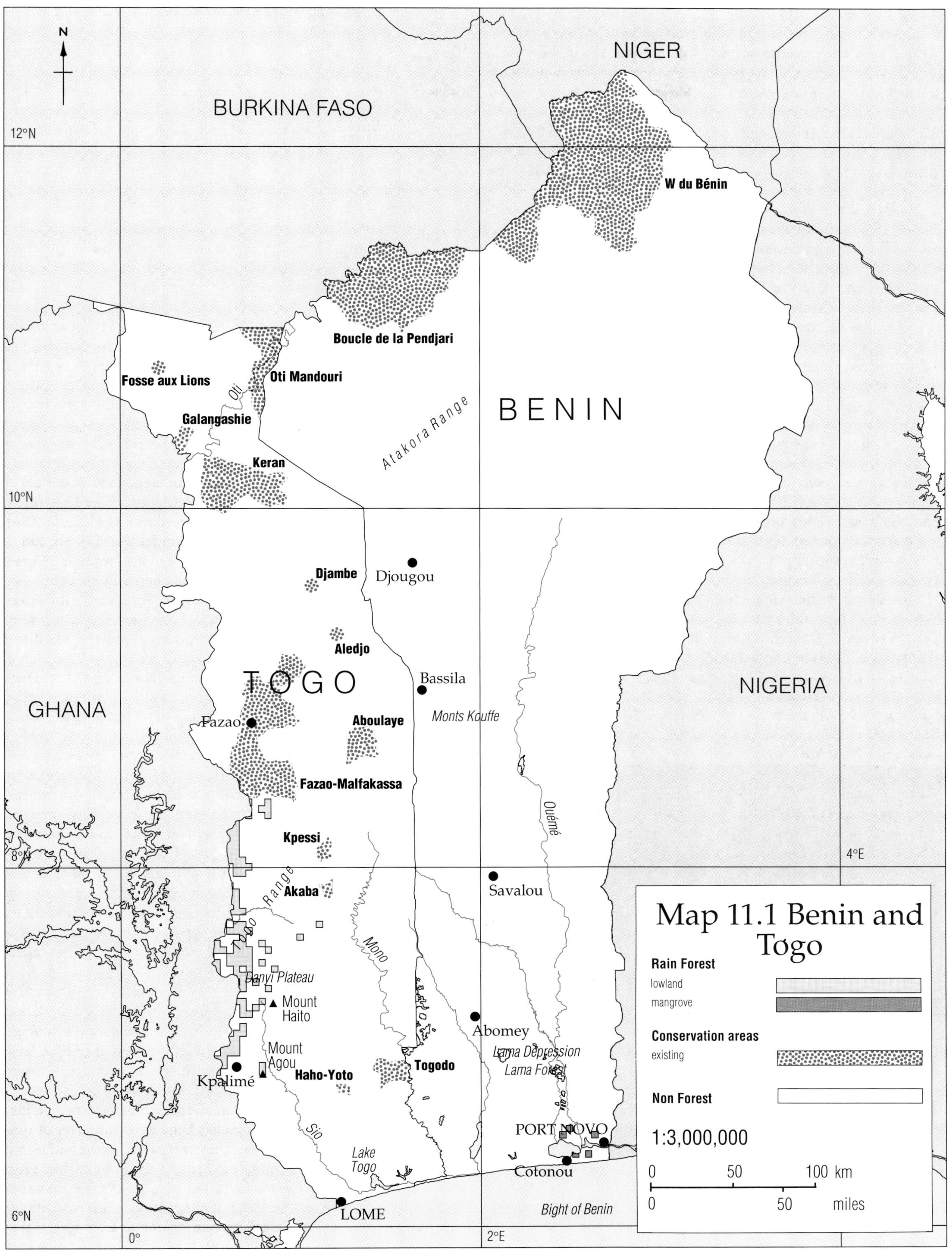
N
NIGER
BURKINA FASO
12°N
W du Bénin
Boucle de la Pendjari
Fosse aux Lions
Oti Mandouri
Oti
Galangashie
Atakora Range
BENIN
Keran
10°N
Djambe
Djougou
Aledjo
TOGO
Bassila
NIGERIA
GHANA
Fazao
Aboulaye
Monts Kouffe
Fazao-Malfakassa
Ouémé
Kpessi
8°N
4°E
Akaba
Savalou
Togo Range
Mono
Danyi Plateau
Mount Haito
Abomey
Mount Agou
Lama Depression
Lama Forest
Kpalimé
Haho-Yoto
Togodo
Sio
PORT NOVO
Lake Togo
Cotonou
LOME
Bight of Benin
6°N
0°
2°E
Map 11.1 Benin and Togo
Rain Forest
lowland
mangrove
Conservation areas
existing
Non Forest
1:3,000,000
0
50
100 km
0
50
miles

are in a natural condition. Considerable amounts of charcoal and firewood are harvested in forests, savannas and in agricultural areas.

Both countries are attempting to strengthen their forest protection and to concentrate timber production in plantations. The seasonal climate and deep, well-drained soils of the south-central parts of both countries are suitable for teak *Tectona grandis* production. Annual yields from plantations in Benin have attained levels of 15–24 cu. m per ha, equal to the best yields obtained in Southeast Asia. Benin had 78 sq. km of teak plantations in 1980 (FAO/UNEP, 1981), managed by a parastatal forestry corporation. Togo had 73 sq. km of teak plantations, some of them dating back to the German colonial period at the beginning of the century. They are dispersed in numerous small plantings, many of which have been poorly maintained and have been encroached upon and degraded. Attempts have ben made to establish plantations of numerous other species in both countries but these have hardly left any traces although some *Cedrela* is grown in association with teak in Benin.

The only natural forests under silvicultural management for timber production are small-scale experimental areas whose management is supported by aid agencies. Projects in Benin supported by FAO and German bilateral assistance have attempted some enrichment planting and more carefully controlled logging on a pilot scale. A similar scheme with German assistance is now operating in Togo. A further German-supported project to protect the central 30 sq. km core of the Lama Forest in Benin and to establish plantations and intensively managed forest in buffer zones in the peripheral part of the forest reserve, is one of the more promising forest conservation and development activities in the two countries.

Deforestation

As early as 1937 it was reported that most of Benin's coastal forests had already disappeared (Aubréville, 1937). For the years from 1981 to 1985, FAO (1988) estimated an annual net deforestation rate of 12 sq. km for Benin and 21 sq. km for Togo.

Agriculture, particularly near the coast, was one of the chief causes of the disappearance of the forests. Those inland, in the more humid regions, initially remained relatively undisturbed. In addition, fire has had a major impact. Vast areas that would once have been covered by dry deciduous forests have been converted to open wooded savannas by centuries of dry-season burning.

The active economy of Togo has generated demand for timber and this has greatly depleted such limited forest resources as the country enjoyed. Togo has been importing timber for two decades. Benin's economy, meanwhile, has not been totally stagnant. Even doctrinaire socialism could not resist the opportunities offered by an excellent port facility and a long, permeable border with densely populated, oil-rich Nigeria. Smuggling liquor and cigarettes to adjacent Moslem states is said to be a significant economic activity of the country. Nigerian cacao moves in the other direction; Benin was once one of Africa's major exporters of this commodity in spite of having an extremely modest domestic production. Timber and other forest products cross the weakly controlled border to meet the needs of the dense populations of adjoining parts of Nigeria. Thus, for totally different reasons the forests of Benin are also sadly depleted.

Manioc is an important crop in Togo, it is frequently grown on land from which forest has been cleared. G. Martin/WWF/BIOS

Biodiversity

There are no comprehensive studies of the fauna or flora of either country. Raynaud and Georgy (1969) describe the species in Benin, but in rather general terms and they focus mainly on the mammals of the savanna zone. Benin has around 2000 species of plants, but the number of endemics is unknown; of the 2300 plants species in Togo at least 20 are found only there (Davis *et al.*, 1986). Numbers of mammals in the two countries are estimated at 187 in Benin and 196 in Togo (Stuart *et al.*, 1990). These include about ten species of primates in each, with the endangered white-throated guenon *Cercopithecus erythrogaster* probably occurring in Benin. There may be as many as 17 species of antelope in each country but forest species such as bongo *Tragelaphus euryceros* and duikers *Cephalophus* spp. are rare and declining. The Fosse aux Lions Forest Reserve in northern Togo has a population of about 150 elephants *Loxodonta africana*, while those in 'W' National Park in Benin form one of the largest remaining elephant populations in West Africa. Sayer and Green (1984) give maps of the distribution of larger mammals in Benin.

There are reported to be 630 bird species in each country (Stuart *et al.*, 1990). Only one threatened species, the white-necked rockfowl *Picathartes gymnocephalus*, is resident in Togo. There are several studies of the birds of the protected areas in the north of Benin (e.g. Green and Sayer, 1979).

The countries do not contain any sites known to be of critical importance for forest biological diversity conservation at a regional level. However, the isolated forest patches in central and southern Benin contain populations of a reasonable variety of primates which would make them important conservation sites at the national level.

Conservation Areas

As in much of Africa, conservation programmes have focused almost exclusively on the savanna areas whose populations of large mammals are a tourist attraction. Both countries have extensive national parks in the dry north (Table 11.2) which are quite well managed. Until recently there were no protected areas whose objective was to conserve natural closed forest. The recent programme to establish a reserve in the Lama Forest in Benin is the first of its kind in either country. Several forest reserves would be important areas for conservation if the laws governing their protection were properly applied. There has been considerable interest over the years in establishing a forest national park in the area

Table 11.2 Conservation areas of Benin and Togo

Classified forests, forest reserves and hunting zones are not included. For data on Biosphere reserves see chapter 9.

	Area (sq. km)
BENIN	
National Parks	
Boucle de la Pendjari	2,755
W du Bénin	5,680
Total	8,435
TOGO	
National Parks	
Fazao-Malfakassa	1,920
Fosse aux Lions	17
Keran	1,636
Faunal Reserves	
Aboulaye	300
Akaba	256
Aledjo	8
Djambe	17
Galangashie	75
Haho-Yoto	180
Kpessi	280
Oti Mandouri	1,478
Togodo	310
Total	6,477

No areas contain significant closed moist forest.
(*Sources:* IUCN, 1990; WCMC, *in litt.*)

of the Monts Kouffé Forest Reserve in central Benin. The Forest Department has never had the resources to pursue this project and the area has continued to be degraded by loggers, poachers and farmers (Green and Sayer, 1978).

Initiatives for Conservation

The European Economic Community is supporting major conservation programmes in the protected savanna areas in the north of both countries. The most important of these projects aims to improve the management of the Pendjari and 'W' National Parks in Benin, including an outlying area of gallery forest adjacent to the Pendjari National Park (see Verschuren *et al.*, 1989).

The Fazao-Malfakassa National Park in central Togo contains some riparian forest and relict patches of forest on steep hillsides. A Swiss-based foundation is supporting the protection of this park. A major project to bring the Keran National Park under management is at present being prepared with support from South Africa.

There are several projects in both countries to establish plantations and promote the use of agroforestry techniques. These will all help to relieve the pressure on natural forests but in only one case is conservation of biological diversity a primary objective. The exception is the German project for the conservation management of the Lama Forest in Benin.

Togo has embarked upon a national Tropical Forestry Action Plan and Benin has indicated its intention of doing so in the near future. Togo is also preparing an Environmental Action Plan with assistance from the World Bank.

References

Aubréville, A. (1937) Les forêts du Dahomey et du Togo. *Bulletin Comité d'Etude Historique et Scientifique d'Afrique occidentale Française* **20**: 1–112.

Davis, S. D., Droop, S. J. M., Gregerson, P., Henson, L., Leon, C. J., Villa-Lobos, J. L., Synge, H., and Zantovska, J. (1986) *Plants in Danger: What do we know?* IUCN, Gland and Cambridge.

FAO (1988) *An Interim Report on the State of Forest Resources in the Developing Countries.* FAO, Rome, Italy. 18 pp.

FAO (1991) *FAO Yearbook of Forest Products 1978–1989.* FAO Forestry Series No. 24 and FAO Statistics Series No. 97. FAO, Rome, Italy.

FAO/UNEP (1981) *Tropical Forest Resources Assessment Project. Forest Resources of Tropical Africa. Part II Country Briefs.* FAO, Rome, Italy.

Green, A. A. and Sayer, J. A. (1978) *La Conservation des Ecosystèmes forestiers dans la Région des Monts Kouffé.* FAO/PNUD BEN 77/011 Document du Travail No. 4, Cotonou. Pp. 1–37 plus annexes.

Green, A. A. and Sayer, J. A. (1979) The birds of Pendjari and Arli National Parks (Benin and Upper Volta). *Malimbus* **1**(1): 14–28.

IUCN (1990) *1989 United Nations List of National Parks and Protected Areas.* IUCN, Gland, Switzerland and Cambridge, UK. 275 pp.

Raynaud, J. and Georgy, G. (1969) *Nature et Chasse au Dahomey.* Sécretariat d'Etat aux Affaires Etrangères, Paris. Pp.1–323.

Robbins, C. B. (1978) The Dahomey Gap – A re-evaluation of its significance as a faunal barrier to West African high forest. *Bulletin of Carnegie Museum of Natural History* **6**: 168–74.

Sayer, J. A. and Green, A. A. (1984) The distribution and status of large mammals in Benin. *Mammal Review* **14**(1): 37–50.

Stuart, S. N., Adams, R. J. and Jenkins, M. D. (1990) *Biodiversity in Sub-Saharan Africa and its Islands: Conservation, Management and Sustainable Use.* Occasional Papers of the IUCN Species Survival Commission No. 6. IUCN, Gland, Switzerland.

Verschuren, J., Heymans, J.-C. and Delvingt, W. (1989) Conservation in Benin. *Oryx* **23**(1): 22–6.

Authorship

Jeff Sayer, IUCN with contributions from Arthur Green, Korup National Park, Cameroon and Denys Bourque, Quebec, Canada.

Map 11.1 Forest cover in Benin and Togo

Data on total forest cover for Togo and mangroves in Benin were taken from 1989–90 UNEP/GRID data, which accompanies an unpublished report *The Methodology Development Project for Tropical Forest Cover Assessment in West Africa* (Päivinen and Witt, 1989). UNEP/GEMS/GRID, with aid from the EEC and FINNIDA, have developed a system to delimit forest/non-forest boundaries in West Africa by mapping and using 1 km resolution NOAA/AVHRR-LAC satellite data. Higher resolution satellite data (Landsat MSS and TM, SPOT) and field data from Ghana, Côte d'Ivoire and Nigeria were also drawn upon. Forest and non-forest data have been categorised into five vegetation types: forest (closed, defined as greater than 40 per cent canopy closure); fallow (mixed agriculture, clear-cut and degraded forest); savanna (includes open forests in the savanna zone and urban areas); mangrove and water. Areas obscured by cloud are also portrayed. The forest and mangrove types and cloud-obscured areas have been mapped in this Atlas.

The UNEP/GRID data set showed no dryland forest in Benin. The information on this vegetation type in Benin shown on Map 11.1 has been taken from the *Ecological Map of the Vegetation Cover of Benin* (FAO, 1979) at a scale of 1:500,000. This map was prepared in 1978 by the Pilot Project on Tropical Forest Cover Monitoring (People's Republic of Benin, UNEP, FAO) from interpretation of Landsat images (recorded between 1973 and 1976) and ground surveys.

Conservation areas for Benin are taken from a 1:600,000 scale map *République Populaire du Bénin* published by the Institut Géographique National, France in 1984 portraying national parks, hunting zones and classified forests. Protected areas for Togo were extracted from two maps: an unpublished blueline map *Forêts Classées du Togo* at a scale of 1:200,000 (nd) and a published map produced by the Institut Géographique National in 1977 *Togo* at a 1:500,000 scale.

12 Burundi and Rwanda

BURUNDI
Land area 25,650 sq. km
Population (mid-1990) 5.6 million
Population growth rate in 1990 3.2 per cent
Population projected to 2020 13.7 million
Gross national product per capita (1988) US$230
Rain forest (see map) 413 sq. km
Closed broadleaved forest (end 1980)* 150 sq. km
Annual deforestation rate (1981–5)* 4 sq. km
Industrial roundwood production† 49,000 cu. m
Industrial roundwood exports† nd
Fuelwood and charcoal production† 4,034,000 cu. m
Processed wood production† 3000 cu. m
Processed wood exports† nd

RWANDA
Land area 24,950 sq. km
Population (mid-1990) 7.3 million
Population growth rate in 1990 3.4 per cent
Population projected to 2020 19.7 million
Gross national product per capita (1988) US$310
Rain forest (see map) 1554 sq. km
Closed broadleaved forest (end 1980)* 1010 sq. km
Annual deforestation rate (1981–5)* 28 sq. km
Industrial roundwood production† 240,000 cu. m
Industrial roundwood exports† nd
Fuelwood and charcoal production† 5,602,000 cu. m
Processed wood production† 15,000 cu. m
Processed wood exports† nd

* FAO (1988)
† 1989 data from FAO (1991)

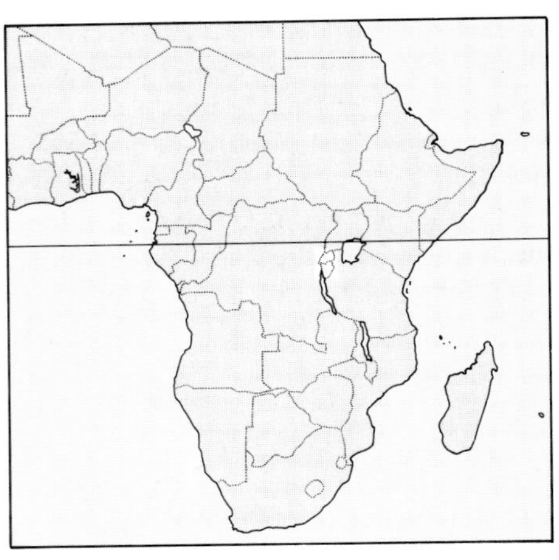

Burundi and Rwanda, located in the equatorial highlands of east-central Africa along the Western Rift, are two of the smallest countries on the continent. They each straddle a major portion of the Zaïre–Nile Divide: a rugged mountainous region of the Rift which is thought to have served as a refugium for moist forest species during dry climatic periods of the Pleistocene. Although interconnected at times in the past, the mountain forests of the region have become disconnected to form an archipelago of forest islands, allowing for the separate evolution of species. These factors, in combination with a large range of topographic, edaphic and climatic characteristics represented in a small area, have resulted in montane forests of unusual species richness as well as high levels of local endemism and species rarity.

These biological attributes are found in a human context of high population density and growth, low per capita supplies of both natural and financial resources and weak institutions. In both countries, 90–95 per cent of the population relies on subsistence farming and virtually all the lands with arable potential have been converted to agriculture. The forest estate has consequently declined to approximately 1.4 per cent in Burundi and 5 per cent in Rwanda. Population pressure is currently driving people to convert marginal lands and seek alternative sources of subsistence goods, especially fuelwood, and commercial products such as timber and gold. This puts additional pressure on the remaining forests.

Despite growing pressures, strong forest conservation programmes have developed. All major montane forests in both countries are now gazetted as either national parks or reserves. Conservation and management of these areas have not always been effective, however. The challenge lies in designing and implementing sustainable management schemes for existing reserves.

Introduction

Burundi This small, landlocked country in east-central Africa stretches 232 km from north to south between 2°20'S and 4°28'S and 203 km from west to east between 29°E and 30°58'E. Zaïre and Lake Tanganyika lie to the west of the country, while Tanzania and Rwanda are on its eastern and northern borders respectively.

There are four natural ecological regions in Burundi (White, 1983a). The Imbo region in the west is a narrow subsidence plain extending along the Rusizi River and the north bank of Lake Tanganyika, lying between altitudes of 780 m and 1000 m. Remnant savanna and a small patch of Guineo-Congolian forest are situated in this region. The elongated, folded ridges of the Zaïre–Nile massif rise some 1500 m above the Imbo plains and then merge into the Central Plateau in the east at around 2100 m. There are four summits higher than 2500 m in this area. Lower montane rain forest remains along the highest reaches of the massif. The Central Plateau is a hilly region above 1500 m, where more than half of Burundi's population lives. Finally, in the east of the country lie the lowlands, including extensive wetlands, of the Mosso.

Burundi has an equatorial montane climate, characterised by mild and stable temperatures and moderate rainfall which varies with altitude. Typically there is a short dry season from December to January, relatively heavy rains from February to May, low precipitation between May and October (June–August may be completely dry), followed by substantial rains from October to December. Precipitation levels vary considerably from region to region (500–2000 mm per annum). Mean annual temperatures on the central plateau are around 21°C with summer maxima of about 33°C and winter minima of approximately 6°C (below 2000 m altitude).

Burundi's original inhabitants were probably the Batwa, a forest-living people who now make up less than 1 per cent of the population. Later, Bahutu agriculturalists and Batutsi pastoralist groups moved into the country and they now constitute 99 per cent of the population. With a mean of 193 persons per sq. km, Burundi's population density is second only to that of Rwanda on the African continent. The current population is estimated to be 5.6 million, growing at 3.2 per cent per year, with 95 per cent of the people living in rural areas (PRB, 1990). Wheat, sorghum, maize, beans, peas, cassava and bananas are the main subsistence crops, while coffee is the most important commercial crop (Wilson, 1990).

Rwanda This hilly country is situated between 1°4'S and 2°51'S and 28°50'E and 30°53'E, along the eastern lip of the Western Rift Valley. Stretching only 185 km north to south and 225 km east to west, it is bounded by Zaïre to the west, Uganda to the north, Tanzania to the east and Burundi to the south.

Ecological zones have been described for Rwanda (Sirven *et al.*, 1974; Delepierre, 1982) and can be summarised as follows. The south-west corner of the country is the lowest in altitude and is an extension of the Imbo region of Burundi. To the east the Kivu slopes run along the western edge of Rwanda, an area known as the Impara. This area consists of extremely rugged hills, averaging 1900 m in altitude, now entirely converted from forest to agriculture. Rising to the east is the Zaïre–Nile Divide, forming a prominent backbone along the length of the country from north to south. This region contains all of Rwanda's remaining montane forests, except for the tiny (less than 0.5 sq. km) Ndiza forest in the district of Gitarama (A. Monfort, *in litt.*). The greater part of the Divide lies above 2000 m and was formed by upthrust during the formation of the Western Rift. The northernmost portion, Virunga, is of volcanic origin and includes three of Africa's higher mountains. Less rugged hills and lava plains descend eastward to the Central Plateau, a highland area (1500–2000 m) dissected by many rivers, where most Rwandans live. Finally, in the eastern and south-central portions of the country lie the more sparsely populated savanna regions of the Akagera basin and the Bugesera. These areas are of lower altitude (generally 1300–1600 m) and end in extensive wetland and papyrus marsh at their respective international borders.

Rwanda experiences an equatorial climate, moderated by its high elevation. Temperatures are generally mild and stable, with annual means between 14°C and 21°C depending on altitude, although frosts may occur in the Zaïre–Nile Divide and snow falls periodically on the highest peaks of the Virunga mountains. Rainfall in Rwanda is related to altitude, with annual means of 2400 mm on the Zaïre–Nile Divide, 1200 mm in the Central Plateau and as little as 650 mm in the eastern savannas. Over most of the country there are four seasons: a major dry season from June to September, a short rainy season from October to December, a relatively dry season from January to February and a long rainy season from March to May (Sirven *et al.*, 1974).

Rwanda's current population is estimated to be 7.3 million or 286 inhabitants per sq. km, which is the highest density on the African continent. Numbers are increasing at approximately 3.4 per cent each year. More than 90 per cent of the people depend primarily on farming for a living and virtually all arable land is now under cultivation (Fossey, 1983; Harcourt, 1986; WRI, 1990). Ethnic composition is similar to that of Burundi: the original inhabitants, the Batwa pygmies, now number less than 1 per cent of the population; the Bahutu, who occupied the country sometime between the 7th and 10th centuries, make up around 90 per cent of the population while the Batutsi, who migrated there in the 14th or 15th century, comprise the remainder.

The Forests

It is estimated that one-third to half of Burundi and Rwanda was originally montane forest (Weber and Vedder, 1984; Runyinya, 1986), which was found mostly in the western highlands of the Zaïre–Nile Divide and the slopes leading to Lake Tanganyika and Kivu. Almost all the remaining forest in the two countries is in the highest reaches of the Zaïre–Nile massif where it is now subdivided into six major discontinuous patches. Apart from these, one very small patch (a few sq. km) of lower altitude closed forest survives at Kigwena along the banks of Lake Tanganyika in Burundi. This is one of the easternmost patches of Guineo-Congolian forest and thus is of special interest.

Burundi The montane forests of Burundi are chiefly closed forest, but they vary with altitude and latitude due to changes in temperature and length of dry season respectively. Bururi forest is located at the southern tip of the Zaïre–Nile Divide. It contains a unique assemblage of species, some more typically found in the savanna regions to the east, some from the lowland forests of Zaïre and others common to other montane forests of the region (Weber and Vedder, 1983; Weber and Vedder, 1984). More than 90 tree species are present, six of which are deciduous due to the long dry season. The lowest lying regions (1600–1900 m) are dominated by *Anthonotha pynaertii, Albizia gummifera, Parinari excelsa, Newtonia buchananii, Croton macrostachyus* and *Tabernaemontana stapfiana* (Project Bururi-Rumonge-Vyanda, 1990). Epiphytes and ferns are numerous. The upper regions (1900–2300 m) include *Albizia* and *Tabernaemontana* as listed above but are characterised by *Chrysophyllum gorungosanum, Symphonia globulifera* and *Entandrophragma excelsum* (Weber and Vedder, 1983). Under favourable conditions in Bururi, *Entandrophragma* individuals emerge from the canopy to reach as high as 65 m.

Kibira forest, along the divide to the north, is at a higher altitude (1000–2660 m) and is more humid. It is contiguous with the Nyungwe forest of Rwanda. Primary forest constitutes about 20 per cent of these areas; the rest is secondary due to such human disturbances as past burning, tree-felling and cattle raising. Bamboo *Arundinaria alpina* is mixed with both the primary and secondary forests. The primary forest is tall closed canopy forest dominated by species such as *Parinari excelsa, Entandrophragma excelsum, Cassipourea ndando, Albizia gummifera* and *Syzygium guineense.* Secondary forest stages include the tree species *Macaranga kilimandscharica, Neoboutonia macrocalyx, Polyscias fulva* and *Hagenia abyssinica* (Trenchard, 1987; Nduwumwami, 1990). In the past, this forest was used as a royal hunting ground and some areas today hold almost magical qualities for the local people and thus are left undisturbed (Wilson, 1990).

The small patch of evergreen forest at Kigwena lies at an altitude of 780–800 m. Exceptionally large (some are over 10 m high) and spectacular *Dracaena steudnesi* plants occur in the forest along with a wide variety of trees including *Maesopis eminii, Newtonia buchananii* and *Pycnanthus angolensis* (Wilson, 1990).

Rwanda Nyungwe forest, located in the south-west, and contiguous with Kibira forest in Burundi, is Rwanda's largest and most diverse montane forest. Dowsett-Lemaire (1990) estimates that it now covers 900 sq. km. Ranging in altitude from 1600 m to 2950 m, it is a dynamic mosaic of closed forest (with *Parinari excelsa, Strombosia scheffleri, Chrysophyllum* sp., *Entandrophragma excelsum, Symphonia* sp., *Newtonia buchananii, Podocarpus* spp. and *Ocotea* spp.), secondary forest, drier forest ridges, swamp forest (dominated by *Carapa grandiflora, Syzygium guineense* and *Anthocleista grandiflora*), large homogeneous stands of bamboo and openings filled with herbaceous plants (J.-P. Van de Weghe, *in litt.*; Bahigiki and Vedder, 1987). Most of the openings and secondary forest are thought to be the consequence of landslides rather than human disturbance.

The Mukura, Gishwati-south and Gishwati-north forests are found further along the Divide. Except that a far greater percentage of the forest cover is in secondary forest (due largely to cattle pasturing) and is generally more impoverished in character, all three are similar to portions of Nyungwe forest at equal altitude (D'Huart, 1983). The dominant secondary species is *Neoboutonia macrocalyx.*

The forests of the Virunga mountains, in the north-west corner of the country, stretch from 2400 m to 3500 m. The lowest slopes of the dormant volcanoes (2500–3200 m) are carpeted with bamboo which covers approximately 35 per cent of the forest. Patches of *Dombeya goetzenii* and *Neoboutonia* are found but the *Prunus africana* forest once located below the bamboo has been converted to agriculture. Higher up, from 2600 m to 3500 m, particularly on the more humid slopes of the east and south, *Hagenia-Hypericum* forest predominates. This open forest, reaching heights of only 10–12 m, covers approximately 30 per cent of the area and is accompanied by lush growth of terrestrial herbaceous vegetation with many grassy, waterlogged clearings. Rocky ridges frequently carry a more diverse tree community and these grade into ericaceous heath leading up from the forest to the sub-alpine zone. This zone is found above 3500 m and is dominated by giant *Lobelia* and *Senecio*. The mountains are topped by *Alchemilla*, sedge and grass meadows in the alpine zone.

Forest Resources and Management

At present, only 360 sq. km (1.4 per cent) of Burundi's land area remains covered in forest, while about 5 per cent of Rwanda is forested. Map 12.1 shows slightly higher forest areas, with 411 sq. km (1.6 per cent) of montane forest remaining in Burundi and 1554 sq. km (6.2 per cent) in Rwanda (Table 12.1). FAO (1988) gives lower figures than suggested here, particularly in Burundi where it estimates only 150 sq. km remains. Map 12.1 shows the extent of Kigwena forest to be only 2.4 sq. km although it is reported to be 5 sq. km in area.

The montane forests of Rwanda and Burundi form an archipelago of high-altitude islands surrounded by a sea of intensive agriculture. This exacerbates both the threats to the forests and the importance of them to their host countries. The value of these forests lies not in the commercial extraction of timber, since as montane forests they harbour only low densities of valuable hardwoods. Instead, their greatest value is realised through the ecological services performed by the forest cover: regulation of the entire region's hydrological system, prevention of soil erosion and reduction of flooding downstream. For instance, the forested areas of the Virungas provide 10 per cent of Rwanda's water catchment area although they cover only 0.6 per cent of its land area (Harcourt, 1986; Weber, 1987).

More easily quantifiable are tangible economic returns from a variety of forest resources. First among these is tourism, which has proved to be an important source of foreign exchange in Rwanda (Weber, 1989). Initially based on visits to mountain gorillas in the Volcanoes National Park (see case study), spin off programmes have begun in the Nyungwe (see case study in chapter 9), Gishwati and Kibira-Teza forests. It is estimated that Rwanda earned US$17 million of foreign exchange in 1989 from its tourist industry (Monfort, 1990). However, the civil war is presently (1991) having a devastating effect on tourist income. Although not tourism *per se*, the establishment of research centres in some of the forests (Volcanoes National Park, Nyungwe and Kibira-Teza) provides benefits in the form of local employment and foreign exchange earnings. More important in the long term, these centres provide the capability to train host-country students and professionals.

Montane forest, note the lichen and moss draping the trees, in the Volcanoes National Park, Rwanda. C. Harcourt

Table 12.1 Estimates of forest extent in Burundi and Rwanda

	Area (sq. km)	*% of land area*
BURUNDI		
Rain forests		
Lowland	2	<0.01
Montane	411	1.6
Totals	413	1.6
RWANDA		
Rain forests		
Montane	1,554	6.2
Totals	1,554	6.2

(Based on analysis of Map 12.1. See Map Legend on p. 109 for details of sources.)

Forest products play an important direct economic role in the lives of local people. Major products include fuel and construction wood, timber, bamboo, honey, medicinal plants, thatch and bushmeat (Bahigki and Vedder, 1987). These resources provide subsistence goods for the people surrounding the forest and some form the major source of income for the harvester. At present, extraction of these products is illegal and frequently involves over-exploitation such that product availability declines.

In order to conserve these resources, each of the remaining forests of Rwanda and Burundi is classified as either national park or reserve. Various management strategies have been adopted in each of these areas, ranging from total protection to multiple use (see Conservation Areas on p. 107).

Deforestation

Although forest conversion began more than 2000 years ago, it was not until the first half of this century that the destruction became significant. During the 1920s the forest edge was pushed back at the rate of nearly one kilometre per year along the entire Zaïre–Nile Divide (Weber and Vedder, 1984). Despite the fact that the

THE MOUNTAIN GORILLA PROJECT

Only six other areas in Africa reach the 4000m altitudes found in the Virunga Volcano region of Rwanda, Uganda and Zaïre. Probably no other forest at these altitudes – the forest stretches up from 2500 m – is as well protected as the montane forest of Rwanda's Volcanoes National Park (Parc National des Volcans) and Zaïre's Virungas National Park (Parc National des Virungas-Sud). Mountain gorillas are the reason. They possess all the attributes required of a species to raise public sympathy, and through them some of the world's rarer forest habitat has been saved.

In the mid-1970s the situation looked bleak for the gorillas and the forest (Harcourt and Fossey, 1981). Repeated censuses indicated that the size of the Virunga gorilla population had not only declined markedly from the estimated 450 of the early 1960s, but that a trade in infants and heads had started. In addition, in the late 1960s, about 20 per cent of the forest had been excised for pyrethrum *Tanacetum cinerariifolium* plantations, with further appropriations planned for cattle ranches. The killing of animals in Dian Fossey's famous study groups in 1978 signalled a nadir. The attacks shocked the world conservation community, and so was born the Mountain Gorilla Project (MGP). Four international conservation organisations (WWF, FFPS, The African Wildlife Foundation and the Peoples Trust for Endangered Species) joined forces and, along with the Belgian foreign aid programme, provided funds, personnel and expertise to help Rwanda protect its Volcanoes National Park and the mountain gorilla. The level of funding required was high – at least US$150,000 per year for the first five years was spent in Rwanda alone.

Starting in 1979, the MGP supplied equipment to the park guard force; it helped the Rwandan Office of Tourism and National Parks (ORTPN) to habituate three gorilla groups to visits by tourists and to establish a strictly controlled tourism programme; and it instituted a country-wide conservation awareness campaign that reached all levels of the community from the local people, through primary and secondary schools, to government officials and diplomats. Until then, the only significant improvement in management of the whole Virunga area had occurred in 1976, when Rwanda increased its guard force and expelled all livestock from the park. Later censuses showed that the improvement in Rwanda's gorilla population stemmed from this date, a momentum maintained and augmented by the Mountain Gorilla Project (Harcourt, 1986).

After ten years' operation, the Mountain Gorilla Project was by any standards a success:

- An eighty-fold increase in park revenue from US$10,000 to more than US$800,000.
- A five-fold increase in number of visitors to the park.
- A doubling of park guards.
- Newly built park headquarters and visitors' accommodation.
- No gorilla killed by poachers since 1983.
- A drop from 50 per cent of the local farmers to less than 20 per cent who think the park should be made available for agriculture (Harcourt *et al.*, 1986).
- About a 15 per cent increase in the gorilla population since the mid-seventies, from around 270 to 310, due to increased immature recruitment to the population (Harcourt, 1986).

A vital reason for the success of the gorilla conservation programme in Rwanda was the previous two decades' biological study of the gorillas in the region. Building on George Schaller's studies in Zaïre in the late 1950s, Dian Fossey established in 1967 in the Volcanoes National Park the field camp that was to become the Karisoke Research Centre. Her work (Fossey, 1983) and that of others there has since provided one of the best compendiums available of knowledge and understanding of the behaviour, ecology and population dynamics of a tropical mammal species. The speed with which Rwanda's new park management policies were formulated and implemented was largely due to this knowledge. In recognition of the importance of biological research to effective management, the Mountain Gorilla Project and ORTPN have always kept separate the research and tourist areas in the park, despite the large financial gain possible from taking groups of tourists to the habituated study groups. The Centre is now funded by the American foundation the Digit Fund, and in 1989 the Rwandan government agreed to a five-year, several hundred thousand dollar USAID project with the Centre to fund research on aspects of the park's ecology other than gorillas (sadly neglected to date) and to strengthen ties between the Centre and other institutions in Rwanda concerned with conservation.

Initially the situation improved only in Rwanda; in Zaïre and Uganda it worsened. In 1985, former personnel of the Mountain Gorilla Project moved to Zaïre to cooperate with the Zaïre Institute for the Conservation of Nature (IZCN) in establishing an equivalent programme there with international funding from WWF and Frankfurt Zoo. The continued health of the gorillas and the forest now owes as much to Zaïre's efforts as to Rwanda's.

If the Virunga gorillas are to continue to multiply and thrive, conservation projects in the area are surely going to have to involve themselves directly in alleviating the ever increasing pressure on the surrounding land from a rapidly growing population and its demands for agricultural expansion. This has happened in Zaïre, with afforestation now an intrinsic part of management of the region. However, the authorities in Rwanda have yet to integrate conservation of the Volcanoes National Park with agricultural development in the community.

Source: Alexander Harcourt

The mountain gorillas have proved an ideal species for attracting tourists and thereby ensuring the protection of their habitat. C. Harcourt

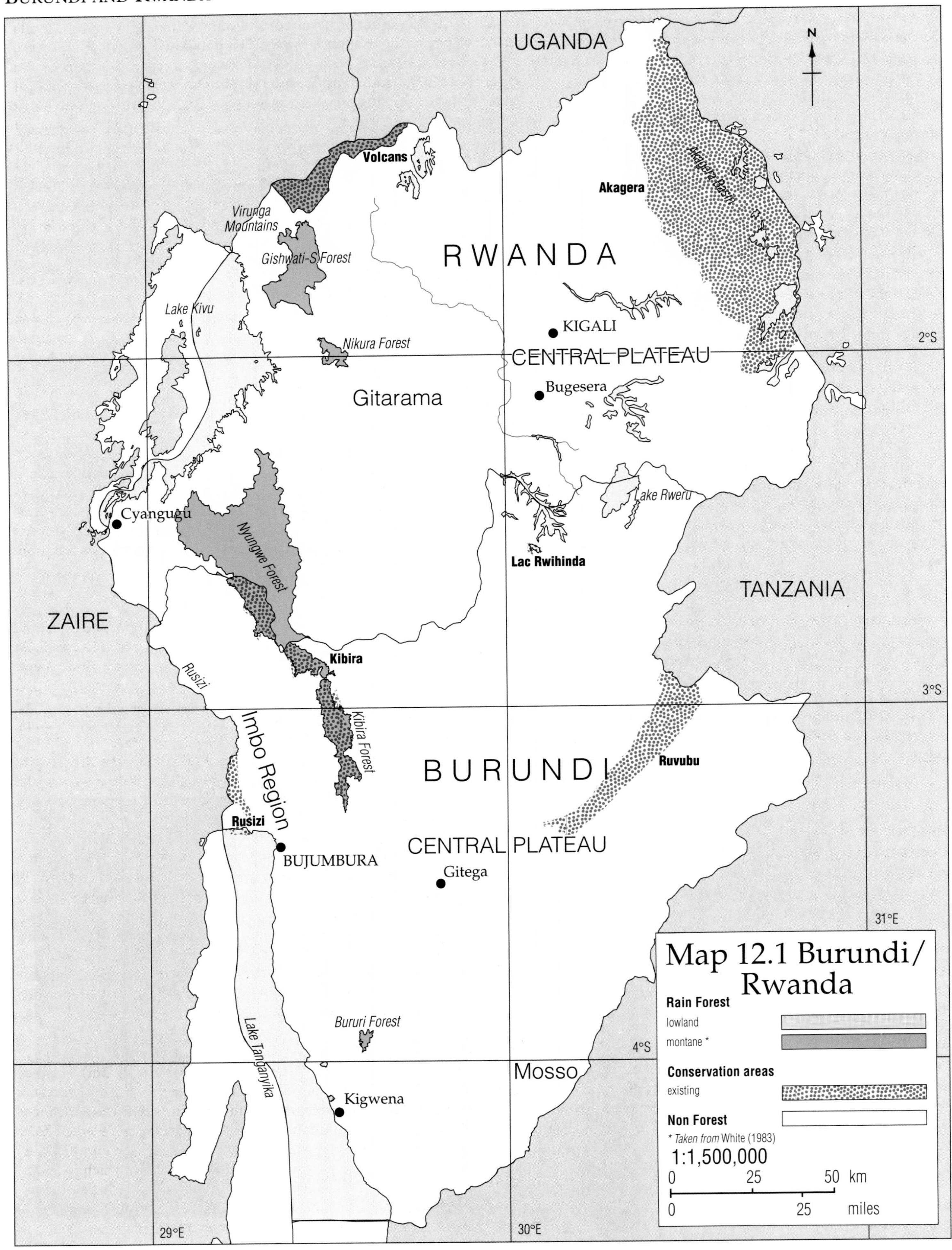
UGANDA
N
Volcans
Akagera
Akagera Basin
Virunga Mountains
Gishwati-S Forest
RWANDA
Lake Kivu
KIGALI
Nikura Forest
2°S
CENTRAL PLATEAU
Bugesera
Gitarama
Lake Rweru
Cyangugu
Nyungwe Forest
Lac Rwihinda
TANZANIA
ZAIRE
Kibira
Rusizi
3°S
Imbo Region
Kibira Forest
BURUNDI
Ruvubu
Rusizi
BUJUMBURA
CENTRAL PLATEAU
Gitega
31°E
Map 12.1 Burundi/ Rwanda
Rain Forest
lowland
montane *
Conservation areas
existing
Non Forest
* Taken from White (1983)
1:1,500,000
0 25 50 km
0 25 miles
Bururi Forest
Lake Tanganyika
4°S
Mosso
Kigwena
29°E
30°E

Rwandan forests were declared national reserves in 1933 by the Belgian colonial authorities, enforcement of the regulations was irregular and often lax. Reserve declaration resulted in relatively stable boundaries for many years, though it rarely prevented some degree of internal degradation. As population densities increased, however, greater pressure was put on the forest reserves, pushing the boundaries back both legally and illegally. For example, 40 per cent of the Volcanoes National Park was excised in the late 1960s for conversion to a large-scale agricultural settlement project. At the same time, the Nyungwe forest reserve declined by 170 sq. km due to encroachment by individual farmers (Harroy, 1981). Similar trends were taking place in Burundi.

By the late 1970s, both countries recognised the problem and took action to stabilise reserve borders. In particular, this involved increased patrols and buffer tree plantations. Subsequently, effort has been placed on controlling illegal activities within the reserves. Notable successes include the exclusion of cattle, the destruction of thousands of snares set for antelope and the control of wood and bamboo cutters in the Volcanoes National Park. Cattle have also been effectively excluded from Kibira National Park, but problems of small-scale hunting and tree-felling continue, as they do in the Nyungwe forest (Nduwumwami, 1990). A comparatively new threat in both Kibira and Nyungwe forests is the invasion of thousands of gold miners who clear forest patches for their camps and denude the stream banks in their search for alluvial deposits. Historically this activity has caused a great reduction in numbers, or even extinction, of large mammals in all the forests (with the excepion of particular regions of the Volcanoes National Park). Most affected are elephant *Loxodonta africana*, buffalo *Syncerus caffer* and duikers (*Cephalophus nigrifrons*, *C. sylvicultor*, *C. weynsi* and possibly *C. monticola*).

Though there has been improved protection of forests in recent years, the Gishwati Reserve is an exception. From 1980 to 1986, more than 100 sq. km of this forest was converted for cattle ranching and timber production, thereby cutting the forest into two isolated sectors (Vedder, 1985).

FAO (1988) estimated deforestation in Rwanda to be 28 sq. km per year and in Burundi to be 4 sq. km per year, a rate of around 2.7 per cent per annum in both countries. However, the present rate of deforestation of the montane forests, in Rwanda at least, is probably zero.

Biodiversity

There are approximately 2500 plant species in Burundi and 2150 in Rwanda (Davis *et al.*, 1986). Many of the animal taxa have not been catalogued, but there are considered to be 633 bird species in Burundi and 669 in Rwanda and 103 mammal species in Burundi (Stuart *et al.*, 1990) and at least 168–174 in Rwanda (A. Monfort, *in litt.*). Wilson (1990) gives information about some of the larger mammals in Burundi. The number of species endemic to each country is undoubtedly low, due largely to the small area in question. However, the montane forests are home to many species endemic to the Kivu-Rwenzori Highlands (or Central African Highlands), which are limited in distribution but are located on the international boundaries of Uganda, Zaïre, Rwanda and Burundi (White, 1983b). The high proportion of endemic species and the unusual species richness, are used as evidence for the claim that this region was part of the most significant Pleistocene forest refugium in Africa (Hamilton, 1982 and see chapter 2).

The best-known Afromontane species are certainly the primates, a number of which are threatened. A well-studied population of mountain gorillas *Gorilla gorilla berengei* is found in the Virunga mountains, shared between Rwanda, Zaïre and Uganda. They number approximately 310 individuals but the population has been rising during the past nine years after a long period of decline (Vedder and Weber, 1990). Other rare or threatened primates include the owl-faced monkey *Cercopithecus hamlyni*, golden monkey *C. mitis kandti*, L'Hoest's monkey *C. lhoesti*, Rwenzori black-and-white colobus *Colobus angolensis ruwenzorii* and chimpanzee *Pan troglodytes* (Storz, 1983). The last four are also found in Burundi. The threatened Uganda red colobus *Procolobus [badius] rufomitratus tephrosceles*, is also present in Burundi.

Both countries contain the threatened Grauer's swamp warbler *Bradypterus graueri*, papyrus yellow warbler *Chloropeta gracilirostris* and Kungwe apalis *Apalis argentea* (Collar and Stuart, 1985). The rare, little-known Albertine owlet *Glaucidium albertinum* is also recorded from Rwanda.

There are 394–400 butterfly species in Rwanda, of which three (*Charaxes turlini*, *Bebearia dowsetti* and *Acraea turlini*) are endemic, while another 12 endemic subspecies are also found only in Rwanda (A. Monfort, *in litt.*). The vulnerable cream-banded swallowtail *Papilio leucotaenia* occurs only in the forests of south-west Uganda, north-east Zaïre, Rwanda and western Burundi. The Nyungwe forest of Rwanda is the single most important locality (Collins and Morris, 1985).

The combination of great species richness, a high proportion of species endemic to the Central African Highlands and significant numbers of rare and threatened forms has caused IUCN to rate the montane forests of the Albertine Rift in the highest priority grouping for conservation in Africa (MacKinnon and MacKinnon, 1986).

Conservation Areas

Burundi It was only in 1980 that a decree allowed for the establishment of national parks and nature reserves in Burundi. At that time two national parks, two nature reserves, three forest reserves and two national monuments were proposed (IUCN, 1987). In 1982, conservation efforts began in the Kibira National Park (379 sq. km) and the Bururi Forest Reserve (16 sq. km) along with several other, non-forested, protected areas. In 1983, a management plan for Bururi was developed under the direction of the Burundian Institute for Nature Conservation with the assistance of USAID. It involved protection of the remaining forest, natural reafforestation of clearings, development of woodlots and agroforestry plots around the reserve and a preliminary study of the area's tourist potential (Weber and Vedder, 1983; Weber and Vedder, 1984). Much of this plan has now been implemented.

Protection activities for the Kibira National Park have included the planting of extensive plantations of exotic trees on the eastern border, organisation of park patrols and a study of resident chimpanzees to assess their conservation status and the possibility of chimp-focused tourism (Trenchard, 1987). Although the area has park status, commercial timber species have been planted in the forest to generate revenues for park management.

Rwanda Only two national parks have been gazetted (Table 12.2), Akagera (2500 sq. km) and the Volcanoes (150 sq. km), yet they occupy over 10 per cent of the nation's territory. Moist forest is found in only the latter. Although small in itself, the Volcanoes National Park is contiguous with the Virunga National Park of Zaïre and the Uganda Gorilla Game Reserve. The Volcanoes National Park was originally part of Albert National Park, which included the entire area of the volcanoes. This was set up in 1925 under the Belgian colonial regime and was the first national park to be gazetted in Africa. The Zaïre and Rwanda sectors were divided in

Table 12.2 Conservation areas of Burundi and Rwanda

Conservation areas are listed below. Forest reserves and natural forest reserves are not included or mapped. For data on Biosphere reserves see chapter 9.

	Area (sq. km)
BURUNDI	
National Parks	
Kibira*	379
Ruvubu	436
Managed Nature Reserves	
Lac Rwihinda	4
Rusizi	52
Total	871
RWANDA	
National Parks	
Akagera	2,500
Volcans (Volcanoes)*	150
Total	2,650

(*Sources:* IUCN, 1990; WCMC, *in litt.*)

* Area with moist forest within its boundaries according to Map 12.2.

1960 when Zaïre became independent. The Volcanoes National Park is currently the best protected of the reserves of Rwanda and Burundi due to the high profile of, and economic return from, mountain gorilla tourism (Weber, 1989; Vedder and Weber, 1990 and see case study). It was declared a Biosphere reserve in 1983.

The forest reserves of Mukura (55 sq. km) and Gishwati (North, 70 sq. km; South, 80 sq. km) are being managed as multiple-use zones, except for a 50 sq. km portion of the latter which is planned as a nature reserve. A buffer of exotic tree plantations is being planted round each and various options for natural forest management are being considered (particularly selective logging and enrichment planting). Further commercial timber plantations are planned inside Gishwati forest (Vedder, 1985).

A general management plan has been prepared for the Nyungwe Forest Reserve (970 sq. km). It designates three zones to be treated in different manners: one to be protected as a wilderness (40 per cent), the second to be cleared and replanted with commercial tree species (10 per cent) and the last to be managed for sustainable timber production and possibly local non-timber product extraction (50 per cent). In addition, a buffer of tree plantations is being established entirely to surround the forest reserve. Plantings of indigenous species may take place in limited areas of the buffer zone, while mixed forestry and cattle-raising activities are being implemented in another limited portion. Finally, a multi-pronged approach to general forest conservation is in progress consisting of:

- An inventory of forest habitats and species.
- Studies of their distribution and abundance.
- Monitoring of human use of the forest.
- Development of a tourism programme centred on visits to selected monkey groups and scenic forest trails (see case study in chapter 9).
- Public awareness discussions with people living around the forest.

Initiatives for Conservation

Since all significant blocks of montane forest in Burundi and Rwanda are within designated conservation areas, the remaining challenge is chiefly that of management. Legislation regarding forest conservation is up to date, but the means effectively to enforce regulations are inadequate. Enforcement capability is severely constrained by the lack of trained professionals, such as biologists, wardens, foresters and managers. Initiatives in both formal and informal training are sorely needed (Weber, 1987; Vedder and Weber, 1990). Although financial self-sufficiency in reserve management is the ultimate objective, external assistance will be necessary, at least for some time to come.

Support for forest conservation efforts has been provided by several international conservation organisations during the past 12 years in Rwanda and five years in Burundi. These organisations include WCI/NYZS, FFPS, USAID, WWF, US Peace Corps, Belgium aid and AWF. Renewal of support from USAID has recently been negotiated in both countries, which will allow for a continuation and expansion of current conservation efforts.

Finally, regional coordination efforts were initiated at the first of a series of workshops on conservation and management of Afromontane forests of Rwanda, Burundi, eastern Zaïre and south-western Uganda. Future coordination is planned in the form of further meetings, site visit exchanges for reserve personnel, exchange of reports and information and a regional training programme.

References

Bahigki, E. and Vedder, A. (1987) *Etude Socio-Economique et Propositions Ecologiques sur la Forêt de Nyungwe*. Unpublished report, World Bank. 213 pp.

Collar, N. J. and Stuart, S. N. (1985) *Threatened Birds of Africa and Related Islands. The ICBP/IUCN Red Data Book Part 1.* ICBP/IUCN, Cambridge, UK. 761 pp.

Collins, N. M. and Morris, M. G. (1985) *Threatened Swallowtail Butterflies of the World. The IUCN Red Data Book.* IUCN, Gland, Switzerland and Cambridge, UK. 401 pp.

Davis, S. D., Droop, S. J. M., Gregerson, P., Henson, L., Leon, C. J., Villa-Lobos, J. L., Synge, H. and Zantovska, J. (1986) *Plants in Danger: What do we know?* IUCN, Gland, Switzerland and Cambridge, UK.

Delepierre, G. (1982) Les régions agroclimatiques en rélation avec l'intensité de l'érosion du sol. *Bulletin Agricole du Rwanda* **15**(2): 87–96.

D'Huart, J. P. (1983) *Conservation et Aménagement des Forêts Naturelles de la Crête Zaïre-Nil au Rwanda.* Unpublished report, IUCN.

Dowsett-Lemaire, F. (1990) Physionomie et végétation de la forêt de Nyungwe, Rwanda. *Turaco Research Report* **3**: 11–30.

FAO (1988) *An Interim Report on the State of Forest Resources in the Developing Countries.* FAO, Rome, Italy. 18 pp.

FAO (1991) *FAO Yearbook of Forest Products 1978–1989.* FAO Forestry Series No. 24 and FAO Statistics Series No. 97. FAO, Rome, Italy.

Fossey, D. (1983) *Gorillas in the Mist.* Hodder and Stoughton, London, UK.

Hamilton, A. C. (1982) *Environmental History of East Africa: A Study of the Quaternary.* Academic Press, London, UK. 328 pp.

Harcourt, A. H. (1986) Gorilla conservation: anatomy of a campaign. In: *Primates: The Road to Self-Sustaining Populations.* Benirshke, K. (ed.), pp. 31–46. Springer Verlag, New York, USA.

Harcourt, A. H. and Fossey, D. (1981) The Virunga gorillas: decline of an 'island' population. *African Journal of Ecology* **19**: 83–97.

Harcourt, A. H., Pennington, H. and Weber, A. W. (1986) Public attitudes to wildlife and conservation in the Third World. *Oryx* **20**(3): 152–4.

Harroy, J. P. (1981) *Evolution Entre 1958 et 1979 du Couvert Forestier.* Assistance International pour le Développement Rural. Brussels, Belgium.

IUCN (1987) *The IUCN Directory of Afrotropical Protected Areas.* IUCN, Gland, Switzerland and Cambridge, UK. xix + 1043 pp.

IUCN (1990) *1989 United Nations List of National Parks and Protected Areas.* IUCN, Gland, Switzerland and Cambridge, UK. 275 pp.

MacKinnon, J. and MacKinnon, K. (1986) *Review of the Protected Areas System in the Afrotropical Realm.* IUCN/UNEP, Gland, Switzerland. 259 pp.

Monfort, A. (1990) *Rapport Annuel du Projet Tourisme et Parcs Nationaux.* ORTPN, Kigali, Rwanda.

Nduwumwami, D. (1990) Protection des forêts de montagne: le Parc National de la Kibira. In: *Proceedings from the First International Workshop for the Conservation and Management of Afromontane Forests: June 19–23, 1989.* WCI/NYZS.

PRB (1990) *1990 World Population Data Sheet.* Population Reference Bureau, Inc., Washington, DC, USA.

Project Bururi-Rumonge-Vyanda (1990) Développement rural: project Bururi-Romonge-Vyanda (Burundi). In: *Proceedings from the First International Workshop for the Conservation and Management of Afromontane Forests: June 19–23, 1989.* WCI/NYZS.

Runyinya, B. (1986) *L'Ecology et Conservation de Massif Forestiers de Rwanda.* PhD thesis, unpublished. Université Libre, Brussels, Belgium.

Sirven, P., Gontanegre, J. F. and Prioul, C. (1974) *Géographie du Rwanda. Editions A.* DeBoeck, Brussels, Belgium.

Storz, M. (1983) *La Forêt Naturelle de Nyungwe et sa Faune.* Kibuye, Projet Pilote Forestier, Rwanda.

Stuart, S. N., Adams, R. J. and Jenkins, M. D. (1990) *Biodiversity in Sub-Saharan Africa and its Islands: Conservation, Management and Sustainable Use.* Occasional Papers of the IUCN Species Survival Commission No. 6. IUCN, Gland, Switzerland.

Trenchard, P. (1987) *Ecology and Conservation of the Kibira National Park, Burundi.* Report to INCN, Wildlife Conservation International/New York Zoological Society, Bujumbura, Burundi. 85 pp.

Vedder, A. (1985) *Rwanda Agro-sylvo-pastoral Project – Phase II: Ecological Aspects of the Project and Natural Forest Conservation.* Report to the World Bank/Direction Generale des Forêts (Minagri). Kigali, Rwanda. 59 pp.

Vedder, A. and Weber, A. W. (1990) Mountain Gorilla Project (Volcanoes National Park) – Rwanda. In: *Living with Wildlife: Wildlife Resource Management with Local Participation in Africa.* Kiss, A. (ed.). World Bank, Washington, DC, USA.

Weber, A. W. (1987) Socioecological factors in the conservation of Afromontane forest reserves. In: *Primate Conservation in the Tropical Rain Forest.* Marsh, C. W. and Mittermeier, R. A. (eds). Monographs in Primatology, Vol. 9, pp. 205–29. Alan R. Liss, New York, USA.

Weber, A. W. (1989) *Conservation and Development on the Zaïre–Nile Divide: an Analysis of Value Conflicts and Convergence in the Management of Afromontane Forests in Rwanda.* Unpublished PhD thesis, University of Wisconsin, Madison, USA.

Weber, A. W. and Vedder, A. (1983) *Socio-ecological Survey of the Bururi Forest Project Area.* Report to USAID. Bujumbura, Burundi. 111 pp.

Weber, B. and Vedder, A. (1984) Forest conservation in Rwanda and Burundi. *Swara* 7(6): 32–5.

White, F. (1983a) *The Vegetation of Africa: a descriptive memoir to accompany the Unesco/AETFAT/UNSO vegetation map of Africa.* Unesco, Paris, France. 356 pp.

White, F. (1983b) Long-distance dispersal and the origins of Afromontane flora. *Sonderbd. Naturwiss. Ver. Hamburg* 7: 87–116.

Wilson, V. J. (1990) *Preliminary Survey of the Duikers and Other Large Mammals of Burundi, East Africa.* Chipangali Wildlife Trust, Bulawayo, Zimbabwe.

WRI (1990) *World Resources 1990–91.* Prepared by the World Resources Institute, UNEP and UNDP. Oxford University Press, Oxford, UK, and New York, USA.

Authorship

Amy Vedder, Biodiversity Program Coordinator, WCI, with contributions from John Hall, School of Agricultural and Forest Sciences, Bangor; Alexander Harcourt, University of California at Davis; Alain Monfort, Belgium and Roger Wilson, FFPS.

Map 12.1 Forest cover in Burundi and Rwanda

Forest and protected area data for Burundi were taken from a tourist map *Burundi* (1984), at a 1:250,000 scale, prepared by the Institut Géographique National – France, Paris, in collaboration with the Institut Géographique du Burundi, Bujumbura. The map was financed by Fonds d'Aide et de Coopération de la République Française (French Aid). The 'Forêt' category and 'Limite de parc ou réserve' were digitised from this map. Vegetation cover data were then overlain on White (1983) to delimit montane and lowland forest, as shown on Map 12.1.

Remaining indigenous forest in Rwanda and national parks were extracted from a published map *République Rwandaise, Carte Administrative et Routière* (nd), at a scale of 1:250,000, published by the Service de Cartographie, Kigali and financed by the Administration Belge de la Coopération au Développement. On Map 12.1 the 'Forêt Naturelle' category has been digitised and mapped as montane forest as categorised by White (1983).

13 Cameroon

Land area 465,400 sq. km
Population (mid-1990) 11.1 million
Population growth rate in 1990 2.6 per cent
Population projected to 2020 23.5 million
Gross national product per capita (1988) US$1010
Rain forest (see map) 155,330 sq. km
Closed broadleaved forest (end 1980)* 179,200 sq. km
Annual deforestation rate (1981–5)* 800 sq. km
Industrial roundwood production† 2,708,000 cu. m
Industrial roundwood exports† 457,000 cu. m
Fuelwood and charcoal production† 10,142,000 cu. m
Processed wood production† 733,000 cu. m
Processed wood exports† 82,000 cu. m
* FAO (1988)
† 1989 data from FAO (1991)

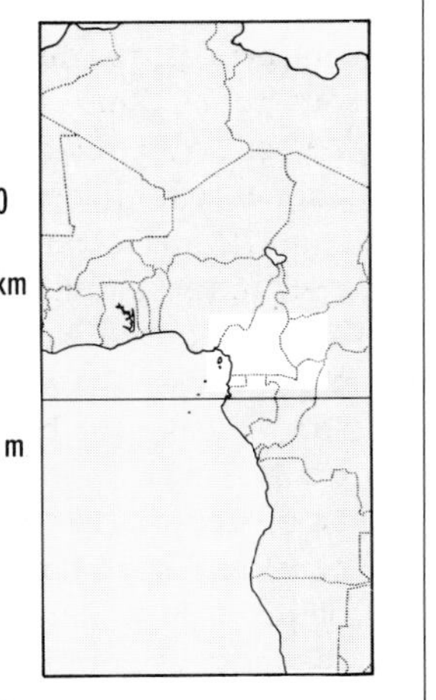

The Republic of Cameroon is one of the most important countries in Africa in terms of the biodiversity of its forests. It contains montane, submontane, lowland evergreen and semi-deciduous forests. The rich biological diversity of the lowland forests is attributable to their very stable existence even during periods of cool, dry weather such as occurred in the Pleistocene during which rain forests were considerably reduced elsewhere. High endemism occurs in the montane forests, which were isolated from one another during these same periods. The different forest types are subject to different pressures. The most highly endangered are the montane, coastal evergreen and semi-deciduous forests. The evergreen forests of the south-east were less endangered until the recent announcement of a government plan to increase logging. Timber is an important component of the Cameroon economy, both for export income and for domestic consumption of building and firewood. Current production of industrial wood is 2.7 million cu. m and the government plans to double this by the year 2000.

Those forests that are legally gazetted (nine different categories including production and protection forests, national parks and faunal reserves) are strictly controlled. Other forests on so-called 'national' land, where most commercial logging occurs, have fewer controls. They are often encroached upon and degraded. There is a need for legal control of the national lands but also a need to provide commercial loggers with enough resources for the sawmills they are obliged to construct. Logging activities, agricultural encroachment (encouraged by the present laws) and fire are the major causes of forest degradation and destruction in Cameroon.

Introduction

Cameroon is roughly triangular in shape with a base of some 700 km and a height of 1200 km and an area of 475,440 sq. km. It lies between latitudes 2° and 13°N and between longitudes 8° and 16°E. It is bounded to the south by Equatorial Guinea, Gabon and Congo, to the north by Chad, to the east by the Central African Republic and to the west by Nigeria and the Atlantic Ocean.

The coastline is 590 km and is highly indented. The coastal plain decreases in width from 100 km in the north to 30 km in the south and hills rise abruptly from this plain to a vast plateau block (500–1000 m above sea level). A mountain ridge, oriented south-west to north-east, continues inland from the oceanic volcanic ridge, extends along the north-western side of the interior plateau, along the Nigeria border and then east across the country in the Massif de l'Adamaoua. The highest point in Cameroon, which is part of the ridge, is Mt Cameroon (4095 m). Drainage is very complex, with nine major river basins. The largest river in Cameroon, the Sanaga, drains much of the central highland region before reaching the sea at a delta on the Bight of Biafra. North of the Sanaga River is the Wouri which rises in the coastal mountains of the great south-west to north-east ridge and reaches the coast at Douala. South of the Sanaga, the principal rivers are the Nyong, Lokoundjé and Ntem. The south-eastern and south-central parts of the plateau drain to the Zaïre River System and much of the north drains into Lake Chad (Hughes and Hughes, 1991).

Many of Cameroon's forests are subject to an equatorial climate with four seasons per year (a long and a short dry season and a long and short rainy season), but the coastal and montane forests tend to have an anomalous climate with only two seasons (a long wet season and a short, albeit often severe, dry season). Most of the coastal plain has more than 4000 mm of precipitation annually and at Debundscha, at the foot of Mt Cameroon, rainfall regularly exceeds 10,000 mm. Mountains receive more rain than lowlands at similar latitudes so that montane forest islands are often surrounded by relatively dry savanna (Génieux, 1961).

Cameroon had, in 1990, a population of 11.1 million people and, if the present annual growth rate of 2.6 per cent is maintained, the population will double in less than 30 years. Fifty per cent of the people are of working age (15–64), while most of the rest are below it. The population is very unevenly distributed with concentrations in the west, south-central and the Sudan savanna zone in the north, while in the Adamaoua plateau and in the south and east densities tend to be low. Mean population density is 24.3 persons per sq. km while the mean rural population density is 13.1 persons per sq. km.

The country is divided into two major religious, social and cultural zones. The people of the humid forest zone in the south of the country are farmers and cultivators. They constitute about 33 per cent of the population and have been profoundly influenced by Christianity and by the European introduction of an externally oriented economy with plantations, commercial agriculture, forestry, railways, urbanisation, some oil production and some industrialisation. The pastoral and sedentary people of the north, on the other hand, are either Muslim (16 per cent) or animist (51 per cent) and have largely retained their traditional ways of life. Consequently the south is much more developed than the north, both economically and socially, although the government has made efforts to reduce this regional disparity.

The capital, Yaoundé, with an estimated population of 750,000, is located in the centre-south while the main port of Douala (population 1,000,000) is located on the Wouri estuary. Much of their recent growth is the result of migration from rural areas. The rural exodus intensified in 1983/4 following severe drought with heavy losses of both food and cash crops. The percentage of rural village dwellers dropped from 71.4 per cent in 1976 to 61.8 per cent in 1986 and continues to fall rapidly (WRI, 1990). This is of great concern to the government for two reasons: first, the loss of peasant farmers who produce most of the cash crops and second, the degradation of the cities with overcrowding, high rates of crime, disease and unemployment.

Before 1977, the economy of Cameroon was based principally on agriculture. The major exports were cocoa, coffee, timber, cotton, rubber, palm oil, bananas, tobacco and tea. By 1980, petroleum was the country's primary export and in 1985 production peaked, bringing in US$1617 million. By 1987, this income was more than halved to US$783 million (Jeune Afrique, 1988). However, this fairly modest production gave a major boost to economic growth, helping to sustain real growth rates of 7–8 per cent between 1980 and 1985. Prudently, the strength of agriculture was maintained and in 1984 an estimated 79 per cent of the working population was engaged in this sector. With the exception of rubber and palm oil, peasant farmers dominate agricultural export production. In contrast, in spite of efforts to build up Cameroonian participation, timber production and export remain dominated by large foreign firms.

The Forests

Comprehensive accounts of the forest vegetation in Cameroon are given by Letouzey (1968, 1985). Much of the information in this section is derived from these publications. Cameroon contains moist forest of two of Africa's four major biogeographical regions: the Afromontane and the Guineo-Congolian (White, 1983). The Afromontane region comprises two major domains, Afro-subalpine grassland and montane forest, both of very limited extent. This region covers about 725 sq. km, or less than 1 per cent of the land area of the country. The Guineo-Congolian region, which includes submontane forest and extensive dense, humid, evergreen forest as well as semi-deciduous forest of middle and lower elevations, covers a total of 267,000 sq. km, or 56 per cent of the land area of the country; about 66 per cent of the region remains forested.

The height of the trees in the montane forest is around 15–25 m, the crown is evergreen, the leaves leathery, and there are few lianes. The understorey tends to be open and lichens and mosses are common. Five species of tree characterise the montane zone: *Nuxia congesta*, *Podocarpus latifolius*, *Prunus africanus*, *Rapanea melanophloeos* and *Syzygium staudtii*. *Arundinaria alpina* also occurs and *Olea hochstetteri* is found in the drier montane forests. Other montane species include *Crassocephalum mannii*, *Hypericum lanceolatum*, *Myrica arborea*, *Philippia mannii* and *Schlefflera abyssinica*. While levels of endemism are fairly high, species diversity is low. It seems likely that this phenomenon may correlate with the severe reduction in the area of the forest which occurred during dry climatic periods (see chapter 2).

The submontane forest zone is found between 800 and 2200 m in the south of the country and from 1200 to 1800 m in the north. It is characterised by floral uniformity and an abundance of plants of the family Guttiferae. It covers about 3775 sq. km, or about 1 per cent of national land. At lower altitudes, the species structure of the forest is similar to that of the adjacent lowland forests; as elevation increases the epiphytic flora, principally orchids and mosses, increases and tree species not found in lowland forests (e.g. *Caloncoba lophocarpa*, *Crotonogynopsis manniana*, *Dasylepis racemosa*, *Erythrococca hispida*, *Prunus africanus* and *Xylopia africana*) begin to appear. The submontane forests are very poorly known biologically compared to both the lowland and montane types.

Medium and low altitude forests are found from sea level to 800 m in the south and from sea level to 1200 m in the north of Cameroon. Within this domain, the dense, humid, semi-deciduous forest is often fragmented and it is seriously endangered by brush fires set during the dry season. This forest type covers around 40,000 sq. km or about 8.6 per cent of national land. The dense humid evergreen forest covers about 27.5 per cent (128,000 sq. km) of the country's land area and is made up of two principal zones: evergreen Cameroon-Congolese forest and evergreen Atlantic forest.

The evergreen Cameroon-Congolese zone of medium altitude forest covers about 81,000 sq. km or 17.4 per cent of the national land. The floristic diversity of this zone tends to be lower than that of the Atlantic coastal forests. Principal affinities are with the Congo basin forests with such species as *Lannea welwitschii*, *Cleistopholis patens*, *Xylopia staudtii*, *Bombax buonopozense*, *Cordia platythyrsa*, *Swartzia fistuloides*, *Irvingia grandifolia* and *Entandrophragma utile*. With the notable exception of *Gilbertiodendron dewevrei* this forest, unlike parts of the Atlantic zone, is not characterised by gregarious Caesalpiniaceae. Associations found within this zone include the swamp forests of the Upper Nyong with *Sterculia subviolacea* and *Macaranga* spp., swamp forests with *Phoenix reclinata* and *Raphia monbuttorum* and flooded forests with *Guibourtia demeusei*.

The evergreen Atlantic (or Nigerio-Cameroon-Gabon) zone of the low and medium altitude forest covers about 47,000 sq. km or 10.1 per cent of national land. The floristic diversity here is very high and there is marked endemism. The flora has affinities with the forests of South America. For instance, the trees *Erismadelphus exsul* and *Sacoglottis gabonensis* belong to families poorly represented in Africa, but which are abundant in South America. *Andira inermis*, which has a very local distribution in this forest zone, is another species that is also found in South America. This zone is the centre of diversity for various plant taxa including the genera *Cola*, *Diospyros*, *Garcinia* and *Dorstenia*. In addition, many narrow endemics occur in the forest including *Hymenostegia bakeri*, *Soyauxia talbotii*, *Deinbollia angustifolia*, *D. saligna*, *Ouratea dusenii* and *Medusandra richardsiana*. The forest shares species with the Ituri forest of eastern Zaïre (e.g. *Diospyros gracilescens*), with the forests of the Congo basin (e.g. *Oubanguia alata*, *Afzelia bipindensis* and *Enantia chlorantha*) and with those of Upper Guinea (e.g. *Diospyros kamerunensis* and *D. piscatoria*). These species shared with other regions are evidence of past connections between the forests.

Mangroves

Map 13.1 indicates that in 1985 there were 2434 sq. km of mangroves remaining in Cameroon. Two major areas, together covering some 2300 sq. km, lie on the coast east and west of Mt Cameroon (SECA/CML, 1987). The red mangrove *Rhizophora racemosa* makes up 90 to 95 per cent of the mangrove area. It can reach 25 m in height while the other two *Rhizophora* species, *R. harrisonii* and *R. mangle* rarely exceed 6 m. The white mangrove *Avicennia nitida* also occurs.

The mangroves and the adjacent coastal waters of up to 50 m in depth nurture and protect a major fishery resource of great economic and nutritive importance for Cameroon. The annual fish production of the Rio del Rey and the Cross River estuary is about 12,800 metric tonnes and at least one-third of this, with a value of US$8 million, comes from the Cameroonian sector. There is currently little information on the actual status of mangroves although there are signs of local damage. For instance, the pesticides and fertilisers used on the large industrial plantations (chiefly of rubber, oil palm and bananas) which are one of the features of the coastal area of Cameroon, drain into the mangroves and have a deleterious effect on them. The fertilisers cause eutrophication and algal growth which interferes with mangrove transpiration, and the pesticides accumulate in the trophic chain. There is also pollution from the offshore oil operations. Very little of Cameroon's mangrove forest is protected apart from a small area in the northern part of the Douala-Edea Faunal Reserve, and even this is threatened with degazettement.

Forest Resources and Management

The most recent comprehensive accounts of the organisation of the forestry sector in Cameroon are given in FAO (1990) and IIED (1987). According to these reports, the forests in Cameroon cover around 175,200 sq. km or 37.6 per cent of the country's land area. The dense humid evergreen and semi-deciduous forests, covering 168,000 sq. km, make up the majority of this. Map 13.1 shows, in 1985, a total rain forest cover of 155,330 sq. km, a slightly lower area than that reported by FAO (1990) and IIED (1987). The types of forest making up this total are shown in Table 13.1.

The Forestry Directorate, under the administrative authority of the Ministry of Agriculture, is charged with the establishment and implementation of forest policy, with the preparation of regulations and the coordination of management plans. It is also responsible for the application of the forestry legislation as it concerns the production and protection forests and the supervision and control of forestry exploitation at both central and regional levels.

The Office National de Régénération des Forêts (ONAREF) has responsibility for forest inventories, the development of management plans, the promotion of wood and wood products, forest regeneration and increasing forest productivity. However, its responsibility for land management and regeneration is limited to state lands. Although ONAREF has carried out an important forest inventory of almost 110,000 sq. km, it has provided management plans for very few of the state forests.

Most of ONAREF's reforestation activities have focused on the creation of plantations (usually of fast-growing exotic species) in savannas and for desertification control. Only 30 sq. km have been reforested annually in recent years and only one-third of that has been in the dense forest zone. On the national lands, where most logging occurs, there is no requirement for management and there is a range of different options for exploitation. These options and the regulations controlling them tend to be minutely detailed but mainly in economic terms. The requirements of the reporting system and the system of log measurements are demanding but in the forest controls are weak. A further problem on national land is a concession licence system which allows locals to log small areas of forest for a three-year period. Concessions are granted without approval from any form of technical committee and there is little or no field supervision of the operations. There is no obligation for the licence holder to construct a sawmill or a wood processing unit and the wood is often sold to existing (expatriate) mills. Indeed, some expatriate companies rely on this source to provide sufficient throughput to operate their mills.

Of the 49 tree species officially recognised as commercial, only about 30 are used and three species (ayous *Triplochiton scleroxylon*, sapele *Entandrophragma cylindricum* and azobé *Lophira alata*) account for almost 60 per cent of production. Ayous (known as obéché or samba in West Africa) is a white wood, while azobé is a hard, heavy, red wood.

Cameroon is currently the seventh largest exporter of tropical timber in the world and third in Africa after Côte d'Ivoire and Gabon. Timber occupies fourth place in order of importance of Cameroon's exports, after petroleum, coffee and cocoa beans. However, the diminishing petroleum resource and the falling world prices for coffee and cocoa combine to put pressure on the forestry sector to make good the difference. Indeed, the present government policy is to increase the amount of logging so that timber production will replace petroleum as the engine that drives the Cameroonian economy. The production target is 4 million cu. m by the year 2000 and 5 million cu. m by the year 2010. In 1988/9, the export of wood represented an income of some US$190 million. The sector engages about 20,000 persons in full-time employment, represents 9 per cent of the total industrial production and provides 4 per cent of the GNP.

The logging industry of Cameroon is under the effective control of foreign companies. In 1987/8 there were 67 foreign exploitants with a total of 54,000 sq. km of concession area and 49 nationals with a total area of only 12,000 sq. km. The smaller national companies tend to concentrate on the more accessible areas. Average yield in Cameroon is about 5 cu. m per hectare, which is low by standards elsewhere in the tropics and indicates that logging is very selective. The volume of timber exports is approximately 1.2 million cu. m per year, of which 62 per cent is raw logs and the rest processed wood. Countries of the EEC, principally Belgium, France, Germany, Greece and Holland, take 85 per cent of the exported logs and 91 per cent of the processed wood. Over half of the production comes from the semi-deciduous and Cameroon-Congolese moist forests in the east of the country.

Government policy is that 60 per cent of logs are processed locally. It is likely that this will rise to 70 per cent with the possibility that eventually the export of unprocessed logs will be banned altogether. One problem with this is the inefficiency with which logs are transformed and the lower prices paid (often as much as 50 per cent less) for the processed logs. The average recovery of timber from raw logs processed for export is about 30 per cent, but can be as low as 20 per cent. Furthermore, it is estimated that as much

Table 13.1 Estimates of forest extent in Cameroon

Rain forests	*Area (sq. km)*	*% of land area*
Lowland	147,480	31.7
Montane	3,186	0.7
Mangrove	2,434	0.5
Swamp	2,230	0.5
Totals	155,330	33.4

(Based on analysis of Map 13.1. See Map Legend on p. 118 for details of sources.)

as 20–35 per cent of each felled tree is lost at the logging site. As a result, the waste from felled tree to sawn product is as high as 65 to 75 per cent. Part of the reason for this is that concessions are granted for a five-year renewable period and, under the current rules, sawmill-based concessions have a working life of only nine years before their licence expires. This does not make it economically viable to invest in expensive, efficient machinery. Instead, old, outdated and inefficient machines tend to be used.

About 10.1 million cu. m of wood are used for fuel, mostly as firewood with only about 10 per cent of this being made into charcoal. Firewood represents a value of more than US$200 million per year. There is little control of this resource and taxes on its harvest are rarely collected.

Concessions have been granted on at least 80,000 sq. km of forest, that is on more than half the land area officially classified as exploitable. By 1992, 50 per cent of production forests will have been logged at least once and some will have been logged three or four times.

Deforestation

Deforestation in Cameroon is difficult to quantify. Clear-felling in the context of logging operations does not occur within the country. The main problem is an insidious and fast-growing degradation of the forests. Logging, agricultural encroachment (which is encouraged by the present legal framework) and fire are the major causes of forest degradation and destruction in Cameroon.

Eliminating seed dispersers such as elephants *Loxodonta africana* and duikers *Cephalophus* spp. from a forest whose tree species have co-evolved with them, will initiate a process of ecological succession towards a forest with a different species composition (see chapter 4). This is occurring in most of Cameroon's ecologically valuable coastal and montane forests. Forest clearing, even in the absence of human settlement, can be followed by invasion of the aggressive weed, *Eupatoria chlorantha*, which suppresses the forest regeneration cycle. Some of the forests have been repeatedly logged, their species composition is essentially secondary and all the mammals, except for a few squirrels, rats and mice, have been eliminated from them. These forests may be further damaged by fire, which is also an increasing threat to the semi-deciduous forests of the Eastern province.

Partly because of the differences in definition and partly because of the difficulties in assessment, estimates of loss of forest cover, or of deforestation, vary considerably. For instance, FAO (1988) gives a figure of 800 sq. km lost per year during 1981–5, while IIED (1987) estimated an annual loss of 1500 sq. km. These figures give an annual loss of forest of 0.5–1 per cent. Of even greater concern is that the rate of forest degradation (for instance, intact forest which has lost one or two key mammal species) is much higher and true primary forests are now virtually restricted to a few areas in the south-east of the country.

Deforestation affects the different forest domains differently. Montane forests, which are usually located on fertile volcanic soils, are seriously threatened by clearance for agriculture and by fire. Indeed, burning has caused much of the natural vegetation to be replaced by secondary grassland. The submontane forests are subject to similar pressures. The coastal Atlantic forests have been heavily logged (often several times), cleared for plantation agriculture, subjected to agricultural encroachment and over-hunted. The semi-deciduous forests on the northern margins of the Congolese forests are threatened by human settlement, heavy hunting pressure and by fire. The Congolese forest is the only forest type of which substantial areas remain intact, but it is being targeted for increased logging.

Most logging occurs on state lands where post-logging protection and management is not mandatory. In fact, on completion of logging, the forest is subject to usage rights by the local people. This directly encourages forest invasion and is an outcome actively supported by the government.

Biodiversity

Cameroon is one of the most ecologically diverse countries in Africa (see for example Gartlan, 1989). The main reason for the high biodiversity of the forests is that they are an ancient and very stable system, particularly in the lowland coastal forests. Present evidence suggests that the coastal forests persisted even in the cool, dry climate of around 18,000 years ago (see chapters 2 and 3) when the forest biome was much reduced. There are 9000 species of plants in the country, with at least 156 endemics including 45 on Mt Cameroon alone. A recent study found more than 200 species of woody plants in a sample site of 0.1 ha, a level of diversity comparable with the highest in the world. Well over 1000 butterfly species have been recorded from the forests of the Bight of Biafra. This area is also a centre of diversity for frogs – eight genera are limited largely to the region. Cameroon contains some 297 species of mammal and 848 species of birds (Stuart *et al.*, 1990).

The Cameroon forests are a major centre of endemism for the ginger and arrowroot family (Zingiberaceae). One species, *Aframomum giganteum*, also found in Gabon, has fronds reaching up to 6 m and is the tallest ginger plant in the world. *Cola lepidota* and *C. pachycarpa*, small trees bearing large, edible fruit, are also endemic to Cameroon. The yam, *Dioscorea*, is indigenous to the forests of the Bight of Biafra. There are several species within the genus but all protect their tubers with toxins. People learnt to destroy these poisons (by peeling, fermenting and cooking the tubers) many centuries ago, thus acquiring a staple crop that enables large numbers of people to live within the forest zone. Cameroon's forests are the centre of dispersion for the world's premier oil-producing plant, the oil palm *Elaeis guineensis*, and its major pollinator, *Elaeidobius kamerunicus*, is endemic to Cameroon. This weevil was exported to Southeast Asia in 1981 where, within one or two years of its introduction, oil production rose by almost 20 per cent.

The montane forests of Cameroon, though not as rich in number of bird species as the lowland forests, are particularly important for the 22 endemic bird species they support (Stuart, 1986). Bannerman's turaco *Tauraco bannermani*, is restricted to the montane forests, while Mt Cameroon has an endemic francolin *Francolinus camerunensis*. The Mount Kupe bush-shrike *Malaconotus kupeensis*, another montane species, is one of the rarest birds in Africa (see case study on ICBP Conservation Projects). Endemism in this area is also high among animals that are poor dispersers such as amphibians and invertebrates.

The lowland forests of Cameroon are of particular importance for the conservation of primates. With 29 primate species, the country is the second richest in Africa in this respect. It contains such rare and threatened species as the drill *Mandrillus leucophaeus* and the mandrill *Mandrillus sphinx*. Other species of conservation concern in the country include the gorilla *Gorilla gorilla* and chimpanzee *Pan troglodytes*, the black colobus *Colobus satanas*, Preuss's guenon *Cercopithecus preussi* and the red-eared guenon *Cercopithecus erythrotis*.

Cameroon is a major squirrel centre in Africa and includes endemics such as *Paraxerus cooperi*, which are restricted to the montane forest of Mt Cameroon. The flightless scaly-tailed squirrel *Zenkerella insignis* is also endemic to Cameroon and is a very rare mammal belonging to the family which, 30 million years ago, contained the dominant rodents in Africa. All other members of the family, the anomalures, are gliders with broad membranes between their legs but *Zenkerella* has no such membrane and is probably close to

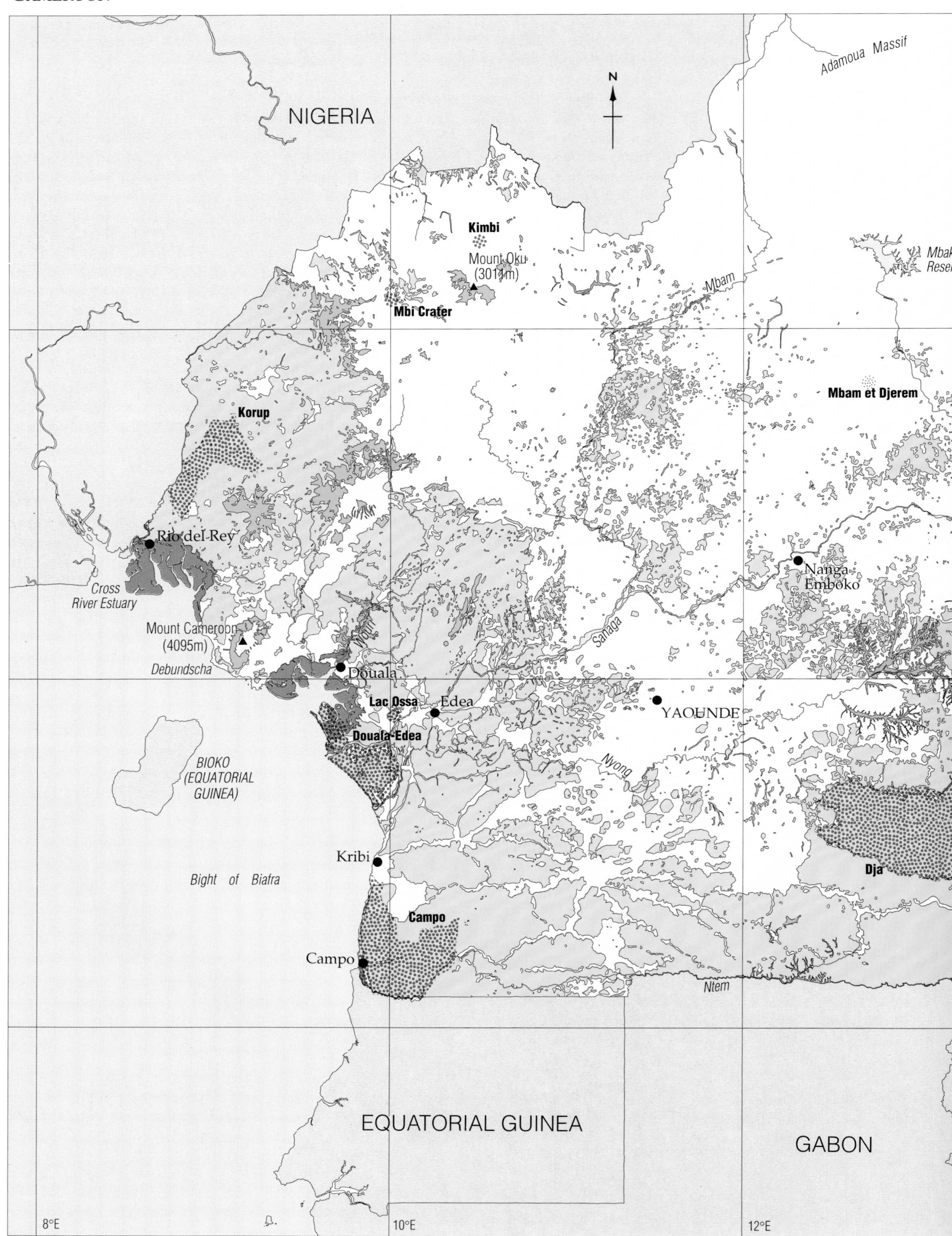
NIGERIA
N
Adamoua Massif
Kimbi
Mount Oku
(3011m)
Mbam
Mbi Crater
Korup
Mbam et Djerem
Rio del Rey
Cross
River Estuary
Nanga
Emboko
Mount Cameroon
(4095m)
Wouri
Sanaga
Debundscha
Douala
Lac Ossa
Edea
YAOUNDE
Douala-Edea
Nyong
BIOKO
(EQUATORIAL
GUINEA)
Kribi
Bight of Biafra
Dja
Campo
Campo
Ntem
EQUATORIAL GUINEA
GABON
8°E
10°E
12°E

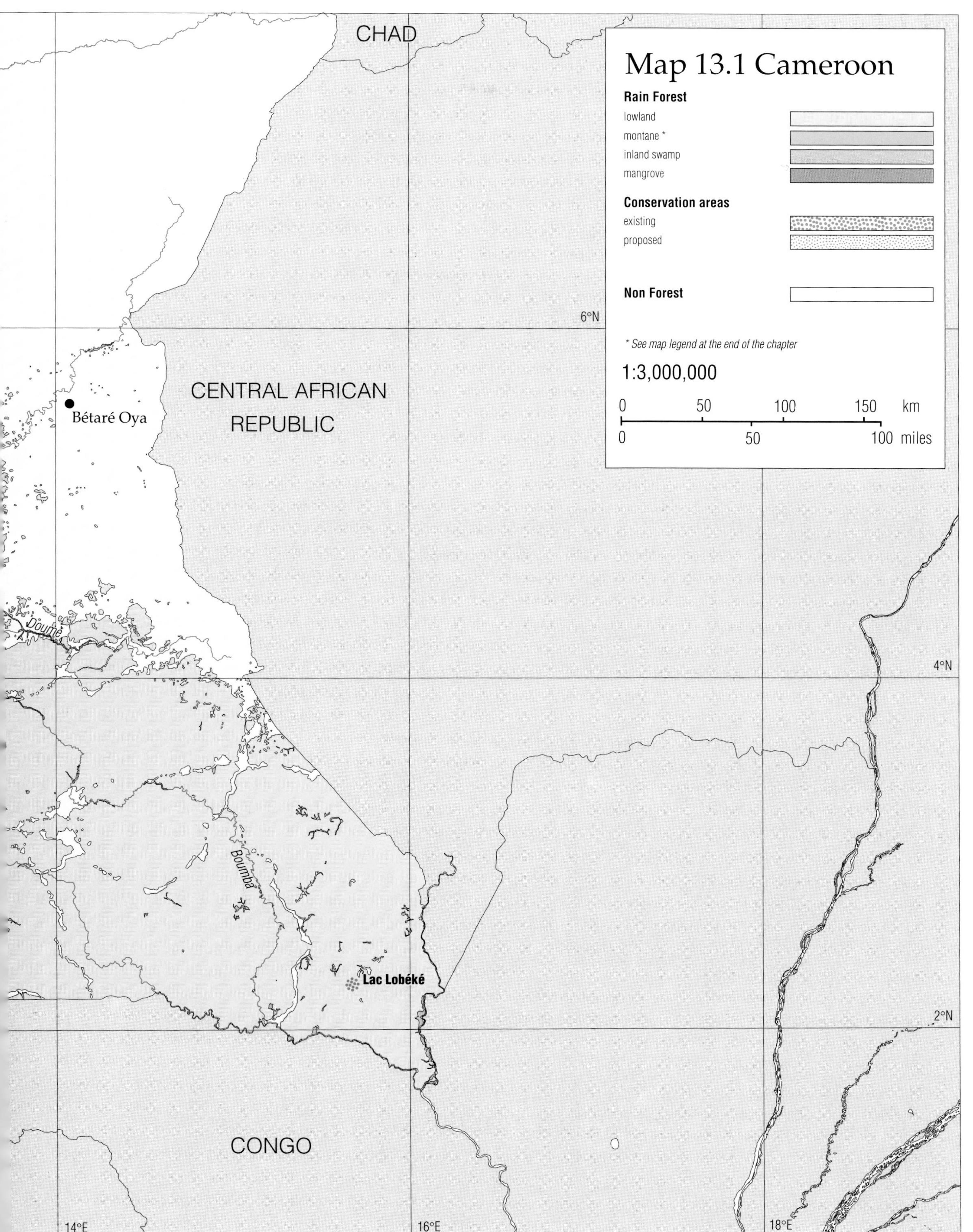
CHAD
Map 13.1 Cameroon
Rain Forest
lowland
montane *
inland swamp
mangrove
Conservation areas
existing
proposed
Non Forest
6°N
* See map legend at the end of the chapter
1:3,000,000
0 50 100 150 km
0 50 100 miles
CENTRAL AFRICAN REPUBLIC
Bétaré Oya
Doumé
4°N
Boumba
Lac Lobéké
2°N
CONGO
14°E
16°E
18°E

ICBP CONSERVATION PROJECTS IN CAMEROON

The montane forests of Cameroon and Bioko island (see chapter 18) and the adjacent region of Nigeria support 22 endemic species of bird (see Figure 13.1), and many other endemic animals and plants. Two of these birds, Bannerman's turaco and the banded wattle-eye *Platysteira laticincta*, are restricted to the Bamenda-Banso Highlands, where they are under serious threat from forest clearance. The only extensive area of forest remaining here is on Mt Oku, where ICBP, together with the government of Cameroon, is running a forest conservation project. The aim is to encourage sustainable use of the forest by the people of the surrounding villages, by helping to market forest products and plant trees to replace those cleared for firewood and timber.

Another endemic bird, the Mount Kupe bush shrike is known only from this one mountain. A recent survey by an ICBP team located eight pairs of these birds and collected valuable information on the habitat of the species. The long-term aim is to develop a new project here, following the Mt Oku model. ICBP is mapping the distributions of all bird species in Africa whose overall range size is estimated at 50,000 sq. km or less. Areas with concentrations of such species are considered priorities for conservation action. *Source:* Michael Crosby

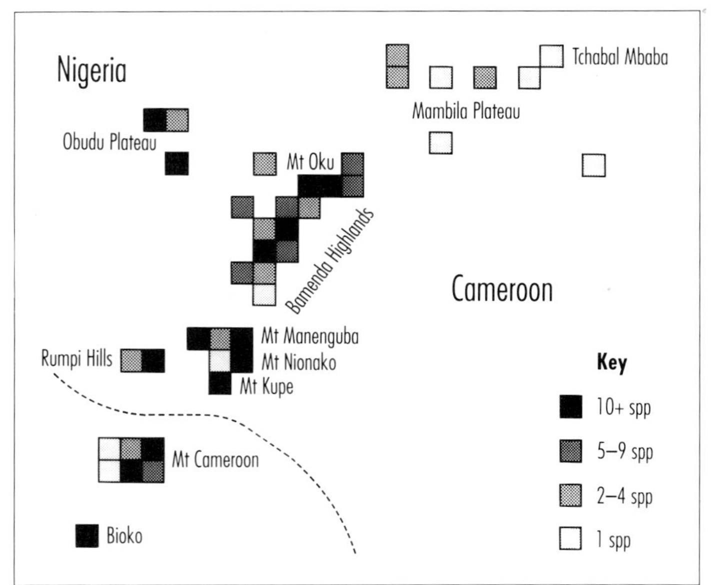

Figure 13.1 The distribution of endemic birds in the Cameroon Highlands (*Source:* Stuart *et al.*, 1990)

the primitive condition. Additional information on the biogeography of Cameroonian mammals is given in Kingdon (1990).

The present diversity, and thus the conservation priority, of the various forests in Cameroon depends on their history. The coastal Biafran forests are a high priority, as are the submontane and the montane forests. The Congolese forests are also important, although marginally less diverse than the coastal forests. They are currently less degraded but are scheduled for logging and are thus in danger. The semi-deciduous forests are the least important for biodiversity.

Conservation Areas

The national protected area system comprises the seven national parks and eleven faunal reserves under the jurisdiction of the Ministry of Tourism (Table 13.2). There are, in addition, approximately 125 forest reserves under the jurisdiction of the Ministry of Agriculture (MacKinnon and MacKinnon, 1986; Gartlan, 1989). The national parks cover an area of 10,319 sq. km, only 1260 sq. km of which is in the dense forest zone, faunal reserves cover more than 10,372 sq. km (there is no data available on the size of the Sanaga Faunal Reserve) and the forest reserves cover an area of around 18,593 sq. km. The present total for the protected area system is thus at least 4.4 per cent of the country's land area (or over 8.4 per cent if the forest reserves are included). The national goal, as set by the law of 1981, is 20 per cent. Apart from the listed areas, there is legal protection of river banks and watercourses. The law requires no environmentally destructive activities for 50 m along each river bank and for 100 m around springs. There is, however, little enforcement of this law.

Under 1981 legislation, most forest reserves on state lands will be reclassified as production forests although a few will be reclassified as protection forests. A protection forest is an area which protects a watershed, a steep slope or other physical feature. Current proposals for protection forests cover in total only about 50 sq. km of the country. It is intended that they will be few in number and rather small.

Most of the protected areas of biological significance are, therefore, the national parks and faunal reserves. The national parks are,

Table 13.2 Conservation areas of Cameroon

Existing and proposed conservation areas are listed below. Hunting reserves (which offer little or no protection) and forest reserves are not listed. For data on World Heritage sites and Biosphere reserves see chapter 9.

	Existing area (sq. km)	*Proposed area (sq. km)*
National Parks		
Benoué†	1,800	
Bouba Ndjidah†	2,200	
Faro†	3,300	
Kalamaloue†	45	
Korup*	1,260	
Mozogo-Gokoro†	14	
Waza†	1,700	
Faunal Reserves		
Bafia†		420
Campo*	2,712	
Dja*	5,260	
Douala-Edea*	1,600	
Kalfou†	40	
Kimbi	56	
Lac Lobéké*	430	
Lac Ossa	40	
Mbam et Djerem		3,532
Mbi Crater*	4	
Nanga Eboke†	160	
Ngoro†		270
Sanaga†	nd	
Santchou†	70	
Totals	20,691	4,222

(*Source:* WCMC, *in litt.*)
* Area with moist forest within its boundaries according to Map 13.2.
† Not mapped – location data not available for this project, or area located in the northern savanna zone of the country.

by and large, found in the savanna zones, while the faunal reserves tend to be in the dense forest or transitional zones. Korup National Park is the only park in the dense forest zone; created in 1986 it covers 1260 sq. km or 12 per cent of the land area under this category of protection (see case study). While the legal protection of the faunal reserves is, theoretically, fairly robust, there have been problems. The Campo Faunal Reserve, established by the French colonial government in 1932 and protecting biologically important forests in the coastal zone, covers an area of 2712 sq. km. In 1968, logging activities covering 2370 sq. km were permitted on a 25-year licence. No special requirements for logging procedures in a faunal reserve were imposed and the current biological value of the reserve must be questioned. It is apparently the intention of the logging company to request a prolongation of the licence when it expires in 1993.

Other faunal reserves have been degraded to the point of disappearance. Examples include the Sanaga River Faunal Reserve, established in colonial times to protect the hippopotamus, but destroyed by dam construction, and the Nanga Eboke Faunal Reserve which has been hunted out. Others are being actively invaded. For instance, 45 per cent of the Santchou Faunal Reserve in the Western Province has been converted to farms and plantations. Yet another problem is the proposed declassification of some areas. For example, the faunal reserve of Douala-Edea has a particular biological importance as it is bisected by the Sanaga River and has different species and subspecies on each bank. Yet development plans to the north of the reserve area are leading to pressure to declassify the northern section. The law requires that, in such a case, an area of equal size be added to the protected area, but it will be impossible to add an area of equal biological value and the danger is that repeated declassification and reclassification will result in completely degraded forest.

A major problem with the faunal reserves has been an inadequate budget and infrastructure. The boundaries are neither cleared nor marked. There are few guards and they are ill-equipped. The faunal reserves have not been developed as tourist attractions and they thus receive a much lower budgetary and infrastructure priority than do the national parks. Realistically, in the dense forest zone, only the Korup National Park, Dja Faunal Reserve (also a Biosphere reserve and a World Heritage site) and the southern section of the Douala-Edea Reserve are protected. These (including all of Douala-Edea) total 8120 sq. km or 1.7 per cent of Cameroon's land area.

The coverage of protected areas is inadequate. Montane, submontane and semi-deciduous forests are barely represented. The coastal forests need additional protection because much of the area has already been logged-over and degraded; parts of the Douala-Edea and the Campo Reserves have been effectively lost. The Congolese forests are under-protected. A scheme for extending protection for adequate coverage of the dense forest zone was presented by IUCN (Gartlan, 1989). It is clear that if the intention is to double the output of timber, then it is even more important to implement an ecologically sound and effective scheme for the protection of all the various types of forest that make up the dense forest zone.

Initiatives for Conservation

A major initiative towards promoting conservation in Cameroon was the opening, in 1990, of a World Wide Fund for Nature (WWF) national office for the country. Cameroon was selected as one of five focal countries in Africa for the development of a national conservation programme. The implementation of such a programme can be expected to make a substantial difference to public awareness of environmental matters. One of the first actions of the country office has been to appoint a national coordinator to develop a programme for environmental education. The WWF Korup Project (see case study) integrates environmental protection and community development. The project has received support from several multilateral and bilateral development agencies. The WWF Mount Kilum Project (on Mt Oku) is designed to protect the highly endangered montane forest. To reduce pressure on the forest it encourages local communities to produce honey and supports farmers' cooperatives.

A scheme to develop the Dja Faunal Reserve as a national park on broadly similar lines to the Kilum and Korup projects was recently initiated by the European Development Fund. This scheme should protect an exceptionally important forest in the Cameroon-Congolese forest type. An EEC-funded buffer zone project will also promote agroforestry around the Dja Biosphere Reserve.

The Ministry of Tourism was created in April 1989 and has the responsibility for the protection and management of the country's national parks and protected areas. The existence in the north of the country, at Garoua, of a school for the training of wildlife technicians for the whole of francophone Africa has been a very useful tool for the dissemination of environmental information throughout much of the continent. Joint projects are being developed between the Wildlife School and WWF.

One of the most encouraging recent developments in Cameroon has been the formation of a number of indigenous NGOs with an interest in the environment. This has been partly a result of internal political developments, but also a reaction to the global situation. The trend for conservation in Cameroon is now more positive, with considerably greater awareness than just a few years ago.

References

FAO (1988) *An Interim Report on the State of Forest Resources in the Developing Countries*. FAO, Rome, Italy. 18 pp.

FAO (1990) *Tropical Forestry Action Plan: Joint Interagency Planning and Review Mission for the Forestry Sector, Cameroon*. FAO, Rome, Italy.

FAO (1991) *FAO Yearbook of Forest Products 1978–1989*. FAO Forestry Series No. 24 and FAO Statistics Series No. 97. FAO, Rome, Italy.

Gartlan, S. (1989) *La Conservation des Ecosystèmes forestiers du Cameroun*. UICN, Gland, Switzerland and Cambridge, UK.

Génieux, M. (1961) *Climatologie du Cameroon. Atlas du Cameroun*. ORSTOM, Yaoundé, Cameroon.

Hughes, R. H. and Hughes, J. S. (1991) *A Directory of Afrotropical Wetlands*. IUCN, Gland, Switzerland and Cambridge, UK/UNEP, Nairobi, Kenya/WCMC, Cambridge, UK.

IIED (1987) *Le Territoire Forestier Camerounais: les Ressources, les Intervenants, les Politiques d'Utilisation*. IIED, London, UK.

Jeune Afrique (1988) *Economie du Cameroun*. Hors série Collection Marchés Nouveaux. Paris, France.

Kingdon, J. S. (1990) *Island Africa*. Collins, London, UK. 287 pp.

Létouzey, R. (1968) *Etude Phytogéographique du Cameroun. Encyclopédie Biologique* LXIX. Paul Lechevalier, Paris, France.

Létouzey, R. (1985) *Notice de la Carte Phytogéographique du Cameroun au 1:500,000*. Institut de la Carte Internationale de la Végétation, Toulouse, France.

MacKinnon, J. and MacKinnon, K. (1986) *Review of the Protected Areas System in the Afrotropical Realm*. IUCN/UNEP, Gland, Switzerland. 259 pp.

SECA/CML (1987) *Mangroves d'Afrique et de Madagascar: les mangroves du Cameroun*. Société d'Eco-aménagement, Marseilles, France and Centre for Environmental Studies, University of Leiden, The Netherlands. Unpublished report to the European Commission, Brussels.

Stuart, S. N. (ed.) (1986) *Conservation of Cameroon Montane Forests*. ICBP, Cambridge, UK.

Stuart, S. N., Adams, J. R. and Jenkins, M. D. (1990) *Biodiversity in Sub-Saharan Africa and its Islands: Conservation Management and Sustainable Use.* Occasional Papers of the IUCN Species Survival Commission No. 6. IUCN, Gland, Switzerland.

White, F. (1983) *The Vegetation of Africa: a descriptive memoir to accompany the Unesco/AETFAT/UNSO vegetation map of Africa.* Unesco, Paris, France. 356 pp.

WRI (1990) *World Resources 1990–91.* Prepared by the World Resources Institute, UNEP and UNDP. Oxford University Press, Oxford, UK, and New York, USA.

Authorship

Steve Gartlan, WWF, Cameroon with contributions from Joseph B. Besong, Forestry Department, Yaoundé, Michael Crosby, ICBP, Cambridge, Charles Doumenge, IUCN, Gland and Clive Wicks, WWF-UK.

Map 13.1 Forest cover in Cameroon

The data showing remaining forest cover in Cameroon were extracted from a published vegetation map *Carte Phytogéographique du Cameroun* (1985), prepared by R. Létouzey for the Institut de la Carte Internationale de la Végétation, Toulouse, France and the Institut de la Recherche Agronomique (Herbier National), Yaoundé, Cameroon. At a scale of 1:500,000 the national map has been separated into six sections and is published in six sheets. Sheet numbers 3, 4, 5 and 6, covering the southern, moist region of Cameroon, were digitised for Map 13.1. Some 267 phytogeographical types have been categorised by Létouzey; of these, classes 159–68, 185–90, 199, 203–7, 215–19, 228–33, 247–50, 266 have been harmonised into the lowland rain forest category portrayed on Map 13.1; classes 108 and 117 into montane forest; classes 155–7, 178–80, 193–7, 211, 223, 224, 242, 243, 256, 259–61 into swamp forest and classes 262–5 into the mangrove category.

Protected areas were taken from a tourist map *Road Map of Cameroon* (1988) at a scale of 1:1,500,000 published by Macmillan and from spatial data held within WCMC files.

Korup Project

The Korup project aims to conserve a unique and biologically important forest through a programme of sustainable development, extension and conservation education which will raise the standard of living of the local people and provide them with a better future. The area managed within the project is situated in the dense evergreen humid Atlantic Biafran coastal forest in the south-west corner of Cameroon. It consists of the Korup National Park, a fully protected area of 1260 sq. km, and a support zone of 3200 sq. km. The national park was created by Presidential decree in 1986 and is currently the only rain forest national park in Cameroon. The project is contiguous with the proposed Oban National Park in Nigeria (see case study in chapter 27).

Most of the original rain forest west of Korup and the Oban Hills has either been destroyed or severely disturbed and the large mammal fauna has essentially disappeared. Indeed, the entire forest block along the 1500 km stretch of coastline between the Niger River and Côte d'Ivoire has virtually vanished. Korup itself is part of a Pleistocene refuge and the forest is over 60 million years old. It has been largely untouched by man, mainly because of the isolation of the forest and its generally poor soils. Korup contains over 3000 species of plants and vertebrate animals.

The Korup project takes the view that no protected area can survive without the active support of the community that lives in or around it. One key principle is that, 'for every restriction imposed by the project in the interest of conservation, an equal opportunity should be provided'. The aim is to reconcile the people and their social and economic development with the protection of the natural resource base.

The main threats to the park came from the expanding population of Nigeria, many of whom cross into Cameroon to farm and hunt. The park is also threatened by the hunters in the six small villages inside the park and the 27 villages containing around 12,000 people that live within 3 km of its boundaries. The villages inside the park are inhabited by approximately 750 people, who depend on hunting and slash and burn agriculture. They kill some 12,000 animals, with a total weight over 140,000 kg, each year. These are sold in towns in Cameroon and Nigeria. Efforts have been made to find alternative sustainable sources of income for these villages. However, as they are situated on very poor acidic soils and are isolated both from each other and from any potential markets for other cash crops, this has proved unsuccessful.

The villagers want the same level of development as in other villages outside the park, including road access to towns, hospitals and schools. As this cannot be provided inside the park, they have agreed to move to more fertile areas outside. Road systems are planned to allow development of these areas of better soil some distance from the park. The local people will still stay within their tribal area and within the Korup forest which has been their home for many years. All resettlement is voluntary and the people are being helped to build their own villages on sites of their own choice.

The objective of the rural development programme is to replace the income and the protein obtained from unsustainable hunting by other sources of income, and also to help the local community to develop sustainable land use systems, including agroforestry. The principle is to develop sustainable farming systems using the minimum amount of imported equipment and materials and thus minimise destruction of the forest.

The Ministry of Tourism is responsible for the project. World Wide Fund for Nature with financial support from the British and the US governments and the EEC has provided the main technical support. Other institutions, including the United Kingdom Natural Resources Institute, Wisconsin Primate Center, Missouri Botanical Garden and Wildlife Conservation International, have provided staff, finance or scientific advice.

With contributions from aid agencies, WWF has already spent over US$2 million on the surveys needed to determine the views of the local people and to prepare a detailed management plan for the area, the implementation of which will cost about US$30 million. In addition, WWF has built the park headquarters and equipped it, an education centre has been established with an education officer employed to run it, a boundary line has been cut round the park and a 120 m suspension bridge has been built over a river into the park. Tourist and scientific camps have been set up. Two rural artisan training centres and a women's institute have been provided with equipment. WWF has also supplied eight vehicles, motor cycles, outboard engines and garage equipment.

The project will be judged a success when the villagers leave the park and are happily settled on good soils outside and when the local people provide the main protection for the park because they are convinced of its importance to them and their successors. *Source:* Clive Wicks

14 Central African Republic

Land area 622,980 sq. km
Population (mid-1990) 2.9 million
Population growth rate in 1990 2.5 per cent
Population projected to 2020 5.9 million
Gross national product per capita (1988) US$390
Rain forest (see map) 52,236 sq. km
Closed broadleaved forest (end 1980)* 35,900 sq. km
Annual deforestation rate (1981–5)* 50 sq. km
Industrial roundwood production† 400,000 cu. m
Industrial roundwood exports† 28,000 cu. m
Fuelwood and charcoal production† 3,055,000 cu. m
Processed wood production† 56,000 cu. m
Processed wood exports† 25,000 cu. m

* FAO (1988)
† 1989 data from FAO (1991)

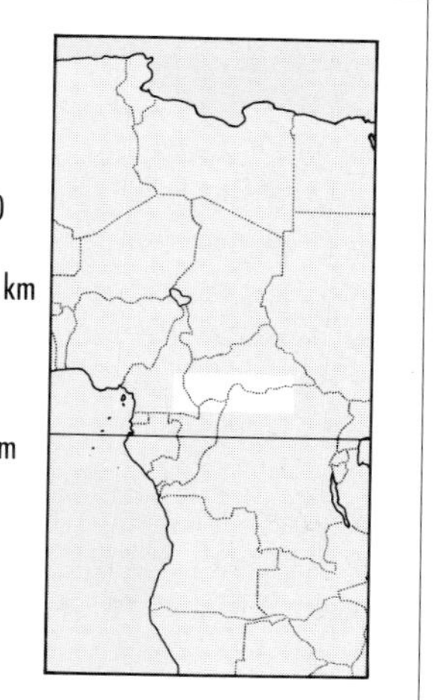

The sparsely populated Central African Republic contained, until recently, some of the most spectacular wilderness areas in Africa. The larger mammals in the north have now been decimated by heavily armed bands of poachers from neighbouring Chad and Sudan, but the forests in the south still shelter a very rich fauna. These forests contain some of Africa's richest stands of valuable hardwoods, protected in part by the enormous cost of transporting the timber to the sea for export. The timber trade has declined in recent years in the face of competition from cheaper timbers from other African and Asian countries and it may continue to do so in spite of the threat posed by the construction of a major highway linking the capital, Bangui, with the Atlantic Ocean.

The dense forests of the south-west are the scene of one of Africa's most interesting forest conservation projects: an attempt by WCI and WWF to achieve forest protection and wise use in an area occupied by Aka pygmies. The project is supported by the World Bank and USAID.

Introduction

The Central African Republic (CAR), as its name suggests, is situated in the heart of the African continent between 2°13' and 11°01'N and 14°25' and 27°27'E. It is bordered by Chad to the north, to the east by Sudan, to the south by Zaïre and the Congo, and by Cameroon to the west.

The country is an immense rolling plain oriented on a south-west to north-east axis. The altitude of this plateau varies between 500–700 m, but it is bordered on the west and east by two mountain ranges. To the west the Massif du Yade and Mont Pana, an extension of the Adamaoua highlands of Cameroon, reach 1410 m in height at Mont Ngaoui. In the east, the Massif des Bongo and the Massif du Dar Chala attain a height of 1330 m. In the south, there are sedimentary formations with sandstones and quartz forming a hilly region with many streams and rivers. In the north near Bria, a floodplain extends toward the Chad basin. Two large watersheds divide the country: one runs from west to east across the centre separating the Zaïre and Chad basins, while the other, along the eastern border of CAR, separates the Chad, Zaïre and Nile basins.

Four principal climatic zones are distinguishable, with rainfall decreasing from south to north. The Congolese Equatorial climate, occurring in the region south of Bayanga and in the forest around Bangassou, is characterised by the absence of a true dry season, although there is less rainfall during January and February. Annual precipitation is over 1500 mm and temperatures here are around 26°C with little seasonal or daily variation. Slightly further north, annual precipitation is still over 1500 mm and there is high humidity throughout the year, but there is a dry season of at least three months each year (subequatorial climatic zone). Temperatures in this region are much more variable: extreme values of 13°C in January and 40°C in March and April have been recorded. The Sudano-Guinean climate is characterised by annual rainfall of around 1400 mm and a three to six month dry season. Lastly, in the far north, with a tropical-Sahelian climate, there is less than 1200 mm precipitation annually and the dry season is more than six months long.

Vegetation broadly follows the climatic zones. In the south is the dense forest zone with both evergreen and deciduous trees. In the Sudanian zone there is dry forest with an upper canopy of deciduous trees and a clear understorey. Elsewhere wooded savanna occurs with abundant grass cover and many trees and shrubs. The savanna-park zone is limited to the alluvial depressions south of the Chad basin and is composed of vast grasslands with small groups of trees. Finally, the Sudano-Sahelian zone contains few trees and much open grassland. Boulvert (1986) provides a more detailed description of the vegetation zones in CAR.

Population density in CAR, at around four people per sq. km, is low compared to that in most other African countries. In the forests, the density is even lower, at only one person per 2 sq. km. Permanent settlements within the forest are few and localised but the entire area is used to a certain degree, particularly by the pygmies (generally referred to as Aka pygmies; see chapter 5) that live there, as well as by Bantu farmers.

The Forests

The common species of the dense evergreen forests are *Pycnanthus angolensis*, *Lophira alata*, *Manilkara mabokeensis*, *Pericopsis elata*, *Trichilia heudelotii*, *Ricinodendron heudelotii*, *Calamus* spp., *Petersianthus macrocarpus*, *Lovoa trichilioides*, *Afzelia bipendensis*, *Guarea cedrata*, *Entandrophragma angolense*, *Monodora myristica*, *Piptadeniastrum africanum* and *Piper guineense*. Throughout there are forests of *Diospyros* spp., and in Mbaéré and Lobaye there are forests of *Heisteria parvifolia* and of *Uapaca guineensis* with *Amphimas pterocarpoides* and *Pentaclethra macrophylla*.

Central African Republic

The semi-deciduous dense forest is characterised by two major types. The first, the forest of *Celtis* spp. and *Triplochiton scleroxylon*, is the most widespread. The other characteristic species of this type are: *Celtis adolfi-friderici*, *C. zenkeri*, *C. philippensis*, *C. mildbraedii*, *Gambeya perpulchra*, *Aningeria altissima*, *Funtumia elastica*, *Mansonia altissima*, *Holoptelea grandis*, *Pterygota macrocarpa*, *Nesogordonia kabingensis*, *Antiaris africana* and *Teclea grandifolia*. The herbaceous plants include species such as *Olyra latifolia*, *Leptaspis cochleata*, *Streptogyna crinita* and *Amorphophallus* spp.

The second type of semi-deciduous forest, the forest of *Aubrevillea kerstingii* and *Khaya grandifoliola*, forms the edge of the forest zone and the forest islands in savannas. *Tripochliton scleroxylon* is always abundant, while other species characteristic of the type include *Afzelia africana*, *Albizia coriaria*, *Parkia filicoidea*, *Berlinia grandiflora*, *Blighia unijugata* and *Chaetacme aristata*. *Mansonia altissima*, *Gambeya perpulchra* and the various *Celtis* species are rare or absent. This forest type is very limited in the Carnot sandstone region, but is extensive in the Lobaye and Ombella-Mpoko areas. It is also common in the Mbomou district.

The forests on the Carnot sandstone are thought to be little disturbed. Characteristics species of secondary forests, such as *Milicia excelsa* and *Ceiba pentandra*, are either absent or occur only in low densities. It is probable that this is because streams in the area are few and far apart and most of them dry up from December to March so villages were never established in the region.

In contrast, the forests of the Haute-Sangha and Basse-Lobaye districts on Precambrian geologic formations appear to be secondary forest. This characteristic is confirmed by the abundance of Mimosaceae (in particular various *Albizia* spp.), as well as *Ricinodendron heudelotii*, *Alstonia boonei* and *Ceiba pentandra*. The most common commercial species in the Haute-Sangha forests are *Tripochliton scleroxylon*, *Terminalia superba* and *Mansonia altissima*.

Forest Resources and Management

The northern limit of the dense forest has receded in the past centuries as a result of a general drying of the climate. According to Aubréville (Sillans, 1958) the former northern limits followed a line connecting Zemio, Bakouma, Bambari and Boda then passed 100 km north of Carnot before crossing the Cameroonian border. Recent Landsat imagery has been used to map the current extent of the forest zone (Boulvert, 1986), as shown on Map 14.1.

Boulvert (1986) estimated the total amount of forest of all types in CAR to be 92,200 sq. km or close to 15 per cent of the country's land area. This was made up of 37,500 sq. km of dense evergreen forest in the west, 10,000 sq. km of dense semi-deciduous forest in the east, 6500 sq. km of dry deciduous forest in the Sudanian zone (woodland in White's 1983 classification) and 38,200 sq. km of semi-humid, dry forests and gallery forests. FAO (1988) estimated that closed broadleaved forest covered only 35,900 sq. km of the country at the end of 1980 but it is not clear which of Boulvert's forest types were included in this figure – presumably only the dense evergreen and semi-deciduous forest. Map 14.1, which has been produced from the map that accompanies Boulvert's report (see Map Legend on p. 124) and includes his dense evergreen and semi-deciduous forest as lowland rain forest (Table 14.1), indicates that there are 47,405 sq. km of this forest type and 4831 sq. km of inland swamp forest in the country.

Commercial forest exploitation in CAR began in 1945 on a small scale, and until 1970 was limited to the Lobaye Prefecture, within a 130 km radius of Bangui. The first forestry permits were allocated in the Haute-Sangha Prefecture in 1967 after a forest inventory had been carried out by the French Centre Technique Forestier Tropical (CTFT, 1967). Indeed, CAR was the only African country to have undertaken a substantial inventory of forest resources before issuing exploitation permits. It was also the first francophone country in Africa to have established detailed forestry regulations and management schemes before beginning logging. In addition, in contrast to other African countries, CAR encouraged the transformation of raw wood; as early as 1970, 72 per cent of the timber was processed within the country.

Table 14.1 Estimates of forest extent in the Central African Republic

	Area (sq. km)	*% of land area*
Rain Forests		
Lowland	47,405	7.6
Swamp	4,831	0.8
Totals	52,236	8.4

(Based on analysis of Map 14.1. See Map Legend on p. 124 for details of sources.)

In 1970, the Central African Republic Forest Service estimated that 80 per cent of the southern dense forest was economically exploitable (24,000 sq. km). In CAR the principal tree species being cut are ayous *Triplochiton scleroxylon*, limba *Terminalia superba*, sipo *Entandrophragma utile*, sapele *E. cylindricum*, tiama *E. angolense*, eyong *Eribroma oblonga*, and dibétou *Lovoa trichilioides*. Ayous comes mostly from the northern area of the forest, where it is found at a density of 80–90 stems (dbh greater than 60 cm – the minimum diameter for felling any of these commercial tree species) on a 25 ha plot. This tree has a white wood and is principally used for plywood. Limba is also cut for plywood, but is found at much lower densities – 15 or more stems per 25 ha plot is considered rich for this species. Sipo, sapele, tiama, and dibétou come from the southern area. Sipo is very rare, on average one stem over 60 cm dbh per 25 ha plot, and is the most expensive timber. Its rich red wood is used in making fine furniture. Sapele, also a red wood, is used mostly for planks and veneers of high quality. Each tree cut represents approximately 12–14 cu. m of timber. The estimated volumes of the principal tree species cut in 1970 are shown in Table 14.2. Germany has provided the greatest market for wood from CAR, followed by Romania, Spain, Portugal and Poland.

There are currently eight forestry companies in the country, four in the Lobaye Prefecture and four in Haute-Sangha and Sangha-Mbaéré Prefectures. Even though the forests of Central Africa have great potential commercial value, production is limited by the distance from the international markets. Transportation from Bangui to the Atlantic Ocean at Pointe Noire is a journey of 1210 km by

Table 14.2 Estimated volumes of tree species cut in CAR in 1970

Species	*Volume (1000 cu.m)*
Limba (*Terminalia superba*)	24,900
Sapele (*Entandrophragma cylindricum*)	23,600
Ayous (*Triplochiton scleroxylon*)	16,900
Bété (*Mansonia altissima*)	2,400
Mukulungu (*Autranella congolensis*)	2,300
Tiama (*Entandrophragma angolense*)	1,700
Iroko (*Milicia excelsa*)	1,700
Kosipo (*Entandrophragma candollei*)	1,500
Sipo (*Entandrophragma utile*)	1,500
Dibétou (*Lovoa trichilioides*)	1,400
Doussié (*Afzelia* spp.)	700
Khaya (*Khaya ivorensis*)	600

river and 515 km by rail. The logging companies in the west of the country (Sangha-Mbaéré Prefecture) are 1250 km from Pointe Noire by the Sangha and Congo rivers and, again, 515 km by rail. There are only a few months of the year when river levels permit exportation and there is a great deal of loss in stockage and transport. Transport costs account for 60 per cent of the final product price and, as fuel costs rise rapidly, many companies have been forced out of business in recent years. Production has, indeed, been declining because of these high overheads (Figure 14.1).

The '4th Parallel' road, which will eventually link Bangui with the Atlantic port of Kribi in Cameroon, has been completed from Bambio to Yamendo (40 km north of Nola) and most companies are turning towards road export of logs by way of Cameroon. Although this road will facilitate the export of logs from CAR, a vast increase in production will be necessary to justify the construction costs of the road. The road could be the salvation of the wood industry in CAR but a disaster for the forest, its wildlife and people. Already there is evidence of unregulated exploitation and lack of control along the road. Logging companies have constructed unauthorised feeder roads and have begun felling and transportation without the consent of the Forestry Department. The only control of exploitation and poaching along the road is by two guards posted at Bambio and they have no means of transport.

Logging is the third largest export industry in CAR after diamonds and coffee. Between 1980 and 1985, its annul contribution to the national economy was approximately US$8 million, that is, between 3.8 per cent and 5.5 per cent of national export earnings. In 1981, logging provided 15.5 per cent of direct employment (forestry is the largest employer in the country) and 12 per cent of all salaries in the 'modern sector'. In spite of its obvious importance to CAR, it is necessary to examine the ecological and social costs associated with commercial logging. Economic analyses of the forestry sector have not looked at the value of the intact forest to the local population who depend on many non-timber forest products for food, medicines and other materials. Little attention has been given either to the consequences of the rapid extermination of forest wildlife or the loss of the ecological benefits of the intact forest. The benefits and losses need to be carefully re-evaluated, taking all these factors into consideration.

Due to the high costs associated with logging, only the most valuable trees are selectively extracted from the forest, removing only a few large trees per hectare. However, considerable disturbance to the forest is caused by the roads constructed to extract this wood. If there is a sufficient density of commercially valuable trees in a sector, a grid system of roads is laid out. This grid consists of a principal road with parallel secondary roads cut each kilometre perpendicular to the principal road. A skidder path is cut between each secondary road (every 500 m), parallel to them, and another series of skidder paths is cut every 250 m perpendicular to the secondary roads and connecting them. The principal road is clear-cut and graded to approximately 50 m in width, while the secondary roads are cut and graded to around 40 m wide. Skidder paths are ungraded, but are clear-cut to approximately 25 m in width. The selective logging itself removes a relatively small number of trees, but many more are cut down in the process of making the roads.

These roads also provide easy access for hunters: the large numbers of transient workers, who come long distances to work in the logging operations, provide a ready market for the meat. Primates, including gorillas *Gorilla gorilla* and chimpanzees *Pan troglodytes*, are prime targets for hunters. The Aka traditionally hunt monkeys with crossbows with poison tipped arrows, but the advent of the shotgun has meant that many more individuals can be killed. Duikers *Cephalophus* spp. are also widely sought after for food by Aka as well as by

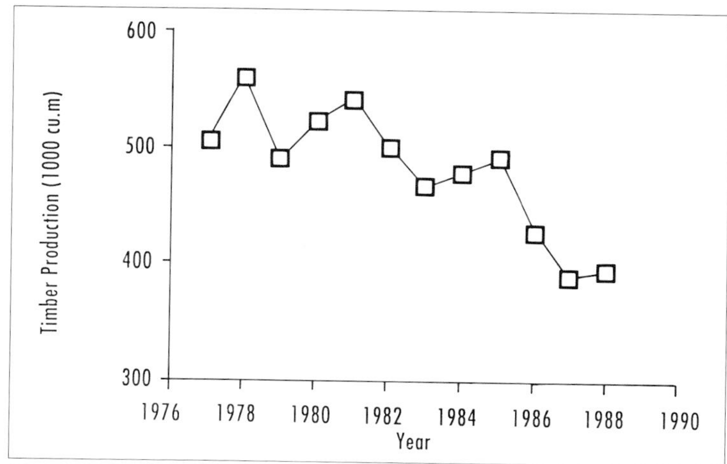

Figure 14.1 Production of industrial roundwood in CAR 1977–88

(*Source:* FAO, 1990)

Bantu hunters and they, particularly the blue duiker *C. monticola*, are the most common species sold in local markets. The principal means of trapping these animals is with snares, even though this is illegal. The snares were originally of rope made from vegetable material, but are now made predominantly from metal cable. The cable is often derived from strands unwound off larger cables discarded by the logging firms, or is made from motor-cycle brake cables. The snares are set along game trails and catch the passing animal by the leg. Duikers can also be driven into large nets strung through the underbrush. Even snakes are eaten and tortoises *Kinixys* spp. are frequently a major food of Aka hunting parties. Hunting for subsistence using traditional methods such as those practised by the Aka (for example, using spears, nets and rope snares) is permitted for all non-protected species.

Deforestation

The only figure for deforestation rate in CAR is that of FAO (1988) which estimated that 50 sq. km of closed broadleaved forests were lost annually between 1981 and 1985. Much more extensive areas of open forest (500 sq. km) are thought to be lost each year. However, there are accounts of a major fire in the forest areas west of Bangui in 1983. The fate of these forests after burning is not known. Rural populations are increasing only slowly as population growth is counteracted by migration to towns, but it is likely that some forest may be at risk from people moving into the forest belt to avoid drought and desertification in the Sahelian and Sudanian zones to the north. Hearsay accounts by specialists with long experience in the country suggest that forests are encroaching on savannas in areas where roads have fallen into disuse and people have moved away. The 4th parallel road will certainly attract people into the forest areas that it crosses and will provoke some deforestation. The overall situation appears to be one of relative equilibrium between localised forest loss in more accessible areas and gain in depopulated forest zones. The major threats come from new infrastructure, large scale population movements and fire in disturbed forest areas. The latter threat may be exacerbated by regional and global climate changes.

Biodiversity

The flora of the Central African Republic is very poorly known. At least 3600 plant species have been recorded but there are probably nearer 5000 in the country (Davis *et al.*, 1986). There is a concentration of endemic species on the hills of the north-east.

There are 19 or 20 species of primate recorded in CAR (Oates, 1986), of which 16 are recorded in the forest. They range in size from the tiny (60 g) dwarf galago *Galagoides demidoff*, to the huge 180 kg gorilla. They include the chimpanzee, six species of

Cercopithecus and the crested and grey-cheeked mangabey, *Cercocebus galeritus* and *C. albigena* respectively. The striking black-and-white colobus monkey *Colobus guereza*, and the red colobus *Procolobus [badius] rufomitratus* are also found in the forests.

Two subspecies of the African elephant occur in CAR, *Loxodonta africana africana* in the northern savanna zone, and *L. a. cyclotis* in the lowland forest. In some forests, the elephants make trails which crisscross the area and these open corridors through the dense undergrowth are the major thoroughfares of the Aka people. However, elephants remain in significant numbers only in the forests of south-western CAR in the newly created Dzanga-Sangha Dense Forest Faunal Reserve which incorporates Dzanga-Ndoki National Park. In most other areas of the forest elephants have been virtually eliminated by ivory poachers. Similarly, the black rhinoceros *Diceros bicornis longipes*, has been reduced almost to extinction. In 1970, it was estimated that there were 3000 rhinos in the country but, by 1984, only 170 remained (Western and Vigne, 1985) and now just a few scattered individuals persist.

Forest antelopes found within CAR include bongo *Tragelaphus euryceros*, sitatunga *T. spekei*, chevrotain *Hyemoschus aquaticus* and six species of duiker. A dwarf form of the African buffalo *Syncerus caffer nanus* is found in the lowland forests. Two species of wild pigs are found in the dense forest, the giant forest hog *Hylochoerus meinertzhageni* and the bushpig *Potamochoerus porcus*.

Approximately 700 species of birds have been recorded for the country (Carroll, 1988), over 400 in the forested regions. The first breeding record of the rare brown nightjar *Caprimulgus binotatus* was made in the Bayanga region (Carroll and Fry, 1987). Only one bird species is listed as threatened in CAR, the shoebill *Balaeniceps rex*, and this inhabits swamps, not forests (Collar and Stuart, 1985).

Several spectacular reptiles are found in the dense forest zone. Most notable are the highly venomous gaboon and rhinoceros vipers (*Bitis gabonica* and *B. nasicornis* respectively). The huge African python *Python sebae* is also found in the forest.

Conservation Areas

An autonomous organisation, the Centre National pour la Protection et l'Aménagement de la Faune (CENPAF) within the Department of Wildlife and National Parks is responsible for protected area management. Reserves cover 10 per cent (Table 14.3) of the country's land area but, in general, management and protection of them is poor. The creation of Dzanga-Sangha Reserve and the Dzanga-Ndoki National Park, December 1990 has added a considerable extent of forest to the protected areas system.

The Biosphere reserve of the Basse-Lobaye, on the border with the Congo, is one of the few other protected areas of forest. It was established for ethnographic studies of the pygmies. Large mammals have been exterminated from it by heavy hunting and parts of this reserve are currently being logged.

Initiatives for Conservation

Following wildlife surveys in the Bayanga region of south-west CAR, conducted by teams supported by Wildlife Conservation International (WCI) and WWF, the Dzanga-Sangha Reserve and the Dzanga-Ndoki National Park were created (see case study). The management of this reserve and park system has been supported by WWF-US/USAID as a 'Wildlands and Human Needs' project, and has received additional support from the World Bank.

As part of a regional conservation programme the EEC will be supporting a project in the Mbaéré-Bodingué-Ngoto forest region. This

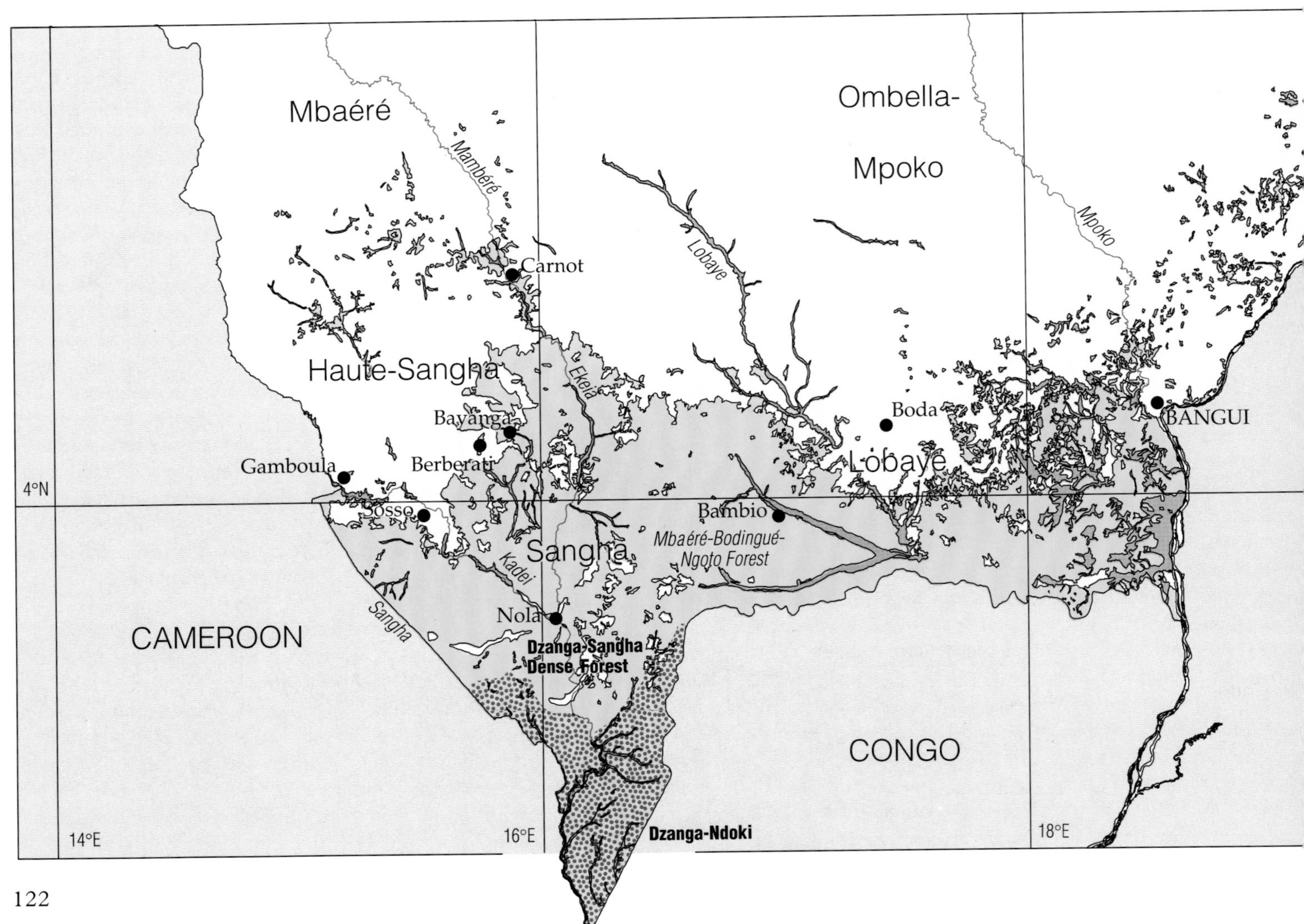

Table 14.3 Conservation areas of the Central African Republic

Existing and proposed conservation areas are listed below. Forest reserves are not listed or mapped. For data on Biosphere reserves and World Heritage sites see chapter 9. Note that only the moist, southern half of the Central African Republic is mapped in this Atlas, therefore Dzanga-Sangha Dense Forest Faunal Reserve is the only conservation area shown on Map 14.1.

	Existing area (sq. km)	Proposed area (sq. km)
National Parks		
André Félix†	1,700	
Bamingui-Bangoran†	10,700	
Manovo-Gounda-Saint Floris†	17,400	
Dzanga-Ndoki*[1]	1,287	
Strict Nature Reserves		
Vassako-Bolo†	860	
Conservation Areas		
Tri-National Rainforest*† (CAR, Cameroon, Congo)		10,500
Special Reserves		
Eastern CAR Elephant†		200,000
Faunal Reserves		
Aouk-Aoukale†	3,300	
Bahr Oulou†		3,200
Gribingui-Bamingui†	4,380	
Koukourou-Bamingui†	1,100	
Nan-Barya†	2,300	
Ouandjia-Vakaga†	1,300	
Yata-Ngaya†	4,200	
Zemongo†	10,100	
Dzanga-Sangha Dense Forest*	4,359	
Private Reserves		
Avakaba Presidential Park†	1,750	
Totals	63,449	213,700

(*Sources:* WCMC, *in litt.*; R. Carroll, pers. comm.)
* Area with moist forest within its boundaries.
† Not mapped – no location data available, or areas located in northern savanna zones of the country.
[1] Dzanga-Ndoki National Park is situated within Dzanga-Sangha Faunal Reserve; its area is therefore not included in the total area.

is an important conservation effort as this forest is traversed by the newly completed 4th parallel road. The project aims to protect the forest after logging and manage it to encourage regeneration of valuable timber trees. A core area will be totally protected as a nature reserve.

At the time of writing, surveys are being undertaken by WWF, WCI and the Experiment in International Living to create a tri-national conservation area centred around the Dzanga-Sangha area in CAR. The aims of the project are to conserve and manage the forest in the contiguous areas of northern Congo, in the Nouabalé region, and in south-eastern Cameroon in the Lake Lobéké forests. This would encompass an area of over 10,000 sq. km of forest with up to 15,000 elephants and a rich forest fauna. It is anticipated that these areas would be managed in an integrated fashion with core areas, multiple use zones and community development zones.

WWF is funding a survey of the forests in the Bangassou region to assess their conservation potential. This survey began in early 1991.

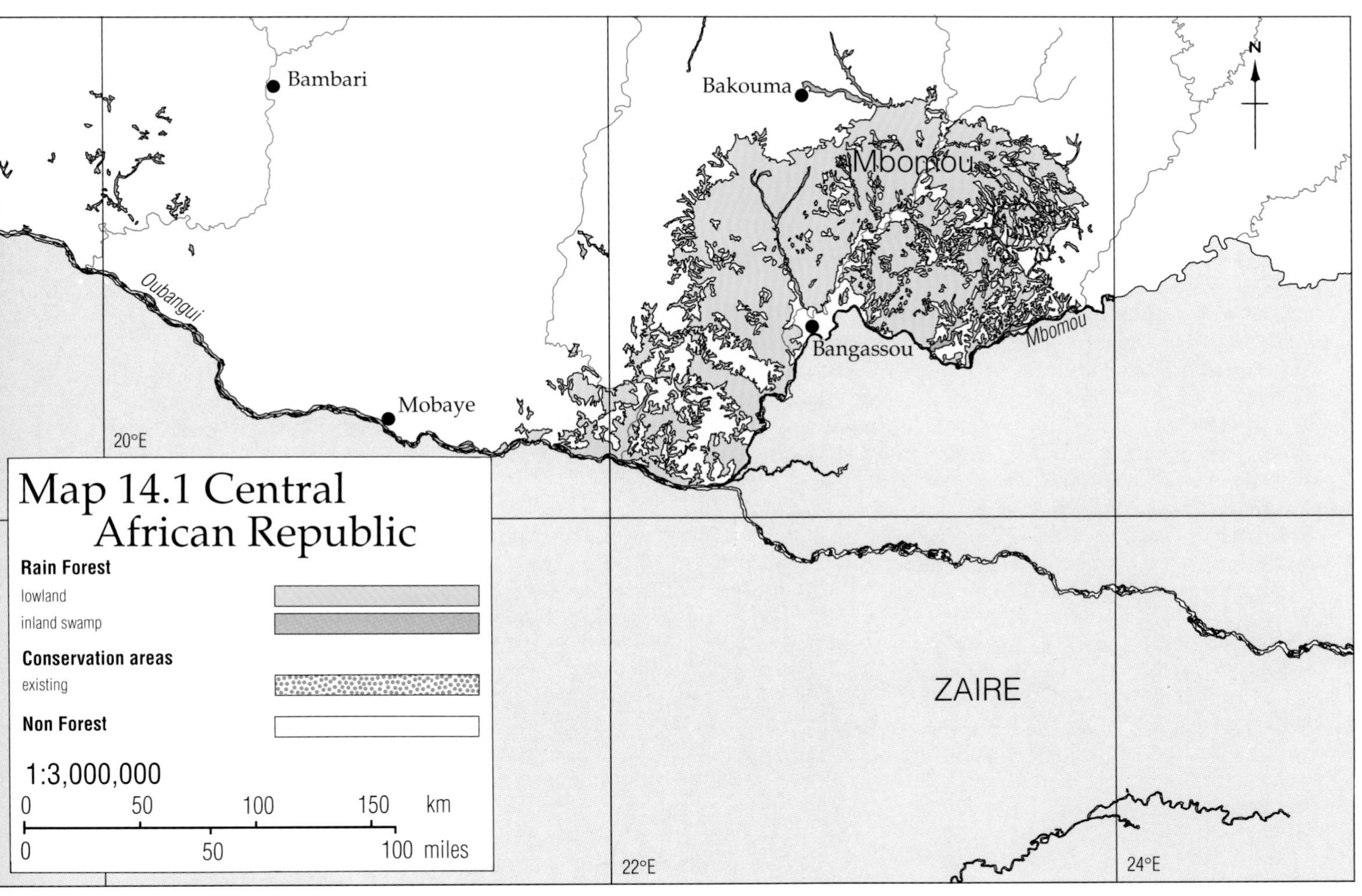

FORESTS OF THE DZANGA-SANGHA REGION

The Dzanga-Sangha Dense Forest Faunal Reserve and the Dzanga-Ndoki National Park system cover approximately 4350 sq. km of dense forest in the southern part of CAR, with 1287 sq. km gazetted as national park and the rest as a multiple use reserve. This system is managed to integrate conservation and regional development. The area contains the last unlogged forest and intact fauna in the country. Elephants reach very high densities (0.9 individuals per sq. km), with major points of concentration around the marshy clearings with saline deposits. One such clearing, Dzanga, means, according to the Aka, the village of elephants. Several other such large clearings occur throughout the forest and are all connected by a network of elephant trails. These clearings offer excellent wildlife viewing and with proper management, have great potential for tourism development.

Lowland gorillas reach population densities of between 0.2 and 11 per sq. km in various habitats, being most frequent in secondary forest and light gaps, but also occurring in primary forest and marshy areas. Chimpanzees and 16 other species of primates are found throughout the park/reserve area. Bongo, dwarf forest buffalo and six species of duiker are common in the Dzanga-Sangha forest.

The aims of the Dzanga-Sangha project are the conservation of the forest and its wildlife, sustainable use of these resources, and the development of options for economic development consistent with the conservation goal. Management of hunting in the reserve allows for sustainable offtake by traditional hunters, while community development activities such as health and sanitation programmes, fisheries development, agroforestry, and small enterprise development attempt to provide alternatives to over-harvesting the wildlife population. Payments from tourism and safari hunting are made directly to the reserve managers and divided between the community and management of the reserve. In this way, the people receive direct financial benefits from the conservation of their wildlife, and therefore realise that it is worth supporting the effort.

Bongo Tragelaphus euryceros, *an elusive forest antelope, at a salt lick in the Dzanga-Sangha National Park.* G. Renson

References

Boulvert, Y. (1986) *République Centrafricaine. Carte Phytogéographique a 1:1,000,000. Notice explicative # 104.* ORSTOM, Paris, France.

Carroll, R. W. (1988) Birds of CAR. *Malimbus* **10**: 177–200.

Carroll, R. W. and Fry, H. (1986) A range extension and probable breeding record of the brown nightjar (*Caprimulgus binotatus* Bonaparte) in south-western CAR. *Malimbus* **9**: 125–7.

Collar, N. J. and Stuart, S. N. (1985) *Threatened Birds of Africa and Related Island. The ICBP/IUCN Red Data Book Part 1.* ICBP/IUCN, Cambridge, UK. 761 pp.

CTFT (1967) *Inventaire forestier dans le Secteur de Nola.* Centre Technique Forestier Tropical, Nogent-sur-Marne, France.

Davis, S. D., Droop, S. J. M., Gregerson, P., Henson, L., Leon, C. J., Villa-Lobos, J. L., Synge, H. and Zantovska, J. (1986) *Plants in Danger: What do we know?* IUCN, Gland, Switzerland and Cambridge, UK.

FAO (1988) *An Interim Report on the State of Forest Resources in the Developing Countries.* FAO, Rome, Italy. 18 pp.

FAO (1990) *FAO Yearbook of Forest Products 1977–1988* FAO Forestry Series No. 23, Statistics Series No. 90. FAO, Rome, Italy.

FAO (1991) *FAO Yearbook of Forest Products 1978–1989.* FAO Forestry Series No. 24, Statistics Series No. 97. FAO, Rome, Italy.

Oates, J. F. (1986) *Action Plan for African Primate Conservation 1986–90.* IUCN/SSC Primate Specialist Group, Stony Brook, New York, USA.

Sillans, R. (1958) *Les Savannes de l'Afrique Centrale. Essai sur la Physiognomonie, la Structure et le Dynamisme des Formations Végétales Ligneuses des Régions Sèches de la RCA.* Lechevalier, Paris, France.

Western, D. and Vigne, L. (1985) The deteriorating status of African rhinos. *Oryx* **19**: 215–20.

Authorship

Richard Carroll, in Dzanga-Sangha, CAR.

Map 14.1 Forest cover in the Central African Republic

Forest data shown on Map 14.1 were extracted from a published map *Carte Phytogéographique de la République Centrafricaine* (1985), at a scale of 1:1 million. The map was prepared for ORSTOM (Institut Française de Recherche Scientifique pour le Développement en Coopération), Bondy, France, by Y. Boulvert and published in association with the Ministère des Relations Extérieures (France) Service de la Coopération et du Développement Fonds d'Aide et de Coopération. The vegetation shown on this map has been categorised into 149 phytogeographical types. Categories *137–143 (IV.B. Secteur Congo-Guinéen de la forêt dense humide)* have been digitised and are depicted on Map 14.1 as lowland rain forest; categories *144 (Forêt ripicole à inondation prolongée à* Uapaca heudelotii *et* Cathormion altissimum) and *145 (Forêt à inondation temporaire)* have been harmonised into the swamp forest category, also shown on Map 14.1.

The mapped conservation area, Dzanga-Sangha Dense Forest Faunal Reserve, has been taken from an unpublished manuscript, WWF-US and New York Zoological Society *Création, Développement, Protection et Aménagement du Sanctuaire de Forêt Dense de Dzanga-Sangha et Parc National Dzanga-Ndoki, République Centrafricaine* (1987).

15 Congo

Land area 341,500 sq. km
Population (mid-1990) 2.2 million
Population growth rate in 1990 3.0 per cent
Population projected to 2020 5.0 million
Gross national product per capita (1988) US$930
Closed broadleaved forest (end 1980)* 213,400 sq. km
Annual deforestation rate (1981–5)* 220 sq. km
Industrial roundwood production† 1,524,000 cu. m
Industrial roundwood exports† 961,000 cu. m
Fuelwood and charcoal production† 1,776,000 cu. m
Processed wood production† 100,000 cu. m
Processed wood exports† 53,000 cu. m

* FAO (1988)
† 1989 data from FAO (1991)

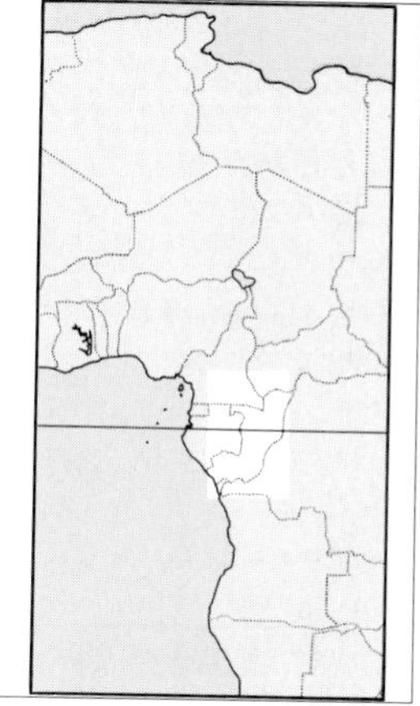

The Congo exhibits a considerable diversity of landscapes and natural environments and hosts an impressive array of both plant and animal species. As in neighbouring Zaïre, the forests have remained relatively intact so that today the Congo is the second most densely forested country in the Afrotropical realm. It contains large areas of both lowland and swamp forest. Up to the 1970s, timber was the main source of revenue for the country, although this has since been surpassed with the discovery of oil. Nonetheless, the forests of the Congo have considerable importance and potential. The authorities recognise the value of forests for the economy and have invested heavily in forestry planning and in several successful plantation schemes. At the time of preparing this Atlas no comprehensive map of the forests existed and the editors have been obliged to use a sketch map prepared by IUCN staff based on local kmowledge.

Introduction

The Congo – or People's Republic of Congo – was formerly one of four territories included in French Equatorial Africa, along with the Central African Republic, Chad and Gabon. The Congo straddles the Equator between latitudes 3°34'N and 5°S and longitudes 11°11'E and 18°35'E, and is bordered by Gabon to the west, Cabinda (Angola) to the south, Zaïre to the south and east, and Cameroon and the Central African Republic to the north. A 200 km Atlantic Ocean coastline in the extreme south-west of the country gives access to the sea.

The coastal plain, about 60 km wide with an altitude of 0–200 m above sea level, is covered in mixed savanna and forest. This rises to low hills and then to the Mayombe massif which crosses the country on a south-east to north-west incline. The low-lying savanna plain of Niari separates this massif, which rises to around 1000 m, and the more northerly Chaillu massif (400–900 m). Both these massifs are still forested. Further north is a central savanna zone with the Batéké and Cataractes plateaux, which are between 400 and 800 m in altitude. In the north-west is the massif of Sangha, another heavily forested region, which rises to 1000 m near Souanké. Finally, in the north-east the 'Cuvette Congolaise' is a vast area of swamp forest at an altitude of between 200 and 400 m above sea level (see Figure 15.1).

Rainfall is generally highest in the north of the country and decreases towards the south. The coastal region receives around 1200–1700 mm of rain per year and has a four or five month dry season between May and September; in the central area annual rainfall is between 1600 and 2000 mm and the dry season is one to three months long. In the north, precipitation is around 1800 mm and there is no totally dry month (UICN, 1990). Mean annual temperatures are between 23° and 27°C.

The Congo is home to more than two million people and has an annual population growth rate of 3 per cent; approximately 67 per cent of the country's inhabitants are less than 20 years of age. The average population density is six people per sq. km but about half the population live in an urban environment, with the vast majority of these living in the capital, Brazzaville, and in Pointe Noire on the coast (UICN, 1990). Indeed, around 82 per cent of the population live in the region between Brazzaville and the coast. Northern Congo is among the most sparsely populated regions of Africa with an average of only 1–3 persons per sq. km. Rural communities tend to concentrate along main access routes, such as rivers, roads or the railway line from Brazzaville to Pointe Noire. Aka and Bakola pygmies live in the forests of northern Congo and there are smaller numbers of Babango pygmies in the south.

Farming is the main activity of rural people and in 1986, 13.6 per cent of the population were engaged in agriculture. Approximately 2000 sq. km are currently cultivated, but it is estimated that some 100,000 sq. km (30 per cent of the country) has potential for traditional forms of farming. Agriculture falls into three categories: traditional, state-operated and private industrial ventures. The traditional farmers exploit about 70 per cent of the existing agricultural land and produce approximately 98 per cent of the country's basic foods, such as manioc, bananas, plantains and maize. However, the difficult life in many village communities and the attractions of employment in the cities, are causing an exodus from rural areas. This is not in keeping with the state's plan to achieve self-sufficiency in food requirements by the year 2000. The state-owned sector exploits only 27 per cent of the cultivated land, but uses 95 per cent of the budget allocated for agriculture and employs just 5 per cent of the country's agricultural work force. The main products of the state sector are palm oil and sugar cane. Offshore fishing is slowly developing as a small industry, while coastal and riverine fishing is conducted largely on a subsistence basis.

As recently as the 1970s, timber was the main export and source of revenue for the Congo. However, this has changed in recent years with the discovery of oil, which is now the main export. Timber and veneer are the next most important exports.

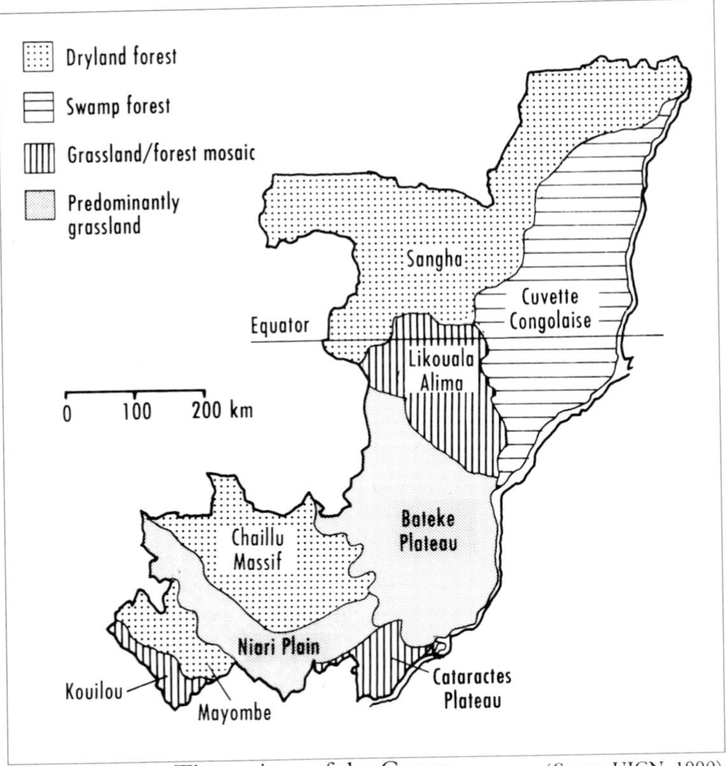

Figure 15.1 The regions of the Congo (*Source:* UICN, 1990)

The Forests

There are two main types of moist forest in the Congo: swamp forest and dry lowland forest. The latter is found mostly on the massifs of the Mayombe and Chaillu in the south and in Sangha in the north-west. The swamp forests are found in the north-east in the Cuvette Congolaise.

In the south, the forests of the Mayombe are mostly semi-deciduous and account for almost 3 per cent of Congo's territory. They are biologically extremely rich, and contain important timber species such as *Aucoumea klaineana*, *Staudtia gabonensis*, *Dacryodes* spp., *Nauclea diderrichii* and *Pycnanthus angolensis*, which are widely distributed. *Terminalia superba*, *Berlinia grandifolia* and *Oxystigma oxyphyllum* are also important, but are more restricted in distribution.

Forests on the Chaillu massif are partially deciduous and comprise 11 per cent of the country. These forests are dominated by *Aucoumea klaineana*, *Terminalia superba* and *Entandrophragma utile*. This region also has a particularly high abundance of *Lovoa trichilioides*.

The rain forests of northern Congo are also partially deciduous and are rich in the following groups of plants: Meliaceae (*Entandrophragma cylindricum*, *E. candollei*, *E. utile*, *E. angolense*, *Khaya anthotheca*, *Guarea* spp.), and Leguminosae (*Piptadeniastrum africanum*, *Pterocarpus soyauxii*, *Erythrophleum* spp.). Other groups, such as Irvingiaceae, are distributed throughout the forest, while species such as *Terminalia superba* (fraké) and *Triplochiton scleroxylon* (ayous) are locally abundant. These forests, which occupy 31 per cent of the country's surface area, present a number of interesting transition zones with the semi-deciduous Sterculiaceae-Ulmaceae forests of Cameroon, the mesophile forests of central Congo and the swamp forests of the Congo basin. Evergreen forests include some pure stands of *Gilbertiodendron dewevrei* while, in other areas, this species is associated with other rainforest and swamp forest species (Bégué, 1967). An understorey dominated by Marantaceae and Zingiberaceae is widespread, for example in the Odzala National Park (IUCN, 1991).

Inland swamp forests are well developed along several rivers in the country, as well as in the Cuvette Congolaise. They are complex forests, characterised by an abundance of such species as *Entandrophragma palustre*, *Uapaca heudelotii*, *Manilkara* spp., *Garcinia* spp., *Sterculia subviolacea* and *Alstonia congensis*. The forests of the Cuvette are similar to those of the central Zaïre Depression. The northern forests have a canopy 25–30 m high with emergents above that, while those in the south tend to have a canopy of about 20 m. *Raphia* palms occupy significant areas within the inland swamp forests. On the banks of certain large rivers and on the islands of northern Congo, there are several stands of *Guibourtia demeusei* (copal tree), which occur on only the highest reaches above the flood plain.

In addition, stretched along the length of the littoral zone there are relict coastal forests which are rich in species such as *Symphonia globulifera*, *Pentaclethra macrophylla*, *Pycnanthus angolensis* and *Chrysobalanaceae* spp. However, in most of the coastal area only thickets, bush and mangroves remain.

There are also dry, semi-evergreen forests, characteristic of regions further north, scattered throughout the savannas in the centre of the country. Principal tree species are *Millettia laurentii*, *Pentaclethra eetveldeana*, *Staudtia gabonensis*, *Petersianthus macrocarpus* and *Trichilia heudelotii*.

Mangroves

The mangrove *Rhizophora racemosa* occurs in association with *Phoenix reclinata* in the Conkouati Wildlife Reserve, in the extreme south-western corner, bordering Gabon. The extent and size of these mangroves has not been determined and the conservation status of mangroves, in general, in the Congo is poorly known; thus they have not been mapped in this Atlas. Mangroves certainly occur elsewhere along the coastline and at the mouths of some of the major rivers, including the Noumbi, Loémé and Kouilou, but there is little information on the extent or status of these. Some of the mangroves have been cleared in forestry operations and the timber is widely used for fuelwood.

Forest Resources and Management

One of the most striking features of the Congo is the sheer extent of its forests. Just over 213,000 sq. km of land, some 65 per cent of the country is still forested (IUCN, 1990). In Africa, these forests are rivalled in extent only by those in Zaïre. The Congo has almost 10 per cent of the continent's closed forest and 12.3 per cent of that of Central Africa (UICN,1990). The lowland forests occupy approximately 45 per cent of the country's total area, while the swamp forests occupy around 20 per cent. Map 15.1 has been digitised from a generalised hand-drawn original (see Map Legend on p. 132) and consequently the figures derived from it may be unreliable. As a result, a figure for rain forest cover has not been quoted at the beginning of this chapter, nor has a table of forest extent been presented here. However, for the reader's interest, Map 15.1 shows 62,521 sq. km of swamp forest and 234,752 sq. km of lowland rain forest.

On the basis of an FAO/UNEP (1981) survey, the forests of the Congo have been broadly divided into various categories, according to their status and potential for future exploitation (Table 15.1). 'Unproductive forests' include savanna scrub, flooded forest, stands of *Raphia* palms, and those in areas where exploitation is not possible because of access problems or dangerous relief. FAO/UNEP (1981) estimated that in 1980 there were 136,900 sq. km of exploitable forest in the Congo: 40 per cent of the country or 61 per cent of the total forested area.

The main species of timber which are exploited include okoumé *Aucoumea klaineana*, sapele *Entandrophragma cylindricum*, sipo *E. utile*, niové *Staudtia gabonensis*, limba *Terminalia superba*, bilinga

Table 15.1 Area of forest cover in the Congo (sq. km)

Forest category	*South Region*	*North/Central Region*	*Total*
Intact closed forest	5,500	97,800	103,300
Exploited closed forest	32,400	1,200	33,600
Total productive forest	37,900	99,000	136,900
Unproductive closed forest			
– for physical reasons	5,500	69,700	75,200
– for legal reasons	0	1,300	1,300
Total unproductive forest	5,500	71,000	76,500
Totals (all forest)	43,400	170,000	213,400

(*Source:* FAO/UNEP, 1981)

Nauclea diderrichii, moabi *Baillonella toxisperma*, iroko *Milicia excelsa*, tiama *Entandrophragma angolense* and longhi *Gambeya* spp.

Forest exploitation in the country is divided among four sectors: the state, national and foreign private sectors, and a mixture of these. The most active operators of this group in recent years have been the private foreign investors who, in 1986, accounted for the production of 58 per cent of the unprocessed logs, 82 per cent of the sawn timber and 35 per cent of veneer exports. State-operated and private national enterprises contribute a very small proportion of total production.

The Congolian output of unprocessed logs has increased gradually since the world timber crisis in 1974, but has not yet recovered to the levels of exploitation achieved in the late 1960s and early 1970s. In 1986 the Congo exported 286,973 cu. m of uncut logs, compared with 205,714 cu. m in 1982. However, due to the increase in the price obtained for the wood, forest products contributed 3.5 per cent of the gross domestic product in 1986, compared with just 1.1 per cent in 1982. The major buyers are Portugal, France, Italy, Germany, Spain, Japan and the Soviet Union.

The first area of forest to be exploited was the Mayombe, chiefly because it was more easily accessible by road and rail from Pointe Noire. The main species extracted there were limba and okoumé. Later, exploitation moved to the dense forests of central and northern Congo, where two different species – sapele and sipo – were heavily exploited. Since 1984, okoumé and sapele have made up between 60 and 65 per cent of the unprocessed logs exported. In general, small and less valuable species are not exploited to any great extent.

There are 25 registered sawmills which produced 53,658 cu. m and 46,113 cu. m of timber during 1988 and 1989, respectively. In recent years, 50–55 per cent of the wood has been locally processed. There are four veneer factories in the Congo: two were fully operational during 1989, another was working at one-third of its normal capacity and the fourth had been closed for the past three years.

The quality and output of timber from the Congo has increased during the past three years, largely as a result of the establishment of logging agencies, the availability of modern logging materials and a reduction in taxes on timber exports and on imported equipment. (See also the case study on Forest Management Units.) However, despite this growth, the future of the timber industry is uncertain, as timber from the Congo is generally more expensive than that, for example, from Southeast Asia. This is mainly because of the high cost of transporting timber from the north of the country to the coast.

The forests of the Congo fulfil an important need for the people of the country, especially the subsistence agriculturalists who are generally unable to obtain enough food from their farmed land

Forest Management Units

The Republic of Congo has more than 200,000 sq. km of forest and a population of only two million people. The forests are relatively rich in commercial timber species and, especially in the north, there is little pressure to clear forests for agriculture. The situation appears suitable for a sustainable and highly profitable timber industry.

With help from FAO, the Congo authorities have divided up the national forest estate into forest management units (Unités Forestières d'Amènagement, UFA), each of sufficient size to support an independent forest industry. In principle, each industry is required to conduct an inventory of its UFA and propose a management plan for ministerial approval. The plans should provide for selection felling on a 25-year cycle with a minimum diameter limit of 60 cm. These calculations were based upon an annual girth increment of 0.8 cm per year observed in trial plots in similar forest types in Gabon. Twenty-five years is the period required for all trees in the 40–60 cm diameter class to pass into the exploitable 60–80 cm class. Extraction is subject to three-year exploitation permits which prescribe the maximum area to be logged and the minimum volume of timber to be produced. Limited selective elimination of non-commercial species was planned in order to encourage the regeneration of valuable species but this has not been applied on an operational scale.

The UFA system developed in the Congo could have provided a sound basis for a sustainable forest industry but various factors have meant that it has never been put into practice properly. First, the government never invested sufficiently in field staff for the forestry service fully to supervise the UFAs. Second, and more important, all forest land is the property of the state and all citizens enjoy constitutional rights to use this land. The Forest Administration does not have the authority to curtail these rights even in an area under management for timber.

Customary rights of subsistence hunting or collecting non-timber forest products do little harm to the forest's timber potential. However, customary rights also allow local people to burn the forest and grow their crops on the cleared land. The result has been that in the more densely populated and accessible south of the country many of the UFAs have been invaded by shifting agriculturalists and potential timber yields have been seriously reduced. In the sparsely populated, inaccessible north, the forests remain largely undisturbed after logging and regenerate well. Valuable timber crops are likely to be available after the 25-year logging rotations.

Thus, in two parts of the same country, subject to the same laws and administration, the potential for sustainable management of the forests for timber is radically different. In the south, sustainability can probably be achieved only in intensively managed plantations taken out of state ownership. In the north, cyclical selective logging appears eminently sustainable and maintains a 'near-natural' forest which is excellent habitat for many wildlife species.

Source: Jeff Sayer

Mangroves (Rhizophora sp.) *in Conkouati Faunal Reserve.* C. Doumenge

and who, therefore, depend on the forest for fruit and roots, as well as for building materials, medicinal plants and bushmeat. The latter is the most important source of protein for a large part of the population. The forests are also important as the home of the pygmies. Small numbers occur throughout the forested parts of the country. They still frequent the forests of the Odzala National Park.

Since the 1970s, there has been an active programme of plantation establishment in the Congo, particularly in the savannas. Near Pointe Noire, for example, the Unité d'Afforestation Industrielle du Congo (UAIC) has created plantations of *Eucalyptus* sp. (320 sq. km) and *Pinus* sp. (10 sq. km). In addition, the Service National de Reboisement (National Reforestation Service) has established more than 100 sq. km of pine and *Eucalyptus* on the savannas and around 80 sq. km of limba plantations in forests. Within the next decade, the UAIC aims to have planted in the region of 1000 sq. km of *Eucalyptus*, mostly in the coastal region. UICN (1990) estimates that between 5000 and 10,000 sq. km of savanna are suitable for the establishment of fast growing plantations. If these are developed they would be an important economic resource for the country.

Deforestation

FAO estimated that 220 sq. km of forest were lost each year in the Congo in 1981–5 (FAO, 1988). No alternative estimates are available. In countries such as the Congo with extensive forests and few people the question of deforestation rates is secondary to that of qualitative changes in the forest cover. Even though deforestation is low, the degradation of forest by accelerating cycles of shifting cultivation and over-hunting of wildlife is a major problem in south Congo. Some forest areas of great biological interest such as the coastal forests in and around the Conkouati Faunal Reserve and on the Chaillu and Mayombe massifs are at special risk.

Sibona (1985) has proposed that within the southern part of the country, each cultivator clears approximately 0.5 ha each year for agriculture. On a countrywide basis, it has been estimated that almost 20,000 sq. km are cleared, mostly in the densely populated south. However, much of the clearance is in secondary scrub and forest regenerating from previous cycles of shifting cultivation. Land exposed to these cycles of shifting agriculture will lose many of the animal and plant species adapted to primary forest and may be exposed to erosion. But when the fallow period is long enough, forest does regenerate and figures for the annual extent of shifting cultivation cannot be equated with deforestation. Unfortunately in the south of the Congo the cycles are often too short to allow adequate regeneration and the fallow phase is dominated by annual weeds and scrub. When shifting agriculture succeeds logging the impact on the residual stand of timber trees is disastrous. Large tracts of the savanna grasslands are burned each year. This prevents recolonisation of this habitat by woody species and fire encroaches on the peripheral zones of the forests. The result is that small islands of grassland or savanna within the forest zone will gradually increase in size as fires penetrate a few metres further into the forest each year. Such fires are usually set by hunters and have a major impact on the forests even in areas where the population density is low. The problem is especially severe in areas with sandy soils, for instance on the Batéké Plateau.

Timber concession areas increased in 1986 with the granting of a further 71,073 sq. km of forest (39,079 sq. km in the north, 31,994 sq. km in the south). This was again increased in July 1988 with the granting of an additional 39,079 sq. km in the north and 37,835 sq. km in the south, representing altogether 34.3 per cent of the total forest cover in the Congo (UICN, 1990). Most logged areas in the north regenerate satisfactorily, but encroachment of logged-over forests by farmers is a major problem in the south (see case study). Nonetheless the Congo is an example of a country where logging has been environmentally benign and in many areas, even in the absence of post-logging management, logging is sustainable and compatible with the conservation of biological diversity.

Biodiversity

The flora of the Congo has been little studied but a few recent surveys have contributed a great deal of information on the range of species that occur in the country. To date, over 4000 plant species have been recorded (Bouquet, 1976), and UICN (1990) estimates that there are probably as many as 6000 species of plants within the country. Much of this diversity is accounted for by the varied topography, tropical climate and the frequent mixing of forest and savanna ecosystems in the southern and central parts of the Congo. As yet, there is insufficient evidence to assess the level of species endemism.

There has been no systematic survey of the fauna of all the Congo, although Dowsett and Dowsett-Lemaire had, in February 1991, just completed intensive surveys of the flora and fauna (birds, mammals, fish, snakes, frogs, lizards and butterflies) in the western Mayombe, the coastal forests and savannas in the country (F. Dowsett-Lemaire, *in litt.*). Dowsett and Dowsett-Lemaire (1989b) give a preliminary list of the larger mammals in the country and they now estimate that there is a total of 132 mammal species in the Congo (F. Dowsett-Lemaire, *in litt.*). Some of these, such as the African wild dog *Lycaon pictus*, bongo *Tragelaphus euryceros*, gorilla *Gorilla gorilla*, lion *Panthera leo*, elephant *Loxodonta africana*, giant pangolin *Manis gigantea* and manatee *Trichechus senegalensis* are fully protected legally but they are rarely able to be protected in reality. The forests of the Congo are home to at least 14 species of antelope, and are particularly rich in primates – 22 species in all. These include one endemic subspecies, Bouvier's red colobus monkey *Procolobus [badius] pennanti bouvieri*, which has been found only in the Lefini Faunal Reserve (Oates, 1986; Lee *et al.*, 1988).

The avifauna within the forests is also diverse. After a study of the literature and museum specimens and completing a short survey in the Mayombe forest, Dowsett-Lemaire and Dowsett (1989) recorded 223 species in the forest, 16 of which were only in the Zaïre section of the Mayombe. While the total number of species inhabiting all the forests is still unknown, it is estimated that there are at least 700 bird species in the Congo (F. Dowsett-Lemaire, *in litt.*). About 500 of these have been listed by Dowsett and Dowsett-Lemaire (1989a). Only one threatened species is recorded from the Congo and that, the black-chinned weaver *Ploceus nigrimentum*, is probably not a forest species.

An investigation of the reptile diversity in the Mayombe forests alone, revealed 45 species (Unesco/PNUD, 1986). Three species of endangered marine turtles – the loggerhead *Caretta caretta*, green *Chelonia mydas*, and the olive ridley *Lepidochelys olivacea* – have been recorded on Congo's coastline (Groombridge, 1982). Three species of crocodile still survive in the country, but are heavily hunted (F. Dowsett-Lemaire, *in litt.*).

Few data exist on the diversity of fish or invertebrates, although two rare species of swallowtail butterfly – the African giant swallowtail *Papilio antimachus* and *Graphium aurivilliusi*, the latter known only from a few specimens – have been recorded from the Congo (Collins and Morris, 1985). In addition, the dragonfly *Aethiothemys watuliki* has been recorded only in Mambili forest (Stuart *et al.*, 1990).

Conservation Areas

The establishment of protected areas in the Congo dates back to 1935, with the creation of the Odzala National Park. At present this is still the only national park in the country and no exploitation is allowed within it. There are also two other kinds of protected areas: faunal reserves, where hunting is totally prohibited but other traditional rights of usage are permitted, and hunting reserves, where hunting is allowed with a big game hunting permit, which limits the number of animals that can be taken. There are two Biosphere reserves in the country, namely Odzala National Park and the Dimonika reserve. Details of the protected areas are given in Table 15.2.

The administration of wildlife and conservation in the Congo is the responsibility of the Ministry of Forest Economy, within which the protected areas are the responsibility of the Hunting and Wildlife Service. In recent years, management of protected areas has been placed under the direction of the Wildlife Inventory and Management Project (DPIAF). DPIAF is seriously understaffed, having only 43 guards to cover the entire network of protected areas, which means that each person is responsible for more than 300 sq. km. IUCN considers 10 sq. km to be the most that can be properly patrolled by an individual guard. In addition, field personnel are poorly equipped and inadequately trained. As an indication of the poor protection of some of these reserves, it has been reported that concessions have been granted for logging and mining operations in Conkouati Faunal Reserve, and that there are many people living in the reserve who are commercial hunters (F. Dowsett-Lemaire, *in litt.*).

At present, there is no programme to protect threatened or endangered flora in the Congo. Insufficient data exist either on a regional or specific taxonomic basis accurately to assess the status of a given species. However, it is certain that many species are in need of protective measures, especially near towns where increasing population pressures result in rising rates of forest clearance for agriculture, fuelwood and building materials. Useful tree species may be depleted. For example, the wood of wengé *Millettia laurentii*, widely used for decoration and wood carving, is now becoming extremely difficult to locate near Brazzaville, although it is still found elsewhere in the Congo.

Table 15.2 Conservation areas of the Congo

For data on Biosphere reserves see chapter 9.

	Area (sq. km)
National Parks	
Odzala*	1,266
Faunal Reserves	
Conkouati*	3,000
Lefini*	6,300
Lékoli-Pandaka*	682
Loudima†	60
Mont Fouari	156
Nyanga Nord	77
Tsoulou	300
Hunting Reserves	
M'boko*	900
Mont Mavoumbou	420
Nyanga Sud	230
Total	13,391

(*Sources:* IUCN, 1990; WCMC, *in litt.*)

* Area with moist forest within its boundaries according to Map 15.1.

† Not mapped.

As in other West African countries, illegal hunting and poaching for bushmeat is widespread, even in 'protected' areas. Hunting provides a lucrative source of income and is also the main protein source for a large section of the population.

Initiatives for Conservation

The Congo has developed conservation policies with support from international organisations such as FAO, EEC, IUCN, WWF and Unesco/UNDP. National organisations, as well as the French Tropical Forestry Centre (CTFT) are conducting reforestation programmes, largely in the southern section of the country. These are, however, mostly of *Eucalyptus*. Additional scientific study is being conducted by the French overseas research agency ORSTOM.

In 1988, a joint Unesco/UNDP project led to the establishment of the Dimonika Biosphere Reserve, the country's first reserve of this kind. A management plan is currently being prepared and, to promote the ideals of this project among local people, the Mayombe Development Committee has been created.

The EEC is currently coordinating an integrated scheme for conservation and rural development among seven Central African countries, of which the Congo is one. Within this scheme, a demonstration project will be established within the Odzala National Park and surrounding areas to illustrate how conservation and the sustainable exploitation of forest products can benefit local people.

USAID and the EEC are supporting the establishment of a large, new protected area in the Nouabalé area on the northern border of the Congo, adjoining the Central African Republic. This project is being coordinated by Wildlife Conservation International, the conservation arm of New York Zoological Society. IUCN is developing a project at Conkouati to improve management of the reserve by working with local communities.

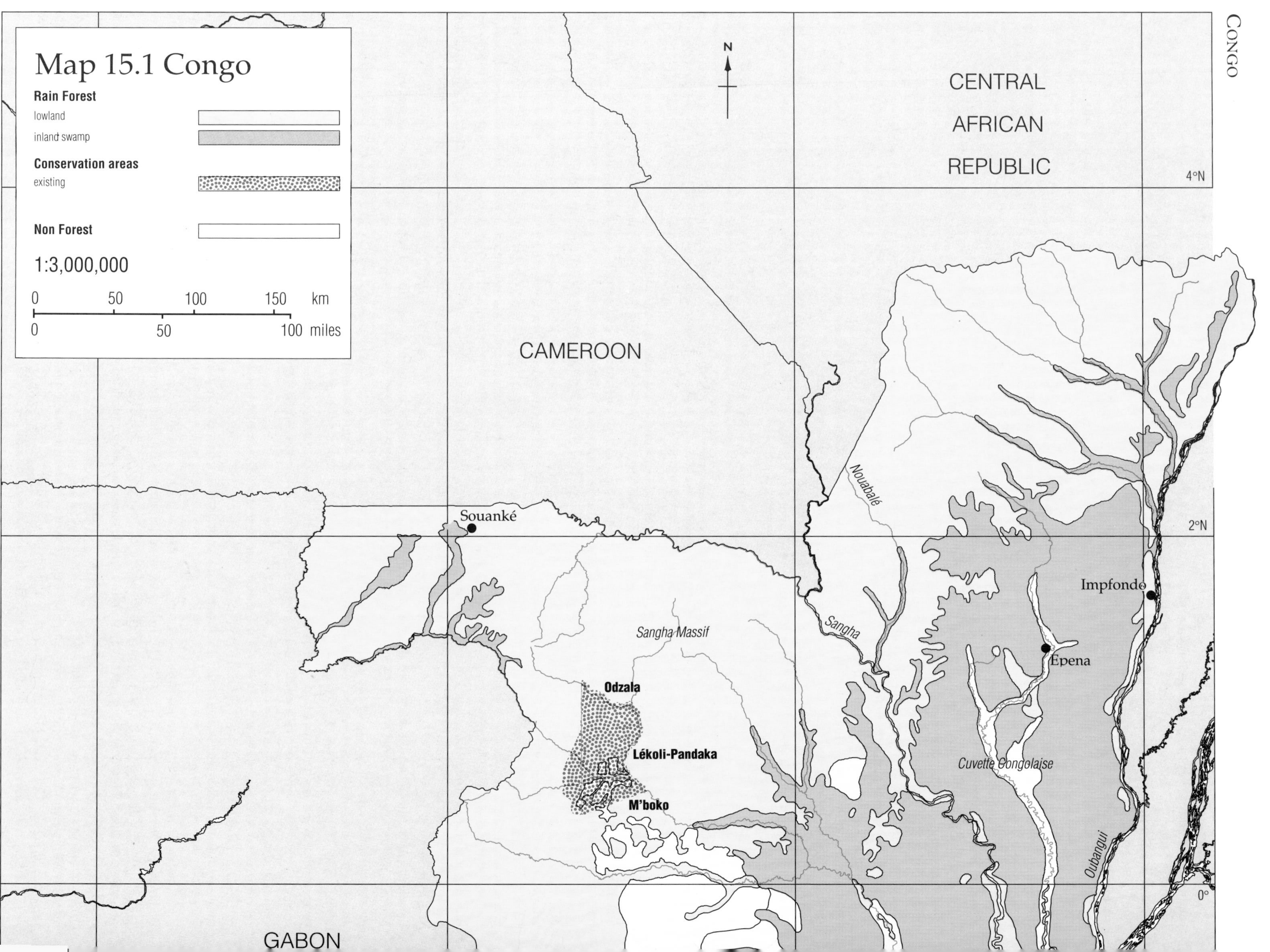

Map 15.1 Congo
Rain Forest
lowland
inland swamp
Conservation areas
existing
Non Forest
1:3,000,000
0 50 100 150 km
0 50 100 miles
N
4°N
2°N
0°
CENTRAL AFRICAN REPUBLIC
CAMEROON
GABON
Souanké
Nouabalé
Sangha
Sangha Massif
Odzala
Lékoli-Pandaka
M'boko
Impfondo
Epena
Cuvette Congolaise
Oubangui

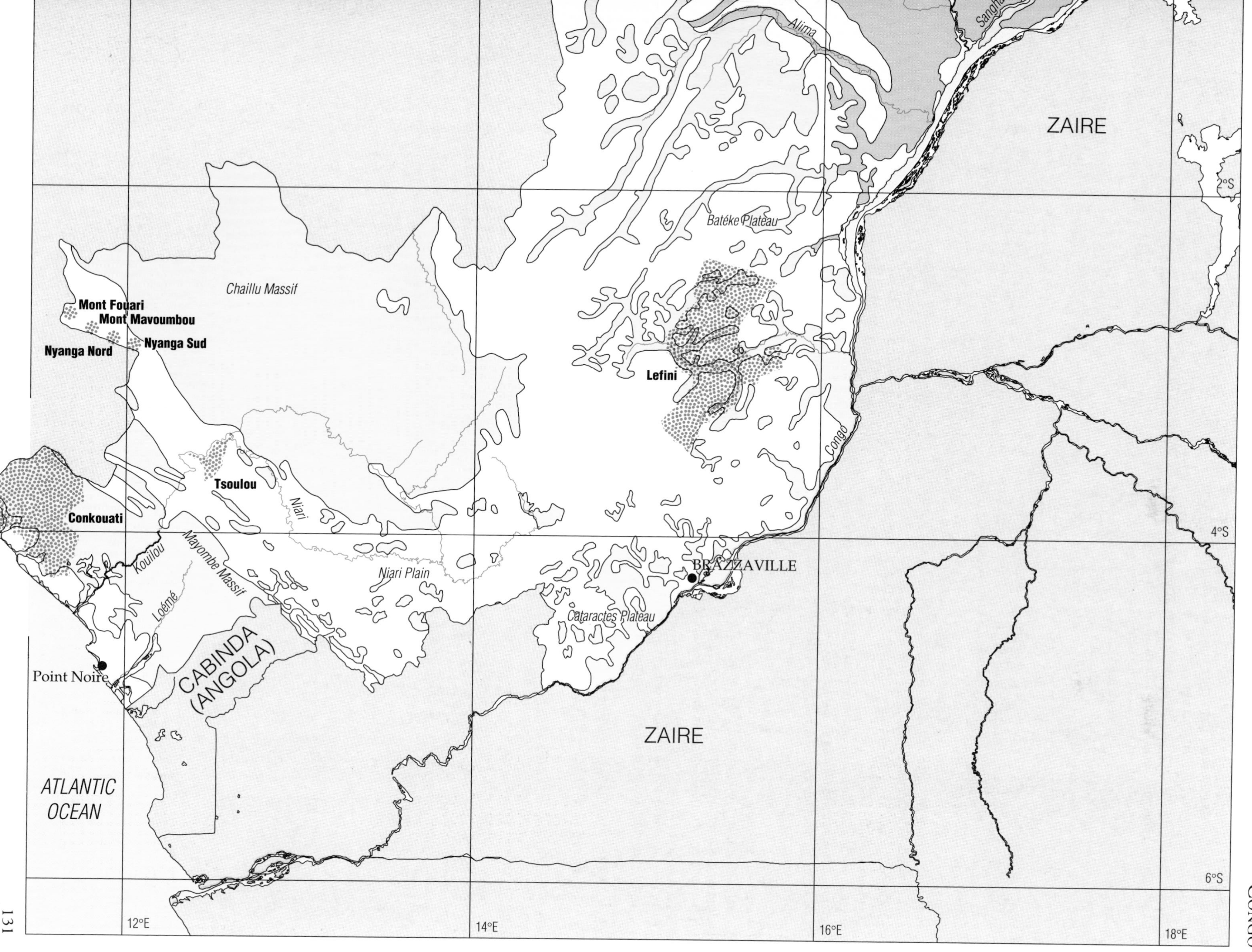
Sangha
Alima
ZAIRE
2°S
Batéke Plateau
Chaillu Massif
Mont Fouari
Mont Mavoumbou
Nyanga Nord
Nyanga Sud
Lefini
Congo
Tsoulou
Niari
Conkouati
4°S
Kouilou
Mayombe Massif
Niari Plain
BRAZZAVILLE
Loémé
Cataractes Plateau
CABINDA
(ANGOLA)
Point Noire
ZAIRE
ATLANTIC
OCEAN
6°S
12°E
14°E
16°E
18°E

References

Bégué, L. (1967) Les forêts du nord de la République du Congo (Brazzaville). *Bois et Forêts des Tropiques* **111**: 63–76.

Bouquet, A. (1976) Etat d'avancement des travaux sur la Flore de Congo-Brazzaville. In: *Comptes Rendus de la VIIIe Réunion de l'AETFAT, 2 vols. Proceedings of the 8th plenary meeting of AETFAT in Geneva, 16–21 September, 1974.* Miége, J. and Stork, A. L. (1975, 1976) (eds). Appendix 1, p. 581.

Collins, N. M. and Morris, M. G. (1985) *Threatened Swallowtail Butterflies of the World. The IUCN Red Data Book.* IUCN, Gland, Switzerland and Cambridge, UK. 401 pp.

Cusset, G. (1989) La flore et la végétation du Mayombe Congolais. Etat des connaissance. In: *Revue des Connaissances sur le Mayombe*, pp. 103–36. Unesco/PNUD, République Populaire du Congo.

Descoings, B. (1969) Esquisse phytogéographique du Congo. In: *Atlas du Congo*. ORSTOM, Paris, France (1 carte couleurs 1:2 million, 2 pages de texte).

Descoings, B. (1975) Les grandes régions naturelles du Congo. *Candollea* **30**: 91–120.

Dowsett, R. J. and Dowsett-Lemaire, F. (1989a) Liste préliminaire des oiseaux du Congo. *Turaco Research Report* **2**: 29–51.

Dowsett, R. J. and Dowsett-Lemaire, F. (1989b) Liste préliminaire des grands mammifères du Congo. *Turaco Research Report* **2**: 20–8.

Dowsett-Lemaire, F. and Dowsett, R. J. (1989) Liste commentée des oiseaux de la forêt du Mayombe (Congo). *Turaco Research Report* **2**: 5–16.

FAO (1988) *An Interim Report on the State of Forest Resources in the Developing Countries.* FAO, Rome, Italy. 18pp.

FAO (1991) *FAO Yearbook of Forest Products 1978–1989.* FAO Forestry Series No. 24 and FAO Statistics Series No. 97. FAO, Rome, Italy.

FAO/UNEP (1981) *Tropical Forest Resources Assessment Project. Forest Resources of Tropical Africa. Part II Country Briefs.* FAO, Rome, Italy.

Groombridge, B. (1982) *The IUCN Amphibia-Reptilia Red Data Book.* IUCN, Gland, Switzerland and Cambridge, UK.

IUCN (1990) *1989 United Nations List of National Parks and Protected Areas.* IUCN, Gland, Switzerland and Cambridge, UK. 275 pp.

Koechlin, J. (1961) *La Végétation des Savanes dans le Sud de la République du Congo (Capitale Brazzaville).* ORSTOM, Paris, France. 310 pp.

Lee, P. C., Thornback, J. and Bennett, E. L. (1988) *Threatened Primates of Africa. The IUCN Red Data Book.* IUCN, Gland, Switzerland and Cambridge, UK.

Oates, J. F. (1986) *Action Plan for African Primate Conservation: 1986–1990.* IUCN/SSC Primate Specialist Group, Stony Brook, New York, USA.

Sibona, F. (1985) *Rapport Final en Sociologie Rurale (Projet 'Développement Forestier Sud-Congo').* FAO, Rome, Italy. 156 pp.

Stuart, S. N., Adam, R. J. and Jenkins, M. D. (1990) *Biodiversity in Sub-Saharan Africa and its Islands: Conservation, Management and Sustainable Use.* Occasional Papers of the IUCN Species Survival Commission No. 6. IUCN, Gland, Switzerland.

UICN (1990) *La Conservation des Ecosystèmes forestiers du Congo.* Basé sur le travail de P. Hecketsweiler. UICN, Gland, Switzerland and Cambridge, UK. 187 pp.

UICN (1991) *Le Pan National d'Odzala, Congo.* Basé sur le travail de Hecketsweiler, P., Doumenge, C. and Makolo Ikonga, J., UICN, Gland Switzerland and Cambridge, UK. xvi + 334 pp.

Unesco/PNUD (1986) *Le Mayombe. Description de la Région et Présentation du Projet sur les Bases Scientifiques de son Développement Intégre.* Plaquette Unesco/PNUD. République Populaire du Congo. Unesco/PNUD, Paris, France. 41 pp.

Authorship

Dominique N'Sosso in Brazzaville and Philippe Hecketsweiler in Montpellier with contributions from Raphael Tsila in Brazzaville, Françoise Dowsett-Lemaire in Brussels, David Stone in Begnins, Switzerland and Jeff Sayer, IUCN.

Map 15.1 Forest cover in the Congo

Vegetation data were digitised from an ONC 1:1 million map on which the extent of the forests had been hand drawn by Philippe Hecketsweiler, one of the authors of this chapter. He was able to produce a map from data gathered from field work during 1989 and 1990 and from numerous reports (Koechlin, 1961; Bégué, 1967; Descoings, 1969, 1975; Cusset, 1989). It shows four vegetation categories: Forêts Denses de Terre Ferme (lowland rain forest), Forêts Inondées Marécageuses (inland swamp forest), Formations Herbeuses Sèches (a dry grassland/bush formation) and Formations Herbeuses Marécageuses (a swamp grassland/bush formation). Of these, lowland rain forest and inland swamp forest are shown on Map 15.1.

Conservation areas are taken from a 1:1 million map *République Populaire du Congo* (1990) produced by Centre de Recherche Géographiques du Congo (CERGEC), Brazzaville and from spatial data held within files at WCMC.

16 Côte d'Ivoire

Land area 318,000 sq. km
Population (mid-1990) 12.6 million
Population growth rate in 1990 3.7 per cent
Population projected to 2020 35.4 million
Gross national product per capita (1988) US$740
Rain forest (see map) 27,464 sq. km
Closed broadleaved forest (end 1980)* 44,580 sq. km
Annual deforestation rate (1981–5)* 2900 sq. km
Industrial roundwood production† 3,413,000 cu. m
Industrial roundwood exports† 550,000 cu. m
Fuelwood and charcoal production† 9,830,000 cu. m
Processed wood production† 1,041,000 cu. m
Processed wood exports† 558,000 cu. m
* FAO (1988a)
† 1989 data from FAO (1991)

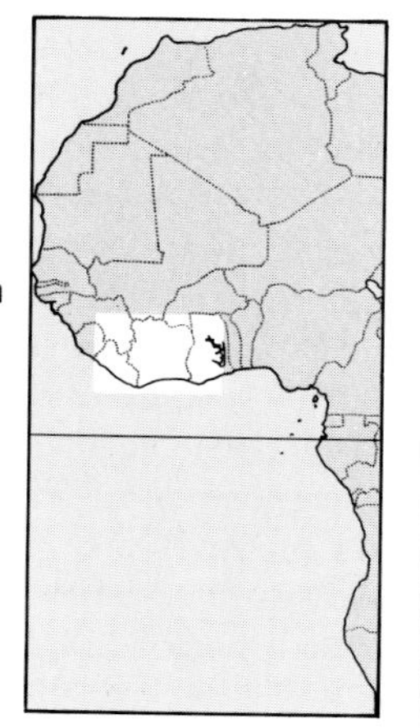

Since independence in 1960, Côte d'Ivoire has shown rapid economic development based mainly on timber, cocoa and coffee exports. For the past 30 years the country has had a remarkable degree of political stability and now has one of the best infrastructures and one of the highest GNP per capita in West Africa. In contrast to this excellent record, deforestation in Côte d'Ivoire has been between 2800 and 3500 sq. km per year for the past 35 years. Indeed, it is predicted that, without stringent protective measures, all the rain forests will have disappeared by the turn of the century.

The country has experienced diminishing total rainfall and irregular rainfall patterns, problems which many people consider to be caused by the severe deforestation. Widespread public concern awoke in 1984 when, after successive dry years, the hydroelectric supply was interrupted and bush fires ravaged the country. It became clear that emergency measures were needed to reduce the rampant deforestation and to avoid the prospect of Côte d'Ivoire becoming a timber importing country. As a result, the 1988–2015 National Forestry Plan was developed, which outlined a long-term strategy of forest resource protection and management and an immediate action plan for rehabilitation of the forestry sector by 1995. President Felix Houphouët Boigny also proclaimed 1988 the National Year of the Forest with the aim of improving public participation in nature conservation, tree planting and fire prevention.

In April 1990, a World Bank loan of US$90 million was agreed for Côte d'Ivoire to implement a project for protection and rational management of the country's remaining tropical forests. For the first time, environmental sustainability seemed to be addressed but northern non- governmental organisations (NGOs) now accuse this project of contributing to further deforestation.

Introduction

Situated between latitudes 4°20' and 10°45'N and longitudes 2°40' and 8°30'W in the central part of the Upper Guinea forest block, Côte d'Ivoire originally must have been predominantly tree covered. Its rolling uniform landscape, intersected by four main rivers running north–south, slowly rises from the Gulf of Guinea to about 400 m above sea level at the northern savanna plateau. The only mountainous region is at the border with Guinea in the Nimba region, where there are peaks of up to 1750 m. Smaller chains of rock outcrops, called inselbergs, occur in the south-western and south-eastern regions, relieving the otherwise monotonous, undulating landscape.

The country has a total area of 322,460 sq. km and it can be divided into two main bioclimatic zones: the tropical moist forest belt in the south, with average rainfall ranging from 1400 to 2500 mm per year and a biannual dry season which does not exceed three to five months in total, and the inland savanna zone which was originally covered with more open, dry forest formations. Temperatures in the forest belt are remarkably constant throughout the year, averaging around the middle to high twenties (°C).

Total population in Côte d'Ivoire has increased from five million in 1970 to almost 13 million in 1990; the annual growth rate, at 3.7 per cent in 1991, is one of the highest in Africa. Mass immigration from drought-stricken Sahelian countries, particularly in the 1970s and early 1980s, is responsible for some of this population increase. About one-third of the people live in the northern savanna regions, while the remaining two-thirds are in the more densely populated forest zone in the south. There are also more people in the eastern and central parts of the southern region than in the western section. The differences in population density are the result of important historical and contemporary movements of the inhabitants (Lena, 1984).

Approximately 55 per cent of the people still live in rural areas, but the general trend of migration to urban areas, recently stimulated by the drop in cocoa and coffee prices and hence farmers' incomes, is likely to continue. Two million people are concentrated in the large urban district of Abidjan and there are some 750,000 people in and around the second largest city, Bouaké. Further important concentrations occur near Korhogo in the north, at Yamoussoukro, which is the newly established capital in the centre of the country, and at Man, Danané and Daloa.

Government policies in the past 25 years have tended to encourage people to move from more densely populated rural regions to relatively unpopulated areas. Farmers are offered incentives to establish new coffee and cacao plantations in the areas to be developed. Some uncontrolled land settlement has been caused by the large numbers of migrants from the Sahel, which has resulted in widespread deforestation and anarchic land occupation in projected industrial plantation areas and forest reserves.

Côte d'Ivoire

The Forests

From south to north of Côte d'Ivoire, a range of successively drier vegetation types can be distinguished. However, gradual changes and strong topographical influences prohibit rigorous definition and exact cartographic representation of each type. Grading from evergreen moist forest in the south-west and south-east, to semi-deciduous forest that fragments into a forest savanna mosaic in the centre of the country (e.g. in Marahoué National Park), the sequence ends up in open Sudan-type savanna land with pockets of dry forest near the border with Mali and Burkina Faso.

The moist forest belt or, more accurately, the fragmented remnants of the moist semi-deciduous and evergreen forest types, shows scattered but omnipresent evidence of early human habitation with pottery fragments and charcoal layers in the soil. As a result, the occurrence of truly virgin forest sites can almost be excluded. Chevalier (1908) suggested their presence in the early years of this century, but if they do still exist they must be limited to some remote areas in the interfluve of the Cavally and Sassandra rivers, for instance in the Taï National Park. However, the degree and extent of vegetation change caused by early human interference remain unknown and now almost impossible to detect due to the subsequent large-scale forest destruction.

Forest composition is associated with soil and climatic conditions. The moderate rainfall regime to which the major part of the moist forest belt is subject (less than 2000 mm rain and a main dry season of at least two to three months) results in forest which is characterised by a high frequency of large trees, towering 55–60 m high, a dense intermediate canopy and a relatively open undergrowth which shows a typically crooked and tortuous habit, induced by drought stress.

These general features contrast with overall differences in tree species composition between evergreen and semi-deciduous forest types. Characteristic tree species of the latter do not occur, or are only sparsely represented, in the former and vice versa. Typical semi-deciduous tree species (cf. Aubréville, 1957) include members of the well represented Malvaceae, Sterculiaceae, Ulmaceae and Moraceae families, for example, *Celtis* spp., *Mansonia altissima*, *Pterygota macrocarpa*, *Nesogordonia papaverifera*, *Sterculia rhinopetala* and *Milicia excelsa*. Less characteristic are *Triplochiton scleroxylon* and *Terminalia* spp. (Combretaceae); these are common in secondary formations of the evergreen forest.

Moist evergreen forest shows a dominance of Mimosaceae and Caesalpiniaceae, including species such as *Piptadeniastrum africanum*, *Parkia bicolor*, *Erythrophleum ivorense* and *Anthonotha* spp. *Parinari excelsa* (Chrysobalanaceae) and *Klainedoxa gabonensis* (Irvingiaceae) are also dominant tree species. The famous African mahoganies or redwoods (*Entandrophragma* spp. and *Khaya* spp.), which attain huge dimensions, can be encountered in both evergreen and semi-deciduous forests. Their relatively high frequency in both formations make these forests extremely attractive for logging.

The boundary between evergreen forest formations and semi-deciduous forest can be traced only approximately. Deciduousness is difficult to assess and local site differences (type of rock, soil and drainage) often play a dominant role in the extent of one forest type or another. For instance, moist formations penetrate along water courses into drier zones and typical elements of drier formations occur on sites with poor water retention capacity inside the evergreen zone. Thus, large transition areas exist where both forest types occur, each in its specific topographical position in the undulating landscape. For mapping purposes, however, an approximate division can be fixed that fairly closely fits the 1700 mm isohyet.

The marked seasonality of climate, with four alternating dry and wet seasons in the moist forest belt and two distinct seasons in the northern savannas, excludes the widespread occurrence of a perhumid forest type. This vegetation type, with strong dominance of the Caesalpiniaceae family in the tree flora (such as *Cynometra ananta* and *Gilbertiodendron* spp.), is limited to the extreme south-west and south-east corners of Côte d'Ivoire where there is more than 2100 mm rainfall per annum. It once covered extensive areas in neighbouring Liberia (Voorhoeve, 1979).

There is a wedge shaped intrusion of savanna woodland (characterised by *Borassus* palms) into the moist forest belt in the central part of the country. This so-called 'Baoulé-V' is most often explained by a rain shadow effect arising from a change in orientation of the West African coastline (Aubréville, 1949). While Liberia is exposed to a frontal arrival of the prevailing south-westerly monsoon winds and hence receives average yearly rainfall of up to 4500 mm, a sudden change in coastline direction to a near parallel orientation occurs after Cape Palmas. This provokes a rather sharp drop in rainfall from 2500 to 1400 mm in the south-western region of Côte d'Ivoire. Recurving of the coastline towards Cape Three Points in Ghana restores higher rainfall (up to 2300 mm) in the south-eastern part of the country.

This central woodland wedge divides the moist forest belt into eastern and western parts. The southern extension of the savannas approaches to within about 150 km of the coast and it probably facilitated deep penetration into the country by expanding savanna tribes (Malinké and Ashanti) invading from the north and east in the 14th century. Unequal population densities in the east and west partly explain today's situation of nearly total deforestation in the east while there is still some forested land remaining in the west.

Forest composition also differs between east and west Côte d'Ivoire in the occurrence of endemic plant species which spread out from their respective Pleistocene refuge areas around Cape Three Points and Cape Palmas. Mangenot (1955) and Guillaumet (1967) described the south-western endemics with the term 'Sassandriennes' after their geographic boundary, the Sassandra river. According to a recent revision by Hall and Swaine (1981), 72 plant species originated in this important centre of diversity.

Another Pleistocene refuge area probably existed in the Nimba mountains. These are nowadays covered by predominantly *Parinari excelsa* forests and submontane woodland (Schnell, 1952).

Mangroves

Mangrove formations, dominated by *Rhizophora* and *Avicennia* spp., originally occupied most of the lagoon shores, river deltas and estuaries. This tidal vegetation can still be encountered along the Ebrié lagoon, which stretches from Abidjan to Grand-Lahou, and in other smaller lagoon areas. The mangroves are being cut for firewood and are declining rapidly as a result. Indeed, Map 16.1 indicates that only 29 sq. km of mangrove remain in Côte d'Ivoire, though it is possible that there are other areas in the country which are too small to appear on a map of this scale.

Forest Resources and Management

An accurate and up-to-date estimate of total remaining forest cover in Côte d'Ivoire is difficult to establish because of the rampant deforestation (Bertrand, 1983) and the low reliability of the available data. Explanation of what is covered by the term 'forest resources' is often lacking or, at best, is ambiguous. Further confusion arises by indiscriminate use of figures for total gazetted forest land (which often includes deforested and non-wooded areas) as an indication of remaining forest cover. As a result, most recent available estimates from reports of the Ministry of Forests (1988) and the FAO (1988b) and data used in reports of the World Bank (1990) show large discrepancies.

Table 16.1 Estimates of forest and woodland cover in Côte d'Ivoire (forest fragments of less than 1 sq. km excepted)

	Moist forest zone	*Savanna zone*	*Total area*
Total land area	132,220	190,145	322,365*
Total estimated forest cover			
(1966)[1]	86,150	88,660	174,810
(1980)[2]	44,580	53,760	98,340
(1987)[3]	21,950	24,720	46,670
(1990)[4]	13,000	18,000	31,000
Permanent Forest Domain (before 1978: 'Forêts Classées')			
(1956)[5]	43,000	25,000	68,000
(1966)[5]	27,840	25,340	53,180
(1974)[5]	28,986	13,000	41,986
(1978)[6]	24,043	12,222	36,265
(1987)[3]	16,000	13,000	29,000
National Parks and Fauna Reserves			
(1974)[5]	5,480	11,750	17,230
(1982)[7]	5,717	13,670	19,387

(Areas in sq. km)

* NB: this figure for country area, supplied by the author, is slightly different from that given by FAO.
1 Guillaumet, J. L. and Adjanohoun, E. (1966). Végétation de la Côte d'Ivoire. Mapping (1:500,000) after aerial photographs of 1965.
2 FAO/UNEP (1981).
3 Ministère des Eaux et Forêts (1988). Plan Directeur Forestier 1988–2015. Annexe 2 (aerial survey 1986) and table 1.
4 Estimate by extrapolation of mean deforestation rates of 3000 sq. km per year in moist forest and 2000 sq. km in savanna.
5 Arnoud and Sournia (1980), pp. 69 and 70. NB: 'Forêts Classées' of 1956 and 1966 include national parks and fauna reserves.
6 Permanent Forestry Domain, gazetted 15 March 1978.
7 Yamoussoukro (1982). Working document, Conference on 'La forêt au service du développement'.

Estimates based on extrapolation of the forest and woodland cover regression rates of the past 30 years (Table 16.1) indicate a remaining total area of 31,000 sq. km (10 per cent of national territory). However, of this only 13,000 sq. km is closed broadleaved forest in the forest zone; the remaining 18,000 sq. km is dry woodland in the savanna zone. It should further be emphasised that most of this is fragmented and timber-depleted forest. Map 16.1 indicates that, in 1987, total remaining moist forest cover (excluding mangroves) was 27,435 sq. km (see Table 16.2), a somewhat higher figure than that calculated from forest regression rates in Table 16.1.

Faced by the spread of slash and burn agriculture and anarchic forest exploitation, the colonial forest service created so-called 'Forêts Classées' (forest reserves) as early as 1926. These were to ensure future timber production. The total gazetted area of 240 forest reserves was as much as 68,000 sq. km in 1956 (Table 16.1), but the reserves lacked distinctively marked boundaries and efficient protection. Consequently, by 1987, agricultural encroachment had caused the area to shrink to only 29,000 sq. km in 147 reserves. Within this area, productive timber stands were thought to occupy no more than 15,000 sq. km (FAO, 1988b). Satellite data are to be compiled for an assessment of the remaining productive area in the 1978 gazetted Permanent Forestry Domain. Preliminary surveys already indicate an alarming 25 per cent further loss of productive timber area due to anarchic land occupation.

The first acajou *Khaya ivorensis* logs were exported from Côte d'Ivoire to France in 1885, to replace rapidly declining *Swietenia* mahogany stocks from the French colonies in the West Indies (Alba, 1956). Contrary to common belief, it was English companies that started trading the precious cabinet timber from the hinterlands of Assinie and Bassam and they continued to control the acajou market until the First World War (Arnoud and Sournia, 1980). Exports of this wood to France, England and Germany steadily increased from 22,850 cu. m in 1909 to some 60,000 cu. m in 1913 (Meniaud, 1922). Acajou remained the only exported timber species until 1920 and the most important one until 1951. In 1920, 50,000 cu. m of acajou was sent overseas, but also 14,000 cu. m of tiama *Entandrophragma angolense*, iroko *Milicia excelsa*, makoré *Tieghemella heckelii*, bossé *Guarea cedrata*, badi *Nauclea diderrichii* and niangon *Heritiera utilis* were shipped. Like other precious cabinet timbers such as sipo *Entandrophragma utile* and makoré, acajou has been logged to such an extent that the species is close to commercial extinction. A forced shift to medium density sawnwood species like framiré *Terminalia ivorensis*, fraké/limba *Terminalia superba* and samba *Triplochiton scleroxylon*, has taken place and these have formed the bulk of timber exports since 1970. The total number of traded species rose from 28 in 1973 to 36 in 1988.

Timber exports increased rapidly after independence in 1960 (Table 16.3), funding the economic development of the country. Subsequently, total log production for industrial use went up almost fivefold and peaked in 1977 at 5.3 million cu. m. The onset of resource exhaustion became evident after 1980 (see Figure 7.1 on p. 59).

Fuelwood and charcoal consumption is difficult to assess, but was estimated by Catinot (1984) to amount to some 6 million cu. m in 1985 while the World Bank (1990) estimated 10 million cu. m for that year and predicted it would reach 14 million cu. m in 1995. Firewood need will further increase with population growth and is already considered to be a major cause of deforestation in the savanna regions and in the vicinity of urban centres. If the official estimates are correct, wood for domestic use (including minor quantities of utensil wood) has outstripped industrial wood consumption since 1980.

Forest legislation, enacted in June 1912, allotted logging rights to private companies after payment of exploration and exploitation taxes. Timber concessions, parcelled out in blocks of 25 sq. km along the major rivers and the railway, extended in 1921 to over 6500 sq. km of acajou-rich forests in the south-eastern hinterland. Besides the annual permit fees, the Supplementary Forest Act of 1920 imposed species-indexed stumpage fees and reforestation taxes. Reforestation, however, rarely occurred because of the reluctance of timber companies to replant and because there was a lack of knowledge of appropriate methods and species requirements (Alba, 1956). At that time, agricultural encroachment was not a problem and the need for reforestation was less apparent.

Table 16.2 Estimates of forest extent in Côte d'Ivoire

	Area (sq. km)	*% of land area*
Rain forests		
Lowland	26,890	8.5
Montane	138	0.04
Swamp	407	0.1
Mangrove	29	<0.01
Totals	27,464	8.64

(Based on analysis of Map 16.1. See Map Legend on p. 142 for details of sources.)

Table 16.3 Changes in volume of timber logged annually for industrial use (volumes in cu. m)

Year	*Total log production*	*Log exports*	*Processed[1] logs*	*Processing percentage*
1900	8,750	8,750	–	–
1910	17,200	17,200	–	–
1920	93,000	78,000	15,000	16
1930	138,800	113,800	25,000	18
1945[2]	80,800	11,300	69,500	86
1950	227,700	133,700	93,600	41
1955	372,600	208,300	164,300	44
1960[3]	1,034,000	823,000	211,000	20
1965	2,560,000	1,905,000	655,000	26
1970	3,548,000	2,511,000	1,037,000	29
1973[4]	5,169,000	3,497,000	1,672,000	32
1975	3,960,000	2,419,000	1,541,000	39
1977[5]	5,321,000	3,335,000	1,986,000	37
1980	4,969,000	3,064,000	1,905,000	38
1985	3,227,000	1,394,000	1,813,000	56
1986	3,020,000	1,020,000	2,000,000	66
1987	3,252,000	1,231,000	2,021,000	62

(*Source:* F. Vooren)

1 Logs delivered at processing plants. End products represent only 50 per cent of this volume. For example, in 1987, 2,021,000 cu. m of logs were processed to 998,000 cu. m of finished wood products (60 per cent for the export market and 40 per cent for domestic use; World Bank, 1990).
2 War recession.
3 Year of independence.
4 Record year of log exports.
5 Record year of total log production.

The revised Forestry Acts of 1935 and 1965 retained the principles of area taxation and volume fees, but stimulated domestic wood processing through a system of variable permit-renewal periods that favoured forest exploitation enterprises with processing facilities. Log exportation quotas linked to volumes of processed wood products were introducd in 1972 further to increase domestic processing (Table 16.3). The export quota system, however, stimulated the maintenance of inefficient processing plants, kept up in order to obtain the required 40 per cent local processing and so continue the more profitable log exports. Currently, fiscal reforms are proposed in an effort to eliminate inefficient and wasteful logging and processing operations (FAO, 1988b; World Bank, 1990).

Past forest management efforts include silvicultural trials with different tree species (e.g. Martineau, 1932) and forest enrichment with line plantation techniques developed by Aubréville in the 1930s (Catinot, 1965). Trials with natural regeneration techniques were conducted in the 1950s but did not prove satisfactory. It was not until 1966 that industrial forest plantations were established. At the same time, a state reforestation service (SODEFOR) was created to pursue large scale reforestation programmes. Up to 1976, 224 sq. km of forested and cleared land had been planted with various tree species (mostly teak *Tectona grandis*), with variable success. Around 170 sq. km of plantation is estimated to remain from this planting period. From 1976, plantations were established mainly on clear-felled lands by removal of residual secondary forests. High yielding, short rotation species such as fraké and cédréla *Cedrela odorata* received priority (Table 16.4). The first yields of timber from plantations were expected in 1990.

Forest management techniques have been developed on a pilot scale over the past ten years (see case study in chapter 8) and consist of progressive elimination, by poison girdling, of undesirable species in order to encourage growth in merchantable trees (Berthault, 1986). This treatment is said to favour regeneration of saleable species and is expected substantially to improve future timber production. The question remains, however, as to whether there are enough production forests left with sufficient residual stocks of saleable trees to make it worthwhile applying the treatment (FAO, 1988b). There are further concerns about the genetic erosion caused by destruction of the other tree species (including potentially merchantable ones) and about the future stem qualities of the trees growing in the open conditions created by the poisoning programme. It is anticipated that this treatment will be initially applied to 5000 sq. km of production forests in the south-western region (World Bank, 1990).

Village tree planting is also planned in the savanna zone, principally to meet the region's future firewood needs. An initial programme of planting 100 sq. km of village woodlots was foreseen within the scope of the 1988 'Year of the Forest' project. Earlier village forestry programmes established some 20 sq. km of teak plantations in the north of Côte d'Ivoire (FAO, 1988b).

Deforestation

Every study of deforestation in the tropics has concluded that Côte d'Ivoire has experienced the most rapid deforestation rate in the world (Gillis, 1988). As elsewhere in West Africa, there is hardly any stretch of natural, unmodified vegetation left in the country. Woodlands have been severely altered or degraded by past and present human activities such as burning, farming and grazing of animals; this probably started as early as 3000 years BP (Sowunmi, 1986). Savanna grasslands with patches of shrub and open woodland remain; these can be considered a type of fire climax. The 11,500 sq. km of Comoé National Park in the extreme north-east constitutes a fine example of such anthropogenic woodlands. A mosaic of small forest patches is all that remains of the moist forest belt in the densely populated south-eastern and central parts of the country.

Intensive logging operations in the rain forest, in search of the highly valued Meliaceae (mahoganies) and other merchantable hardwood species, began shortly after 1950. Log exports rose to a peak of 3.5 million cu. m in 1973 and brought Côte d'Ivoire into fourth place among tropical timber exporting countries. There was no attempt to manage the land to obtain a sustainable yield of trees. Instead, the opening up of timber concessions was generally followed by total conversion of forested land by invading slash and

Table 16.4 Industrial plantations by SODEFOR

Species	*1966–76*	*1977–81*	*Total ha*
Teak	8,851	8,432	17,283
Niangon, okoumé	2,295	48	2,343
Cédréla	403	8,886	9,289
Framiré, samba	4,535	4,681	9,216
Fraké	1,869	20,112	21,981
Sipo, acajou, makoré	4,185	1	4,186
Others	241	1,984	2,225
Total ha	22,379	44,144	66,523

(*Source:* adapted from FAO, 1988b)

Now an unusual sight in Côte d'Ivoire, forest as far as the eye can see. This is a view from Mt Nienokoue over the Hana River and Taï National Park. F. Lauginie/WWF

burn agriculturalists. Smallholder coffee and cacao plantations were also a major cause of large-scale forest clearance. Indeed, throughout this period, government policies actively encouraged the conversion of the forest to these plantations.

In the mid-1970s, the process accelerated through a cacao and coffee boom and affected the still largely forested south-western part of the country. Mass immigration from impoverished and drought-stricken Sahelian countries further contributed to a rapid spread of cultivated areas. South-west Côte d'Ivoire, however, still possesses around 10,000 sq. km of forest in gazetted reserves and this includes the 3500 sq. km Taï National Park. Conservation efforts should be directed at preserving this Upper Guinea Refuge Area as it is exceptionally rich in endemic plant and animal species.

Biodiversity

Côte d'Ivoire contains some of the most important sites for biological diversity in West Africa. Of particular importance is the Taï National Park and the contiguous N'Zo Faunal Reserve. These include the largest area of undisturbed lowland rain forest in West Africa and are an important centre of diversity for many species of plants and animals. Around 4700 species of plant occur in the country, including 90 or so endemics (Davis *et al.*, 1986). Mt Nimba, with over 2000 species, is particularly rich, while some 1300 plant species have been recorded in Taï.

There are 17 species of primate in Côte d'Ivoire (Oates, 1986) and five of them are listed as threatened (Lee *et al.*, 1988). These include the chimpanzee *Pan troglodytes* of which there is a well-known tool-using population in Taï National Park. Other threatened primates are the red-capped mangabey *Cercocebus torquatus*, the olive colobus *Procolobus verus*, the western red colobus *Procolobus [badius] badius badius* and the diana monkey *Cercopithecus diana*.

Côte d'Ivoire holds important populations of a number of large mammals, including around 3600 elephants *Loxodonta africana*, and other threatened species such as the manatee *Trichechus senegalensis* and pygmy hippopotamus *Choeropsis liberiensis*. Two rare viverrids, Johnston's genet *Genetta johnstoni* and Leighton's linsang *Poiana richardsoni liberiensis* occur in the forest zone. The threatened zebra and Jentink's duikers (*Cephalophus zebra* and *C. jentinki*), found in the south-west of the country, are among the 19 antelope species that have been recorded in Côte d'Ivoire (Roth and Hoppe-Dominik, 1990).

Of the 668 species of bird recorded for Côte d'Ivoire (Stuart *et al.*, 1990), seven are listed as threatened. All of these are forest species (Collar and Stuart, 1985) and all have been recorded in Taï National Park. This park, listed by Unesco as a Biosphere reserve, may be one of the last strongholds for the white-breasted guineafowl *Agelastes meleagrides*.

Two endemic amphibians, *Bufo danielae* and *Kassina lamottei*, are restricted to the forests in the south-west. There are eight amphibian species of conservation concern in the country

Conservation Areas and Initiatives

Between 1968 and 1974, Côte d'Ivoire issued a series of decrees to conserve some of its biotic diversity in national parks and nature reserves (Table 16.5). Six per cent of the land area is covered by such conservation areas. Considerable efforts are made to set aside a balanced sample of the major vegetation types and thus to preserve some of the country's genetic capital. Biodiversity in plant and animal communities locally attains relatively important levels through the fact that some major Pleistocene refuge areas were located near or on the country's territory. Mention has already been made of the Cape Palmas and Cape Three Point refuges and the Nimba mountain areas. Also of special value are some biotopes that can be found in the lagoon district, in the forest-savanna ecotone and in the arid savannas.

Zebra duiker Cephalophus zebra, *is a threatened forest species found only in Côte d'Ivoire, Liberia and Sierra Leone.* F. Lauginie/WWF

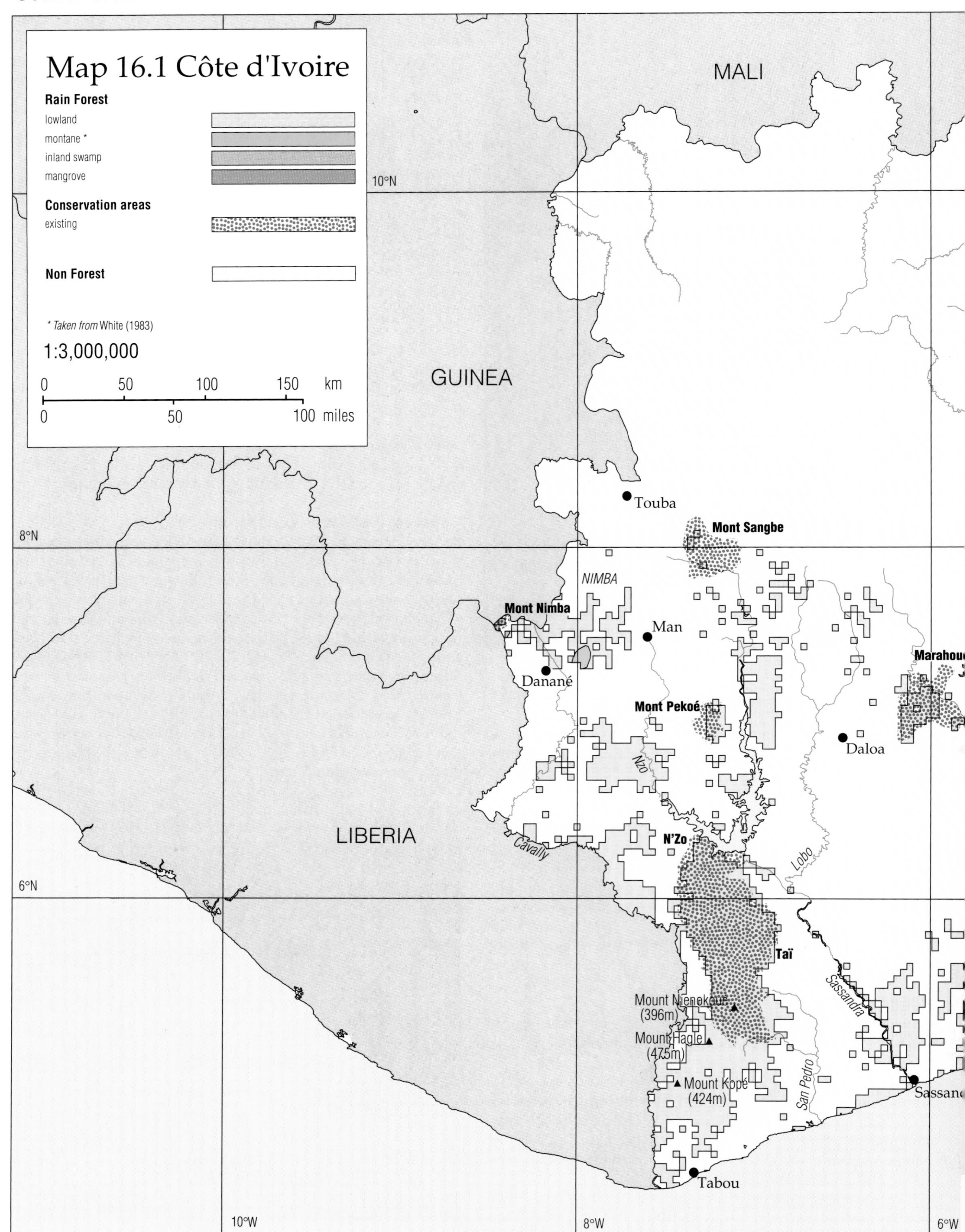
Map 16.1 Côte d'Ivoire
Rain Forest
lowland
montane *
inland swamp
mangrove
Conservation areas
existing
Non Forest
* Taken from White (1983)
1:3,000,000
0 50 100 150 km
0 50 100 miles
MALI
GUINEA
LIBERIA
10°N
8°N
6°N
10°W
8°W
6°W
Touba
Mont Sangbe
NIMBA
Mont Nimba
Man
Danané
Mont Pekoé
Daloa
Marahou
Nzo
N'Zo
Cavally
Lobo
Taï
Sassandra
Mount Nienokoue (396m)
Mount Hagle (475m)
Mount Kopé (424m)
San Pedro
Tabou

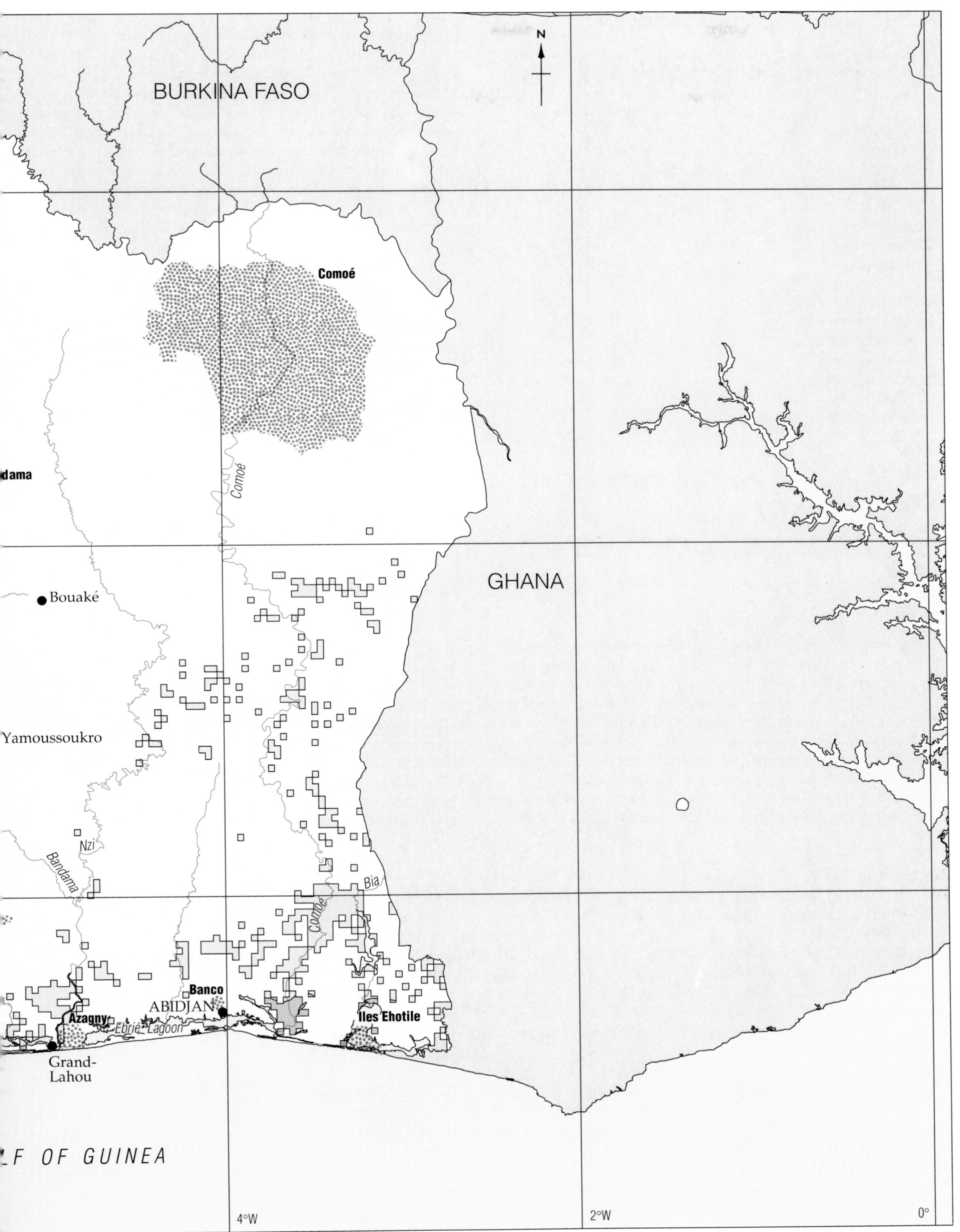

N
BURKINA FASO
Comoé
dama
Comoé
GHANA
Bouaké
Yamoussoukro
Nzi
Bandama
Bia
Comoé
Banco
ABIDJAN
Azagny
Ebrié Lagoon
Iles Ehotile
Grand-
Lahou
LF OF GUINEA
4°W
2°W
0°

Côte D'Ivoire

Table 16.5 Conservation areas of Côte d'Ivoire

Forest reserves are not included or mapped. For data on Biosphere reserves and World Heritage sites see chapter 9.

	Area (sq. km)
National Parks	
Azagny*	190
Banco†	30
Comoé	11,500
Iles Ehotile*	105
Marahoué*	1,010
Mont Pekoe*	340
Mont Sangbe*	950
Taï*	3,500
Strict Nature Reserves	
Mont Nimba*	50
Fauna and Flora Reserves	
Haut Bandama	1,230
Botanical Reserves	
Divo	74
Partial Faunal Reserve	
N'Zo[1]*	950
Total	19,929

(*Sources:* IUCN, 1990; WCMC, *in litt.*)

* Area with moist forest within its boundaries according to Map 16.1.
† Banco National Park does contain some moist forest but not enough to appear on Map 16.1.
[1] 180 sq. km of N'Zo reserve has now been inundated by Buyo reservoir.

Conservation areas are, however, under severe human pressure and they still lack efficient protection and management. Furthermore, it has become increasingly evident that conservation strategies that rely heavily on legal measures and police action have little or no effect. Active involvement of the local populations in, for instance, buffer zone management (currently being investigated for Taï (see case study) and Marahoué national parks), may offer better possibilities to limit poaching and illegal settlement.

Hunting was officially closed in 1974 by a presidential order and a ban on logging activities in the entire savanna region was decreed in 1982. Notwithstanding these measures, elephant populations have been reduced from 10,000 animals in 1950, to 5000 in 1980, to current estimates of 3600 individuals (see chapter 4); savanna woodlands continue to be cut down for fuelwood needs. Hunting activities are also still widespread. Indeed, bushmeat from forest duikers *Cephalophus* spp., antelopes, monkeys, bushpigs *Potamochoerus porcus* and smaller game such as porcupines *Hystrix cristata* and cane rats *Thryonomys swinderianus*, is highly appreciated and is traded at local markets and served in numerous popular restaurants in Abidjan. An attempt to undermine bushmeat traffic was initiated in 1988 by the creation of the Abokouamékro game farm near Yamoussoukro. This project aims to combine wildlife tourism with bushmeat production.

A World Bank loan recently contributed to improving the infrastructure and tourist potential of Comoé and Azagny national parks. A further proposed loan will support rehabilitation and management operations in Taï, Marahoué and Banco national parks. Besides wildlife and nature tourism, scientific research (for instance, on the still unexploited potential of their genetic resources) and education have been deemed as major objectives for these conservation areas.

Recent conservation initiatives include an African Development Bank study of a fairly undisturbed forest region in the coastal area of Fresco. Some 450 sq. km of mangrove habitats – containing major populations of wetland birds, nesting sea turtles and manatees, together with fauna-rich inland forests – are proposed for protection in a regional park.

The hinterland of Tabou in the south-western region deserves further attention. This was once covered by the most humid type of evergreen forest in Côte d'Ivoire and, as a Pleistocene refuge, is exceptionally rich in endemic plant and animal species (Aubréville, 1957; Guillaumet, 1967). Most of the area is now lost to agro-industrial oil palm and rubber plantations, active timber exploitation and slash and burn agriculture. A last, virtually intact, but unprotected, remainder of this unique biome can be found in the Mounts Kopé-Haglé hilltop range, located near the Taï National Park. Rare or unique tree species such as *Brachystegia* spp., *Didelotia unifoliolata* and *Cynometra ananta* can still be found on these inaccessible hilltops where forest exploitation has not occurred. However, slash and burn cultivators, forced uphill by the expanding oil palm plantations, are taking possession of the steep hillsides and threaten the continued existence of this last per-humid forest sanctuary in Côte d'Ivoire.

References

Alba, P. (1956) Le développement de la foresterie en Afrique Occidentale Française. *Journal of the West African Science Association* **2**: 158–71.

Arnoud, J. C. and Sournia, G. (1980) Les forêts de Côte d'Ivoire. Essai de synthèse géographique. *Annals of the University of Abidjan, series G (Géography)* **IX**: 5–93.

Aubréville, A. (1949) *Climats, Forêts et Désertification de l'Afrique Tropicale.* Société d'Editions Géographiques, Maritimes Coloniales, Paris, France.

Aubréville, A. (1957) A la recherche de la forêt en Côte d'Ivoire. *Bois et Forêts des Tropiques* **56**: 17–32; **57**: 12–27.

Berthault, J. G. (1986) *Etude de l'Effet d'Interventions Sylvicoles sur la Régénération Naturelle au Sein d'un Périmètre Expérimental d'Aménagement en Forêt Dense Humide.* Unpublished thesis, University of Nancy, France.

Bertrand, A. (1983) La déforestation en zone de forêt de Côte d'Ivoire. *Bois et Forêts des Tropiques* **202**: 3–18.

Catinot, R. (1965) Sylviculture tropicale en forêt dense africaine. *Bois et Forêts des Tropiques* **100**: 5–18; **101**: 3–6; **102**: 3–16; **103**: 3–16; **104**: 17–31.

Catinot, R. (1984) *Appui à Sodefor pour l'Implantation d'un Programme de Protection Contre les Incendies de Forêts.* Rapport en sylviculture et aménagement FO: TCP/IVC/2304 (T). FAO, Rome, Italy.

Chevalier, A. (1908) La forêt vierge de la Côte d'Ivoire. *La Géographie* **17**: 201–10.

Collar, N. J. and Stuart, S. N. (1985) *Threatened Birds of Africa and Related Islands. The ICBP/IUCN Red Data Book Part 1.* ICBP/IUCN, Cambridge, UK. 761 pp.

Davis, S. D., Droop, S. J. M., Gregerson, P., Henson, L., Leon, C. J., Villa-Lobos, J. L., Synge, H. and Zantovska, J. (1986) *Plants in Danger: What do we know?* IUCN, Gland, Switzerland and Cambridge, UK.

FAO (1988a) *An Interim Report on the State of Forest Resources in the Developing Countries.* FAO, Rome, Italy. 18 pp.

TAI NATIONAL PARK BUFFER ZONE

The Taï National Park covers an area of 3300 sq. km in south-western Côte d'Ivoire. It was established as a national park in 1972 and declared a World Heritage site in 1982. It is the largest fully protected area in the Upper Guinea forest block and pessimists would say that it is the only area that is sufficiently large and secure to guarantee the survival of the numerous animal and plant species endemic to this region. Species such as the pygmy hippopotamus, Jentink's and zebra duikers and chimpanzees, rare elsewhere in the Upper Guinea zone, are comparatively numerous in Taï. IUCN's review of the protected area systems of the Afrotropical realm ranked the Taï National Park as the single highest priority for rain forest conservation in West Africa. The area has been the subject of many long term ecological studies (summarised in Guillaumet *et al.*, 1984). The Taï is now a focal area for the Dutch Tropenbos scheme which promotes ecological research for wise use of rain forests throughout the tropics. WWF also supports conservation through the provision of equipment for the staff, the preparation of a management plan and help with the demarcation of the boundaries. However, the Taï National Park suffers from acute problems.

The Taï region has experienced spectacular population growth in the last few decades. Baoulé and Dioula peoples from the north of the country have moved into the forest to grow coffee and cacao, initially with encouragement from the government. This deprived the traditional forest dwelling Gur and Oubi peoples of their community forests and forced many of them into illegal land settlement and poaching. Forced resettlement of various peoples in response to changing buffer zone boundaries has built up considerable resentment among the local populations. At present any agricultural activity even in the Taï buffer zone is regarded by the authorities as a direct threat to the integrity of the protected area.

The problems of the buffer zones of Taï National Park have been particularly contentious. Some people feel that the buffer zone should be totally protected and should be a *de facto* extension of the park itself. Others see that sustainable use of the buffer zone forest to meet the needs of local communities is the highest priority, so as to relieve pressure on the central area.

One of the more interesting schemes proposed for the buffer zone by the Dutch Tropenbos programme is to promote the sustained yield management of populations of the African giant snail *Achatina achatina*. These snails provide a protein rich food which is highly appreciated by the local population. It is estimated that 8000 tons were sold in Côte d'Ivoire in 1986. The snails are easy to catch, can be kept alive for up to a week, and come ready packed for transport in their own shells. Each snail provides 100–300 grams of meat with no waste. The snails produce up to 200 eggs per year and grow rapidly. They live in forested areas and in regenerating second growth around farms. The Dutch believe that by introducing measures to limit the offtake of these snails and to promote their reproduction, they could become a major economic justification for the local people to maintain forest in the buffer zone around Taï.

Source: Jeff Sayer

Giant snails Achatina sp. *are widely harvested for food in West Africa. There is a proposal to promote their use in the buffer zone around Taï National Park.* I. and D. Gordon

FAO (1988b) *Côte d'Ivoire. Programme Sectoriel Forestier.* Rapport de Préparation. Volumes 1 and 2 14/88 CP-IVC 22. FAO, Rome, Italy.

FAO (1991) *FAO Yearbook of Forest Products 1978–1989.* FAO Forestry Series No. 24 and FAO Statistics Series No. 97. FAO, Rome, Italy.

FAO/UNEP (1981) *Tropical Forest Resources Assessment Project. Part 2: Country Briefs.* FAO, Rome, Italy.

Gillis, M. (1988) West Africa: resource management policies and the tropical forest. In: *Public Policies and the Misuse of Forest Resources.* Repetto, R. and Gillis, M. (eds), pp. 299–351. Cambridge University Press, Cambridge, UK.

Guillaumet, J. L. (1967) Recherches sur la végétation et la flore de la région du Bas-Cavally (Côte d'Ivoire). *Mémoire ORSTOM* **20**: 1–247.

Guillaumet, J. L., Conturier, G. and Dosso, H. (eds) (1984) *Recherche et Aménagement en Milieu forestier Tropical Humide: le Projet Taï de Côte d'Ivoire.* MAB Technical Notes No. 15. Unesco, Paris, France.

Hall, J. B. and Swaine, M. D. (1981) Distribution and ecology of vascular plants in a tropical rain forest. Forest vegetation in Ghana. *Geobotany* **1**.

Lee, P. J., Thornback, J. and Bennett, E. L. (1988) *Threatened Primates of Africa. The IUCN Red Data Book.* IUCN, Gland, Switzerland and Cambridge, UK.

Lena, P. (1984) Le développement des activités humaines. In: *Recherche et Aménagement en Milieu forestier Tropical Humide: le Project Taï de Côte d'Ivoire.* Guillaumet, J. L., Couturier, G. and Dosso, H. (eds), pp. 58–112. MAB Technical Notes No. 15. Unesco, Paris, France.

Mangenot, G. (1955) Etudes sur les forêts de plaines et plateaux de Côte d'Ivoire. IFAN, *Etudes Eburnéennes* **4**: 5–61.

Martineau, A. (1932) Etudes sur l'accroissement des arbres en Côte d'Ivoire. *Actes et Comptes Rendus de l'Association Colonies-Sciences* **9**: 183–7.

Meniaud, J. (1922) *La Forêt de la Côte d'Ivoire et son Exploitation.* Publications Africaines, Paris, France.

Ministère des Eaux et Forêts (1988) *Plan Directeur Forestier 1988–2015.* République de Côte d'Ivoire.

Oates, J. F. (1986) *Action Plan for African Primate Conservation 1986–1990.* IUCN/SSC Primate Specialist Group, Stony Brook, New York, USA.

Päivinen, R. and Witt, R. (1989) *The Methodology Development Project for Tropical Forest Cover Assessment in West Africa.* Unpublished report. UNEP/GRID, Geneva, Switzerland.

Roth, H. H. and Hoppe-Dominik, B. (1990) Ivory Coast. In: *Antelopes Global Survey and Regional Action Plans. Part 3: West and Central Africa.* East, R. (ed.). IUCN, Gland, Switzerland.

Schnell, R. (1952) *Végétation et Flore de la Région Montagneuse des Monts Nimba.* IFAN, 22, Dakar, Senegal.

Sowunmi, M. A. (1986) Change of vegetation with time. In: *Plant Ecology in West Africa.* Lawson, G. W. (ed.), pp. 273–307. John Wiley and Sons, New York, USA.

Stuart, S. N., Adams, R. J. and Jenkins, M. D. (1990) *Biodiversity in Sub-Saharan Africa and its Islands: Conservation, Management and Sustainable Use.* Occasional Papers of the IUCN Species Survival Commission No. 6. IUCN, Gland, Switzerland.

Voorhoeve, A. G. (1979) *Liberian High Forest Trees.* 2nd Edition. Pudoc, Wageningen, The Netherlands.

World Bank (1990) *Forestry Sector Project, Republic of Côte d'Ivoire.* Staff appraisal report No. 7421-RCI.

Authorship

Fred Vooren, Forestry Department, University of Wageningen, The Netherlands with contributions from Jeff Sayer, IUCN.

Map 16.1 Forest cover in Côte d'Ivoire

Information on Côte d'Ivoire's remaining moist forests was digitally extracted from 1989–90 UNEP/GRID data which accompanies an unpublished report *The Methodology Development Project for Tropical Forest Cover Assessment in West Africa* (Päivinen and Witt, 1989). Forest and non-forest boundaries for West Africa have been mapped by UNEP/GEMS/GRID. Together with the EEC and FINNIDA, they have delimited these boundaries using 1 km resolution NOAA/AVHRR-LAC satellite data. These data have been generalised for this Atlas to show 2 × 2 km squares which are predominantly covered in forest. Higher resolution satellite data (Landsat MSS and TM, SPOT) and field data from Ghana, Côte d'Ivoire and Nigeria were also used. Forest and non-forest data have been categorised into five vegetation types: forest (closed, defined as greater than 40 per cent canopy closure); fallow (mixed agriculture, clear-cut and degraded forest); savanna (includes open forests in the savanna zone and urban areas); mangrove and water. This dataset also portrays areas obscured by cloud. The 'forest' and 'mangrove' classification and cloud obscured areas have been mapped here. Montane, lowland rain forest, mangrove and swamp forest, as shown on Map 16.1, have been demarcated by overlaying White's vegetation map (1983) on to the UNEP/GEMS/GRID 'forest' and 'mangrove' categories.

Protected area spatial data are taken from two sources, namely a blue-line 1:1 million map (no title, nd) showing the delimitation of forest classes, which includes conservation areas, and a 1:800,000 scale map *Côte d'Ivoire* (1988) published by Michelin, Paris showing national parks and reserves.

17 Eastern Africa

SUDAN
Land area 2,376,000 sq. km
Population (mid-1990) 25.2 million
Population growth rate in 1990 2.9 per cent
Population projected to 2020 54.6 million
Gross national product per capita (1988) US$340
Closed broadleaved forest (end 1980)* 6,400 sq. km
Annual deforestation rate (1981–5)* 40 sq. km
Industrial roundwood production† 2,087,000 cu. m
Industrial roundwood exports† nd
Fuelwood and charcoal production† 20,112,000 cu. m
Processed wood production† 15,000 cu. m
Processed wood exports† nd

ETHIOPIA
Land area 1,101,000 sq. km
Population (mid-1990) 51.7 million
Population growth rate in 1990 2.0 per cent
Population projected to 2020 126 million
Gross national product per capita (1988) US$120
Rain forest (see Map 17.1) 47,256 sq. km
Closed broadleaved forest (end 1980)* 27,500 sq. km‡
Annual deforestation rate (1981–5)* 60 sq. km
Industrial roundwood production† 1,756,000 cu. m
Industrial roundwood exports† nd
Fuelwood and charcoal production† 37,884,000 cu. m
Processed wood production† 49,000 cu. m
Processed wood exports† nd

SOMALIA
Land area 627,340 sq. km
Population (mid-1990) 8.4 million
Population growth rate in 1990 3.1 per cent
Population projected to 2020 18.7 million
Gross national product per capita (1988) US$170
Closed broadleaved forest (end 1980)* 14,800 sq. km
Annual deforestation rate (1981–5)* 30 sq. km
Industrial roundwood production† 90,000 cu. m
Industrial roundwood exports† nd
Fuelwood and charcoal production† 6,896,000 cu. m
Processed wood production† 14,000 cu. m
Processed wood exports† nd

DJIBOUTI
Land area 23,180 sq. km
Population (mid-1990) 0.4 million
Population growth rate in 1990 3.0 per cent
Population projected to 2020 1 million
Gross national product per capita (1988) nd
Closed broadleaved forest (end 1980)* 10 sq. km
Annual deforestation rate (1981–5)* nd
Industrial roundwood production† nd
Industrial roundwood exports† nd
Fuelwood and charcoal production† nd
Processed wood production† nd
Processed wood exports† nd

KENYA
Land area 569,690 sq. km
Population (mid-1990) 24.6 million
Population growth rate in 1990 3.8 per cent
Population projected to 2020 60.5 million
Gross national product per capita (1988) US$360
Closed broadleaved forest (end 1980)* 6900 sq. km
Annual deforestation rate (1981–5)* 110 sq. km
Industrial roundwood production† 1,766,000 cu. m
Industrial roundwood exports† nd
Fuelwood and charcoal production† 33,884,000 cu. m
Processed wood production† 237,000 cu. m
Processed wood exports† 1000 cu. m

TANZANIA
Land area 886,040 sq. km
Population (mid-1990) 26 million
Population growth rate in 1990 3.7 per cent
Population projected to 2020 68.8 million
Gross national product per capita (1988) US$160
Closed broadleaved forest (end 1980)* 14,400 sq. km
Annual deforestation rate (1981–5)* 100 sq. km
Industrial roundwood production† 1,989,000 sq. km
Industrial roundwood exports† nd
Fuelwood and charcoal production† 31,114,000 cu. m
Processed wood production† 171,000 cu. m
Processed wood exports† 6000 cu. m

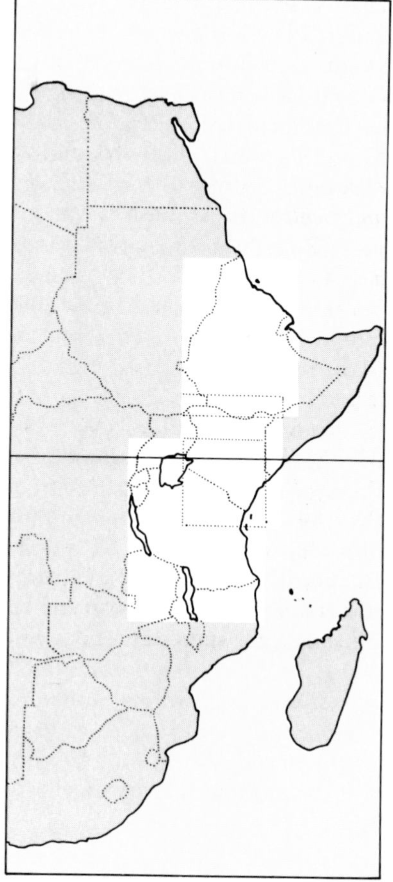

* FAO (1988a)
† 1989 data from FAO (1991)
‡ Figure includes coniferous and bamboo forest, estimated by FAO (1988a) to cover 16,000 sq. km in 1980.

The East African nations form a cohesive unit in discussions on moist forests as they share the problems of small, fragmented areas of forest under extreme pressure of encroachment and exploitation. Yet these forest patches, in most countries making up less than 2 per cent of the land area, have great significance for both water catchment and for the conservation of biological diversity. While East Africa has long been noted for its excellent network of protected areas, these were usually developed for savanna animals – not for forest biota. Despite high diversity and endemism among plants, birds, mammals and other taxa, few forest areas are included in the national parks network. Forest reserve status, while no longer allowing clear-felling, does permit heavy levels of exploitation and often cannot prevent encroachment. With the deteriorating economic situation there has been a reduction in management capability and it is likely that the biodiversity of East Africa's forests will be increasingly at risk in the future.

Introduction

The six East African nations of Sudan, Ethiopia, Somalia, Djibouti, Kenya and Tanzania share many similarities with respect to their forest cover. For instance, their very restricted closed canopy moist forests have all suffered severe deforestation in the past hundred years. The remaining forest patches are often isolated, biologically diverse and contain numerous local endemics, often of less mobile taxa. Nevertheless, the forests have surprising levels of similarity in structure and species composition. In addition, most of the forests are under intense pressure from high and growing populations, all of whom need land, fuel, poles and other resources. Lastly, forest products are not major export items in any of these countries.

Forests

In chapter 1 the limitations of the forest types considered within this Atlas are described. In East Africa it becomes particularly important to appreciate these limits, since there is a wide variety of forest and bushland types, some of which will not be included even though they may be dense, to the point of having a closed canopy. The maps in this chapter indicate lowland and montane rain forest, swamp forest and mangrove, but exclude those closed and open canopy formations that occur in drylands (such as *Acacia-Commiphora* woodlands).

Table 17.1 is a summary of the formations occurring in eastern Africa that lie within the scope of this Atlas, with some indication of their original extent. As may be seen, lowland rain forests occur only in Kenya – except for relict inland forest mosaics with grassland in Sudan, Kenya and Tanzania – and as coastal forest mosaics in Somalia, Kenya and Tanzania. Swamp forests occur only in Tanzania. Montane rain forests are extensive as mosaics with other vegetation in all countries except Djibouti, while mangrove is represented in all countries.

East African forests usually require at least 800 mm rainfall for full development, and do not grow above 3000 m altitude (Lind and Morrison, 1974). There is similarity in forest structure and, to some extent, species composition across geographically wide areas in the region. For example, the Imatongs of Sudan, Gambella of Ethiopia, Cheranganis of Kenya and Mt Meru of Tanzania have similar forest communities at corresponding altitudes. Local conditions of exposure, geology, soil depth, mist frequency and seasonality of rainfall all lead to a finer pattern of community differentiation than White (1983) was able to portray. No standardised detailed classification is available for eastern Africa and communities are best described by dominant species and environmental features.

The mangrove flora of East Africa is richer and totally different from that of West Africa. Nine tree species occur, all with wide Indo-Pacific ranges; all are found in Kenya and Tanzania, but only three reach the Red Sea coasts of Sudan and Ethiopia (White, 1983). Mangrove distribution is very fragmented, with concentrations at the mouths of larger rivers such as the Rufiji.

Forest Resources and Management

Forestry has long been a major land use activity in eastern Africa. In Kenya and Tanzania, the forestry services were established, many reserves gazetted and resources documented by 1905 (Rodgers, in press). Reservation was for maintaining timber value and, in many regions, for water catchment. Early forestry was not conscious of the need for biological diversity conservation, and the development of the national park concept focused almost exclusively on the large mammal faunas of the savanna woodlands and grasslands, not the forests.

However, forestry has changed in its attitude to natural forest resources. Until the 1960s much natural forest, after selective logging, was converted to large scale exotic monoculture plantations (*Pinus patula* and *Cupressus lusitanica* were common choices). This was due to the perceived low proportion of favoured indigenous species, such as *Ocotea*, and inadequate regeneration. However, all nations have now formally renounced such practices and the remaining forests are seen as providing a multiplicity of benefits: climate buffering, water, timber (from a much greater mix of species) and genetic resource conservation.

High rainfall forest land is also important in providing environmental services in support of agriculture. But forest lands can produce high income cash crops such as tea and coffee and support dense populations. In Kenya only 12 per cent of the land is suitable for rain-fed agriculture and one-quarter of this is forest reserve. The human population is largely dependent on agriculture (88 per cent) and is growing at 3.8 per cent per annum. The pressures on forest land are thus immense and increasing. This is mirrored throughout East Africa (Rodgers, in press; Hallsworth, 1982).

These pressures will, in the long term, materially affect the ability of the forest sector to maintain goods and services. Encroachment for agricultural land remains a major threat, as described by Kokwaro (1988) for Kakamega Forest in Kenya.

Table 17.1 Moist forest types and their distribution in Eastern Africa, with figures giving the estimated original extent of each type (in sq. km)

Forest type	*Sudan*	*Ethiopia*	*Somalia*	*Kenya*	*Tanzania*	*Djibouti*
Lowland dryland rain forests						
2. Guineo-Congolian rain forests: drier types				2,900		
11a. Guineo-Congolian mosaic of rain forest and grassland	22,800			11,500	12,700	
16. East African coastal mosaic:						
(a) Zanzibar Inhambane			18,200	28,400	97,700	
(b) forest patches				4,600	600	
17. Cultivation and secondary grassland replacing upland and montane forest		12,700				
Lowland wetland rain forests						
8. Swamp					800	
Montane rain forests						
19a. Afromontane vegetation	4,200	213,400	1,200	28,900	50,300	
65. Altimontane vegetation		23,200		1,800	1,200	
Mangrove						
77. Mangrove			1,800	3,100	5,300	300

(*Sources:* forest types: from White, 1983; forest extent data: MacKinnon and MacKinnon, 1986)

Garnett's bushbaby Galago garnettii, *is a nocturnal primate that is common in Diani forest on the coast of Kenya.* C. Harcourt

Slowly forestry departments are recognising that protection by force cannot maintain forest cover in the long run. In many areas protection measures have failed, largely as a result of economic problems. Tanzania, for example, no longer has the field capability to protect forests. This is reflected in the lack of convictions for forest offences. Governments, often assisted by international aid agencies, are attempting to develop sustainable land use practices in key forest areas so as to reduce pressure on forest resources. Projects usually involve a combination of community agroforestry techniques for fuelwood production, improved agricultural productivity, conservation awareness and increased forest protection.

Tanzania, Kenya and Ethiopia have either completed or have begun a Tropical Forestry Action Plan (TFAP) and a National Conservation Strategy (NCS). The intention of such plans has been to reinforce the long-term strategic importance of natural forests in agriculture, water and energy. The TFAP views commercial use as a secondary activity wherever there are environmental concerns, and calls for industrial wood needs to be met from plantations on non-forest land. Such programmes have, however, met with mixed success. The Tanzanian TFAP forest sector review, for example, contrasts markedly with the Kenya review in its detailed coverage of environmental issues.

Detailed planning is still inadequate and few natural forests have working or management plans. Mapping is incomplete, resource inventories and documentation usually poor. There is a need for much greater investment in the natural forest if it is to survive and continue to provide resources in the face of growing land pressure.

Surveys of forest status can be found in FAO/UNEP (1981) and are further summarised in FAO (1988a). The World Bank has provided status reports on the forestry sector for many countries, although these do not always comment on natural forest or biodiversity. MacKinnon and MacKinnon (1986) provided a comprehensive review of conservation needs for protected areas of Africa based on White's vegetation types. Descriptions of existing protected areas – regrettably, these are rarely forests – are found in IUCN (1987a, b). Patterns of biodiversity are detailed in Stuart *et al.* (1990).

SUDAN

Rain forests in Sudan are restricted to forest and grassland mosaics on the Imatong, Dongotona and Didinga mountains on the southern border of the country, and to the wetter south-western border with Zaïre (see Figure 17.1; it has not been possible to obtain a recent, accurate vegetation map of Sudan showing 'actual' closed moist forest). Mangroves occur only in small and scattered patches on the Rad Sea coast, the principal species being *Avicennia marina*, *Rhizophora* and *Bruguiera*.

The Southern Mountains are in three main blocks, Imatong, Dongotona and Didinga, comprising a total mountain area of 60,000 sq. km. The highest point is 3187 m at Kinyeti on the Imatongs, Mt Lotuke is 2963 m on Didinga and Dongtona reaches 2800 m (Jackson, 1956). Receiving an annual rainfall of 1500 mm the mountains have great catchment value and support cash crops such as tea, a growing agricultural population, and both natural forest and plantation logging industries. The area above 2500 m has never been cultivated (Jenkins *et al.*, 1977). The Imatongs are by far the largest mountains, and Jenkins *et al.* (1977) described lowland, lower montane and upper montane forest zones on them. The lowland forests are scattered patches of *Milicia*, *Khaya* and *Albizia* trees, while the lower montane forests are more diverse with *Croton*, *Olea*, *Ocotea*, *Podocarpus*, *Prunus*, *Syzygium* and *Teclea*. The upper montane forests have *Hagenia*, *Maesa* and *Podocarpus* species merging into *Gnidia-Hypericum* heath. Upper montane forest is found on Didinga, which has only tiny patches of this type, while the Dongotona massif has a 17 sq. km relict of it.

Tropical rain forest (the drier type of Guineo-Congolian rain forest) exists in small patches in the west and south of Sudan, e.g. around Talanga, Lott, Lamboni at the foot of the Imatongs, Azza in Meridi District and Yambio on the Aloma plateau. Annual rainfall exceeds 1300 mm and emergent trees can reach 50 m. *Celtis*, *Chrysophyllum*

Figure 17.1 Approximate ecological zones of Sudan. Rain forests are restricted to the south and south-west (*Source:* Hillman, 1985)

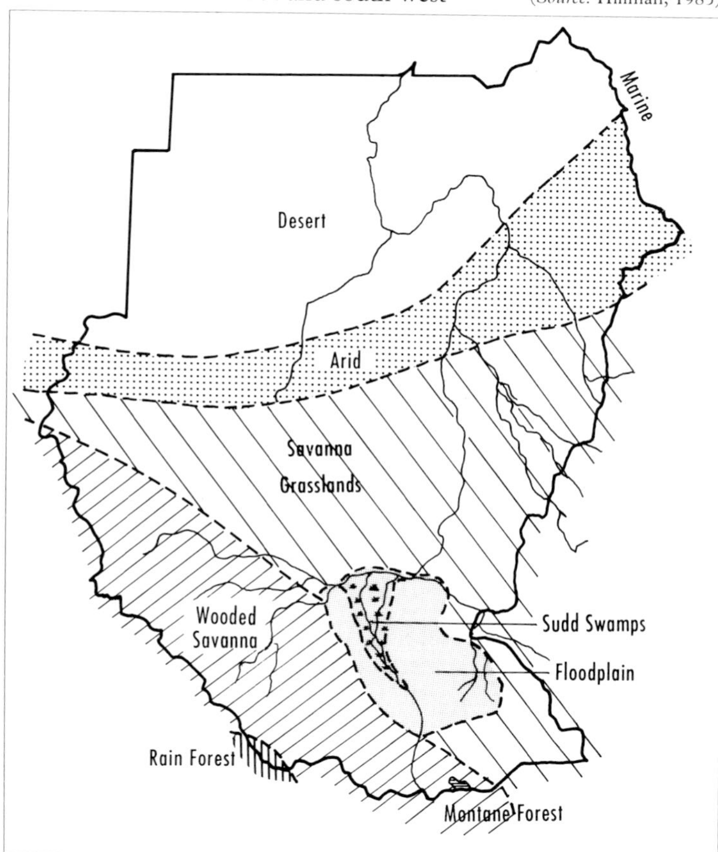

Table 17.2 Major forest protected areas of eastern Africa

SUDAN

Existing Protected Areas	*Size (sq. km)*	*Protected Area Formally Proposed by Government*	*Size (sq. km)*
Bengangai GR	170	Imatong Mountains NCA	1,000
Bire Kpatuos GR	5	Jebel Marra Massif NCA	1,500

Further Requirements: Red Sea Hills; upgrade forest GRs and NCAs to full NP

ETHIOPIA

Bale Mountains NP	2,471		
Gambella NP	5,061		
Simen Mountains NP	Small forest area		
Nechisar NP	Small forest area		

Further Requirements: Upgrade Gambella GR to NP; Illubabor forests (lower forests); Gara Ades Mount (dry montane forest); Lakes Asaita and Gargori (riverine forest); Belleta Sai and Gore to Tepi forests (for coffee and for upland forest); Tiro Boter (medium altitude forest); Neghelli-Arero forest patches

SOMALIA AND DJIBOUTI

Forêt du Day NP	100	Daalo Forest NP	2,510
		Gaan Libaah NP	500

Further Requirements: Libaah Xeela Helleh as a national park; Ahl Mescat Mts and Wagger Mts as a reserve; Juba and Scebeli rivers (riparian forest); Jowhar-Warshek (Mogadishu GR)

KENYA

Aberdare NP	766 (part forest)	Marsabit NP	360
Mt Elgon NP	169 (part forest)	Diani Marine NP Complex (may have forest)	
Mt Kenya NP	715 (part forest)		
Kakamega NP†	45 (part forest)		
Nairobi NP	117 (part forest)		
Arabuko-Sokoke NR	43		
South-Western Mau NR	430		
Nandi North NR	34		
Boni National NaR	1,339 (little forest)		
Marsabit NaR	21 (little forest)		
Masai Mara NaR	1,510 (little forest)		
Shimba Hills NaR	193 (little forest)		
Tana River Primate NaR	169 (little forest)		

Further Requirements: Nature reserve or park status for: part of Cheranganis, much more of Kakamega, part of Ngurimans, Taita Hills, more forest for Mt Kenya

TANZANIA*

Arusha NP	<50	Uzungwa NP	500
Gombe NP	<10		
Kilimanjaro NP	<20		
Mahale Mountain NP	<200		
Rubondo	180		
Ruhaha	20		
Mt Meru	1000		
Mikumi NP	<10		
Ngorongoro CA	<200		
Selous GR	<100 (+1000 thicket)		

Further Requirements: Itigi Thicket; Southern Highlands; Minziro; Uluguru Mts; E and W Usambara Mts; Rufiji Delta mangroves

† Exists as a NaR, proposed as a NP.
* Sizes given here are estimates of closed forest extent within existing and proposed protected areas

GR: Game Reserve
NP: National Park
NR: Nature Reserve
NaR: National Reserve
CA: Conservation Area
NCA: Nature Conservation Area.

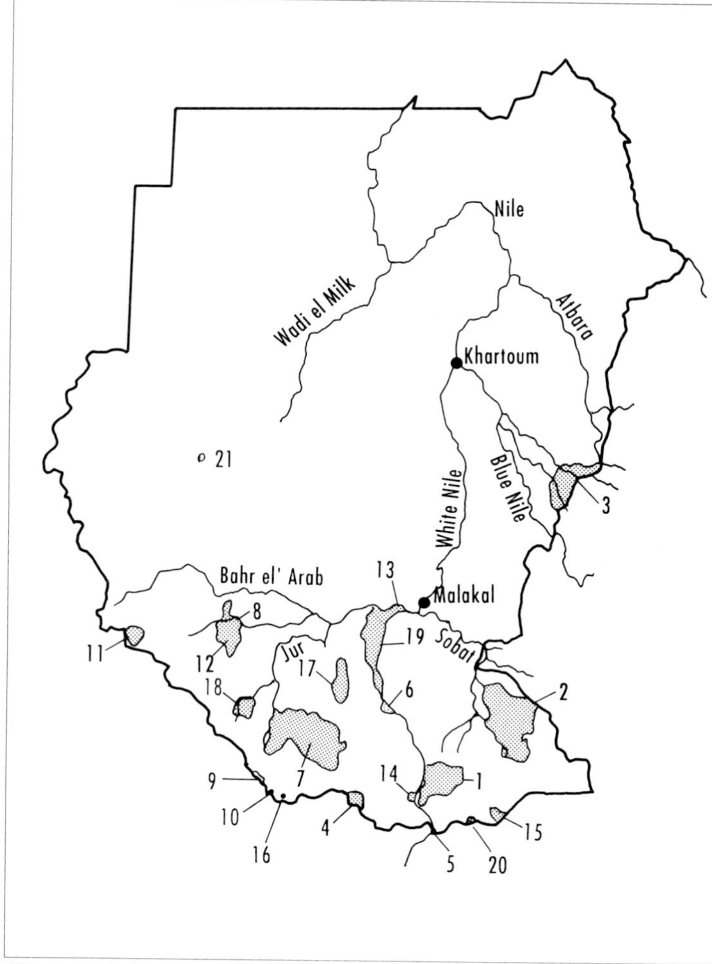

Figure 17.2 Conservation areas of Sudan

(*Source:* Hillman, 1985)

albidum, *Erythrophleum*, *Entandrophragma angolense*, *Holoptelea grandis*, *Khaya*, *Maesopsis eminii* and *Milicia* are typical canopy species and genera.

There are no recent reliable estimates of forest cover in Sudan. The World Bank (1986) gives an overall forest area estimate of 940,000 sq. km, with 16,200 sq. km of tropical high forest, all in south Equatoria Province. FAO (1988a) estimates a total of 6400 sq. km of closed forest, based on data in FAO/UNEP (1981) from a survey in the 1970s. Persson (1975) quoted figures of 12,000 sq. km of gazetted forest reserve, of which 3000 sq. km was in high forest. Imatong Central Forest Reserve alone is 1032 sq. km (Jenkins *et al.*, 1977). Deforestation is said to be rapid, estimated at 40 sq. km per annum for the years 1981–5 (FAO, 1988a). The World Bank (1986) stressed the severe consequences of deforestation for agriculture.

Biodiversity

There is little information on the biodiversity of Sudan's forest. Brenan's (1978) review of plant diversity and endemism in Africa, suggests overall diversity is low: about 3200 species of which fewer than 50 (that is, less than 1.5 per cent of the total) are endemic. These are mainly dryland species and include 11 endemics of the Jebel Marra massif (Wickens, 1976), an example of which is *Kickxia dibolophylla*. The Imatong Mountains are recognised as being of outstanding biological importance, on account of their geographical position and the variety of plant and animal life in the forest and non-forest habitats. A subspecies of spotted ground-thrush *Turdus fischeri maxis* was recently named from the Lotti Forest on the south-west Imatongs (Collar and Stuart, 1985).

Table 17.3 Conservation areas of Sudan

Existing and proposed areas are listed below. Marine national parks and bird sanctuaries are not listed. For data on Biosphere reserves see chapter 9. For locations see Figure 17.2.

	Existing area (sq. km)	*Proposed area (sq. km)*	*Number*
National Parks			
Bandingilo	16,500		1
Boma	22,800		2
Dinder	8,900		3
Lantoto		760	4
Nimule	410		5
Radom†	12,500		
Shambe	620		6
Southern	23,000		7
Wildlife Sanctuaries			
Arkawit†	820		
Arkawit-Sinkat†	120		
Khartoum Sunt Forest†	15		
Game Reserves			
Abroch†		nd	
Ashana	900		8
Bengangai	170		9
Bire Kpatuos	5		10
Boro		1,500	11
Chelkou	5,500		12
Fanyikango Island	480		13
Juba	200		14
Kidepo	1,200		15
Machar†		nd	
Mbarizunga	10		16
Meshra		4,500	17
Mongalla†	75		
Numatina	2,100		18
Rahad†	3,500		
Sabaloka†	1,160		
Tokar†	6,300		
Wadi Howar†		nd	
Zeraf	9,700		19
Nature Conservation Areas			
Imatong Mountains		1,000	20
Jebel Elba†		4,800	
Jebel Marra Massif		1,500	21
Lake Ambadi†		1,500	
Lake No†		nd	
Totals	116,985	15,560	

(*Source:* IUCN, 1990b; WCMC *in litt.*)

† Not mapped – no location data available to this project.
Areas with moist forest are listed in Table 17.2.

Conservation Areas

While Sudan already has a large area gazetted as national parks, wildlife sanctuaries and game reserves (see Figure 17.2), with others proposed, there is little protection for the forests. The Imatongs and other southern mountain forests are forest reserves and so still open to legal exploitation. Their gazettement as full national parks is a major priority (MacKinnon and MacKinnon, 1986). The Red Sea Hills and Jebel Marra also need protection.

Two game reserves, Bengangai and Bire Kpatuos totalling 175 sq. km, protect mammals in the south-western lowland forest, but no information is available on the effectiveness of this protection for the total forest ecosystem. Details are given in Tables 17.2 and 17.3.

ETHIOPIA

Ethiopia has extensive and varied montane rain forests (Table 17.1, Map 17.1). Nine distinct associations each with several constituent communities, are recognised (Friis and Tadesse, 1988). Humid, mixed forests occur in southern Ethiopia and Harerge province, with genera of the species *Podocarpus*, *Croton*, *Olea* and *Schefflera*, and *Hagenia* at higher altitudes. In the south-west, broadleaved forests with Ugandan affinities are found. *Aningeria adolfi-friedericii* is the main emergent, reaching a height of 40 m. In addition, some riparian forests occur along water courses in the lowlands. Small areas of mangrove (mainly *Avicennia*) occur in shallow bays on the Eritrea coast.

Historical estimates suggest some 87 per cent of the Ethiopian highlands had forest and woodland cover, but this was reduced to 40 per cent by 1950 and just 5.6 per cent by 1980 (IUCN, 1990a). Forest in the entire country declined from an original cover of 35 per cent to 16 per cent by 1952, 3.6 per cent by 1980, 2.7 per cent by 1987 and an estimated 2.4 per cent in 1990. Deforestation rates estimated from Pohjonen and Pukkala (1990) are shown in Table 17.4 but these figures are for woodland and thicket as well as 'forest' as defined in this Atlas. This forest loss and the subsequent marginal agriculture have led to some 270,000 sq. km of the 540,000 sq. km plateau area having moderate or serious erosion, causing a soil loss of 1.5 billion tonnes per annum.

Forests are destroyed for fuelwood, and an estimated 10,000 sq. km of fast growing plantations are needed to prevent continued loss of natural forests. Today there are 3100 sq. km of plantation, but some of these contain slow growing species such as *Cupressus lusitanica*, and many are poorly stocked. If villagers were to change from cow dung fuel to wood, and so restore soil fertility, then some 20,000–30,000 sq. km of plantations would be needed. If the population reaches 70 million people by the year 2000 (almost 52 million today and 2 per cent growth) then 30,000–40,000 sq. km would be needed. Continuous deforestation, therefore, seems inevitable (Pohjonen and Pukkala, 1990).

Map 17.1 shows 47,256 sq. km of closed canopy forest, all of it montane (for sources see legend for Map 17.1 on p. 160). In comparison with the estimates made by FAO (1988a) this figure is generous and is taken to include not only broadleaved forest (the FAO figure is 27,500 sq. km) but also coniferous forest (FAO 1980, 8000 sq. km) and bamboo. Most of the estimated 27,500 sq. km of closed broadleaved forest is in the south-west, and 60 per cent is in the Illubabor Administrative Region. This area has some 140 million cu. m of standing timber of which 50–60 per cent was considered to be merchantable species (Chaffey, 1979). Wood use is estimated at 24 million cu. m per annum, which is 60 per cent more than the sustainable cut of 15 million cu. m which could be achieved with good management; therefore there is a loss of 1600–2000 sq. km of forest and woodland per annum (IUCN, 1990a).

Forestry and wildlife matters are within the State Forest Conservation and Development Department in the Ministry of Agriculture. Traditionally much of the forest was not on government land and so there has not been a long history of control and management planning. The preliminary Tropical Forestry Action Plan for Ethiopia (FAO, 1988b) identified 37 priority forest areas totalling 36,000 sq. km, of which 16,000 sq. km are in the Illubabor Region. All were in urgent need of protection and management.

Table 17.4 Deforestation rates for the Ethiopian Highlands

Year	*Forest area (sq. km)*	*Deforestation Loss rate per annum (sq. km)*
1900	530,000	–
1950	210,000	6,400
1965	90,000	8,000
1985	35,000	2,750

(*Source:* Pohjonen and Pukkala, 1990)

Table 17.5 Conservation areas of Ethiopia

Conservation areas are listed below. Marine national parks and controlled hunting areas are not included or mapped. For data on World Heritage sites see chapter 9.

	Area (sq. km)
National Parks	
Abijatta-Shalla Lakes	887
Awash[1]	756
Bale Mountains*	2,471
Gambella	5,061
Mago	2,162
Nechisar	514
Omo*	4,068
Simen Mountains[1]	179
Yangudi Rassa	4,731
Wildlife Reserves	
Alledeghi	1,832
Awash West	1,781
Bale*	1,766
Chew Bahr	4,212
Gash Setit	709
Gewane	2,439
Mille Sardo	8,766
Nakfa	1,639
Shire	753
Tama*	3,269
Yob	2,658
Sanctuaries	
Babile Elephant	6,982
Senkelle Swayne's Hartebeest	54
Yavello	2,537
Totals	60,226

(*Sources:* IUCN, 1990b; WCMC, *in litt.*)

* Area with moist forest within its boundaries according to Map 17.1.
[1] Awash and Simen are the only two legally gazetted national parks. The remainder are 'managed' as national parks but are not yet gazetted.

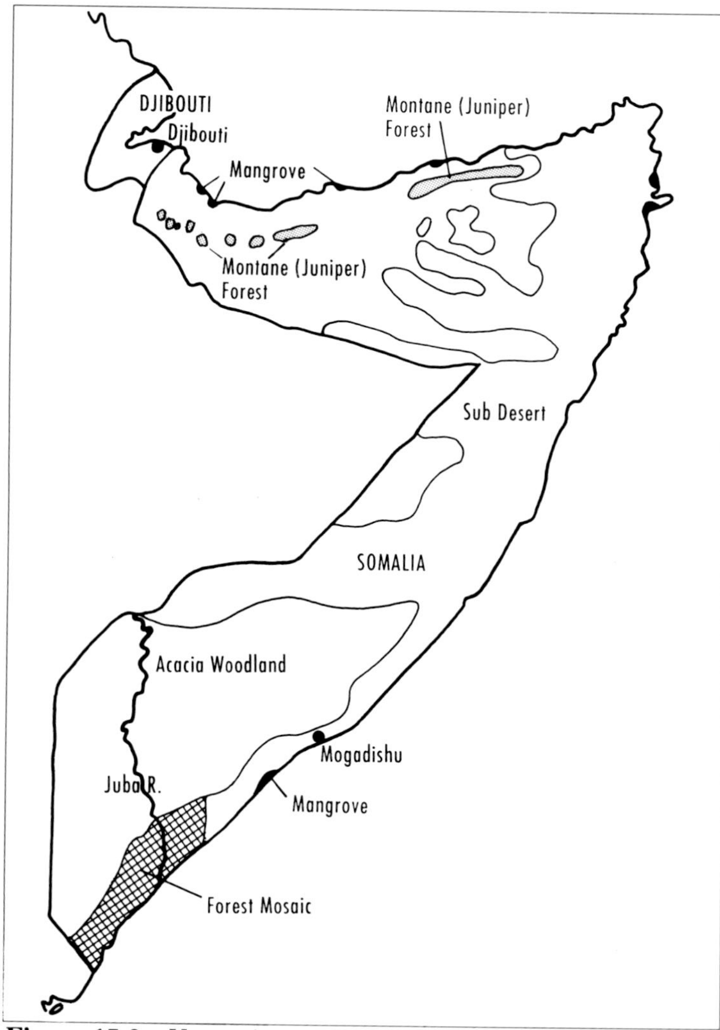

Figure 17.3 Vegetation types of Somalia. Moist forests include the montane forests in the north, the coastal mosaic in the south and coastal mangrove patches

(Map compiled from various sources on file at WCMC. No map of actual forest extent is available.)

Biodiversity

IUCN (1989) reviews overall biodiversity levels in Ethiopia, which are relatively high in many animal groups and plants. Hamilton (1989) considers Ethiopia to be a minor core area for endemism and biodiversity, while Brenan (1978) suggested plant endemism, based on a part sample of the flora, to be over 20 per cent. More recent work, for instance by Friis (1982), puts the figure between 10 and 20 per cent. One major centre of endemism is the Ogaden desert, with several endemic genera, but there are also four recognisable centres of endemism in the high mountain regions. Egziabher (1990) discusses the importance of the south-western forests of Tepi, Didessa, Gore and Harenna for conserving rapidly dwindling wild gene-pools of *Coffea arabica*.

Of the mammals, 22 of Ethiopia's 242 species (9 per cent) are endemic, though few are forest forms. Twenty-seven of 847 bird species are endemic (3.2 per cent) including several forest species, the highest level of any mainland country of Africa. One such species is Prince Ruspoli's turaco *Tauraco ruspolii*, which is endemic to forest patches near Neghelli (Collar and Stuart, 1985). Several reptile and amphibian species, especially tree-frogs, are localised endemics in different forest blocks.

Conservation Areas

Table 17.2 lists four national parks with closed forest; three of them, Bale, Gambella and Nechisar, do not have full legal protection. Consolidation of Gambella National Park and creation of further protected areas in the Illubabor and Gore to Tepi forests must be the highest forest conservation priorities in Ethiopia. Beals (1968) discussed conservation needs in detail, while proposed protected areas are summarised in MacKinnon and MacKinnon (1986) and Table 17.5 lists existing and proposed conservation areas in the country.

SOMALIA AND DJIBOUTI

Somalia has coastal mosaic forest in the extreme south, small areas of montane forest in the northern hills and some coastal mangrove patches (see Figure 17.3). The montane forests are in ranges rising to 2500 m, with evergreen thicket (mainly *Buxus hildebrandtii*) and relict patches of juniper forest above 1300 m. Deforestation has been severe and the vegetation is grossly overgrazed. In the south, the northernmost limits of White's (1983) coastal mosaic forests occur in the Kismayo depression, known as the Holawajir Forest (Douthwaite, 1987). Riparian forests on the Juba and Scebeli rivers are of importance but are rapidly being felled.

There are no recent reliable estimates of closed forest cover or deforestation rates. For Somalia FAO (1988a) estimates 14,800 sq. km of closed forest, although much of this is not forest in the sense used here. Forest reserves total 4000 sq. km, but a good deal of this area is believed to be scrub and woodland.

Biodiversity

The northern forests do have some bird and plant species communities of interest. Collar and Stuart (1988) and IUCN (1987b) mention the Day Forest of the Goda Mountains (1983 m above sea level) in Djibouti, which is ostensibly a national park, but little managed. This has the endemic Djibouti francolin *Francolinus ochropectus* and near-endemic palm *Livistona carinensis*; both are very rare. These documents stress the value of Daalo (Daloh) Forest Reserve, which is in good *Juniperus-Olea* forest in north Somalia. It has an endemic bird, the Warsangali linnet *Carduelis johannis* and an isolated leopard *Panthera pardus* population.

Conservation Areas

Little attention has been paid to forest relicts in Somalia; indeed, Madgwick (1989) is exceptional in detailing the decline and destruction of riverine forests. Table 17.2 lists two proposed parks in the northern mountains, Gaan Libaa and Daalo, and suggests that three more, plus patches of riparian forest, are needed to protect forest values. The Mogadishu Game Reserve, which should have been included in the proposed national park of Jowhar-Warshek (IUCN, 1987b), reputedly no longer exists. Parker (1987) describes patches of relatively undisturbed forest on the Juba River at Shoonto and Barako, but Arbowerow forests on the lower Scebeli River have disappeared. All were suggested for strict nature reserve status by Sale (1989). The Forêt du Day National Park of Djibouti has little protection, and as a consequence forest cover is degrading rapidly. Table 17.6 lists proposed and existing conservation areas in Somalia and Djibouti: Figure 17.4 shows those protected areas for which data were available.

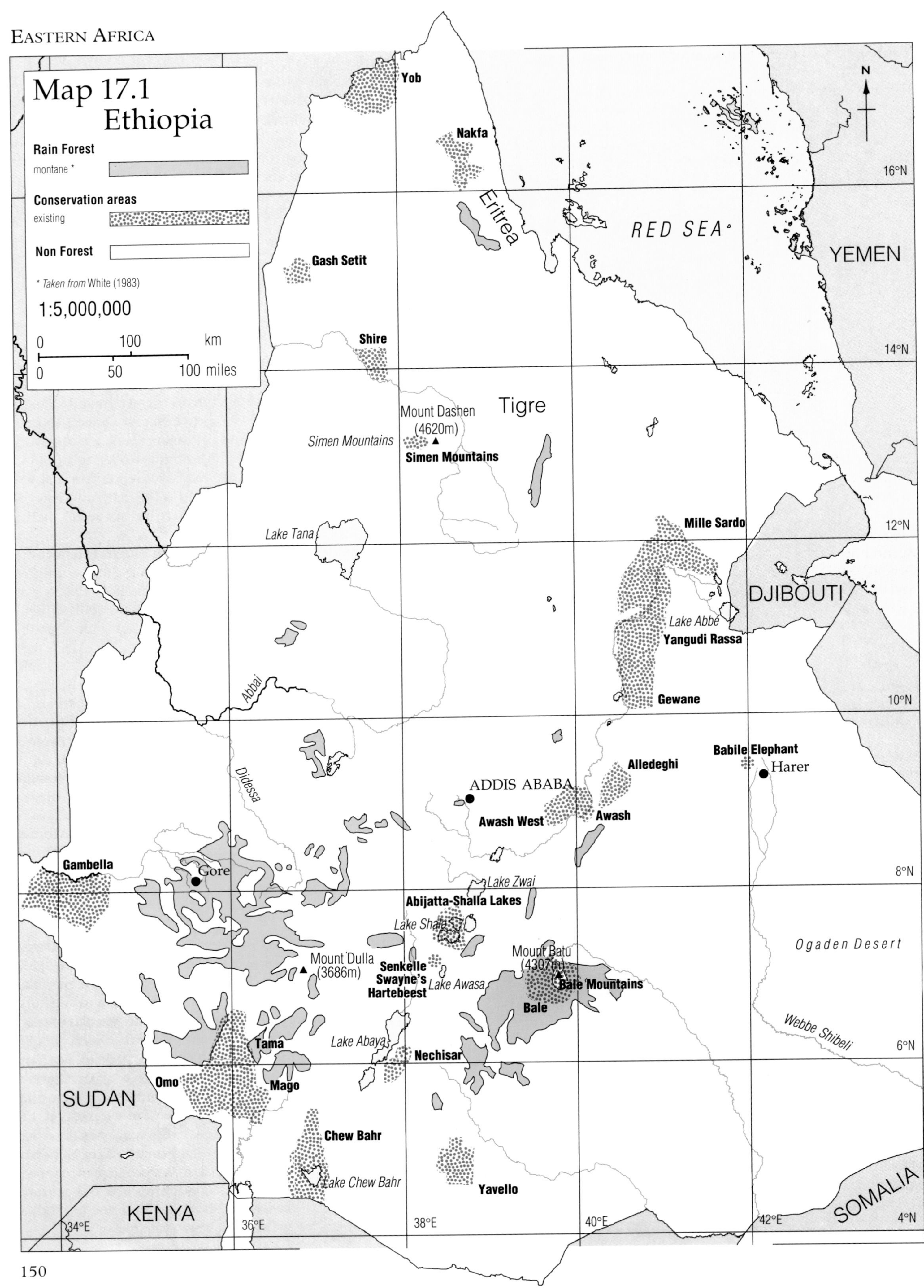
Map 17.1
Ethiopia
Rain Forest
montane *
Conservation areas
existing
Non Forest
* Taken from White (1983)
1:5,000,000
0 100 km
0 50 100 miles
N
16°N
14°N
12°N
10°N
8°N
6°N
4°N
34°E
36°E
38°E
40°E
42°E
Yob
Nakfa
Eritrea
RED SEA
YEMEN
Gash Setit
Shire
Mount Dashen
(4620m)
Tigre
Simen Mountains
Simen Mountains
Lake Tana
Mille Sardo
DJIBOUTI
Lake Abbe
Yangudi Rassa
Abbai
Gewane
Didessa
Babile Elephant
Alledeghi
Harer
ADDIS ABABA
Awash West
Awash
Gambella
Gore
Lake Zwai
Abijatta-Shalla Lakes
Lake Shala
Mount Dulla
(3686m)
Senkelle
Swayne's
Hartebeest
Lake Awasa
Mount Batu
(4307m)
Bale Mountains
Bale
Ogaden Desert
Webbe Shibeli
Tama
Lake Abaya
Nechisar
Omo
Mago
SUDAN
Chew Bahr
Lake Chew Bahr
Yavello
KENYA
SOMALIA

Table 17.6 Conservation areas of Somalia and Djibouti

Existing and proposed conservation areas are listed below. Controlled hunting areas are not recorded. For locations see Figure 17.4.

SOMALIA	*Existing area (sq. km)*	*Proposed area (sq. km)*	*Number*
National Parks			
Angole-Farbiddu		nd	1
Awdhegle-Gandershe		800	2
Daalo Forest		2,510	3
Gaan Libaah		500	4
Gezira Lagoon†		50	
Har Yiblane†		nd	
Jowhar-Warshek		2,200	5
Lag Badana-Bushbush		3,340	6
Lag Dere		5,000	7
Las Anod-Taleh-El Chebet		8,000	8
Rus Guba		nd	9
Nature Reserves			
Alifuuto (Arbowerow)		1,800	10
Balcad†		2	
Game Reserves			
Bushbush†	3,340		
Geedkabehleh	104		11
Mandera†	nd		
Mogadishu	nd		12
Partial Game Reserves			
Belet Wein	nd		13
Bulo Burti†	nd		
Jowhar†	nd		
Oddur†	nd		
Wildlife Reserves			
Boja Swamps		1,100	14
Eji-Oobale		nd	15
El Hammure		4,000	16
Far Libah†		nd	
Far Wamo		1,400	17
Haradere-Awale Rugno		2,500	18
Harqan Dalandoole		8,000	19
Hobyo		2,500	20
Qurajo†		nd	
Ras Hajun		nd	21
Zeila		4,000	22
Totals	5,246	45,900	

DJIBOUTI	*Existing area (sq. km)*	*Proposed area (sq. km)*	*Number*
National Parks			
Forêt du Day	100		23
Integral Reserves			
Maskali Sud†	nd		
Parks			
Musha Territorial†	nd		
Total	100		

(*Source:* IUCN, 1990b)

* Area with moist forest are listed in Table 17.2.
† Not mapped – no spatial data available to this project.
nd no data

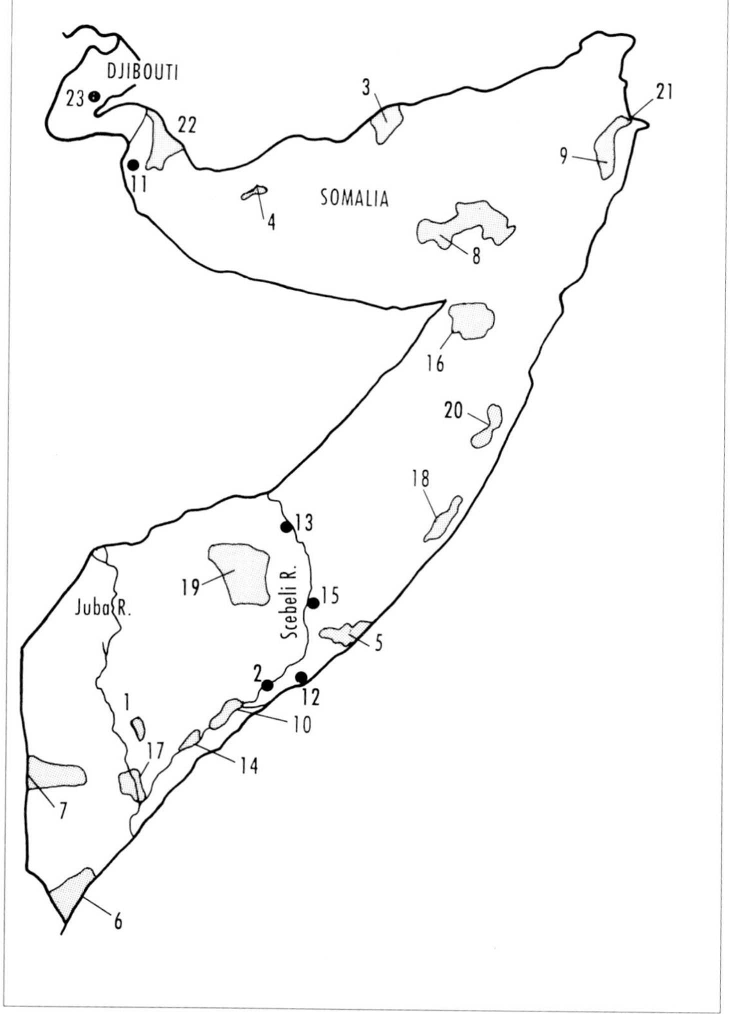

Figure 17.4 Conservation areas (existing and proposed) of Somalia and Djibouti

(*Sources:* Kingdon, 1990; Stuart *et al.*, 1990)

KENYA

Kenya has the most diverse forests in East Africa, with lowland rain forest in western Kenya, montane forest in the central and western highlands and on higher hills and mountains along the southern border. In addition, there are some coastal mosaic forests and fairly extensive mangroves, particularly at Lamu and at the mouth of the Tana River.

Despite the long history of managed forestry in Kenya there has been little published synthesis of the variety of forest vegetation types until recently. The major forest blocks of Mt Kenya, the Aberdares and Mau have sparse descriptions of vegetation and floristics. Lind and Morrison (1974) give a general account of East African forests including those of Kenya; Hamilton and Perrott (1981) describe Mt Elgon in detail; Coetzee (1967) and Zamierowski (1975) give an introduction to Mt Kenya; Beentje (1990) provides an overview, recognising 21 forest types (excluding mangroves).

Map 17.2 shows closed forest to be widely but sparsely distributed across most of Kenya except the north-east. Forests can be summarised in six main blocks:

1 *The volcanic mountains*: Elgon, Kenya and associated high ranges of the Aberdares, Cherangani, Mau;
2 *The western plateau*: Kabarnet, Kakamega, Nandi, Trans-Mara;
3 *The northern mountains*: Ndotos, Mathews, Leroghi, Kulal, Marsabit;
4 *The coastal forests*: Arabuko-Sokoke, Tana, Kayas, coral rag;
5 *The southern hills*: Taita, Taveta and Shimba, Nguruman, and
6 *The riverine forests*: Tana and tributaries, Ewaso-ngiso, Kerio, Turkwell.

The most widespread montane associations are the moist *Ocotea-Polyscias* and drier *Podocarpus-Cassipourea* forests. *Juniperus-Olea* dominate upper slopes, while the lowland forests are extremely diverse.

All forests are under pressure, being the only lands suitable for expansion of rain-fed agriculture (Young, 1984; Polhill, 1988; Beentje, 1990). The World Bank (1988) recognises the central problem of watershed protection, the failure of which threatens agriculture and hydroelectric schemes and, consequently, the national economy.

Until a Presidential Directive in 1984 reversed the practice, indigenous forest was cut for plantations of exotic species. Kenya's forests are being over-exploited by excessive legal harvesting as well as by illegal pitsawyers (Young, 1984; World Bank, 1988). The Mau forest block, which is the largest single block of forest in East and Central Africa with some 2440 sq. km, has lost 30 per cent of its forest cover to illegal encroachment in recent years.

Kokwaro (1988), describing in detail the biologically important Kakamega Forest, reports a decrease from 238 sq. km to 100 sq. km in the past 30 years due to serious encroachment and continued overuse for timber, charcoal, firewood, cash crops and forest plantation. The adjacent Nandi Reserve decreased by 7 per cent in the 1970s. The impact of a further Presidential Directive banning the cutting of natural trees, in 1988, remains to be seen.

Estimates of closed forest vary, depending on definition. FAO's 1980 data describe 6900 sq. km of broadleaved closed canopy forest, 2500 sq. km of coniferous forest and 1650 sq. km of bamboo, a total of 11,050 sq. km, or some 2 per cent of the country. Beentje's estimate of evergreen forest, as opposed to deciduous open forest or woodland, is 5856 sq. km (Beentje, 1990), virtually all of which is in Kenya's 206 gazetted forest reserves. These reserves total 17,000 sq. km, but not all are forest or even woodland, and some are plantation. Part of this closed forest resource is further protected as nature reserves (526 sq. km), national reserves or national parks (less than 180 sq. km). The country's protected forest areas are listed in Table 17.2.

KAYA FORESTS OF THE KENYAN COAST

The Kaya forests of the Kenyan coast are relict patches of the once extensive and diverse Zanzibar-Inhambane lowland forests of eastern Africa. The word 'Kaya' means a homestead in several Bantu languages and historically these forest patches sheltered fortified villages or Kayas which were set up by the Mijikenda people who were fleeing from enemy groups in the north. During the last century, the villages moved outside the forest patches and Kayas have come to mean the forest patches which have survived and been protected by the traditions and customs of elders who used the old Kaya clearings for ceremonies. Over the past few decades an increasing disregard for traditional values and a decline in respect for the elders has led to damage to these small forests and associated sacred groves.

All areas of unprotected forest and woodland in coastal Kenya are under extreme threat because of the rising population and an ever increasing need for more land on which to grow food and the demand for more building poles and fuelwood. The rapidly expanding tourist industry has also given rise to a demand for wood for hotel construction, furniture and carvings. This is met partly by tree poaching from gazetted forest reserves, which are inadequately protected by the overstretched and underfunded Forestry Department, and partly by unlicensed tree cutting in ungazetted areas. There are also serious threats from developing commercial activities such as lead ore mining at Kaya Kauma, marble quarrying at Pangani and Kambe, a proposed lime factory at Pangani and the planned reopening of the rare earth mines on Mrima Hill.

The Kayas and groves are often on hill tops protecting water catchments and they are also important as representative forest remnants supporting a diverse flora and often containing many rare plants. Those on Jurassic limestone outcrops are particularly interesting: in one or two of these the endemic African violet *Saintpaulia rupicola* is just surviving.

There has been increasing national and international concern over these forests, particularly as it became known that the village elders were worried over their inability to care for their sacred places and sources of medicinal plants. The National Museums of Kenya (NMK) and WWF have supported studies of the Kayas. A survey carried out in 1986/7 listed about 35 Kayas and important sacred groves, the largest being no more than 1.5 sq. km, and information on a few more has been obtained since then. They occur in the two southern coastal districts, Kilifi and Kwale, scattered in the 30–40 km wide coastal strip from just north of the Sabaki River and Malindi to the border with Tanzania.

After some debate over whether the Kayas should be protected as National Monuments under NMK, involving the elders and the local community in their protection and rehabilitation, or whether they should be forest reserves, it was decided that the former is more appropriate. A few of the Kayas in Kwale District occur in existing forest reserves and are thus theoretically protected, although legal and illegal tree cutting takes place. As a result, the elders are denied their traditional use of these Kayas. NMK will now be involved with the protection of these Kayas in Kwale District and donor funding is being sought to support this challenge in conservation. *Source:* Anne Robertson

The area around Mau forest in Kenya is being cleared to make a buffer zone of tea plantations. It is hoped that this will stop further encroachment into the forest, seen on the left side of this photograph in the distance. C. Harcourt

The forests depicted on Map 17.2 cover approximately 19,141 sq. km. The main source is a 1983 land-use map produced by the Kenya Rangeland Ecological Monitoring Unit. A degree of uncertainty was introduced by the difficulty in identifying plantation forests from the Landsat imagery used (Young, 1984). The map and statistics are generally considered to over-estimate the present extent of natural moist forest. Although the vegetation category *Dense Natural Forest* from the source map has been mapped on Map 17.2, more open formations may also be included in this category. FAO estimated a total for closed and open broadleaved forest to be 19,450 sq. km in 1980.

Biodiversity

Many of Kenya's largest forested mountain blocks are of recent volcanic origin and hence are relatively species poor (Rodgers and Homewood, 1982). Highest diversities are in the coastal forests, the western plateau forests such as Kakamega (Kokwaro, 1988) and especially the tiny, geologically older mountains at the northern end of the Eastern Arc of block mountains – the Taita and Taveta Hills (Beentje, 1988a). Table 17.7 summarises Kenya's biological diversity.

The figures for numbers of species and endemism are very much lower than those for Tanzania, which has the bulk of the Eastern Arc forests (Lovett, 1988). Details of plant endemism are given by Brenan (1978) who highlights the importance of the coastal forests. Beentje (1988b) analysed the distribution of rare trees in Kenya. The distribution and proportion of endemics are shown in Table 17.8 which again stresses the value of coastal and Eastern Arc forests for biodiversity.

The Taita Hills have less than 3 sq. km of closed forest, rising to 2200 m asl, but have 13 totally endemic plant species including four woody plants and one of the African violets, *Saintpaulia teitensis*. There are three butterfly endemics; several reptiles, including *Amblyodipsas teitana*; some amphibia, e.g. *Afrocaecilia teitana*; and three or four bird species (depending on taxonomy), of which the Taita thrush *Turdus helleri* is one. The area has been documented recently (Beentje, 1988a) and is desperately in need of strong conservation programmes.

Kakamega Forest is considered to be the easternmost outlier of the Guinea-Congolian forest, and has many species not found elsewhere in Kenya including L'Hoest's monkey *Cercopithecus lhoesti*. Sixty-two birds are restricted to this area in Kenya. Turner's eremomela *Eremomela turneri* and Chapin's flycatcher *Muscicapa lendu* are two globally threatened species (Collar and Stuart, 1985, 1988) which occur there. However, floristic diversity and endemism are relatively low.

The coastal forests are diverse and not easy to classify. There are several endemics, and birds, mammals and plants all have thorough studies devoted to them. Collar and Stuart (1988) consider Sokoke the second most important forest for birds on continental Africa. Two bird species are endemic: the Sokoke scops owl *Otus ireneae* and Clarke's weaver *Ploceus golandi* and six species are rare or threatened. The golden-rumped elephant-shrew *Rhynchocyon chrysopygus* is an almost endemic small mammal.

Other significant coastal areas are: the Shimba Hills, where there are wetter forests with endemic plants (e.g. *Dichapetalum fructuosum*) and frogs (e.g. *Afrixalus sylvaticus*); the 'Kayas', which are

Table 17.7 Biological diversity in Kenya

	Species	*Endemics*
Plants	6,500	265
Mammals	307	8
Birds	860	9
Snakes	106	4
Amphibia	97	4

(*Source:* WCMC, 1988)

Table 17.8 Distribution and number of rare trees in Kenya and number of endemics

Area	*Total no. of rare species*	*No. which are endemic*
Coastal	32	12
Taita Hills (Eastern Arc)	10	7
Central Dry Forest	9	6
Central Moist Forest	3	3
Central Riverine	5	3
Non-forest	29	2

(*Source:* Beentje, 1988b)

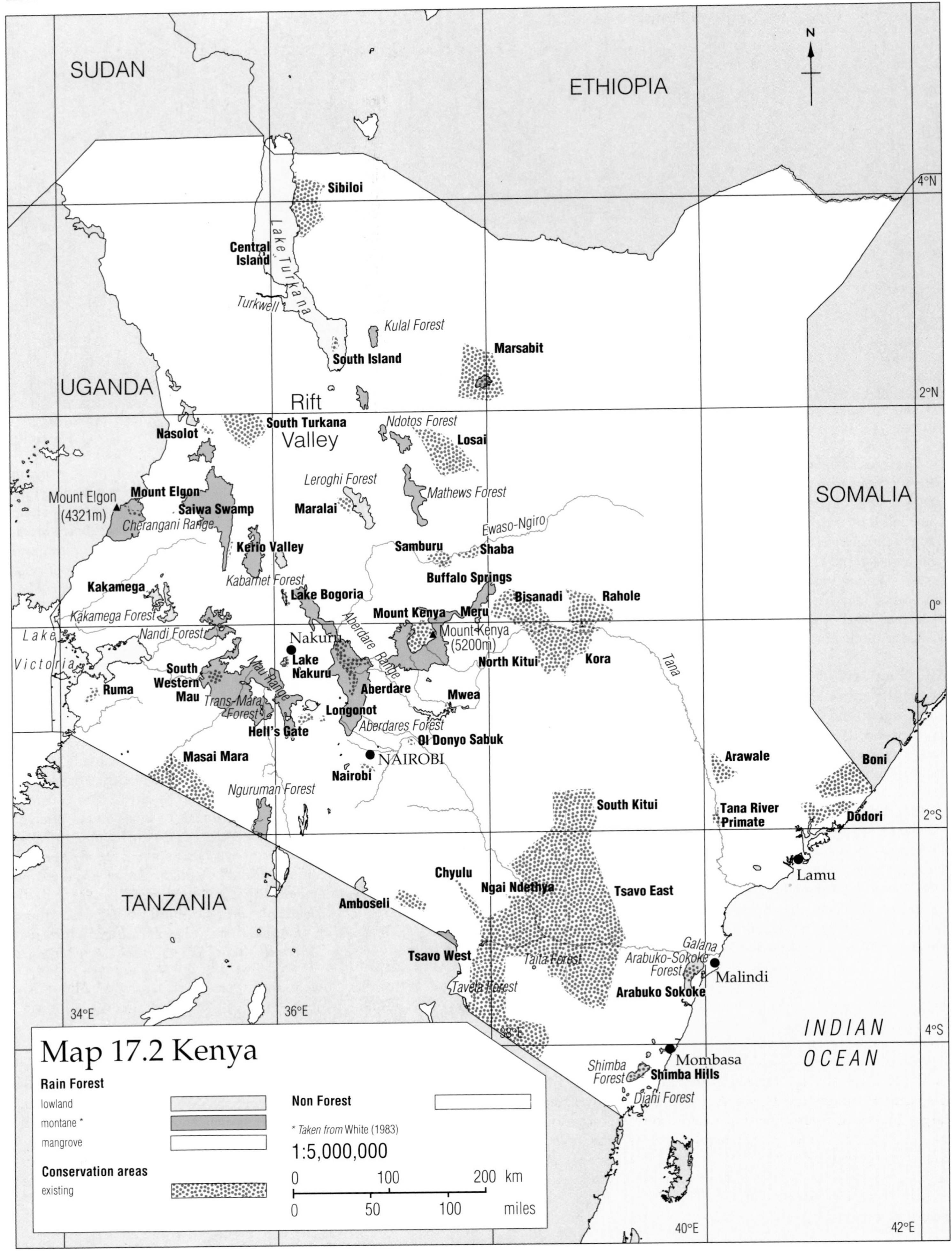
SUDAN
ETHIOPIA
N
UGANDA
SOMALIA
TANZANIA
INDIAN OCEAN
Rift Valley
Lake Turkana
Sibiloi
Central Island
Turkwell
Kulal Forest
Marsabit
South Island
South Turkana
Nasolot
Ndotos Forest
Losai
Leroghi Forest
Mathews Forest
Mount Elgon (4321m)
Mount Elgon
Saiwa Swamp
Cherangani Range
Maralai
Ewaso-Ngiro
Kerio Valley
Samburu
Shaba
Kabarnet Forest
Buffalo Springs
Kakamega
Lake Bogoria
Bisanadi
Rahole
Kakamega Forest
Mount Kenya
Meru
Lake Victoria
Nandi Forest
Nakuru
Aberdare Range
Mount Kenya (5200m)
Lake Nakuru
North Kitui
Kora
Tana
South Western Mau
Mau Range
Aberdare
Mwea
Ruma
Trans-Mara Forest
Longonot
Hell's Gate
Aberdares Forest
Ol Donyo Sabuk
Masai Mara
NAIROBI
Arawale
Boni
Nairobi
Nguruman Forest
South Kitui
Tana River Primate
Dodori
Lamu
Chyulu
Ngai Ndethya
Tsavo East
Amboseli
Galana
Tsavo West
Taita Forest
Arabuko-Sokoke Forest
Malindi
Taveta Forest
Arabuko Sokoke
Mombasa
Shimba Forest
Shimba Hills
Diani Forest
4°N
2°N
0°
2°S
4°S
34°E
36°E
38°E
40°E
42°E
Map 17.2 Kenya
Rain Forest
lowland
montane *
mangrove
Non Forest
* Taken from White (1983)
Conservation areas
existing
1:5,000,000
0 100 200 km
0 50 100 miles

Table 17.9 Conservation areas of Kenya

Existing and proposed conservation areas are listed below. Forest reserves and marine parks are not included or mapped. For data on Biosphere reserves see chapter 9.

	Existing area (sq. km)	*Proposed area (sq. km)*
National Parks		
Aberdare★	766	
Amboseli	392	
Arabuko-Sokoke	6	360[1]
Central Island	5	
Chyulu	471	
Hell's Gate	68	
Kora	1,788	
Lake Nakuru	188	
Longonot	52	
Malka Mari†	876	
Meru	870	
Mt Elgon★	169	
Mt Kenya★	715	
Nairobi	117	
Ndere Island†	4	
Ol Donyo Sabuk	18	
Ruma	120	
Saiwa Swamp	2	
Sibiloi	1,571	
South Island	39	
Tsavo East	11,747	
Tsavo West★	9,065	
Nature Reserves		
Arabuko-Sokoke★	43	
Cheptugen-Kapchemutwa†	<1	
Kaimosi Forest†	<1	
Kaptagat Forest†	nd	
Karura†	1	
Katimok Kabarnet†	1	
Langata	1	
Mbololo†	nd	
Nandi North†	34	
South-Western Mau★[2]	430	
Uaso Narok†	16	
National Reserves		
Arawale	533	
Bisanadi	606	
Boni	1,339	
Buffalo Springs	131	
Dodori	877	
Kakamega★	45	
Kamnarok†	88	
Kerio Valley	66	
Lake Bogoria★	107	
Losai	1,806	
Marsabit★[3]	2,088	
Masai Mara	1,510	
Mwea	68	
Nasolot	92	
Ngai Ndethya	212	
North Kitui	745	
Rahole	1,270	
Samburu	165	
Shaba	239	
Shimba Hills★	193	
South Kitui	1,833	
South Turkana	1,091	
Tana River Primate	169	
Game Sanctuaries		
Maralai	5	
Totals	44,855	360

(*Sources:* IUCN, 1990b; WCMC, *in litt.*)

★ Area with moist forest within its boundaries according to Map 17.2..

† Not mapped – no spatial data available to this project. The percentage of forest included in protected areas cannot readily be calculated for Kenya. Although closed forest occurs in several protected areas, it is often restricted to tiny fragments within the total reserve or national park.

nd no data

1 The 360 sq. km Arabuko-Sokoke Forest Reserve, encompassing the already existing nature reserve (43 sq. km) and national park (6 sq. km) is now under proposal for national park status.

2 Area proposed as a national park.

3 360 sq. km of the national reserve has been proposed as a national park.

relict forests on limestone (see case study); and the coral rag coastal margin forests of Diani-Jadini.

Lower Tana River forests have important primate habitats, with two endemic primate subspecies (Tana River red colobus *Procolobus [badius] rufomitratus rufomitratus* and Tana River mangabey *Cercocebus galeritus galeritus*), six rare bird species, and a near endemic poplar (*Populus ilicifolia*) (Collar and Stuart, 1988).

Conservation Areas

Kenya has an extensive protected area network (Table 17.9), mainly established for large mammal preservation; but several areas contain closed forests (see Table 17.2). The protected area legislation is complex, with several categories besides national park. Nature reserves are gazetted and administered by the Forest Department under the Forest Act and are usually areas within larger blocks of forest reserves, where exploitation is forbidden. National reserves may permit regulated uses such as grazing or fuelwood harvesting. IUCN (1987a) lists, in addition, game sanctuaries, reserves, sanctuaries, nature parks and Biosphere reserves.

Key areas for protecting forest diversity are mainly the nature reserves, especially Arabuko-Sokoke, South-Western Mau and Nandi and a few of the national reserves, such as Boni, Marsabit, Kakamega and Shimba Hills. The Aberdares, Mt Kenya and Mt Elgon national parks have some forest areas. Park status is imminent for Kakamega and Arabuko-Sokoke but doubt exists as to whether all key areas will be included in these parks.

Recent (mid-1991) initiatives between Kenya Wildlife Services (which administers the national parks) and the Kenya Forest Department will allow for joint management of forest areas considered of conservation importance, for instance the Aberdares. British Aid is making an inventory of key forest areas for biodiversity and preparing conservation management plans.

There are still many gaps; several key forests and vegetation types have only reserve forest or trust forest status. This is the case for Taita, Kulal and Kerio riverine forest, Diani and other patches of coastal forest, Cherangani and most of Mt Kenya. The Taita Hills may become a World Heritage site, although this in itself will not provide sufficient protection. In addition, the sacred Kaya forests (see case study) are to be given legal national monument status. IUCN (1987b) and Stuart *et al.* (1990) present general overviews but a more comprehensive study of forest conservation priorities is urgently needed.

TANZANIA

Tanzania is rich in vegetation types, including lowland inland and coastal mosaic forests, swamp forest, extensive montane forests and mangroves. The mangroves are the most extensive in the region, principally on the Rufiji Delta, but also at Tanga, Kilwa and on eastern Zanzibar and Mafia islands. The most significant forests are the montane rain forests which lie in an arc in the east of the country, stretching from the Pare and Usambara Mountains in the north to the Southern Highlands (see Map 17.3). Apart from the recent volcanic extrusions of Kilimanjaro, Mt Meru and Ngorongoro, most of these hills are ancient, and thus of exceptional value and biological richness.

Swamp forest occurs on the Tanzania-Uganda border, the main location being Minziro Forest Reserve, with 250 sq. km. This is an important site for primates and birds as well as for plants. Guineo-Congolian rain forest is now very restricted. Rubondo Island National Park in Lake Victoria has 180 sq. km of little disturbed forest, while Gombe National Park on the shores of Lake Tanganyika has several small forest patches with some affinity to those of Zaïre and Burundi. The coastal mosaic forests are extensive, variable and very important biologically, but are now greatly depleted, fragmented and degraded.

Table 17.10 Conservation areas of Tanzania

Existing and proposed protected areas are listed below. Forest reserves and marine reserves are not included. For data on World Heritage sites and Biosphere reserves see chapter 9.

	Existing area (sq. km)	*Proposed area (sq. km)*
National Parks		
Arusha*	137	
Gombe	52	
Katavi	2,253	
Kilimanjaro*	756	
Lake Manyara	320	
Mahale Mountain	1,613	
Mikumi*	3,230	
Ruaha*	12,950	
Rubondo*	457	
Serengeti	14,763	
Tarangire	2,600	
Uzungwa		1,000
Game Reserves		
Biharamulo	1,300	
Burigi	2,200	
Ibanda	200	
Kilimanjaro*	890	
Kizigo	4,000	
Maswa	2,200	
Mkomazi	1,000	
Mt Meru*	300	
Moyowosi	6,000	
Rumanyika	800	
Rungwa	9,000	
Sadani	300	
Selous*	50,000	
Ugalla	5,000	
Umba	1,500	
Uwanda	5,000	
Conservation Areas		
Ngorongoro*	8,288	
Unclassified		
Mafia Islands*	nd	
Totals	137,109	1,000

* Area with moist forest within its boundaries. Moist forests occur only as small fragments within these protected areas.
(*Sources:* IUCN, 1990b; WCMC, *in litt.*)

Estimates of closed forest vary considerably. Kessy (1982 in Rodgers *et al.*, 1985) suggested 9360 sq. km; FAO (1988) reported 14,400 sq. km and Rodgers *et al.* (1985) estimated a higher figure of 16,185 sq. km of natural closed high forest. This is about 1.5 per cent of Tanzania's land area. The legal forest reserves cover some 133,500 sq. km, but of this only 9519 sq. km is high forest and a further 800 sq. km is mangrove. Map 17.3 shows about 1975 sq. km of mangrove, 4990 sq. km of montane forest and 11,140 sq. km of lowland forest, giving a total of 18,105 sq. km of moist forest in the country. However, due to the outdated maps and somewhat sketchy information available these figures are not considered reliable and have not been quoted at the head of this chapter.

Forest reserves are classified as 'production' forest and 'protective' forest; the latter is estimated as 16,000 sq. km including grassland and woodland, and is of importance for water catchment protection. There are 115,000 sq. km of production forest, most of it miombo woodland; and 7100 sq. km of plantation, most of which is softwood, used in industry or for fuelwood. There are still large areas of forest outside the forest reserves. Some are in the game reserve category, like the Selous Game Reserve, but large areas such as the Kichi-Matumbi Hills near the Rufiji Delta are not protected at all. The national parks include little forest (see Table 17.2). In addition to the high forest, Rodgers *et al.* (1985) map some 2868 sq. km of 'itigi' thicket in central Tanzania and 2145 sq. km of coastal thicket (these thicket formations are not mapped in Map 17.3). Both are outside the formal reserve network.

Most forest has been exploited to some degree in the recent past. 'Protection' or 'Catchment' status is ill-defined and can readily be changed by the Forestry Division. The biologically rich forests of the East Usambara Mountains were heavily exploited using unsuitable machinery in the early 1980s. This caused considerable permanent ecological damage including total deforestation of some sites (Hamilton and Bensted-Smith, 1989; Rodgers, in press). In some areas, intensive pitsawing has replaced mechanical logging. Economic problems and a weak forest service have led to a relaxation in the regulation of forest exploitation (FAO, 1989; Rodgers, in press). Encroachment, illegal harvesting and burning are all major problems. In 1988, the national wood-based industry consumed some 730,000 cu. m of logs, over 70 per cent of which were supplied by softwood plantations. The output from closed natural forests, however, exceeds regrowth and is affecting water catchment function (Mbwana in Rodgers, in press). The demand for fuelwood is by far the greatest drain on forest resources, annual consumption being about 25 million cu. m for domestic supply and 5 million cu. m for commercial use. Most comes from the woodland areas, but natural forests are heavily exploited, especially in the poorly protected lowland areas.

There are no reliable figures for deforestation in the closed forest areas. FAO (1988a) estimated an annual deforestation rate of closed moist forest of 100 sq. km from 1981–5, but nationally, deforestation is thought to exceed 4000 sq. km per year; desertification is a recognised consequence in the drier regions of north Tanzania (Tanzania Forest Division, *Daily News* 30/10/1989).

Biodiversity

Tanzania has long been acknowledged as one of the most important nations in Africa for conservation. However, in continental African terms, Tanzania is an extremely important nation for forest biodiversity: an overview is given in Stuart *et al.* (1990). The geographical and historical reasons for this are explained in Rodgers and Homewood (1982), Hamilton (1989) and Lovett and Wasser (in press).

Brenan (1978) estimated that 1100 plant species are endemic, 10 per cent of the flora in Tanzania (compare this with the 30

FOREST CONSERVATION IN THE USAMBARA MOUNTAINS

The chain of mountains in Eastern Tanzania, known as the Eastern Arc has existed since the Oligocene, 100 million years ago. The mountains have been forested for much of this period and at least in their recent history have been isolated from the Guineo-Congolian rain forests. Forests on the Eastern Arc mountains persisted through the periods of dry climate associated with the Pleistocene glaciations. The persistence of these forests through this long period of isolation has resulted in the evolution of a highly endemic flora. Of the 2000 species of plants known from the Eastern Arc mountains, 25–30 per cent are endemic. The forests are also the home to several endemic mammals, birds, reptiles and a remarkably large number of amphibians. One of the most extensive areas of forest now remaining in the Eastern Arc mountains, is on the East Usambaras mountains in Tanga region in northern Tanzania. At least 100 species of plants are strict endemics to the Usambaras and the forests are noteworthy for several rare and near endemic species of birds. Collar and Stuart (1985) list six threatened and three near-threatened bird species from the area.

Much of the forest was cleared in colonial time, first for coffee and later for tea cultivation. Migrant workers were brought to the area from various parts of East Africa to work on the estates. Recently the tea estates have suffered from poor management. Many people have abandoned work on them and have cleared forest land to practise low-grade agriculture. A particular problem has come with the growing of cardamon. The ground storey of the forest is cleared to cultivate the cardamon under the canopy. This cultivation prevents regeneration of the forest trees and when, after seven or eight years, the yields of the cardamon begin to decline the people clear the canopy trees to plant maize, sugar cane and manioc. Large areas of forest are being destroyed in this way. Further problems have come from pitsawyers who illegally cut timber in the forests for sale, both in nearby markets and across the border in Kenya. Industrial logging in the forest was stopped when it became clear that the logging was unsustainable. However, the past few years have seen a tremendous increase in legal and illegal pitsawing, which has exceeded industrial output. Furthermore, Tanzania's economic decline has led to decreased forest management capability.

Since 1986 an IUCN/EEC project in the Usambaras has attempted to give more effective protection to the forest while helping the local people to develop agricultural systems which will be less destructive. The project is based on consultation with local communities. A locally recruited staff member is assigned to a village in the East Usambaras to encourage the formation of a village development committee. Through the village coordinators the project is able to engage in a dialogue with the villagers, both to ascertain their needs and aspirations and to explain the long-term environmental problems that will result if the forests are lost. The village committees are encouraged to develop their own solutions to the problem, and if these ideas are thought to be viable, support is available to help implement them, through the provision of agricultural tools, tree seeds and seedlings and help in transporting building materials and agricultural produce.

Attempts have been made to bring pitsawing of timber under proper control. The intention was to license villagers to pitsaw timber in forests adjacent to their village land. So far the difficulties in preventing outsiders from logging the forests have been so great that it has been found necessary to put a total ban on all timber extraction. Nonetheless, there is no doubt that the forest could support a moderate yield of valuable timber – much in demand in local towns – and this could provide an incentive to the villagers to maintain forest areas. In its first three years of operation the villagers working with the project have planted boundary strips around the forest reserves and 30 ha of communal plantations. Central and village tree nurseries have been established. The project has helped villagers put in contour strips on a thousand farms. More importantly there have been changes in the attitudes of the farmers, who are adopting contour terracing on a wide scale and planting cloves, pepper and coffee as a source of cash. The forests of the East Usambaras are still very much under threat. The number of people in the area is simply too great for the land resources available. However, by working with the villagers, the project has made them much more sensitive to conservation problems.

A new FINNIDA project to strengthen forest management is to be coordinated with the IUCN/EEC initiatives. This new project recognises the importance of forest for water catchment and biodiversity as well as village level resource exploitation. The problems of an inadequate research base for the fragmented forest resource is of concern both for monitoring biological values and improving management. The project will consider higher conservation status, including forest parks and Biosphere reserves for these exceptionally valuable forests. *Source:* Jeff Sayer

species or 0.3 per cent in Uganda). More than 93 per cent of the endemics are from the mountains of the Eastern Arc. It is probable that some 25 per cent of Tanzania's forest species are endemic. Some plantgroups are especially rich in endemics: the Annonaceae, Caesalpiniaceae and Rubiaceae are three outstanding examples.

Forest loss and fragmentation has often led to the concentration of endemic species in very restricted areas. Kimboza Forest Reserve of 4 sq. km has 17 endemic plant taxa including six tree species, of which *Baphia pauloi* is one. Magombera Forest Reserve, now less than 8 sq. km and included in the Selous Game Reserve, has three endemic species, including *Polyalthia verdcourti*. The Rondo forests in south-east Tanzania have at least ten endemic species.

Diversity and endemism within animal taxa is well documented: for mammals by Kingdon (1990), for birds by Stuart (1985) and for a variety of other taxa in Rodgers and Homewood (1982) and Lovett and Wasser (in press). The Uzungwas have two endemic primates – Gordon's red colobus *Procolobus [badius] gordonorum* and the Sanje mangabey *Cercocebus galeritus sanjei*. IUCN's Primate Action Plan recommends an increased conservation effort in this area (Oates, 1985). The Eastern Arc mountains are a major centre of endemism for birds, with 14 endemic species, many confined to small forest blocks. The Usambaras have two strict endemics, the Ulugurus two and the Uzungwas one.

Conservation Areas

Tanzania has 11 national parks, a conservation area, and 16 game reserves (Table 17.10) as well as an extensive forest reserve network. Most parks and game reserves were established for large mammals or, as in the case of Kilimanjaro National Park, exceptional landscapes. As a result few of them contain closed forest. Exceptions are given in Table 17.2.

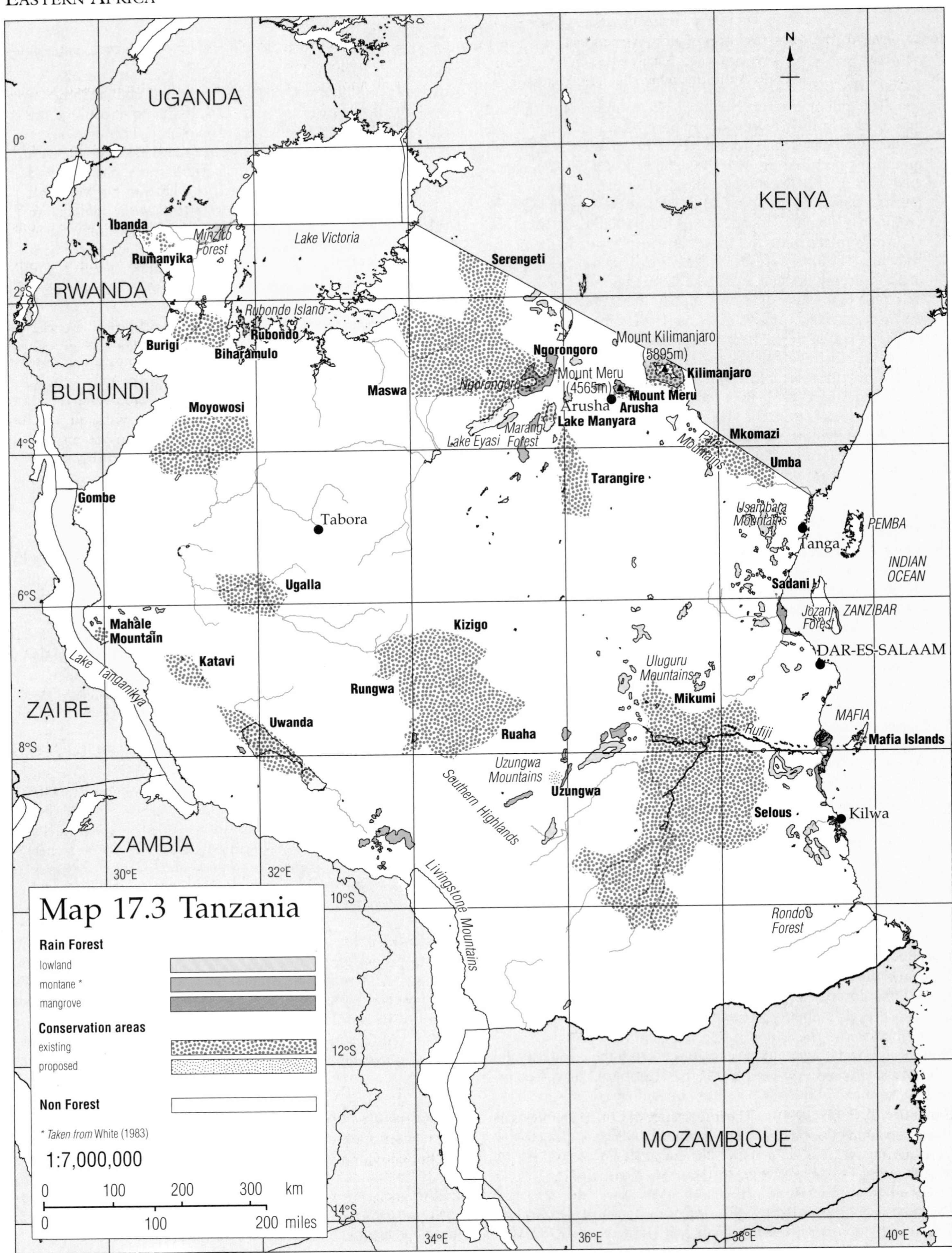
UGANDA
KENYA
RWANDA
BURUNDI
ZAIRE
ZAMBIA
MOZAMBIQUE
N
0°
2°S
4°S
6°S
8°S
10°S
12°S
14°S
30°E
32°E
34°E
36°E
38°E
40°E
Lake Victoria
Ibanda
Rumanyika
Minziro Forest
Rubondo Island
Rubondo
Burigi
Biharamulo
Serengeti
Ngorongoro
Mount Kilimanjaro (5895m)
Kilimanjaro
Mount Meru (4565m)
Mount Meru
Arusha
Lake Manyara
Marang Forest
Lake Eyasi
Maswa
Moyowosi
Pare Mountains
Mkomazi
Umba
Tarangire
Gombe
Tabora
Usambara Mountains
Tanga
PEMBA
INDIAN OCEAN
Ugalla
Sadani
Jozani Forest
ZANZIBAR
Mahale Mountain
Kizigo
Katavi
Lake Tanganikya
DAR-ES-SALAAM
Uluguru Mountains
Mikumi
Rungwa
Uwanda
Ruaha
Rufiji
MAFIA
Mafia Islands
Uzungwa Mountains
Uzungwa
Southern Highlands
Selous
Kilwa
Livingstone Mountains
Rondo Forest
Map 17.3 Tanzania
Rain Forest
lowland
montane *
mangrove
Conservation areas
existing
proposed
Non Forest
* Taken from White (1983)
1:7,000,000
0 100 200 300 km
0 100 200 miles

Forest reserve status does not offer an adequate level of protection for biodiversity. Biological resources of such international importance demand the highest conservation status legally available. At present, however, the National Parks Authority, which has a major task managing and funding existing parks, is not keen to acquire new closed forest areas, while the Forest Division is not keen to lose more areas. A new legal category of forest national park, or an arrangement whereby forest areas can be legally created as parks but managed by the Forest Division, needs to be developed.

The 1000 sq. km Uzungwa National Park may soon be gazetted. Lake Manyara National Park may be extended to include the Marang Forest and Kilimanjaro National Park may take in more of the lower forest areas. Zanzibar is considering national park status for Jozani Forest and parts of Ngezi Forest. Major decisions need to be taken over the higher forests on the Uluguru Mountains, which are of crucial catchment and biological value and are compact and not heavily affected by settlement pressure. More difficult is the case of the fragmented forest patches of the East Usambaras (see Hamilton and Bensted-Smith, 1989 for details, and case study). However, there is no reason why the key patches cannot be given total protection and managed by the Forest Department, thus forming the core area of a larger Man and the Biosphere (MAB) reserve. Further protected areas are needed for patches of itigi thicket, coastal forest, high altitude forest and grassland mosaics in the Mporoto and Livingstone Mountains and for Minziro Forest.

Conclusion

This brief account of eastern Africa's closed forests has documented four major facts:

- The small and still decreasing extent of natural closed forests.
- The great variety of forests with high internal diversity and endemism.
- The total inadequacy of the existing national park network for conserving forests, and the need for more parks.
- African countries are only now realising the significant role of closed forests in many development sectors, not only for timber and biological resources, but also for sustaining water, soils, climate, energy, industry and agriculture.

The problem of how to incorporate forests within the protected area network has not been solved. Many forest areas are at risk even within the protected area system as management is either inadequate (e.g. Kakamega in Kenya) or lacking altogether (e.g. Day National Park in Djibouti). Forests must be conserved through integrated programmes addressing issues of catchment policy, sustainable development for local people, levels of permissible exploitation and protection of biodiversity. While protected area managers have learned to deal with some of the threats facing savanna and large mammal ecosystems, these problems have not been addressed in forests, and this needs to be rectified.

References

Banyikwa, F. F. (1986) The geographical distribution of mangrove forests along the East African coast. In: *Status and Utilization of Mangroves. Proceedings of a Workshop on 'Save the Mangrove Ecosystems in Tanzania' 21–22 February 1986, Dar es Salaam*. Published by the Faculty of Science, University of Dar es Salaam.

Beals, E. W. (1968) Conservation of vegetation in Ethiopia. *Acta Phytogeographica Suecica* **54**: 137–40.

Beentje, H. J. (1988a) An ecological and floristic study of the forests of the Taita Hills, Kenya. *Utafiti – the Occasional Papers of National Museums of Kenya* **1**(2): 23–66.

Beentje, H. J. (1988b) Atlas of rare trees of Kenya. *Utafiti – the Occasional Papers of National Museums of Kenya* **1**(3): 71–123.

Beentje, H. J. (1990) The forests of Kenya. *Mitteilungen aus dem Institüt für allgemeine Botanik in Hamburg* **23a**: 265–8.

Brenan, J. P. M. (1978) Some aspects of the phytogeography of Tropical Africa. *Annals of the Missouri Botanic Gardens* **65**: 437–78.

Chaffey, D. R. (1979) *South-west Ethiopia Forest Inventory Project*. Land Resources Project Report 31. Ministry of Overseas Development, UK.

Coetzee, J. A. (1967) The East African Mountains. *Palaeoecology of Africa* **3**: 1–146.

Collar, N. J. and Stuart, S. N. (1985) *Threatened Birds of Africa and Related Islands. The ICBP/IUCN Red Data Book Part 1.* ICBP/IUCN, Cambridge, UK. 761 pp.

Collar, N. J. and Stuart, S. N. (1988) *Key Forests for Threatened Birds in Africa. ICBP Monograph* 3. ICBP, Cambridge, UK.

Douthwaite, R. J. (1987) Lowland forest resources and their conservation in Southern Somalia. *Environmental Conservation* **14**(1): 29–33.

Egziabher, T. B. G. (1990) The importance of Ethiopian forests in conservation of arabica coffee pools. In: *Proceedings of 12th Plenary Meeting of AETFAT*. Peters, C. R. and Lejoly, J. (eds). Vol. 23 A/B, Mitteilungen aus dem Institüt für allgemeine Botanik in Hamburg.

FAO (1988a) *An Interim Report on the State of Forest Resources in the Developing Countries*. FAO, Rome, Italy. 18 pp.

FAO (1988b) *Report of the Mission to Ethiopia on Tropical Forestry Action Plan. July 1988*. FAO, Rome, Italy.

FAO (1989) *Tropical Forestry Action Plan*. United Republic of Tanzania. FAO, Rome, Italy.

FAO (1991) *FAO Yearbook of Forest Products 1978–1989*. FAO Forestry Series No. 24 and FAO Statistics Series No. 97. FAO, Rome, Italy.

FAO/UNEP (1981) *Tropical Forest Resources Assessment Project. Forest Resources of Tropical Africa. Part II: Country Briefs*. FAO, Rome, Italy.

Friis, I. (1982) Studies in the flora and vegetation of south-west Ethiopia. *Opera Botanica* **63**: 1–76.

Friis, I. and Tadesse, M. (1988) The Evergreen Forests of Tropical North East Africa. In: *Floristic Inventory of Tropical Countries*, pp. 218–31. Campbell, D. G. and Hammond, H. D. (eds). Science Publishing Department, New York Botanical Gardens, New York, USA.

Hallsworth, E. G. (1982) *Socio-economic Effects and Constraints in Tropical Forest Management*. John Wiley, London, UK.

Hamilton, A. (1989) African forests. In: *Tropical Rain Forest Ecosystems*, pp. 155–82. Lieth, H. and Werger, M. J. A. (eds). Vol. 14B of 'Ecosystems of the World'. Springer Verlag.

Hamilton, A. C. and Bensted-Smith, R. (eds) (1989) *Forest Conservation in the East Usambara Mountains, Tanzania*. IUCN, Gland, Switzerland and Cambridge, UK. 392 pp.

Hamilton, A. C. and Perrott, R. A. (1981) A study of altitudinal zonation in the montane forest belt of Mount Elgon. *Vegetatio* **45**: 107–25.

Hillman, J. C. (1985) *Seminar on Wildlife Conservation and Management in the Sudan*. Topic III: Wildlife Research in Relation to Conservation and Management. Deutsche Gesellschaft für technische Zusammenarbeit.

IUCN (1987a) *IUCN Directory of Afrotropical Protected Areas*. IUCN, Gland, Switzerland and Cambridge, UK. xix + 1043 pp.

IUCN (1987b) *Action Strategy for Protected Areas in the Afrotropical Realm*. CNPPA. IUCN, Gland, Switzerland.

Eastern Africa

IUCN (1989) *The IUCN Sahel Studies*. IUCN, Gland, Switzerland.

IUCN (1990a) *Ethiopia National Conservation Strategy Phase One*. March 1990, Gland, Switzerland.

IUCN (1990b) *1990 United Nations List of National Parks and Protected Areas*. IUCN, Gland, Switzerland and Cambridge, UK. 275 pp.

Jackson, J. K. (1956) The vegetation of the Imatong Mts, Sudan. *Journal of Ecology* **44**(2): 341–74.

Jenkins, R. N. *et al.* (1977) *Forest Development Prospects in Imatong Central Forest Reserve, S. Sudan*. Vols. 1 & 2. Land Resources Division, Ministry of Overseas Development, UK.

Kingdon, J. (1990) *Island Africa*. Collins, London, UK.

Kokwaro, J. O. (1988) Conservation status of the Kakamega Forest in Kenya. *Monograph Systematic Botany Missouri Botanic Gardens* **25**: 471–89.

Lind, E. M. and Morrison, M. (1974) *Vegetation of East Africa*. Longmans, Nairobi, Kenya.

Lovett, J. C. (1988) Tanzania. In: *Floristic Inventory of Tropical Countries*, pp. 232–5. Campbell, D. G. and Hammond, H. D. (eds). Science Publishing Department, New York Botanical Gardens, New York, USA.

Lovett, J. C. and Wasser, S. (in press) *Ecology and Biogeography of the Forests of Eastern Africa*. Cambridge University Press, Cambridge, UK.

MacKinnon, J. and MacKinnon, K. (1986) *Review of the Protected Areas System in the Afrotropical Realm*. IUCN/UNEP, Gland, Switzerland. 259 pp.

Madgwick, J. (1989) Somalia's threatened forests. *Oryx* **23**(2): 94–101.

Oates, J. F. (1985) *Action Plan for African Primate Conservation 1986–1991*. IUCN/SSC Primate Specialist Group. IUCN, Stony Brook, New York, USA.

Parker, T. (1987) *Report on an Aerial Reconnaissance of Current and Potential Reserves in Southern and Central Somalia*. WWF and Somalia Ecological Society.

Persson, R. (1975) *Forest Resources of Africa*. Royal College of Forestry, Stockholm, Sweden.

Pohjonen, V. and Pukkala, T. (1990) *Eucalyptus globulus* in Ethiopia. *Forest Ecology and Management* **36**: 19–31.

Polhill, R. M. (1988) East Africa (Kenya, Tanzania, Uganda). In: *Floristic Inventory of Tropical Countries*, pp. 218–31. Campbell, D. G. and Hammond, H. D. (eds). Science Publishing Department, New York Botanical Gardens, New York, USA.

Rodgers, W. A. (in press) The conservation of the forest resources of Eastern Africa: Past influences, present practices and future needs. In: *The Ecology and Biogeography of the Forests of Eastern Africa*. Lovett, J. and Wasser, S. (eds). Cambridge University Press, Cambridge, UK.

Rodgers, W. A. and Homewood, K. M. (1982) Species richness and endemism in the Usambara Mountain forests. *Biological Journal of the Linnean Society* **18**: 197–242.

Rodgers, W. A., Mziray, W. and Shishira, E. K. (1985) The extent of forest cover in Tanzania using satellite imagery. *Institute of Resources Assessment Research Paper* 12, University of Dar es Salaam.

Sale, J. B. (1989) *Forestry Sector Support and Training: Somalia. Wildlife Resources Management*. UNDP/FAO, Rome, Italy.

Stuart, S. N. (1985) Rare Forest Birds and their Conservation in Eastern Africa. In: *Conservation of Tropical Forest Birds*. Diamond, A. W. and Lovejoy, T. E. (eds). ICBP Technical Publication No. 4. ICBP, Cambridge, UK.

Stuart, S. N., Adams, R. J. and Jenkins, M. D. (1990) *Biodiversity in Sub-Saharan Africa and its Islands: iConservation, Management and Sustainable Use*. IUCN, Cambridge, UK and Gland, Switzerland.

WCMC (1988) *Kenya. Conservation of Biological Diversity*. World Conservation Monitoring Centre, Cambridge, UK. 17 pp.

White, F. (1983) *The Vegetation of Africa: a descriptive memoir to accompany the Unesco/AETFAT/UNSO vegetation map of Africa*. Unesco, Paris, France. 356 pp.

Wickens, G. E. (1976) The flora of Jebel Marra (Sudan Republic) and its geographical affinities. *Kew Bulletin, Additional Series* **V**: 1–368.

World Bank (1986) *Forest Sector Review for Sudan*. Report 5911-Su. Washington, DC, USA.

World Bank (1988) *Forestry Sector Review: Kenya, Volume I*. World Bank, Washington, DC, USA.

Young, T. P. (1984) *Kenya's Indigenous Forests. Status, Threats and Prospects for Conservation Action*. WWF/IUCN. East Africa Office, Nairobi.

Zamierowski, E. E. (1975) Leaching losses of minerals from leaves of trees in montane forest in Kenya. *Journal of Ecology* **63**(2): 679–87.

Authorship

Alan Rodgers, Cambridge with contributions from Anne Robertson, National Museums of Kenya (NMK) and Jeff Sayer, IUCN.

Map 17.1 Forest cover in Ethiopia

Data on the extent of natural closed forest in Ethiopia were taken from *National Atlas of Ethiopia* (1988) compiled by the Ethiopian Mapping Authority, Addis Ababa. The *High Forest* category from the eight land use categories shown, was mapped to depict montane rain forest on Map 17.1. Mangroves for Ethiopia are not shown in this Atlas as data are unavailable. Conservation areas have been extracted from various sources on file at WCMC.

Map 17.2 Forest cover in Kenya

A detailed *Land Use Map of Kenya* was published in 1983 by the Kenya Rangeland Ecological Monitoring Unit (KREMU), Ministry of Environment and Natural Resources, at a scale of 1:1 million. This shows 32 land use classes, one of which has been mapped to produce the forest cover shown on Map 17.2, namely, *Dense Natural Forest*. The source map was compiled from remote sensed data of 1972–80 by KREMU, Ministry of Environment and Natural Resources. Mangroves are not depicted on this map and have, therefore, not been shown on Map 17.2. More recent satellite imagery from between 1986 and 1989 will be published by KREMU but was unavailable for this project. More recent and detailed information is available for south-western Kenya from *Vegetation and Climate Maps of SW Kenya* (1987), produced at a 1:250,000 scale, compiled by C. G. Trapnell and M. A. Brunt for the Land Resources Development Centre, Overseas Development Administration but, as these maps do not cover the whole country, they were not used in this Atlas.

Protected areas for Kenya have been selected from a topographical map *Kenya* (nd), at a scale of 1:1,250,000 published by Bartholomew & Son.

Map 17.3 Forest cover in Tanzania

Forest cover data for Tanzania have been taken from a number of sources. Dense closed moist forests, not including mangrove, have been extracted from a 1:2 million scale map *Forest Cover in Tanzania* which accompanies a report evaluating the extent of forest cover in Tanzania using 1973–9 satellite imagery (Rodgers *et al.*, 1985). The distribution of five vegetation types have been demarcated on this source map. Only the *Natural Forest* category is shown on Map 17.3. Mangrove forests were not included in this survey and coastal and *itigi thicket* (thicket formations) have not been mapped. Forest cover was then overlain with White's vegetation types. An attempt has been made to map the mangrove formations from various sketch and topographical maps, but no precise data were available. Location of mangroves illustrated on a 1:1 million ONC map (*M-5* [1969] and *N-5* [1973]) were used as a guide and then updated with reference to Banyikwa (1986) and a map of forest reserves (WCMC, *in litt.*). Reference has also been made to a 1:2 million scale map *Tanzania Vegetation Cover Types* (1974), prepared by the Forest Division, Ministry of Lands, Natural Resources and Tourism, Dar es Salaam.

The system of protected areas for Tanzania has been extracted from a map published by the Ministry of Natural Resources and Tourism at a scale of 1:2 million *Tanzania* (1974).

18 Equatorial Guinea

Land area 28,050 sq. km
Population (mid-1990) 400,000
Population growth rate in 1990 2.6 per cent
Population projected to 2020 800,000
Gross national product per capita (1988) US$350
Rain forest including degraded forest (see map) 17,004 sq. km
Closed broadleaved forest (end 1980)* 12,950 sq. km
Annual deforestation rate (1981–5)* 30 sq. km
Industrial roundwood production† 160,000 cu. m
Industrial roundwood exports† 120,000 cu. m
Fuelwood and charcoal production† 447,000 cu. m
Processed wood production† 61,000 cu. m
Processed wood exports† 13,000 cu. m
* FAO (1988)
† 1989 data from FAO (1991)

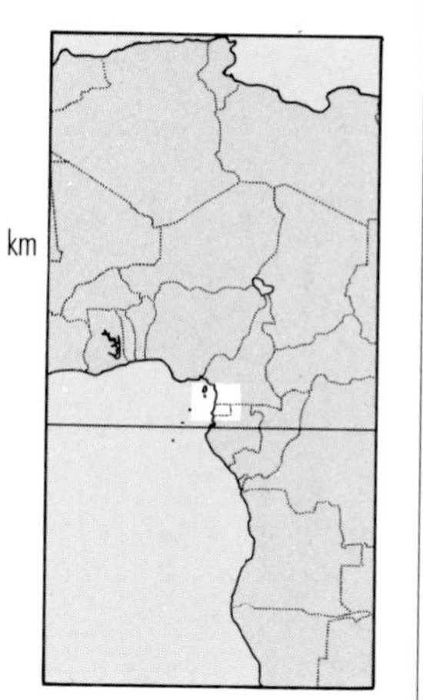

Equatorial Guinea, the only Spanish-speaking nation in sub-Saharan Africa, is one of the least developed countries on the continent. In the early years of this century it was very prosperous as a result of its flourishing cacao and coffee trade, but it now faces severe economic difficulties. Forestry is now a major contributor to the economy. The country has few other natural resources and industry is insignificant. Timber is seen as a panacea and commercial logging is likely to extend into all of the country's remaining primary forest in the next two decades (Fa, 1988; UICN, 1991). Extraction rates will increase and, since the forest management capacity of the government is weak, there are fears that serious degradation of the forest may follow (Castroviejo *et al.*, 1986; UICN, 1991). About 78 per cent of the population are rural and the demand for agricultural land in much of the country will exert pressure on forests opened up by logging (Castroviejo *et al.*, 1986; Fa, 1988).

In 1986 Castroviejo and a team supported by Spanish development aid proposed the establishment of several protected areas, both on the mainland and on the islands. These proposals have now been adopted by the government but little progress has been made in implementing them. The richness of Equatorial Guinea's flora and fauna makes the effective management of these areas an urgent priority.

Introduction

Equatorial Guinea consists of the small mainland territory of Rio Muni on the coast of West Africa and several islands in the Gulf of Guinea. The most important of the islands are Bioko (formerly Fernando Poo) and Annobon (formerly Pagalu). Rio Muni, with a land area of 26,000 sq. km, is bordered by Gabon to the south and east and by Cameroon to the north. Its relief is complex: the land rises from the coastal plain through a series of stepped tablelands into the interior, where there are a number of granitic inselbergs. The highest mountains reach 1250 m above sea level.

Bioko, 2017 sq. km in area and lying only 32 km from Africa's coast, is of volcanic origin but was connected to the mainland in the recent past (10,000–11,000 years ago) when sea levels were lower. It has a rugged topography, dominated by two mountain masses. In the north, Pico Basilé is the main feature, reaching a height of 3011 m, while the southern third of the island consists of a jagged plateau where the two highest peaks are Pico Biao at 2009 m and the Gran Caldera de Luba at 2261 m. The plateau falls steeply to the sea on the southern coast. Around much of the rest of the coast there is a relatively flat belt of land stretching about two kilometres inland.

Annobon is a much smaller (17 sq. km) volcanic island lying 340 km from the mainland. It has always been isolated from the continent and also from the islands of São Tomé and Príncipe, which lie between it and Bioko. It rises precipitously from the sea to several distinct peaks, the highest of which is 613 m (Castro and Calle, 1985).

Bioko has a typically equatorial climate, with a mean annual temperature of around 25°C (maximum in February of 26.2°C and a minimum in September of 24°C), high humidity and an enormous rainfall. The southern coast of Bioko, with a mean annual precipitation of 10,900 mm, is one of the wettest places in the world; on the north coast, this falls to a mean of 1930 mm. The wet season is from April to October. Rio Muni's climate is similar, although mean annual precipitation is considerably less, between 1800 and 3800 mm, most of which falls between September and December. Annobon, too, has less rain than Bioko, about 1016 mm per year falling mainly from November to April (Fry, 1961). Temperatures during the day at this time are between 19.5°C and 30.5°C.

The population on the mainland portion of Equatorial Guinea is predominantly rural, with only about 18 per cent of the people in the capital, Bata. Rural population densities are relatively high (more than ten inhabitants per sq. km) in some regions such as the Ebebiyin, Micomeseng and Mongomo (Castroviejo *et al.*, 1986; UICN, 1991). However, a large portion of Rio Muni is still unpopulated, with less than five individuals per sq. km in more than half the country (e.g. in the Evinayong, Acurenam and Nsoc regions). A census in 1983 recorded 230,000 inhabitants, an increase of 72 per cent from 1932. On Bioko, meanwhile, the 1983 census revealed 59,196 people, approximately the same as in 1969. It is estimated that the island lost 30–50 per cent of its people after independence under the murderous regime of Macias Nguema. Many were executed or fled the country during the ten years (1969–79) when Bioko suffered under one the world's most destructive dictatorships. More than 60 per cent of the population are based in and around the capital city of Malabo, another 13,000 or so live in the towns of Baney, Rebola, Luba and Riaba, while most of the remainder live along the east, north and west coasts (Butynski and

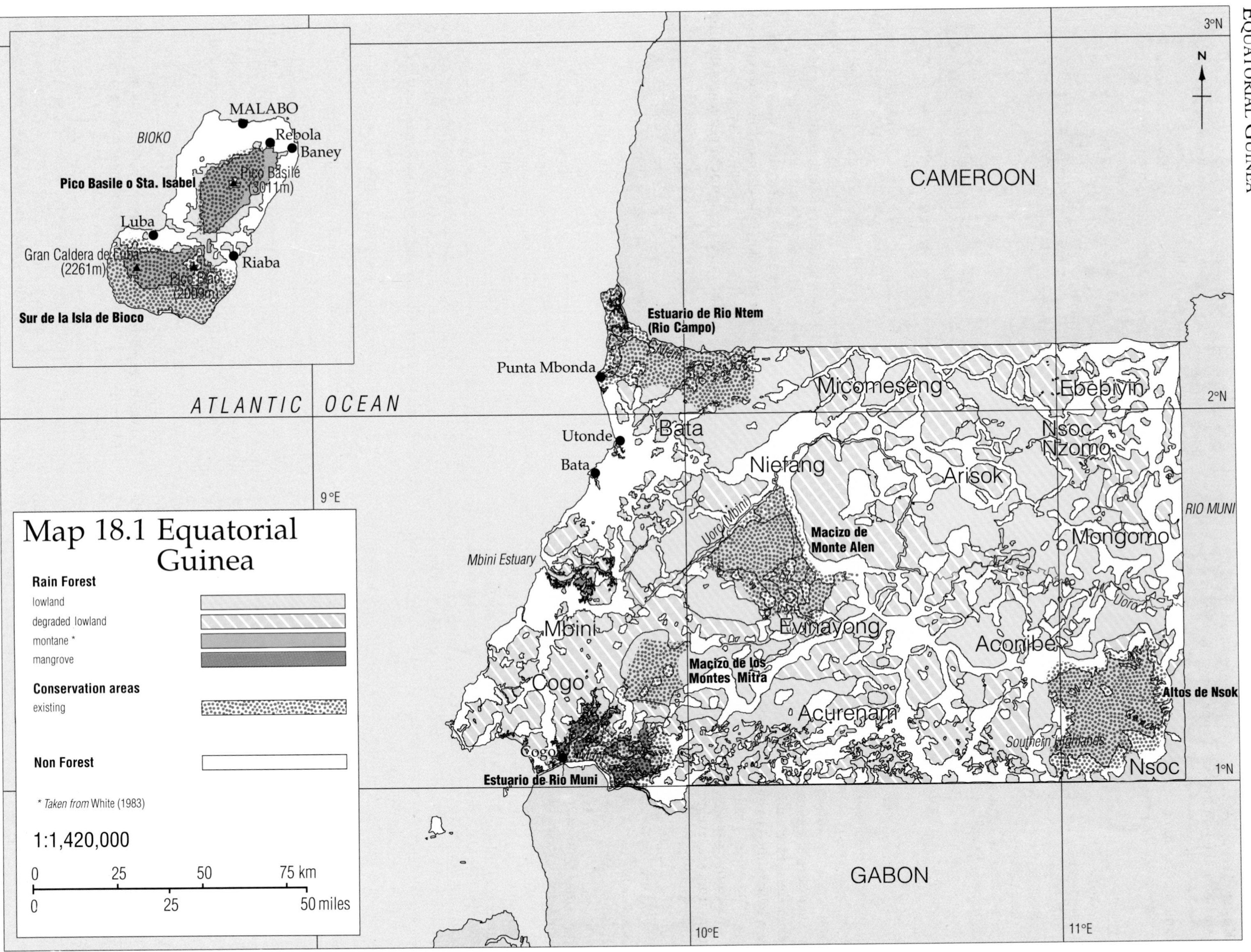

Map 18.1 Equatorial Guinea

Koster, 1989a). About half of Bioko is uninhabited. Finally, on Annobon, the population of approximately 1500 is concentrated in the only town of San Antonio de Palé; the number of inhabitants here has remained largely unchanged from the figure reported early this century (Harrison, 1990).

The Forests

Tropical rain forest is found a few kilometres inland from the coast in Rio Muni and it still covers most of the country (Alers and Blom, 1988). Before the forest is reached, there are *Terminalia* and palm trees along the coast and then a belt of savanna no more than 2 km wide. The forest canopy reaches a height of 35–40 m with emergents as tall as 70 m. Floristic composition is variable depending on soil conditions and exposure, but around 70–80 common tree species are typical. The most abundant of these are abina *Petersianthus macrocarpus*, ilomba *Pycnanthus angolensis*, okoumé *Aucoumea klaineana*, dahoma *Piptadeniastrum africanum*, ébé *Pentaclethra macrophylla* and tali *Erythrophleum ivorense*.

Guinea (1946) distinguished four vegetation formations in Rio Muni: one in the Utamboni basin and the Evinayong–Acurenam–Nsoc corridor with floristic affinities with Gabon's forests; another in the Rio Ntem–Ebebiyin corridor similar to southern Cameroonian coastal forests; a third, a flooded formation, in the north-east in the Bimvile basin and Abia and the fourth, the coastal forests between Utonde, Punta Mbonda and Pemba.

Maps by Guinea (1968) and by Ocaña Garcia (1962) show that, from sea level to 800 m, the major vegetation type on Bioko was lowland rain forest. However, much of this has been cut and the land planted with cacao so now patches of primary forest remain only between 600 and 800 m. These are mainly on Pico Basilé and in the southern third of the island (Butynski and Koster, 1989a). Montane rain forest is found at altitudes between 800 m and 1400 m. Approximately 19 per cent (393 sq. km) of the island was originally covered with this vegetation, and now there are two separate rings of it, one round Pico Basilé and one round the Southern Highlands (Butynski and Koster, 1989a). Higher up (1400–2600 m), there is *Schefflera* forest, scrub and ultimately subalpine and alpine meadows on the summits of Bioko's volcanoes. The decline of the cacao plantations from a peak of 410 sq. km to the present 46 sq. km of cultivation (World Bank, 1985) has meant that considerable secondary vegetation is regenerating on the island (Butynski and Koster, 1989a). The rich soils and abundant rainfall are particularly favourable to cacao and many of the old plantations are now being rehabilitated. After timber, cacao is likely to be the next highest foreign exchange earner for the country.

On the island of Annobon, moist forest (similar in composition to the mist forests on Príncipe) is found only above an altitude of 500 m (Exell, 1952/3). Some of the principal species are: *Agelaea* spp., *Cassipourea annobonensis*, *Craterispermum montanum*, *Heisteria parviflora*, *Rubus pinnatus*, *Schefflera mannii*, *Strombosia* sp., *Pavetta monticola*, *Discoclaoxylon occidentale* and *Ficus clarencensis*. There is a profuse development of epiphytes in this forest.

Mangroves

The extensive development of mangroves is confined to three estuaries in mainland Equatorial Guinea (Hughes and Hughes, 1991). Mangrove stands begin 1 km up the estuary of the Ntem River and extend inland for a distance of 15 km in belts 500–2500 m wide. Here tidal forests have an area of 18 sq. km. Other small patches of mangroves occur at the mouth of the Mbia River (1.5 sq. km) and the Ekuko River (10 sq. km). There are four major blocks of mangroves in the Mbini estuary, of about 53 sq. km in total, while the other major area for mangroves is the Muni estuary. Here they extend from just inside the mouth to the head of the estuary, 17 km inland, and up the tributary rivers. The tidal forests there occupy about 65 sq. km and there are also some extensive patches of swamp forest in the area. Map 18.1 shows 257 sq. km of mangrove remaining around the coastline of Rio Muni.

Bioko has only a few small patches of mangroves on its coast. The mangroves have suffered little degradation although traditionally they have been exploited for firewood and building materials.

Forest Resources and Management

Lepitre *et al.* (1988) estimated that approximately 59 per cent of Rio Muni is still covered by relatively intact forest. Most of the pristine primary forest is located in the Centre-South province, while the Litoral province has a large expanse of secondary forest. No good vegetation map of Equatorial Guinea is available; the one used to generate the forest cover of Rio Muni shown on Map 18.1 was produced in 1960. The source map (see Map Legend on p. 167) distinguished between *bosque claro* and *bosque dense* and on Map 18.1 the former is shown as degraded forest. Table 18.1 gives areas, measured from Map 18.1, for the different forest types in Rio Muni and Bioko together. The map shows 7704 sq. km of dense rain forest and 7945 sq. km of the open (degraded) rain forest on the mainland as well as 257 sq. km of mangroves, which gives a total forest cover of 15,906 sq. km or 61 per cent of the land area of Rio Muni.

Butynski and Koster (1989a) give a detailed breakdown of the forest cover remaining on Bioko. There are 572 sq. km (28 per cent) of lowland rain forest, 375 sq. km (19 per cent) of montane forest, 202 sq. km (10 per cent) of *Schefflera* forest and another 388 sq. km (19 per cent) of secondary forest. The vegetation cover in Bioko shown on Map 18.1 has been taken from a map within Butynski and Koster's (1989a) report and, as expected, the rain forest cover measured from it is similar to that given above. It shows 552 sq. km of lowland rainforest and 546 sq. km of montane forest (no distinction has been made between the *Schefflera* forest and the lower montane forest).

There are no data for the extent of the forests remaining on Annobon, but they cannot be extensive; they are mostly in the south of the island (Harrison, 1990).

At present there is no management framework for the forests of Equatorial Guinea. Three ordinances relate to natural resources: a forestry law enacted on 15 January 1985 for the protection of forest products and the regulation of their use; another, enacted in 1989, for the protection of wildlife and the third, which took effect

Table 18.1 Estimates of forest extent in Rio Muni and Bioko

	Area (sq. km)	*% of land area*
Rain forests		
Lowland (closed)	8,256	29.4
Lowland (degraded)	7,945	28.3
Montane	546	1.9
Mangrove	257	0.9
Totals	17,004	60.5

(Based on analysis of Map 18.1. See Map Legend on p. 167 for details of sources.)

in January 1990, concerned with protected areas. The government department in charge of production and export of timber and the use of the forests is the Secretaría de Aguas y Bosques y Repoblación Forestal. The Dirección General de Aguas y Bosques is the ultimate authority responsible for the administration, exploitation and conservation of forest resources in Equatorial Guinea. Its offices in Malabo grant felling concessions and deal with other matters relating to the management of forest land. However, staffing levels are low and few employees are trained in either forestry or conservation.

It is difficult to estimate potential timber resources in Equatorial Guinea due to the lack of accurate inventory data. The available information comes primarily from FAO/UNEP (1981) and Malleux (1987). Both estimate that, although commercially valuable timber is found throughout most of the country, in Rio Muni its potential for exploitation is restricted to 85 per cent of the territory and in Bioko to less than 20 per cent. Only about 1 per cent of Equatorial Guinea's commercial forest is found on Bioko (FAO/UNEP, 1981). Indeed, it was estimated in 1985 that there were only 4 sq. km of unexploited, commercially productive forest land on the island (Butynski and Koster, 1989a). Recent reports of industrial logging in forests on Bioko's mountain slopes are therefore rather alarming.

Timber has been extracted in the country since the early 1920s, its use coinciding largely with the development of the plywood market and the demand for veneer logs. Between 1930 and 1939, an average of 60,941 cu. m of timber was exported annually, a large percentage (50–99 per cent) of which was *Aucoumea klaineana* (Capdeveielle, 1969). There was a drop in exports between 1939 and 1963 followed by a recovery which continued until independence in 1968. Between 1963 and 1970 a total of 2,543,000 cu. m of wood was exported and at least 1017 sq. km of forest was logged. Jacobson (1968), Vannière (1969) and FAO (1972) all reported that exploitation in Equatorial Guinea, at 25 cu. m of timber per ha of forest, was more intense at that time than in any other African country. As a result, large individuals of valuable species such as *Aucoumea klaineana* have virtually disappeared from the western forests. From independence until 1981, exportation of wood was limited to 50,000 cu. m per year. Thereafter the volume extracted has been increasing so that 133,000 cu. m has been exported each year from 1986–8 (FAO, 1990). Veneer logs have been the major product, making up 90–100 per cent of the exported wood.

Timber exploitation is carried out entirely by private logging companies to which the government gives concessions for selective logging. The government can grant land through direct contract or public auction, and government inspectors work with the companies on the ground to verify volume extracted. Timber is subject to high export taxes and there are also taxes on petrol and handling. This high taxation and the excessive costs of road building – vehicles transporting wood are not allowed to use public roads – are considered to be the main obstacles to the expansion of the timber industry in the country. However, the National Forestry Programme aims to promote the expansion of forestry in Equatorial Guinea (to reach 450,000 cu. m) by improving and enlarging the forestry road network in Rio Muni to accommodate the increase in timber traffic (Lepitre *et al.*, 1988). In addition, it is proposed that a second port be opened at Cogo; at present timber is exported exclusively from the port of Bata.

Wildlife is under heavy hunting pressure north of the Uoro River, but the situation is much better south of the river (Castroviejo *et al.*, 1986; Fa, 1988; UICN, 1991). There are few guns in the country, so animals are usually trapped with wire snares. Bushmeat, particularly from duikers, primates and the giant forest rat *Cricetomys gambianus* is a major source of protein in human diet. Most of the meat is consumed locally, but some is sold on the roadside or taken to the market in nearby large towns (Castroviejo *et al.*, 1986; Fa, 1988; UICN, 1991).

Equatorial Guinea's National Forestry Programme aims to expand the logging industry in the country. J. Fa

Deforestation

Most deforestation in Equatorial Guinea is a result of shifting agriculture (Vannière, 1969). Forest is cleared, usually incompletely, the debris is burnt and the land cultivated for less than five years before being allowed to revert to forest or other secondary vegetation. FAO estimated that 140 sq. km of forest in Rio Muni and 10 sq. km in Bioko were being cleared for agriculture every year from 1980 to 1985, but this was largely in already exploited areas (FAO/UNEP, 1981). Lepitre *et al.* (1988) calculated that 1425 sq. km of land was affected by agriculturalists and that 80 per cent of this was forest. In Rio Muni, severe forest destruction has occurred in the Micomeseng-Ebebiyin-Mongomo region. Cacao plantations occupied much of the land before independence, but they now constitute only 3 per cent of their former size. There are, however, plans to rehabilitate the cacao industry and this may increase pressure on the forests.

A considerable amount of forest destruction on Bioko must have taken place between 1860 and 1902 since, in the former year, Gustav Mann found it totally unexploited (Hooker, 1862), but when Alexander (1903) visited in 1902 he reported a belt of cultivation more than 3 km wide around most of the island. By 1911, Mildbraed (1922) observed that most of the tropical lowland forest had been felled. Nosti reported that almost 97 per cent of the tropical zone was under cultivation in 1947, but, in contrast, the vegetation maps of Guinea (1968) showed that about 20 per cent of the lowland forest remained in that year. In short, it appears that during the 100 years from 1860–1959 about 50 per cent of Bioko's natural vegetation was destroyed with at least 80 per cent of the lowland tropical forest and 55 per cent of the subtropical montane forest being removed (Butynski, 1985). Until recently, though, there was little destruction of the remaining lowland rain forest on the island. Indeed, the abandonment of the plantations meant that much of the forested land that had been cleared for agriculture was regenerating to bush and secondary forest. However, the recent investments in rehabilitating the cacao industry are reversing these trends.

Biodiversity

Given the paucity of information on the plants and animals present in Equatorial Guinea, only a cursory view of the major aspects of the country's biodiversity is possible at present. However, the importance of Bioko as a locality of unique insular life forms and Rio Muni's position within a Pleistocene refuge (see chapter 2), makes Equatorial Guinea a significant country for biological diversity. For the country as a whole, it is reported that there are 141 mammal species and 392 bird species (Stuart *et al.*, 1990), but these are probably not complete lists.

The floristic affinities of Rio Muni are Guineo-Congolian (White, 1983). There are no figures available for the number of plant species there, but the flora is likely to be rich (Davis *et al.*, 1986). Little is known about any of the fauna on continental Equatorial Guinea. However, it is certain that Rio Muni contains important populations of large mammals such as gorilla *Gorilla gorilla*, chimpanzee *Pan troglodytes*, elephant *Loxodonta africana*, buffalo *Syncerus caffer nanus* and forest duikers (Castroviejo *et al.*, 1986; Fa, 1988; UICN, 1991). The primates of particular conservation importance on the mainland are the red-capped mangabey *Cercocebus torquatus*, mandrill *Mandrillus sphinx*, black colobus *Colobus satanas*, chimpanzee and gorilla (Oates, 1986).

Of the recorded 1105 plant species on Bioko, 49 are endemic (Davis *et al.*, 1986). Endemism is low for vertebrate species on this island. Indeed, the only known endemic vertebrates are a bird, the Fernando Po speirops *Speirops brunneus* (Amadon, 1953), a skink, *Scelotes poensis*, and a caecilian, *Schistometopum garzonheydti*. However, 28 per cent of the 65 mammals and 32 per cent of the 144 resident birds are endemic subspecies (Amadon, 1953; Eisentraut, 1973). A total of 32 amphibians and 52 reptiles have been recorded on the island (Butynski and Koster, 1989a). A large number of the plants and animals on Bioko have a very restricted distribution on the mainland, many of them being found only on Mt Cameroon (references in Butynski and Koster, 1989a). As a result, the populations of these species are frequently small and highly vulnerable to extinction (Butynski and Koster, 1989a). Rare primates on the island are an endemic subspecies of drill *Mandrillus leucophaeus poensis*, Preuss's guenon *Cercopithecus preussi insularis*, red-eared guenon *Cercopithecus erythrotis erythrotis* and Pennant's red colobus *Procolobus [badius] pennanti pennanti*.

Green and hawksbill turtles (*Chelonia mydas* and *Eretmochelys imbricata*) both nest on Bioko's beaches and are exploited – and probably depleted – by the local people for food and tortoiseshell (Butynski and Koster, 1989b). Fishes have been the focus of a study by Roman (1971), but more extensive work, being carried out by a Spanish Cooperation Programme, has resulted in the discovery of several new and previously unrecorded species.

Annobon has 17 single island endemic species of plant out of a total indigenous flora of 208 species (Exell, 1944, 1973). Only two species of reptile have been reported on the island, the endemic gecko, *Hemidactylus newtoni*, and *Lygodactylus thomensis*, a gecko that is found also on São Tomé and Príncipe. There are only nine species of resident land birds on Annobon, two are endemic subspecies, one confined to Annobon and the other shared with São Tomé and Príncipe.

Conservation Areas

Although the Monte Alén Partial Reserve and several other parks and reserves were initially proposed before 1970, they have only recently been officially protected. The newly protected areas are those proposed by Castroviejo *et al.* (1986) and UICN (1991). The network is composed of two areas on Bioko and five on Rio Muni, as well as the island of Annobon in its entirety (Table 18.2).

On Bioko, the protected area on Pico Basilé extends from the moist lowland forests to subalpine heaths and meadows. It is here that the endemic Fernando Po speirops is found, as well as nine of Bioko's ten primate species (Butynski and Koster, 1989a). There are no people living in this area, but hunting is common around the lower parts of the peak and along the road to the summit

Table 18.2 Conservation areas of Equatorial Guinea

Conservation Areas	*Area (sq. km)*
Altos de Nsok*	400
Estuario de Rio Muni*	700
Estuario de Rio Ntem (Rio Campo)*	200
Macizo de Monte Alén*	800
Macizo de los Montes Mitra*	300
Pico Basilé o Sta. Isabel (Bioko)*	150
Sur de la Isla de Bioco (Bioko)*	600
Isla de Annobon†	17
Total	3,167

(*Sources:* Castroviejo *et al.*, 1986; J. Castroviejo, pers. comm.)
* Area with moist forest within its boundaries according to Map 18.1, including some areas mapped as 'degraded forest'.
† Not mapped

(Castroviejo *et al.*, 1986; Fa, 1988; UICN, 1991). The other protected area, in the south of the island, contains Bioko's last extensive stands of primary lowland forest, with montane forest and subalpine heaths at higher altitudes (Castroviejo *et al.*, 1986; Fa, 1988; UICN, 1991). The beaches in this region are particularly important for nesting marine turtles (Butynski and Koster, 1989b). Hunting occurs here in only a few easily accessible areas and there is no human settlement.

The largest of the protected areas on Rio Muni, and one of the most important for large mammals, is Macizo de Monte Alén in the north of the Niefang mountain range. This site is an important water catchment area for the Uolo River and it contains intact primary forest which is unlikely to be exploited because of the rugged topography of the region (Castroviejo *et al.*, 1986; UICN, 1991). There is some secondary vegetation and agricultural land within the site and a low population density (less than 2.5 individuals per sq. km). Although animals are trapped, particularly around villages, there are still considerable populations of elephant, buffalo, chimpanzees, gorillas and leopards *Panthera pardus* in the area (Castroviejo *et al.*, 1986; Fa, 1988). The threatened mandrill is also present in high densities (Castroviejo *et al.*, 1986).

Castroviejo *et al.* (1986, 1990) consider that strict protection and exclusion of human activity are not feasible in the protected areas. Instead, they and Fa (1988) suggest that the areas should be managed for the sustainable use of natural resources; logging, cultivation and hunting should be kept to a minimum within core areas and rare animal species should be protected from exploitation, while there could be sustained-yield offtakes of the commoner species. However, Butynski and Koster (1989a) suggest that, on Bioko at least, a minimum of 60 per cent of each protected area should be designated as strictly protected core zones and that only the remaining 40 per cent in the buffer zone should be used for sustainable, multiple-use land practices.

Initiatives for Conservation

Conservation-related projects in Equatorial Guinea are a very recent phenomenon although some measures for protection of certain areas were implemented during the colonial period. Since independence in 1968, FAO has been the focal point for institutional support to the forestry sector in the country. The earlier projects focused on the need for forestry inventories (Vannière, 1969), but recent emphasis has been placed on evaluating the growth potential of the forestry sector (Catinot, 1980; Lepitre, 1986). A 1986 mission concentrated on the training of personnel for reforestation (Troensegaard, 1986). However, the most recent effort, begun in 1988, is a two and a half year project to culminate in a 1:200,000 forest resources map of Rio Muni, detailed inventories of tree resources, management of a pilot study and demonstration area of 160 sq. km for agroforestry, and initiation of action for a more rational use and administration of the country's forest resources. The latter includes the start of a data bank containing details of all forestry concessions, a management plan for the forests and the training of forestry personnel.

A Spanish supported project which started in 1985, focuses on biological research and nature conservation (Castroviejo *et al.*, 1986). It is funded by the Cooperation Española and undertaken by scientists from Doñana in Seville, Spain. They have worked on an inventory of the flora and fauna of Equatorial Guinea, recommended areas for conservation and have been training local people in research techniques.

The EEC, in collaboration with IUCN, has developed a regional project in Central Africa for the rational use and conservation of forest resources (Fa, 1988; UICN, 1991). This project will support Equatorial Guinea's National Forestry Programme by assisting training, ecological studies and practical conservation management. It will also help establish the conservation area at Monte Alén.

References

Alers, M. P. T. and Blom, A. (1988) *Elephants and Apes of Rio Muni: report of a first mission to Rio Muni (Equatorial Guinea).* Unpublished report to Wildlife Conservation International.

Alexander, B. (1903) On the birds of Fernando Po. *Ibis*: 330–403.

Amadon, D. (1953) Avian systematics and evolution in the Gulf of Guinea. *Bulletin of the American Museum of Natural History* **100**: 395–451.

Butynski, T. M. (1985) *Survey of the Rain Forests and Primates on Bioko Island (Fernando Poo), Equatorial Guinea.* A research proposal submitted to the New York Zoological Society. Unpublished.

Butynski, T. M. and Koster, S. H. (1989a) *The Status and Conservation of Forests and Primates on Bioko Island (Fernando Poo), Equatorial Guinea.* WWF Unpublished Report, Washington, DC, USA.

Butynski, T. M. and Koster, S. H. (1989b) *Marine Turtles on Bioko Island (Fernando Poo), Equatorial Guinea: A Call for Research and Conservation.* WWF Unpublished Report, Washington, DC, USA.

Capdeveielle, J. M. (1969) *Tres Estudios y un Ensayo sobre Temas Forestales de la Guinea Continental Española.* Instituto de Estudios Africanos, Consejo Superior de Investigaciones Científicas, Madrid, Spain.

Castro, M. and Calle, M. (1985) *Geografia de Guinea Ecuatorial. Programa de Colaboración Education con Guinea Ecuatorial.* Secretaria General Tecnica. Ministerio de Educacion y Ciencia, Madrid, Spain.

Castroviejo, J., Juste, J. and Castelo, R. (1986) *Proyecto de Investigación y Conservación de la Naturaleza en Guinea Ecuatorial.* Ministério de Asuntos Exteriores, España, Secretaría de Estado para Cooperación Internacional y para Iberoamérica Oficina de Cooperación con Guinea Ecuatorial.

Castroviejo, J., Blom, A. and Alers, M. P. T. (1990) Equatorial Guinea. In: *Antelopes Global Survey and Regional Action Plans. Part 3: West and Central Africa.* East, R. (ed.) IUCN/SSC Antelope Specialist Group. IUCN, Gland, Switzerland.

Catinot, R. (1980) *Perspectives de Développement forestier en Guinée Equatoriale et Programme d'Action.* FAO, Rome, Italy.

Davis, S. D., Droop, S. J. M., Gregerson, P., Henson, L., Leon, C. J., Villa-Lobos, J. L., Synge, H. and Zantovska, J. (1986) *Plants in Danger: What do we know?* IUCN, Gland, Switzerland and Cambridge, UK.

Eisentraut, M. (1973) Die Wirbeltierfauna von Fernando Poo und Westkamerun. *Bonner Zoologische Monographien* **3**: 1–428.

Exell, A. W. (1944) *Catalogue of the Vascular Plants of São Tomé (with Príncipe and Annobon).* British Museum (Natural History), London. 428 pp.

Exell, A. W. (1952/3) The vegetation of the islands of the Gulf of Guinea. *Lejeunica* **16**: 57–66.

Exell, A. W. (1973) Angiosperms of the islands of the Gulf of Guinea (Fernando Po, Príncipe, São Tomé and Annobon). *Bulletin of the British Museum (Natural History), Botany* **4**(8): 325–411.

Fa, J. E. (1988) *Equatorial Guinea: Wildlife Conservation and*

Rational Use of Natural Resources. Universidad Nacional Autonoma de Mexico, Mexico.

FAO (1972) *Forestry Report on Equatorial Guinea.* Based on assignment FAO/EQG/70/001 of A. G. Forester. FAO, Rome, Italy.

FAO (1988) *An Interim Report on the State of Forest Resources in the Developing Countries.* FAO, Rome, Italy. 18pp.

FAO (1991) *FAO Yearbook of Forest Products 1978–1989.* FAO Forestry Series No. 24 and FAO Statistics Series No. 97. FAO, Rome, Italy.

FAO/UNEP (1981) *Tropical Forest Resources Assessment Project. Forest Resources of Tropical Africa. Part II Country Briefs.* FAO, Rome, Italy.

Fry, C. H. (1961) Notes on the birds of Annobon and other islands in the Gulf of Guinea. *Ibis* **103**: 267–76.

Guinea, E. (1946) *Ensayo Geobotanico de la Guinea Ecuatorial.* Instituto de Estudios Africanos, Consejo Superior de Investigaciones Científicas, Madrid, Spain.

Guinea, E. (1968) Fernando Po. In: *Conservation of Vegetation in Africa South of the Sahara.* Hedberg, I. and Hedberg, O. (eds). Acta Phytogeographica Suecica **54**: 130–2.

Harrison, M. J. S. (1990) A recent survey of the birds of Pagalu (Annobon). *Malimbus* **11**: 135–43.

Hooker, J. D. (1862) On the vegetation of Clarence Peak, Fernando Po; with descriptions of the plants collected by Mr Gustav Mann on the higher parts of that mountain. *Proceedings of the Linnean Society of London* **6**: 1–30.

Hughes, R. H. and Hughes, J. S. (1991) *A Directory of Afrotropical Wetlands.* IUCN, Gland, Switzerland and Cambridge, UK/UNEP, Nairobi, Kenya/WCMC, Cambridge, UK.

Jacobson, C. A. (1968) *Logging in the Rio Muni (Guinea Ecuatorial).* Internal FAO document, Rome, Italy.

Lepitre, C. (1986) *Exploitation Forestière en Guinée Equatoriale.* Ministère de la Coopération Francaise, Centre Technique Forestier Tropical, Paris, France.

Lepitre, C., Mille, G. and de Royer, G. (1988) *Programme d'Appui au Secteur Forestière en Guinée Equatoriale.* Fond Européen de Développement, Brussels, Belgium.

Malleux, J. (1987) *Informe de la Misión de Inventario y Ordenación Forestal.* FAO, Rome, Italy.

Mildbraed, J. (1922) Fernando Poo in Wissenschaftlicke Ergebnisse der Zweiten Deutschen Zentral-Africa-Expedition, 1910–1911. *Botanik, Leipzig* **2**: 164–95.

Nosti, J. (1947) El bosque en Fernando Poo. *Africa* año VI, núms 66–7.

Oates, J. F. (1986) *Action Plan for African Primate Conservation: 1986–1990.* IUCN/SSC Primate Specialist Group, Stony Brook, New York, USA.

Ocaña Garcia, M. (1962) Factores que influencian la distribución de la vegetación en Fernando Poo. *Archivos del Instituto de Estudios Africanos* XIV **55**: 67–85.

Roman, B. (1971) *Peces de Rio Muni, Guinea Ecuatorial (aguas dulces y salobres).* Fundación la Salle de Ciencias Naturales, Caracas, Venezuela.

Stuart, S. N., Adams, R. J. and Jenkins, M. D. (1990) *Biodiversity in Sub-Saharan Africa and its Islands: Conservation, Management and Sustainable Use.* Occasional Papers of the IUCN Species Survival Comission No. 6. IUCN, Gland, Switzerland.

Troensegaard, J. (1986) *Formulación del TCP/s-Y-EQG-20, Capacitación en reforestación.* Informe de Misión. FAO, Rome, Italy.

UICN (1991) *Conservacíon de los Ecosistemas Forestales de Guinea Ecuatorial.* Basado en el trabajo de John E. Fa. UICN, Gland, Switzerland and Cambridge, UK. 221 pp.

Vannière, B. (1969) *Mission d'Inventaire Forestier, 20 Juin–29 Juillet 1969, Guinée Equatoriale.* FAO, Rome, Italy.

White, F. (1983) *The Vegetation of Africa: a descriptive memoir to accompany the Unesco/AETFAT/UNSO vegetation map of Africa.* Unesco, Paris, France. 356 pp.

World Bank (1985) *Staff Appraisal Report Equatorial Guinea: Cocoa Rehabilitation Project.* Western Africa Projects Department, World Bank, Washington, DC, USA.

Authorship

John Fa, Gibraltar with contributions from Javier Castroviejo Bolibar, Coto Doñana, Spain.

Map 18.1 Forest cover in Equatorial Guinea

Forest cover for Rio Muni shown on Map 18.1 has been extracted from a national forest map (no title), produced in 1960 and prepared for the Servicio Geográfico del Ejército. There is no recent vegetation map of Equatorial Guinea available. The source map has been published at a scale of 1:100,000 in 15 sheets but does not cover the island of Bioko. Three forest types have been extracted, namely, *Bosque denso* shown as lowland rain forest in this Atlas, *Bosque claro* mapped as degraded lowland rain forest and *Mangle* covering the distribution of mangrove. Although the source map is over 30 years old, it is believed that there have been no major changes in forest cover since then and therefore Map 18.1 provides a fairly accurate account of the distribution of closed and disturbed forest within Rio Muni.

Vegetation cover for Bioko has been taken from a sketch map accompanying a report written by Butynski and Koster (1989a), on which the approximate present distribution of cultivation, secondary forest and the four main natural vegetation types on Bioko Island are depicted. *Lowland forest* and *Montane forest* vegetation types have been taken from this map. There are no spatial data for the extent of the forests remaining on Annobon.

Spatial data for conservation areas have been taken from various sketch maps and data on file at WCMC.

19 Gabon

Land area 257,670 sq. km
Population (mid-1990) 1.2 million
Population growth rate in 1990 2.2 per cent
Population projected to 2020 2.6 million
Gross national product per capita (1988) US$2970
Rain forest (see map) 235,445 sq. km‡
Closed broadleaved forest (end 1980)* 205,000 sq. km
Annual deforestation rate (1981–5)* 150 sq. km
Industrial roundwood production† 1,222,000 cu. m
Industrial roundwood exports† 913,000 cu. m
Fuelwood and charcoal production† 2,478,000 cu. m
Processed wood production† 354,000 cu. m
Processed wood exports† 53,000 cu. m

* FAO (1988)
† 1989 data from FAO (1991)
‡ Figure almost certainly a considerable overestimate resulting from the poor resolution of the source map used, which does not distinguish agricultural enclaves within the remaining forest and extensive areas of degraded forest.

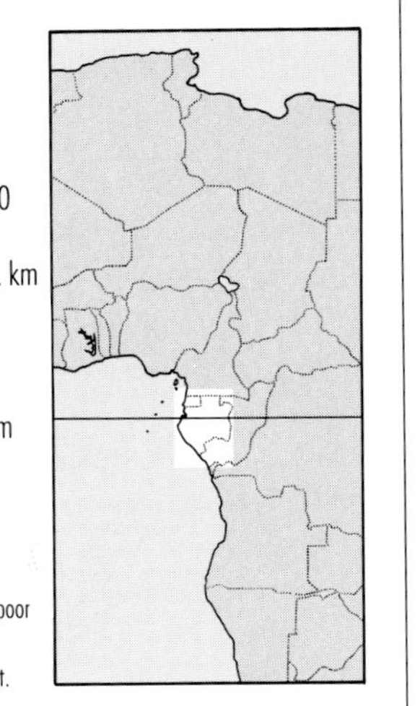

Gabon, on the west coast of equatorial Africa is not a country typical of the region. Tropical forest still covers 88 per cent of its land area, deforestation rates are low and large populations of elephants, gorillas and chimpanzees, species severely threatened in most countries, still exist.

Historically, the tropical forest ecosystem of the country has been naturally protected by the low human population density, limited transport routes in the interior and by many of the people having found productive employment in development associated with the extraction of minerals such as oil, manganese and uranium. By the early 1980s Gabon had the second highest per capita income in continental Africa, after Libya. Many large development projects were undertaken, the biggest being the construction of a railway, completed in 1987, linking the capital Libreville, on the coast, to Franceville 650 km inland. The decline in oil prices reduced Gabon's revenues by almost 50 per cent and the country now has the second largest debt per capita in Africa. These two factors have introduced a real threat to the country's forests as the railway has allowed selective logging to spread to central and eastern regions and exploitation of all natural resources is being accelerated to generate export revenue. In the scramble to raise revenues, the standard of logging practices is said to be declining.

Five protected areas exist which together account for 6.9 per cent of the country's land area; however, the forests of all but the smallest have been selectively logged. At present no forest area in Gabon is protected from selective logging although the forests of the north-east which do not contain okoumé (an important timber tree) are not immediately threatened. There are now plans to extend the network of protected areas and to exclude logging from some parts of the forest within existing reserves. The long-term conservation of Gabon's forests requires not only that new reserves be created but also that the selective logging practices be improved with a view to ensuring sustainability.

Introduction

Gabon has an area of 267,667 sq. km, straddling the equator on the west coast of Africa (2°12'N to 3°55'S and 8°20'–14°40'E). Its Atlantic coastline stretches for 750 km and it borders Equatorial Guinea and Cameroon to the north and Congo to the east and south.

The climate is equatorial with mean temperatures varying between 21°C and 27°C. Rainfall is relatively low, ranging from 1500 mm in the north-east and in the savanna areas, to 3300 mm in the north-west. There are two wet seasons (March–June and October–December) and two dry seasons (January and July–September). Humidity is high even during the long dry season when evaporation rates and temperatures are low due to persistent cloud cover (Hladik, 1973).

The Ogooué river drains about two-thirds of Gabon's area. It originates in the south-east close to the border with Congo, and flows north and then west to reach the Atlantic through a complex delta. Three geomorphic zones are recognised: the sedimentary coastal basin which is relatively flat with altitudes rising only a few hundred metres above sea level; the north-south ranges of mountains (Monts de Cristal in the north and Massif du Chaillu in the south) with rugged terrain rising to 900 m; and the eastern plateaux intersected by deep valleys. The highest point in the country is Mont Sassamongo (1001 m) in the north-east.

The human population density is low and Gabon is in the low fertility crescent of the Congo basin. The official estimate of the population, based on a census in 1980, is 1,232,000 but other sources suggest a lower figure of about 900,000. The figure given for 1990 by the Population Reference Bureau is 1.2 million. At least 50 per cent of the population are urban, living in the coastal towns of Libreville and Port Gentil, the mining towns of the south-east, Moanda and Mounana, or in the two other sizeable towns, Franceville and Oyem. Rural population density does not exceed two inhabitants per sq. km and vast areas of the interior are unpopulated. A resettlement programme in the 1950s led to all villages being sited on roads or navigable rivers. Meanwhile, a few indigenous hunter-gatherers, the pygmies, maintain their traditional way of life in parts of the north-east and south-west.

Rural land is state owned. Forest management is overseen by the Ministry of Water and Forests which has 11 divisions, three of which are concerned with forestry (Inventories, Forest Exploitation and Reforestation) and one with faunal management (Wildlife and Hunting).

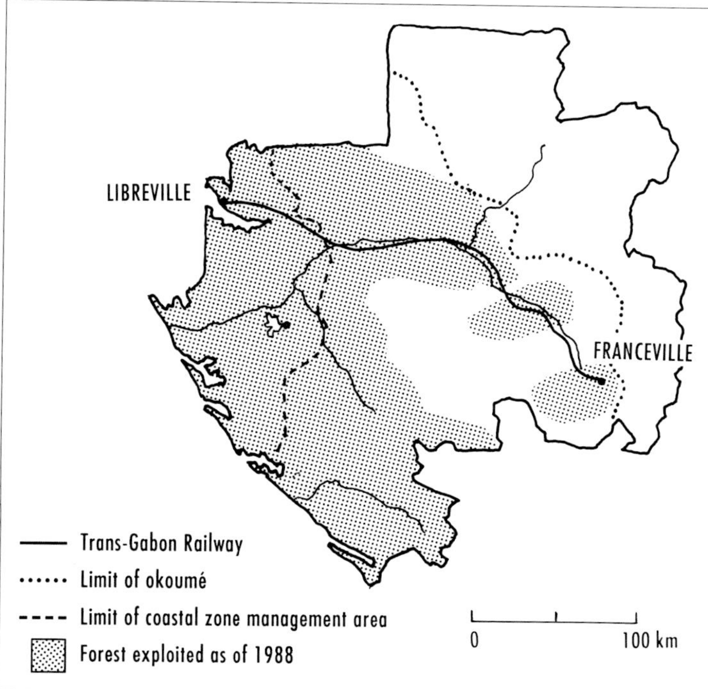

Figure 19.1 The distribution of okoumé in Gabon (it occurs to the west but not to the east of the country) *(Source:* UICN, 1990)

The Forests

Gabon falls within the Guineo-Congolian regional centre of endemism (White, 1983) and the diversity of plants is very high (Hladik, 1986; , Reitsma, 1988; UICN, 1990). The forests are dominated by trees of the Caesalpiniaceae family, but other well represented families are the Burseraceae, Euphorbiaceae and Olacaceae (Reitsma, 1988). Subdivision of the forest into categories is not easy as data are incomplete, but useful distinctions can be made both from an economic point of view and in relation to geomorphology. The former approach distinguishes between forests with the important timber tree okoumé *Aucoumea klaineana* and those without; thus Gabon is divided into a larger western zone, covering about 70 per cent of the country, in which okoumé occurs and is usually dominant, and a smaller zone in the east where okoumé is absent (Figure 19.1). The second approach to forest classification distinguishes three major types (Caballé, 1978; UICN, 1990):

- Forests of the sedimentary basin where *Sacoglottis gabonensis* dominates and okoumé, *Desbordesia glaucescens* and *Dacryodes buettneri* are common.
- Forests of the mountainous zones where *Sacoglottis* disappears while okoumé, *Desbordesia* and *Dacryodes* are abundant and *Tetraberlinia polyphylla* and *Monopetalanthus* spp. become common. The mountainous zone of the Monts du Cristal has a high level of species endemism.
- Forests of the eastern plateaux, where *Desbordesia* becomes rare while *Scyphocephalium ochocoa* and *Paraberlinea bifoliolata* are abundant. In the north-east, okoumé is absent.

Grass savannas interspersed with gallery forests cover a total area of about 35,000 sq. km in the south-west, south-east and centre of the country. It has been suggested that these savannas are man-made, dating from the migrations of Bantu tribes in the 15th and 16th centuries (Descoings, 1974; Fontes, 1978), but recent research suggests that the savannas of central Gabon may date at least from the Pleistocene and would be colonised by forest if left unburnt (Oslisly and Peyrot, pers. comm.) In fact, both human interference and natural climatic variations are probably responsible for the existing vegetation patterns.

Mangroves

Mangrove forests characterised by *Rhizophora racemosa* cover about 3000 sq. km in the Gabon estuary and in the estuary of the river Ogooué alone. The area of mangroves shown on Map 19.1 is a little over 6000 sq. km (Table 19.1). They are not subject to any commercial exploitation at present.

Forest Resources and Management

FAO (1988) estimated that there were 205,000 sq. km of closed broadleaved forest at the end of 1980. This is only 79.6 per cent of Gabon's land area. A later estimate (UICN, 1990) suggests that 88 per cent of the land, or around 227,500 sq. km, is still forested. Table 19.1, derived from Map 19.1, suggests that as much as 91 per cent of Gabon is forested. However the source map (see Map Legend on p. 174) is generalised and undoubtedly overestimates the extent of forest cover.

Considerable areas of the forest has been either logged or cleared for shifting agriculture at some time in the past. This is especially true in the north-central region, close to the border with Equatorial Guinea where much of the forest is secondary. The apparent contrast in the state of the forest on the two sides of the frontier (see Maps 18.1 and 19.1) is evidence of the very generalised nature of the maps available for Gabon. The southern forests of the central zone are also substantially degraded (Figure 19.2).

Selective logging began in Gabon at the turn of the 20th century. Until the discovery of oil reserves in the 1950s, wood represented almost 90 per cent of the country's exports but, by 1985, this had fallen to 6 per cent. The figure for 1987 was 12 per cent, not due to an increase in exports of wood, but to the declining value of oil production.

Okoumé is a species of Burseraceae found only in Gabon, Equatorial Guinea, Cameroon and Congo and it is by far Gabon's most important timber species, responsible for at least 60 per cent of exports in 1987. Okoumé is a light wood (density 0.6) which is valuable for peeling for plywood and it dominates the selective logging industry in Gabon. Forest inventories indicate that there are 100 million cu. m of okoumé in Gabon's forests and trees sufficiently large to be harvested (that is, at least 70 cm dbh) are found at densities of 1–3 per hectare. Annual production peaked at 1.6–1.8 million cu. m between 1969 and 1973 (Barret, 1983), but declined because of reduced demand from Europe in 1974, and has since remained stable at just under one million cu. m per year. The stated aim of the government is to raise annual production of okoumé to 2.1 million cu. m (Diop, 1989). Fifty-five other species are logged but only 14 of these accounted for more than 5000 cu. m of exported wood in 1987.

Table 19.1 Estimates of forest extent in Gabon

	Area (sq. km)	*% of land area*
Rain forests		
Lowland	225,276	87.4
Swamp	4,040	1.6
Mangrove	6,129	2.4
Totals	235,445	91.4

(Based on analysis of Map 19.1. These figures and Map 19.1 from which they are derived undoubtedly overestimate forest cover in Gabon. See Map Legend on p. 174 for details of sources.)

Gabon

Forest management practices are defined by a law dating from July 1982 (No. 1/82/PR), but legislation relating to selective logging is not yet complete. Minimum diameters, between 55 and 70 cm, are fixed for the species of trees which are commonly logged and a minimum interval of 20 years should elapse between successive logging of any area of forest, but neither of these stipulations is strictly enforced.

For management purposes, forests in which okoumé occurs are divided into two zones covering about 90 per cent of the geographical range of okoumé. The first is the coastal zone where licences are issued only to Gabonese nationals, while the second zone, encompassing the inland forests, is open to all-comers. Exploitation in this zone is closely linked to the recently completed railway line that runs from Libreville to Franceville in the south-east. The 650 km railway has made selective logging an economically feasible activity by providing the means of transporting logs to the coast. Licences for concessions near the track were offered in advance of the completion of the railway and the fees contributed to the construction costs. Currently, approximately 50 companies are involved in selective logging.

In 1988 it was estimated that 46 per cent (105,000 sq. km) of Gabon's forest had already been selectively logged at least once and that, each year about 2500 sq. km of forest is logged, 60 per cent of which is primary forest (UICN, 1990). Extraction rates for okoumé are low, averaging 1.5 trees per hectare, and an estimation of the direct and indirect damage caused by logging to the forest is that between 10 and 20 per cent of the canopy is destroyed.

Okoumé is marketed by the state company Société Nationale des Bois du Gabon (SNBG) which has a monopoly on sales of okoumé and ozigo *Dacryodes buettneri*. Other species can be sold directly by forestry companies. The SNBG has experienced difficulties maintaining an export market for okoumé and, in early 1989, imposed strict quotas on the amount of okoumé each logging company could produce. The quotas represented a reduction in output of okoumé compared to 1988 levels and contradicted the government policy, announced six months earlier, to increase timber production in order to help compensate for reduced oil revenues. Severe financial problems ensued for most of the logging companies as they had invested in equipment with the aim of increasing production in 1989. To help recoup this outlay they were forced to diversify their activities, increasing the logging of other species to compensate for lost revenues from okoumé. The change is dramatic and, although statistics are not yet available, okoumé's dominance of the logging industry in Gabon is now much reduced. This has important implications for forest management as extraction rates have risen and forests that had been recently logged for okoumé are being relogged to extract the other timber species for which a market, negotiable independently by each logging company, exists.

Reforestation projects are in progress at six sites in Gabon but, at present, involve a total of only 310 sq. km (UICN, 1990). Planting and husbandry of okoumé in logged forest is a promising way of increasing future production. While attempts to plant okoumé in single-species plantations have not always been successful, artificially increasing okoumé density is predicted to give annual production levels of 15–18 cu. m per hectare after 35–40 years (Nicoll and Langrand, 1986). The current rate of reforestation is about 10 sq. km per year – equivalent to the area of forest that is selectively logged each day in Gabon.

Deforestation

Deforestation is rare in Gabon compared to most other tropical forest countries. The low human population density in rural areas limits forest clearance for plantations. Most logging companies work from isolated bases and thus clearance for shifting cultivation is rarely associated with selective logging.

Figure 19.2 Floristic regions of Gabon

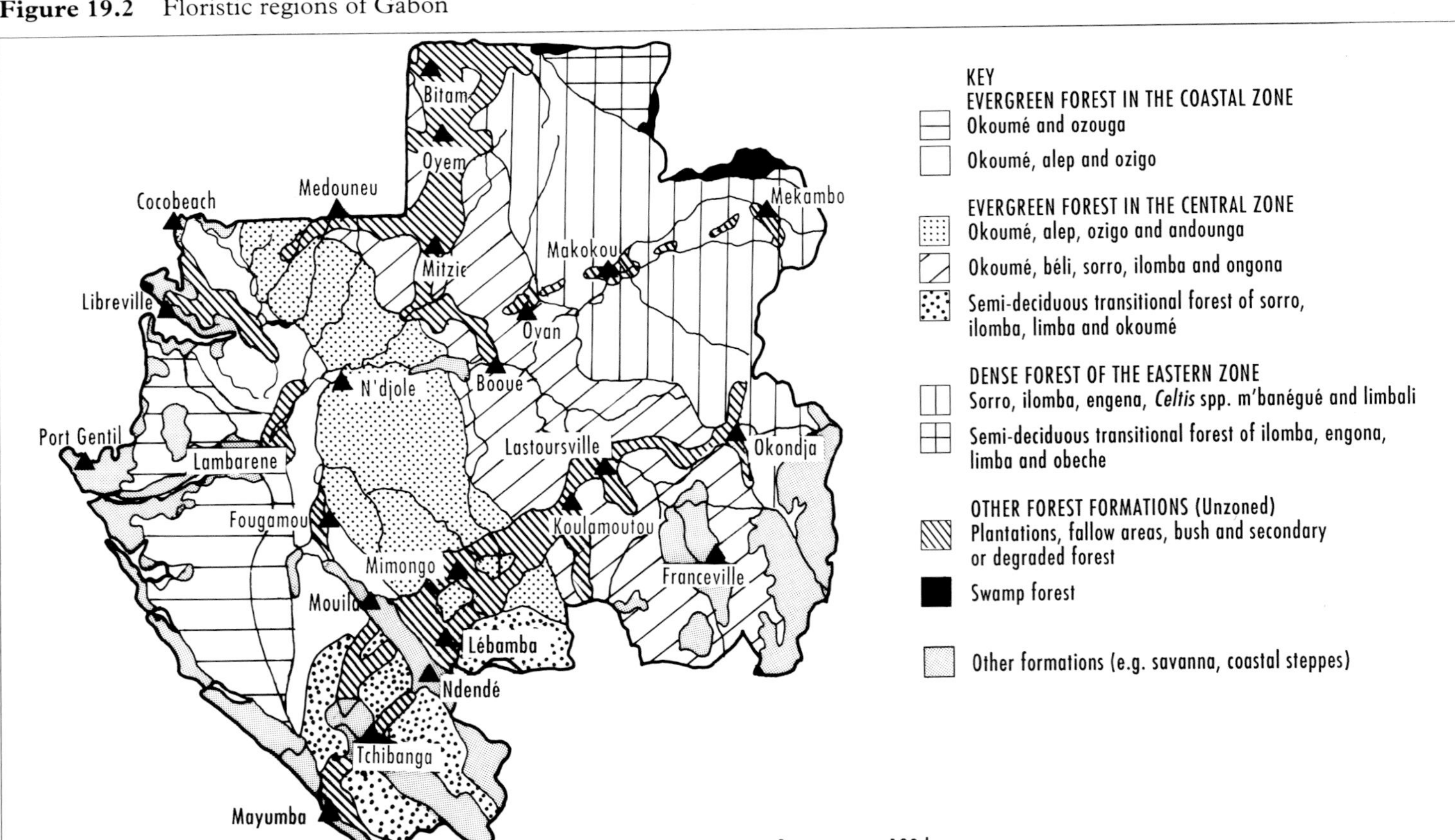

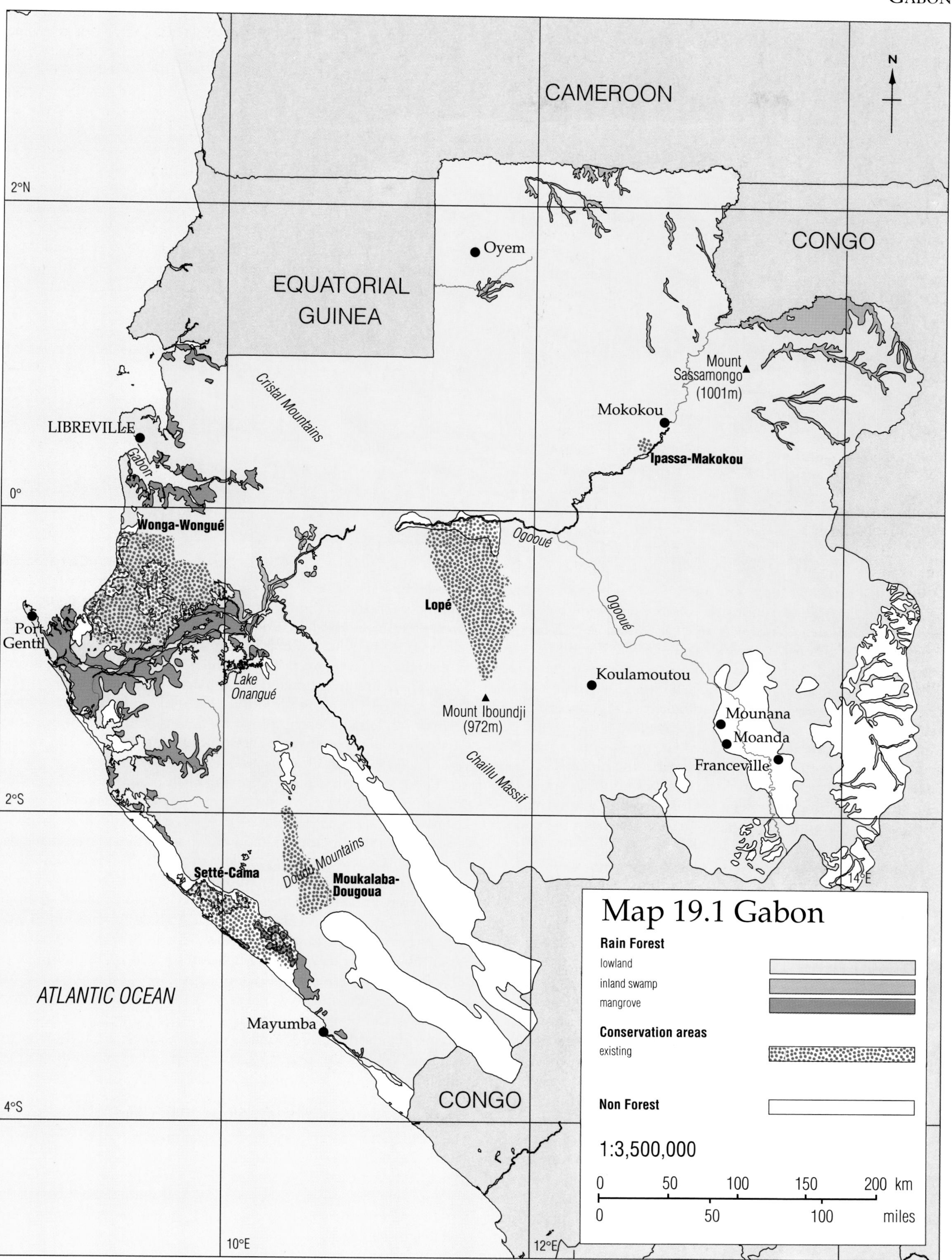
CAMEROON
EQUATORIAL GUINEA
CONGO
Oyem
Mount Sassamongo (1001m)
Mokokou
Ipassa-Makokou
LIBREVILLE
Gabon
Cristal Mountains
2°N
0°
2°S
4°S
10°E
12°E
14°E
Wonga-Wongué
Ogooué
Lopé
Port Gentil
Lake Onangué
Mount Iboundji (972m)
Koulamoutou
Mounana
Moanda
Franceville
Chaillu Massif
Doudou Mountains
Setté-Cama
Moukalaba-Dougoua
ATLANTIC OCEAN
Mayumba
CONGO
N
Map 19.1 Gabon
Rain Forest
lowland
inland swamp
mangrove
Conservation areas
existing
Non Forest
1:3,500,000
0 50 100 150 200 km
0 50 100 miles

A hut in north-eastern Gabon made entirely from forest products by Bakola people. C. Doumenge

In 1988 a total of 130 sq. km (0.05 per cent of the forest area) had been clear felled for industrial plantations of oil palms, rubber, coffee and cacao. A project to clear 227 sq. km of forest for *Eucalyptus* plantations was abandoned in 1982. Current development plans include expansion of commercial crops of basic foods such as bananas, manioc and rice, with the eventual aim that Gabon become agriculturally self-sufficient. The country is already self-sufficient for sugar, palm oil, pineapples, chickens and eggs. Some of the plantations are located in the savanna zone of the south-east, as are the chicken farms and cattle ranches. The area covered by subsistence plantations was about 600 sq. km in 1988. The area cleared for these each year is currently diminishing as a result of the migration of people to urban areas from the countryside.

Although selective logging in Gabon does not lead to deforestation, it is far from clear whether current logging practices are sustainable. Research on this topic is urgently needed as there are reasons to believe that the country's forests are particularly fragile given the generally low rainfall (Tutin and Fernandez, 1987). With the increased diversity of tree species being logged, extraction rates have risen and this obviously increases the percentage of the canopy eliminated. Forests in coastal areas that have been logged several times differ conspicuously in appearance from unlogged forests, with lower canopy and dense herbaceous undergrowth which appears to limit the regeneration of many trees (Nicoll and Langrand, 1986).

While shifting cultivation does not pose a great threat in Gabon, hunting certainly increases as a result of logging. The roads opened by loggers allow access to new hunting areas and the trans-Gabon railway provides the means for the products of large-scale hunting to be transported to urban areas. Subsistence hunting by small rural communities has a limited impact, but professional hunters who provide meat for larger towns can decimate the fauna of a particular area in a short time.

Biodiversity

Gabon's forests are home to a diverse array of flora and fauna. Recent studies (Hladik, 1986; Reitsma, 1988) have shown plant diversity to be equal to that of tropical forests of South America although these were previously considered unrivalled in their diversity. There are an estimated 8000 plant species and of the 1900 already described in the 'Flore du Gabon', an impressive 19 per cent are endemic. Description of the flora is far from complete and discoveries of new species are almost commonplace (see, for example, Hallé, 1987 and Floret *et al.*, 1989).

There is a diverse range of mammalian fauna with 130 species recorded from the best-studied area, close to Makokou in the north-east (CENAREST, 1979). At least 20 species of primate occur in Gabon, including important populations of western lowland gorilla *Gorilla g. gorilla*, chimpanzee *Pan t. troglodytes*, black colobus *Colobus satanas* and mandrill *Mandrillus sphinx* and a newly discovered endemic species of monkey, the sun-tailed guenon *Cercopithecus solatus* (Harrison, 1988). A recent census found elephant *Loxodonta africana* to be widely distributed and estimated a total population of 74,000 (Michelmore *et al.*, 1989). Manatee *Trichechus senegalensis* and hippopotamus *Hippopotamus amphibius* occur in lagoons and coastal rivers. Water chevrotain *Hyemoschus aquaticus* and otter shrew *Potamogale velox* occur throughout the country at low densities. Bongo *Tragelaphus euryceros* and giant forest hog *Hylochoerus meinertzhageni* appear to be limited to parts of the north-east and are probably rare. Seven species of forest duiker exist including an endemic sub-species, the white-legged duiker *Cephalophus ogilbyi crusalbum* (Blom *et al.*, 1990). Several species become common in forests close to savanna zones, including bushbuck *Tragelaphus scriptus*, buffalo *Syncerus caffer nanus* and yellow-backed duiker *Cephalophus sylvicultor*. Lion *Panthera leo* previously occurred in the savanna zones but had been eliminated by the early 1970s. Leopard *Panthera pardus* and golden cat *Felis aurata* are widely distributed.

Table 19.2 Conservation areas of Gabon

	Area (sq. km)
Strict Nature Reserves	
Ipassa-Makokou*	100
Presidential Reserves	
Wonga-Wongué*	4,800
Wildlife Management Areas	
Lopé*[1]	5,000
Setté-Cama*[2]	7,000
Unclassified	
Moukalaba-Dougoua*	1,000
Others	
Sibang Experimental Forest Station†	<1
Total	17,900

(*Sources:* UICN, 1990; WCMC, *in litt.*)

* Area with moist forest within its boundaries according to Map 19.1..

† Not mapped – no location data available.

[1] Lopé Reserve incorporates the following zones: Réserve de faune de l'Offoué-Okanda and the Domaine de chasse de la Lopé-Okanda.

[2] Setté-Cama Reserve incorporates the following zones: Réserve de faune du Petit Loango, the Réserve de faune de la plaine Ouango, the Domaine de chasse d'Iguéla, the Domaine de chasse de Ngoué-N'Dogo and the Domaine de chasse de Setté-Cama.

The avifauna of Gabon is also diverse with 618 species recorded to date (P. Christy, pers. comm.). Several threatened or poorly known species occur, such as the Loango slender-billed weaver *Ploceus subpersonatus*, Damara tern *Sterna balaenarum*, grey-necked rockfowl *Picathartes oreas* and Dja river warbler *Bradypterus grandis*. The majority of birds in Gabon are sedentary forest species but savanna species typical of the West African region are also well represented. Migratory birds include at least 90 palearctic species and about 50 inter-African species.

In the Makokou area 65 species of land and fresh water reptiles have been recorded (CENAREST, 1979) but systematic data are lacking for other parts of the country. Two species of small crocodile (*Crocodylus cataphractus* and *Osteolaemus tetraspis*) are widely distributed, whereas the Nile crocodile *Crocodylus niloticus* has disappeared from many rivers as a result of hunting. Gabon is an important nesting area for leather-backed turtle *Dermochelys coriacea* but human predation on eggs is high. Approximately 100 species of amphibians have been recorded. Systematic research has been conducted on neither fish nor invertebrates but the rare giant African swallowtail butterfly *Papilio antimachus* has been recorded in the Lopé Reserve.

The low human population density, the isolation of much of the interior and the mineral resources which have reduced the need to exploit timber, have combined to provide natural protection of huge areas. These historical factors explain why Gabon shelters larger numbers of elephants, gorillas and chimpanzees than any other country (Tutin and Fernandez, 1984; Michelmore *et al.*, 1989; WCI, 1989). However, the situation is rapidly changing, principally since the completion of the railway in 1987, as this has opened up previously inaccessible regions to selective logging and hunting. The current economic problems are also increasing pressure on the forest. Gabon's external debt crisis has led to the current policy of maximising expoitation of the country's natural resources despite the fact that demand for these products on international markets is weak.

Conservation Areas and Initiatives

Protected areas in Gabon currently total 17,900 sq. km (Table 19.2) which represents 6.9 per cent of the land area. No national parks exist and selective logging has affected the forests of four of the five reserves; indeed, the four large reserves all contain sizeable areas of savanna. Three faunal reserves are administered by the Wildlife Department: the Lopé in central Gabon, Moukalaba in the south-west, and Setté-Cama on the south-west coast. The other two protected areas are Wonga-Wongué, the presidential reserve on the coast south of Libreville, and Ipassa-Makokou in the north-east, a Biosphere reserve run by the National Centre for Scientific and Technical Research (CENAREST) which has an ecological field station.

There are problems with all the existing reserves: none is legally protected from selective logging; none has a management plan and all are critically understaffed and underfunded. A planned EEC-financed project for the Lopé Reserve would resolve a number of these problems but implementation is urgent as three logging companies are currently working within the reserve. Research on the socio-ecology of chimpanzees and gorillas and on the impact of selective logging on populations of large mammals is being conducted at the field station in this reserve. Few mammals remain in the small Ipassa-Makokou Reserve which has been very heavily poached, due to its proximity to the town of Makokou and to a total lack of policing (Lahm, pers. comm.).

A recent EEC/IUCN report (UICN, 1990) recommends that inventories and more studies of the forests be made, that the forests are managed for sustainable use of the timber, that there is better control of commercial hunting and that legislation and education about the forests is improved. The report also proposes the creation of several additional reserves and extensions of some of the existing protected areas (Figure 19.3) as the present network does not include

Figure 19.3 Areas suggested by IUCN for protection (existing conservation areas are shown but not numbered) (*Source:* UICN, 1990)

1 Monts Doudou (this area would protect the endemic white-legged duiker and would extend Setté-Cama to make the largest reserve in Gabon). **2** Soungou-Milonda. **3** Mont Iboundji (this is an extension of Lopé Reserve). **4** Forêt des Abeilles (suggested specifically to protect the endemic sun-tailed guenon and is an extension of the Lopé Reserve). **5** Mingouli. **6** Monts de Belinga. **7** Grottes de Belinga (an important area for bats). **8** Djoua. **9** Minkébé. **10** Tchimbélé. **11** Akanda. **12** Mondah. **13** Ozouri. **14** Lac Onangué. **15** Leconi.

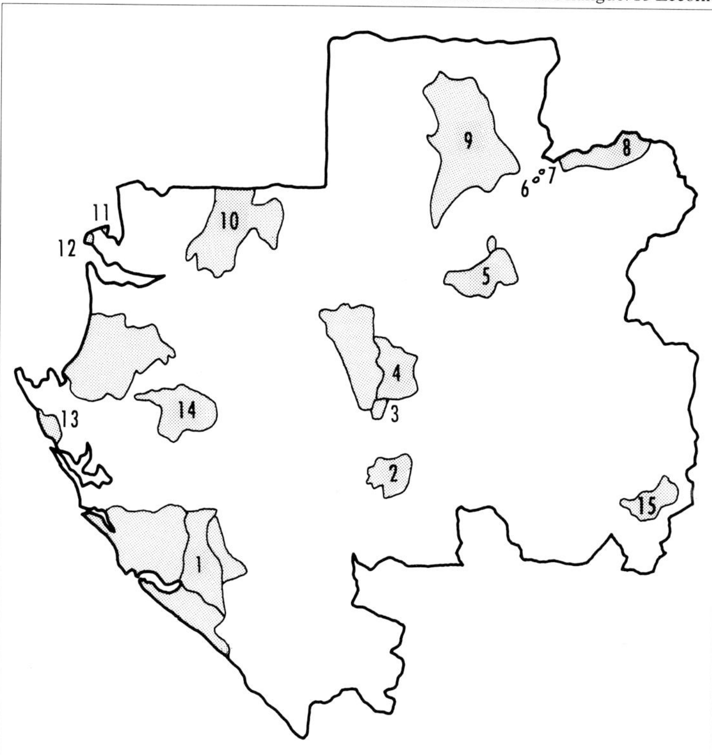

the full range of diversity of Gabon's flora and fauna. One of the suggestions in the report was that Moukalaba Reserve and the Setté-Cama Reserve are amalgamated which, with the inclusion of the Monts Doudou region in between, would create the country's largest protected area (Gamba Reserve) of approximately 9000 sq. km.

A survey funded by WWF is presently being undertaken in one of the largest of the proposed areas, Minkébé (7000 sq. km) in the north-east of the country. Minkébé is outside the limits of okoumé distribution and thus less threatened by selective logging and would constitute a very significant addition to Gabon's protected area network (McShane-Caluzi and McShane, 1990). It has been estimated that creation of a new reserve at Minkébé and improvement of conditions at Petit Loango (part of the Setté-Cama Reserve) would assure the protection of 40,000 elephants and this has been recommended by the CITES' Working Group on African Elephants (Anon., 1989). Other important new reserves are designed to protect the endemic sun-tailed guenon, the white-legged duiker, centres of floristic endemism in the Monts de Cristal and Mont Iboundji and areas of mangrove close to Libreville.

There are encouraging signs that both national and international bodies are concerned with conservation issues in Gabon. An Ecology Party is one of the 40 political groups represented in the newly formed National Assembly Party. WWF is active in conservation projects while large development schemes with a strong conservation element are planned by the EEC and the World Bank. The potential for ecosystem conservation still exists in Gabon (McShane, 1990), but if there is to be 'conservation before the crisis', action must not be too long delayed.

References

Anon. (1989) *Plan d'Action pour l'Eléphant. Document du Groupe de Travail de la CITES sur l'Eléphant d'Afrique.*

Barret, G. (1983) L'exploitation forestière. In: *Géographie et Cartographie du Gabon*, pp. 58–63. Paris, France.

Blom, A., Alers, M. P. T. and Barnes, R. F. W. (1990) Gabon. In: *Antelopes Global Survey and Regional Action Plans. Part 3: West and Central Africa.* East, R. (ed.). IUCN/SSC Antelope Specialist Group. IUCN, Gland, Switzerland.

Caballé, G. (1978) Essai sur la géographie forestière du Gabon. *Adansonia Séries 2* **17**(4): 425–40.

Caballé, G. (1983) Végétation. In: *Géographie et Cartographie du Gabon*, pp. 34–7. Atlas Illustré. EDICEF, Paris, France.

CENAREST (1979) *Liste des Vertébrés de la Région de Makokou, Gabon.*

Descoings, B. (1974) *Les Savanes du Moyen-Ogooué, Région de Booué, Gabon.* CNRS/Centre d'Etudes Phytosociologiques et Ecologiques, Document 69, Montpellier, France.

Diop, M. (1989) La Forêt Gabonaise. *Marchés Tropicaux* 15 décembre 1989.

FAO (1988) *An Interim Report on the State of Forest Resources in the Developing Countries.* FAO Rome, Italy. 18 pp.

FAO (1991) *FAO Yearbook of Forest Products 1978–1989.* FAO Forestry Series No. 24 and FAO Statistics Series No. 97. FAO, Rome, Italy.

Floret, J.-J., Louis, A. M. and Reitsma, J. M. (1989) Un *Microdesmis* (Pandaceae) à dix étamines découvert en Afrique: *M. afrodecandra* sp. nov. *Bulletin du Muséum National d'Histoire Naturelle, Paris 4th serie* **11**: 103–15.

Fontes, J. (1978) Les formations herbeuse du Gabon. *Annales de l'Université Nationale du Gabon* **2**: 127–53.

Hallé, N. (1987) *Cola lizae* N. Hallé (Sterculiaceae): Nouvelle espèce du Moyen Ogooué (Gabon). *Adansonia* **3**: 229–37.

Harrison, M. J. S. (1988) A new species of guenon (genus *Cercopithecus*) from Gabon. *Journal of Zoology, London* **215**: 561–75.

Hladik, A. (1986) Données comparatives sur la richesse spécifique et les structures des peuplements des forêts tropicales d'Afrique et d'Amérique. In: *Vertébrés et Forêts Tropicales Humides d'Afrique et d'Amérique.* Gasc, J. P. (ed.). Muséum National d'Histoire Naturelle, Paris, France.

Hladik, C. M. (1973) Alimentation et activité d'un groupe de chimpanzés réintroduit en forêt Gabonaise. *La Terre et la Vie* **27**: 343–413.

McShane, T. O. (1990) Conservation before the crisis – an opportunity in Gabon. *Oryx* **24**: 9–14.

McShane-Caluzi, E. and McShane, T. O. (1990) *Conservation Avant la Crise: Strategie pour la Conservation au Gabon.* WWF, Washington, DC, USA and Gland, Switzerland.

Michelmore, F., Beardsley, K., Barnes, R. and Douglas-Hamilton, I. (1989) Elephant population estimates for the Central African forests. In: *The Ivory Trade and the Future of the African Elephant.* Cobb, S. (ed.). Ivory Trade Review Group, International Development Centre, Oxford, UK.

Nicoll, M. and Langrand, O. (1986) *Conservation et Utilisation Rationelle des Ecosystèmes Forestiers du Gabon.* IUCN/WWF, Gland, Switzerland.

Reitsma, J. M. (1988) Forest vegetation of Gabon. *Tropenbos Technical Series 1.* The Tropenbos Foundation, Ede, The Netherlands. 142 pp.

Tutin, C. E. G. and Fernandez, M. (1984) Nationwide census of gorilla (*Gorilla g. gorilla*) and chimpanzee (*Pan t. troglodytes*) populations in Gabon. *American Journal of Primatology* **6**: 313–36.

Tutin, C. E. G. and Fernandez, M. (1987) Gabon: a fragile sanctuary. *Primate Conservation* **8**: 160–1.

UICN (1990) *La Conservation des Ecosystèmes forestiers du Gabon.* Basé sur le travail de C. Wilks. UICN, Gland, Switzerland and Cambridge, UK. 215 pp.

WCI (1989) The status of elephants in the forests of central Africa: results of a reconnaissance survey. In: *The Ivory Trade and the Future of the African Elephant.* Cobb, S. (ed.). Ivory Trade Review Group, International Development Centre, Oxford, UK.

White, F. (1983) *The Vegetation of Africa: a descriptive memoir to accompany the Unesco/AETFAT/UNSO vegetation map of Africa.* Unesco, Paris, France. 356 pp.

Authorship

Caroline Tutin, Lopé Reserve, Gabon, with contributions from Charles Doumenge, IUCN, Gland, G. Caballé, Montpellier, France Tom McShane, Erica McShane-Caluzi, WWF-US, A. Blom, WWF-Zaïre and Richard Barnes, University of California.

Map 19.1 Forest cover in Gabon

There are no detailed published maps of existing vegetation for Gabon. Forest cover data and protected areas were taken from a 1:1 million generalised published map Gabon (1987) which was prepared by the Institut Géographique National – France, Paris, in collaboration with the Institut National de Cartographie, Libreville. This map has been adapted to show areas of degraded forest as indicated on a sketch map in Caballé (1983). Nevertheless the vegetation data on this map are general and the forest cover is likely to be more fragmented than shown on Map 19.1, hence caution is necessary when quoting the statistics of forest extent which are derived from this map.

20 The Gambia and Senegal

THE GAMBIA
Land area 10,000 sq. km
Population (mid-1990) 0.9 million
Population growth rate in 1990 2.6 per cent
Population projected to 2020 1.7 million
Gross national product per capita (1988) US$220
Rain forest (see map) 497 sq. km
Closed broadleaved forest (end 1980)* 650 sq. km
Annual deforestation rate (1981–5)* 22 sq. km
Industrial roundwood production† 21,000 cu. m
Industrial roundwood exports† nd
Fuelwood and charcoal production† 901,000 cu. m
Processed wood production† 1000 cu. m
Processed wood exports† nd

SENEGAL
Land area 192,530 sq. km
Population (mid-1990) 7.4 million
Population growth rate in 1990 2.7 per cent
Population projected to 2020 15.2 million
Gross national product per capita (1988) US$510
Rain forest (see map) 2045 sq. km
Closed broadleaved forest (end 1980)* 2200 sq. km
Annual deforestation rate (1981–5)* 25 sq. km
Industrial roundwood production† 605,000 cu. m
Industrial roundwood exports† nd
Fuelwood and charcoal production† 3,786,000 cu. m
Processed wood production† 11,000 cu. m
Processed wood exports† nd

* FAO (1988)
† 1989 data from FAO (1991)

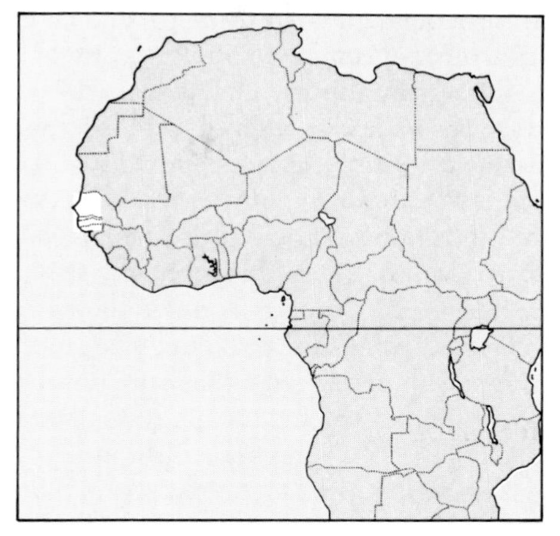

The Gambia is a tiny semi-enclave within the much larger country of Senegal on the west coast of Africa. In both countries very little closed forest remains and that which does still exist is rapidly being lost or modified through a combination of increased population, bushfires, desertification, uncontrolled grazing and extensive forest exploitation. The landscape is dominated by open savanna, with small islands of relict closed forests in the moister south-western and western regions. The mangroves in The Gambia are the least disturbed remaining natural forest within that country.

The small size of The Gambia, the intensity of cultivation there and the expansion in both the human population and livestock, exclude the possibility of establishing any large conservation areas in the country. At present only 1 per cent of land is protected. In Senegal, deforestation, fires and increasing desertification place natural ecosystems under great stress. The country possesses a varied fauna but outside the protected areas the ability of the land to support wildlife has been severely impaired.

INTRODUCTION

Senegal and The Gambia are the most westerly countries in Africa and cover a total land area of 202,530 sq. km (country areas are 196,720 and 11,300 sq. km respectively). Senegal is bordered by arid Mauritania in the north and Mali in the east, and the more humid countries of Guinea-Bissau and Guinea in the south. The Gambia is a narrow strip of land on each bank of the River Gambia and is entirely enclosed by Senegal apart from its coastline on the Atlantic. It extends 320 km inland from the coast where its maximum width of 45 km shrinks to only 7–15 km either side of the river. Senegal stretches from 12°30'–16°30'N and 11°20'–17°10'W and The Gambia runs along 13°30'N from 13°53'–16°50'W. Before the Trans-Gambia Highway was opened in 1958, The Gambia and its river had isolated the Casamance region in the south of Senegal from the rest of the country.

The Gambia is essentially the valley of the navigable Gambia River. The coast is one of submergence and the river valley is on a low sandstone plain – no part of the country exceeds 100 m asl. Away from the river in the west, there are sandy hills separated by flat, sandy plains, while in the east higher land prevails between tributaries of the river (Hughes and Hughes, 1991). Senegal comprises monotonous plains, the floodplains of ancient river systems, with most of the land below 100 m. Only small areas in the east and extreme south-east reach 500 m.

Both countries are characterised by a tropical monsoon climate, with a long and severe dry season from November to May – dominated by the dry Harmattan wind from the Sahara – and a wet season from June to October. Annual precipitation decreases from south to north. For instance, in Senegal, 200 mm fall per year in the Bakel region in the north and 1500–1600 mm per year at Ziguinchor in the south. Precipitation comes mainly from the south-westerly monsoon. Consequently, in comparison to The Gambia, Senegal has a contracted rainy season and reduced precipitation in the northern arid zone and a longer rainy season in the south (May–October). Duration of the wet season and annual precipitation are variable, with the present trend being a reduction in rainfall. In The Gambia in 1965, when there was still good forest cover, precipitation in the capital city of Banjul was 1240 mm; in the past 20 years the mean level (measured between 1982 and 1988) has almost halved to 650 mm. This has had obvious effects on the ecology of the region. Mean monthly temperatures are generally between 25°C and 30°C while temperatures inland may exceed 40°C in the hottest months of April and May.

The population of Senegal and The Gambia is mainly rural – 60 per cent of Senegal's population and 82 per cent of The Gambia's live in the countryside. With a population of around 900,000, The Gambia is one of the most densely populated countries in Africa. Most people live in small villages fairly evenly scattered throughout the country. The population is greatest on the south bank of the river and this area has better developed communications and infrastructure and larger settlements. Population density in Senegal is less than half that of The Gambia, but the rate of growth of the urban population in Senegal is one of the highest on the continent (FAO/UNEP, 1981). In Senegal the inhabitants are spread unevenly across the country: there are very few people in the east, while Casamance, Louga, Fleuve and Sine Saloum provinces are very heavily populated.

Senegal was the first country in the region to introduce groundnut cultivation and, by independence in 1960, the country was heavily dependent on the crop. France helped to support this pattern in its former colony, by providing price subsidies for Senegalese groundnuts. They remain the most important crop and are grown in about 40 per cent of the cultivated land area. However, exports, which accounted for 50 per cent of export earnings in 1976, now account for less than 20 per cent. The government is promoting agricultural cooperatives as a means of improving production. Such an approach seems to be acceptable in Senegal because property is traditionally owned by the tribe, with the chief assigning land to a family or group of families depending on their needs and ability. There are now about 2200 cooperatives in operation, of which 1700 are engaged in groundnut cultivation (USAID, 1982). Large-scale clearance of indigenous vegetation for groundnut and millet cultivation is thought by some people to be a major factor in the increasing aridity of the north, a trend that is aggravated by tree felling for timber and charcoal burning. Tourism, cotton and marine fisheries are now becoming important to the economy.

In The Gambia, the principal crop is again the groundnut which furnishes 95 per cent of export revenues. Rice is the staple food crop, with maize, manioc, fruits and vegetables cultivated on a small scale.

The Forests

The rain forests of southern Senegal and The Gambia are at the north-western limit of this biome in Africa, as the long dry season militates against the development of typical rain forest. However, because of high rainfall during the wet season and a high water-table during the dry season, particularly in the coastal plain of Basse Casamance, Guineo-Congolian forest species have been able to extend their range into this area (White, 1983). Sudanian savanna woodland predominates, with small areas of evergreen lowland forests in the west of The Gambia and south and west of Senegal.

The Gambia Most of the closed moist forest in The Gambia is mangrove, which penetrates deep along the Gambia River and its tributaries. In addition, there were some scattered riparian forests along streams, above the tidal waters of the river. These contained commercial species such as mahogany *Khaya senegalensis* and iroko *Chlorophora regia* as well as *Parinari excelsa, Detarium senegalense, Dialium guineense* and *Erythrophleum guineense* (FAO/UNEP, 1981). In the past 15 years, most of the broadleaved trees have been lost. The Abuko Nature Reserve, in the west, houses The Gambia's last remnant of riparian forest (45 ha) with more than 50 tree species protected there. The most notable are the cabbage tree *Anthocleista procera, Piliostigma bilboa, Ficus* spp., *Parkia biglobosa, Parinari excelsa,* oil palm *Elaeis guineensis, Raphia* palm and rattan palm *Calamus deeratus.* Lianas are well developed under the canopy.

Senegal Most of Senegal is savanna country. The transition from arid to more humid climate vegetation occurs at roughly the 750 mm isohyet (USAID, 1982). On both sides of that line, however, the existence of forest species is determined mainly by the presence or absence of depressions where water can accumulate, and by the amount of grazing activity. There are some remnants of lowland tropical rain forest in the southernmost region, Casamance, which give a picture of the once more extensive moist forests. *Parinari excelsa* was abundant in swamp forest in depressions and in drier forest on better-drained soils; drier forest species include *Erythrophleum suaveolens, Detarium senegalense, Afzelia africana* and *Khaya senegalensis.* Among the rarer species are *Albizia adianthifolia, A. ferruginea, A. zygia, Antiaris toxicaria, Chlorophora regia, Cola cordifolia, Daniellia ogea, Dialium guineense, Morus mesozygia, Schrebera arborea* and *Sterculia tragacantha* (White, 1983). The canopy of the Basse Casamance forest is 18–20 m high and is made up of large trees with abundant lianas. This is the wettest part of the country and in the past there were large areas of moist seasonal riparian forest along the banks of the Casamance River. These have all but disappeared and are reduced to degraded copses of mature trees. The Basse Casamance National Park (50 sq. km) contained the best example of seasonal moist forest, and still contains mangrove, the last vestiges of moist forest and climax Guinea-woodland.

Mangroves

Mangroves are present in both countries, but are of particular note in The Gambia, where they are the most pristine remaining natural habitat in that country. FAO estimated that 600 sq. km occurred in The Gambia and 1620 sq. km in Senegal at the end of 1980 (FAO/UNEP, 1981). Map 20.1 gives the slightly lower area of 497 sq. km in The Gambia and the higher figure of 1853 sq. km in Senegal.

The Gambia A nearly continuous belt of mangroves exists along The Gambia River from its mouth at Banjul to 150 km inland. The dominant species are red mangrove *Rhizophora racemosa,* white mangrove *Avicennia nitida* and *Laguncularia racemosa.* These reach a height of at most only five metres for the first 25 km upstream but, further up river, they can grow as tall as 15–20 m, supporting nesting colonies of many large bird species. The mangroves are best developed at the mouths of small tributaries, while upstream the tallest forest, comprising *Rhizophora* spp., occurs as a narrow strip backed by low open woodland of *Avicennia africana.* On the north bank of the Gambia River an extensive mangrove block of some 80 sq. km occurs in the Salikene district. Tall mangroves occur at the mouths of the Jurunkku and Mini Minium Bolons in a continuous forest block which spreads upstream for 30 km.

Away from the capital most mangrove stands are largely intact. However, at present the only protected area on the Gambia River is the Gambia River National Park. Here a group of five islands is protected. Some mangrove swamp and lagoons are also protected in the Kiang National Park on the south bank of the Gambia River. Hunting occurs in the riverine wetlands and many species are taken; the Nile crocodile *Crocodylus niloticus* and hippopotamus *Hippopotamus amphibius* have been hunted almost to the point of extinction and are both endangered in The Gambia.

Senegal Mangroves in Senegal occupy the estuaries and channels of seawater which penetrate deep into the territory all along the coast. These do not differ from other mangroves of West Africa, the main species being *Rhizophora mangle, R. racemosa, R. harrisonii, Avicennia africana, Laguncularia racemosa* and *Conocarpus erectus.* A large area of mangrove, 620 sq. km (UN, 1987), exists on the south-west coast. The Saloum River and its delta are also important sites for mangroves. In total the delta wetlands cover 1500 sq. km, some 70 per cent of which is mangrove forest (Hughes and Hughes, 1991). The Saloum Delta National Park occupies 720 sq. km of the delta. The delta stretches along the coast for a distance of 72.5 km and reaches 35 km inland, while mangrove swamps extend almost 70 km upstream to Kaolack.

South of the main river channel there is a network of cross-connecting streams weaving between mangrove-covered mud islands, with only small areas above high tide level. Marine turtles are present and Campbell's monkey *Cercopithecus campbelli* and colobus monkeys are found in the forest trees.

There are plans for industrial development and expansion of rice in the area. In the past 15 years several anti-salt barrages have been constructed along the coast to prevent high-tide inundation. As a result large areas of mangrove have died. The building of roads across the mouths of mangrove creeks has had the same effect in many areas of Casamance. In both Senegal and The Gambia the intensification of agriculture and the destruction of natural riverine wetlands have brought the problem of increased erosion and siltation.

Forest Resources and Management

The Gambia In The Gambia, at the end of 1980, FAO estimated that 650 sq. km of closed broadleaved forest remained (FAO/UNEP, 1981). This figure is somewhat higher than that given by Map 20.1, where all remaining forest is shown as mangrove. There are no substantial areas of closed dryland forest remaining. Management of the forests is the responsibility of the Forestry Department, by way of land gazetted as 'forest parks'. All other forest land is communal, for there is no privately owned forest. However, the government may make regulations for the protection, control and management of any forest park and make regulations for areas outside such parks. Prohibited activities on land outside forest parks include: the cutting of firewood; the collection of fibre, rubber, palm nuts, palm kernels or gum; conversion of wood to charcoal; the extraction of palm wine, and quarrying. Traditional chiefs are responsible for protecting land under their jurisdiction from the ravages of bush or forest fires, but fire remains a serious problem.

The Gambia's 66 forest parks cover a total of 340 sq. km and management, in the form of boundary maintenance, fire protection and early burning, is the responsibility of Area Councils (FAO/UNEP, 1981). Parker (1973) suggested that the forest parks should be considered as potential wildlife conservation units, so that further land is not put out of bounds to the population; they could form the basis of a conservation system that would ensure the survival of representative habitats. This, however, has not occurred. Some of the forest parks are protection forests, many are surrounded by barbed wire, and felling, burning, grazing and hunting are all prohibited within them. Despite this, there appears to be a thriving trade in the haulage of charcoal and wood both ways across the borders and a limited amount of fuelwood collection, hunting and grazing does occur. Forest plantations are also established within The Gambia's forest parks. Management objectives include production of timber trees (mostly *Gmelina*), bamboo *Oxytenanthera abyssinica* and palms *Borassus aethiopum*.

FAO estimated that 21,000 cu. m of industrial roundwood was produced in 1989, none of which was exported (FAO, 1991). Trees are rarely felled for local fuel use (dead wood is collected by women and children), but they may be commercially poached on a small scale. Charcoal remains the principal domestic fuel although its manufacture is now prohibited; FAO estimated (1991) that in 1989, 901,000 cu. m of charcoal and fuelwood were used.

Senegal According to FAO/UNEP (1981), closed broadleaved forest covered 2200 sq. km of Senegal at the end of 1980 (1.1 per cent of the land area). Just 130 sq. km of this total was dense forest (20 sq. km protected in Basse Casamance National Park), 450 sq. km was riparian forest (170 sq. km protected in Niokolo-Koba National Park) and 1620 sq. km was mangrove. Map 20.1 shows a slightly lower total area of closed forest but this figure does not include riparian forest. It indicates 192 sq. km of dryland forest and 1853 sq. km of mangroves, giving a total of 2045 sq. km. There is no moist forest north of the River Gambia, apart from mangrove. Most of the broadleaved forests have now been felled to provide timber, fuel or new agricultural land. Extraction has been selective in some areas, leaving, for instance, the commercially important oil palm *Elaeis guineensis* in place. There are also vast areas of open palm plantations with cash crops underplanted in the wet season. There has been some reforestation in the Casamance region with small-scale teak plantations of a few hundred to a thousand hectares, and between the Gambian border and the Casamance River there are commercial stands of *Eucalyptus*.

Table 20.1 Estimates of forest extent in The Gambia and Senegal

	Area (sq. km)	*% of land area*
GAMBIA		
Rain forests		
Mangrove	497	5.0
Totals	497	5.0
SENEGAL		
Rain forests		
Lowland	192	0.1
Mangrove	1,853	1.0
Totals	2,045	1.1

(Based on analysis of Map 20.1. See Map Legend on p. 182 for details of sources.)

There are 197 'classified' forests in Senegal covering an area of 39,000 sq. km, or 20 per cent of the country. The government forestry programme has five aims:

- Forest protection and management, including bushfire control, reserve protection, management of existing forests and creation of a botanical garden.
- Agrarian and pastoral land reform, including reforestation in the Senegal Delta and management of grazing in forest areas.
- Reforestation for production of teak, *Gmelina* and cashew *Anacardium occidentale* and gum trees.
- Reforestation for environmental protection and fuelwood production, by stabilising and protecting sand dunes, establishing plantations for fuelwood, encouraging reforestation around urban centres, creating village woodlots and plantations along roads.
- Establishing a wildlife programme, including providing equipment to support the management of national parks and hunting zones (USAID, 1982).

Six categories of forest have been distinguished (FAO/UNEP, 1981): forest reserves (forêts domaniales classées) where hunting and forest exploitation is prohibited; managed forests (forêts domaniales aménagées) for the production of fuelwood and charcoal; the Noflaye Botanical Reserve where there is total protection of flora and fauna; wildlife reserves (réserves de faune), where hunting is prohibited, and finally, hunting reserves (zones d'intérêt cynégétique), where hunting is allowed. However, grazing and felling are widespread in all these categories.

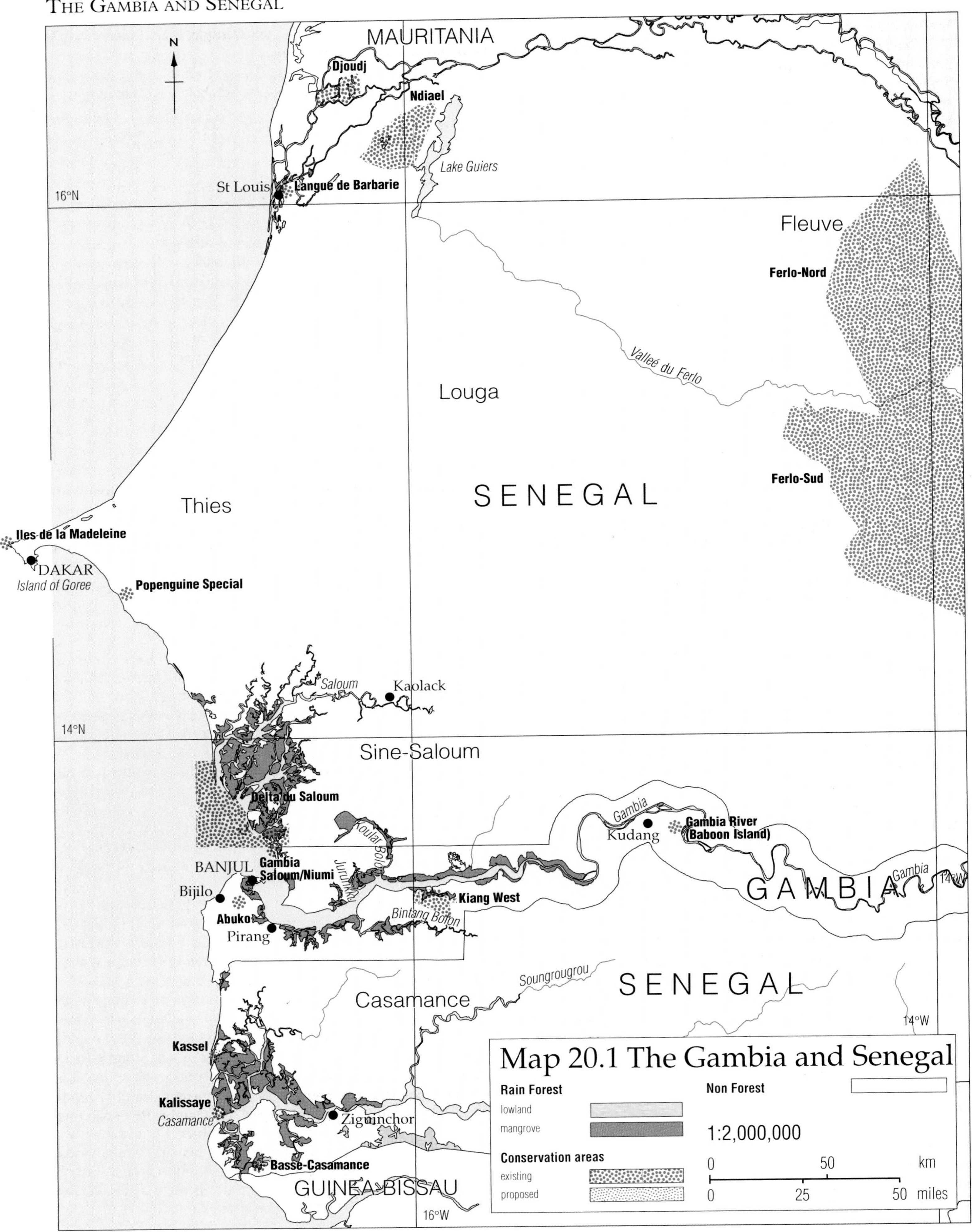

MAURITANIA
Djoudj
Ndiael
Lake Guiers
St Louis
Langue de Barbarie
16°N
Fleuve
Ferlo-Nord
Valleé du Ferlo
Louga
SENEGAL
Ferlo-Sud
Thies
Iles de la Madeleine
DAKAR
Island of Goree
Popenguine Special
Saloum
Kaolack
14°N
Sine-Saloum
Delta du Saloum
Koular Bolon
Gambia
Kudang
Gambia River (Baboon Island)
BANJUL
Gambia Saloum/Niumi
Jurunku
Bijilo
Abuko
Pirang
Bintang Bolon
Kiang West
GAMBIA
14°W
Soungrougrou
SENEGAL
Casamance
Kassel
Kalissaye
Casamance
Ziguinchor
Basse-Casamance
GUINEA-BISSAU
16°W
Map 20.1 The Gambia and Senegal
Rain Forest
lowland
mangrove
Non Forest
1:2,000,000
Conservation areas
existing
proposed
0
50
km
0
25
50
miles

Deforestation

The Gambia There are no areas of closed forest which have not been modified by humans. Most of The Gambia's closed broadleaved forests have disappeared over the past 15 years or so and the country is now mainly degraded open wooded savanna. Virtually all the original climax riparian forest has been cleared to the east of Kudang, although large areas of mangrove are still intact. Other than mangroves, dense forests are now restricted to riparian forests along the banks of the main steams, with these disappearing at a rate of 2 sq. km per year at the end of 1980 (FAO/UNEP, 1981). Overall, during 1981–5, FAO estimated annual deforestation for closed broadleaved forests to be 22 sq. km (3.4 per cent) (FAO, 1988). Bushfires are a problem (Starin, 1989), particularly as post-harvest farm burns are often poorly controlled and during the Harmattan season these frequently spread beyond village boundaries. Overgrazing by domestic stock and a lowered water table prevents regeneration, especially of species adapted to high water tables such as oil palm, swamp palm and the cabbage tree. Many of the largest riparian trees and the oil palms in Abuko Nature Reserve have died in the past five years; this is probably due to the drop in the water table.

Senegal The increase in population and demand for fuelwood and other wood products in Senegal has led to clearance of forested land, and indeed the moist forest is largely reduced to degraded copses of mature trees. It has been replaced by huge areas of open palm savanna, underplanted with cash crops. In 1988 the Food and Agriculture Organisation of the United Nations (FAO) estimated that deforestation of broadleaved closed forest had occurred at an annual rate of 25 sq. km (2.5 per cent) for the period 1981–5. Bushfires have accelerated the process of deforestation. When these occur, the grasses and fire-resistant species survive, whereas the original forest trees are killed. Grasses are especially well adapted to colonise sandy soils and are more tolerant of dry conditions than the forest species. As a result of the fires and of shifting agriculture, extensive grasslands have become established in the once forested regions. These are dominated by *Panicum maximum*, *Pennisetum purpureum* and *Imperata cylindrica*. Human migration is also a serious problem. The Casamance region is the wettest and most fertile part of the country and earns revenue from tourism as well as from agriculture. People with different cultures and agricultural practices from the north of Senegal and from Mali have migrated to this region, a movement that has increased the demand for land and had a deleterious effect on the surviving forests.

Biodiversity

The great range of habitats in Senegal, from semi-desert in the north to degraded moist forest, mangrove and deltaic formations in the south, explains the richness of the country's flora compared to other Sahelian states. Senegal has a flora of approximately 2200 species (Berhaut, 1976) of which 26 are endemic (Davis *et al.*, 1986). The Gambia has around 530 plant species, including three endemics (Davis *et al.*, 1986).

Most larger mammals have disappeared from The Gambia but approximately 108 species of mammals still occur (Stuart *et al.*, 1991). These include the hippopotamus, three species of duiker *Cephalophus* spp., bushbuck *Tragelaphus scriptus*, Temminck's red colobus monkey *Procolobus [badius] badius temminckii*, red patas monkey *Erythrocebus patas*, Campbell's monkey and serval *Felis serval*. None is endemic. Some tropical forest species such as forest genet *Genetta pardina* are found at the northernmost limit of their distribution. Several species do not breed in the country but occasionally visit from Senegal: for example, lion *Panthera leo* and roan antelope *Hippotragus equinus*.

Large mammals in Senegal are confined mainly to the national parks. There are very few endemics or globally threatened species. Some extinctions have occurred within the country: for instance, the giraffe *Giraffa camelopardalis* no longer occurs. Approximately 169 mammal taxa are recorded here. In Senegal, in addition to species listed for The Gambia, other noteworthy mammals include leopard *Panthera pardus*, hunting dog *Lycaon pictus*, common eland *Tragelaphus oryx*, western black-and-white colobus *Colobus polykomos* and chimpanzee *Pan troglodytes* (Stuart *et al.*, 1991). The elephant *Loxodonta africana* population in Senegal was reduced by poachers to about 20–40 in 1989, and in fact wildlife rangers predicted their extinction during the next rainy season because, at this time of the year, protection from poachers cannot be provided. Smaller mammals present in the country include Beecroft's flying squirrel *Animalurops beecrofti*, Libyan striped weasel *Poecilictis libyca*, dark mongoose *Crossarchus obscurus* and giant ground pangolin *Manis gigantea*.

Around 490 species of birds have been recorded in The Gambia and 625 species for Senegal (Stuart *et al.*, 1991). No threatened bird species are listed in Collar and Stuart (1985) for either country. In Senegal, the martial eagle *Polemaetus bellicosus*, bateleur eagle *Terathopius ecaudatus*, greater flamingo *Phoenicopterus ruber*, tree ducks *Dendrocygna* spp. and the royal tern *Sterna maxima* are noteworthy (Serle *et al.*, 1977). The wetland and mangrove areas in both countries have very rich avifaunas.

Forty-six reptiles are known from The Gambia and 57 from Senegal. In The Gambia the slender-snouted crocodile *Crocodylus cataphractus* has disappeared in the past decade and the dwarf crocodile *Osteolaemus tetraspis* survives only at Abuko Nature Reserve (see case study). The coastal skink *Chalcides armitagei* is the only vertebrate endemic to The Gambia although it is likely to occur in similar habitats in Senegal. Several species of snake recorded at Abuko (Hakansson, 1981) are not otherwise known north of the forests of Guinea. In Senegal, all three African crocodiles are still present, although both the dwarf crocodile and slender-snouted crocodile are close to extinction. All of the amphibian species recorded in The Gambia (26) and Senegal (29), belong to the savanna faunal assemblage of Schiotz (1967); none is a closed forest animal.

Conservation Areas

Two departments deal with conservation in The Gambia: the Wildlife Conservation Department which is responsible for park and reserve administration, and the Forestry Department which protects the forest parks. In Senegal, the National Parks Directorate is the responsibility of the Ministry of Tourism and Nature Protection, while areas other than national parks are administered by the Direction des Eaux, Forêts et Chasses (IUCN, 1987). The conservation areas of the two countries are listed in Table 20.2.

The Gambia The protected areas system here dates back to 1916 when the main part of Abuko Nature Reserve was protected as a water catchment area. Abuko was given nature reserve status in 1968 and extended by about 29 ha to its current size (107 ha) in 1978 (Edberg, 1982). In that same year, the Gambia River National Park was created, while the other two protected areas, namely Gambia Saloum/Niumi and Kiang West National Parks, were designated in 1987. Now a total of 1 per cent of the country is incorporated into protected areas. All 66 of the country's forest parks were created in 1955.

Table 20.2 Conservation areas for The Gambia and Senegal

Existing and proposed areas are listed below. Forest parks and classified forests are not included. For data on Biosphere reserves and World Heritage sites see chapter 9.

	Existing area (sq. km)	*Proposed area (sq. km)*
THE GAMBIA		
National Parks		
Gambia River (Baboon Island)	6	
Gambia Saloum/Niumi	20	
Kiang West*	100	
Nature Reserves		
Abuko	1	
Total	127	
SENEGAL		
National Parks		
Basse-Casamance*	50	
Delta du Saloum*	760	
Djoudj	160	
Iles de la Madeleine	5	
Langue de Barbarie*	20	
Niokolo-Koba†	9,130	
Faunal Reserves		
Dindefello**		1
Ferlo-Nord	4,870	
Ferlo-Sud	6,337	
Gueumbeul Special**	8	
Ndiael	466	
Popenguine Special	10	
Special Reserves		
Kalissaye*	16	
Kassel*		1
Hunting Reserves		
Maka-Diama†	600	
Others		
Elephants du Fleuve**		nd
Faleme**		nd
Senegambien**		nd
Totals	22,432	2

(*Sources:* IUCN, 1990; WCMC, *in litt.*)
* Area with moist forest within its boundary according to Map 20.1.
† Area not shown on Map 20.1 but given on Figure 20.1..
nd No data available.
** Area not mapped as no location data available.

The Abuko Nature Reserve supports the last few hectares of riparian forest, and houses many species of plants and animals that have all but disappeared from the rest of the country (see case study). The reserve comprises remnant gallery forest and a recent extension of regenerating open wooded savanna. The water table at Abuko has fallen markedly in the past decade. This has resulted in the reduction of surface water available for the wildlife and many of the largest gallery forest trees and oil palms have died. It is unlikely that this last remaining stand of forest will survive for long, in spite of its complete protection. The River Gambia National Park, a group of five islands, also supports a mature riparian forest fringe and is the site of a long-term (about 15 years) chimpanzee rehabilitation project. There are already 40 chimpanzees on the islands and this is the probable carrying capacity of their habitat. The effect of such a high ape density on the islands' plant and other animal life is not known (Starin, 1989). Kiang West, south of the River Gambia, contains mangroves, seasonal marsh and open wooded savanna.

In addition, a degree of protection is provided by the forest parks in The Gambia although the prohibitions preventing felling, burning, grazing and hunting are not always adhered to. The most closely studied forest park in The Gambia is Pirang which is only 64 ha. This, and the coastal forest at Bijilo (50 ha), are the sole representatives of original climax vegetation outside protected areas. Pirang acts as an important refuge for species such as Temminck's red colobus, vervet *Cercopithecus aethiops*, and red patas monkey and is the only place where the Guinea baboon *Papio papio* is safe from hunters. The better maintained parks, nearer the capital, are mostly converted to *Gmelina* plantations.

Senegal In 1925 Niokolo-Koba National Park was established and since then another five parks have been gazetted (Dupuy and Verschuren, 1977). Approximately 12 per cent of Senegal's land area is protected. Its conservation areas include Biosphere reserves at

Apart from mangroves, The Gambia contains very little closed moist forest . I. and D. Gordon

DWARF CROCODILE PROJECT IN THE ABUKO NATURE RESERVE

The Abuko Nature Reserve contains the last population of the endangered dwarf crocodile *Osteolaemus tetraspis* in The Gambia, which is the northernmost limit of its distribution. The virtual disappearance of *O. tetraspis*, down to 12–15 adults in 1988, is mainly due to forest loss and the resultant depletion of their breeding habitat, pools in gallery forest. The Gambian Wildlife Conservation Department has initiated a programme to halt the decline of the crocodiles with the development of an intricate irrigation scheme.

Since 1981–2, decline in the water table had prevented the normal pattern of forest floor flooding which the dwarf crocodile requires in order to breed. Bristol University carried out a census of the population in 1988/9 and all the crocodiles recorded were mature adults, born before 1981–2 and there had been no apparent recruitment since then. At the request of the Gambian Wildlife Conservation Department, the scientists from Bristol University devised an irrigation scheme to increase the surface water available to the breeding crocodiles. This involved the construction of a deep-bore pump to reach the lowered water table and extensive subsurface piping throughout the reserve to irrigate pools constructed or renovated for the crocodiles. In addition each new pool had a 3–4 m long, 45 cm diameter tunnel dug leading down from one bank to act as refuges for the crocodiles.

Osteolaemus adopted the irrigated natural and artificial pools at the start of the 1989 wet season, three months after construction. There was a combination of unaltered natural pools, new pools and dry pools partially filled with the extra water. Irrigated pools, already well established with dead leaves, tree roots and plants colonising the banks, were in use approximately six weeks before unaltered pools had filled. The first successful breeding took place during the 1989 wet season and a census in 1990 revealed that a minimum of seven juveniles had survived their first year.

Further suitable habitat for the West African dwarf crocodile Osteolaemus tetraspis *is being created at Abuko Nature Reserve, which is the only place in Gambia where this species still occurs.* E. Dragesco/WWF

The irrigation scheme also helped take the pressure off other species within the reserve, as the loss of small bodies of surface water had forced mammals and birds to drink and bathe at the larger permanent pools inhabited by Nile crocodiles, probably leading to the local increase in the predation on species such as sitatunga *Tragelophus spekii*. Over 100 vertebrate species use the smaller pools including red colobus, vervet and red patas monkey, north African crested porcupine *Hystrix cristata*, turacos, monitor lizards and forest cobras *Naja melanoleuca*.

It seems that the irrigation scheme has been successful for *O. tetraspis*, at least in the short term. By slightly modifying the crocodiles' natural environment to reinstate former hydrological cycles, breeding has recommenced after an eight-year break.

Niokola-Koba National Park, Samba Dia Classified Forest and Saloum Delta National Park and three World Heritage sites: Djoudj National Bird Sanctuary, the Island of Goree and Niokolo-Koba National Park. Four of the protected areas contain some moist forest (see Table 20.2).

The 50 sq. km Basse-Casamance National Park, established in 1970, contains the best example of seasonally moist forest. Located in the extreme south of the country, the park incorporates mangroves at its west end and Guinea-woodland, savanna and secondary forest throughout the rest of its area. However, many of the largest emergent trees have died in recent years and the surrounding buffer zone has been modified by felling and grazing. As a result most of the park's larger mammal species seem to have declined. More than 50 species of mammal have been recorded including leopard, buffalo *Syncerus caffer*, Campbell's monkey, Demidoff's galago *Galagoides demidoff*, Temminck's red colobus, giant pangolin and the threatened manatee *Trichechus senegalensis* (IUCN/UNEP, 1987).

Niokolo-Koba National Park is the other protected area in the country containing moist forest, but this is all riparian forest and not, therefore, included in the forest statistics. Its position is indicated on Figure 20.1 because Map 20.1 does not cover the eastern section of Senegal. The park is mostly covered by savanna, but also supports seasonal swamp, bamboo, patches of climax Guinea-woodland, riparian forest with lianas and tree species such as *Raphia sudanica*, *Baissea multiflora*, and *Dalbergia saxatilis*. The park is watered by the River Gambia and its Niokolo-Koba and Koulountou tributaries, which support well developed but narrow stands of gallery forest.

Figure 20.1 Location of Nikola-Koba National Park in eastern Senegal. (*Source:* WCMC, *in litt.*)

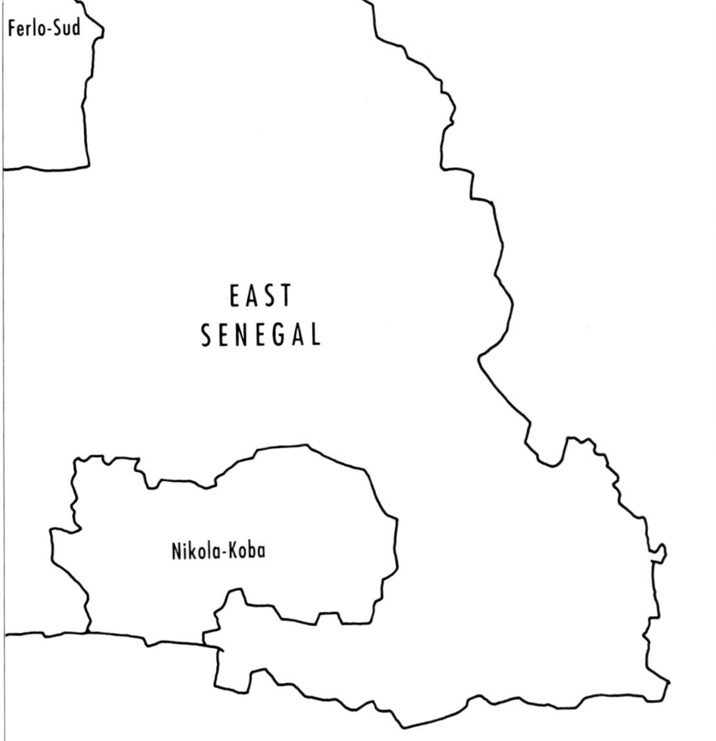

Initiatives for Conservation

Both countries have received considerable amounts of aid in connection with various anti-desertification programmes that focused on village-level forestry and associated rural development activities. In addition, in the Casamance region of Senegal, Canada has supported projects to bring forests under sustained yield management. An interesting initiative is an attempt to involve local people in the management of the Bakov and Mahen Forest Reserves which together cover 1000 sq. km. Canada has also supported a major programme to control bushfires in Casamance over the past decade.

IUCN has supported wetland conservation activities in Senegal by providing advice on the management of the Djoudj and Saloum Delta National Parks and by conducting training courses. A National Conservation Strategy was initiated in Senegal several years ago and there are now plans to complete the strategy for adoption by the government. A forest sector study is under preparation as a preliminary to a Tropical Forestry Action Plan (TFAP).

Various small-scale conservation initiatives have been supported by NGOs in The Gambia. There is also interest in an Environmental Action Plan which would be supported by the World Bank and The Gambian authorities are negotiating with FAO to develop a TFAP.

References

Berhaut, J. (1976) *La flore illustrée du Senegal*. XI volumes. Dakar.

Collar, N. J. and Stuart, S. N. (1985) *Threatened Birds of Africa and Related Islands. The ICBP/IUCN Red Data Book Part 1.* ICBP/IUCN, Cambridge, UK. 761 pp.

Davis, S. D., Droop, S. J. M., Gregerson, P., Henson, L., Leon, C. J., Villa-Lobos, J. L., Synge, H. and Zantovska, J. (1986) *Plants in Danger: What do we know?* IUCN, Gland, Switzerland and Cambridge, UK. 461 pp.

Dupuy, A. R. and Verschuren, J. (1977) Wildlife and parks in Senegal. *Oryx* **XIV**: 36–46.

Edberg, E. (1982) *A Naturalists' Guide to The Gambia*. J. G. Sanders. 46 pp.

FAO (1988) *An Interim Report on the State of Forest Resources in the Developing Countries*. FAO, Rome, Italy. 18 pp.

FAO (1991) *FAO Yearbook of Forest Products 1978–1989*. FAO Forestry Series No. 24 and FAO Statistics Series No. 97. FAO, Rome, Italy.

FAO/UNEP (1981) *Tropical Forest Resources Assessment Project. Forest Resources of Tropical Africa. Part II Country Briefs*. FAO, Rome, Italy. 586 pp.

Hakansson, N. T. (1981) An annotated checklist of reptiles known to occur in The Gambia. *Journal of Herpetology* **15**: 155–61.

Hughes, R. H. and Hughes, J. S. (1991) *A Directory of Afrotropical Wetlands*. IUCN, Gland, Switzerland and Cambridge, UK/UNEP, Nairobi, Kenya/WCMC, Cambridge, UK.

IUCN (1987) *IUCN Directory of Afrotropical Protected Areas*. IUCN, Gland, Switzerland and Cambridge, UK. xix + 1043 pp.

IUCN (1990) *1990 United Nations List of National Parks and Protected Areas*. IUCN, Gland, Switzerland and Cambridge, UK. 275 pp.

Parker, I. S. C. (1973) *Prospects of Wildlife Conservation in The Gambia*. A Consultant Report to the Foreign and Commonwealth Office. ODA's Project for a Land Resource of The Gambia. Unpublished manuscript distributed within The Gambia.

Schiotz, A. (1967) The treefrogs (Rhacopharidae) of West Africa. *Spolia zooligica Musei Hauniensis* **25**: 1–346.

Serle, W., Morel, G. J. and Hartwig, W. (1977) *A Field Guide to the Birds of West Africa*. Collins, London, UK.

Starin, E. D. (1989) Threats to the monkeys of The Gambia. *Oryx* **23**: 208–14.

Stuart, S. N., Adams, R. J. and Jenkins, M. D. (1991) *Biodiversity of Sub-Saharan Africa and its Islands: Conservation, Management and Sustainable Use*. Occasional Papers of the IUCN Species Survival Commission No. 6. IUCN, Gland, Switzerland.

UN (1987) *Inventaire de Forêts en Casamance et en Senegal Oriental et Aménagement de Forêts Classées*. FO:DP/SEN/82/027. FAO, Rome, Italy.

USAID (1982) *Gambia–Senegal: A Country Profile*. Country Profiles – USAID (Office of Foreign Disaster Assistance). 123 pp.

White, F. (1983) *The Vegetation of Africa: a descriptive memoir to accompany the Unesco/AETFAT/UNSO vegetation map of Africa*. Unesco, Paris, France. 356 pp.

Authorship

Scott Jones, Bristol University.

Map 20.1 Forest cover in The Gambia and Senegal

Information on rain forests in The Gambia and Senegal is taken from a digital map entitled *Range and Forest Resources of Senegal* at a 1:1 million scale (n.d.). The map was prepared for the US Agency for International Development (USAID) by the US Geological Survey, National Mapping Division, EROS Data Center. The digital map is based on a generalisation of the *Map of Vegetation Cover* from the report entitled *Mapping and Remote Sensing of the Resources of the Republic of Senegal* (1986) prepared by the Remote Sensing Institute, South Dakota State University under contract to USAID, for Senegal's Direction of Land Use Planning, Ministry of Interior (unseen). The map was compiled from the interpretation of Landsat imagery of different dates and from extensive ground surveys. In this Atlas only the 'Forests' and 'Mangroves' datasets were digitised out of the 31 rangeland and forest types shown. This map was kindly made available to the project by the EDC International Projects department of the EROS Data Center.

Protected areas for The Gambia are mapped from a 1:350,000 (c.) Tourist Information and Guide Map *The Gambia* (1987). Protected areas for Senegal are digitised from a 1:1 million Institut Géographique National map *Sénégal* and from spatial data held within files at WCMC.

21 Ghana

Land area 230,020 sq. km
Population (mid-1990) 15 million
Population growth rate in 1990 3.1 per cent
Population projected to 2020 33.9 million
Gross national product per capita (1988) US$400
Rain forest (see map) 15,842 sq. km
Closed broadleaved forest (end 1980)* 17,180 sq. km
Annual deforestation rate (1981–5)* 220 sq. km
Industrial roundwood production† 1,101,000 cu. m
Industrial roundwood exports† 201,000 cu. m
Fuelwood and charcoal production† 16,068,000 cu. m
Processed wood production† 590,000 cu. m
Processed wood exports† 170,000 cu. m
* FAO (1988)
† 1989 data from FAO (1991)

Present-day Ghana covers the area of the ancient Ashanti kingdoms of pre-colonial Africa. They were among the most developed civilisations of the forest belt and traded gold for salt produced in mines in the Sahara to the north. Ghana's forests were thus probably subject to relatively intense human influences several centuries before those of other parts of Africa. Ghana attained independence on 6 March 1957 but suffered serious economic decline in the post-independence period. The total collapse of Ghana's economy in the late 1970s was followed by the Economic Recovery Programme of 1983 and a period of increased foreign investment. This led to an upturn in timber exports – traditionally a major source of foreign revenue – sawmills and logging operations were modernised and ports for log exports were rebuilt resulting in the recovery of the timber trade which had been declining. The timber industry is now contributing to improved economic conditions in the country but there must be concern that this will be at the expense of Ghana's forests.

Introduction

The Republic of Ghana lies between 4°45'N and 11°11'N latitude and between 1°14'E and 3°07'W longitude. The rectangular-shaped country, with a total area of 238,540 sq. km, extends 675 km inland with a coastline of 567 km. Côte d'Ivoire is on the western border, Burkina Faso to the north and Togo to the east.

The topography of Ghana is undulating with prominent scarps seldom exceeding 600 m, occurring at Akwapim, Kwahu, Mampon, Ejura and Gambaga. The highest hills (880 m) run in a north-east to south-west direction between the Volta River and the Togo border (FAO/UNEP, 1981). To the west of this range is Lake Volta, formed in 1964 through the damming of the Volta River. This is the largest artificial lake in Africa with an area of 8500 sq. km (Owusu *et al.*, 1989). Prominent rivers found west of the Volta, which drain the wetter south, are the Ankobra, Pra and Tano (FAO/UNEP, 1981).

The two predominant ecological zones in Ghana are the closed forest zone, covering an area of 81,342 sq. km (or 35 per cent of the land area) and the savanna zone, including Lake Volta, covering 156,300 sq. km (Hall and Swaine, 1981; Silviconsult, 1985). The closed forest zone is found mainly in the south-western third of the country with a section extending into the northern part of the Volta Region. Rainfall in this zone follows a bimodal pattern with peaks in June and October, and an annual range of 1000–2100 mm. The forest zone, comprising seven vegetation types, supports two-thirds of the country's population and the majority of its economic activities, including timber, cacao, oil palm, rubber, cola and mineral production (Foggie, 1951; Owusu *et al.*, 1989).

The savanna zone covers the northern two-thirds of the country and there is also a savanna coastal band which extends from the Accra area to the Togo border (Owusu *et al.*, 1989). Three vegetation types are recognised within this zone: coastal savanna, interior (Guinea) savanna and north-east (Sudan) savanna. Major economic activities of this zone include livestock production and annual crops, such as maize, millet, cassava, groundnuts, bambarra nuts and cotton (Foggie, 1951; World Bank, 1988).

Temperatures in the country range from minima of 15–24°C to maxima of 33–43°C, with August being the coolest month in the forest zone. Harmattan winds from the Sahara blow between December and February, lowering humidity and temperatures and bringing dust to the northern parts of the country.

Agriculture, both subsistence and cash crop, is the major economic activity in Ghana. It has been estimated that about half of the economically active population is involved in agriculture and lives in rural settlements of less than 5000 people (FAO/UNEP, 1981). Cacao, first introduced in 1878, is the main export crop and is the leading Ghanaian commodity export, although output has fallen considerably since the 1970s due to poor prices and the ageing of trees (Asibey and Owusu, 1982). Other commercial crops such as copra, oil palm, coffee and citrus, and subsistence crops including maize, plantain, cassava, yams and millet are cultivated mostly by individual farmers (FAO/UNEP, 1981).

The forestry and logging sector has, since the early 1970s, accounted for 5–6 per cent of total GDP and ranks third behind cacao and the minerals bauxite, diamonds and gold among commodity exports (World Bank, 1988; Abbiw, 1989). In 1987, export earnings for this sector were approximately US$100 million and it employed 70,000 people (World Bank, 1988). The forests not only produce timber, they also provide 75 per cent of the country's energy requirements.

The people of Ghana are distinguished mainly by language and, to a lesser degree, by their political, social and cultural institutions. The Akan, centred around Kumasi, makes up more than half the country's population. In the forest zone, population densities (1984) range from 30 people per sq. km in the Brong Ahafo Region to 117 people per sq. km in the Central Region. The Upper East Region is one of the most densely populated areas in the savanna zone, with 87 people per sq. km (World Bank, 1987). The capital of Ghana is Accra, while other major cities are Kumasi and Takoradi in the south, and Tamale, Wa and Bolgatanga in the north.

The Forests

The two main vegetation types in Ghana are the closed forest and the savanna ecosystem. The latter is characterised by an open canopy of trees and shrubs with a distinct ground layer of grass. The Ghanaian forests are part of the Guineo-Congolean phytogeographical region; the flora has strong affinities with the forests of Côte d'Ivoire, Liberia and Sierra Leone and a lesser affinity with the Nigerian rain forests from which they are separated by the arid Dahomey Gap (W. Hawthorne, pers comm. and see chapter 11).

The classification by Hall and Swaine (1981) recognises seven vegetation types within the closed forest (Table 21.1 and Figure 21.1), each with distinct associations of plant species and corresponding rainfall and soil conditions.

Wet evergreen forest occurs in the extreme south-western corner of the country and enjoys the highest annual rainfall (1500–2100 mm). It is floristically rich and closely corresponds to Taylor's *Lophira-Tarrietia-Cynometra* 'Rainforest' Association (Hall and Swaine, 1981). Timber species logged from this forest type include dahoma *Piptadeniastrum africanum*, kaku *Lophira alata*, walnut *Lovoa trichilioides* and niangon *Heritiera utilis* (Owusu *et al.*, 1989).

The moist evergreen forest is transitional between the wet evergreen and moist semi-deciduous forests. Although not as rich as wet evergreen, this type has great floristic diversity and a greater number of commercial timber species (Hall and Swaine, 1981). Annual rainfall is 1500–1700 mm.

The moist semi-deciduous forest is subdivided into the north-west and south-east subtypes and has the Kwahu and Mampon scarps and hills of western Ashanti as prominent topographical features. Rich forest soils and annual rainfall of 1200–1800 mm give rise to tree heights which often exceed 50 m and are the tallest in the Ghanaian forest (Hall and Swaine, 1981). Occupying 40 per cent of the closed forest zone, the moist semi-deciduous forest is considered to be the most important for timber production and is characterised by such species as utile *Entandrophragma utile*, African mahogany *Khaya ivorensis* and wawa *Triplochiton scleroxylon* (Owusu *et al.*, 1989). The north-west subtype harbours more elephants *Loxodonta africana* than any other part of Ghana's forest (Hall and Swaine, 1981).

The dry semi-deciduous forest type is found as a peripheral band around the moister forest types to the south and is adjacent to the Guinea savanna zone to the north (Hall and Swaine, 1981). This forest type is widespread in West Africa, has lower tree heights than those found in the moist semi-deciduous forest and receives rainfall in the order of 1250–1500 mm per annum. It is comprised of a wetter subtype, characterised by species such as *Hymenostegia afzelii*, and a drier subtype containing species such as *Diospyros mespiliformis* and *Anogeissus leiocarpus* (Owusu *et al.*, 1989)

The upland evergreen forests occur as outliers from the main evergreen forest block on the high ground (500–750 m) of the Atewa Range, Tano Ofin Forest Reserve and Mt Ejuanema near Nkawkaw. Although surrounded by moist semi-deciduous forest, they are floristically similar to the moist evergreen type. The most characteristic species are herbaceous rather than woody, with epiphytes and ground ferns being both abundant and diverse (Hall and Swaine, 1981). Deciduous trees are relatively rare and the soils tend to be bauxitic in composition (Owusu *et al.*, 1989).

The southern marginal forest type is found as a narrow band from west of Cape Coast to Akosombo. Precipitation is relatively low at 1000–1250 mm per annum and the soils tend to be shallow. Forests occur in isolated patches as most of the land has been converted to

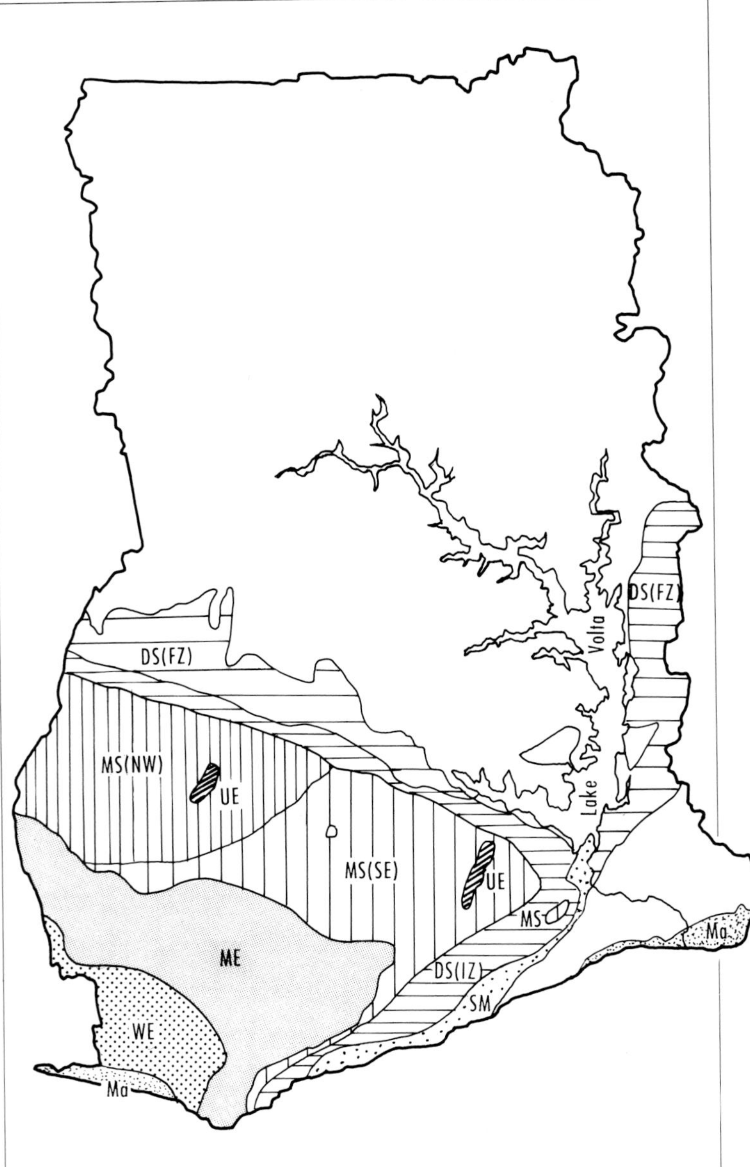

Figure 21.1 Distribution of forest types

(*Sources:* Hall and Swaine, 1981; Forest Resources Management Project, Kumasi, *in litt.*) **WE** wet evergreen. **ME** moist evergreen. **MS** moist semi-deciduous. (**NW** north-west subtype, **SE** south-east subtype). **DS** dry semi-deciduous (**FZ** fire zone subtype, **IZ** inner zone subtype). **UE** upland evergreen. **SM** southern marginal. **Ma** mangrove (source for mangrove zone: Forest Resources Management Project, Kumasi). Forest is stippled.

Table 21.1 Forest zone types and their coverage in Ghana

Type	*Area (sq. km)*	*Percentage of total forest zone area*
Wet evergreen	6,570	8.1
Moist evergreen	17,770	21.8
Moist semi-deciduous	32,890	40.4
Dry semi-deciduous	21,440	26.4
Upland evergreen	292	0.3
Southern marginal	2,360	2.9
South-east outliers	20	0.02
Total	81,342	

(*Sources:* Hall and Swaine, 1981; Silviconsult, 1985)

thickets, farms and savanna (Hall and Swaine, 1981). Trees are sparse and small in stature, while species such as *Hildegardia barteri* and *Talbotiella gentii* show a high degree of gregariousness.

The south-east outliers represent the driest of forest types (rainfall of 750–1000 mm per annum) and are the least extensive, occupying an area of approximately 20 sq. km in small scattered patches (Hall and Swaine, 1981). South-east outliers are found predominantly on the Accra plain, one notable example being at Shai Hills Game Production Reserve. This forest type is characterised by a low floral diversity and trees with low canopies. Typical species include *Millettia thonningii*, *Talbotiella gentii* and *Drypetes parvifolia* (Hall and Swaine, 1981; Owusu *et al.*, 1989). Within this forest type, there are several very rare tree species and few commercial timber species.

Mangroves

Along the coastline of Ghana, mangrove stands are best developed on the western coast between Côte d'Ivoire and Cape Three Points. These stands are restricted in area and are usually found as thickets at river mouths and on some lagoons (Hughes and Hughes, 1991). Map 21.1 shows an area of only 3 sq. km of mangrove remaining in the country, but this is because other patches are too small to be mapped at this scale. The total remaining area is unknown, though it is certainly not large. Typically, *Laguncularia racemosa* and *Rhizophora racemosa* are found on the seaward side of saline lagoons while *Avicennia nitida* occurs on the landward side of the swamps (FAO/UNEP, 1981).

Ghanaians use mangroves in a number of ways. Oysters, crustaceans such as crabs and prawns, as well as birds, reptiles and mammals found in the mangrove habitat are all collected for subsistence purposes, while wood is commonly used for dock pilings and as building poles. The wood of *Rhizophora* and *Avicennia* is used for firewood or converted to charcoal for domestic purposes (Fiselier, 1990) and large quantities are also used in a salt extraction process (SECA/CML, 1987). This process has led to the almost complete disappearance of mangroves in Ghana (SECA/CML, 1987). Salt extraction occurs predominantly in Accra and the surrounding area (Toth and Toth, 1974). Other threats to the mangrove and wetland system include plastics and oil pollution, urban landfill (Toth and Toth, 1974) and reclamation of land as rice paddies (Hughes and Hughes, 1991).

Forest Resources and Management

At the turn of the century, it was estimated that Ghana had 88,000 sq. km of forest. By 1950, this had fallen to 42,000 sq. km and by 1980 it was estimated at 19,000 sq. km (Frimpong-Mensah, 1989). This corresponds closely to the FAO (1988) estimate of 17,180 sq. km for 1980. The current area of intact closed forest is about 15,000 sq. km. Map 21.1 shows 15,839 sq. km of lowland rain forest remaining in the country (Table 21.2), a figure that agrees well with other estimates and shows a reduction, in seven years, of less than 1500 sq. km from other figure estimates. This reduction also accords well with FAO's annual deforestation estimate of 220 sq. km (FAO, 1988).

Concern for the effects of deforestation and its impact on the environment of Ghana stretches back to the beginning of this century with the passing of the Timber Protection Ordinance in 1907 and an assessment of the forest estate by H. N. Thompson. This was followed by the establishment of the Forestry Department in 1909, the passing of the Forest Ordinance (Cap. 157) in 1927, introduction of the taungya system for reforestation in 1928 and the systematic selection, demarcation and reservation of forest in the 1920s and 1930s to create permanent forest estate. At that time, forest reserves there were established to meet local needs for forest products, to create a suitable climate for agriculture, and to prevent environmental deterioration (Foggie, 1951; Hall and Swaine, 1981). With emphasis placed on timber extraction after the Second World War, reserves became increasingly important for maintaining the viability of the timber industry (Taylor, 1960). The latter half of the 1980s and early 1990s have marked a return to the sustainable use and conservation of forest within reserves.

Table 21.2 Estimates of forest extent in Ghana

Rain forests	*Area (sq. km)*	*% of land area*
Lowland	15,839	6.9
Mangrove	3	<0.01
Totals	15,842	6.9

(Based on analysis of Map 21.1. See Map Legend on p. 192 for details of sources.)

Today, there are approximately 280 forest reserves, 100 of which serve a protection function. In addition, within most of the production forest reserves there are areas serving a predominantly conservation function where logging is supposedly prohibited (Ghartey, 1989). These areas, or protection working circles, make up 30 per cent of the forest reserves. Two-thirds of all forest reserves are located in the closed forest zone, representing 21 per cent of the area in that zone (Owusu *et al.*, 1989). Approximately 14,100 sq. km of intact closed forest is currently protected by forest reserves, with as little as 1000 sq. km being found outside them. Hawthorne (1990) has estimated that less than 1 per cent of forest cover is found outside forest reserves, much of it in small, scattered patches in swamps and sacred groves (see case study on the Boabeng-Fiema Monkey Sanctuary). An unspecified amount of

Boabeng-Fiema Monkey Sanctuary

Conservation of forests and wildlife in Ghana has a long tradition and has expressed itself through local customs, practices and taboos. These include protection of snails *Helix* sp., tree and plant species in sacred groves, protection of the Nile crocodile on Katorgor Pond and the establishment of a monkey sanctuary at Boabeng-Fiema in Brong-Ahafo Region.

The Boabeng-Fiema Monkey Sanctuary is 2.6 sq. km in area and represents the driest extreme of forests, being dominated in part by *Khaya grandifoliola* and *Aubrevillea kerstingii* (Hawthorne, 1989; Owusu *et al.*, 1989). Although representing an isolated stand in the savanna, where fire and other disturbances are prevalent, this area remains intact as a consequence of the relationship that local residents have with their environment. A stream and the surrounding forest are considered sacred due to the presence of the god Abujo and the stream spirit Dawaro (Nuhu, 1986). The mona monkey *Cercopithecus mona* and western black-and-white colobus *Colobus polykomos* are said to be children of Abujo and Dawaro and are revered and protected as such. These species cannot be hunted, access to and use of their habitat is restricted and offences are dealt with by the traditional council (Nuhu, 1986). Upon death, these monkeys are afforded the same funeral rites as human community members.

This harmonious relationship between man and his environment has resulted in a stable ecosystem and an area of growing interest to national and international visitors alike. In recognition of the cultural and biotic significance of Boabeng-Fiema, it was given legal protection in 1974 under Section 52 of the Local Government Act of 1971.

secondary forest is developing throughout much of southern Ghana, making for a complex mosaic of various land cover types.

The export of timber from Ghana began in 1888 when African mahoganies were shipped overseas to a number of foreign markets. The mahoganies accounted for 98 per cent of all timber exports until the Second World War (Asibey, 1978; François, 1987). From that time until the 1970s, eight timber species were sold abroad: white mahogany *Khaya anthotheca*, *K. ivorensis*, gedu nohor *Entandrophragma angolense*, sapele *E. cylindricum*, *E. utile*, afrormosia *Pericopsis elata*, baku *Tieghemella heckelii* and *Triplochiton scleroxylon*. Together with cacao, these timbers accounted for 75 per cent of the country's overseas earnings. Today, of the approximately 126 forest tree species which grow to timber size, 50 are considered merchantable, 23 of which are commercially important for logs, sawn timber or for processing into veneers and plywoods, or furniture (François, 1987; Frimpong-Mensah, 1989). Of these, ten or so species account for around 75 per cent of sawlog production (Addo-Ashong, 1989).

Since the late 1940s, more than 90 per cent of the country's closed forest have been logged (Asibey and Owusu, 1982). It has been estimated that the gross national standing volume of timber in the closed forest is 188 million cu. m, of which 102 million cu. m is in trees greater than 70 cm dbh (Ghartey, 1989).

Timber cutting is permitted through long-term concessions and short-term licences. The structure of the industry is such that there are no fewer than 500 logging companies, 85 sawmills, 13 veneer slicing plants and in excess of 200 furniture firms (World Bank, 1988; Frimpong-Mensah, 1989). Logging and wood-processing are centred at Kumasi, while Takoradi is the main port of export. In 1987, 320,000 cu. m of logs and 200,000 cu. m of processed wood were exported, mainly to Britain and West Germany, at a value of approximately US$100 million (World Bank, 1988). The industry provides employment for 1400 people in the industrial operations and many others in the rural sector.

Following a peak in the timber industry in the 1970s, there was considerable decline in all sectors with the collapse of Ghana's economy at the end of that decade. Under the Economic Recovery Programme (ERP) of 1983, the timber industry has been revitalised and timber production has grown dramatically. The production of logs increased by approximately 59 per cent from 560,000 cu. m in 1983 to 890,000 cu. m in 1986, while sawn timber production rose by about 23 per cent from 189,000 cu. m to 232,000 cu. m over the same period (Frimpong-Mensah, 1989).

Over-exploitation of a limited number of timber species led to a ban on export, in log form, of 14 primary species in 1979 and an additional four in 1987. Among these are odum *Milicia excelsa*, *Khaya ivorensis*, emeri *Terminalia ivorensis*, and afrormosia. This has been complemented by the increased use of secondary species such as kyenkyen *Antiaris africana*, oprono *Mansonia altissima* and kyerere *Pterygota macrocarpa* (Friar, 1987). Initiatives aimed at long-term sustained yield management have included the Ghana Forest Simulation Model (GHAFOSIM) which provides for assessment of current and alternative exploitation practices, and the introduction of a 40-year felling cycle (Ghartey, 1990). Proposals include rationalisation of working plans and stock maps for areas to be logged, tighter control of logging activities and a reduction in wastage of logs through stronger forest management under the Forest Resources Management Project (World Bank, 1988).

The development of plantations is one alternative for providing domestic and industrial wood requirements. In Ghana, plantations date back to the first decade of this century when they were situated in the Guinea-savanna woodland (FAO/UNEP, 1981). Between 1948 and 1961, plantation activities were well organised with the establishment of permanent nurseries (FAO/UNEP, 1981; Bennuah, 1987). In 1960, the FAO proposed a national forest plantation estate of 59,000 sq. km, commencing with the planting of 50 sq. km per annum in 1968 (FAO/UNEP, 1981). This was reviewed by the Land Use Planning Committee in 1979 with a recommendation that 110 sq. km be planted annually to meet domestic and industrial demand by the year 2030. Between 1968 and 1977, 400 sq. km were planted through the conversion of natural forests in logged-over forest reserves. As a consequence of poor funding and management, however, current planting has fallen to just 10–20 sq. km per annum, much of which is used to rehabilitate failed plantations (World Bank, 1987; Owusu *et al.*, 1989). In 1980, it was estimated that there was a total of 263 sq. km covered by industrial plantations and 490 sq. km covered by fuelwood plantations (FAO/UNEP, 1981). A second estimate lists a total plantation area of 760 sq. km, 520 sq. km of which is used for the production of sawn timber, while the other 240 sq. km are woodlots and plantations in the savanna zone (Silviconsult, 1985).

Indigenous tree species planted include *Terminalia ivorensis*, *Heritiera utilis* and *Khaya ivorensis*. Exotics planted include teak *Tectona grandis*, *Cedrela odorata*, *Eucalyptus* spp. and *Pinus* spp. (FAO/UNEP, 1981). The choice of species planted has been dependent upon end use, with *Triplochiton scleroxylon* planted for timber production, *Gmelina arborea* for pulp and paper and teak for fuelwood and, increasingly, for telephone and construction poles (FAO/UNEP, 1981; Friar, 1987).

At the present time, little systematic management of plantations occurs. Under the auspices of the Forest Resources Management Project, the approximately 300 sq. km of industrial forest plantations is to be surveyed and rehabilitated. A long-term plan for industrial plantations outside reserves is also to be drawn up, the goal of which is to alleviate pressure on the natural forest. In addition, considerable support is being provided for the establishment of district and local village nurseries to help meet domestic wood demand (World Bank, 1988).

Of the minor forest products used in Ghana (see case study), bushmeat is one of the most important. It has been estimated that 75 per cent of the country's population relies on bushmeat for protein and that in some rural areas, local residents rely exclusively on fish (mostly dried) and bushmeat for their animal protein (Asibey, 1974; MacKinnon and MacKinnon, 1986; World Bank, 1988). In the closed forest zone, all species of mammal, including primates, are eaten. The favourite is probably the cane rat *Thryonomys swinderianus*. In one survey, it was found that more than 12 million cedis (or US$202,000) worth of bushmeat was sold from a single market in Accra in 1985. Three-quarters of this trade involved the cane rat (Ntiamoa-Baidu, 1987). In another six-year survey of one Accra market, Asibey (1987) reported that at least 79,000 kg of bushmeat was traded every year. Apart from subsistence use and local trade, there is a lucrative export market for wildlife species. In 1985, approximately 21,000 live animals were exported at a value of US$344,032 (Ntiamoa-Baidu, 1987).

In Ghana, there is a strong tradition of group hunting, the exercise of which is often tied to customary rites and practices. The Aboakyer (bushbuck) festival of the Efutus in Winneba is an annual hunt to determine whether the next year will be one of peace and abundance or war and famine (Akyempo, n.d.). The outcome is determined by which of two hunting parties is the first to capture a bushbuck and present it live to the local chief. From its inception, this festival has moved from sacrificing a human being, to a leopard, then to the bushbuck of today (Akyempo, n.d.).

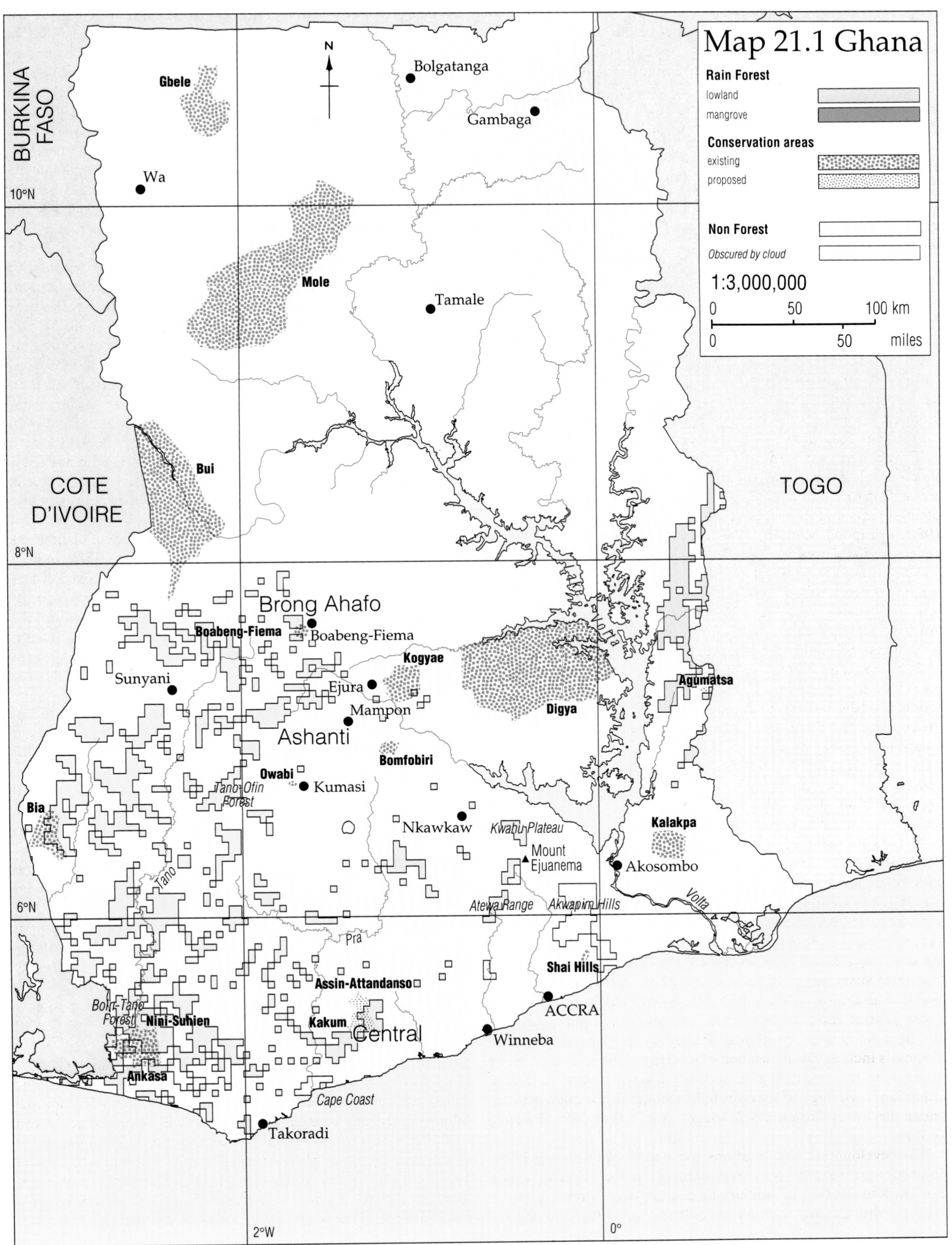
Map 21.1 Ghana
Rain Forest
lowland
mangrove
Conservation areas
existing
proposed
Non Forest
Obscured by cloud
1:3,000,000
0 50 100 km
0 50 miles
BURKINA FASO
COTE D'IVOIRE
TOGO
10°N
8°N
6°N
2°W
0°
Gbele
Wa
Bolgatanga
Gambaga
Mole
Tamale
Bui
Lake Volta
Brong Ahafo
Boabeng-Fiema
Kogyae
Sunyani
Ejura
Mampon
Digya
Agumatsa
Ashanti
Bomfobiri
Owabi
Kumasi
Tano-Ofin Forest
Bia
Nkawkaw
Kwahu Plateau
Mount Ejuanema
Kalakpa
Akosombo
Tano
Volta
Atewa Range
Akwapim Hills
Pra
Shai Hills
Assin-Attandanso
ACCRA
Boin-Tano Forest
Nini-Suhien
Kakum
Central
Winneba
Ankasa
Cape Coast
Takoradi

MINOR FOREST PRODUCTS

The use of minor forest products has had a long history in Ghana. As early as 1850, oil palm was collected and exported, followed by gum copal *Daniellia ogea* in the 1870s. By the 1920s, trade in cola nuts *Cola nitida* exceeded 10,000 tons, the majority of which was collected from wild trees (IUCN, 1988). Today, close to 100 per cent of rural people and 60 per cent of the urban population rely on traditional medicine as their main source of health care. Most of this medicine is derived from forest plants (World Bank, 1988). Traditional medicine is so important that research centres have been established such as that at Mampon.

Many small-scale rural industries depend on minor forest products for their existence. Carvers of drums and utensils, weavers of baskets and cane furniture, and manufacturers of tools and musical instruments are typical examples. The trade in these items also provides a livelihood for thousands of residents throughout the country. It has been estimated that there are roughly 700 people employed full-time in the forest product trade in Kumasi's central market alone (Falconer, 1990).

Several tree species are used in the production of canoes. African pear wood *Manilkara obovata*, asoma *Parkia bicolor* and onyina *Ceiba pentandra* are commonly used in making fresh-water boats, while *Triplochiton scleroxylon* is used for sea-going vessels (World Bank, 1988; N. O'Neill, pers. comm.).

Hundreds of different plant and animal species are used by the local people. Approximately 300 forest species provide wild fruit, while plant beverages include tea, coffee, cocoa and palm wine (Abbiw, 1989). *Thaumatococcus daniellii*, a rhizomatous herb, is reported to be 20,000 times sweeter than ordinary sugar. Between 1975 and 1980, 288,800 kg of this herb in the form of fresh fruit was exported for use in countries such as the UK (Enti, n.d.). Ahensaw *Momordica angustifolia* is a forest climber that is used as a bathing sponge, demmere *Calamus deeratus*, a climbing palm, is used to make baskets, while *Garcinia* spp. are sought after as chewing sticks and the seeds of *Griffonia simplicifolia* are taken to reduce high blood pressure (Abbiw, 1989; Enti, n.d.). Animal species, such as the pangolin *Manis* sp. and genet *Genetta* sp. are valued in the preparation of various medicines as well as providing substantial quantities of meat (World Bank, 1988).

Deforestation

The World Bank (1988) has estimated that closed forest has been reduced at an annual rate of 750 sq. km since the turn of the century, while FAO (1988) reported a deforestation rate of 220 sq. km per annum for the period 1981–5. The current deforestation rate is probably negligible as very little closed forest remains outside the reserve network. The major causes of deforestation are fire damage, over-logging, shifting cultivation and an ever increasing demand for fuelwood. These, coupled with a rapidly increasing population, are issues of national concern.

Fire damage There are reports that fire damage in Ghana, following the drought of 1982–3, altered the structure and composition of about 30 per cent of the forest remaining in the semi-deciduous forest zone, and led to the loss of 4 million cu. m of high quality timber. Indeed, this is now the greatest single threat to the long-term survival of forest in the country. In recent years, fire damage has progressively expanded southwards and heavily logged and previously burnt areas are at a higher risk from subsequent fires .

Over-logging Particularly when coupled with fire, over-logging is a contributing factor to deforestation. It has been predicted that an annual allowable cut of 1.1 million cu. m for commercial timber species could be sustained for the foreseeable future (World Bank, 1988). This would comprise 720,000 cu. m from reserved forests, 120,000 cu. m from unreserved forests and 260,000 cu. m from plantations. However, with log production in the region of 1.35 million cu. m per annum and wastage of merchantable wood at 25–50 per cent, it has been estimated that the actual cut is in the range of 2.0 to 2.7 million cu. m per annum, or 1.6 to 2.5 times the sustainable cut (World Bank, 1988). Compounding the problems caused by this unacceptably high extraction rate, the timber industry has concentrated on a limited number of tree species. It is believed that unless a more balanced approach to exploitation is undertaken, species such as odum, afrormosia and sapele will become virtually extinct as commercial timber species within two or three decades (Alder, 1989).

Shifting cultivation Traditionally, shifting cultivation has accounted for up to 70 per cent of deforestation. The most serious effects have been felt in areas outside legally protected reserves (Agyeman and Brookman-Amissah, 1987). Furthermore, the land tenure system permits the renting of land and encourages an attitude of maximising short-term returns; natural forest has been replaced by plantation and cash crops, while high densities of such domestic stock as cattle, sheep and goats exist in the forest zone. All these factors have contributed to both deforestation and land degradation (World Bank, 1987).

Fuelwood demand Woodfuels, in the form of both fuelwood and charcoal, account for more than 75 per cent of all energy consumed in Ghana and an even higher percentage of energy for household cooking and water heating in rural and urban areas alike (Owusu *et al.*, 1989). In one study, it was found that approximately 84 per cent

Bundles of firewood for sale by the roadside near Kumasi. D. and I. Gordon

of urban households sampled were either dependent on charcoal alone, fuelwood alone (14.2 per cent) or a combination of the two (5 per cent) (Nketiah *et al.*, 1988). In 1985, consumption of woodfuels was roughly 12 million cu. m. By 1988 this figure had reached 15.9 million cu. m, and it is predicted to reach 17 million cu. m by the turn of the century (World Bank, 1988; FAO, 1990).

Between the years 1986 and 2000, according to a World Bank (1988) estimate, fuelwood consumption will grow by approximately 2.8 per cent as against a decline in wood availability of 0.7 per cent per annum. This difference will result in a fuelwood deficit of 11.6 million cu. m by the year 2000. Since fuelwood comes almost exclusively from natural ecosystems, with very little from plantations and woodlots, wood resources will become increasingly scarce in areas outside reserves, while pressure for wood within forest reserves will continue to intensify (Owusu *et al.*, 1989). In the Accra area and parts of the northern regions, some local residents have taken to burning roots and cassava stems to satisfy their fuelwood requirements.

Localised causes of deforestation include extensive cultivation resulting from migration of cacao farmers and the mining of diamonds, gold and manganese in the south-west because these rely heavily on the use of timber, poles and firewood (World Bank, 1987). In addition, high population density in closed forest areas such as the Central Region is putting increasing stress on both wood and land resources (World Bank, 1987).

Biodiversity

The closed forest and savanna zones of Ghana support a wide diversity of plants and animals. More than 3600 plant species have been identified, over 2100 being found in the forest zone (Lebrun, 1976; World Bank, 1988). Within this zone, 125 plant families have been identified and a species diversity of about 300 plants has been recorded in a single hectare (World Bank, 1988). Of the 43 endemic plant species in the country, 23 are known to exist in the forest zone. Seven are found only in the wet evergreen forest (Brenan, 1978; Hall and Swaine, 1981). These include *Hymenostegia gracilipes*, *Cola umbratilis* and *Alsodeiopsis chippi* (IUCN, 1988). In total, 730 tree species, of which 680 attain a dimension of 5 cm or more at breast height, have been recorded from the closed forests (Hawthorne, 1989).

The wet evergreen forest is the most prolific in its floral diversity (Hall and Swaine, 1981). In contrast, the much drier southern marginal and south-east outlier forest is species-poor, with 90 per cent of the vegetation attributed to a single species in some of the southern outlier forest plots. Nevertheless, five endemic or near endemic species, including *Talbotiella gentii*, *Dalbergia setifera* and *Turraea ghanensis*, are found in these forest types.

The mammal fauna of the closed forest zone is biotically diverse and includes over 200 species, many of which are rare or endangered. Ungulate species include Maxwell's duiker *Cephalophus maxwelli*, bushbuck *Tragelaphus scriptus*, buffalo *Syncerus caffer*, bongo *Tragelaphus euryceros* and the rare Ogilby's duiker *Cephalophus ogilbyi* (Ankudey and Ofori-Frimpong, 1990). Carnivores are represented by species such as leopard *Panthera pardus* and golden cat *Felis aurata* which are rare, African civet *Viverra civetta* and several species of mongoose (World Bank, 1988; Mensah-Ntiamoa, 1989). Of the 16 primates recorded in the country, many, including the western black-and-white colobus monkey *Colobus polykomos*, spot-nosed monkey *Cercopithecus petaurista*, white-collared mangabey *Cercocebus Otys*, bushbaby *Galago senegalensis*, Bosman's potto *Perodicticus potto* and chimpanzee *Pan troglodytes* are found in the forest zone. Eight primates occur in Bia and Nini-Suhien national parks (Asibey and Owusu, 1982). Other species of this zone include forest elephant *L. a. cyclotis* and the increasingly rare pygmy hippopotamus *Choeropsis liberiensis* (Mensah-Ntiamoa, 1989).

This black scorpion Pandinus imnperator *found in Ghana's forests, can live up to 25 years and get as large as 23cm.* M. Spaulding

The closed forest zone also supports 74 species of bats, 37 rodents, three species of flying squirrel and a variety of reptiles including African python *Python sebae*, Bosc's monitor *Varanus exanthemathicus*, common hinged tortoise *Kinixys* sp. and Nile crocodile *Crocodylus niloticus* (World Bank, 1988; Mensah-Ntiamoa, 1989).

Some 200 species of birds, out of a total of 721 listed for the country, have been recorded within the forest zone, with 80 species being restricted to primary forest (IUCN, 1988; WRI, 1990). This includes six species of hornbills, the African grey parrot *Psittacus erithacus* and the endangered white-breasted guineafowl *Agelastes meleagrides*, which was seen in Boin-Tano Forest Reserve in 1989 (Nash, 1990). Amphibian and fish species are yet to be systematically surveyed in the forest zone.

Conservation Areas

The Wild Animals Preservation Act No. 43 of 1961, Legislative Instrument 710 of 1971 and the National Wildlife Conservation Policy of 1974 provide the legislative authority and the guidelines for the conservation of wildlife and the establishment of conservation areas. The administration of wildlife legislation is the responsibility of the Department of Game and Wildlife within the Ministry of Lands and Natural Resources.

The total reserved forest in Ghana is approximately 38,400 sq. km, which corresponds to 16.7 per cent of land area. This is divided into 26,300 sq. km of forest reserves and 12,105 sq. km of wildlife sector reserves. The latter represent a little over 5 per cent of Ghana's land area (IUCN, 1988).

Conservation areas fall within four categories: national parks, strict nature reserves, wildlife sanctuaries and game production reserves, of which 93.5 per cent by area, are located in the savanna zone. Threats to conservation areas include the poaching of wildlife, unresolved resettlement issues with local residents and an inadequate knowledge of ecological systems upon which to base sound management decisions (Nuhu, 1986; Owusu *et al.*, 1989). To date, only one management plan, that for the Bia Conservation Area (Martin, 1982), has been proposed.

Ghana

A faunal survey of the closed forest led to the conversion of Bia Tributaries South Forest Reserve to Ghana's first closed forest national park in 1974. Bia National Park was followed by the creation of Nini-Suhien National Park and a number of other reserves (Asibey and Owusu, 1982). In addition to these two national parks there are two wildlife sanctuaries and three game production reserves in the forest zone (IUCN, 1988). Almost 1 per cent of the forest zone is included in these categories of protected area.

Three additional conservation areas are proposed in the forest zone: Kakum National Park and Assin-Attandanso Game Production Reserve (being developed under the auspices of the Central Region Integrated Development Project) and Agumatsa Wildlife Sanctuary (Cloutier and Dufresne, 1991). Existing and proposed conservation areas are listed in Table 21.3.

Two forest habitats will still lack protected area coverage: upland evergreen, which has a number of rare plant species and mangroves which are important bird nesting sites and nursery areas for fish and prawns (IUCN, 1988). Furthermore, the inner zone of the dry semi-deciduous forest type is minimally protected and conservation areas at present do not cover any large tracts of this forest. Several conservation areas are too small to maintain viable populations of animal and plant species in the long term (Hall and Swaine, 1981; IUCN, 1988).

Initiatives for Conservation

Recently, several conservation initiatives have been undertaken by the Environmental Protection Council (EPC) in addition to the efforts of the Department of Game and Wildlife and the Forestry Department. Through the Environmental Outreach Programme, the EPC has been working with district assemblies in the preparation of environmental guidelines for local area development. The EPC also liaises with environmental clubs throughout the country in promoting conservation education, organising field trips to protected areas and sites of ecological significance and conducting preliminary research into sacred groves and the role they play in conservation. The EPC is currently supporting the development of a National Conservation Strategy for Ghana (Benneh, 1987).

Currently, there are no reserve areas along the coast of Ghana and none of the mangroves is protected. However, this situation is being reviewed by the Department of Game and Wildlife as many of these coastline areas have historical (castles and forts), recreational (beaches) and ecological significance. One conservation initiative currently in progress is a joint venture between the UK Royal Society for the Protection of Birds (RSPB), International Council for Bird Preservation (ICBP) and the government of Ghana for the protection of seabirds and shorebirds and their habitats (Hepburn, 1987). Entitled *Save the Seashore Bird Project*, this is one step towards ensuring the protection and sustainable use of the coastal wetland ecosystem. In addition to playing a leading role in this project, the Department of Game and Wildlife has a conservation education officer who works with wildlife clubs, local community groups and schools in promoting conservation, and is planning to extend its protected areas network in the closed forest.

The Forestry Department is currently involved in a multiplicity of management and conservation initiatives under the auspices of the Forest Inventory and Forest Resources Management projects. The Forest Inventory Project is funded by the UK Overseas Development Administration (ODA) while the Forest Resources Management Project is supported by the government of Ghana, World Bank, ODA and the Danish Ministry of Foreign Affairs' Department of International Development (DANIDA) (Howard, 1989). The Forest Inventory Project is surveying closed forest in forest reserves as a basis for sustained yield management (Wyatt, 1989). By November 1988, 5460 sq. km, covering 43 forest reserves, had been inventoried (Ghartey, 1989). The project is also surveying protection working circles within production forest reserves with a view to protecting vulnerable ecosystems.

Table 21.3 Conservation areas of Ghana

Existing and proposed areas are listed below. Forest reserves are not included or mapped. For data on Biosphere reserves and Ramsar sites see chapter 9.

	Existing area (sq. km)	*Proposed area (sq. km)*
National Parks		
Bia*	78	
Bui	2,074	
Digya	3,126	
Kakum*		213
Mole	4,914	
Nini-Suhien*	106	
Strict Nature Reserves		
Kogyae*	324	
Wildlife Sanctuaries		
Agumatsa*		12
Bomfobiri	52	
Owabi	73	
Game Production Reserves		
Ankasa*	207	
Assin-Attandanso*		154
Bia*	228	
Gbele	547	
Kalakpa	324	
Shai Hills	54	
Monkey Sanctuaries[1]		
Boabeng-Fiema*	3	
Totals	12,110	379

(*Sources:* IUCN, 1990; WCMC, *in litt.*)

* Area with moist forest within its boundaries as shown on Map 21.1.
[1] Protected by a local by-law only.

In 1986, the World Bank, with support from FAO, the Canadian International Development Agency (CIDA) and ODA, conducted a forestry sector review for Ghana. As a result, the US$64.6 million Forest Resources Management Project was launched in 1989, under which the forestry and wildlife sectors are being reviewed and working plans will be prepared for all the closed forest reserves. Furthermore, the project will strengthen the present network of conservation areas and improve game management in the areas outside parks and reserves (World Bank, 1988). Institutions such as the Ministry of Lands and Natural Resources, the Forestry Department, the Department of Game and Wildlife, the Forest Products Inspection Bureau and the Timber Export Development Board will be strengthened, as will such training bodies as the Institute of Renewable Natural Resources (IRNR) at Kumasi, and School of Forestry, Sunyani.

Agroforestry is being promoted throughout the country by the establishment of a unit within the Crop Services Division of the Ministry of Agriculture. A National Agroforestry Committee will advise and coordinate research, and support community-based agroforestry (Owusu *et al.* 1989).

A part of the Forest Resources Management Project is the Forestry Commission's revision of the national forest policy which was in effect before Ghana's independence. The policy will focus on the overall management of the forest estate, seek to protect critical areas, provide support for forest-based industries (fuelwood and charcoal) and encourage private and community forestry (EPC, 1989). Actions stemming from the policy will promote a reduction in logging waste, the development of economic uses for wood residues and encourage the development and use of wood preservation techniques.

The EEC is contributing to a study on protected area development in south-western Ghana and the preparation of a regional West African programme of environmental awareness (F. W. Nagel, pers. comm.). Friends of the Earth-Ghana (FoE-Ghana), established in January 1986, has initiated a number of tree planting, environmental education and research programmes in the closed forest zone (FoE, n.d.). Agroforestry schemes have been set up in ten Ashanti villages, with a number of other villages expressing an interest in setting up similar projects with the assistance of Friends of the Earth.

The Ghana Association for the Conservation of Nature (GACON), in conjunction with the Harrogate Conservation Volunteers (UK), was established in June 1988. One of the objectives of GACON is the protection of 'sacred groves', burial grounds and watersheds which are to be managed by local communities. One notable example has been the establishment of Jachie Conservation Area in the Ashanti Region. The declaration of this land as sacred fulfilled the dual purpose of providing an important burial ground for the citizens of Jachie and serving as a refuge for local plant and animal species. Other local wildlife reserves are being supported at Kokobiriko, Asiempong and Santasi (K. Frimpong-Mensah, pers. comm.).

References

Abbiw, D. K. (1989) Non-wood forest products (minor forest products). *Ghana Forest Inventory Project Seminar Proceedings*, pp. 79–88. Forestry Department, Accra, Ghana.

Alder, D. (1989) Natural forest increment, growth and yield. *Ghana Forest Inventory Project Seminar Proceedings*, pp. 47–53. Forestry Department, Accra, Ghana.

Addo-Ashong, F. W. (1989) *Timber Industry Development Within the Tropical Forest Ecosystem – the Ghanaian Experience*. Paper presented at the first ASEAN Forest Products Conference, 4–7 September 1989, Kuala Lumpur, Malaysia.

Agyeman, V. K. and Brookman-Amissah, J. (1987) Agroforestry as a sustainable agricultural practice in Ghana. *National Conference on Resource Conservation for Ghana's Sustainable Development*: Volume 2 – conference papers, pp. 131–40. Environmental Protection Council/Forestry Commission/European Economic Community, Accra.

Akyempo, K. (n.d.) *Aboakyer: Deer Hunt Festival of the Efutus*. Anowua Educational Publications, Accra, Ghana. 43 pp.

Ankudey, N. K. and Ofori-Frimpong, B. Y. (1990) Ghana. In: *Antelopes Global Survey and Regional Action Plans. Part 3: West and Central Africa*. East, R. (ed.) IUCN, Gland, Switzerland.

Asibey, E. O. A. (1974) Wildlife as a source of protein in Africa south of the Sahara. *Biological Conservation* **6**(1): 32–9.

Asibey, E. O. A. (1978) Primate Conservation in Ghana. In: *Recent Advances in Primatology, Vol. 2 – Conservation*. Chivers D. J. and Lane-Petter, W. (eds), pp. 55–74. Academic Press, London, UK. xiii and 312 pp. illustr.

Asibey, E. O. A. (1987) *The Grasscutter*. Unpublished paper. Prepared for the FAO Regional Office, Accra, Ghana.

Asibey, E. O. A. and Owusu, J. G. K. (1982) The case for high-forest national parks in Ghana. *Environmental Conservation* **9**(4): 293–304.

Benneh, G. (1987) Approaches towards a national conservation strategy. In: *State of Ghana's Natural Resources*, pp. 43–7. Environmental Protection Council, Accra, Ghana.

Bennuah, S. (1987) *Development of Forestry in Ghana*. Unpublished BSc thesis, Institute of Renewable Natural Resources, Kumasi, Ghana. 68 pp.

Brenan, J. P. M. (1978) Some aspects of the phytogeography of tropical Africa. *Annals of Missouri Botanical Garden* **65**(2): 437–78.

Cloutier, A. and Dufresne, A. (1991) *Appraisal of the Protected Areas System of Ghana*. IUCN, Gland, Switzerland. 55 pp.

Enti, A. A. (n.d.) *International Trade in Non-traditional Forest Produce*. Unpublished. Forestry Department, Accra, Ghana. 7 pp.

EPC (1989) *Environmental Protection Council Action Plan*. EPC, Accra, Ghana. Draft. Pp. 1–9.

Falconer, J. (1990) *The Major Significance of 'Minor' Forest Products: The Local Use and Value in the West African Humid Forest Zone*. FAO Community Forestry Note 6. FAO, Rome, Italy. 232 pp.

FAO (1988) *An Interim Report on the State of Forest Resources in the Developing Countries*. FAO, Rome, Italy. 18 pp.

FAO (1991) *FAO Yearbook of Forest Products 1978–1989*. FAO Forestry Series No. 24 and FAO Statistics Series No. 97. FAO Rome, Italy.

FAO/UNEP (1981) *Tropical Forest Resources Assessment Project. Forest Resources of Tropical Africa. Part II: Country Briefs*. FAO, Rome, Italy. 586 pp.

Fiselier, J. L. (1990) *Living Off the Tides*. Environmental Database on Wetland Interventions, Leiden, The Netherlands. 119 pp.

FoE (n.d.) Background information brochure. Friends of the Earth, Ghana. 2 pp.

Foggie, A. (1951) Management and Conservation of Vegetation in Africa. In: *Bulletin No. 41, Commonwealth Bureau of Pastures and Field Crops*, pp. 80–92. Penglais, Aberytswyth, Wales. C. J. Cousland and Sons Ltd, Edinburgh, UK.

François, J. H. (1987) Timber resources: demands and management approaches. *National Conference on Resource Conservation for Ghana's Sustainable Development*: Volume 2 – conference papers, pp. 151–5. Environmental Protection Council/Forestry Commission/European Economic Community, Accra.

Friar, F. A. (1987) The state of the resources of Ghana. In: *State of Ghana's Natural Resources*, pp. 16–25. Environmental Protection Council, Accra, Ghana.

Frimpong-Mensah, K. (1989) Requirement of the timber industry. *Ghana Forest Inventory Project Seminar Proceedings*, pp. 70–9. Forestry Department, Accra, Ghana.

Forestry Department (1987) *Forestry Department Annual Report*. Forestry Department, Accra, Ghana.

Ghartey, K. K. F. (1989) Results of the Inventory. *Ghana Forest Inventory Project Seminar Proceedings*, pp. 32–46. Forestry Department Accra, Ghana.

Ghartey, K. K. F. (1990) *The Evolution of Forest Management in the Tropical High Forest of Ghana*. Paper presented at Conférence sur la Conservation et Utilisation Rationnelle de la Forêt Dense d'Afrique Centrale et de l'Ouest. 5–9 November 1990. World Bank/African Development Bank/IUCN. Abidjan, Côte d'Ivoire. 11 pp.

Hall, J. B. and Swaine, M. D. (1981) *Distribution and Ecology of Vascular Plants in a Tropical Rain Forest: Forest Vegetation in Ghana*. Junk, The Hague, The Netherlands. 383 pp.

Hawthorne, W. D. (1989) The flora and vegetation of Ghana's forests. *Ghana Forest Inventory Project Seminar Proceedings*, pp. 8–14. Forestry Department, Accra, Ghana.

Hawthorne, W. D. (1990) Knowledge of plant species in the forest zone of Ghana. *Proceedings of the Twelfth Plenary Meeting of AETFAT. Symposium II Mitteilungen aus dem Institüt für allgemeine Botanik in Hamburg Band 23a*. Pp. 177–86.

Hepburn, J. R. (1987) Conservation of wader habitats in coastal West Africa. In: *The Conservation of International Flyway Populations of Waders*. Davidson, N. C. and Pienkowski, M. W. (eds). Wader Study Group Bulletin No. 49, Supplement/IWRB Special Publication No. 7. Slimbridge, UK.

Howard, W. (1989) The Forest Resources Management Project (World Bank/ODA/DANIDA). *Ghana Forest Inventory Project Seminar Proceedings*, pp. 59–62. Forestry Department, Accra, Ghana.

Hughes, R. H. and Hughes, J. S. (1991) *A Directory of Afrotropical Wetlands*. IUCN, Gland, Switzerland and Cambridge, UK/UNEP, Nairobi, Kenya/WCMC, Cambridge, UK.

IUCN (1988) *Ghana: Conservation of Biological Diversity*. Draft. IUCN Tropical Forestry Programme. World Conservation Monitoring Centre, Cambridge, UK. 17 pp.

Lebrun, J. P. (1976) Richesses spécifiques de la flore vasculaire des divers pays ou régions d'Afrique. *Candollea* **31**: 11–15.

MacKinnon, J. and MacKinnon, K. (1986) *Review of the Protected Areas System in the Afrotropical Realm*. IUCN/UNEP, Gland, Switzerland. 259 pp.

Martin, C. (1982) *Management Plan for the Bia Wildlife Conservation Areas*. Prepared for Ghana Forestry Commission/IUCN/WWF (IUCN/WWF Project 1251). Final Report. 152 pp.

Mensah-Ntiamoa, A. Y. (1989) *Pre-feasibility Study on Wildlife Potentials in the Kakum and Assin-Attandanso Forest Reserves – Central Region – Ghana*. Unpublished. Department of Game and Wildlife, Accra, Ghana. 60 pp.

Nash, S. (ed.) (1990) *Project: GREEN – Ghana Rainforest Expedition Eighty-Nine*. 30 June 1989–12 September 1989. Final Report. Rose Hill Lodge, Dorking, UK. 87 pp.

Nketiah, K. S., Hagan, E. B., Addo, S. T. (1988) *The Charcoal Cycle in Ghana – a Baseline Study*. Final Report. UNDP, Accra, Ghana. 184 pp.

Ntiamoa-Baida, Y. (1987) West African wildlife: a resource in jeopardy. *Unasylva* **156**: 139.

Nuhu, V. A. N. (1986) *Wildlife Conservation in Ghana: Pre- and Post-colonial Era*. BSc thesis, unpublished. Institute of Renewable Natural Resources, Kumasi, Ghana. 67 pp.

Owusu, J. G. K., Manu, C. K., Ofosu, G. K. and Ntiamoa-Baidu, Y. (1989) *Report of the Working Group on Forestry and Wildlife*. Revised version. Prepared for the Environmental Protection Council, Accra, Ghana.

SECA/CML (1987) *Mangroves of Africa and Madagascar, Conservation and Reclamation*. Société d'Eco-aménagement, Marseilles, France and Centre for Environmental Studies, University of Leiden, The Netherlands. Unpublished report to the European Commission, Brussels.

Silviconsult (1985) *The Forest Department Review*. Consultancy Report, World Bank Export Rehabilitation Programme.

Taylor, C. J. (1960) *Synecology and Silviculture in Ghana*. Nelson, Edinburgh, UK.

Toth, E. F. and Toth, K. A. (1974) *Coastal National Park Site Selection Survey*. Unpublished. Department of Game and Wildlife, Accra, Ghana. 8 pp.

World Bank (1987) *Ghana: Forestry Sector Review*. Washington, DC, USA. 35 pp.

World Bank (1988) *Staff Appraisal Report: Ghana Forest Resources Management Project (No. 7295-GH)*. World Bank, Washington, DC, USA. 119 pp.

WRI (1990) *World Resources 1990–1991: a guide to the global environment*. World Resources Institute, New York, USA. 383 pp.

Wyatt, A. (1989) Opening remarks. *Ghana Forest Inventory Project Seminar Proceedings*, pp. 5–6. Forestry Department, Accra, Ghana.

Authorship

Donald Gordon of WCMC, Cambridge with contributions from William Hawthorne, N. M. Bird and J. E. F. Falconer of the ODA/Forest Resources Management Project in Kumasi, N. O'Neill, UK, E. O. A. Asibey of the World Bank, Washington, J. H. François, K. T. Boateng and K. Ghartey of the Forestry Department and K. Tufour and R. K. Bamfo of the Forestry Commission, Accra, G. Pungese of the Department of Game and Wildlife, Accra and K. Frimpong-Mensah, J. G. K. Owusu and S. J. Quashie-Sam from the Institute of Renewable Natural Resources, Kumasi, D. S. Amlalo and K. Omasi from the EPC.

Map 21.1 Forest cover in Ghana

Information on forest cover in Ghana has been extracted from 1989–90 UNEP/GRID data which accompany an unpublished report *The Methodology Development Project for Tropical Forest Cover Assessment in West Africa* (Päivinen and Witt, 1989). Forest/non-forest boundaries in West Africa have been mapped by UNEP/GEMS/GRID, who, together with the EEC and FINNIDA, have developed a system using 1 km resolution NOAA/AVHRR-LAC satellite data to delimit these boundaries. These data have been generalised for this Atlas to show 2 × 2 km squares which are predominantly covered in forest. Higher resolution satellite data (Landsat MSS and TM, SPOT) and field data from Ghana, Côte d'Ivoire and Nigeria have also been used. Forest and non-forest data have been graded into five vegetation types: forest (closed, defined as greater than 40 per cent canopy closure); fallow (mixed agriculture, clear-cut and degraded forest); savanna (includes open forests in the savanna zone and urban areas); mangrove and water. In addition this dataset shows areas obscured by cloud. In this Atlas, UNEP/GRID's 'forest' and 'mangrove' classifications have been mapped. Delimitation of 'types' of forest shown on Map 21.1 have been made by overlaying White's vegetation map (1983) on to the UNEP/GRID dataset.

Reference has also been made to a blueline map 1:500,000 scale *Map of Forest Reserves in Ghana* illustrating forest reserves and distribution of forest zones (Hall and Swaine, 1981) and to a 1:2 million scale map *Ghana* compiled by the Survey of Ghana in 1969 showing the main vegetation zones, reserved and unreserved forests.

Conservation areas were drawn from a 1:1 million unpublished map *Ghana* (1989) showing district assembly areas and established protected areas prepared by the Town and Country Planning Department, Ghana.

22 Guinea

Land area 245,857 sq. km
Population (mid-1990) 7.3 million
Population growth rate in 1990 2.5 per cent
Population projected to 2020 14.4 million
Gross national product per capita (1988) US$350
Rain forest (see map) 7655 sq. km
Closed broadleaved forest (end 1980)* 20,500 sq. km
Annual deforestation rate (1981–5)* 360 sq. km
Industrial roundwood production† 647,000 sq. km
Industrial roundwood exports† 8000 cu. m
Fuelwood and charcoal production† 4,022,000 cu. m
Processed wood production† 90,000 cu. m
Processed wood exports† nd
* FAO (1988)
† 1989 data from FAO (1991)

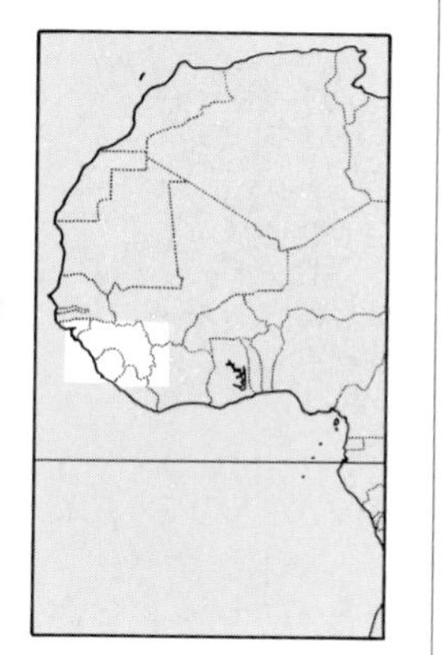

The moist forests of Guinea are severely reduced in area and deforestation is continuing. It is expected that all moist forest outside reserves will be lost in the near future, except for some degraded relicts protected by terrain or tradition. Nonetheless, the remaining forest tracts have a diverse flora and fauna, characteristic of the Upper Guinea centre of endemism, and have international importance for biodiversity conservation.

The Forestry Department, which in Guinea is responsible for the conservation of wildlife and all renewable natural resources, is at present in a state of reorganisation and reorientation after a period of inactivity in the early 1980s. The National Forest Policy, drawn up within the Tropical Forestry Action Plan, emphasises conservation and wise use of all forest resources for the benefit of rural populations and the maintenance of environmental quality.

National capacity to implement forestry and conservation programmes remains limited, but assistance is being provided by the World Bank, the European Development Fund, Germany and France.

INTRODUCTION

The Republic of Guinea, lying between 7°05'–12°51'N and 7°30'–15°10'W, is bounded on the west by the Atlantic, in the south by Sierra Leone, Liberia and Côte d'Ivoire, and in the north by Guinea-Bissau, Senegal and Mali. It divides into four natural regions: Guinée Maritime, Moyenne Guinée or Fouta Djallon, Haute Guinée and Guinée Forestière. The country becomes increasingly dry towards the north and east. It straddles the Upper Guinea Forest, the transitional savanna/forest mosaic and the dry Sudanian vegetation zones (White, 1983). The headwaters of many of the major rivers of West Africa (for example, The Gambia and Senegal) lie in the highlands of the Fouta Djallon (culminating in Mt Loura, 1538 m) in the central part of the country. The 'Dorsale Guinéenne' constitutes a second highland chain traversing the south-east on a south-east to north-west axis which includes Mt Nimba (1752 m), Pic de Fon (1656 m) and the Ziama massif (1387 m). The 'dorsale' forms the watershed between the Niger drainage system and the major rivers flowing to the Atlantic coast through Guinea's neighbouring countries to the south.

The population was estimated at 7.3 million in 1990 and it is expected to be almost double this size by the year 2020. Population density is highly variable according to region, with 44 people per sq. km on the coast, 23 people per sq. km in the moist forest region in the south-east and only 12 people per sq. km in the dry north-east.

Guinea is relatively rich in natural resources, particularly minerals, and has a high agricultural potential in certain regions. However, the economy declined in the 1960s and 1970s and this is now one of the least developed countries in Africa. A radical economic reform programme is under way with a policy framework designed to create a more vigorous, market-based economy.

The Forests

The type of forest cover varies across the country, following the general trend of increasing dryness towards the north and east. Three principal formations (including mangroves) occur and the following descriptions are largely based upon those given in the National Forestry Policy and Action Plan (République de Guinée, 1987).

Much of the northern half of Guinea, with the exception of the montane section of Fouta Djallon, was originally covered with dry forest. In this area, rainfall declines from 1600 mm to 1250 mm per year and the dry season lengthens from three to seven months between Faranah in the south-west of the region and Siguiri in the north-east. To the west the forest was mostly dominated by *Parkia biglobosa* and *Pterocarpus erinaceus*, with much bamboo *Oxytenanthera abyssincia* in the underbrush. In the north, *Afzelia africana* dominated on sandy soils, while the eastern foothills of the Fouta Djallon carried a mixed forest in which *Erythrophleum guineense* was prominent. To the south of the dry forest zone, principal elements included *Isoberlinia doka*, *I. dalzielli* and *Uapaca somon*. Human activity has now degraded the dry forests of Guinea into more open, wooded savannah formations. Dense *Isoberlinia* forest is still reported, however, from the region between Faranah and Kouroussa.

The moist forests include evergreen forest in the extreme south-east of the country, higher-altitude submontane forests in the Fouta Djallon and on the higher peaks in the south-east and semi-deciduous forests elsewhere. The evergreen moist forests are similar to those found in the neighbouring regions of Côte d'Ivoire and Liberia. They are mixed forests with no clearly dominant species although *Piptadeniastrum africanum*, *Parkia bicolor*, *Heritiera utilis*, *Entandrophragma* spp. and *Lophira alata* are important constituents. Valley-bottoms often contain *Raphia* and *Uapaca* dominated swamp forest, while considerable areas within the remaining large tracts are prone to waterlogging during the wet season.

These forests appear typical of the southern part of Guinée Forestière, up to the area around Macenta and Dièckè. Further north a drier forest, tending to become semi-deciduous, appears in which a number of species that are also present in the south, but that are more typical of secondary conditions, become more prominent.

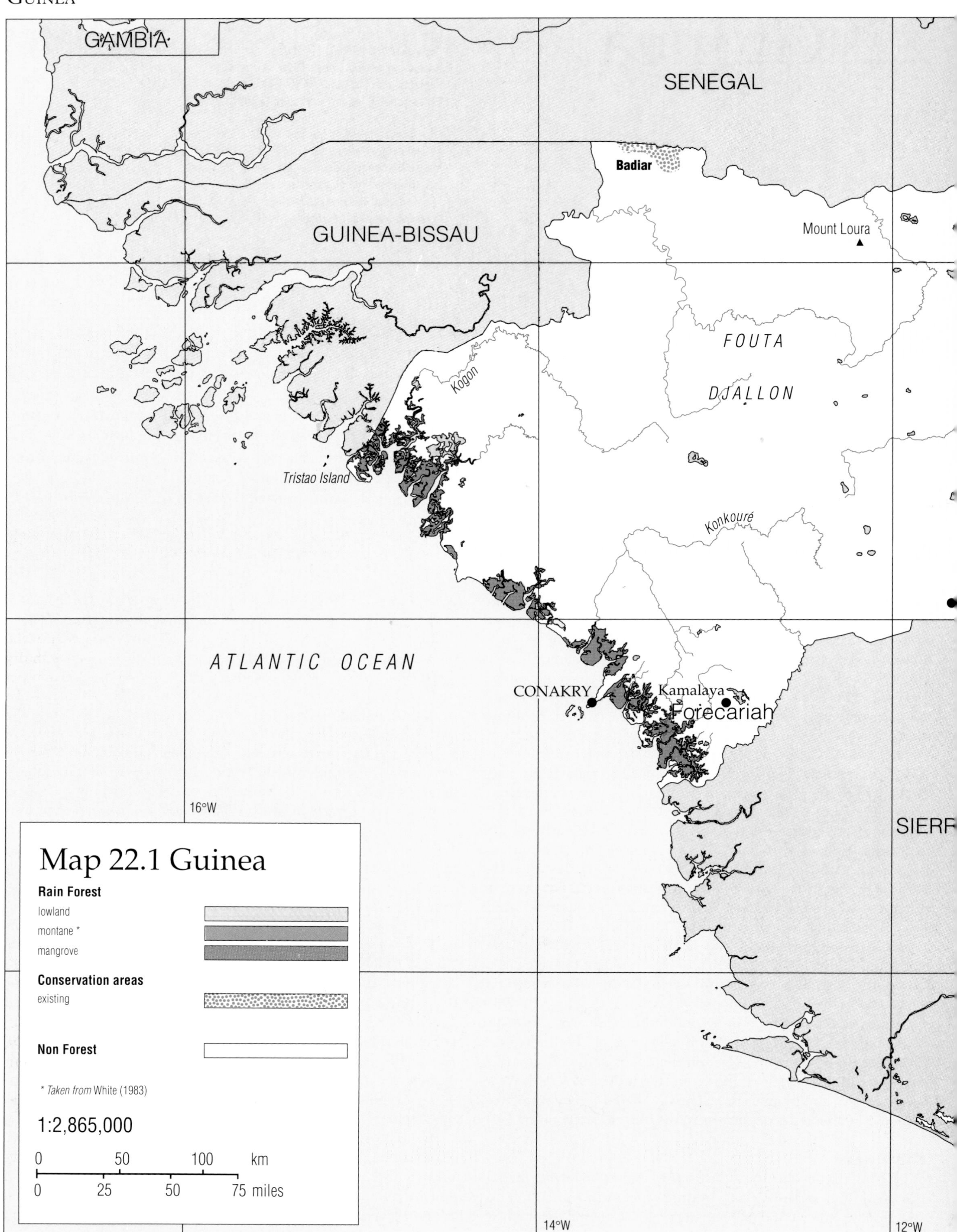
GAMBIA
SENEGAL
Badiar
GUINEA-BISSAU
Mount Loura
FOUTA
DJALLON
Kogon
Tristao Island
Konkouré
ATLANTIC OCEAN
CONAKRY
Kamalaya
Forecariah
16°W
14°W
12°W
SIERR
Map 22.1 Guinea
Rain Forest
lowland
montane *
mangrove
Conservation areas
existing
Non Forest
* Taken from White (1983)
1:2,865,000
0 50 100 km
0 25 50 75 miles

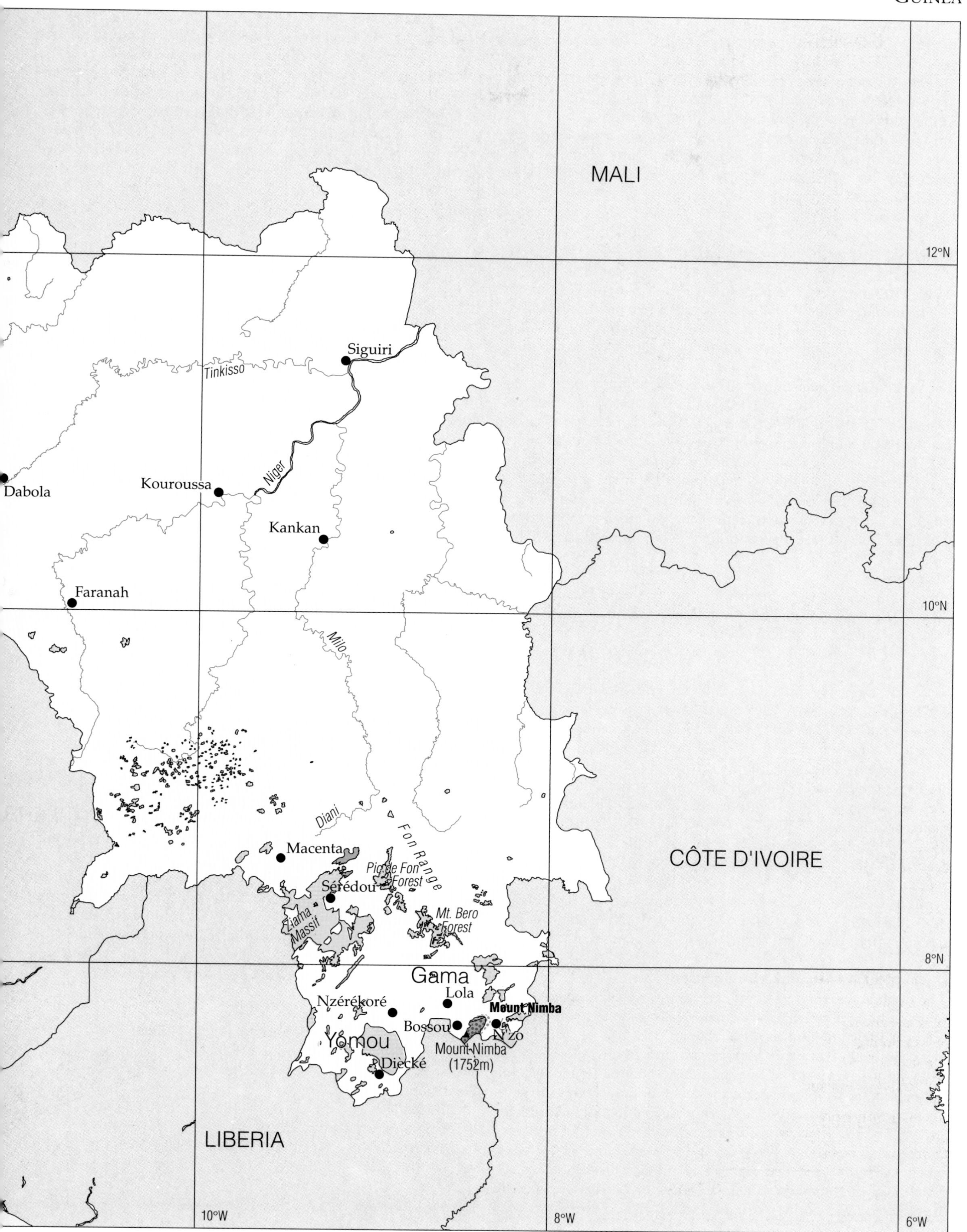
MALI
12°N
Siguiri
Tinkisso
Niger
Dabola
Kouroussa
Kankan
Faranah
10°N
Milo
Diani
Fon Range
Macenta
Pic de Fon Forest
Sérédou
Liama Massif
Mt. Bero Forest
CÔTE D'IVOIRE
8°N
Gama
Lola
Nzérékoré
Mount Nimba
Bossou
N'zo
Yomou
Mount Nimba (1752m)
Dièckè
LIBERIA
10°W
8°W
6°W

These include *Triplochiton scleroxylon*, *Terminalia ivorensis*, *T. superba*, *Chlorophora regia* and *Antiaris excelsa*. With increasingly dry conditions, species such as *Khaya grandifoliola* and *Afzelia* spp. appear. Rainfall in the evergreen forest zone varies from 2700 mm in the south, with a dry season of less than two months, to 1700 mm in the north, where the dry season is three months.

Much of the remaining area of southern Guinea was originally covered with semi-deciduous moist forest. It was similar to that described for the northern evergreen forest but it lacked *Triplochiton* and contained *Khaya senegalensis* and *Erythrophleum*. On the coast the forests were richer in *Parinari excelsa*, were without *Erythrophleum* but exhibited an abundance of *Canarium*, *Aningeria*, *Antiaris*, *Carapa* and *Terminalia ivorensis*. The coastal climate has sharply contrasting dry and wet seasons, rainfall of 2500–4500 mm, and a dry season of five to six months' duration.

The uplands of the Fouta Djallon, with a five-month dry season during which there are frequent mists, relatively high rainfall, cooler temperatures and high relative humidity, carried a dense submontane forest. *Parinari excelsa* was a dominant species, and *Parkia biglobosa* a prominent constituent. The higher altitude areas of the Dorsale Guinéenne such as Nimba, the Simandou range and the Ziama massif, also bear forests in which *Parinari* is common, becoming dominant with increasing altitude. In the Ziama massif above Sérédou, *Parinari* becomes frequent and tree-ferns (*Cyathea* sp.) appear in valleys at about 700 m altitude. This corresponds to the level at which mist cover becomes regular (Bourque and Wilson, 1990). A similar progression, with slight variations in the levels at which the changes take place, is found on the slopes of Mt Nimba.

Mangroves

Areas of mangrove occur along the entire coast of Guinea, except on rocky promontories, and particularly around its many river estuaries. Species include *Rhizophora harrisonii* (the commonest species), *R. racemosa*, *R. mangle*, *Avicennia africana* and *Laguncularia racemosa*.

Present cover is estimated at 2500 sq. km (Altenburg and van der Kamp, 1989), while the original cover was reckoned to be in the order of 3000 sq. km (MacKinnon and MacKinnon, 1986). Map 22.1 shows an area of 2963 sq. km of mangroves remaining in the country. The main human activities in the mangrove areas are fishing, firewood cutting and rice production (Altenburg and van der Kamp, 1989). A considerable amount of wood from the mangroves is used as fuel for salt extraction. It is believed that about 780 sq. km has been cleared for rice fields alone, which suggests that MacKinnon and MacKinnon's figure for original cover was an underestimate. The government of Guinea is very concerned about the degradation and loss of mangroves (République de Guinée, 1987) as they help sustain numerous economic activities and afford protection to the coast and its hinterland.

Forest Resources and Management

The original moist forest cover (evergreen, semi-deciduous, submontane and the moist forest element of the savanna/forest mosaic, but excluding mangroves) was around 182,800 sq. km or 74 per cent of the country (MacKinnon and MacKinnon, 1986). Much of this was lost long ago, particularly on the Fouta Djallon, although considerable areas of semi-deciduous forest may still have been present at the turn of the century (République de Guinée, 1987). The semi-deciduous, drier, moist evergreen and northern submontane forests are now reduced to scattered and degraded relicts. Nonetheless a good example of coastal semi-deciduous forest remains at Kamalaya in the Prefecture of Forecariah where it is protected by its position in a deep valley. In addition, fragments of the Fouta Djallon forests persist in a number of forest reserves. However, the only extensive forest is now concentrated in Guinée Forestière in the moist evergreen zone of the south-east.

Various figures of forest cover have been published, but methodological differences and the rapidly changing situation mean that the figures are not consistent. FAO (1988) estimated that, in 1980, as much as 20,500 sq. km of 'closed broadleaved forest' remained in the country. However, République de Guinée (1987) reported that only 10,750 sq. km of 'forêt dense' existed at that time. More recent estimates are 5887 sq. km of forest with dense or medium canopy cover (Atlanta, 1989) and 3970 sq. km of 'forêt dense' (Esteve *et al.*, 1986). Map 22.1 indicates a figure intermediate between these two estimates: 4482 sq. km of lowland rain forest are shown and 210 sq. km of montane forest, giving a total of 4692 sq. km of dryland closed forest (Table 22.1).

The largest remaining areas, and the only ones containing extensive primary forest, are the forest reserves of Ziama (1123 sq. km, of which *c.* 750 sq. km is high forest) and Diècké (556 sq. km of which 520 sq. km is high forest). Other forest areas are found on Mt Nimba, at the Côte d'Ivoire frontier near Nzo (Forest of Déré) and within the Pic de Fon and Mt Bero Forest Reserves. Smaller relicts persist elsewhere, including on the Diani River and on the Liberian frontier in the Prefecture of Yomou and in the Gama region.

Forest management and wildlife conservation is the responsibility of the Direction Nationale des Forêts et Chasses (DNFC), within the Ministère de l'Agriculture et des Ressources Animales. The basis of the forest estate is the network of forest reserves (Forêts Classées) largely created before independence and covering some 11,000 sq. km in total. Areas protected for wildlife conservation are also under the aegis of the DNFC; indeed, in broad terms, it has responsibility for the management of all forest resources and land not converted for agriculture or other uses.

After independence the exploitation of natural forest concentrated upon the evergreen forest zone of Guinée Forestière where valuable timber species were found within reach of good access to the sea, by way of Côte d'Ivoire and Liberia. Timber was extracted from the forests of eastern Guinée Forestière at Lola, Nzo and Gama, while the more accessible northern sectors of the Diècké Reserve were managed to supply a sawmill and plywood factory at Nzérékoré. Part of a 300 sq. km concession at Maluetta in the Ziama Reserve was also exploited, with a sawmill (attached to the Forestry Centre) and chipboard factory at Sérédou. By the early 1980s all these state enterprises had effectively ceased to function. Both the chipboard and plywood factories had closed down completely and the sawmills operated at a very reduced rate, adequate only to keep them in working order and to provide a living for the workforce. The Sérédou and the Nzérékoré sawmills have both now been sold into private ownership. Although in 1989, FAO (1991) recorded exports of 8000 cu. m of industrial roundwood from the country, log and timber exports are now prohibited. However, reliable data on the forest industry are scarce and published figures are contradictory.

Table 22.1 Estimates of forest extent in Guinea

	Area (sq. km)	*% of land area*
Rain forests		
Lowland	4,482	1.8
Montane	210	<0.1
Mangrove	2,963	1.2
Totals	7,655	3.1

(Based on an analysis of Map 22.1. See Map Legend on p. 199 for details of sources.)

Several inventories and studies were undertaken in the second half of the 1980s (e.g. Atlanta, 1989; CTFT, 1989) to assess the forest resource as a basis for resuscitating the moribund forestry sector. At the same time a national forest policy was developed, the 'Politique Forestière et Plan d'Action' (République de Guinée, 1987), within the framework of the Tropical Forestry Action Plan. This document is wide-ranging in scope and places emphasis firmly on the necessity to conserve forest cover in order to safeguard environmental quality. The policy recognises the need for a protected area network, for firm measures to arrest forest loss and to rehabilitate degraded areas and the management of the full array of forest values and services both for local people and the national economy.

As a first step towards implementation of the Forest Policy, the forest service is to be restructured and its staff strengthened. Forest administration is to be decentralised with technical staff of the DNFC attached to local administrations at the level of prefectures, regions and sectors.

The inventories have shown that the exploitable timber reserves of Guinée Forestière now stand at an estimated 10.1 million cu. m over an area of 12,500 sq. km, of which the greater part is dispersed in open woodlands (Atlanta, 1989). The forests of eastern Guinée Forestière are depleted to such an extent that they are no longer worthy of attention and the relicts are being swiftly cleared. The emphasis has therefore shifted to the last two extensive tracts of forest, Ziama and Diècké, and a World Bank/KfW-funded project (PROGERFOR) to manage these forests commenced in September 1991 (see section on conservation initiatives below).

Deforestation

The major cause of forest loss is the traditional agricultural and pastoral practice in which land is cleared by fire. The Fouta Djallon was deforested in this manner in historic times and the drier forest types have proved particularly susceptible during this century. The main area of loss is now Guinée Forestière where it is a continuing and accelerating process. Depletion of forest cover has been estimated at 260 sq. km per year (République de Guinée, 1987). The general consensus is that all natural forest outside forest reserves, excepting fragments protected by terrain or tradition, is liable to be lost in the near future. Even the forest reserves are subject to encroachment.

The loss of forest to agriculture is driven by population growth, estimated at 2.5 per cent a year, and exacerbated by immigration from the north following drought and environmental degradation in the Sahelian zone during the 1970s and 1980s. More recently, the civil war in Liberia has led to an influx of refugees from the south, of whom a proportion may be expected to remain. Slash and burn agriculture is the norm, with its cycle of forest clearance for crops and decreasing fallow periods resulting in loss of productivity and the subsequent need to clear fresh forest land. Cash-crop planting, principally of coffee, cacao and cola under tree cover, is now being promoted as part of the economic revival of the agricultural sector. This has the effect of putting land under permanent crops and removing it from the food-producing domain, so increasing the need for new land for subsistence cultivation.

Logging has obviously led to alteration of forest composition and structure and has undoubtedly reduced the area of primary forest in the country, but has not of itself resulted in total forest loss. Its main impact upon forest cover has been to open the way for agricultural settlement. In December 1984, it was estimated that 290 sq. km of forest remained in the Lola region (Lola, Gama, Nzo) but farmers were dispersed throughout the area and deforestation was taking place at a rate of 4 sq. km per year. The Gama forests were already reduced to degraded fragments and all the forests are expected to be lost completely during the 1990s (Esteve *et al.*, 1986).

Permanent settlements within Diècké Forest Reserve, developed in contravention of regulations. R. Wilson

Parts of the Ziama and Diècké Forest Reserves were exploited under the taungya system whereby, after the timber trees have been removed from an area, the local people are allowed to grow crops among replanted tree seedlings. The farmers thus keep the seedlings clear of weeds for their early years. After three years, they are supposed to move on and leave the forest to regrow. However, control was inadequate and the land became permanently occupied, not only where the taungya system was applied but in large adjoining areas as well. Fields were still being cut out of the forest in these reserves in early 1991, indicating continuing management deficiencies at a local level. Rates of forest loss are estimated at 1.5 sq. km and 9.6 sq. km per year in Diècké and Ziama respectively (Atlanta, 1989).

Forest reserves are, in theory, totally protected against agricultural development. However, encroachment by agro-industrial concerns has been tolerated on a small scale. A 3 sq. km quinine plantation was established in the Ziama massif in the 1940s and has recently passed into private ownership. In 1990 permission was being sought for a concession on a further 25 sq. km in the reserve. Another private company occupied land within the forest reserve boundary (although on land already cleared of forest) and apparently intended to plant up to 15 sq. km of a variety of crops including rice, coffee, bananas and papaya. A third company, SOGUIPAH, is developing oil palm and rubber plantations in valley-bottom land on the edge of the Diècké Reserve near Diècké town, and may have encroached inadvertently upon the reserve boundary. There issues are now being resolved by PROGERFOR.

Mt Nimba contains valuable iron ore deposits. These are already exploited in the Liberian sector of the Nimba range and plans are being advanced for a similar development in Guinea. If these are realised, one may foresee increased pressure for land for a growing population attracted by the mining operations, increased demands for fuel and construction wood, and direct damage resulting from the mining itself or its associated pollution. The forests of the area are thus under permanent threat of degradation and loss. The Guinean government is acutely aware of the environmental implications of mining development on Nimba, which includes a Man and the Biosphere (MAB) reserve and a World Heritage site. Although the revised mining plans will definitely involve the loss of about 150 ha of submontane forest, the remaining 11,850 ha on the massif will be contained within the World Heritage site and the MAB core zone, which has been enlarged to include the Forest of Déré. A management plan for the MAB reserve has been drawn up, which provides for improved conservation management. It is hoped that these measures will minimise or offset the impacts of the mine and lead to better protection overall.

The hydro-electric dam scheme on the Diani River will, if carried through, create a 470 sq. km lake, covering about 74 sq. km of Ziama Forest Reserve. Around 50 sq. km of dense and medium tree cover is likely to be flooded (Atlanta, 1989) of which 7 sq. km, all within the reserve, would be closed canopy forest.

The overall situation, then, is one of heavy deforestation that has resulted in the disappearance of 96 per cent of Guinea's original forest cover. Unless firm measures are adopted without delay, and vigorously implemented, the remainder may well not survive.

Biodiversity

The forests of Guinea form part of the Upper Guinea forest block, isolated from the rest of the Guineo-Congolian forests by the more arid Dahomey Gap. The Upper Guinea forests contain a distinctive flora and fauna and constitute the 'Upper Guinea centre of endemism'. Although the forests within the Republic of Guinea are now greatly reduced in total extent, important tracts remain and are valuable for the conservation of biodiversity.

Only Nimba has been thoroughly studied and proved to be of outstanding importance. New information from Ziama and Dièckè (Bourque and Wilson, 1990) shows that these forests are also among the most important in the region. A preliminary reworking of the priority scores used by Collar and Stuart (1988) to assess the relative importance of forests for the conservation of African birds, shows Nimba to be the third most important in the region, Ziama the fourth and Dièckè the seventh. Threatened and 'near-threatened' bird species in these forests include the western wattled cuckoo-shrike *Campephaga lobata*, yellow-throated olive greenbul *Criniger olivaceus*, white-necked rockfowl *Picathartes gymnocephalus*, Nimba flycatcher *Melaenornis annamarulae* and black-headed stream warbler *Bathmocercus cerviniventris* (Collar and Stuart, 1985; Bourque and Wilson, 1990). A further six threatened and near-threatened species are known to occur just over the border in the non-Guinean sectors of Nimba. The white-breasted guineafowl *Agelastes meleagrides* is now believed to be extinct in the region (Collar and Stuart, 1985).

The following threatened mammal species are known, or strongly suspected to occur, in the moist forests of Guinea: diana monkey *Cercopithecus diana*, red colobus *Procolobus [badius] badius*, olive colobus *Procolobus verus*, chimpanzee *Pan troglodytes*, leopard *Panthera pardus*, African elephant *Loxondonta africana*, and pygmy hippopotamus *Choeropsis liberiensis* (IUCN, 1987; Bourque and Wilson, 1990). The manatee *Trichechus senegalensis* occurs in the coastal regions in the north of the country (IUCN, 1988). A number of little-known and rare species have been recorded, including Johnston's genet *Genetta johnstoni*, lesser otter shrew *Micropotamogale lamottei*, slender-tailed giant squirrel *Protoxerus aubinnii* and golden cat *Felis aurata* (IUCN, 1987; Schreiber *et al.*, 1989; Bourque and Wilson, 1990). Others, notably the Liberian mongoose *Liberiictis kuhnii*, are known from the Guinean frontier regions in neighbouring countries (Schreiber *et al.*, 1989).

Among the reptiles, the threatened West African dwarf crocodile *Osteolaemus tetraspis* is found in Ziama and Dièckè (Bourque and Wilson, 1990), while several other rare and endemic reptiles have also been recorded. Eleven species of endemic or near endemic amphibians occur (WCMC, 1987). Of particular interest, although it is not a forest species, is the viviparous toad *Nectophrynoides occidentalis*, known from only the Guinean and Côte d'Ivoire sectors of Mt Nimba. Little is known about the invertebrates, but it is certain that some of these too will be rare and endemic.

Some 99 plant species or subspecies are believed to be endemic to Guinea, of which 37, including those of forests and forest relicts, are considered threatened (IUCN, 1988). The important centres of plant endemism are the highland areas of the Dorsale Guinéenne (particularly Nimba) and the Fouta Djallon. A number of the high value timber trees, such as *Entandrophragma* spp., *Khaya* spp., and *Milicia excelsa*, are considered endangered or priorities for genetic resource conservation, or both (FAO, 1986; Read, 1990). Some wild species, notably *Coffea* spp., may have value for crop improvement.

The fauna and flora of Guinea's forests, with the exception of Nimba, remain poorly known and their full importance for biodiversity conservation may well prove to be greater than the considerable value that can already be attributed to them. Given the rates of deforestation and the level of hunting now taking place in the remaining tracts, however, all forest-dependent species must be considered vulnerable, if not endangered on a national scale.

Conservation Areas

The strict nature reserve on Guinean Mt Nimba (140 sq. km) was created in 1944, declared a Biosphere reserve in 1980 and designated a World Heritage site in 1981. Boundary changes to the Biosphere reserve and a renomination of the World Heritage site took place in 1991, to take proper account of the mining proposals that postdated the earlier designations. The Biosphere reserve core zones now cover the whole of the Guinean sector of the Nimba massif with the exception of 800 ha directly affected by the planned mine, and include some 11,800 ha of forest. They also cover the Forest of Déré (8920 ha) and a small forest at Bossou. The non-forested areas on Mt Nimba are clothed in savannas and upland grassland formations. The exceptional biological diversity of Mt Nimba can, in great measure, be attributed to the variety of distinct vegetation types and their ecotones.

The Ziama Forest Reserve was declared a Biosphere reserve in 1980, covering 1162 sq. km. In principle all forest reserves in Guinea constitute conservation areas for wildlife, with prohibitions on hunting and other unauthorised activities. The most recent Guinean protected area to be created, the Badiar National Park, is not a moist forest site. (See Table 22.2 for a list of conservation areas.)

Hunting and agricultural encroachment occur within the boundaries of all Guinean forest conservation areas. The Mt Nimba Strict Nature Reserve is relatively well protected from these pressures but as noted above its future is threatened by the possibility of full-scale mining operations.

Table 22.2 Conservation areas of Guinea. Forest reserves are not included or mapped. For data on Biosphere reserves and World Heritage sites see chapter 9.

	Existing area (sq. km)	*Proposed area (sq. km)*
National Parks		
Badiar	382	
Strict Nature Reserves		
Mt Nimba*	140	
Bird Reserves		
Alkatraz†		nd
Tristao†		nd
Totals	522	

(*Sources:* IUCN, 1990; WCMC, *in litt.*)

* Area with moist forest within its boundaries according to Map 22.1.

† Area not mapped in this Atlas.

Initiatives for Conservation

Under the Sekou Touré regime between the 1960s and the early 1980s nature conservation was given a very low priority in Guinea (MacKinnon and MacKinnon, 1986). This situation has altered and conservation is now a matter of great government concern. Although conditions on the ground still reflect previous neglect, a range of initiatives has been launched with the potential to achieve real improvements.

The key policy statement upon which this revitalisation is based is the National Forest Policy, while the process of developing a national Environmental Action Plan was begun in 1989. Aspects of the Forest Policy relating to subsistence and sport hunting and to the creation of protected areas are given a legal framework in the new 'Code de la Protection de la Faune Sauvage et de la Réglementation de la Chasse', drafted in 1988 and now gazetted (République de Guinée, 1988).

The Badiar National Park is already established and a number of other sites have been identified as potential protected areas. These include Bossou (near Nimba) and Ouré-Kaba in the moist forest region, and Iles Tristao, a coastal site with mangrove and considerable ornithological interest. However, except for Bossou and Tristao, the identification of the sites is based largely on anecdotal information and there is a recognised need for proper survey and inventory work upon which to base the further development of the protected area programme.

At present it seems that moist forests are under-represented, given their national and international importance, in the array of sites under consideration for protected area status. This makes it all the more important that the forest reserves should be well managed. A project for the management of the Ziama and Diècké forest reserves, the strengthening of the management capability of DNFC and the clarification of land rights in the vicinity of the two reserves, is seen as an important step in this respect. This project began, with World Bank and German funding, in September 1991.

An assessment of the special problems of Nimba is being undertaken by a pilot project initiated in 1989 by UNDP and Unesco. This project, which has a strong fundamental research element, is being executed by the Ministry of Natural Resources, Energy and Environment (which is also responsible for mining affairs) and the Directorate of Scientific and Technical Research.

The policy framework for forest conservation in Guinea has largely been established. The issues now revolve around the translation of policy into appropriate practical action. The results remain to be seen but the next decade is a critical one in which the remaining forests of Guinea will either be adequately protected or irrevocably degraded and lost.

References

Altenburg, W. and van der Kamp, J. (1989) *Etude Ornithologique Préliminaire de la Zone Cotière du Nord-ouest de la Guinée.* ICBP Study Report No. 30. ICBP, Cambridge, UK.

Atlanta Consult Industrie-und Unternehmensberatung GmbH (1989) *Inventaire Forestier de la Guinée Forestière* (2 vols. Rapport de Synthèse and Rapport Technique).

Bourque, J. D. and Wilson, R. (1990) *Rapport de l'Etude d'Impact Ecologique d'un Projet d'Aménagement Forestier Concernant les Forêts Classées de Ziama et de Dièckè en République de Guinée.* Unpublished report to IUCN, Gland, Switzerland.

Collar, N. J. and Stuart, S. N. (1985) *Threatened Birds of Africa and Related Islands. The ICBP/IUCN Red Data Book Part 1.* ICBP/IUCN, Cambridge, UK. 761 pp.

Collar, N. J. and Stuart, S. N. (1988) *Key Forests for Threatened Birds in Africa.* ICBP Monograph No. 3. ICBP, Cambridge, UK.

CTFT (1989) *Potentialitiés et Possibilités de Relance L'Activité Forestière.* Synthèse Regionale et National.

Esteve, J., Labrousse, R. and Laurent, D. (1986) *Guinée Forestière: Potentialités et Possibilités de Relance de l'Activité Forestière.* Secrétariat d'Etat aux Eaux et Forêts, Ministère des Ressources Naturelles, de l'Energie et de l'Environnement, République de Guinée.

FAO (1986) *Databook on Endangered Tree and Shrub Species and Provenances.* FAO Forestry Paper 77. FAO, Rome, Italy.

FAO (1988) *An Interim Report on the State of Forest Resources in the Developing Countries.* FAO, Rome, Italy. 18 pp.

FAO (1991) *FAO Yearbook of Forest Products 1978–1989.* FAO Forestry Series No. 24 and FAO Statistics Series No. 97. FAO, Rome, Italy.

IUCN (1987) *The IUCN Directory of Afrotropical Protected Areas.* IUCN, Gland, Switzerland and Cambridge, UK. xix + 1043 pp.

IUCN (1988) *Guinea: Conservation of Biological Diversity and Forest Ecosystems.* Briefing document for IUCN, Gland, Switzerland. 13 pp.

MacKinnon, J. and MacKinnon, K. (1986) *A Review of the Protected Area System in the Afrotropical Realm.* IUCN/UNEP, Gland, Switzerland. 259 pp.

Read, M. (1990) *Mahogany: Forests or Furniture?* FFPS, Brighton, UK.

République de Guinée (1987) *Politique Forestière et Plan d'Action.* Plan d'Action Forestier Tropical. Conakry, Guinea.

République de Guinée (1988) *Code de la Protection de la Faune Sauvage et Réglementation de la Chasse.* Conakry, Guinea.

Schreiber, A., Wirth, R., Riffel, M. and Van Rompaey, H. (1989) *Weasels, Civets, Mongooses and their Relatives. An Action Plan for the Conservation of Mustelids and Viverrids.* IUCN/SSC Mustelid and Viverrid Specialist Group. IUCN, Gland, Switzerland.

WCMC (1987) *The Amphibians of Africa.* Unpublished report.

White, F. (1983) *The Vegetation of Africa: a descriptive memoir to accompany the Unesco/AETFAT/UNSO vegetation map of Africa.* Unesco, Paris, France. 356 pp..

Authorship

Roger Wilson, FFPS, London with contributions from H. F. Maitre, CTFT, Nogent-sur-Marne, France.

Map 22.1 Forest cover in Guinea

Rain forests and protected areas for Guinea are taken from a vegetation map which accompanies the report entitled *Potentialités et Possibilités de Relance de l'Activité Forestière*, CTFT (1989); Synthèse Régionale et Nationale. This is a series of detailed maps covering each prefecture within Guinea. The land use map, drawn by CTFT in 1989 at a scale of 1:700,000 is a synthesis of work carried out in 1985 (south-east: forest zone), 1986 (west) and 1987 (centre and north-east: Upper Guinea). The data are derived from 1979–80 aerial photography taken by the Japan International Cooperation Agency (JICA) and updated using Landsat MSS 1984–1985–1986 imagery. Vegetation has been classified into 29 different categories (A1–7 through to E1–7, R, M) within six biogeographical regions. To compile Map 22.1 the following categories have been extracted: A2 ('Forêt dense humide' – Guinée Forestière), B3 ('Forêt d'altitude' – Fouta Djalon et Contre-Forts), B6 ('Reliques de forêt dense humide en voie défrichement' – Guinée Forestière) and M ('Mangroves – dégradées' – Guinée Occidentale et Maritime). Lowland and montane rain forest and mangroves are shown on Map 22.1. Badiar National Park and Mt Nimba Strict Nature Reserve are also delimited by the CTFT map.

23 Guinea-Bissau

Land area 28,120 sq. km
Population (mid-1990) 1 million
Population growth rate in 1990 2.1 per cent
Population projected to 2020 2 million
Gross national product per capita (1988) US$160
Closed broadleaved forest (end 1980)* 6600 sq. km
Annual deforestation rate (1981–5)* 170 sq. km
Industrial roundwood production† 145,000 cu. m
Industrial roundwood exports† nd
Fuelwood and charcoal production† 422,000 cu. m
Processed wood production† 16,000 cu. m
Processed wood exports† 2000 cu. m
* FAO (1988)
† 1989 data from FAO (1991)

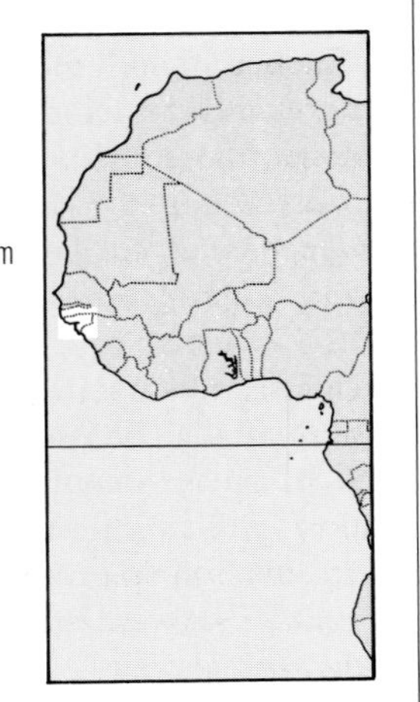

Guinea-Bissau, formerly Portuguese Guinea, is one of the smallest and poorest countries in Africa. Little is known about the natural history of the country. Once covered by a mosaic of lowland forest and woodland, the original vegetation has now largely been cleared and secondary grasslands and savannas dominate most of the land. The main factors that contributed to the destruction of the vegetation were the growing of groundnuts during the colonial era, the bad planning of dam construction for rice growing during the 1970s, development of cashew plantations (mostly provoked by the policies of liberalism supported by the World Bank and the IMF) and forest exploitation from the 1980s onwards.

Guinea-Bissau's coastal zone is one of the most important in Africa. Vast expanses of mangrove along the coast and on the offshore islands provide a valuable forest resource encompassing important fishing grounds and a haven for wildlife, especially for aquatic mammals such as manatees, otters, dolphins and large populations of migrating palaearctic birds. A protected area system has not yet been established, but a number of important sites, mainly in the extensive coastal and estuarine mangroves and intertidal mudflats, have been selected for conservation. Major initiatives are under way to survey these and other critical ecosystems. In addition, a National Conservation Strategy is being prepared with special emphasis on coastal zone management and the development of the protected areas network.

Introduction

Guinea-Bissau is situated on the Atlantic coast between 10°55'–12°40'N and 13°38'–16°43'W, bounded by Senegal to the north and Guinea to the south and east. The topography is generally low-lying, rising from sea-level in the west to low mountains in the east. These mountains represent a continuation of the Fouta Djallon in Guinea, the highest point in Guinea-Bissau being only 262 m. Most of the country comprises a wide coastal plain characterised by mangrove-stabilised estuaries and meandering, sluggish rivers that flow roughly south-west towards the coast. Over the centuries, the continuously rising sea level has created a large number of creeks with higher mainland tongues in between. The Geba-Corubal, Cacheu, Grande de Buba, Cacine and Mansoa are the largest rivers in the region. Wide intertidal mud flats almost link the Bijagos Archipelago (comprising 50 islands of varying sizes and about 30 small islets) with the other coastal islands (Jeta, Pecixe, Bolama and Melo) and the continent. A low plateau succeeds the coastal plain towards the higher ground in the east.

The tropical climate is seasonal, with a wet season from June to October, and a drier season from November to May dominated by the dry Harmattan wind. Most of the country receives 1500–2000 mm of rain per year, but in the south-east, annual precipitation averages more than 2000 mm. Over the past ten years, precipitation has declined with a shortening of the rainy season and the appearance of brief dry periods during the wet season (IUCN, 1985). Mean temperatures in the capital, Bissau, are 25°C in the coolest month and 28°C in the hottest.

This country is one of the world's poorest with over 70 per cent of the people living in rural areas, usually in small villages. The mean population density is around 27 people per sq. km with approximately 60 per cent occupying the coastal region and 40 per cent the plateau area (WCMC, 1991); the central eastern region is very sparsely populated (FAO/UNEP, 1981). Even though the country is small, there are at least 20 ethnic groups, the main ones being the Balante (27 per cent of the population in 1979), the Fulani (23 per cent), the Mandjako (11 per cent) and the Malinké (12 per cent) (Paxton, 1990).

Guinea-Bissau has a subsistence economy but is not self-sufficient in food and has to rely heavily on international aid. Eighty per cent of the population worked in the agricultural sector in 1986 (including forestry and fishing). The staple food is rice, with about 70 per cent of the rice production taking place in the region of Tombali in the south. Other crops include groundnuts, sorghum, cassava, maize, beans, coconut, cashews, millet, palm nuts and sweet potato. Over the past decade the government has encouraged the establishment of cashew plantations to earn foreign currency. For this purpose, it has granted concessions to private producers on 45 per cent of the territory, causing numerous land conflicts over the whole country.

Forests

Closed broadleaved forests occur on the lowland plain and along the coast. The latter was once totally covered by mangrove, flooded savannas and coastal shrub savannas. The lowland forests of the Guinea-Congolian/Sudanian transition zone (White, 1983) are closely related to those of coastal Guinea, Liberia and Sierra Leone. They still cover small areas of the

Tombali and Quinara regions to the south of the country and of the Cacheu region in the north-west. The canopy height is 30 m or more with the principal species being *Afzelia africana*, *Alstonia congensis*, *Anisophyllea lamina*, *Antiaris africana*, *Ceiba pentandra*, *Detarium senegalense*, *Dialium guineense*, *Elaeis guineensis*, *Erythrophleum guineense*, *Ficus* spp., *Milicia excelsa* and *Parinari excelsa*. Lianas, *Raphia* and rattans are also well developed (SCET International, 1978). In the north of the country there are significant areas of palm groves (*Elaeis* spp.) and of more scattered *Borassus aethiopum*.

'Semi-dry' broadleaved forests predominate in the centre of the country (SCET International, 1978). Again, as in the lowland forests, the principal tree species are *Afzelia africana*, *Erythrophleum guineense*, *Parinari excelsa* together with the African mahogany *Khaya senegalensis*.

The interior is dominated by a mosaic of open forest and tree savanna, characterised by a continuous grass-cover and some gallery forest along streams. These secondary vegetation types of woodland and tree savanna, covering more than 10,000 sq. km in 1975 (FAO/UNEP, 1981), have developed from the original closed forests as a result of repeated fires.

Mangroves

The country still supports important areas of mangroves; in fact the mosaic of mangroves and coastal flats in Guinea-Bissau is the largest of this habitat type in Africa (FAO/UNEP, 1981). Historically 11 per cent of the country was covered with mangroves (IUCN, 1988). They clothe the coast and estuarine shores and penetrate deep inland up the tidal waters of the six major estuaries. For instance, mangrove habitat reaches 100 km inland following the Cacheu River, before grading into palm swamps and fresh water swamp forest. However these mangroves are being cleared and transformed into flooded grassy areas, colonised by palm groves at the edges.

The entire coast is mangrove covered except for areas south of Cape Roxo and north of the Cacheu River mouth, a 3 km strip south of Point Cabaciera and a 15 km strip along Varela Bay. Mangroves also occur on numerous offshore mudflats along the coast. The offshore islands are only partly fringed by mangroves, as the islands of the Bijagos Archipelago are old hilltops and, in places, they rise steeply from the sea. In the Bijagos, mangroves cover approximately 30 per cent of the area.

According to EDWIN (1987), mangroves cover an area of approximately 2360 sq. km (8 per cent of the country). Once they were more extensive; along the Cacheu River system alone there were 1110 sq. km of mangrove forest and 130 sq. km of fresh water swamp forest (Hughes and Hughes, 1991). It has been suggested that in 1975 there were approximately 2500 sq. km of tidal forest in Guinea-Bissau but that by 1986 15–20 per cent had been cleared for conversion of the land to rice farming (Hughes and Hughes, 1991). Reclamation of mangrove areas for rice cultivation by traditional techniques involves the construction of 1.5–2 m high dykes along tidal creeks, so that the land behind the dykes is no longer inundated at high tide.

Vegetation is dominated by the genus *Rhizophora* (*racemosa*, *mangle* and *harrisonii*) and *Avicennia africana*, while in the south of the country *Laguncularia racemosa* and *Conocarpus erectus* are found. The higher sandy islands in the mangroves are colonised by the oil palm *Elaeis guineensis* which is often present in almost pure stands in low lying areas behind the mangrove fringe.

Mangrove areas are now experiencing large rainfall deficits and increasing salinity. In an attempt to improve conditions for rice cultivation, anti-salt barriers have been constructed. These unfortunately have had adverse effects on the environment because of a lack of understanding of soil science and hydrodynamics, as well as of social considerations, land regimes and migration. Entire creeks have been isolated from the sea, cutting off the life-line of the remaining mangrove ecosystem. More than 40 of these anti-salt barrages have been constructed (EDWIN, 1987).

Mangroves cleared to make way for rice fields in Guinea-Bissau. The dyke prevents sea water flooding into the fields. J. Pierot/IUCN

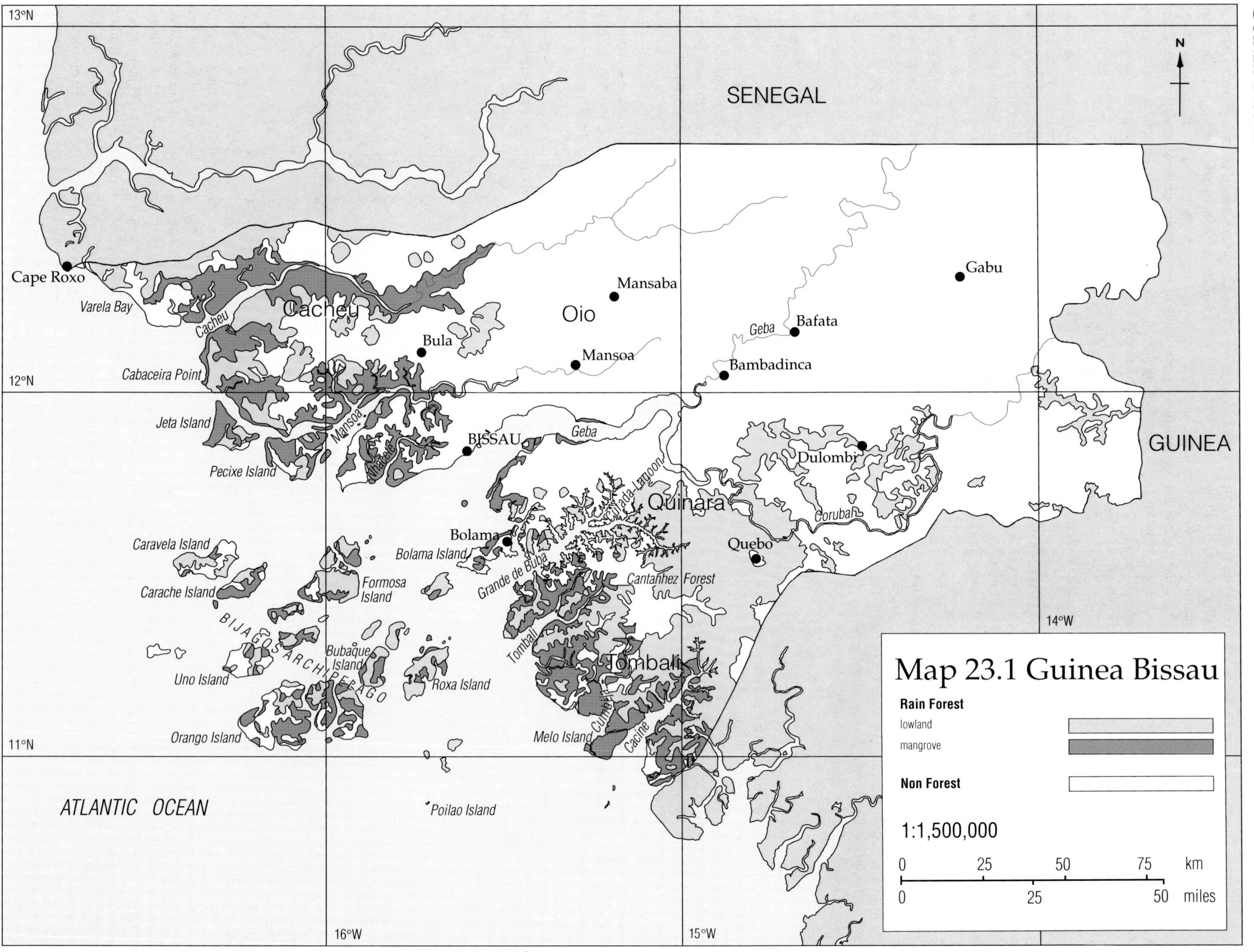
Map 23.1 Guinea Bissau
Rain Forest
lowland
mangrove
Non Forest
1:1,500,000
0 25 50 75 km
0 25 50 miles
N
SENEGAL
GUINEA
ATLANTIC OCEAN
13°N
12°N
11°N
14°W
15°W
16°W
Cape Roxo
Varela Bay
Cacheu
Cacheu
Cabaceira Point
Jeta Island
Pecixe Island
Mansoa
Bula
Mansaba
Oio
Mansoa
Gabu
Geba
Bafata
Bambadinca
BISSAU
Geba
Dulombi
Quinara
Corubal
Quebo
Bolama
Bolama Island
Grande de Buba
Cantanhez Forest
Tombali
Tombali
Cumbija
Cacine
Melo Island
Caravela Island
Carache Island
Formosa Island
BIJAGOS ARCHIPELAGO
Bubaque Island
Uno Island
Roxa Island
Orango Island
Poilao Island

As in other countries, the mangroves are an important source of fuelwood, timber for construction and medicinal products. They also play a vital role as nursery and breeding areas for fish and shellfish and are an important habitat for other wildlife, particularly for palaearctic migratory birds. Conversion of mangroves has disrupted fresh water supplies and caused soil acidification. IUCN (1985) has argued that government agricultural schemes must be coordinated with coastal lowland conservation measures.

IUCN is collaborating with the Ministry of Rural Development to plan future development in the coastal zone, including the identification of conservation areas. The establishment of three protected areas in the coastal zone has been recommended. These are Cantanhez Forest in the south, the Lagoa de Cufada (the first Ramsar site in Guinea-Bissau) (Scott and Pineau, 1990) and mangroves of the Cacheu River. In addition, it is proposed that a Biosphere reserve incorporating all the Bijagos islands be set up. The project in the coastal zone will be the first to be developed under the National Conservation Strategy for Guinea-Bissau. It builds upon a study, initiated in 1986 by WWF, that examined the threat to the wildlife resources of the mangroves posed by excessive ricefield development, and which also identified sites of special interest that would merit protection in the form of reserves or parks (WWF, 1987).

Forest Resources and Management

The FAO/UNEP Tropical Forest Resources Assessment Project at the end of 1980, estimated closed broadleaved forest cover to be 6600 sq. km (FAO, 1988) or 23.5 per cent of the country. A government study in 1985 estimated dense humid forest at 2209 sq. km, moderately dense forest at 1133 sq. km and 'semi-dry' dense forest at 1285 sq. km. It is likely that this area has been further depleted since then. Map 23.1 has been generated from a hand drawn original and it appears likely that both the mangroves and the dryland forest in the south are more fragmented than shown here. Because of the unreliability of the source map, the forest area statistics measured from Map 23.1 must also be regarded with caution and hence are not quoted at the beginning of this chapter nor presented as a table. However, for the readers' interest, Map 23.1 shows 5368 sq. km of lowland rain forest and 3491 sq. km of mangrove.

Since the early 1980s, forests in the Tombali area have suffered heavy pressure from the development of fruit farming and rice culture. A decrease in the size of the semi-dry forests of Quinara is attributable to bush fires, expansion of cashew plantations and timber exploitation. Huge tracts of forest between Mansoa, Bafata and Mansaba have been cleared and almost no moist forest now exists between Bula in the west and Gabu in the east.

The Ministry of Rural Development and Agriculture is responsible for the management of forests and wildlife through the Directorate General of Forestry. No forest reserves have been created by the government, but over the entire country small forest zones, commonly called 'sacred forests', are protected by different ethnic groups for religious reasons. Timber exploitation is prohibited but hunting is permitted in some of these forests.

The government is promoting timber extraction as a main source of foreign revenue. Nine timber species are exploited. The primary commercial species is *Khaya senegalensis* with about 10,000 cu. m of rough sawn timber produced per year, while afzelia *Afzelia africana* and iroko *Milicia excelsa* have a combined annual yield of around 2500 cu. m of timber (FAO/UNEP, 1981). *Prosopis africana* is an important source of fuelwood and charcoal in the north of the country. In 1989 FAO estimated industrial roundwood production to be 145,000 cu. m and fuelwood and charcoal production to be 422,000 cu. m (FAO, 1991).

Other non-timber products are obtained from the forest. Bamboo is very versatile and is valued for basketwork, fencing and furniture-making while palms are used to make wine and oil. Many different plant species are collected for use in traditional medicine such as the leaves and the roots of *Combretum* and the bark of the *Khaya* mahogany. Numerous fruits are also collected from the forest.

Deforestation

FAO (1988) estimates that 170 sq. km of closed broadleaved forest are lost each year. The principal causes of deforestation relate not only to population increase and resulting pressures for more land, but also to bushfires and the development of cashew and groundnut cultivation, fruit farming, rice culture and timber exploitation.

Large areas of former closed forest around Mansoa and Bissau were cleared for cashew plantations in the 1970s and 1980s, while clearance for groundnut, millet and timber has resulted in the loss of all closed forest north of a line between Bissau and Gabu. The three largest districts, Oio, Bafata and Gabu in the north and east, which cover more than half the country, have lost virtually all their original sub-humid forest and a large part of the dense semi-dry forests in these areas has also been eliminated. The rate of shifting cultivation has increased, and soil exhaustion and erosion are serious problems, particularly in these three large districts, as well as in the Quinara region where soils prone to erosion are also found.

As commercially valuable trees are felled, the forest degrades and becomes drier; bushfires are more extensive and these hinder regeneration. In spite of the existence of vast areas of uncultivated land in the north of the country, the moist forests in the south-west are still being cleared for agriculture. Between Bambadinca and Quebo (Bafata-Tombali Districts), European Development Fund schemes are clearing large areas of primary and secondary moist forest, primarily for groundnut cultivation, although some timber is also being extracted. With the loss of the trees in the more accessible north, the timber industry is increasingly active in more remote areas, in particular the ecologically important south-west peninsula.

Road construction has opened up forest lands to loggers, agriculturalists and settlers. In addition, closed forests are threatened by rice growing in the western part of the Oio region, the Bafata region and the centre and north-west of Gabu.

Biodiversity

Very little information is available on the biological diversity of Guinea-Bissau. There are no complete faunal or botanical reference collections. The Portuguese started a survey of the flora in the late 1960s but this was never completed. About 1000 species of plants with 12 endemics are recorded (Davis *et al.*, 1986), but there are no reliable estimates of numbers of threatened or rare plants. However, a survey of the large mammals was undertaken in 1988–9 by the Direction Générale des Fôrets et de la Chasse/Centre Canadien d'Etudes et de Coopération International (DGFC/CECI/UICN, 1989, 1990a, b).

Stuart *et al.* (1990) record 109 mammal species, but no endemics, in the country. Eleven primates are found including the threatened Temminck's red colobus *Procolobus [badius] badius temminckii*, the western black-and-white colobus *Colobus polykomos*, red patas monkey *Erythrocebus patas* and Campbell's monkey *Cercopithecus campbelli*. Ungulates include hippo *Hippopotamus amphibius*, bushbuck *Tragelaphus scriptus* and several duikers *Cephalophus* spp. Leopards *Panthera pardus* still survive and the markets at Bissau usually have skins of golden cats *Felis aurata* and serval *F. serval*. The chimpanzee *Pan troglodytes* is probably extinct along with the giant eland *Tragelaphus derbianus* and bongo *Tragelaphus euryceros* although all occurred until recently. The status of manatees *Trichechus senegalensis* is of concern, and small numbers of dolphins are caught in coastal fisheries (Stuart *et al.*, 1990). Elephants *Loxodonta africana* have been reduced to tiny numbers: a herd of 40 was reported as recently as 1988 (Douglas-Hamilton, 1988). Because hunting is mostly for subsistence rather than commerce, there is relatively little trade in wildlife or wildlife products. Nevertheless, bushmeat is one of the most important sources of protein for rural people and over-hunting has almost certainly contributed to the depletion of stocks of large mammals. The fauna suffered from excessive hunting during the war of liberation, as well as from habitat destruction. As a result, rural populations are increasingly depending on marine resources for subsistence .

The precise number of bird species in Guinea-Bissau is unknown, and none is listed as threatened by Collar and Stuart (1985). Birds of Guinea-Bissau, mainly water birds of the mangroves, mudflats and ricefields in the south-west, were surveyed by Altenburg and van der Kamp (1985). The mangroves and wetlands are very important for migratory species, huge numbers of waders occurring on the inter-tidal mudflats. Wintering species of palaearctic waders which were recorded from coastal mudflats and mangroves include curlew sandpiper *Calidris ferruginea*, knot *C. canutus*, redshank *Tringa totanus*, whimbrel *Numenius phaeopus*, grey plover *Pluvialus squatarola* and bar-tailed godwit *Limosa lapponica*. In 1982 WWF (1983/4) carried out a study of these species and reported that their population on the coastal wetlands of Guinea-Bissau totalled over 1 million. This amounts to 12 per cent of all the estimated 8 million waders that migrate down the west Atlantic seaboard (Altenburg, 1987).

The reptile fauna is not well documented. It is likely that all three species of crocodile (Nile crocodile *Crocodylus niloticus*, slender-snouted crocodile *C. cataphractus* and dwarf crocodile *Osteolaemus tetraspis*) occur in several areas, although the slender-snouted species, which naturally occurs in low densities, may be extinct. The royal and rock pythons (*Python regius* and *P. sebae*) are killed for their skins and meat, while other snakes include the common tree snake *Boaedon fuliginosus*, olive grass snake *Psammophis sibilans*, green tree mamba *Dendroaspis viridis*, and half-banded garter snake *Elapsoidea semiannulata*. The islands to the south of the Archipelago probably constitute the most important breeding ground for green turtles *Chelonia mydas* on the West African coast. Other turtles to have been recorded are the loggerhead *Caretta caretta*, olive ridley *Lepidochelys olivacea*, hawksbill *Eretmochelys imbricata* and leather-backed *Dermochelys coriacea*. One rare amphibian species occurs, *Pseudhymenochirus merlini*, which can also be found in Guinea and Sierra Leone. The dragonfly *Brachythemis liberiensis* is known only from Guinea-Bissau, though there are no recent records.

Threats to biological diversity arising from habitat degradation are evident everywhere in Guinea-Bissau. Shifting cultivation, bushfires, inappropriate use of new technology and failure to take ecological constraints into account in the National Development Plan have resulted in destruction of ecosystems (UNSO/Unesco, 1984).

Conservation Areas

At present Guinea-Bissau has no protected areas other than six reserves where hunting is permanently prohibited; the country needs to establish a protected area system as a matter of urgency. In 1981 it was reported that the government was trying to establish a national parks programme (IUCN, 1987) but little progress has been made to date.

Guinea-Bissau has the chance of incorporating sound conservation policies into its national development plans with the elaboration of its National Conservation Strategy (NCS). As part of the NCS process, a faunal inventory is under way to identify sites of conservation interest in the interior of the country.

IUCN (1985) has identified six critical sites for palaearctic birds. These are:

- The Bijagos Archipelago, a group of 18 inhabited islands with extensive oil palm, mangrove, mudflats and climax woodland.
- The Cacheu Peninsula.
- The Basin of the Mansoa and Nhacete rivers including Jeta and Pecixe islands with 505 sq. km of mangrove.
- The Basin of the Geba River and the Corubal River including Bolama island.
- The Basin of the Grande de Buba River where estuarine mangrove (170 sq. km) and mudflats meander into patches of dense forest.
- The Basin of the Tombali, Cumbija and Cacine rivers – biologically the richest area of the country – with extensive mangroves (785 sq. km), mudflats and dense closed moist forest providing important habitat for the golden cat, cape clawless otter *Aonyx capensis* and western black-and-white colobus monkeys. Forest elephants have recently disappeared from this site.

Initiatives for Conservation

An assessment of the environmental situation and of the exploitation of natural resources has been undertaken in the coastal zone by IUCN in collaboration with the General Directorate of Forests and Wildlife (DGFC) and the National Research Institute (INEP). The study has concluded that conservation of mangroves is urgently required because the rich marine resources are being used increasingly by the local population and are also being exploited by an uncontrolled fishing industry. The process of privatisation of land brings with it over-exploitation of forest resources, including deforestation, and causes social fragmentation of the population and their exodus to urban centres. There is a need to create an efficient environmental protection service as well as a protected areas network which could involve ecotourism. Areas that should be protected include the Cacheu River mangroves, to maintain the ecological processes necessary for fishing and for the conservation of species such as sitatunga *Tragelaphus spekei*; the Cufada lagoons and their dense open forest mosaics; the remaining patches of sub-humid forests of Cantanhez in the south of the country; and the zone of Dulombi in the east, for the conservation of large mammals. As for the Archipelago, preservation zones corresponding to the core areas of the Biosphere reserve have been recommended on the island of Poilão, for nesting marine turtles; on the islands of the Orango complex, for manatees, otters and crocodiles and in the bay of Caravela, as well as other as yet imprecisely defined marine zones.

The DGFC collaborates with IUCN, INEP, CECI and the Centre International Pour L'Exploitation des Oceans (CIEO) with the support of UNDP on the proposal to make the Archipelago a Biosphere reserve. DGFC and CECI are preparing the management plan for the Dulombi conservation zone and DGFC and the Portuguese Parks Service are doing the same for the proposed Cufada National Park. The DGFC and IUCN have established preliminary contacts with the Swedish International Development Agency (SIDA) for the management of resources of the Grande de Buba River and in the Cacheu mangrove conservation zone. A Tropical Forestry Action Plan (TFAP) is being prepared with EEC support. The elaboration of a National Environmental Strategy has begun under the control of the National Environment Commission, which includes seven ministers under the presidency of the Head of State.

References

Altenburg, W. (1987) Waterfowl in West African Coastal Wetlands: A summary of current knowledge. International Working Group on Waterfowl and Wader Research, Leiden, The Netherlands. *WIWO Report* **15**.

Altenburg, W, and van der Kamp, J. (1985) *Oiseaux d'Eau dans les Rizières de la Guinée-Bissau, Résultats Préliminaires d'un Récensement entre Mi-novembre et Mi-décembre 1983.* IUCN/WWF project 3096. IUCN/WWF, Gland, Switzerland.

Collar, N. J. and Stuart, S. N. (1985) *Threatened Birds of Africa and Related Islands. The ICBP/IUCN Red Data Book Part 1.* ICBP/IUCN, Cambridge, UK. 761 pp.

Davis, S. D., Droop, S. J. M., Gregerson, P., Henson, L., Leon, C. J., Villa-Lobos, J. L., Synge, H. and Zantovska, J. (1986) *Plants in Danger: What do we know?* IUCN, Gland, Switzerland and Cambridge, UK.

DGFC/CECI/UICN (1989) *Résultats de l'Inventaire Faunique au Niveau National et Propositions de Modifications de la Loi sur la Chasse.* Bissau. 144 pp.

DGFC/CECI/UICN (1990a) *Propositions d'un Réseau d'Aires Protégées en Guinée-Bissau (zone continentale).* Bissau. 154 pp.

DGFC/CECI/UICN (1990b) *Utilisation et Perception de la Faune et du Milieu Naturel en Guinée-Bissau.* Bissau. 106 pp.

Douglas-Hamilton, I. (1988) *African Elephant Population Study.* EEC/WWF/Global Environment Monitoring Centre. Nairobi, Kenya.

EDWIN (1987) *West Africa Review. Assessment of Environmental Impacts of Water Management Projects on Wetlands.* EDWIN Report No. 1. Compiled by van Ketel, A., Marchand, M. and Rodenburg, W. F. Centre for Environmental Studies, Leiden University, The Netherlands.

FAO (1988) *An Interim Report on the State of Forest Resources in the Developing Countries.* FAO, Rome, Italy. 18 pp.

FAO (1991) *FAO Yearbook of Forest Products 1978–1989.* FAO Forestry Series No. 24 and FAO Statistics Series No. 97. FAO, Rome, Italy.

FAO/UNEP (1981) *Tropical Forest Resources Assessment Project. Forest Resources of Tropical Africa. Part II: Country Briefs.* FAO, Rome, Italy.

Hughes, R. H. and Hughes, J. S. (1991) *A Directory of Afrotropical Wetlands.* IUCN, Gland, Switzerland and Cambridge, UK/UNEP, Nairobi, Kenya/WCMC, Cambridge, UK.

IUCN (1985) *Guinea-Bissau. Vers l'Elaboration d'une Stratègie Nationale de Conservation des Ressources Naturelles.* Rapport de Mission, CDC, Mai 1985. IUCN, Gland, Switzerland.

IUCN (1987) *IUCN Directory of Afrotropical Protected Areas.* IUCN, Gland, Switzerland and Cambridge, UK. xix + 1043 pp.

IUCN (1988) *Consérvation du Milieu et Utilisation durable des Réssources Naturelles dans la Zone Cotière de la Guinée-Bissau.* Rapport d'Activité. IUCN and Le Ministère du Développement Rural et de l'Agriculture République de Guinée-Bissau. IUCN, Gland, Switzerland.

Paxton, J. (ed.) (1990) *The Statesman's Yearbook 1989–1990.* Macmillan Reference Books, London, UK.

SCET International (1978) *Potentialité Agricole, Forestières et Pastorals de la Guinée-Bissau.* 3 vol. Fonds d'Aide et de Coopération de la République Française, Commissariat d'Etat à l'Agriculture et à l'Elevage, Commissariat aux Ressources Naturelles. Bissau.

Scott, D. A. and Pineau, O. (1990) *Promotion de la Convention de Ramsar et Inventaire de la Lagon de Cufada, Guinée-Bissau.* 26 pp.

Stuart, S. N., Adams, R. J. and Jenkins, M. (1990) *Biodiversity in Sub-Saharan Africa and its Islands. Conservation, Management and Sustainable Use.* IUCN, Gland, Switzerland and Cambridge, UK.

UNSO/Unesco (1984) *République de Guinée-Bissau. Plan National d'Action pour Lutter contre la Dégradation du Milieu Naturel en Guinée-Bissau.* Compiled by Bartolucci, I. J. and Lepape, M.-C. FMR/SC/ECO/84/216 (UNSO). Paris, France.

WCMC (1991) *Guia da Biodiversidade de Guiné Bissau.* Prepardo pelo Centro Mundial de Monitoramento para a Concervação da Natureza, Cambridge, UK. 15 pp.

White, F. (1983) *The Vegetation of Africa: a descriptive memoir to accompany the Unesco/AETFAT/UNSO vegetation map of Africa.* Unesco, Paris, France. 356 pp.

WWF (1983/4) *WWF Yearbook.* WWF, Gland, Switzerland.

WWF (1987) *WWF List of Approved Projects. Africa and Madagascar.* Vol 4. WWF, Gland, Switzerland.

Authorship

Scott Jones of Bristol University with contributions from Pierre Campredon, IUCN, Guinea-Bissau and Pat Dugan, IUCN Wetlands Programme, Switzerland.

Map 23.1 Forest cover in Guinea-Bissau

Information on rain forest cover is taken from a generalised map (*c.* 1:1 million) hand drawn by the author of the chapter, Scott Jones (1990). It is based on his personal experience of the region and shows mangrove and lowland rain forest. His map is based on an earlier 1:500,000 Instituto Geográfico Nacional (1981) land use chart *Guiné Bissau* which has been updated to reveal the devastating forest loss in the northern part of the country. The forest areas shown on the hand drawn map in the south-east and west of the country include enclaves of cultivation and degraded forest and therefore present an overly optimistic view of the forest resources of the country. Consequently, statistics for forest extent derived from Map 23.1 are considered to be unreliable. There are no protected areas in Guinea-Bissau.

24 Indian Ocean Islands

COMOROS
Land area 2230 sq. km
Population (mid-1990) 0.5 million
Population growth rate in 1990 3.4 per cent
Population projected to 2020 1.3 million
Gross national product per capita (1988) US$440
Closed broadleaved forest (end 1980)* 160 sq. km
Annual deforestation rate (1981–5)* 5 sq. km
Industrial roundwood production† nd
Industrial roundwood exports† nd
Fuelwood and charcoal production† 308,190 cu. m
Processed wood production† nd
Processed wood exports† nd

MAURITIUS
Land area 1850 sq. km
Population (mid-1990) 1.1 million
Population growth rate in 1990 1.3 per cent
Population projected to 2020 1.3 million
Gross national product per capita (1988) US$1810
Closed broadleaved forest (end 1980)* 30 sq. km
Annual deforestation rate (1981–5)* 1 sq. km
Industrial roundwood production† 13,000 cu. m
Industrial roundwood exports† nd
Fuelwood and charcoal production† 21,000 cu. m
Processed wood production† 4000 cu. m
Processed wood exports† nd

REUNION
Land area 2500 sq. km
Population (mid-1990) 600,000
Population growth rate in 1990 1.8 per cent
Population projected to 2020 800,000
Gross national product per capita (1988) nd
Closed broadleaved forest (end 1980)* 820 sq. km
Annual deforestation rate (1981–5)* nd
Industrial roundwood production† 2000 cu. m
Industrial roundwood exports† nd
Fuelwood and charcoal production† 31,000 cu. m
Processed wood production† 2000 cu. m
Processed wood exports† nd

SEYCHELLES
Land area 270 sq. km
Population (mid-1990) 100,000
Population growth rate in 1990 1.7 per cent
Population projected to 2020 173,000
Gross national product per capita (1988) US$3800
Closed broadleaved forest (end 1980)* 30 sq. km
Annual deforestation rate (1981–5)* nd
Industrial roundwood production† nd
Industrial roundwood exports† nd
Fuelwood and charcoal production† nd
Processed wood production† nd
Processed wood exports† nd

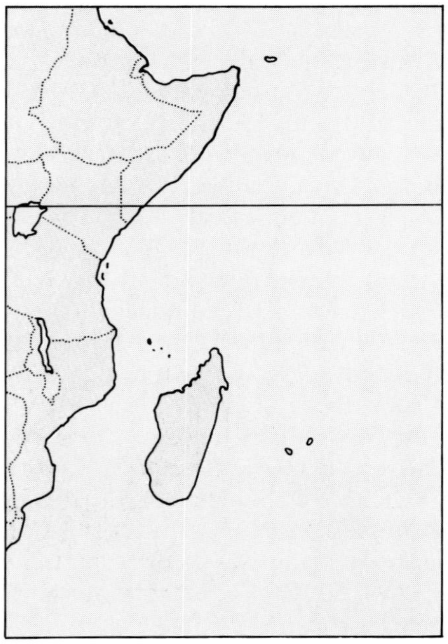

* FAO (1988)
† 1989 data from FAO (1991)

This chapter deals with the groups of islands lying around Madagascar in the Indian Ocean, namely the Comoros, the Mascarenes and the Seychelles. (Note that the statistics given at the head of this chapter treat the Comoros group and the Seychelles each as a single entity, whereas two islands of the Mascarene group, Mauritius and Réunion, are dealt with separately.) Although these islands were all completely forested when first discovered a few centuries ago, none now has any extensive forest cover. Logging and clearing for agriculture have been the main causes of deforestation. Those forest patches which do remain frequently contain species that are unique to the island concerned. They are of particular interest to ornithologists with 37 per cent of the African bird species listed as endangered or vulnerable occurring on these islands.

Introduction – Comoros

The four Comoro islands lie between Africa and the northern tip of Madagascar, between 11°20' and 13°04'S latitude and between 43°11' and 45°19'E longitude. Grande Comore (Ngazidja) is the largest with an area of 1024 sq. km. Anjouan (Ndzuani) at 424 sq. km is slightly larger than Mayotte (Maore) which is 385 sq. km in area. Moheli (Mwali) at 211 sq. km is the smallest of the islands. The Comoros were formed volcanically; they appeared well after Madagascar separated from Africa and have never been linked to a continent. Sea depths of 2000–3000 m separate the islands (Battistini and Vérin, 1984). Mayotte, at 15 million years old, is the most ancient of the islands, while Grande Comore, formed in the Quaternary Period, is the youngest. Grande Comore has no rivers or valleys but reaches the highest altitude, 2361 m at the summit of Karthala Volcano. This volcano is still active, last erupting in 1977.

The climate is tropical with a dry, cool season from April to October. Temperatures during this time average 24°C, warming up from July onwards. Precipitation exceeds 1000 mm per year throughout the archipelago and there is considerable local variation. For instance, on the west flank of Karthala Volcano on Grande Comore the mean annual rainfall is 5600 mm. Devastating cyclones are reported to occur about once every ten years or so (Thorpe *et al.*, 1988).

Although geographically the archipelago forms an entity, three of the islands became independent in 1975 and make up the Republic of the Comoros, but Mayotte remained a dependent territory of France. The population of the islands is expanding rapidly with, in 1980, 47 per cent of the population under 15 years old (Battistini and Vérin, 1984). Population density on the four islands varies from 100–450 per sq. km (Weightman, 1987; Carroll and Thorpe, 1991). Only 23 per cent of the inhabitants live in urban areas (PRB, 1990).

The Forests

Before the Comoros were colonised by man, the higher areas were covered by sub-equatorial rain forest and the lower slopes, where rainfall is less, were clothed with drought-resistant forests. Now the forest on Grande Comore remains only on the steep slopes (Figure 24.1). The forests there have been totally removed between sea level and 500 m for cash crops and coconut plantations; in some places crops reach 1400 m asl (Louette *et al.*, 1988). In higher areas, subsistence farming and grazing by cattle and goats have eliminated the forests and *Philippia* heath. On Moheli, forest is still present above 400 m in one continuous stretch along the ridge

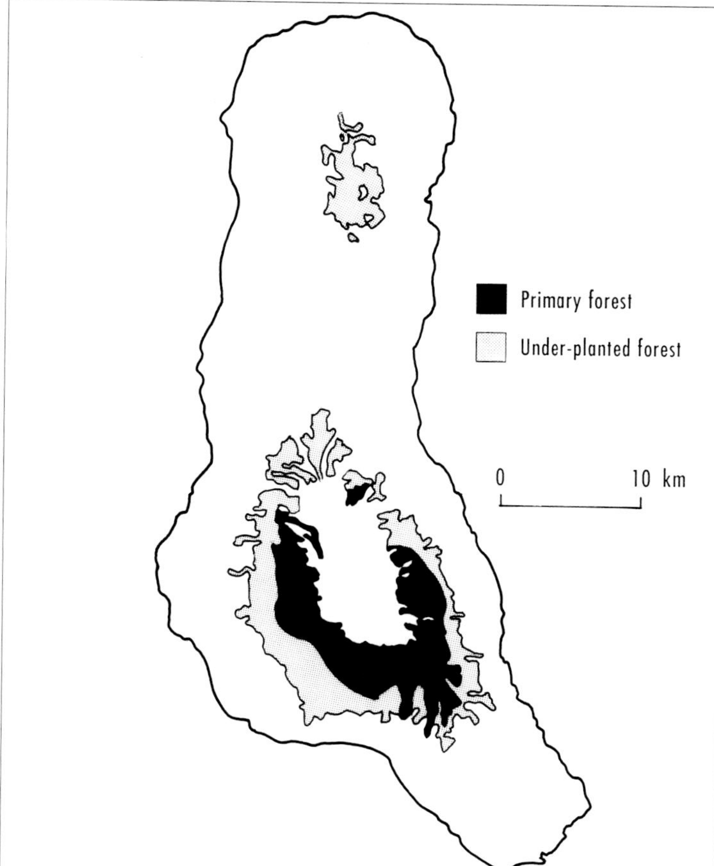

Figure 24.1 Forest boundaries on Grande Comore in 1983

(*Source:* Thorpe *et al.*, 1988)

(Figure 24.2), but this forest is thought to be entirely secondary (Gachet, 1957). On Anjouan, small patches of primary forest remain on the steepest slopes of the central peaks (Figure 24.3). Benson (1960) reports that Mayotte was completely deforested in the 19th century. Now some secondary forest is to be seen on the highest points of the island.

Figure 24.2 Forest boundaries on Moheli in 1983

(*Source:* Thorpe *et al.*, 1988)

The dense humid forest on the Comoros reaches a height of around 20–30 m. Dominant species are *Ocotea comorensis*, *Khaya comorensis*, which sometimes form up to 80 per cent of the canopy in the Karthala forest, *Olea* sp., *Chrysophyllum boivinianum*, *Prunus africana* and *Filicium decipiens* (White, 1983). Epiphytes are common and tree ferns, such as *Cyathea similis*, are also present (Thorpe *et al.*, 1988). Recent lava flows are colonised by *Nuxia pseudodentata*, *Breonia* sp., *Weinmannia* sp., *Apodytes dimidiata* and *Olea* sp. (White, 1983).

Only small and isolated patches of mangrove are found on the Comoros. They are most common on the island of Mayotte. The mangroves are generally floristically poor; plant species present include *Rhizophora mucronata*, *Bruguiera gymnorrhiza*, *Avicennia marina* and *Sonneratia alba*.

Forest Resources and Management

There is an urgent need for new aerial surveys of the islands to obtain accurate measures of the remaining forest extent. In 1949, forest cover was estimated at 314 sq. km (Deville, 1974) but by 1981, this was thought to have diminished to 185 sq. km (Louette *et al.*, 1988). FAO (1988) reported only 160 sq. km remaining in 1980.

Half of the forest is government owned but, on Anjouan, forest land can be claimed by villagers if they plant it with bananas or other crops. There is a system of individual permits for tree felling, but control is difficult (Louette *et al.*, 1988). Most of the timber is for domestic use, with *Takamaka comorensis* the most prized wood for making into furniture and traditional carved boxes. Both the harvesting and export of timber was banned in 1987. Comparatively little timber is removed commercially. For instance, there is a sawmill at Nioumbadjou on Grand Comore which, until 1987, produced about 1000 cu. m (net) of timber from a privately owned forest of 50 sq. km on the south-western flank of Mt Karthala. Allowing for gross extraction of 2300 cu. m, this represented about 0.5 cu. m per hectare per year (Louette *et al.*, 1988). Although only certain tree species are extracted – hence the low annual volume removed per hectare – many others are damaged in the process. As a result, the forest has been degraded and frequently then cleared for agriculture.

Figure 24.3 Forest boundaries on Anjouan in 1983

(*Source:* Thorpe *et al.*, 1988)

In 1971/2, it was estimated that there were about 83 sq. km of primary forest on Anjouan, but by 1983/4 only 22 sq. km remained. By 1987 only 11 sq. km of forest was left on the island and all of this was growing on the steepest slopes (Weightman, 1987). Even this is not secure – significant areas on the margins of this forest have now been cleared or underplanted. There is only limited planting of exotic trees (*Eucalyptus* and pines) for wood and timber on the Comoros.

Deforestation

The increasing human population on the Comoros is causing massive demand for land and fuelwood. As a result, deforestation on the islands is becoming a severe problem. Erosion is extensive and previously perennial streams are drying up on the three smaller islands. On Anjouan, the most degraded of the four islands, erosion from the deforested areas is so bad that, in the rainy season, the sea turns red with the soil washed off the land.

The major cause of loss of primary forest is underplanting with crops such as bananas, maize and peas. At higher altitudes, manioc and sweet potatoes are grown within the forest. The amount of forest destruction is very variable: some areas are underplanted with a low density of bananas and retain an intact canopy and dense lower strata, while in others there is a sparse canopy, no lower strata remain and areas of grassland are formed by overgrazing. Landslides and cyclone damage pose other threats to the remaining forest on Anjouan's peaks.

Biodiversity

The fauna on the Comoros is very depauperate. There are no native mammals other than bats on the islands. Two of these, *Pteropus livingstonii* and *Rousettus obliviosus*, are endemic to the Comoros, and the former is one of the world's rarest bats. A number of mammal species have been introduced from Madagascar including *Tenrec ecaudatus*, *Lemur mongoz* and *Lemur fulvus mayottensis*. The latter was considered to be an endemic subspecies but is now generally thought to be merely a form of *Lemur fulvus fulvus*.

There are 99 bird species on the Comoros, of which 16 are endemic to one or more of the islands (Louette, 1988). A number of these endemic bird species are confined to the slopes of Mt Karthala on Grande Comore. These are the Grande Comoro scops owl *Otus pauliani*, the Grande Comoro flycatcher *Humblotia flavirostris*, the Mount Karthala white-eye *Zosterops mouroniensis* and the Grande Comoro drongo *Dicrurus fuscipennis*.

The reptiles on the islands are mostly poorly known; indeed, two new species of gecko have been described from the Comoros in the last decade. There are 25 species of native snakes and lizards of which at least 11 are endemic to the islands (Stuart *et al.*, 1990).

Little is known about the invertebrates, but there are two threatened species of swallowtail butterfly: one, *Graphium levassori*, found only on Mt Karthala and the other, *Papilio aristophontes*, found in the forests of all the islands except Mayotte.

There are 935 species of plants on the Comoros of which 416 are indigenous and 136 are endemic (Davis *et al.*, 1986).

Conservation Areas and Initiatives

No protected areas have been designated on any of the Comoro islands. Louette *et al.* (1988) proposed that a national park be gazetted on Mt Karthala and that nature reserves should be set up to protect the forests remaining on the other islands.

A National Committee for the Environment was established in 1990 and, in 1991, a regional director for the environment was appointed on Anjouan. A National Conservation Strategy will be written for Anjouan initially and then will be extended to other islands of the Federal Republic of Comoros. In addition, an inventory of primary forest species on Anjouan is to be compiled.

INTRODUCTION – MASCARENES

This group of three volcanic islands, namely, Mauritius, Rodriguez and Réunion, is situated on the southernmost part of the Seychelles–Mauritius ridge. The nearest large land mass is Madagascar. Although all three islands used to be covered in forest, none now has extensive areas remaining. Indeed, the 109 sq. km island of Rodriguez today has no intact patches of native forest left (Gade, 1985; Strahm, 1989) and has not, therefore, been included in this chapter.

Mauritius is 62 km long and 46 km wide. The land rises from the coastal plains to a rugged and precipitous central plateau (305–730 m) on which there are several extinct craters and peaks which reach an altitude of 826 m. Réunion, part of metropolitan France, is 75 km long and 70 km wide and 200 km south-west of Mauritius. The central massif of the island culminates in the Piton des Neiges at 3069 m.

On the coastal plains of Mauritius, mean annual rainfall varies from 890 mm on the leeward side of the island to 1905 mm on the south-west. On the central plateau it ranges from 2540 mm to 4445 mm. Cyclones are frequent and devastate crops but do little damage to the indigenous forests, possibly because of the efficient anchorage of the trees (White, 1983). Réunion has a dry, cool season from May to October and a wet, warm season from November to April. Annual precipitation varies from less than 1000 mm on the west coast to over 9000 mm on the east side of the island. Mean annual temperatures are generally between 12°C and 20°C over most of the island but reach around 24°C on the coast (Doumenge and Renard, 1989).

On Mauritius, sugar cane is the largest source of foreign exchange earnings; it is grown on 90 per cent of the island's cultivable land (45 per cent of total land area) and employs 20 per cent of the island's labour force (World Bank, 1990). Similarly on Réunion, sugar cane is the major export earner and occupies 65 per cent of the cultivable land.

The Forests

Man arrived on Mauritius 400 years ago and found the island covered with tropical vegetation. On the lowlands below about 220 m and on the fertile parts of the uplands, this vegetation consisted of dense tropical evergreen forest. The most common tree species in the original forest were *Diospyros tesselaria*, *Elaeodendron orientale*, *Stadmannia sideroxylon* and *Foetidia mauritiana* (Procter and Salm, 1974). Lowland forest remains only on the western slopes of the mountains between Mt du Rempart and Chamarel with the best examples of forest being found at the base of Trois Mammelles and Mt Brise Fer and at the mouth of the Tamarin Gorge.

The upland forests of Mauritius used to cover the fertile uplands of the Plaine Wilhelms and the south. Today these forests are restricted to an area in the south-west around the Black River Gorges. Some trees exceed 20 m in height, though most are about 17 m high. The upper canopy consists of *Mimusops maxima*, *M. petiolaris*, *Diospyros tesselaria*, *Calophyllum eputamen*, *Canarium* sp. and *Calvaria* sp. A dense lower canopy at about 15 m is formed by tree genera such as *Tabernaemontana*, *Aphloia*, *Eugenia*, *Tambourissa*, *Erythrospermum*, *Antirhea* and *Nuxia*. Lianes, orchids and ferns are common in this layer. A third layer is formed at around 5 m in height where *Colea*, *Psathura* and *Cyathea* are some of the more characteristic plant genera.

Forest types on Réunion were varied but now very little remains of the lowland forests. The moist montane forests are the least disturbed and characteristically contain species such as *Bertiera rufa*, *Dombeya ficulnea*, *D. punctata*, *D. reclinata*, *Forgesia borbonica* and *Monimia rotundifolia* (Domenge and Renard, 1989). Although largely invaded by exotic species and otherwise degraded, Réunion still retains a lot more native woody vegetation than any other of the small Indian Ocean islands (Figure 24.4).

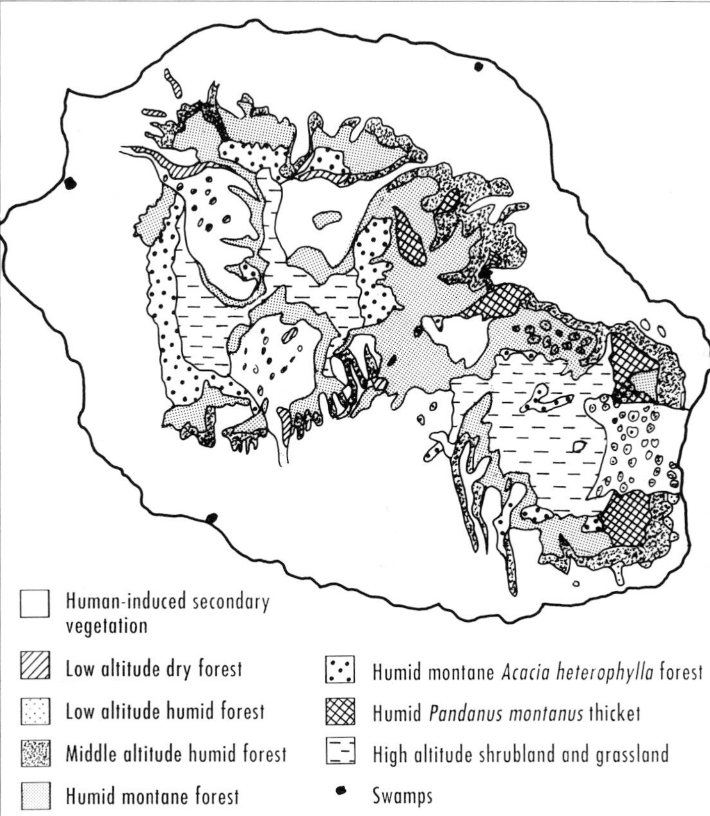

Figure 24.4 Vegetation map of Réunion

(*Sources:* Doumenge and Renard, 1989, after Cadet, 1980 and modified by J. Dupont)

Mangroves are found on the east, south-east and north-east coasts of Mauritius and are relatively abundant in the region of Ile d'Ambre and Ile aux Cerfs. They have disappeared almost completely from the west coast, largely destroyed by clearing for development or by logging (World Bank, 1990). The government has made it a matter of priority to reestablish mangroves where they have been eliminated (MAFNR, 1985).

Forest Resources and Management

The present extent of forest cover on either island appears to be unknown. Procter and Salm (1974) reported that there were a few hundred hectares of indigenous lowland and 16 sq. km of upland forest in Mauritius. FAO estimated 30 sq. km remaining in 1980. No vegetation map, or more recent reports of forest cover on Mauritius have been found for this Atlas. Réunion was reported to contain 820 sq. km of closed forest in 1980 (FAO, 1988); again, no other information on forest cover has been obtained for this island.

The Mauritius Forestry Service is one of the oldest in the Commonwealth. It has been in formal existence since 1777 although it did not become fully autonomous until 1883. As early as 1671, when the Chief Woodcutter was appointed Governor, it was evident that forests were thought to be important. He was later replaced because his lack of education made him an inadequate administrator! Today, the forestry policy of Mauritius provides for the safeguarding of water supplies and soil fertility and the prevention of damage to rivers and agricultural land by flooding and erosion.

A report by the Forestry Service states that it manages about 116 sq. km of forest, for the exploitation of timber, poles and firewood, but these are managed plantations, not natural forests. Most of the timber is exploited by a sole concessionaire, Grewals (Mauritius) Ltd, but any timber of more than 18 cm diameter that this firm leaves, and all timber below 18 cm diameter, is sold to other sawmillers and woodmerchants.

The Forestry Service sells fuel in the form of firewood and charcoal, the amount sold in 1983 (22,146 cu. m) being almost double that sold in 1980 (11,612 cu. m). In addition, a considerable amount of firewood and fodder is collected daily from the forests by the public.

On Réunion, clearfelling of native tamarind (*Acacia heterophylla*) forest to encourage regeneration of straight-boled, even-aged stands, has been the object of intense debate among local conservationists and foresters. The unmanaged forests contain many old and windblown trees and have a rich epiphytic flora. The managed forests are species-poor and are prone to invasion by exotic trees and shrubs. The extent of management of these forests is now to be scaled down. Elsewhere on Réunion the total area of native forest is now reasonably stable but there remains a serious threat to the indigenous biological diversity from invasion by introduced exotic species.

The French National Forestry Office (ONF) is responsible for all forest management and conservation on Réunion. It controls a total of 1002 sq. km of land, 40 per cent of the island, but only 104 sq. km is sufficiently wooded to have any forestry potential. Some 555 sq. km of natural vegetation exists but much of this is high altitude dwarf scrub. There are areas of natural forest in small fragments throughout the island but most have been degraded by grazing, fuelwood and timber harvesting, clearance for agriculture and especially by the invasion of exotic species. The fragmentation and degradation of the natural forest make it virtually impossible to come up with a meaningful figure for natural forest cover. Forty-nine sq. km of production forests are under management. These consist of 17 sq. km of indigenous *Acacia heterophylla*, 4 sq. km of native mixed hardwoods and 28 sq. km of plantations of introduced *Cryptomeria*.

Deforestation

The Dutch began the deforestation of Mauritius by logging it for ebony *Diospyros* sp. in the 17th century; in the following century the French were cutting timber for shipbuilding and clearing land for agriculture, and in the 19th century the British were planting extensive areas of sugar cane, building settlements in the upland areas and cutting wood as fuel for trains and sugar factories (Cheke, 1987). Further clearing this century for sugar, tea and pine plantations and logging has reduced the remaining forest to a few patches and even these are being steadily degraded by illegal wood cutting and invaded by introduced, exotic plants and animals (Cheke, 1987). In addition, the torrential rains that accompany cyclones frequently cause landslides in the montane forests, which are rapidly colonised by exotic plant species.

Similarly on Réunion, deforestation has been a result of logging, clearing for agriculture and the collection of fuelwood and forest products such as palm hearts (Doumenge and Renard, 1989).

Biodiversity

On both islands, the list of plants and animals that have become extinct is almost longer than those that survive. Indeed, the most famous of recent extinctions, that of the dodo *Raphus cucullatus* , occurred on Mauritius. Loss of natural habitat and the introduction of alien species are the main causes of the loss of biodiversity on both Mauritius and Réunion.

About 800–900 plant species occur on Mauritius and around one-third of these are endemic. There are eight endemic genera. Of the estimated 500 species of seed plant found on Réunion, 30 per cent are endemic (Doumenge and Renard, 1989). Several of the surviving species are severely threatened. For instance, only 100 plants of *Crinum mauritianum* are known to survive, *Tetrataxis salicifolia* is reduced to seven individuals on Mauritius and *Drypetes caustica* only just survives with two plants on Mauritius and 12 on Réunion (Stuart *et al.*, 1990). The most important area for threatened endemic species – of both plants and birds – is the Black River Gorge in the Macchabee-Bel Ombre Nature Reserve. Indeed, this forest ranks first in the priority rating used by Collar and Stuart (1988) for 75 key forests for threatened birds in Africa.

As in the Comoros, the only native mammals on Mauritius and Réunion are bats. Mauritius had four species of which only three survive. Two (*Tarida acetabulosa* and *Taphozous mauritianus*) of an original five species still live on Réunion.

Twenty-six per cent (11 of 43) of the entire African total of birds considered to be endangered or vulnerable are from the Mascarenes (Collar and Stuart, 1985). There are ten species of birds remaining on Mauritius and seven of these, including the pink pigeon *Nesoenas mayeri*, the Mauritius kestrel *Falco punctatus*, and the echo parakeet *Psittacula eques*, are threatened and are more or less restricted to the Macchabee-Bel Ombre Nature Reserve (Collar and Stuart, 1985). Over half (19 of 33) of the indigenous birds on Réunion are already extinct (Stuart *et al.*, 1990). Of those remaining, the Réunion cuckoo-shrike *Coracina newtoni*, which is confined to forests in the north-west of the island, is considered to be vulnerable to extinction; while the Mascarene black petrel *Pterodroma aterrima*, a seabird, is endangered and may even be extinct (Collar and Stuart, 1985).

Four endemic reptiles survive on mainland Mauritius while a further six exist on islets; all six are either endangered or rare (Cheke, 1987). Three of Réunion's five indigenous reptiles are extinct, only two geckos, *Phelsuma borbonica* and *P. ornata inexpectata*, survive.

Both islands have an endemic butterfly, *Papilio manlius* in Mauritius and the threatened *Papilio phorbanta* in Réunion (Collins and Morris, 1985).

Conservation Areas and Initiatives

The Mauritius Forestry Service, a department in the Ministry of Agriculture, Fisheries and Natural Resources, is responsible for managing the state-owned nature reserves. There are nine nature reserves on mainland Mauritius (Table 24.1, Figure 24.5) of which Macchabee-Bel Ombre at 36 sq. km is the largest. The total of the others is barely 4 sq. km. Seven further reserves protect offshore islands.

On Réunion, there are five reserves totalling 59 sq. km (Table 24.2, Figure 24.6), but plans exist for a much more extensive system of protected areas (Bosser, 1982; Doumenge and Renard, 1989). At present all reserves are under the management of the ONF but there are plans to give the French Ministry of the Environment a more prominent role in nature conservation programmes on the island.

A National Environmental Action Plan has been written jointly by the Government of Mauritius and the World Bank and this is now at the implementation stage (World Bank, 1990). Other initiatives are taking place. WWF is helping to preserve the flora of Mauritius by fencing some areas, to keep out deer and pigs, and weeding out exotics to give the indigenous vegetation a chance to regenerate. A number of projects to breed some of the rare plants and animals have also been undertaken by WWF in conjunction with the Jersey Wildlife Preservation Trust and the Mauritius Wildlife Appeal Fund (WWF, 1989). Indeed, some of the endangered kestrels and pink pigeons bred in captivity have now been released back on to Mauritius.

In the mid-1980s, the Mauritius Wildlife Appeal Fund was able to lease the 25 ha Ile aux Aigrettes from the government and the fund is using its resources and those from international aid agencies to protect it, both from illegal woodcutting and from invasion by exotic species. This island contains the sole surviving remnant of coastal forest (Parnell *et al.*, 1989).

Bosser (1982) and Doumenge and Renard (1989) give detailed proposals for the reorganisation of conservation programmes on Réunion and for a considerable expansion of the protected area network. Information is given on 13 additional proposed reserves. The Ministry of the Environment and ONF are proceeding with a programme to establish reserves largely along the lines of the proposals in Doumenge and Renard (1989). Another conservation scheme is the Botanic Garden and Conservatoire for the Mascarenes which was established in the 1980s. It undertakes programmes to study, collect and propagate the indigenous flora of Réunion and has plans to restore natural vegetation to experimental areas.

Table 24.1 Conservation areas of Mauritius

Marine and fishing reserves are not covered below. For data on Biosphere reserves see chapter 9.

	Area (sq. km)
Nature Reserves	
Bois Sec	1
Cabinet	<1
Coin de Mire (Gunner's Quoin)	1
Combo	2
Corps de Garde	<1
Gouly Pere	<1
Ile Plate	3
Ile aux Aigrettes	3
Ile aux Serpents	<1
Ilot Gabriel	4
Ilot Marianne	<1
Les Mares	1
Macchabee-Bel Ombre	36
Perrier	<1
Pouce	<1
Round Island	2
Total	60

(*Sources:* Stuart *et al.*, 1990; WCMC, *in litt.*)

Figure 24.5 Protected areas in Mauritius (*Source:* WCMC, *in litt.*)

1 Combo. **2** Bois Sec. **3** Gouly Pere. **4** Les Mares. **5** Macchabee-Bel Ombre. **6** Perrier. **7** Cabinet. **8** Corps de Garde. **9** Pouce. **10** Coin de Mire (Gunner's Quoin). **11** Ile Plate. **12** Ilot Gabriel. **13** Round Island. **14** Ile aux Serpents. **15** Ilot Marianne. **16** Ile aux Aigrettes.

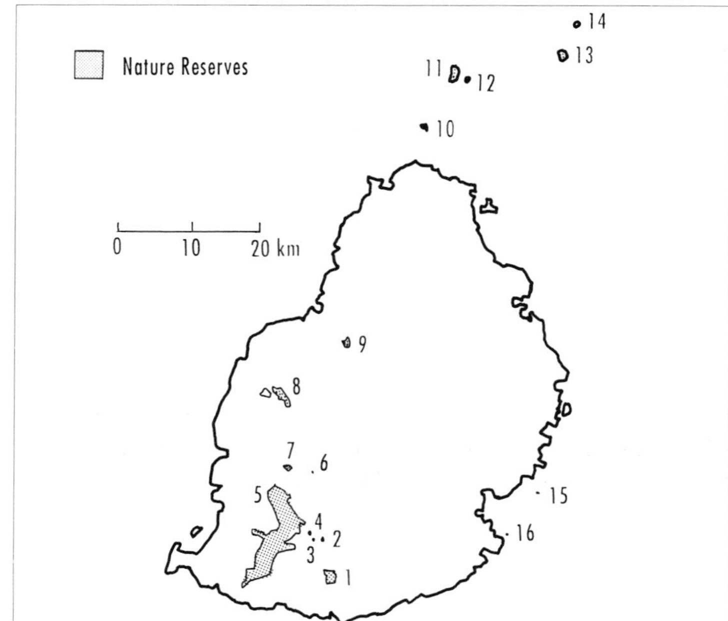

Table 24.2 Conservation areas of Réunion

The list below shows existing and proposed sites. Data for Réunion are poor.

Reserves	*Existing area (sq. km)*	*Proposed area (sq. km)*
Bebour		100
Bois de Bon Accueil		nd
Bord a Martin-Bras Bemale		nd
Col de Bellevue		<1
Colorado		nd
Etang de Saint Paul		nd
Etang du Gol		nd
Hauts du Bois de Nefles	2	
Hauts de St Phillipe	35	
L'Ilot de Patience		5
La Plaine des Chicots		nd
Lieu dit les Palmistes		<1
Mare Longue	1	
Mares et du Sommet de l'Enclos	3	
Mazerin	18	
Notre-Dame de la Paix		2
Ravine de la Grande Chaloupe		nd
Rempart de Cilaos		nd
Totals	59	109

(*Sources:* Doumenge and Renard, 1989; Stuart *et al.*, 1990; WCMC, *in litt.*)

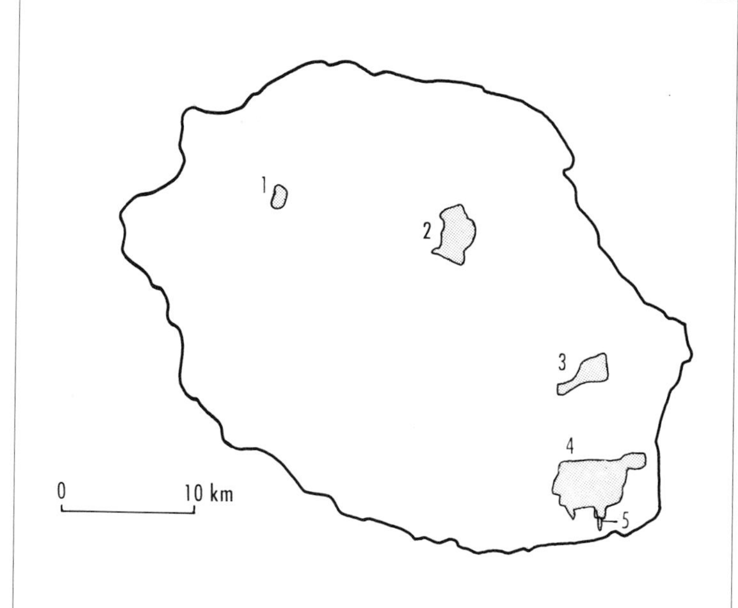

Figure 24.6 Protected areas in Réunion

(*Source:* adapted from Doumenge and Renard, 1989)

1 Hauts de Bon Accueil. **2** Mazerin. **3** Mares et du Sommet de l'Enclos. **4** Hauts de St Phillipe. **5** Mare Longue.

INTRODUCTION – SEYCHELLES

The Seychelles archipelago is made up of 115 granitic and coralline islands spread along a 1000 km arc in the western Indian Ocean. The granitic Seychelles group consists of 41 islands and islets situated between 4–5°S and 55–56°E, rising from the Seychelles Bank, a shoal area of some 31,000 sq. km with water depths of less than 60 m, surrounded by ocean 4–5 km deep. The largest island is Mahé, with a land area of 154 sq. km, rising to 914 m in Morne Seychellois, the highest point of the group. Praslin, 35 sq. km in area, rises to 427 m and Silhouette, which covers 19.6 sq. km, ascends to 867 m. These islands are characterised by rugged granitic mountains, often with smooth bare-rock faces, surrounded by low areas of coastal sand flat and marsh. In addition to the granite islands, there are two other types of islands. One consists of low sand cays on sea-level coral reefs and the other comprises islands of the reef limestone now slightly elevated above the sea. The sand cays are usually less than 5 m above sea level and the raised reef islands less than 8 m.

The great interest of the granitic Seychelles lies in the fact that, alone of isolated oceanic islands, they are built of continental rocks rather than oceanic basalts or reef limestones. They are part of the Gondwanaland super-continent that has never been submerged.

All the islands experience a humid tropical climate with annual rainfall exceeding 700–800 mm and mean monthly temperatures that are always above 20°C (Walsh, 1984) but there are important climatic differences between the islands. Mean annual rainfall on both Mahé and Praslin is above 2000 mm and on the highest parts of Mahé may reach 5000 mm. Average temperature at sea level is around 27°C, and seasonal and diurnal ranges are very low.

The islands were thought to have been sighted by Vasco da Gama in 1502, but the first recorded landing was in 1609. Nobody lived there for a further 160 years. In 1785, the population of the Seychelles was seven Europeans and 123 slaves (EMP, 1990). Originally claimed by the French in 1756, the British took them over in 1814 until independence in 1976. The Seychellois people are of mixed race, primarily of African, Malagasy and French origin with some Indians and Chinese. Today's population of 100,000 people (about one-third of whom are under the age of 15) are concentrated on the three largest granitic islands of Mahé, Praslin and La Digue.

The Forests

The tall lowland forests of the granitic islands, dominated by massive Bois de fer *Vateria sechellarum*, have long since disappeared. Native woodland now exists only in inaccessible inland and upland localities. The diversity of species, at least among canopy forming trees, is much lower than in typical rain forests (Procter, 1984). The development of a palm forest type in Praslin, characterised by the curious endemic Coco-de-Mer *Lodoicea maldivica*, is a unique feature of the Seychelles vegetation (Procter, 1984). There are about 766 species of flowering plants and 85 ferns and fern allies on the granitic islands (Procter, 1984) only 222 of these are indigenous and 69 are endemic.

Formerly very extensive and well developed on the east coast, true mangrove swamps are now found mostly in a few places on the west coast, the best developed being in the Port Glaud area. Small patches of mangrove do, however, occur throughout the islands. The typical species of the East African mangrove swamps are all present.

Forest Resources and Management

No statistics or maps of the forest extent on the Seychelles have been found for this Atlas. The only estimate of closed forest is that of FAO (1988) for 1980 although the Environmental Management Plan (EMP, 1990) does give some figures for remaining productive or exploitable forest on the islands. It estimates that on Mahé, 40 sq. km are currently under forest or shrub of which about 14.5 sq. km are considered as productive forest land, while there are 2 sq. km of exploitable forest on Silhouette, 9.5 sq. km on Praslin and 2 sq. km on La Digue. On the coralline outer islands there are about 4.4 sq.

Palm forest, mainly Coco-de-Mer, in Vallée de Mai, Praslin, Seychelles. R. Wilson

km of productive forest, mostly *Casuarina*. This species, along with *Albizia* and mahogany, has been grown in plantations that were started in the 1950s and now cover about 9 sq. km.

Forest plantations have on the whole proved to be uneconomic and it is cheaper to import wood. As a consequence, current forest management objectives are mainly focused on the conservation of existing forest cover and on erosion control (EMP, 1990).

Deforestation

When man first set foot on the Seychelles in the early 17th century the islands were covered in forest (Procter, 1970). The first settlers in 1770 began the rapid process of clearing the forests and started to cultivate nutmeg, cinnamon, cloves and maize (Stoddart, 1984). For a considerable period, timber was exported and logging was one of the main industries. The timber was used largely for ship building. As the timber trees were depleted, the gaps in the forests were filled, not by endemics but by fast-growing introduced species, chiefly cinnamon *Cinnamomum verum*, *Albizia falcata* and *Adenanthera pavonina* (Procter, 1970).

The lowland forest is now almost completely replaced by coconut plantations or housing, while farmland has taken over in the intermediate forest area between the lowland and mountain zones.

Biodiversity

Only 250 indigenous plant species are known to occur on the Seychelles although 1139 species of seed plant have been recorded there. The plants of the granitic Seychelles include many of great interest. *Medusagyne oppositifolia* in the monotypic endemic family Medusagynaceae was long thought to be extinct and has only recently been rediscovered on Mahé. There are endemic palms of the monotypic genera *Roscheria*, *Verschaffeltia*, *Nephrosperma*, *Deckenia*, *Phoenicophorium* and *Lodoicea* (Fauvel, 1915; Bailey, 1942; Sauer, 1965); three other endemic genera of flowering plants, *Protarum*, *Geopanax* and *Gastonia* (Procter, 1974), and striking endemic pandans, *Pandanus hornei* and *P. seychellarum* (St John, 1961, 1967).

The invertebrates of the Seychelles are varied and distinctive with perhaps as many as 51 per cent (1200–1300 species) of the insects being endemic to the islands (Scott, 1936). The best known of the vertebrates are the birds, of which there are 40 species on the Seychelles as a whole (Benson, 1984). Endemic species of the forests are the Seychelles scops owl *Otus insularis*, Seychelles swiftlet *Collocalia elaphra*, Seychelles white-eye *Zosterops modestus*, and Seychelles kestrel *Falco araea*.

Native mammals are restricted to bats with two species on the granitic islands, *Pteropus seychellensis seychellensis* and the very rare sheath-tailed bat *Coleura seychellensis*.

Conservation Areas and Initiatives

Protection of wildlife and forest resources in the Seychelles dates back to the mid-19th century when two areas of Praslin were purchased as Coco-de-Mer reserves. The legal basis of the network of parks and reserves in the country is the National Park and Nature Conservancy Ordinance of 1969. This provides for the creation of four categories of protected areas: strict nature reserves, special reserves, national parks and areas of outstanding natural beauty.

The Seychelles possess two World Heritage sites, Aldabra atoll (which is wholly set aside for conservation and scientific research) and the Vallée de Mai, which is within Praslin National Park. There is in addition a comprehensive network of marine and land-based terrestrial national parks and reserves (Table 24.3 lists the terrestrial protected areas).

The government of the Seychelles, with the assistance of UNEP and UNDP, has published an Environmental Management Plan for the years 1990–2000 (EMP, 1990). Included in this is a major project to prepare and implement a national forest management plan in 1991. Under this plan it is intended that all the forests will be surveyed and assessed for erosion control and water retention, for preservation of biodiversity, for wood production and landscaping. Each forest area will have its management objectives determined and a course of action will be undertaken to meet these objectives. A policy for timber production will also be produced. Fires present the major threat to forests in the Seychelles, particularly on Praslin and Curieuse islands where the palms are highly prone to burning. Combating fire is recognised as an immediate priority for the forestry sector.

Table 24.3 Conservation areas of the Seychelles

Existing and proposed conservation areas are listed below. Marine national parks that have no land within them have been excluded from this list. For details of World Heritage sites and Biosphere reserves see chapter 9.

	Existing area (sq. km)	*Proposed area (sq. km)*
National Parks		
Morne Seychellois	30	
Praslin†	7[1]	
Strict Nature Reserves		
Aldabra	208[2]	
Marine National Parks		
Curieuse	3[3]	
St Anne	nd[4]	
Reserves		
Aride Island Special	<1	
Cousin Island Special	<1	
La Digue Veuve	<1	
African Banks Special		nd
Desnoeufs Special		nd
Bird and Dennis Islands		nd
Frégate		nd
Totals	251	

(*Source:* WCMC, *in litt.*)

1 The 18 ha Vallée de Mai Nature Reserve is within this national park.
2 This is the area of land and mangrove in the reserve, there are an additional 142 sq. km of sea included.
3 This is the area of the island, the whole park is 15 sq. km.
4 There are six small islands within this park but none is of any conservation significance.

References

Bailey, L. H. (1942) Palmae Sechellarum. *Gentes Herbarum* **6**: 3–48.

Battistini, R. and Vérin, P. (1984) *Géographie des Comores*. Agence de Cooperation Culturelle et Technique, Paris, France.

Benson, C. W. (1960) The Birds of the Comoro Islands: Results of the British Ornithologists' Union Centenary Expedition 1958. *Ibis* **103**b: 5–106.

Benson, C. W. (1984) Origins of Seychelles' land birds. In: *Biogeography and Ecology of the Seychelles Islands*, pp. 469–86. Stoddart, D. R. (ed.). Junk, The Hague, The Netherlands.

Bosser, J. (1982) *Projet de Constitution de Réserves Biologiques dans le Domain forestiers à La Réunion. Rapport de mission*. ORSTOM, Paris, France. 35 pp.

Cadet, T. (1980) *La Végétation de l'île de la Réunion. Etude Phytoécologique et Phytosociologique*. Cazal, Saint Denis, Ile de la Réunion.

Carroll, J. B. and Thorpe, I. C. (1991) The conservation of Livingstone's fruit bat *Pteropus livingstonii* Gray 1866: A report on an expedition to the Comores in 1990. *Dodo, J. Jersey Wildl. Preserv. Trust* **27**: 26–40.

Cheke, A. (1987) The legacy of the dodo – conservation in Mauritius. *Oryx* **21**(1): 29–36.

Collar, N. J. and Stuart, S. N. (1985) *Threatened Birds of Africa and Related Islands. The ICBP/IUCN Red Data Book Part 1*. ICBP/IUCN, Cambridge, UK. 761 pp.

Collar, N. J. and Stuart, S. N. (1988) *Key Forests for Threatened Birds in Africa*. ICBP Monograph No. 3. ICBP, Cambridge, UK.

Collins, N. M. and Morris, M. G. (1985) *Threatened Swallowtail Butterflies of the World. The IUCN Red Data Book*. IUCN, Gland, Switzerland and Cambridge, UK.

Davis, S. D., Droop, S. J. M., Gregerson, P., Henson, L., Leon, C. J., Villa-Lobos, J. L., Synge, H. and Zantovska, J. (1986) *Plants in Danger: What do we know?* IUCN, Gland, Switzerland and Cambridge, UK.

Deville, A. (1974) *Projet de Programme Pour le Futur Développement dans l'Archipel des Comores*. Rapport de mission PNUD/FAO, Tananarive, Madagascar.

Doumenge, C. and Renard, Y. (1989) *La Conservation des Ecosystèmes forestiers de l'île de la Réunion*. IUCN, Gland, Switzerland and Cambridge, UK.

EMP (1990) *Environmental Management Plan of the Seychelles 1990–2000*. Report prepared with the advice and assistance of UNEP, UNDP and the World Bank. Government of Seychelles, Victoria.

FAO (1988) *An Interim Report on the State of Forest Resources in the Developing Countries*. FAO, Rome, Italy. 18 pp.

FAO (1990) *FAO Yearbook of Forest Products 1977–1988*. FAO Forestry Series No. 23 and FAO Statistics Series No. 90. FAO, Rome, Italy.

Fauvel, A. A. (1915) Le cocotier de mer des îles Seychelles (*Lodoicea sechallarum*). *Annals Musée colon. Marseille* **3**(3): 169–307.

Gachet, C. (1957) *Rapport de Mission aux Comores*. Secteur Generale Eaux et Forêts de Madagascar, Tananarive.

Gade, D. W. (1985) Man and nature on Rodrigues: tragedy of an island common. *Environmental Conservation* **12**(3): 207–16.

Louette, M. (1988) *Les Oiseaux des Comores*. Musée Royal de L'Afrique Centrale, Tervuren, Belgium.

Louette, M., Stevens, J., Bijnens, L. and Janssens, L. (1988) *A Survey of the Endemic Avifauna of the Comoro Islands*. ICBP Study Report No. 25. ICBP, Cambridge, UK.

MAFNR (1985) *White Paper for a National Conservation Strategy*. Ministry of Agriculture, Fisheries and Natural Resources, Government of Mauritius, Port-Louis. 24 pp.

Parnell, J. A. N., Cronk, Q., Wyse Jackson, P. and Strahm, W. (1989) A study of the ecological history, vegetation and conservation management of Ile aux Aigrettes, Mauritius. *Journal of Tropical Ecology* **5**: 355–74.

PRB (1990) *1990 World Population Data Sheet*. Population Reference Bureau, Inc. Washington, DC, USA.

Procter, J. (1970) *Conservation in the Seychelles. Report of the Conservation Advisor 1970*. Government Printer, Mahé, Seychelles.

Procter, J. (1974) The endemic flowering plants of the Seychelles: an annotated list. *Candollea* **29**(2): 345–87.

Procter, J. (1984) Vegetation of the granitic islands of the Seychelles. In: *Biogeography and Ecology of the Seychelles Islands*, pp. 193–207. Stoddart, D. R. (ed.). Junk, The Hague, The Netherlands.

Procter, J. and Salm, R. (1974) *Conservation in Mauritius 1974*. IUCN/WWF Report to the Government of Mauritius. Unpublished.

Sauer, J. D. (1965) The Seychelles archipelago and its palms. *Bulletin Missouri Botanic Garden* **53**(9): 8–15.

Scott, H. (1936) General conclusions regarding the insect fauna of the Seychelles and adjacent islands. *Transactions Linnaeus Society London* **19**(2): 307–91.

St John, H. (1961) *Pandanus* of the Maldive Islands and the Seychelles Islands. *Pacific Science* **15**: 328–46.

St John, H. (1967) Revision of the genus *Pandanus* Strickman. Part 24 Seychellea, a new section from the Seychelles Islands. *Pacific Science* **21**: 531.

Stoddart, D. R. (1984) Impact of man in the Seychelles. In: *Biogeography and Ecology of the Seychelles Islands*, pp. 641–54. Stoddart, D. R. (ed.). Junk, The Hague, The Netherlands.

Strahm, W. (1989) *Plant Red Data Book for Rodriguez*. Koeltz Scientific Books, Konigstein, Germany.

Stuart, S. N., Adams, R. J. and Jenkins, M. D. (1990). *Biodiversity in Sub-Saharan Africa and its Islands: Conservation, Management and Sustainable Use*. Occasional Papers of the IUCN Species Survival Commission No. 6. IUCN, Gland, Switzerland.

Thorpe, I. C., Gilby, L., Waters, D. and Turner, P. A. (1988) *University of East Anglia Comoro Island Expedition*. Final Report. UEA, Norwich, UK.

Walsh, R. D. (1984) Climate of the Seychelles. In: *Biogeography and Ecology of the Seychelles Islands*, pp. 39–62. Stoddart, D. R. (ed.). Junk, The Hague, The Netherlands.

Weightman, B. L. (1987) *The Crisis on Anjouan*. Report to FAO.

White, F. (1983) *The Vegetation of Africa: a descriptive memoir to accompany the Unesco/AETFAT/UNSO vegetation map of Africa*. Unesco, Paris, France. 356 pp.

World Bank (1990) *National Environment Action Plan for Mauritius*. World Bank, Washington, DC, USA.

WWF (1989) Mauritius – a paradise being lost. *WWF Reports*. **February/March**: 2–6. Gland, Switzerland.

Authorship

Caroline Harcourt for the sections on the Mascarenes and Seychelles, Ian Thorpe for the section on the Comores, with contributions from M. Louette, Musée Royal de l'Afrique Centrale, Belgium and Charles Doumenge, IUCN, Gland.

25 Liberia

Land area 96,320 sq. km
Population (mid-1990) 2.6 million
Population growth rate in 1990 3.2 per cent
Population projected to 2020 6.5 million
Gross national product per capita (1988) US$450
Rain forest (see map) 41,238 sq. km
Closed broadleaved forest (end 1980)* 20,000 sq. km
Annual deforestation rate (1981–5)* 460 sq. km
Industrial roundwood production† 1,160,000 cu. m
Industrial roundwood exports† 701,000 cu. m
Fuelwood and charcoal production† 4,800,000 cu. m
Processed wood production† 416,000 cu. m
Processed wood exports† 28,000 cu. m

* FAO (1988)
† 1989 data from FAO (1991)

Unlike all other countries in West Africa, Liberia lies wholly within a belt which has the soil and climatic conditions suitable for rain forest. Liberia is small in area, but still has the two most substantial blocks of closed canopy rain forest in the Upper Guinean Forest. The forest fauna of this zone is particularly distinctive, with very few close relatives east of the arid Dahomey Gap in Benin. A lower human population density than other countries in the region, combined with a late start for logging, may explain the survival of these eminently exploitable forests. Today, however, Liberia is suffering similar pressures to those of its deforested neighbours.

Exploitation of the forest is primarily for timber, fuelwood, bushmeat and farming. The foreign-owned timber industry operates within an adequate legislative framework and is more highly taxed than in most other African countries, yet in practice the political-economic situation has allowed the loggers to make their own rules.

Bushmeat is a key part of the diet of most Liberians and a large variety of species is consumed both from bushland in the agricultural areas and from the rain forests. There is an urgent need for improved management systems to protect threatened forest species while still allowing sustainable cropping of other species.

Each year, clearance for shifting agriculture destroys about 2 per cent of the remaining rain forest. Forest resources in these areas are devastated as vegetation is burnt in order to begin the farming cycle with upland rice. This system is well adapted to the local constraints of ecology, economics and labour supply in the sparsely populated Liberian forest areas, but becomes unsustainable at population densities over 25 persons per sq. km. The crucial factor, however, is the farmers' frequent preference for clearing forest land rather than old bush fallow. The opening up of the forests by timber companies further speeds this process.

National forests, which are nominally protected against encroachment by farmers and hunters, cover about 30 per cent of Liberia's rain forest and all but one have logging concessions within them. The only significant area of forest without a concession agreement is Sapo National Park which has been successfully protected since its establishment in 1983. Basic prerequisites for restraining deforestation in Liberia are the capacity to enforce genuine control on the timber industry and a new approach to forest protection which integrates conservation with efforts to find solutions to the problems of shifting agriculture and hunting.

Introduction

Liberia occupies the wet south-western corner of West Africa. Looking inland from the almost unbroken coastal sand strip, it is characterised topographically by three bands: coastal lowlands, central rolling hills with inland plateaux and the highlands of the Wologizi range (up to 1380 m) and Nimba range (up to 1385 m before iron ore extraction started). The climate is hot and humid with a mean annual temperature of 26°C and annual rainfall ranging from 4600 mm in the Monrovia area to about 2000 mm in the centre and north of the country. The major rainy season is from May to September, but there is a short dry spell in July–August which is more or less marked depending on the region. The period between October and April is relatively dry.

Demographically, the pattern is typical of developing nations with a high proportion of children under 14 years (41 per cent in 1981), low average density (27 per sq. km), a high population in the capital Monrovia (800,000 estimated in 1988) and a high migration from country areas into the towns. Although Liberia appears in world markets as pre-eminently an exporter of iron ore, rubber and timber (64 per cent, 18 per cent and 6 per cent of exports in 1985 respectively), it is basically an agricultural country. Seventy-two per cent of the working population were employed on the land in 1980. Upland rice is the staple crop of the shifting agriculturalists.

Liberia's recent history is unique in that it was never directly colonised. It became an independent republic in 1847 and until the coup of 1980 it was ruled by an elite group who were mostly the descendants of freed slaves from the southern USA. There are 16 indigenous African tribes, making up 93 per cent of the population, while the Liberians descended from freed slaves form 2 per cent of the inhabitants.

During 1990, civil war erupted in Liberia, heralding a complete breakdown of state structures and the deaths of thousands of Liberians. President Doe himself was killed in September of that

Table 25.1 Vegetation of Liberia 1979–82

Vegetation type	*Area (sq. km)*	*% Liberian lands*
Total high forest	47,900	49.8
High forest slope 0–15%	34,430	
High forest slope 15–30%	7,050	
High forest slope >30%	6,200	
Swamp forest (fresh water)	220	
Mangrove	190	0.2
Plantations	1,670	1.7
Farmland and regrowth	45,410	47.2
Non-forested swamp	80	0.1
Savanna	360	0.4
Grassland	210	0.2
Other	360	0.4
Subtotal – Land Area	96,180	100.0
Subtotal – Inland Water Area	760	
Grand Total	96,940	

(*Source:* FDA/IDA, 1985)

Definitions: **High forest** is defined as 'forest of a primary or old secondary nature occurring on drier sites with a closed or almost closed canopy exceeding 30 metres in height'. **Swamp forest** is 'high forest but located on swampy or periodically inundated sites'. **Mangrove** includes 'all areas covered by mangrove irrespective of canopy or tree height'. **Plantation** includes 'forest, oil-palm and rubber plantations'. **Farmland and regrowth** includes 'presently farmed lands for subsistence cropping and regrowth resulting from such previously farmed areas. In addition it includes regrowth arising from transitional changes to grassland and low palm/tree cover of coastal formations.' **Swamp** includes 'all lands permanently or seasonally inundated with water other than those supporting swamp forest or mangrove'. **Savanna** includes 'mixed tree/grassland formations with a continuous dense grass layer'. **Grassland** includes 'natural grass areas that contain less than 10 per cent woody vegetation cover'. **Other** includes urban areas, bare ground, non-wooded coastal dunes, mining areas, etc. *NB:* This report does not mention any montane forest present in the country, while Map 25.1 shows this vegetation type but no swamp forest remaining.

year. A peacekeeping force was formed from the armed forces of several West African states with the objective of facilitating the installation of an interim government. The future for the forests, as for the people of Liberia, remains uncertain at the time of writing.

The Forests

Liberia was originally almost completely covered in tropical moist forest, the exception being an area of about 1000 sq. km of Guinea savanna to the north of Mt Wituvi in the north-west. Today, most rain forest is found in two regions, one in the north-west and one in the east and south-east of the country. Two types of rain forest can be broadly distinguished: the evergreen forests of the wetter south and central areas and the moist semi-deciduous forests found in the drier north-western parts of the country, chiefly in upper Lofa County. In addition, a small area of montane forest is still present on Mt Nimba.

Characteristic tree species of the mixed evergreen forests are ekki *Lophira alata*, niangon *Heritiera utilis*, *Sacoglottis gabonensis*, *Calpocalyx aubrevillei* and *Dialium* spp. Often there is a dominance of a single species in one or all storeys such as *Gilbertiodendron preussii*, *Tetraberlinia tubmaniana* or *Parinari excelsa*. The moist semi-deciduous forests have a greater abundance of the Meliaceae such as *Entandrophragma* spp. and *Khaya* spp. although these are increasingly logged out.

It appears that there was a period around 300 years ago when the country was more densely populated than at present and there was considerably less forest cover. It may be appropriate therefore to consider the majority of existing closed forest in Liberia as late secondary forest (Voorhoeve, 1965).

Mangroves

Liberia does not have extensive wetlands along its coast, but mangroves do occur at all river mouths and the Forest Development Authority (FDA/IDA, 1985) estimated the total area of these to be 190 sq. km in 1979–82. The Mafa River has some mangroves 7 km upstream from the sea and these are separated from those at the mouth by a small block of terrestrial forest. There are other areas of mangrove around Lake Piso, which is actually a large open lagoon on the coast rather than a lake. The mangroves are best developed at creek mouths and are backed by *Raphia* swamps. An extensive area of mangrove of some 60 sq. km exists around the Mensurado Creek close to Monrovia, although this has suffered considerably in recent years from road-building, land-fill and fuelwood collection. Other fairly extensive areas of mangrove are at the confluence of the Bo and Junk rivers, along the Mechlin and Benson rivers, at the mouth of the Joda River and on the Decoris, Cestos and Senkwen rivers (Hughes and Hughes, 1991). Note that Map 25.1 indicates only 6 sq. km of mangrove; the area around Mensurado Creek appears to be mostly a savanna or urban area now. Other regions may be too small to show on the scale used for this map. No fresh water swamp forest can be distinguished on the map.

All three species of crocodile occur in the swamps; the Nile and African dwarf crocodile (*Crocodylus niloticus* and *Osteolaemus tetraspis*) are found in the mangroves, although they are not common, while the slender-snouted crocodile *Crocodylus cataphractus* occurs in fresh water. The African manatee *Trichechus senegalensis* is also found in very small numbers in some estuaries. Hunting and fishing pressures are intense in the mangroves as is their exploitation for fuelwood, while clearing of the forests for rice cultivation and settlements is increasing. At present none of the mangroves is protected although the proposed nature reserve at Cape Mont will include this habitat.

Forest Resources and Management

Figures for forest cover in Liberia varied widely until 1985 when the government's Forestry Development Authority (FDA), with support from the International Development Authority of the World Bank, mapped forest cover from aerial photographs taken between 1979 and 1982 (FDA/IDA, 1985). These data, by far the most comprehensive and accurate up to that date, gave a total dryland forest extent of 47,900 sq. km or 49.8 per cent of Liberian lands (Table 25.1). Map 25.1 shows lowland rain forest covering 41,177 sq. km and montane forest covering 55 sq. km (Table 25.2), giving a total dryland moist forest coverage of 41,232 sq. km, or 43 per cent of land area, remaining in early 1987. The figure estimated by FAO (1988) for the end of 1980 was based on only partial aerial photograph coverage and, as a result, underestimated the forest cover.

All but an insignificant fraction of forest lands are under state ownership. About 30 per cent of this forest is currently accorded legal protection as national forests or national parks, but several

Table 25.2 Estimates of forest extent in Liberia

	Area (sq. km)	*% of land area*
Rain forests		
Lowland	41,177	43
Montane	55	<0.1
Mangrove	6	<0.01
Totals	41,238	43

(Based on analysis of Map 25.1. See Map Legend on p. 220 for details of sources.)

A log collecting area in a Liberian forest. J. Sayer

significant blocks remain which are essentially unprotected. Virtually all national forest land and unprotected forest is subject to concession agreements with commercial logging companies.

In the late 1980s, with declining revenues from Liberia's economic mainstay of iron ore, pressure was exerted on the forestry sector to generate foreign exchange to help alleviate Liberia's foreign debt which stood at one billion dollars by 1988. Timber represented 15 per cent of exports in 1986–7 but this had risen to 25 per cent in 1989.

The production of logs increased tenfold between the early 1960s and the early 1970s and by 1979 total production stood at 800,000 cu. m. Production declined in the first half of the 1980s but rose rapidly again in the latter half to the official record of 1,186,000 cu. m in 1989. In the same year, of the 68 registered timber companies, 36 were actively operating under various types of lease agreement with the government. Almost all companies are foreign-owned; the largest is Israeli, several are Lebanese and most of the remainder are European. In 1989, 14 companies accounted for 82 per cent of total timber production. The major destinations for log exports in 1989 were France (34 per cent), Portugal (14 per cent), Italy (13 per cent) and Germany (9.2 per cent).

Currently, eight species account for 70 per cent of total production with niangon alone accounting for almost 30 per cent. New regulations in 1987 created some incentives for the use of 'lesser known species' and made it mandatory for logging companies to convert a greater proportion of their production into sawn timber and for export shipments to contain a minimum of 5 per cent processed wood. However, most investments in the processing industry are seen by timber exploiters as necessary overheads so that they can continue their extremely profitable log exports (Repetto and Gillis, 1988). Thus, despite an increase to 28 sawmills and four veneer or plywood plants, the increase in processed wood has been comparatively slight. There are hearsay accounts of some logging and timber exports continuing in spite of the civil unrest in the country. However, this must be at a very reduced level.

The 1953 Forest Act provides for the creation of forest reserves but does not distinguish between production and protection forest. The 1976 act creating the Forestry Development Authority detail the objectives of the FDA which include: 'To devote all publicly-owned forest lands to their most productive use for the permanent good of the whole people considering both direct and indirect values.' In general terms, the rules for sustained-yield management of the forests are present in the legislation; the problem lies in how they are applied.

The FDA's management of timber exploitation is based on a country-wide forest inventory carried out in the 1960s by the Bureau of Forests and Wildlife Conservation (BFWC) in cooperation with the German Forestry Mission (GFM) to Liberia (Sachtler and Hamer, 1967). GFM has a small silvicultural research programme in the Gola National Forest in the west (Jordan, 1983; Poelker and Wolf, 1989) and in the Grebo, Cavally and Krahn Bassa areas in the east (Woell, 1986; Zwuen, 1988). Their findings point to the need for a felling cycle minimum of 40 years and suggest certain silvicultural practices which might assist in achieving a sustained yield of timber.

However, in reality timber exploitation in Liberia has had little respect for the demands of silviculture. For example, the felling cycle is set at 25 years which therefore gives an annual coupe of 4 per cent of the area within any one concession. It is estimated that there are about 26,000 sq. km of loggable forest in Liberia, yet the total area of land leased to logging companies in 1989 was about 53,000 sq. km. Part of the explanation for this lies in the proliferation in recent years of salvage permits for non-forest areas. There are also a number of timber concession areas in which more than 50 per cent is inaccessible forest or farmland. As the annual coupe is calculated on the total area of the concession, rather than the area of productive forest within it, this means that the annual 4 per cent portion of the concession can be a much higher percentage of the productive forest (Reitbergen, 1989). It appears then that at a national level the felling cycle is in practice reduced to about 12 years.

In addition, the output per hectare is low. In 1989 there was an estimated average output of 6.3 cu. m per hectare (FDA, 1990), a decrease from the 1974 level of 7.5 cu. m per hectare (FAO/UNEP, 1981). This fall occurred despite the supposed increase in processing facilities and reduction of the forest area available. It is estimated that 1260 sq. km of land were logged-over each year between 1984 and 1989.

Logging taxes are high compared with other countries in the region and are collected directly by the FDA. The total assessed forest fees in 1989 were estimated at US$20 million. These levies include a 'conservation tax' which has generated revenue for the recent remarking of the boundaries of some of the national forests. It appears, though, that the timber companies lessen their tax burden by illegally lowering the declared value of the timber exported. For example, the actual quantity of timber loaded on to a ship may be under-declared; the recorded quality or species value of the timber may be downgraded; or the prices quoted to the Liberian government may be less than those negotiated with the foreign buyer. The net effect of these practices is a huge loss of revenue to Liberia and corresponding profits for the timber companies. For instance, the value of logs leaving Liberia for EEC countries (about 80 per cent of all log exports from Liberia) in 1989 was recorded by the FDA as approximately US$80 million – their value assessed on arrival in Europe was nearer US$200 million.

In short, the FDA in the 1980s manifested the contradictions inherent in a government body responsible for both forest conservation and generation of maximum revenue from timber extraction. Under the regime of President Samuel Doe, the timber companies controlled an industry that was beyond jurisdiction and fundamentally corrupt.

Forest wildlife is a crucial food source for the majority of rural Liberians. A large variety of bush and forest animals are hunted for meat. The preferred species are antelopes, especially duikers, bush-pigs *Potamochoerus porcus* and primates (Anstey, 1991). Despite a ban on all hunting announced by the President in 1988 and the passing of a more pragmatic Wildlife and National Parks Act later that year, hunting continues essentially unchecked in many forest areas. A decrease in hunting of large game animals occurred in the 1980s, but this was more a result of the decline in their populations and military restrictions on the availability of guns and cartridges than enforcement of wildlife regulations.

In early 1990, a questionnaire survey of perceptions about conservation, wildlife and bushmeat was carried out in Monrovia by the WWF/FDA Wildlife Survey and the Society for the Conservation of Nature of Liberia. Extrapolating from respondents' estimates of their expenditure, the survey found that Monrovia spends a minimum of US$12 million annually on bushmeat. For the whole of Liberia, a conservative estimate (given the greater dependence of rural people on bushmeat) would give the annual sale value of bushmeat at US$66 million (Anstey, 1991). Clearly, the value of the forests to Liberia is much more than the sum of timber revenues obtained from them. It remains to be seen whether a new government in Monrovia will approach the management of the forests with this in mind.

Deforestation

FAO estimated the annual rate of deforestation for the years 1981–5 to be 460 sq. km. This estimate appears to be based on extrapolating from a sample of air photographs. The FDA/IDA (1985) figures together with unpublished estimates from the author J. Mayers, give an annual rate of deforestation of 946 sq. km for the years 1983–9, which is an annual forest loss of about 2 per cent. This figure accords well with the decline in amount of forest from the 1979–82 information of FDA/IDA (1985) to that shown on the 1987 satellite images from which Map 25.1 has been drawn.

The principal source of deforestation in Liberia has historically been smallholder agriculture. This accounted for over 95 per cent of all national forest clearing in the survey of 1982 (FDA/IDA, 1985). Shifting agriculture includes a variety of practices, some of them quite complex associations of crops and trees, but most commonly involving the cutting and burning of all vegetation in small areas (tens of hectares) before planting upland rice. The rice is cultivated for one or two years and is followed by the planting of root crops, especially cassava. The land is then left fallow for 8–15 years before the cycle begins again.

Liberia's population is low and unevenly distributed. Although the national growth rate is currently about 3.2 per cent, the rural growth rate is expected to average only about 1.7 per cent until 1999. Moreover, existing rural population densities in the three counties (Grand Gedeh, Sinoe and Lofa) with the greatest remaining blocks of relatively undisturbed forest average only about 6.5 people per sq. km at the present time. Deforestation rates in these areas are therefore low.

Road development, especially construction of logging roads, has had a major but variable impact on deforestation in Liberia. In regions characterised by high economic opportunity and high net immigration potential (e.g. the Monrovia–Nimba corridor), such roads have led to an increase in deforestation rate by providing better access to forest lands. Conversely, in areas characterised by low economic opportunity and where more people are leaving than moving in (e.g. Grand Gedeh and Sinoe), the impact of such roads appears to be one of redistributing the regional pressure so that deforestation increases in road corridors but reduces in non-corridor areas.

Logging operations are inextricably linked to the deforestation ultimately caused by shifting agriculture. Agriculture gains a foothold in the logging camps, which could be anything between 30 and 500 houses in size, as about 50 per cent of workers move their families into the camps (MPEA, 1983) and continue to rely on shifting agriculture and hunting for subsistence. Many of the families settle permanently after the logging companies move on. In addition, farmers appear to prefer logged forest for clearing since some of the larger trees have already been removed. This will inevitably influence the species composition of the regrowth forest.

Biodiversity

During the Pleistocene ice ages Africa's westernmost rain forest refuge was probably centred in present-day Liberia (Diamond and Hamilton, 1980; see chapter 2). Some of the forest flora (Kunkel, 1965; Voorhoeve, 1965), and many of the rain forest animals in Ghana, Côte d'Ivoire, Sierra Leone and Liberia are derived from this Upper Guinean refuge population. Considering the extent of deforestation in the other countries of the region, the forests of Liberia could be regarded as present-day refugia and represent the most substantial existing areas of the Upper Guinean biogeographical zone (MacKinnon and MacKinnon, 1986).

The bird list for Liberia currently stands at 590 species (Gatter, 1989), a large proportion of which are found in rain forest. Nine of these are regarded as threatened, while 13 are endemic to the Upper Guinean forest (Thiollay, 1985). These include the white-breasted guineafowl *Agelastes meleagrides* which is 'one of the most threatened birds in continental Africa' (Collar and Stuart, 1985), the Gola malimbe *Malimbus ballmanni* and the white-necked rockfowl *Picathartes gymnocephalus* (Collar and Stuart, 1985; Carter, 1987).

Liberian mammals mentioned in the IUCN Red List of Threatened Animals (IUCN, 1990) include Jentink's duiker *Cephalophus jentinki*, which is the world's rarest duiker, and another strikingly marked duiker endemic to the Upper Guinean Forest, the zebra duiker *Cephalophus zebra*. Also endemic to this zone are the pygmy hippopotamus *Choeropsis liberiensis*, diana monkey *Cercopithecus diana* and the Liberian mongoose *Liberiictus kuhnii*. The latter is known only from a few locations in Liberia and from the Taï National Park, Côte d'Ivoire (Taylor, 1988). Elephant *Loxodonta africana* probably number less than 2000, but there are good populations of giant forest hog *Hylochoerus meinertzhageni*, chimpanzee *Pan troglodytes* and the forest dependent primates, red colobus *Procolobus [badius] badius* and western black-and-white colobus *Colobus polykomos*. Bongo *Tragelaphus euryceros*, leopard *Panthera pardus* and golden cat *Felis aurata celidogaster* are also sparsely but widely distributed.

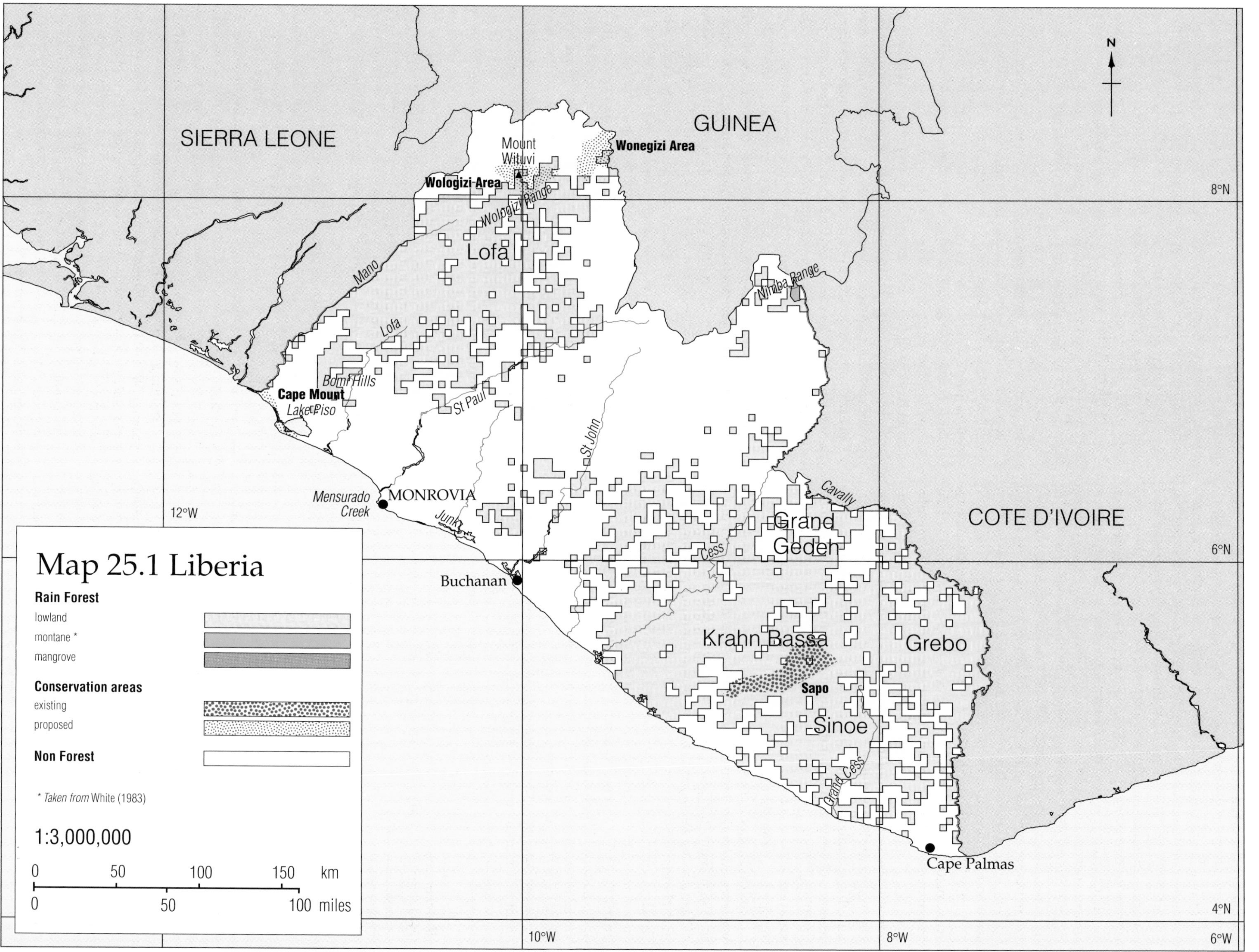
SIERRA LEONE
GUINEA
COTE D'IVOIRE
Mount Wituvi
Wonegizi Area
Wologizi Area
Wologizi Range
Lofa
Mano
Nimba Range
Lofa
Bomi Hills
Cape Mount
Lake Piso
St Paul
St John
Cavally
Mensurado Creek
MONROVIA
Junk
Grand Gedeh
Cess
Buchanan
Krahn Bassa
Grebo
Sapo
Sinoe
Grand Cess
Cape Palmas
N
8°N
6°N
4°N
12°W
10°W
8°W
6°W
Map 25.1 Liberia
Rain Forest
lowland
montane *
mangrove
Conservation areas
existing
proposed
Non Forest
* Taken from White (1983)
1:3,000,000
0 50 100 150 km
0 50 100 miles

Conservation Areas

Liberia had 12 nominally protected national forests, but two have been completely deforested. These cover an area of 14,358 sq. km but have not been shown on Map 25.1 as they are not totally protected. There is only one national park (Table 25.3). The national forests were created by Presidential decree in the late 1950s and early 1960s in areas identified by the then Bureau of Forest and Wildlife Conservation. The 1976 act creating the FDA reaffirmed that the national forests were 'a permanent forest estate made up of reserved areas upon which scientific forestry will be practised'. The boundaries declared at the time of creation excluded all human settlement and agriculture was banned. On the ground, the national forests were demarcated by a cleared line three metres wide and patrolled by forest guards.

In the 1970s and early 1980s, however, the protected area system existed on paper only (Verschuren, 1983). In 1987, the FDA began a major effort to reestablish the national forests using funds generated by the new conservation tax on logging. This realised about US$2 million in 1988. By the end of 1989, the resurvey and cleaning of the boundary lines was complete for the Gola, Gbi, East and West Nimba, Kpelle, Gio, Lorma and North Lorma national forests. A Forest Protection Section was set up in 1987 which in 1989 had 110 forest guards, or one guard for every 127 sq. km of national forest. These guards are under-trained and poorly equipped for the policing job expected of them.

Sapo National Park (see case study), in south-eastern Liberia, was created by the People's Redemption Council Decree No. 73 in 1983. Since the production of a management plan in 1986 (FDA/IUCN, 1986), WWF has been working with the Wildlife and National Parks section of the FDA in the management and development of Sapo National Park and surrounding areas.

Table 25.3 Conservation areas of Liberia

	Existing area (sq. km)	*Proposed area (sq. km)*
National Parks		
Cape Mount[1]*		224
Sapo*	1,308	
Wologizi Area[2]*		202
Wonegizi Area[2]*		261
Totals	1,308	687

(*Source:* WCMC *in litt.*)

* Area with moist forest within its boundaries according to Map 25.1.
[1] Reported to contain mangroves, but these are not shown on Map 25.1.
[2] It is proposed that these three areas, with the 435 sq. km Lorma National Forest form the Loma Conservation Area; the planned status of this area is uncertain at present (S. Anstey, pers. comm.)

Initiatives for Conservation

If the national forests could be effectively conserved, and if the proposed reserves were gazetted, Liberia would have a reasonable network of protected habitats.

In the late 1980s there was some recognition at various government and non-governmental levels in Liberia that survival of the forests required more than just controlling logging. The policing by forest guards simply antagonised other forest users. Forest conservation had to be part of a broader strategy for national land use.

In early 1990 this led the FDA to embark upon a Tropical Forestry Action Plan for Liberia. Among the objectives identified in this exercise were, first, the need for local land-use management units, linked to national forests, which would formulate and implement management plans with the FDA; and, second, the need for substantial improvement of training programmes at all levels to foster conservation expertise. It was felt that much could be learned from the Sapo National Park project. The Society for the Conservation of Nature (SCN) was formed in 1986 and has played a key role in questioning the ways in which Liberia has been exploiting its forests. The SCN now implements field projects and is an effective advocate for conservation.

A nationwide survey of large mammal populations and their use was initiated by WWF and FDA in 1989. The aim of this survey was to produce information on which to base decisions about the revision of hunting regulations and the creation of further protected or managed areas. Initial results on distributions and hunting were presented (Anstey, 1991) before the survey – and most other work in the Liberian forests – was curtailed by the civil war.

References

Anstey, S. G. (1991) *Wildlife Utilisation in Liberia*. Unpublished report to WWF.

Carter, M. F. (1987) *Initial Avifaunal Survey of Sapo National Park, Liberia*. Report to WWF and FDA. Forestry Development Authority, Monrovia. 34 pp.

Collar, N. J. and Stuart, S. N. (1985) *Threatened Birds of Africa and Related Islands. The ICBP/IUCN Red Data Book Part 1*. ICBP/IUCN, Cambridge, UK. 761 pp.

Diamond, A. W. and Hamilton, A. C. (1980) The distribution of forest passerine birds and Quaternary climatic change in Africa. *Journal of Zoology* **191**: 379–402.

FAO (1988) *An Interim Report on the State of Forest Resources in the Developing Countries*. FAO, Rome, Italy. 18 pp.

FAO (1991) *FAO Yearbook of Forest Products 1978–1989*. FAO Forestry Series No. 24 and FAO Statistics Series No. 97. FAO, Rome, Italy.

FAO/UNEP (1981) *Tropical Forest Resources Assessment Project. Forest Resources of Tropical Africa. Part II: Country Briefs*. FAO, Rome, Italy.

FDA (1990) *Annual Report 1989*. Forestry Development Authority, Monrovia, Liberia. 32 pp.

FDA/IDA (1985) *Project Report on Forest Resource Mapping of Liberia*. Report to the Forestry Development Authority. FDA, Monrovia. 18 pp.

FDA/IUCN (1986) *Integrated Management and Development Plan for Sapo National Park and Surrounding Areas in Liberia*. IUCN/WWF, Gland, Switzerland. 66 pp.

Gatter, W. (1989) *The Birds of Liberia: a preliminary check list with status and open questions*. Deutsche Gesellschaft für Technische Zusammenarbeit, Eschborn, Germany.

Hughes, R. H. and Hughes, J. S. (1991) *A Directory of Afrotropical Wetlands*. IUCN, Gland, Switzerland and Cambridge, UK/UNEP, Nairobi, Kenya/WCMC, Cambridge, UK.

IUCN (1990) *1990 IUCN Red List of Threatened Animals*. IUCN, Cambridge, UK and Gland, Switzerland.

Jordan, J. W. (1983) *Fundamentals of Natural Forest Management in the Gola Forest (Liberia)*. Unpublished Masters thesis, University of Goettingen, Germany.

Kunkel, G. (1965) *The Trees of Liberia; Field Notes on the More Important Trees of the Liberian Forests, and a Field Identification Key*. Report No. 3 of the German Forestry Mission to Liberia. Bayerischer Landwirtschafts Verlag, Munich. 270 pp.

SAPO NATIONAL PARK

Sapo National Park (1308 sq. km), in south-eastern Liberia, consists almost entirely of intact rain forest. Characterised by a predominance of the overstorey timber tree *Tetraberlinia tubmaniana*, the forest contains viable populations of most of the forest mammals of Liberia. Sapo is located in Sinoe County which has the lowest population density (six persons per sq. km compared with the national density of 27 people per sq. km) and growth rate (1.6 per cent) in the country (MPEA, 1983). The park itself contains no human settlements, so that pressure on it is mostly brought on by the logging concessions which surround it – through the increased hunting and farming activities necessary to provide for the labour force.

Given the reckless pace of logging in Liberia during the latter half of the 1980s and the political pressure that is at the heart of the concession system, it is remarkable that Sapo has survived intact. This is due to the commitment of a few local and foreign conservationists. Because Sapo is the only national park in Liberia it has been possible for the Wildlife and National Parks section of the Forestry Development Authority to concentrate all its efforts on Sapo. The 30 or so FDA staff have been successful in protecting the area from encroachment. This has been achieved by emphasising extension and education rather than by rigorous policing and with steadily increasing benefits perceived by the people as originating from the existence of the park.

Some villages to the west of the park benefit from a development fund generated from canoe tours for visitors on the Sinoe River organised by the Sapo staff. But the major generation of local support for the park since 1987 has come from the Sapo Agriculture Project which has been supported by WWF and the Society for the Conservation of Nature of Liberia. This project involves village and farm-level groups or individuals in cash crop and food crop tree nurseries and bean farms from which benefits accrue according to labour put in. The project aims, through farmer-managed research, to improve living standards by generating local solutions to the environmental problems of shifting agriculture. By 1990 there were 23 villages around Sapo participating in the tree nursery and bean farm project, some of them receiving no material inputs from the project team – a good sign that this approach is beginning to catch on.

The research camp in Sapo National Park. A. Dunn

MacKinnon, J. and MacKinnon, K. (1986) *Review of the Protected Areas System in the Afrotropical Realm.* IUCN/UNEP, Gland, Switzerland. 259 pp.

MPEA (1983) *Republic of Liberia Planning and Development Atlas.* Ministry of Planning and Economic Affairs, Monrovia.

Poelker, D. A. and Wolf, R. (1989) *Annual Report No. 1 Forest Management Unit.* German Forestry Mission. FDA, Monrovia.

Reitbergen, S. (1989) Africa. In: *No Timber Without Trees, Sustainability in the Tropical Forest.* Poore, D. (ed.). Earthscan Publications, London, UK.

Repetto, R. and Gillis, M. (eds) (1988) *Public Policy and Misuse of Forest Resources.* Cambridge University Press, Cambridge, UK.

Sachtler, M. and Hamer, K. (1967) *Inventory of Krahn-Bassa National Forest and Sapo National Forest.* Technical Report No. 7 of the German Forestry Mission to Liberia. Deutsche Gesellschaft für Technische Zusammenarbeit, Eschborn, Germany.

Taylor, M. E. (1988) Searching for a mongoose. *Species* **11**: 17.

Thiollay, J. M. (1985) *The West African Forest Avifauna: A Review.* ICBP Technical Publication No. 4. ICBP, Cambridge, UK.

Verschuren, J. (1983) *Conservation of Tropical Rainforest in Liberia; Recommendations for Wildlife Conservation and National Parks.* IUCN, Gland, Switzerland. 47 pp.

Voorhoeve, A. G. (1965) *Liberian High Forest Trees.* Centre for Agricultural Publications and Documentation, Wageningen, The Netherlands.

Woell, H. (1986) *Formation of the Forest Management Unit in 'Bomi wood' Concession.* Deutsche Gesellschaft für Technische Zusammenarbeit, Eschborn, Germany.

Zwuen, G. (1988) *Results of the Permanent Sample Plots in Rainforest in Southeastern Liberia.* Research Report No. 2. Forestry Development Authority, Monrovia.

Authorship

James Mayers, WWF with contributions from Simon Anstey, WWF-International and Alexander Peal, FDA, Liberia.

Map 25.1 Forest cover in Liberia

Remaining rain forests of Liberia were digitally extracted from 1989–90 UNEP/GRID data which accompany an unpublished report *The Methodology Development Project for Tropical Forest Cover Assessment in West Africa* (Päivinen and Witt, 1989). UNEP/GEMS/GRID together with the EEC and FINNIDA have developed a methodology to map and use 1 km resolution NOAA/AVHRR-LAC satellite data to delimit forest/non-forest boundaries in West Africa. These data have been generalised for this Atlas to show 2 × 2 km squares which are predominantly covered by forest. The study has also made use of higher resolution satellite data (Landsat MSS and TM, SPOT) and field data from Ghana, Côte d'Ivoire and Nigeria. Forest and non-forest data have been categorised into five vegetation types: forest (closed, defined as greater than 40 per cent canopy closure); fallow (mixed agriculture, clear-cut and degraded forest); savanna (includes open forests in the savanna zone and urban areas); mangrove and water. In addition this dataset shows areas obscured by cloud. In this Atlas, UNEP/GRID's 'forest' and 'mangrove' classifications have been mapped. Delimitation of 'types' of forest shown on Map 25.1 have been made by overlaying White's vegetation map (1983) on to the UNEP/GRID dataset.

Existing and proposed protected areas are taken from an unpublished sketch map produced by the Forestry Development Authority *National Forest and Parks (Liberia)* at a scale of 1:2 million and from spatial data held within files at WCMC.

26 Madagascar

Land area 581,540 sq. km
Population (mid-1990) 12 million
Population growth rate in 1990 3.2 per cent
Population projected to 2020 29.6 million
Gross national product per capita (1988) US$180
Rain forest 41,715 sq. km‡
Closed broadleaved forest (end 1980)* 103,000 sq. km
Annual deforestation rate (1981–5)* 1500 sq. km
Industrial roundwood production† 807,000 cu. m
Industrial roundwood exports† 2000 cu. m
Fuelwood and charcoal production† 7,049,000 cu. m
Processed wood production† 239,000 cu. m
Processed wood exports† nd

* FAO (1988)
† 1989 data from FAO (1991)
‡ Figure partly derived from Map 26.1 and partly from information in Green and Sussman (1990).

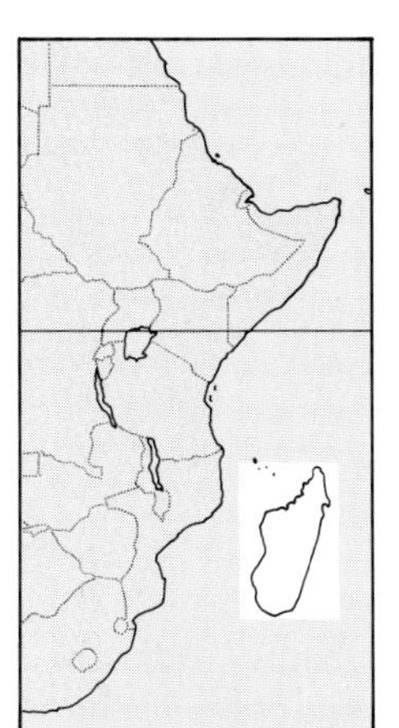

Once described as the 'naturalists' promised land', Madagascar is now considered by some to be the world's most threatened natural area. This is not only because of the extreme richness and unusual diversity of its flora and fauna, most of which is endemic to the island, but because of the acute problem caused by increasing levels of forest clearance. Rain forest is now confined to the east and north of the island, with some small patches remaining in the Sambirano region in the north-west. It is calculated that, within the next 35 years, all but the most inaccessible patches of this forest will have been lost. Deforestation in the country is mostly a result of shifting or 'tavy' cultivation.

In recognition of the importance of Madagascar, there is now considerable conservation activity within the country. The government has developed a National Conservation Strategy and an Environmental Action Plan is being implemented. In support of this, substantial funds have been provided by outside donors to help ensure that Madagascar's unique biological diversity is not lost.

Introduction

With a length of 1600 km and a maximum width of 580 km, Madagascar is the world's fourth largest island, after Greenland, New Guinea and Borneo. It lies in the southern Indian Ocean, separated from mainland Africa by at least 300 km of the Mozambique Channel and extends from 11°57'S to 25°35'S and from 43°14'E to 50°27'E. Because of its large size and diverse climate, geology and vegetation and especially because the island lies on its own continental plate, Madagascar is often regarded as a microcontinent. The origins of the country are still disputed (see Tattersall, 1982 for a brief review). It may have broken away from Africa as long as 165 million years ago and arrived in its present position around 121 million years ago (Rabinowitz *et al.*, 1983).

Broadly, the central part of the island consists of an elevated plateau with a mean altitude of around 1200 m. This has been deeply eroded in some places and built up by volcanic activity in others. The highest peak, at 2876 m, is Mt Maromokotra in the Tsaratanana massif. The rugged central plateau falls off sharply to the east where a strongly eroded escarpment gives way to a narrow (mostly less than 30 km wide) but continuous coastal plain. The slope to the west is generally less steep and this region is composed of numerous sedimentary plateaux that range in elevation from sea level to about 800 m.

Madagascar's climate is highly variable, though predominantly tropical (Donque, 1972). In the east and the Sambirano district of the north-west, annual rainfall is around 2000 mm (but may reach over 5000 mm in places), with little seasonal variation except that September and October may be slightly drier. Mean annual temperature in the lowland areas of these regions is 26°C, though this generally declines from north to south. The Central Plateau has a lower annual rainfall, around 1500 mm, and lower mean temperatures (16–20°C). In addition, for areas above 1300 m, a drop in temperature of 0.6°C for every 100 m rise in altitude occurs. Annual rainfall in the western part of the island declines from around 1500 mm in the north to 500–1000 mm in the south and most of this, as in the centre of the island, falls between October and April. Temperatures are generally a little higher on the west coast than at the same latitude on the east. The extreme south of the island is hot and semi-arid with infrequent rainfall averaging around 300–500 mm per year.

The prevalence of cyclones is of considerable environmental importance in Madagascar. The great majority occur between mid-January and mid-March, with most hitting the island along the north-eastern coast. They can cause great devastation with winds up to 300 km per hour and rainfall of 600–700 mm falling in four or five days. Massive destruction of crops and forests, as well as large scale flooding, are often the result of these cyclones.

The origins of the Malagasy people are complex and incompletely understood. They appear to have arrived as recently as 2000 years ago in waves of migration from both Indonesia and Africa. There have been Arab influences in the country since the 12th century and contact with Europeans since the 16th century. The population of Madagascar is estimated to be increasing at a rate of 3.2 per cent; it has more than doubled in 30 years, from 5.4 million in 1960 to 12 million in 1990 and it is calculated that there will be around 30 million people in the country by 2020. Distribution of this population is very uneven, with a density of more than 100 inhabitants per sq. km on the east coast, approximately 60 people per sq. km on the central highlands (although many of these are concentrated in urban areas) and falling to only five or so per sq. km on the west coast. Around 80 per cent of the population live in rural areas though the capital, Antananarivo, contains over one million inhabitants.

The Forests

Opinions are divided as to whether Madagascar was totally covered in forest when man arrived on the island. It appears likely that there was an area of savanna and woodland in the Central Plateau (MacPhee *et al.*, 1985) but some authors, following Perrier de la Bathie (1921) and Humbert (1927), consider that the entire surface of the island was more or less completely covered in forest as recently as 2000 years ago. Very briefly, the country can be divided into two major floristic zones, a moister Eastern Region and a dry Western area (Perrier de la Bathie, 1936; Humbert, 1959; White, 1983) and within these is a wide range of habitats. The Eastern Region covers just over half the island, extending westwards from the east coast to cover the central highlands; it also includes a small enclave, the Sambirano Domain, on the north-west coast. This unit, which contains the only forest type to be considered in this Atlas, was probably originally all forested, but much of it has now been replaced by a mosaic of cultivation and secondary or degraded formations. The Western Region extends from the flat plains on the west coast eastwards up to about 800 m. Within this area are the dry deciduous forests, less dense and with a lower, more open canopy than in most of the moister eastern forests (Figure 26.1). Lusher forests grow alongside rivers. Included within the Western Region biogeographical zone is the semi-arid Southern Domain. This is characterised by thickets or forests of bushy, drought-resistant vegetation, with Euphorbiaceae and the endemic Didiereaceae families predominating; it is frequently referred to as 'spiny forest' (Figure 26.1).

In the east there is low altitude dense rain forest (originally from sea level to 800 m) with a canopy at around 25–30 m or less. Large emergent trees are uncommon. The most widely represented families here are the Euphorbiaceae, Rubiaceae, Sapindaceae, Anacardiaceae, Arecaceae, Fabaceae, Flacourtiaceae and Lauraceae (Koechlin, 1972). There is also a middle stratum of small trees and large shrubs composed of families such as Ochnaceae, Araliaceae, Violaceae and Tiliaceae. Epiphytes, particularly ferns and orchids, are common. Generally, the forests in this region are very rich in species and no individual plant species dominates. The structure of the forest in the Sambirano region is similar though the floristic composition is slightly different; there are, in particular, numerous Sarcolaenaceae (Koechlin, 1972).

The medium altitude (or moist montane, to use White's 1983 category) rain forest grows chiefly between 800 and 1300 m asl but may reach 2000 m in sheltered places. This is a dense formation composed of a large number of species most of which are evergreen. The canopy at 20–25 m in height is not as high as in the lowland forest and there is an undergrowth with more plentiful shrubs and herbaceous plants. In the top stratum, *Tambourissa*, *Weinmannia*, *Symphonia*, *Dalbergia* and *Vernonia* are among the best represented genera (Koechlin, 1972). Epiphytes, including mosses, lichens, ferns and orchids, are abundant and are a striking feature of this forest type.

At higher altitudes (generally 1300–2000 m or above), sclerophyllous montane forest predominates. Composed of small-leaved, very twisted trees of only 10–13 m in height, its structure is intermediate between forest and thicket. Mosses and lichens coat the trees and carpet the ground. *Dicoryphe viticoides*, *Tina isoneura*, *Alberta minor* and *Rhus taratana* are some of the species typical of this region (Koechlin, 1972). Gymnosperms (*Podocarpus* spp.) and bamboo may form pure stands in some places. This type of forest is only slightly less susceptible to fire than the bushy vegetation at higher altitudes. This sclerophyllous montane vegetation has not been differentiated from the moist montane forest on Map 26.1.

The montane vegetation, above 2000 m, on large isolated massifs such as Tsaratanana, Marojejy, Ankaratra and Andringitra, is generally composed of ericoid bush or herbaceous plants which are adapted to the extremes of humidity and temperature that they experience at this altitude.

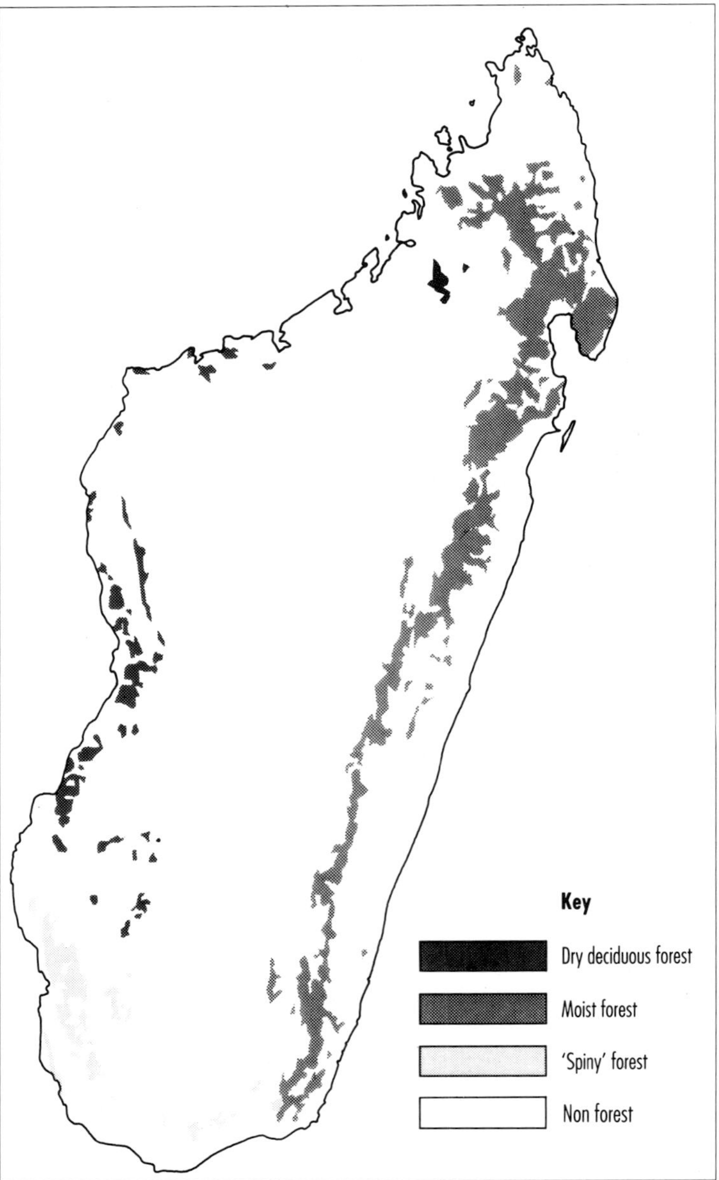

Figure 26.1 Vegetation types in Madagascar showing the western dry deciduous forest and the southern spiny forest as well as the moist forest. Mangrove has been excluded.

The eastern rain forest in this figure has been derived from Green and Sussman (1990), while other vegetation types are from Faramalala (n.d., see Map Legend). Her vegetation categories no. 7: 'Forêts denses sèches' (dry deciduous forests) and no. 27: 'Forêt dense sèche à *Didierea* et *Euphorbia*' (spiny forest) are shown here.

Mangroves

Of the estimated 3300 sq. km of mangrove (Keiner, 1972) in Madagascar, most (97 per cent) are found along the west coast. They are located mainly between Morombe and Antsiranana with some sites in the south-west around Toliara and in the north-west near Iharana. The largest area, of about 460 sq. km, is in the Bay of Bombetoka. The Mangoky, Tsiribihina and Mahajamba rivers also have large areas of mangroves at their mouths. Along the east coast, there are only 11 mangrove sites of any size, the largest, of around 22.2 sq. km, being in Rodo Bay (SECA/CML, 1987). Map 26.1 shows 3315 sq. km of mangrove on the west coast but almost none remaining on the east coast. However, on the source map for mangroves (see Map Legend on p. 229), most of the east coast was obscured by cloud.

The mangroves in Madagascar are floristically more diverse than those in continental Africa. Six plant genera are widespread within the mangrove areas: *Rhizophora*, *Bruguiera*, *Ceriops*, *Avicennia*, *Sonneratia*, and *Carapa* (SECA/CML, 1987). The mangroves are important for inshore fisheries, serving as nursery areas for many species of fish and crustacea. They are also resting and nesting areas for a number of bird species including the Madagascar fish eagle *Haliaeetus vociferoides* and the swift tern *Sterna bergii* (Nicoll and Langrand, 1989). The largest of Madagascar's fruit bats, *Pteropus rufus*, forms large colonies within the mangroves.

Madagascar's mangroves, unlike those of mainland Africa, are not at present exploited to any great extent (SECA/CML, 1987; Nicoll and Langrand, 1989). The greatest threat is that they are becoming silted-up as the extensive inland erosion is resulting in large quantities of soil being carried down the rivers into the mangroves. A shortage of fuelwood may mean that the mangroves are cut for this purpose, especially in areas of moderate to high population density. None of the mangrove sites is protected.

Forest Resources and Management

Even now, there are no accurate figures for the extent of surviving tree cover in Madagascar, though it is usually said that approximately 20 per cent of the island is covered with forest. This is the figure estimated by Guichon (1960), based chiefly on aerial surveys made in the late 1940s. Chauvet (1972), deriving his figures from Guichon's work, calculated that there were 61,500 sq. km of 'eastern type' forest remaining at that time. In 1981, FAO/UNEP gave a figure of 69,550 sq. km of forest left in the east and Sambirano region. The much higher figure given by FAO (1988) of 103,00 sq. km includes all closed broadleaved forest. It thus includes dry deciduous forest in the west and gallery forest throughout the country, as well as the moist forests. The only recent figure for remaining forest area is from a study by Green and Sussman (1990) covering the eastern forests but excluding the Sambirano. Using satellite imagery, they estimated that, in 1985, only 38,000 sq. km of forest remained in the east of Madagascar. Including the Sambirano would not increase this figure to any great extent as recent reports indicate that the forests of this region do not exceed 350–400 sq. km in area and that they are fragmented and rapidly decreasing in size (see, for example, Nicoll and Langrand, 1989; Andrews, 1990).

The eastern rain forests shown on Map 26.1 were digitised from a copy of a map in Green and Sussman (1990). The map indicates 25,820 sq. km of lowland rain forest and 23,000 sq. km of montane forest, giving a total of 48,820 sq. km of dryland forest remaining in 1985. However, this figure is some 28.5 per cent higher than that which Green and Sussman derived from the same data. The discrepancy results from the poor quality of the source map and the consequent generalisation of the forest cover that occurred when it was digitised. Due to the unreliability of the data, the statistic for forest area shown at the beginning of this chapter is only approximate, and there is no table of extent of rain forest types within the chapter. A figure of 41,715 sq. km of moist forest remaining in the country has been calculated for Tables 9.6 and 10.1 by adding to Green and Sussman's (1990) figure of 38,000 sq. km of eastern forest, an estimated area of 400 sq. km of forest around Sambirano and 3315 sq. km of mangroves measured from Map 26.1.

Legal protection of forests in Madagascar began around 200 years ago. In King Andrianapoinimerina's reign (1787–1810) those deforesting an area were liable to be punished. Later, under the ancient Hova Kingdom, anybody found cutting down trees was condemned to be chained in irons. The Direction des Eaux et Forêts is now responsible for the management of the forests and generally employs less drastic punishments.

Reforestation projects have been undertaken for several decades, especially on the Central Plateau. One of the earliest projects was the planting of *Eucalyptus* along the railway from Antananarivo in 1910 to provide fuelwood for the locomotives. A 1981 FAO/UNEP report estimated that there were 1120 sq. km of plantations, composed mainly of pines, with the trees being used for industrial timber. In addition, the report calculated that there were 1540 sq. km of plantations where the timber was used mostly for firewood; these were frequently composed of broadleaved trees. There has been no significant replanting of native trees.

Deforestation

In their recent report on the eastern rain forests, Green and Sussman (1990) calculated that between the years of 1950 and 1985 the rate of deforestation in this region alone has been 1110 sq. km per year (see Figure 26.2). The estimated figure given by FAO, 1500 sq. km per year, is for all closed broadleaved forests and is, therefore, probably an underestimate. Green and Sussman (1990) consider that, if destruction of forests continues at the present rate, only those on the steepest slopes will survive the next 35 years.

Figure 26.2 Distribution of rain forest in Madagascar originally, in 1950 and in 1985 (*Source:* Green and Sussman, 1990)

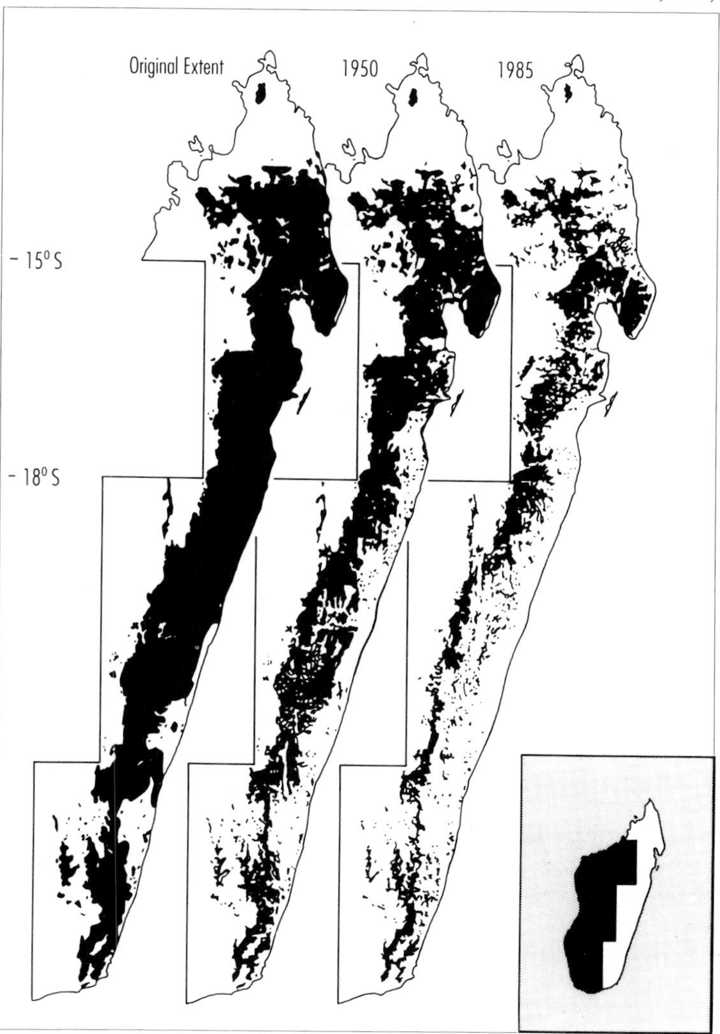

These logs were all cut by hand and individually carried out of the steep rain forest near Ranomafana, from an area inaccessible to vehicles. C.Harcourt

The principal agent of destruction of the eastern moist forests is 'tavy' or slash and burn agriculture. Madagascar's population is still mostly rural and the people depend on agriculture for their survival. To obtain more land, forest areas are cleared, the vegetation is allowed to dry and is burned some months later. Dryland rice is the most common crop, but maize, manioc and others are also cultivated. These are grown for a year or two and then the land is left fallow and the process repeated elsewhere. Degraded vegetation types regrow on the deserted plots which used to be recleared after an interval of ten years or more. Tavy has been practised for centuries, but the increase in the population has put greater pressure on the land and the fallow periods often now last only three or four years. As a result of progressive deterioration of the soil structure and nutrient content, the land degrades to unproductive grassland or becomes weed infested. Tavy cultivation is often practised on steep slopes where there is great risk of erosion and, consequently, the land becomes unusable even more rapidly. Deforestation has been swiftest in areas with low relief and high population density (Green and Sussman, 1990).

Some timber is removed by logging companies but this is not currently a major threat in Madagascar as much of the remaining forest is in steep, isolated areas inaccessible to heavy machinery. Of the 7,856,000 cu. m of roundwood estimated to be cut in Madagascar each year, only 807,000 cu. m are for industrial use, while the majority is for fuel and charcoal (FAO, 1991). These figures are for wood cut from all over the country, not just that taken from the moist forests.

FAO/UNEP (1981) suggests that there has been little deforestation of the montane regions above 2000 m. However, although relatively safe from use by local people, these areas are highly susceptible to fire damage (Jenkins, 1987). The montane forests are now very limited in extent and require special conservation measures.

Biodiversity

Madagascar has been separated from continental Africa for many millions of years and, therefore, until the arrival of man 1500–2000 years ago, its flora and fauna had an independent evolutionary history. As a result, they are diverse and unique, being characterised by high levels of endemism at species and higher taxonomic levels. For the size of the country, Madagascar's flora is one of the richest in the world with at least 8500 plant species (White, 1983) of which around 80 per cent are endemic (Humbert, 1959). Some faunal groups are comparatively impoverished, but the degree of endemism is again very high. There is high species diversity of both plants and animals in the rain forests of eastern Madagascar. Conspicuous examples of endangered Malagasy plants are the tree ferns (Cyatheaceae), which are used as containers for potted plants. The Malagasy ebonies *Diospyros perrieri* and *D. microrhombus* and palissandre (*Dalbergia* spp.) have been heavily exploited for their wood so that there are few large trees of these species remaining (Dorr *et al.*, 1989).

Madagascar's mammalian fauna is relatively impoverished. Indeed, it has 108 mammal species, fewer than any other African country of comparable size. The native living land mammals belong to only five orders: Primates, Carnivora, Rodentia, Insectivora and Chiroptera. Bushpig *Potamochoerus porcus* occurs but was probably introduced and dugong *Dugong dugon* is found in coastal waters. All 30 species of primates are lemurs, none of which is found outside Madagascar and the Comoros (the two species on the Comoros were almost certainly introduced by man from Madagascar). There were at least 14 other lemur species on the island, all of which have become extinct since humans arrived there. Most of those surviving today are considered to be in danger (Harcourt and Thornback, 1990). Habitat destruction is the main threat to the lemurs but hunting also occurs. There are eight species of carnivore, all unique to the island and all belonging to the Viverridae; only one of these is not considered to be threatened (Schreiber *et al.*, 1989). There are between 10 and 16 species of rodents, depending upon the taxonomy employed, and all are endemic. In the most recent classification of Madagascar's insectivores, there are considered to be 24 species of endemic tenrecs (18 of which are rare or threatened species) and two species of

A family group, male, female and infant, of the nocturnal lemur Avahi laniger laniger, *in Madagascar's eastern rain forest.* C. Harcourt

shrews (Nicoll and Rathbun, 1990). There are 28 bat species in the country, of which nine are endemic. There used to be two or three species of hippopotamus and a large termite-eating tenrec, resembling an aardvark, but these became extinct sometime after man's arrival on the island.

Similarly, the number of bird species found on Madagascar is low. There are 250 species of which 198 are non-introduced resident species and 53 per cent of these are endemic (Langrand, 1990). The huge, flightless elephant bird *Aepyornis maximus*, used to occur on the island but this too, along with other flightless species, has vanished within the last few hundred years.

In contrast to the mammals and birds, the reptile and amphibian fauna is rich and over 90 per cent of these groups are endemic (Jenkins, 1987). There are about 260 reptile species which include one crocodile, 13 tortoises and turtles, 60 snakes and approximately 180 lizards. Madagascar contains two-thirds of the world's chameleon species, including the smallest (thumbnail size) and the largest (60 cm in length). Around 150 amphibian species (all anurians) are to be found in the country.

As in many other countries, the invertebrate fauna has not been studied in detail but there are known to be at least 260 species of butterfly in the country and 182 of these are endemic (Jenkins, 1987). Species richness and endemicity tend to be higher in the montane and forested areas than in the lowland and arid regions (Jenkins, 1987).

Conservation Areas

Madagascar has one of the oldest protected area networks in the African region, with ten strict nature reserves (Réserves Naturelles Intégrales) dating back to 1927. There are now 11 nature reserves, six national parks (one of these is a marine park and not mapped or shown on Table 26.1) and 23 special reserves (Réserves Spéciales) in the country. This system of protected areas (Table 26.1) is quite comprehensive, covering around 11,200 sq. km (approximately 1.9 per cent of Madagascar's land area) and containing a good, although incomplete, cross-section of key ecosystems. Unfortunately, much of this network exists only on paper. A shortage of trained personnel and finances has meant that it is impossible to ensure that the reserves and parks are adequately safeguarded. In addition, many areas are too small to be viable and are surrounded by degraded or agricultural land.

Six different categories of protected areas are legally recognised but only the three mentioned above have been used to protect natural ecosystems or threatened species. Most strictly protected, on paper, are the 11 nature reserves (a twelfth strict nature reserve, on Masoala Peninsula, was degazetted in 1964). Access to these is forbidden other than for scientific purposes. Entry to the national parks is controlled, but in Isalo National Park rights are accorded to neighbouring villagers for the collection of *Boroceras* sp. cocoons and *Uapaca bojeri* fruits. The 23 special reserves have generally been set up to protect particular plant or animal species. Although access is unlimited, hunting, fishing, pasturing of livestock, collection of natural products and introduction of any plants or animals is forbidden.

In the forest reserves (Forêts Classées), of which there are 158, all forest exploitation is forbidden but local people can collect some products such as honey and raffia. These forests may be exploited in the future: they are protected for economic rather than conservation purposes. The Reafforestation and Restoration Zones (Périmètres de Reboisement et de Restauration) are intended primarily to protect watersheds and prevent erosion. There are 77 such zones at present. Land-use is regulated within these areas so that management of pasture, tree planting and anti-erosion measures are carried out. There are also four hunting reserves (Réserves de Chasse) where hunting is seasonally closed but the public has free access.

The reserves are essential for conservation of the country's rich biological diversity, as was recognised by the Malagasy government during its 1985 Conference on the Conservation of Madagascar's Natural Resources for Development (Rakotovao *et al.*, 1988). Administration of the protected areas is now the responsibility of the National Association for Management of Protected Areas. This has been established with support from the World Bank-sponsored Environmental Action Plan.

Table 26.1 Conservation areas of Madagascar

Conservation areas are listed below. Forest reserves and marine parks are not included or mapped. For data on Biosphere reserves and World Heritage sites see chapter 9.

	Area (sq. km)
National Parks	
Isalo	815
Mananara*	230
Mantadia*	100
Montagne d'Ambre*	182
Ranomafana*	400
Strict Nature Reserves	
Andohahela*	760
Andringitra*	312
Ankarafantsika	605
Betampona*	22
Lokobe	7
Marojejy*	602
Tsaratanana*	486
Tsimanampetsotsa	432
Tsingy de Bemaraha	1,520
Tsingy de Namoroka	217
Zahamena*	732
Special Reserves	
Ambatovaky*	601
Ambohijanahary	248
Ambohitantely	56
Analamazaotra (Périnet)	8
Analameran	347
Andranomena	64
Anjanaharibe-Sud*	321
Ankarana	182
Bemarivo	116
Beza-Mahafaly	6
Bora	48
Cap Sainte Marie	18
Forêt d'Ambre*	48
Kalambatritra	283
Kasijy	188
Mangerivola*	119
Maningozo	79
Manombo*	50
Manongarivo*	353
Marotandrano*	422
Nosy Mangabe*	5
Pic d'Ivohibe*	35
Tampoketsa d'Analamaitso	172
Total	11,191

(*Sources:* Nicoll and Langrand, 1989; WCMC, *in litt.*)

* Area containing moist forest within its boundaries as shown on Map 26.1.
† Not mapped – data not available to this project.

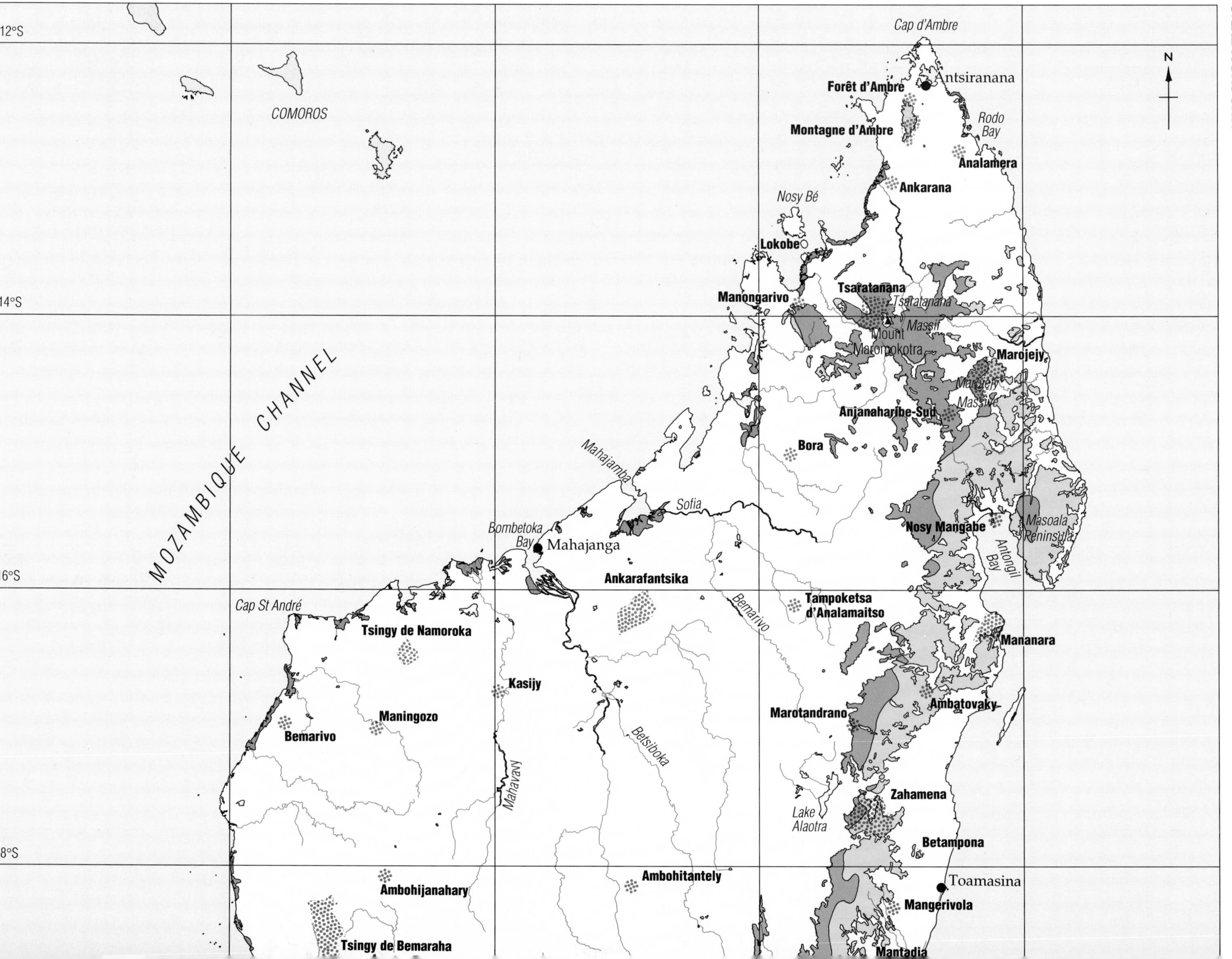
N
12°S
14°S
16°S
18°S
COMOROS
MOZAMBIQUE CHANNEL
Cap d'Ambre
Antsiranana
Forêt d'Ambre
Montagne d'Ambre
Rodo Bay
Analamera
Ankarana
Nosy Bé
Lokobe
Manongarivo
Tsaratanana
Tsaratanana Massif
Mount Maromokotra
Marojejy
Marojejy Massif
Anjanaharibe-Sud
Bora
Mahajamba
Sofia
Nosy Mangabe
Masoala Peninsula
Antongil Bay
Bombetoka Bay
Mahajanga
Ankarafantsika
Cap St André
Tsingy de Namoroka
Tampoketsa d'Analamaitso
Bemarivo
Mananara
Kasijy
Maningozo
Bemarivo
Marotandrano
Ambatovaky
Betsiboka
Mahavavy
Zahamena
Lake Alaotra
Betampona
Ambohijanahary
Ambohitantely
Toamasina
Mangerivola
Tsingy de Bemaraha
Mantadia

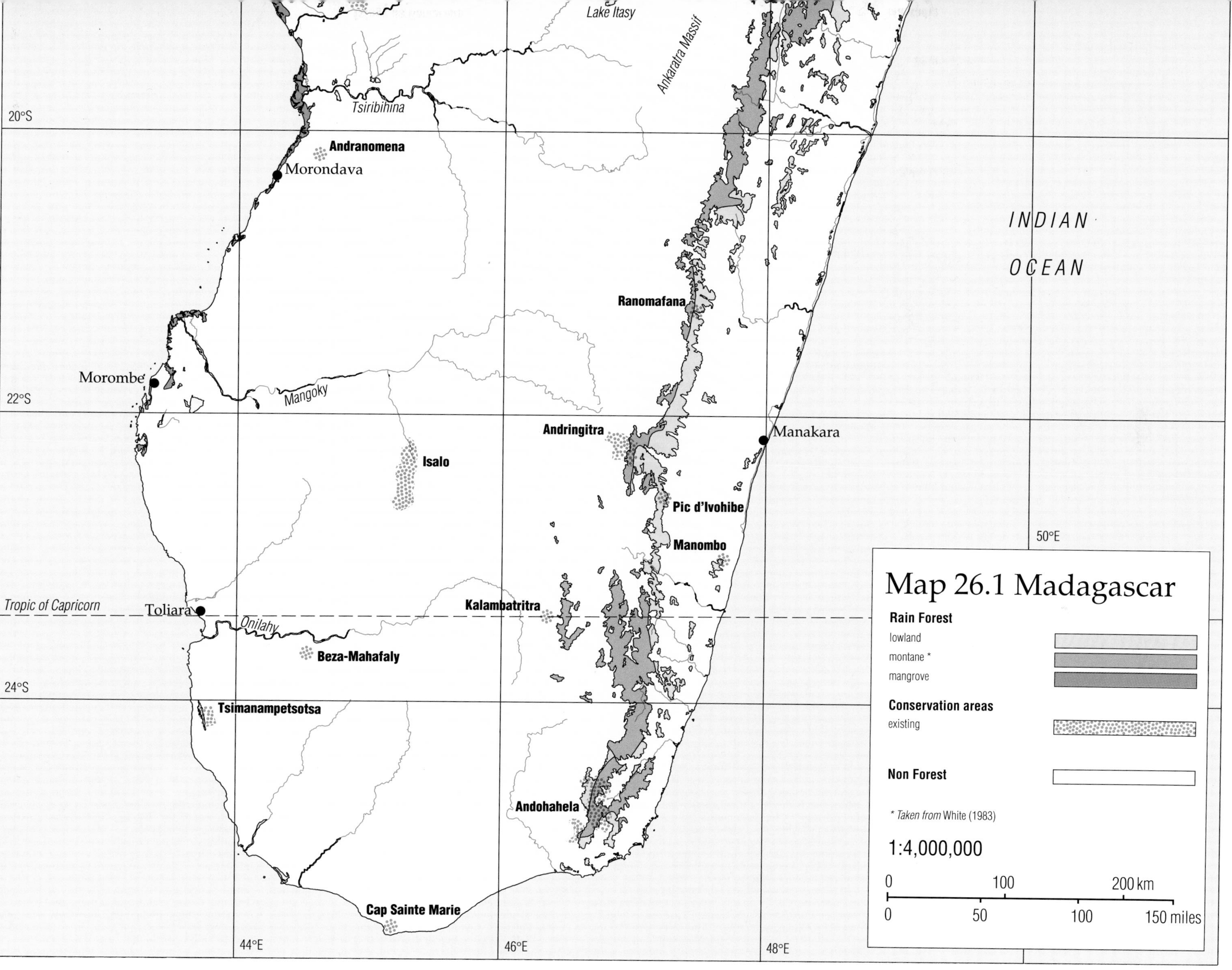
Lake Itasy
Ankaratra Massif
Tsiribihina
20°S
Andranomena
Morondava
INDIAN
OCEAN
Ranomafana
Morombe
Mangoky
22°S
Andringitra
Manakara
Isalo
Pic d'Ivohibe
50°E
Manombo
Tropic of Capricorn
Toliara
Kalambatritra
Onilahy
Beza-Mahafaly
24°S
Tsimanampetsotsa
Andohahela
Cap Sainte Marie
44°E
46°E
48°E
Map 26.1 Madagascar
Rain Forest
lowland
montane *
mangrove
Conservation areas
existing
Non Forest
* Taken from White (1983)
1:4,000,000
0 100 200 km
0 50 100 150 miles

Madagascar

Initiatives for Conservation

The government of Madagascar developed a National Strategy for Conservation for Development in 1984 and now, with the help of several international agencies, including the World Bank, it has prepared an Environmental Action Plan which is presently being implemented. This aims to develop human resources in the country, to conserve and manage the biological diversity, to improve rural and urban living conditions and to promote sustainable development by better management of natural resources (World Bank, 1988). Over US$100 million has been provided for this plan by outside donors.

In 1986, a project to survey the protected areas was set up through WWF's programme in Madagascar in collaboration with a number of government ministries. The aims of this project were to:

- Evaluate the existing protected areas
- Develop and implement management plans for priority protected areas
- Recommend the establishment of new protected areas in key regions, and
- Train Malagasy counterparts in protected area management and conservation biology.

The review of the protected areas and the recommendations to improve the existing reserves and to set up new ones are outlined in a WWF publication by Nicoll and Langrand (1989). Four new national parks have been gazetted since this book was published, which is an indication of the Malagasy government's commitment to conserve its country.

WWF also funds a number of other projects including environmental education, a programme of conservation in southern Madagascar (involving Beza-Mahafaly and Andohahela reserves) and in the north (Mt d'Ambre National Park), a recovery programme for the ploughshare tortoise *Geochelone yniphora*, a study of the ecology and conservation of the diademed sifaka *Propithecus diadema* and the captive breeding of lemurs at Ivolina, near Toamasina. Other non-Malagasy organisations involved in one or more of these projects are Yale University, Washington University, Duke University Primate Center, the Smithsonian Institution, USAID, Unesco and Jersey Wildlife Preservation Trust (WWF, 1990). A Biodiversity Planning Centre is to be set up in the country (Smith *et al.*, 1990). USAID has agreed to fund the Geographic Information System (GIS) to be used in the Centre.

Conservation International has recently set up an office in Antananarivo and intends to become involved with numerous conservation projects in the country (Conservation International, 1990). These will include biological inventories, vegetation mapping, land-use planning, research on particular species and data management, as well as support for local NGOs, training of Malagasy scientists and education programmes for local people.

Duke University and the Duke Primate Center have been very involved with the gazetting of Ranomafana as a national park and in setting up a long-term study site there. Missouri Botanical Garden is particularly concerned with ensuring that a protected area is gazetted on the Masoala Peninsula (see case study). Other projects that seek to integrate protection of natural resources and biodiversity with rural development, are located at Ranomafana National Park, Montagne d'Ambre National Park, the Mananara North Biosphere Reserve and at Beza-Mahafaly and Andohahela Nature Reserves.

The World Bank is funding a project for the management and protection of forests (Gestion et Protection des Forêts – GPF). This project is supporting improved management of several protected areas and the training of reserve staff. Unesco and UNDP are collaborating with the GPF, especially in the development of tourism and rural development in buffer zones. The Ankarafantsika Nature Reserve is one of the more important sites coming under this project.

The forests between Andranomena and the Tsiribihina River on the west coast of Madagascar near Morondava are being managed by a project supported by Swiss development aid. The forests are being managed for sustained yield timber production. Agricultural and agroforestry activities in the buffer zones aim to reduce pressure on the forests from people in surrounding areas.

Madagascar is one of the few countries that has a successful debt-for-nature swap programme, initiated in 1989 by the Malagasy Central Bank in collaboration with WWF, USAID, and the Ministry of Agricultural Production, Water and Forests (MPAEF).

Masoala Peninsula

The Masoala Peninsula in the north-east of Madagascar contains some of the country's largest remaining tracts of undisturbed eastern rain forest, including perhaps the only stands extending down to sea level. The region appears to have an exceptionally high level of biodiversity, with many plant and animal species occurring only there. For instance, on a recent expedition to the peninsula two genera and seven species of previously unknown palms were discovered in just three weeks. In addition it is here, and only here, that the red-ruffed lemur *Varecia variegata rubra*, is found. At present none of the peninsula is protected, but there are plans to gazette a national park in the area which would include at least 2000 sq. km of forest. Missouri Botanical Garden and the Agricultural Development Department of the Malagasy Lutheran Church (SA.FA.FI.) in conjunction with MPAEF, have put forward a number of proposals for the conservation of the peninsula's forests (Missouri Botanical Garden and SA.FA.FI., n.d.).

They plan to implement a rural development programme that is adapted to local needs and conditions but is also linked to natural resource conservation. Local people will be trained in agriculture, land-use and conservation management so that they can carry out the proposed projects. The preparation of a comprehensive inventory of the region's biological resources is under way. They hope that the work on the Masoala Peninsula will provide an example of the successful integration of development and conservation, following the model that is being widely used in other protected areas.

The primary benefits expected from the project are:

- Increased productivity and efficiency in local agricultural production, which will help provide food self-sufficiency.
- Improved fisheries exploitation in coastal areas.
- Availability of basic health care and health education.
- Reduction in the rate of degradation of the area's forests.
- Increased public awareness of the immediate and long-term relationship between local economic conditions and land-use practices.
- Increased employment and stimulation of the local economy.

References

Andrews, J. R. (1990) *A Preliminary Survey of Black Lemurs,* Lemur macaco, *in North West Madagascar.* Unpublished final report of the Black Lemur Survey 1988.

Chauvet, B. (1972) The forests of Madagascar. In: *Biogeography and Ecology in Madagascar. Monographiae Biologicae 21,* pp. 191–9. Battistini, R. and Richard-Vindard, G. (eds). Junk, The Hague, The Netherlands.

Conservation International (1990) *A Program of Conservation for Madagascar.* Unpublished proposal. Conservation International, Washington, DC, USA. 25 pp.

Donque, G. (1972) The Climatology of Madagascar. In: *Biogeography and Ecology in Madagascar. Monographiae Biologicae 21,* pp. 87–144. Battistini, R. and Richard-Vindard, G. (eds). Junk, The Hague, The Netherlands.

Dorr, L. J., Barnett, L. C. and Rakotozafy, A. (1989) Madagascar. In: *Floristic Inventory of Tropical Countries.* Campbell, D. G. and Hammond, D. (eds). New York Botanical Garden, New York, USA.

FAO (1988) *An Interim Report on the State of Forest Resources in the Developing Countries.* FAO, Rome, Italy. 18 pp.

FAO (1991) *FAO Yearbook of Forest Products 1978–1989.* FAO Forestry Series No. 24 and FAO Statistics Series No. 97. FAO, Rome, Italy.

FAO/UNEP (1981) *Tropical Forest Resources Assessment Project. Forest Resources of Tropical Africa. Part II: Country Briefs.* FAO, Rome, Italy.

Green, G. M. and Sussman, R. W. (1990) Deforestation history of the eastern rain forests of Madagascar from satellite images. *Science* **248**: 212–15.

Guichon, A. (1960) La superficie des formations forestières à Madagascar. *Revue Forestière Française* **6**: 408–11.

Harcourt, C. and Thornback, J. (1990) *Lemurs of Madagascar and the Comoros. The IUCN Red Data Book.* IUCN, Gland, Switzerland and Cambridge, UK.

Humbert, H. (1927) Déstruction d'une flore insulaire par le feu. Principaux aspects de la végétation à Madagascar. *Mémoires de L'Académie Malagache* **V**: 1–80.

Humbert, H. (1959) Origines présumées et affinities de la flore de Madagascar. *Mémoire d'Institut Science Madagascar Séries b (Biologique et Végétation)* **9**: 149–87.

Jenkins, M. D. (1987) *Madagascar: an Environmental Profile.* IUCN/UNEP/WWF. IUCN, Gland, Switzerland and Cambridge, UK. 374 pp.

Keiner, A. (1972) Ecologie, biologie et possibilités de mise en valeur des mangroves malagaches. *Bulletin Madagascar* **308**: 49–84.

Koechlin, J. (1972) Flora and vegetation of Madagascar. In: *Biogeography and Ecology in Madagascar. Monographiae Biologicae 21,* pp. 227–59. Battistini, R. and Richard-Vindard, G. (eds). Junk, The Hague, The Netherlands.

Langrand, O. (1990) *Guide to the Birds of Madagascar.* Yale University Press, New Haven, USA.

MacPhee, R. D. E., Burney, D. and Wells, N. A. (1985) Early Holocene chronology and environment of Ampasambazimba, a Madagascar subfossil lemur site. *International Journal of Primatology* **6**(5): 463–89.

Missouri Botanical Garden and SA.FA.FI. (n.d.) *An Integrated Project for Rural Development, Conservation and Biological Inventory of the Masoala Peninsula, Northeastern Madagascar.* Unpublished report.

Nicoll, M. E. and Langrand, O. (1989) *Madagascar: Revue de la Conservation et des Aires Protégées.* WWF, Gland, Switzerland.

Nicoll, M. E. and Rathbun, G. B. (1990) *African Insectivores and Elephant-Shrews: An Action Plan for their Conservation.* IUCN/SSC Insectivore, Tree-Shrew and Elephant-Shrew Specialist Group. IUCN, Gland, Switzerland.

Perrier de la Bathie, H. (1921) La végétation malagache. *Annales du Musée Colonial de Marseille (3 séries)* **9**: 1–268.

Perrier de la Bathie, H. (1936) *Biogéographie des Plantes de Madagascar.* Paris, France.

Rabinowitz, P. D., Coffin, M. F. and Falvey, D. (1983) The Separation of Madagascar and Africa. *Science* **220**: 67–9.

Rakotovao, L., Barre, V. and Sayer, J. (1988) *L'Equilibre des Ecosystèmes forestiers à Madagascar: Actes d'un Séminaire International.* IUCN, Gland, Switzerland and Cambridge, UK. 344 pp.

Schreiber, A., Wirth, R., Riffel, M. and Van Rompaey, H. (1989) *Weasels, Civets, Mongooses and their Relatives: An Action Plan for the Conservation of Mustelids and Viverrids.* IUCN/SSC Mustelid and Viverrid Specialist Group. IUCN, Gland, Switzerland.

SECA/CML (1987) *Mangroves of Africa and Madagascar: the Mangroves of Madagascar.* Sociétée d'Eco-aménagement, Marseilles, France and Centre for Environmental Studies, University of Leiden, The Netherlands. Unpublished report to the European Commission, Brussels.

Smith, A. P., Horning, N., Oliverieri, S. and Andrianifahnana, L. (1990) *Feasibility Study for Establishment of a Biodiversity Planning Service in Madagascar.* University of New England, Armidale, Australia.

Tattersall, I. (1982) *The Primates of Madagascar.* Columbia University Press, New York, USA.

White, F. (1983) *The Vegetation of Africa: a descriptive memoir to accompany the Unesco/AETFAT/UNSO vegetation map of Africa.* Unesco, Paris, France. 356 pp.

World Bank (1988) *Madagascar Environmental Action Plan. Volume 1: General Synthesis and Proposed Actions.* Preliminary report.

WWF (1990) *WWF List of Approved Projects. Volume V Africa and Madagascar.* WWF, Gland, Switzerland.

Authorship

Caroline Harcourt in Cambridge with contributions from Martin Jenkins, Cambridge; Olivier Langrand, WWF-Madagascar; David Stone, Gland, Switzerland and Peter Lowry, Missouri Botanical Garden, USA.

Map 26.1 Forest cover in Madagascar

Vegetation data are taken from two sources. Mangroves are extracted from *Carte des Formations Végétales de Madagascar*, a 1:1 million unpublished map prepared by Faramalala Miadana Harisoa – I.C.I.V. (n.d.). Vegetation category no. 12 'Zone à mangrove' has been included on Map 26.1. The eastern rain forests are from a map compiled from 1985 satellite imagery which accompanies Green and Sussman (1990). A photocopy of the map in this report was supplied by G. Green and this was used to show the extent of the eastern rain forest on Map 26.1. It indicates a total forest cover about 28.5 per cent greater than that reported by Green and Sussman from the same data. The discrepancy apparently results from the poor quality of the photocopy and the generalisation of the map that occurred when it was digitised. Vegetation types have been harmonised with White's (1983) vegetation map.

Conservation areas are taken from a map series *Carte de Madagascar* in 12 sheets at a scale of 1:500,000 (1963), published by the Institut Géographique National, and from spatial data held within files at WCMC.

27 Nigeria

Land area 910,770 sq. km
Population (mid-1990) 118.8 million
Population growth rate in 1990 2.9 per cent
Population projected to 2020 273.2 million
Gross national product per capita (1988) US$290
Rain forest (see map) 38,620 sq. km
Closed broadleaved forest (end 1980)* 59,500 sq. km
Annual deforestation rate (1981–5)* 3000 sq. km
Industrial roundwood production† 7,868,000 cu. m
Industrial roundwood exports† 16,000 cu. m
Fuelwood and charcoal production† 100,430,000 cu. m
Processed wood production† 2,945,000 cu. m
Processed wood exports† 1000 cu. m
* FAO (1988)
† 1989 data from FAO (1991)

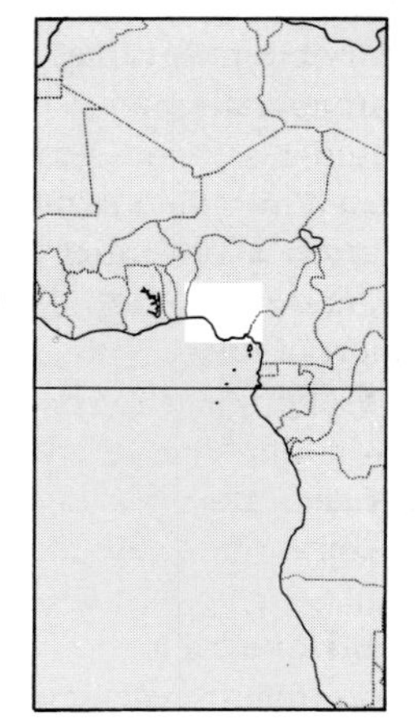

The moist forests of Nigeria are now largely restricted to forest reserves. The forest area has been reduced considerably, with over half having disappeared within living memory. Although natural forests were carefully managed in the early part of the century, they have since been severely over-exploited and it is estimated that at the present rate of deforestation, the country's timber resources will be exhausted by 1995.

The main problem is the pressure put on the land by Nigeria's growing population; the country is already as densely populated as Western Europe and the numbers continue to increase. Nigeria contains a quarter of Africa's total population and accounts for half of the continent's economic production. Subsistence farming and cash cropping have destroyed the forest outside the reserves and there is now considerable encroachment into these areas which are mostly unprotected. In addition, the reserves are now the main source of fuelwood in the country and, with domestic consumption exceeding sustainable yield from these areas by at least four times, there is little hope for their conservation without massive planting programmes.

Today, there is a growing awareness of the urgent need to manage the remaining forests. Management plans are being implemented for several of the protected areas and more are being drafted. Ultimately, though, only changes in the efficiency of agriculture, in systems of land tenure and in people's attitudes to the use of natural resources can rescue the situation.

Introduction

With a country area of 923,770 sq. km Nigeria is by far the largest country in tropical West Africa. It extends between 4°16'N and 13°52'N and between 2°49'E and 14°37'E and is bounded by Cameroon and Chad to the east, Niger to the north and Benin to the west. The southern coastline is dominated by the delta of the River Niger. The general relief of the country is that of a plateau dissected into three parts (south-west, north and south-east) by the Y-shaped Niger/Benue rivers system. The central Jos Plateau has an altitude of about 1200 m with higher peaks such as Wadi Hill at 1698 m and Shere Hill at 1781 m; whereas the general height of the Biu Plateau, in the east, is 600 m with one of its peaks reaching 795 m. The south-eastern border with Cameroon is mountainous, and Chappal Wade at 2419 m is the highest point in Nigeria. Volcanic formations are also found in these regions.

Rainfall decreases, both in amount and duration, from south to north, and it tends to be wetter in the east than in the west. At Calabar in the south-east, the mean annual precipitation is 3070 mm whereas at Ikeja in the south-west it averages 1700 mm; and at Maiduguri in the north-east the average is only 600 mm. The annual variability in rainfall has generally increased in recent years, especially in the north. The dry season is brief in the delta region, lasting one or two months during January and February, whereas in the far north the duration is of eight or nine months from October to May. In the north, the mean maximum temperature is about 35°C, and in the south it averages around 31°C; the mean minimum temperature is about 18°C in the north and 22°C in the south. Thus, the daily range of temperature is much greater in the north, where temperatures are also lower in the dry season when the dust-laden Harmattan wind blows from the desert, causing a haze which obscures the sun.

Although only the twelfth largest country in Africa, Nigeria contains a quarter of the continent's people and a greater population than any other African country. People have congregated on more fertile soils, for example in the southern parts of the Sudan savanna and along the northern fringes of the high forest. In some areas, particularly in the south-eastern states of Akwa Ibom and Imo, rural populations may exceed 1000 persons per sq. km, while in Kano State in the north they reach around 500 persons per sq. km. The Nigerian population has expanded from perhaps 10 million in 1900 to almost 120 million in 1990, although these figures are approximate due to the lack of an accurate census. Cities have existed in Nigeria for many centuries, even in the forest zone, and today some 30 per cent of the total population live in towns, mostly in the south-west.

Food production depends mainly on subsistence farmers, crops being grown on holdings of usually no more than 2 ha. Cash crops, such as groundnuts and cotton grown in the savanna zones and cacao and rubber in the forest zone, were grown mainly for export. However, as a result of the need to feed larger urban populations, and because of rising food prices, it has now become remunerative for farmers to grow food crops for cash. The population growth in the forest zone and the spread of cash cropping, have resulted in almost total destruction of natural high forest outside the reserves.

Nigeria was formerly an important producer of tropical timber, and in 1960 exported a roundwood equivalent of 773,000 cu. m;

domestic consumption was thought to be of a similar level. During the oil-boom decade of the 1970s, vastly increased government and commercial activity, combined with rising standards of living, caused a tremendous growth in demand for timber to make furniture and provide building materials. In consequence, the domestic market for timber began to predominate over the export market, and in 1976 the federal government was obliged to forbid all export of logs or semi-processed timber. It should be noted, however, that FAO (1988) still reports appreciable exports of industrial roundwood from the country. A result of the burgeoning demand has been increased pressure on, and progressive over-exploitation of, the remaining forests.

The Forests

Three major vegetation types may be recognised in Nigeria: the swamp forests (including mangroves) of the Niger Delta and coastal belt, the lowland forests of the humid south, which reach about 250 km inland, and the savannas of the subhumid central area and the drier north (Charter, 1978). Forest outliers may be found, particularly in the savanna zone along watercourses and also in areas of higher rainfall such as on the south-western slopes of the Jos Plateau.

The main closed forest blocks in the country are in the states of Ondo (south-west), Bendel (south-centre) and Cross River (south-east) and in the Niger Delta (River State). The forest in Cross River State has affinities with those of Cameroon and Gabon and has apparently been continuously linked with them (Hall, 1977). Indeed, it is estimated that during the Pleistocene, the Oban Forest in Cross River State remained the only major forest refuge in Nigeria. In contrast, in more westerly forests, because of intermittent dry phases, savanna has periodically reached the coast (Hall, 1977).

White (1983) classifies Nigerian lowland forest as Guineo-Congolian, only distinguishing a drier subtype in the south-west and south-centre from a wetter subtype in the south-east. There have been several studies of the vegetation types (e.g. Rosevear, 1954a, b; Keay, 1965; Redhead, 1971; Charter, 1978; Hall, 1977, 1981). Drier forest occurs towards the northern margins of the forest zone, in which dominant trees generally belong to the Sterculiaceae (*Cola* spp., *Mansonia altissima*, *Nesogordonia papaverifera*, *Pterygota* spp., *Sterculia* spp., *Triplochiton scleroxylon*), to the Moraceae (*Antiaris africana*, *Ficus* spp., *Milicia excelsa*) and to the Ulmaceae (*Celtis* spp., *Holoptelea grandis*). The moister forests are characterised by members of the Leguminosae (*Brachystegia* spp., *Cylicodiscus gabunensis*, *Gossweilerodendron balsamiferum*, *Piptadeniastrum africanum*) and of the Meliaceae (*Entandrophragma* spp., *Guarea* spp., *Khaya ivorensis*, *Lovoa trichilioides*); and the forests in the high rainfall areas by the species *Klainedoxa gabonensis*, *Lophira alata*, *Nauclea diderrichii* and *Pycnanthus angolensis*, and by the presence of climbing palms. However, the boundaries between these forest types are not distinct, and soil differences determined by geological parent material also influence the species composition.

In the lowland forest there is no single canopy layer, but crowns exist at all levels giving an irregular structure to the forest, with large emergents projecting above the level of other crowns. In mature forest, the undergrowth is fairly open, except where climber tangles fill gaps left by fallen emergents.

Nigeria's montane forest, occurring above about 1500 m, is limited in extent and confined mainly to the Mambila and Obudu plateaux which border Cameroon. These forests mostly occupy ravines and steep-sided valleys, between extensive rolling grasslands. The forest has a characteristic appearance, with festoons of mosses, orchids and begonias hanging from tree branches and boles; tree ferns (*Cyathea* spp.) are also present. Typical tree species are *Carapa procera*, *Bridelia speciosa*, *Cephaelis mannii*, *Eriocoelum macrocarpum*, *Sapium ellipticum*, *Symphonia globulifera*, *Syzygium staudtii* and *Tabernaemontana ventricosa* (Hall and Medler, 1975).

Freshwater swamp forest is an important component of the vegetation in the Delta, and also fringes the numerous rivers and estuaries that flow into the creeks and lagoons. There are fewer species than in dryland forests, although large emergents such as *Alstonia boonei*, *Mitragyna ledermannii*, *Symphonia globulifera* and (where better drained) *Lophira alata* are present. *Mitragyna* is a useful timber tree and may be exploited wherever it is accessible. The main canopy is formed of smaller species, such as *Oxystigma mannii*, *Anthostema aubryanum* and *Nauclea pobeguinii*. A member of the Meliaceae, *Carapa procera*, is common though this species does not grow to a large size in Nigeria. The palm *Raphia hookeri* is usually abundant, and there may be dense colonies of *Pandanus candelabrum* especially along margins of water courses. Tree species more associated with dryland forests may also occur on hummocks or other raised areas. The swamps are flooded in the rainy season, but more or less dry out in the dry season.

Mangroves

The Nigerian mangroves are extensive; they were reported to cover over 9700 sq. km in 1980 (FAO/UNEP, 1981). The mangroves are associated with the lagoon systems in the west and particularly with the Niger Delta and the Cross estuary in the east. Along the Niger Delta alone, they are reported to cover an area of 5400 sq. km (SECA/CML, 1987). The distribution of mangroves and swamp forest in this area is shown in Figure 27.1. Map 27.1 shows a total

Figure 27.1 The main ecological zones of the Niger Delta
(*Sources:* SECA/CLM, 1987; from Ibiele *et al.*, 1983; Allen, 1965; Burke, 1972; NEDECO, 1961)

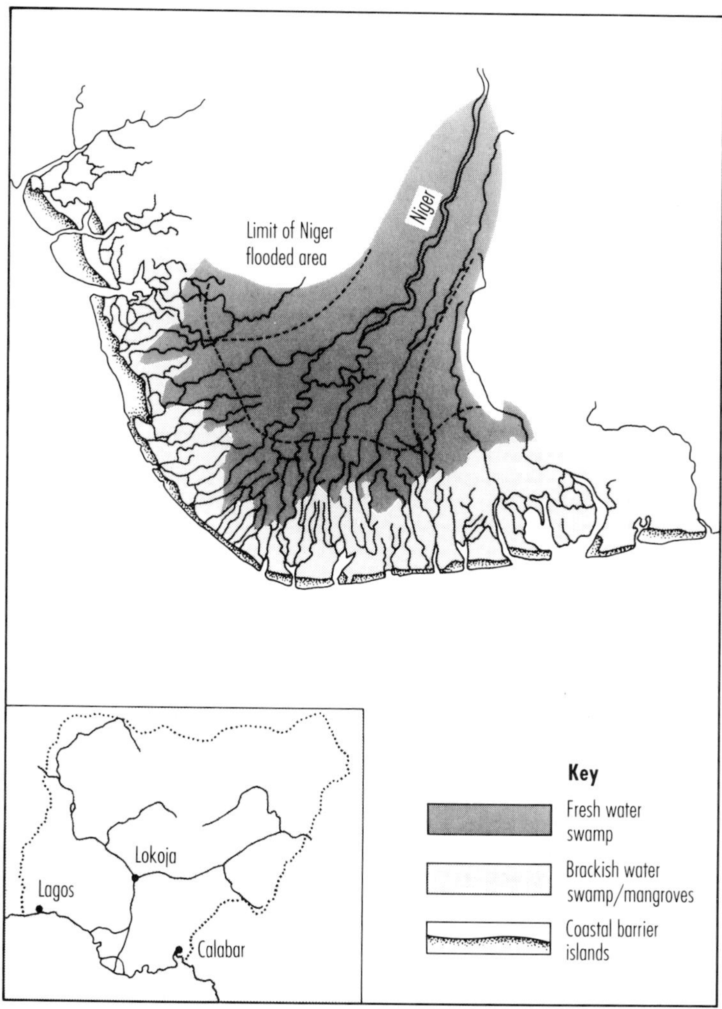

of 10,515 sq. km of mangrove within the country. The most common species is *Rhizophora racemosa*, but *R. harrisonii*, *R. mangle*, *Avicennia africana* and *Laguncularia racemosa* also occur (Rosevear, 1947; Keay, 1989). The trees form a dense tangle about 10 m high, although some trees can reach more than 40 m in height. In many places where mangroves are cleared, the secondary vegetation becomes dominated by the sedge *Cyperus articulatus*, the grass *Paspalum vaginatum* or the fern *Acrostichum aureum*.

Waterbirds are abundant in the mangrove swamps (Rosevear, 1947); the grey parrot *Psittacus erithacus* and palm nut vulture *Gypohierax angolensis* are also common. Among the animals present in the swamps are the sitatunga *Tragelaphus spekei*, the marsh mongoose *Atilax paludinosus* and the spotted-necked otter *Lutra maculicollis*. The mona monkey *Cercopithecus mona* is the only common monkey. Two species of crocodile are present: the Nile *Crocodylus niloticus* and the slender-snouted *Crocodylus cataphractus*.

These forests have long provided fuelwood and building materials to coastal towns, and are being destroyed where they are accessible to urban centres, such as around Lagos, Port Harcourt and Warri. In the Niger Delta the wetlands are being criss-crossed by oil and gas pipelines and considerable areas are disturbed as a result of oil exploration and pollution. In addition, the exotic palm *Nypa fruticans* imported from Southeast Asia is invading the mangrove edges and may have a detrimental effect on them. None of the mangrove areas is presently protected.

Forest Resources and Management

In 1897, Nigeria was thought to contain about 600,000 sq. km of natural vegetation (Stebbing, 1935). By 1951, it was estimated that only 360,000 sq. km remained (IUCN/WWF, 1987), and FAO reported that there were only 140,750 sq. km of forest, of which less than half (59,500 sq. km) was closed forest (FAO/UNEP, 1981). Map 27.1 and Table 27.1 show that now only 38,620 sq. km of forest remains in the country, of which almost a third is mangrove. It was not possible to distinguish between swamp and lowland forest from the UNEP/GRID dataset, but Figure 27.1 indicates the extent of the swamp forest along the Niger Delta where it is reported to cover about 11,700 sq. km (SECA/CML, 1987).

Exploitation of the forests for timber began in the 1880s, by extracting valuable trees such as the African mahoganies (*Entandrophragma* and *Khaya* spp.) that were accessible from river banks and could be transported by boat for export. The Bendel forests in the south-centre were the most easily reached, as they were penetrated by rivers and the terrain was relatively flat and free from surface streams. Tramways were constructed in the forest to facilitate removal of logs that were more distant from rivers. Initially timber for local use was cut by pitsawyers, but in 1901 two small sawmills were established in Lagos. The improvement of roads after the Second World War facilitated expansion into large blocks of forest in Ogun and Ondo states in the south-west. The Cross River State forests in the south-east have been less exploited, owing both to the rugged terrain and to the fact that until 1973 there were no bridges over the Cross River to allow the easy movement of timber to other parts of Nigeria.

Forest management began early in Nigeria. In 1901 the First Forestry Ordinance was promulgated to regulate the size of timber concessions, to impose forestry fees, and to oblige concessionaires to plant 20 economic tree seedlings at each stump site. Duties were also exacted on exported logs. In 1919, the first Governor General of Nigeria, Sir Frederick Lugard, spelled out Nigerian government policy on forestry, which remained valid until the economy was transformed by the oil-boom decade in the 1970s. He stated that, for a well-populated country, it was desirable to reserve for forestry one-third of the total land area; and for Nigeria he proposed an absolute minimum target of 25 per cent.

The constitution of forest reserves required elaborate consultation procedures between government departments as well as with local communities. In creating the reserves, preference was given to blocks of vacant land, especially those containing valuable forest, but also including protection reserves for watersheds and unstable terrain. As communal land tenure was usual, ownership of the reserves became vested in local authorities. Reserve boundaries were marked on the ground, and as survey and demarcation proceeded, existing rights practised in the reserve were investigated by village meetings. Rights were admitted which did not seem likely to conflict with the proposed management objectives of the reserve, and included traditional activities such as hunting, fishing and gathering for food, medicines and sometimes materials for house construction and domestic use.

In the high forest areas, forest reservation was substantially completed during the 1920s and 1930s, although there has been some reservation in the Rivers State during the 1960s and 1970s. In the savanna areas of the north, much of the reservation took place during the 1940s–1960s. Forest reserves now cover about 93,000 sq. km, or 10 per cent of land area (Federal Department of Forestry, 1984), but at least three-quarters of this is in savanna areas (Allen and Shinde, 1981).

In 1960, in order to control exploitation and regeneration of the forests, nearly 10,000 sq. km of the high forest reserves were being managed under working plans prepared by the Forestry Department. The forests were worked on a 100-year felling cycle and a quarter of the area was given out as concessions. Royalties were paid to the local authorities by calculating payments at so much per unit volume of timber removed. Only trees larger than prescribed girths, 60–90 cm dbh depending on species, could be felled. Total log removals normally averaged between 20 and 50 cu. m per ha, whereas the total standing bole volume of the forest is between 150 and 450 cu. m per ha. The timber company was expected to remove all sound trees of listed species above the prescribed girth limits and, if it failed to do so, royalties could be charged on the unfelled trees. However, these plans have now fallen into disuse and new plans have never been written. In addition, the state governments now control forest reserves and the Federal Department of Forestry (FDF) has no executive authority for their management.

After 1960, the tendency was to favour indigenous entrepreneurs, and to use forest reserves as a source of government patronage. Concessions were given to political clients who might not have facilities to exploit them, and who sub-leased to established companies. Concessions were for shorter periods; working plans were allowed to fall into abeyance, or were ignored, and new plans were not prepared. The method of paying royalties was changed to a fee based on coupe area, which removed control of revenue collection from local staff. In 1962, in order to make larger

Table 27.1 Estimates of forest extent in Nigeria

	Area (sq. km)	*% of land area*
Rain forests		
Lowland (and swamp)	28,040	3.1
Montane	65	<0.01
Mangrove	10,515	1.2
Totals	38,620	4.2

(Based on analysis of Map 27.1. See Map Legend on p. 239 for details of sources.)

Forest land cleared for agriculture on the southern escarpment of the Jos Plateau. Only small relict forests remain in this area. R. Wilkinson

areas available for exploitation, the felling cycle for natural forest was reduced to 50 years. In 1970 the original concessions began to expire and, as subsequent blocks had already been given to fresh concessionaires, it became necessary for some agencies to re-exploit compartments that had already been logged 25 years previously. By the 1980s forest exploitation had become virtually unregulated and timber was being removed from the reserves on a massive scale, both legally and illegally.

In 1974–8 the FAO collaborated with the Federal Department of Forestry to carry out land-use surveys and mapping of the whole of Nigeria by means of side-look radar. At the same time, an inventory was conducted through Nigeria's high forests of all trees over 20 cm dbh (Sutter, 1979). This led FAO (1979) to predict that by 1995 forest reserves in the south-west and south-centre of the country would be exhausted of timber. Therefore it was recommended that a plantation programme should be initiated that would establish 10,000 sq. km of forestry plantations in the high forest zone, and another 8000 sq. km in the savanna zones during a 20-year period – ideally before the year 2000.

The forest reserves are capable of yielding perhaps 2 million cu. m of timber annually on a sustained yield basis. However, increasing population, particularly in the towns, and rising standards of living have caused domestic consumption of timber (in 1990) to exceed this sustainable yield fourfold. It is estimated that by the year 2000, the demand for timber in Nigeria will be about 10 million cu. m (roundwood equivalent), not including poles and fuelwood. To save the natural forests would require either importing timber on a considerable scale or a massive planting programme, particularly within the forest zone outside forest reserves. This would need to be of the order of 500 sq. km annually to satisfy present consumption and to meet subsequent growth in demand. The most the government forestry services have attained per year is about 200 sq. km within high forest reserves (in 1975); but for more than a decade they have not exceeded 50 sq. km annually.

When the far greater demand for fuelwood is taken into account, it is apparent that the extent of planting needed is beyond the reach of the government alone and will require participation by private companies and individuals. Indeed, several commercial enterprises, including a pulpmill at Jebba, an integrated mill at Sapele, and a match factory at Ibadan, are already beginning to establish their own tree plantations in the hope of securing future wood supplies for their factories. However, they have to plant within forest reserves to get security of tenure over sufficiently large blocks of land, and they tend to plant fast-growing species on short rotations, mainly exotics such as *Gmelina* or *Eucalyptus*. Establishing tree plantations, whether by natural or artificial regeneration, requires the certainty that in due course the owner will be able to reap the benefit from his investment. Planting trees outside forest reserves, by private individuals or organisations, is discouraged by the land tenure situation and by state or local governments charging royalties for exploiting trees on lands outside forest reserves. Placing a moratorium on felling certain species (such as *Milicia excelsa* and *Triplochiton scleroxylon* in Oyo State) may also prove counterproductive. The situation can be resolved only by ensuring secure and transferable long-term titles to land and by providing inducements for planting timber species with longer rotations.

If the productive capacity of the natural high forests is not to be destroyed, then it is essential that all forest reserves are brought under management, and exploitation is strictly regulated on a sustained yield basis. As the remaining natural forests cannot produce enough timber to satisfy Nigeria's needs, planting timber trees must be undertaken within the forest zone. If this is to include significant areas of land outside forest reserves, then land tenure reform may be inescapable. However, as an interim measure, planting timber crops in forest reserves will need to be included in current working plans, in order to help reduce the pressure on remaining natural forests. In addition, if the existing genetic variety of the natural forests is to be maintained, then silvicultural treatments that destroy currently uneconomic species may have to be abandoned.

Timber is not the only valuable resource found in the forest. Bushmeat is an important source of protein, particularly for rural people for whom it constitutes about 20 per cent of their diets (Charter, 1973). It is also regarded as a delicacy and the more or less unregulated hunting provides a significant source of income for rural communities. Many hunters in the south-east of the country use breech-loading shotguns which are being manufactured

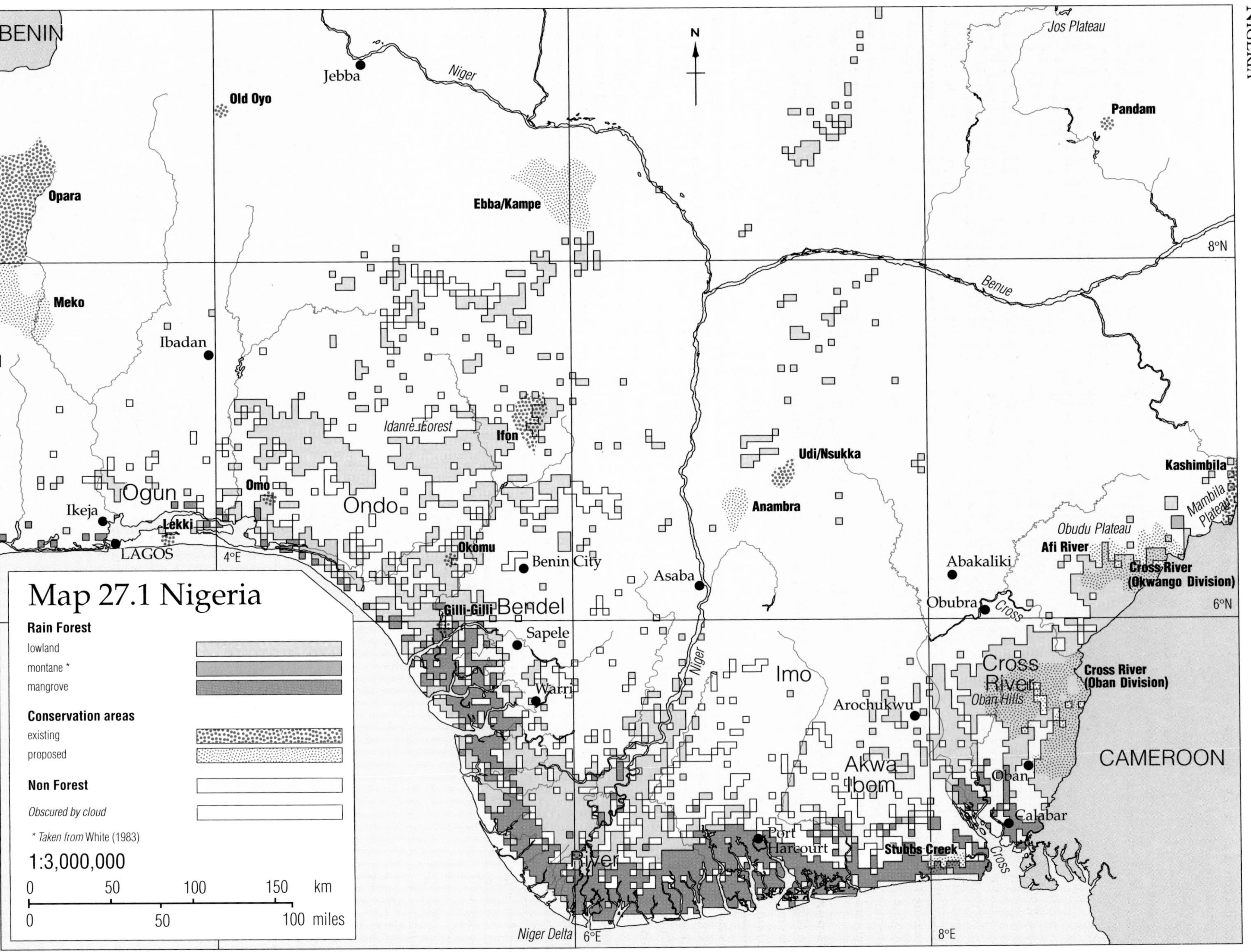
BENIN
Jebba
Niger
Old Oyo
Opara
Ebba/Kampe
Meko
Ibadan
Idanre Forest
Ifon
Udi/Nsukka
Anambra
Ogun
Omo
Ondo
Ikeja
Lekki
LAGOS
4°E
Okomu
Benin City
Asaba
Bendel
Gilli-Gilli
Sapele
Warri
Niger
Imo
Arochukwu
Akwa Ibom
Port Harcourt
Stubbs Creek
Rivers
Niger Delta
6°E
8°E
Benue
8°N
6°N
Jos Plateau
Pandam
Kashimbila
Mambila Plateau
Obudu Plateau
Afi River
Cross River (Okwango Division)
Abakaliki
Obubra
Cross
Cross River
Oban Hills
Cross River (Oban Division)
CAMEROON
Oban
Calabar
N
Map 27.1 Nigeria
Rain Forest
lowland
montane *
mangrove
Conservation areas
existing
proposed
Non Forest
Obscured by cloud
* Taken from White (1983)
1:3,000,000
0 50 100 150 km
0 50 100 miles

locally, although the ammunition is imported. In the south-west, traditional hunters still rely mainly on muzzle-loading guns. Favourite quarry are antelopes, large rodents, bushpigs, pangolins, tortoises, hawks and other large birds. Monkeys are also shot. Crocodiles, pythons and monitor lizards are killed both for their meat and their decorative skins, which are used to manufacture handbags, shoes, and the like. The giant snail *Achatina marginata suturalis* is also regarded as a delicacy, and is gathered in the forest; black snails are preferred, and near villages and towns this causes powerful selection for albinism in the species (Reid, 1989). Another delicacy is the larva of the rhinoceros beetle, *Oryctes* sp., which lives in rotting palm trunks. Wild animals are also used in magical and medicinal concoctions.

A further problem besides unregulated hunting is the gathering of secondary forest products by rural people, mainly for sale in the towns. These include wrapping leaves, chew sticks, peppers, spices, medicines, and so on. They provide a valuable source of income for the people and one that is available virtually throughout the year; but this trade must also be controlled if the resources are not to be destroyed and the forests and their bounty are to remain for future generations.

Deforestation

In Nigeria, deforestation has been caused mainly by two factors: first, the increase in area of subsistence farming to feed a growing population; and second, the spread of cash cropping by peasant farmers. Formerly, cash cropping replaced forest cover with tree crops, such as cacao, cola, oil palm and rubber, the produce of which was mainly exported. However, today, as a result of rising prices for foodstuffs, these tree crops are often themselves displaced by arable crops. When tree crops replace natural vegetation, they do not cause marked site degradation, but this is not true for arable crops.

The successive conversion to tree crops and arable farming has mostly eliminated timber on lands outside forest reserves. Vast tracts are now devoid of forest and in areas where, before 1970, forest came to the road edges (for example, from Benin City to Asaba, Obubra to Agoi and Arochukwu to Abakaliki), it has now disappeared from sight.

Forest reserves are important in maintaining timber supplies and in assisting in the moderation of climate, yet they are not protected adequately. The boundaries are not being maintained and patrolled as before and they are being encroached for farming. Local and state governments often have more pressing calls on their limited finances. Because of the tradition of communal land tenure, local people may consider that the land has been stolen from them, and connive with illegal fellers. Ijaiye Forest Reserve has notably suffered in this way. In contrast, in other major reserves in the south-west, such as Omo, Oluwa and Idanre, timber crop plantations established along the roads have protected the natural forests behind. However, in Bendel State in the south-centre, from 1975 to 1987 the forestry service was organising clearing of about 5 sq. km per year of forest reserves (and previously much more than this), ostensibly for taungya farming, but without establishing a tree crop; meanwhile older taungya plantations were devastated by illegal fellers.

The pressure of deforestation could be reduced if the productivity of arable areas were enhanced, so that less land would be needed for raising food crops. This will require either eliminating the fallow period, or putting it to intensive use, such as the growing of fodder for livestock. However, a feature of peasant agriculture is that every community grows most of the food for its own subsistence. In contrast, the chief consideration for commercial agriculture is to raise the most profitable crop for each locality, which in much of the forest zone could probably be tree crops, while arable farming might be mainly consigned to savanna areas. A problem is that profitable systems of commercial farming are poorly developed in Nigeria and the absence of a free market in land means that land is not used in the most efficient way. Furthermore, there is little alternative to subsistence farming for large numbers of people; this can only be resolved if industry absorbs people off the land, an option made difficult by high population growth. Moreover, even industrialised countries require some 30 per cent of the land area under forest if they are to be self-sufficient in wood (Lowe, 1986).

Biodiversity

Nigeria has a considerable diversity of habitats, from arid thorn savanna in the extreme north to freshwater swamp forest in the south. Associated with this is a wide range of plant and animal species. Lebrun (1967) lists 4614 plant species in the country. There are 205 endemic species (Davies *et al.*, 1986), with the highest degree of endemism occurring in the lowlands of south-east Nigeria, particularly round Oban (Brenan, 1978).

Happold (1987) gives a checklist and distribution data for mammals in the various vegetation zones of Nigeria. He lists a total of 248 species, of which 125 are found in the forests (Table 27.2). Stuart *et al.* (1990) report that there are 274 mammal species in the country.

There is a high diversity of primates, of which at least half are of conservation concern (Lee *et al.*, 1988). Two of them, the white-throated guenon *Cercopithecus erythrogaster* and Sclater's guenon *Cercopithecus sclateri*, are probably endemic although the former may also occur in Benin (Oates, 1986). The drill *Mandrillus leucophaeus* and the gorilla *Gorilla gorilla* still exist, but are endangered. Harcourt *et al.* (1989) estimate that at least 50 per cent more gorillas are killed in Nigeria than are born each year. The chimpanzee *Pan troglodytes* survives only in relict populations. Almost all the surviving species of antelope are considered to be threatened, mainly due to over-hunting (Anadu and Green, 1990). Leopards *Panthera pardus*, once common, are now rare, particularly because they are killed for their skins. There are less than 2000 elephants remaining in Nigeria, the populations are fragmented and declining due to poaching and their range is being restricted by deforestation and the spread of agriculture. Hippos *Hippopotamus amphibius* are threatened by hunting and poaching, and their destructive habits do not endear them to local farmers. Manatees *Trichechus senegalensis* are also threatened by overhunting.

Table 27.2 Number of mammals and their distribution

	All spp.	*Forest*	*Savanna*	
			Guinea	*Sudan*
Insectivora	26	11	10	8
Chiroptera	71	38	40	34
Primates	21	11	11	4
Pholidota	2	2	1	
Lagomorpha	2	1	1	
Rodentia	54	35	29	20
Carnivora	33	11	23	21
Sirenia	1			—
Tubulidentata	1	1	1	
Proboscidea	1	1	1	
Hyracoidea	2	1	1	
Perissodactyla	2		1	
Artiodactyla	32	13	20	18
Totals	248	125	139	105

(*Source:* Happold, 1987)

Nigeria

There are 839 bird species recorded in Nigeria (Elgood, 1965, 1982; Ash and Sharland, 1986), six of which, including three from montane rain forest, are thought to be threatened (Collar and Stuart, 1985). There are two endemics: the Ibadan malimbe *Malimbus ibadanensis*, which is found on forest fringes, and the Anambra waxbill *Estrilda poliopareia*, a grassland species.

Despite being protected by wildlife legislation for several decades, grey parrots and canaries *Serinus mozambicus* are caught as pets, an activity which, in the case of parrots, threatens their survival in the wild. For waterbirds, the draining of wetlands for cultivation or for irrigation schemes are the main hazards. Loss of habitat also puts the montane species at risk.

Little is known about the status and distribution of reptiles in the country. Butler and Reid (1990) list 105 species of snakes, including 56 forest snakes of which one (*Mehelya egbensis*) is endemic. The Nile, African dwarf (*Osteolaemus tetraspis*) and slender-snouted crocodiles all survive in Nigeria, although their numbers are much reduced. There are 18 rare amphibian species found only in Nigeria and Cameroon, and a toad *Bufo perreti* that is confined to Nigeria (Stuart *et al.*, 1990). Reid *et al.* (1990) give the frog fauna of Cross River State alone as numbering about 75 species. There is a rich invertebrate fauna, as may be expected in a country with such a wide range of ecosystems, but little is known about species numbers or degree of threat.

Conservation Areas

Currently some 3.5 per cent of Nigeria is under some sort of management for nature conservation (Table 27.3), but very little of this is forested land. Three categories of protected areas exist: strict nature reserves, game reserves and national parks, all of which were formerly reserved forest. The strict nature reserves are mostly relatively small and are intended to conserve various examples of primary vegetation; they are within forest reserves and under the aegis of the Forestry Research Institute of Nigeria (FRIN), but are not protected by specific legislation. Game reserves, which are controlled by the states, incorporate areas where hunting is supposed to be strictly regulated, habitat protected and wildlife conserved and managed (Oates and Anadu,

Table 27.3 Conservation areas of Nigeria

Existing and proposed conservation areas for Nigeria are listed below; however, only parks in the southern two-thirds of the country are mapped. For data on Biosphere reserves see chapter 9. Forest reserves and safari parks (zoo parks) are not listed or mapped.

	Existing area (sq. km)	*Proposed area (sq. km)*
National Parks		
Cross River (Okwangwo Division)*		950
Cross River (Oban Division)*		2,800
Kainji Lake‡	5,341	
Old Oyo	2,512	
Strict Nature Reserves		
Akure‡	< 1	
Bam Ngelzarma‡	1	
Bonu‡	1	
Lekki*	1	
Milliken Hill[1]‡	< 1	
Omo*	5	
Ribako‡	2	
Urhonigbe‡	1	
Game Reserves		
Afi River*		*c.* 100
Akpaka‡		194
Alawa‡	296	
Anambra[1]		145
Ankwe River‡		nd
Bedde†‡	354	
Chingurmi/Duguma†‡	354	
Dagida‡	294	
Dagona†‡	1	
Damper Sanctuary‡		nd
Ebba/Kampe		1,217
Falgore (Kogin Kano)‡	923	
Gashaka Gumti[4]‡	6,670	
Gilli-Gilli*	362	
Hadejia (Baturiya) Wetlands†‡	297	
Ibi‡	1,540	
Ifon*	282	
Iri-Ada-Obi‡		nd
Kambari‡	414	
Kamuko‡		1,127
Kashimbila*	1,396	
Kwale‡	3	
Kwiambana‡	2,614	
Lake Chad‡	388	
Lame/Burra‡	2,060	
Margadu-Kabak Wetlands†‡	100	
Meko		966
Nguru/Adiani Wetlands†‡	75	
Ohosu‡		471
Okeleuse‡		114
Okomu*	112	
Ologbo‡	194	
Opanda‡		105
Opara[1]	1,100	
Orle[2]‡	54	
Pai River[3]‡	700	
Pandam	224	
Sambisa‡	518	
Stubbs Creek*		210
Taylor Creek (including Nun River)		300
Udi/Nsukka[1]	56	
Wase Rock‡	1	
Yankari[4]‡	2,244	
Totals	31,492	8,699

(*Sources:* WCMC, *in litt.*; J. Caldecott, pers. comm.)

† Part of the proposed Chad Basin National Park.
* Areas with moist forest within their boundaries according to Map 27.1.
nd no data
‡ Not mapped – location data not available to this project or protected area is located to the north of the country which has not been included in this Atlas.
[1] These were once protected natural areas but have now been converted to other land use such as *Melina* plantation. They maintain their legal status, though in reality they no longer represent natural habitats.
[2] Half of the game reserve has been converted to growing wheat.
[3] Mostly degazetted.
[4] Proposed as National parks

THE CROSS RIVER NATIONAL PARK PROJECT

The Cross River National Park (CRNP) Project incorporates the design, establishment and development of a new moist forest conservation area in south-east Nigeria. The park is in two divisions, the Oban (about 2800 sq. km) and the Okwangwo (about 950 sq. km), which are approximately 40 km apart on either side of the Cross River. The project began in late 1988 with a WWF feasibility study for the Oban Division, and was extended to the Okwangwo Division in early 1990; a seven-year, US$70 million programme of work has now been defined and budgeted (Caldecott *et al.*, 1989, 1990). The forests which the project is designed to protect were identified as worthy of special conservation measures in three IUCN publications (MacKinnon and MacKinnon, 1986; IUCN, 1987a, b). These reports emphasised the extreme biological richness of the areas, their relatively intact status and the increasing threat to their integrity as a result of uncontrolled farming, logging and hunting.

The park will be the largest protection forest in the moist forest zone of Nigeria, where more than 90 per cent of the original forest has already been lost or badly degraded. As such it represents one of the country's most important natural resource assets, supporting fisheries, protecting watersheds and climatic stability, providing tourism opportunities and preserving genetic resources. In addition, by constituting the area as a park it is intended to preserve the majority of Nigeria's moist forest species of flora and fauna and the values associated with them. The park will make a valuable contribution to the preservation of the world's biodiversity.

The project will define the boundaries and management zones of the park. It will provide development support for villages around the park and the infrastructure, equipment, staff and expertise needed to manage the park and its buffer zones. About 76,000 people live close to the park, and partly depend on it for income and subsistence from hunting, fishing and collection of forest produce. They live by shifting cultivation and will destroy the biological integrity of the park area unless radical changes in land-use occur: this is the central problem that the project is trying to address.

The mechanism to be used in the CRNP project is the Support Zone Development Programme (SZDP). Each target village and its farmland and communal forests is defined as a Support Zone, within which incentives and disincentives will be applied to encourage the local people to participate actively in the protection and development of the park and to adopt more sustainable, agroforestry-based land-use. The incentives will be in the form of directed credit, advice, grants, planting materials and so on which will be supplied on condition that the boundaries of the park are respected by the beneficiaries.

Non-compliance by SZDP villages will incur a variety of possible sanctions, including: reduction or suspension of regular grant allocations, suspension of SZDP registration, refusal of loans to individuals and withdrawal of village-specific privileges to make use of parts of the park for traditional economic activities. The SZDP and the park management will thus work closely together to monitor village behaviour and to administer appropriate rewards and sanctions. The direct linkage between the two will be achieved through a public relations system with a village liaison officer employed at every village in order to maintain intimate communication between each community and the project authorities. *Source:* Julian Caldecott

1982). The first national park to be created was at Kainji Lake, which was gazetted in 1975 and is operated under a management plan. In 1988 the Council of Ministers approved creation of the Old Oyo National Park, and three more parks (Hadejia-Nguru Wetlands, Gashaka Gumti and Cross River) were approved in 1989 but by mid-1991 had still not been gazetted. A National Parks Decree has been drafted and awaits promulgation. A preliminary management plan has been prepared by the Federal Wildlife Division for Old Oyo National Park.

At the national level, wildlife conservation and protected area management are the responsibility of the Federal Department of Forestry in the Ministry of Agriculture, Water Resources and Rural Development. At the regional level, protected areas are managed by the state departments of forestry in the Ministry of Agriculture and Natural Resources (Caldecott *et al.*, 1989). The Wildlife and Conservation Division of the Federal Department of Forestry is responsible for the establishment and development of national parks and for the enforcement of international wildlife conventions (Anadu, 1987). Each state forestry department is responsible for the establishment and management of forest reserves, game reserves and strict nature reserves.

To assist in formulating official policy and in coordinating development programmes for the country, a Natural Resources Conservation Council was created by Decree No. 50 of 1989. The Council appoints members of a National Advisory Committee on Conservation of Renewable Resources, a government office responsible for promoting, monitoring and mobilising finances for conservation. This will replace the National Wildlife Conservation Committee, which was an adjunct of the National Forestry Development Committee.

However, because of financial stringencies that have affected Nigeria for most of the 1980s, government activities in wildlife conservation have tended to stagnate; staff generally lack funds, transport and resources to do their work. Threats to the protected areas include shifting cultivation, illegal grazing, drought, fuelwood and timber demand, poaching, uncontrolled bushfires, expansion of road networks, oil exploration and extraction and local irrigation and damming schemes (e.g. see Afolayan, 1980; Oates and Anadu, 1982; Ola-Adams, 1983; Anadu, 1987; Osemeobo, 1988).

Initiatives for Conservation

There are two main non-governmental organisations in Nigeria concerned with conservation: the Nigerian Field Society founded in 1930, which publishes an internationally respected journal and organises meetings and lectures; and the Nigerian Conservation Foundation (NCF) established in 1982, which is active in organising and funding conservation projects and in promoting conservation awareness. The activities of the NCF and the attention shown by international organisations and voluntary bodies in conserving natural environments, have begun to regalvanise interest in protecting the remaining flora and fauna. But conservation costs money, and the economic difficulties of the 1980s have discouraged government spending on wildlife. Nevertheless the decade has not been without progress.

The NCF and WWF, with financial assistance from the British Overseas Development Administration (ODA), have prepared an integrated rural development plan for Cross River National Park (Caldecott *et al.*, 1989, 1990; and see case study), the initial phases of which are now being implemented. WWF, funded by the EEC and the ODA, has also prepared a management plan for the conservation of the Boshi Okwangwo forests of Cross River State (Caldecott *et al.*, 1990). Active conservation to date includes protection of gorillas and the cessation of logging in the proposed park area. In addition, WWF has prepared a proposal to develop the newly approved Gashaka Gumti National Park and is seeking funds for this project. This area contains considerable areas of relict lowland moist forest, gallery forest and montane forest and is a vital watershed for the River Benue.

The NCF, in coordination with the Forestry Division of Bendel State, is actively involved in the conservation of Okomu Forest Reserve, especially in preventing poaching and incursions by shifting agriculturalists. ODA and WWF funding is enabling NCF to establish an education centre and develop conservation education programmes in the Okomu area. The farmers in this area are being trained in techniques of sustainable agriculture. The Foundation also intends to develop Stubbs Creek, in Akwa Ibom State, as a game reserve which will protect coastal lowland forest. Mobil Ltd will provide the funding for this project.

There is a growing general awareness of the urgent need to prepare management plans for exploitation and regeneration of forest reserves. There is increasing pressure to confront the impediment to agricultural and forestry development that land tenure provides. The World Bank is negotiating a major loan to support government environmental conservation programmes and a Tropical Forestry Action Plan is being prepared with assistance from the World Bank, British bilateral aid and the EEC.

References

Afolayan, T. A. (1980) A synopsis of wildlife conservation in Nigeria. *Environmental Conservation* 7: 207–12.

Allen, J. R. L. (1965) Coastal Geomorphology of Eastern Nigeria; Beach ridge barrier islands and vegetated tidal flats. *Geologie en Mijnbouw*: **44**.

Allen, P. E. T. and Shinde, N. N. (1981) *Land-Use Area Data for Nigeria.* FO:NIR/77/009, Federal Department of Forestry, Lagos. 85 pp.

Anadu, P. A. (1987) Progress in the conservation of Nigeria's wildlife. *Biological Conservation* **41**: 237–51.

Anadu, P. A. and Green, A. A. (1990) Nigeria. In: *Antelopes Global Survey and Regional Action Plans. Part 3: Central and West Africa.* East, R. (ed.). IUCN, Gland, Switzerland.

Ash, J. S. and Sharland, R. E. (1986) *Nigeria: Assessment of Bird Conservation Priorities.* ICBP Study Report No. 11. ICBP, Cambridge, UK.

Brenan, J. P. M. (1978) Some aspects of the phytogeography of tropical Africa. *Annals Missouri Botanical Garden* **65**(2): 437–78.

Burke, K. (1972) Longshore drift, submarine canyons and submarine fans in development of the Niger Delta. *American Association of Petroleum and Geology Bulletin*: **56**(10).

Butler, J. A. and Reid, J. C. (1990) Records of snakes from Nigeria. *Nigerian Field* **55**: 19–40.

Caldecott, J. O., Bennett, J. G. and Ruitenbeek, H. J. (1989) *Cross River National Park, Oban Division: Plan for Developing the Park and its Support Zone.* Unpublished report to WWF-UK, Godalming, UK.

Caldecott, J. O., Oates, J. F. and Ruitenbeek, H. J. (1990) *Cross River National Park, Okwangwo Division: Plan for Developing the Park and its Support Zone.* WWF-UK, Godalming, UK.

Charter, J. R. (1973) *The Economic Value of Wildlife in Nigeria.* Research Paper (Forestry Series) No. 19. Federal Department of Forestry Research, Ibadan, Nigeria. 11 pp.

Charter, J. R. (1978) Vegetation: ecological zones. In: *The National Atlas of the Federal Republic of Nigeria.* Federal Surveys, Lagos.

Collar, N. J. and Stuart, S. N. (1985) *Threatened Birds of Africa and Related Islands. The ICBP/IUCN Red Data Book Part 1.* ICBP/IUCN, Cambridge, UK. 761 pp.

Davis, S. D., Droop, S. J. M., Gregerson, P., Henson, L., Leon, C. J., Villa-Lobos, J. L., Synge, H. and Zantovska, J. (1986) *Plants in Danger: What do we know?* IUCN, Gland, Switzerland and Cambridge, UK. 461 pp.

Elgood, J. H. (1965) The birds of the Obudu Plateau. *Nigerian Field* **30**(2): 60–8.

Elgood, J. H. (1982) *The Birds of Nigeria.* BOU, London, UK.

FAO (1979) *Forest Development in Nigeria: Project Findings and Recommendations.* FO:DP/NIR/71/546 Terminal Report, FAO, Rome, Italy. 88 pp.

FAO (1988) *An Interim Report on the State of Forest Resources in the Developing Countries.* FAO, Rome, Italy. 18 pp.

FAO (1991) *FAO Yearbook of Forest Products 1978–1989.* FAO Forestry Series No. 24 and FAO Statistics Series No. 97. FAO, Rome, Italy.

FAO/UNEP (1981) *Tropical Forest Resources Assessment Project. Forest Resources of Tropical Africa. Part II: Country Briefs.* FAO, Rome, Italy.

Federal Department of Forestry (1984) *Land Use Planning: Country Report to the Commonwealth Training Workshop on Land Use Planning.* Bangladesh, India.

Hall, J. B. (1977) Forest types in Nigeria: an analysis of pre-exploitation forest enumeration data. *Journal of Ecology* **65**: 187–9.

Hall, J. B. (1981) Ecological islands in south-eastern Nigeria. *African Journal of Ecology* **19**: 55–72.

Hall, J. B. and Medler, J. A. (1975) Highland vegetation in south-eastern Nigeria and its affinities. *Vegetatio* **29**(3): 191–8.

Happold, D. C. D. (1987) *The Mammals of Nigeria.* Clarendon Press, Oxford, UK. xv + 402 pp.

Harcourt, A. H., Stewart, K. J. and Inaharo, I. M. (1989) Gorilla quest in Nigeria. *Oryx* **23**(1): 7–13.

Ibiele, D. D., Powell, C. B., Isoun, M., Selema, M. D., Shou, P. H. and Murday, M. (1983) Establishment of baseline data for complete monitoring of petroleum related aquatic pollution in Nigeria. In: *The Petroleum Industry and the Nigerian Environment.* Proceedings of the 1983 International Seminar Nigerian National Petroleum Corporation.

IUCN (1987a) *Action Strategy for Protected Areas in the Afrotropical Realm.* IUCN, Gland, Switzerland and Cambridge, UK. 56 pp.

IUCN (1987b) *IUCN Directory of Afrotropical Protected Areas.* IUCN, Gland, Switzerland and Cambridge, UK. xix + 1043 pp.

IUCN/WWF (1987) *Centres of Plant Diversity. A Guide and Strategy for Their Conservation.* IUCN Threatened Plants Unit/WWF Plants Conservation Programme, Kew, London, UK.

Keay, R. W. J. (1965) *An Outline of Nigerian Vegetation.* Federal Ministry of Information, Lagos, Nigeria. 46 pp. + 1 map.

Keay, R. W. J. (1989) *Trees of Nigeria.* Clarendon Press, Oxford, UK. viii + 476 pp.

Lebrun, J. P. (1967) Richesses spécifiques de la flore vasculaire des divers pays ou régions d'Afrique. *Candollea* **31**: 11–15.

Lee, P. C., Thornback, J. and Bennett, E. L. (1988) *Threatened Primates of Africa. The IUCN Red Data Book.* IUCN, Gland, Switzerland and Cambridge, UK.

Lowe, R. G. (1986) *Agricultural Revolution in Africa?* Macmillan, London, UK. 295 pp.

MacKinnon, J. and MacKinnon, K. (1986) *Review of the Protected Areas System in the Afrotropical Realm.* IUCN/UNEP, Gland, Switzerland. 259 pp.

NEDECO (1961) *The Waters of the Niger Delta.* Federation of Nigeria, The Hague, The Netherlands.

Oates, J. F. (1986) *Action Plan for African Primate Conservation 1986–1990.* IUCN/SSC Primate Specialist Group. Stony Brook, New York, USA.

Oates, J. F. and Anadu, P. A. (1982) *The Status of Wildlife in Bendel State, Nigeria, with Recommendations for its Conservation.* WWF/IUCN Project 1613. Unpublished report to WWF, Gland, Switzerland.

Ola-Adams, B. A. (1983) *Effects of Logging on the Residual Stands of a Lowland Rain Forest at Omo Forest Reserve, Nigeria.* Paper presented at the 24th Annual Conference of the Science Association. Ibadan, Nigeria.

Osemeobo, G. J. (1988) The human causes of forest depletion in Nigeria. *Environmental Conservation* **15**: 17–28.

Redhead, J. F. (1971) The timber resources of Nigeria. *Nigerian Journal of Forestry* **1**: 7–11.

Reid, J. C. (1989) Albinism in molluscs. *Nigerian Field* **54**: 82–3.

Reid, J. C., Owens, A. and Laney, R. (1990) Records of frogs and toads from Akwa Ibom State. *Nigerian Field* **55**: 113–28.

Rosevear, D. R. (1947) Mangrove swamps. *Farm and Forest* **8**: 23–30.

Rosevear, D. R. (1954a) Vegetation. In: *The Nigeria Handbook*, pp. 139–73 + 1 map. Crown Agents, London, UK.

Rosevear, D. R. (1954b) Forestry. In: *The Nigeria Handbook*, pp. 174–205 + 1 map. Crown Agents, London, UK.

SECA/CML (1987) *Mangroves of Africa and Madagascar: The Mangroves of Nigeria.* Société d'Eco-aménagement, Marseilles, France and Centre for Environmental Studies, University of Leiden, The Netherlands. Unpublished report to the European Commission, Brussels.

Stebbing (1935) The encroaching Sahara. *Geography Journal* **85**(6).

Stuart, S. N., Adams, R. J. and Jenkins, M. D. (1990) *Biodiversity in Sub-Saharan Africa and its Islands: Conservation, Management and Sustainable Use.* Occasional Papers of the IUCN Species Survival Commission No. 6. IUCN, Gland, Switzerland.

Sutter, H. (1979) *High Forest Development, Nigeria: the Indicative Inventory of Reserved High Forest in Southern Nigeria 1973–77.* FO:NIR/71/546 Technical Report 1, Federal Department of Forestry, Ibadan. 381 pp.

White, F. (1983) *The Vegetation of Africa: a descriptive memoir to accompany the Unesco/AETFAT/UNSO vegetation map of Africa.* Unesco, Paris, France. 356 pp. + 2 maps.

Authorship

Richard Lowe, University of Ibadan with contributions from Julian Caldecott and Richard Barnwell, WWF-UK, and Ronald Keay, UK.

Map 27.1 Forest cover for Nigeria

Lowland and montane rain forests and mangroves are shown on Map 27.1. Forest cover has been extracted from 1989–90 UNEP/GRID data which accompany an unpublished report *The Methodology Development Project for Tropical Forest Cover Assessment in West Africa* (Päivinen and Witt, 1989) and montane forest has been delimited by overlaying White's vegetation map (1983) on to this data. It was not possible to use White's map to differentiate between lowland and swamp forest.

UNEP/GEMS/GRID together with the EEC and FINNIDA have developed a methodology to map and use 1 km resolution NOAA/AVHRR-LAC satellite data to delimit forest/non-forest boundaries in West Africa. These data have been generalised for this Atlas to show 2 × 2 km squares which are predominantly covered by forest. Higher resolution satellite data (Landsat MSS and TM, SPOT) and field data from Ghana, Côte d'Ivoire and Nigeria have also been used. Forest and non-forest data have been categorised into five vegetation types: forest (closed, defined as greater than 40 per cent canopy closure); fallow (mixed agriculture, clear-cut and degraded forest); savanna (includes open forests in the savanna zone and urban areas); mangrove and water. In addition this dataset shows areas obscured by cloud. The forest and mangrove types and cloud obscured areas have been mapped in this Atlas.

Existing and proposed protected areas spatial data have been provided by the Nigerian Conservation Foundation with additions from Julian Caldecott, WWF-UK.

28 São Tomé and Príncipe

Land area 960 sq. km
Population (mid-1990) 100,000
Population growth rate in 1990 2.7 per cent
Population projected to 2020 300,000
Gross national product per capita (1988) US$280
Rain forest 299 sq. km
Closed broadleaved forest (end 1980)* 560 sq. km
Annual deforestation rate (1981–5)* nd
Industrial roundwood production† 9000 cu. m
Industrial roundwood exports† nd
Fuelwood and charcoal production† nd
Processed wood production† 5000 cu. m
Processed wood exports† nd

* FAO (1988)
† 1989 data from FAO (1991)

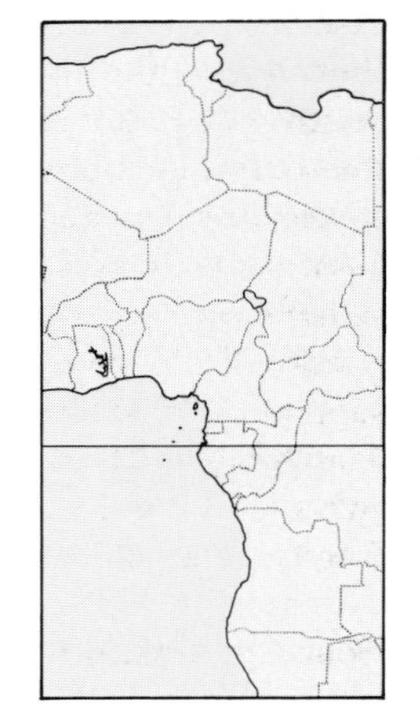

São Tomé and Príncipe are two small oceanic islands in the Gulf of Guinea. Both are mountainous and deeply eroded by high rainfall. Undisturbed primary rain forest remains in the wettest areas of the south-west of each island on steep slopes in virtually inaccessible terrain. Although they have low overall biotic diversity relative to continental Africa, the islands harbour numerous endemic species as a result of their isolation: 109 plants, 26 birds, seven reptiles, seven amphibians and two species of mammal. Many of the terrestrial vertebrates survive well in habitats disturbed by man but others, especially some birds, are confined to pristine forest.

Clearance of lowland forest to plant sugar cane, cacao and other crops began four centuries ago and the only remaining forest is inaccessible and unsuitable for agriculture. Until recently there was no threat to this primary forest as the demand for wood is met from the regenerating secondary forest on abandoned plantations. All the remaining undisturbed forest on each island has been proposed as a *zona ecológica* which will receive complete protection.

Rain forest protection and wildlife conservation are the responsibility of the Commission for Forestry Coordination and promoted through a newly-formed NGO, the Association of Friends of Nature.

INTRODUCTION

São Tomé and Príncipe are a pair of volcanic islands situated between Bioko (formerly Fernando Po) and Annobon (or Pagalu) in the Gulf of Guinea. Príncipe lies 220 km west of the African mainland between 1°32'N and 1°43'N latitude and 7°20'E and 7°28'E longitude, while São Tomé (0°25'N–0°01'S, 6°28'E–6°45' E) is 255 km west of Gabon and has the equator passing through its southern tip. These two islands and Annobon are separated by seas over 1800 m deep and, unlike Bioko, have never been connected to the mainland, nor to each other.

São Tomé is approximately 45 km by 25 km with a surface area of 857 sq. km. It is mountainous with at least ten peaks of over 1000 m in the west-central and southern sections. In the north it rises along a sharp ridge to 2024 m at the Pico de São Tomé, while it falls sharply to the sea in the west. The eastern and north-eastern sides slope less steeply and there are extensive areas of flatter land suitable for cultivation. The main rivers, none navigable, are the Contador, Lembá, Xufexufe, Quija, Mussucavú, Caué, Ana Chaves and Io Grande.

Príncipe is only 139 sq. km in area, about 17 km by 8 km. It too is very mountainous, rising to 948 m (Pico de Príncipe) at its southern edge with a large plug (Pico Papagaio) 680 m high, further north. The Papagaio is the only river of any size.

The climate on the two islands is very similar (Bredero *et al.*, 1977). Both have two long wet seasons with drier periods separating them. The main dry period is between June and September, while the months from December to February have less rain but are not truly dry. Rainfall in the south-western parts probably exceeds 7000 mm annually in São Tomé and 5000 mm on Príncipe and these areas, along with the high interiors of both islands, are wet throughout the year. Temperatures at sea level average 22–33°C, though minimum temperatures at higher altitudes can drop to 9°C or less.

The Forests

Exell (1944, 1956, 1973), who gave the fullest account of the vegetation of São Tomé and Príncipe, made extensive collections from both islands in 1932–3. He considered that, apart from some small areas of mangrove and sand dune on the coast, São Tomé was originally covered in rain forest from sea level almost to the summit of the Pico. Lowland forest extended from the coast to 800 m. Endemic trees there include *Rinorea chevalieri*, *Zanthoxylum thomense*, *Drypetes glabra*, *Anisophyllea cabole* and *Sorindeia grandifolia*.

This region has been brought almost entirely under cultivation and it is only in the catchment areas of the major rivers flowing to the south-east and south-west that untouched lowland forest remains. However, secondary forest, which is at times difficult to distinguish from the virgin forest, has reclaimed extensive areas of abandoned plantations in the low altitudes in the south.

The montane forest region extends from 800–1400 m. The trees here are tall, forming a dense, high canopy and are covered in lianas and epiphytic mosses, ferns, orchids and begonias. The endemic trees include *Trichilia grandiflora*, *Pauridiantha insularis*, *Pavetta monticola*, *Erythrococca molleri* and *Tabernaemontana stenosiphon*. Ferns are especially abundant and diverse. Most of this vegetation zone appears to be still intact, mainly because it is too high for successful plantation.

Mist forest, mapped in this Atlas as part of the montane forest category, extends from 1400 m to the Pico at 2024 m, though at the highest altitudes the trees are small and the canopy open. Epiphytes are even more abundant and ferns remain an important element of the flora. There is no montane grassland. Endemic trees include *Podocarpus mannii*, *Balthasaria mannii*, *Psychotria guerkeana* and *P. nubicola*.

Like São Tomé, Príncipe was also once completely covered in forest but most of the accessible regions of the island have now been converted to agriculture. The remaining virgin rain forest in Príncipe, in the south, resembles the lowland rain forest of São Tomé but is relatively impoverished. Endemic trees include *Rinorea insularis*, *Ouratea nutans*, *Casearia mannii*, *Croton stelluliferus* and *Erythrococca columnaris*. Towards the summit of the Pico de Príncipe the forest assumes a slightly more montane character but the altitude is too low to support the mist forest vegetation seen in São Tomé.

Mangroves are scarce on both islands and are consequently not shown on Figure 28.1. They are cut by the local people for fuel and are severely threatened by such use.

Forest Resources and Management

The total area of primary forest is estimated at about 284 sq. km (Interforest AB, 1990), with 240 sq. km on São Tomé and about 40 sq. km on Príncipe. These figures are somewhat lower than those reported for 1980 by FAO (1988) but are in good agreement with the areas shown on Figure 28.1 (see Table 28.1). The forest cover in this figure was obtained from a photocopy of a map prepared by the Bureau pour la Développement de la Production Agricole (BDPA, 1985).

The present total production of timber and fuelwood is estimated at 168,000 cu. m per year (Interforest AB, 1990), but only a negligible amount, or none at all, of this comes from primary forest. There are 15 sawmills producing 4500 cu. m of sawn wood annually, corresponding to about 9000 cu. m of sawlogs. Fuelwood production, at around 157,500 cu. m per year, is much greater than sawnwood production. Most fuelwood is used domestically for cooking, perhaps 137,000 cu. m, with a further 6200 cu. m used by small industries, schools and bakeries. FAO (1991) gives no estimates of fuelwood and charcoal production on the islands. The biggest industrial demand is for drying cacao, 7000 t of which were produced in 1988, requiring about 14,300 cu. m of wood for drying. The World Bank scheme to rehabilitate the cacao plantations will produce a higher demand for wood. Total wood consumption is forecast to rise to 208,000 cu. m per annum by the year 2000 as cacao plantations are rehabilitated. The regenerating secondary forests and plantations of fast-growing tree species will be used to meet the shortfall in firewood demand (Interforest AB, 1990).

Table 28.1 Estimates of forest extent in São Tomé and Príncipe

Rain forests	*Area (sq. km)*	*% of land area*
Lowland	224	23.3
Montane	75	7.8
Totals	299	31.1

(Based on analysis of maps used to produce Figure 28.1.)

Deforestation

São Tomé and Príncipe were discovered by the Portuguese around 1470 and there is no evidence of occupation by man before that time. Sugarcane cultivation began in the 1480s and it had become an important crop in the north of São Tomé by 1550, reaching its peak production in 1578. As a result, much of the forest at lower altitudes in the north was destroyed at an early date. During the 17th century, sugarcane production declined so, for a while, little more forest was cut down. However, in 1800 and 1802 coffee was introduced to São Tomé and Príncipe respectively. Coffee farms were established up to 1200 m in São Tomé, causing more forest destruction. By 1870 coffee production declined but the deforested areas were replanted with cacao and more trees were cut to extend the plantations. Some indigenous trees were left to provide shade for the cacao and several exotic tree species were planted. After 1975, when São Tomé and Príncipe became independent, many of the plantations were abandoned and forest is regrowing on the cleared land.

Most of the remaining primary forest has survived because of its inaccessibility on steep slopes in the wettest, most inhospitable parts of the islands which are unsuitable for either cultivation or human habitation. However, a World Bank team recently reported that firewood collectors were beginning to encroach on the natural forests in the extreme south-west of São Tomé.

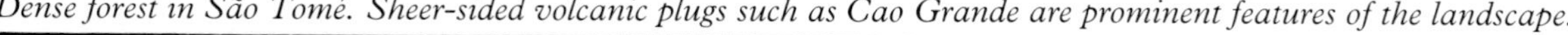

Dense forest in São Tomé. Sheer-sided volcanic plugs such as Cao Grande are prominent features of the landscape. P. Steele/ICCE

São Tomé and Principe

Biodiversity

The rain forests of São Tomé and Príncipe are remarkable for the high levels of endemism found in both their flora and fauna. There are 601 plant species known from São Tomé and 314 from Príncipe. São Tomé has one endemic genus and 87 single-island endemic species (14.5 per cent of the indigenous flora), while Príncipe has one endemic genus and 32 single-island endemic species (10.2 per cent of the total). Of the 194 plant species endemic to the Gulf of Guinea islands, only 16 occur on more than one island. This emphasises the high degree of isolation under which these floras evolved and suggests that each island received its flora separately from the mainland.

Birds are the most numerous and important group of terrestrial vertebrates. There are five endemic genera and 26 endemic species. Many survive well in human-altered habitats, but others are confined to undisturbed primary forest. Seven species are listed by Collar and Stuart (1985) as being threatened and others may give cause for concern (Jones and Tye, 1988). Indeed, the forests of south-west São Tomé have been ranked second in importance in a list of 75 key forests for threatened birds in Africa (Collar and Stuart, 1988). The threatened species include the maroon pigeon *Columba thomensis*, São Tomé scops owl *Otus hartlaubi*, the dwarf olive ibis *Bostrychia bocagei* and the São Tomé white-eye *Zosterops ficedulinus*.

The only indigenous mammals on both islands are bats and shrews. Of the five species of bat recorded, one, the São Tomé little collared fruit-bat *Myonycteris brachycephala*, is an endemic species and two are endemic subspecies. The São Tomé white-toothed shrew *Crocidura thomensis* is endemic to that island, while there is an endemic subspecies of a mainland shrew on Príncipe. All the larger mammals have been

Figure 28.1 Forest cover on São Tomé and Príncipe

(*Source:* BDPA, 1985)

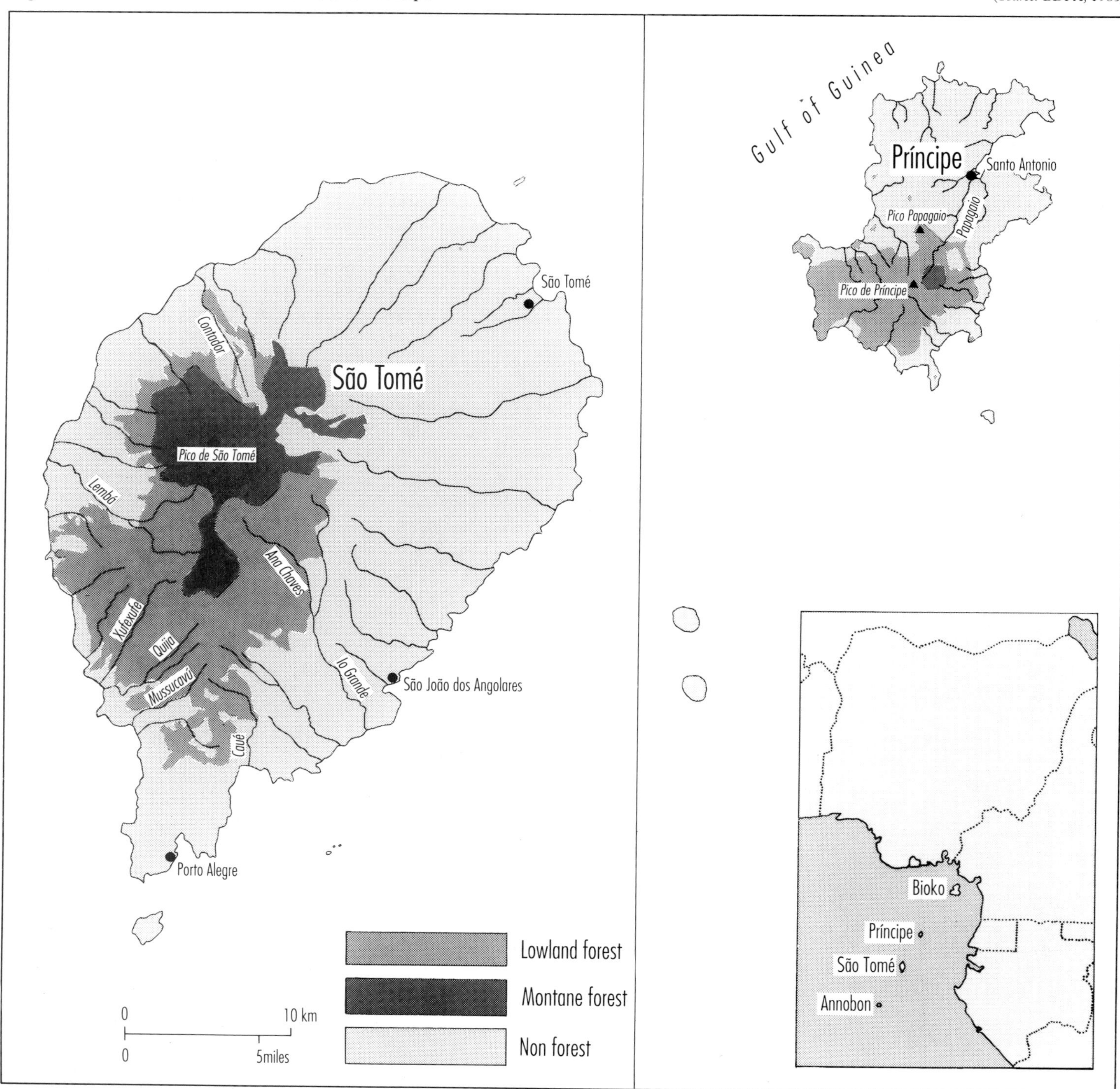

introduced since the islands were discovered. Now both islands have populations of the mona monkey *Cercopithecus mona*, feral cats and pigs, black and brown rats and house mice. In addition, the African civet *Viverra civetta* and the weasel *Mustela nivalis* are found on São Tomé.

There are 14 terrestrial reptiles on São Tomé and 11 on Príncipe of which two are endemic to the latter island (the legless skink *Feylinia polyepis* and the burrowing snake *Typhlops elegans*); four are endemic to both islands – a gecko, *Hemidactylus greefi*, a skink, *Panaspis africana* and two burrowing snakes, *Rhinotyphlops feae* and *R. newtoni*. Another endemic gecko, *Lygodactylus thomensis*, is shared by Annobon as well as the other two islands.

Five species of amphibian are known to occur on São Tomé. Four are single-island endemics: the two tree frogs *Hyperolius molleri* and *H. thomensis*, the ranid frog *Ptychadena newtoni* and the caecilian *Schistometopum* sp. The occurrence of a caecilian on São Tomé is unique for a volcanic, oceanic island. Two of the three amphibians that have been recorded on Príncipe are single-island endemics – a tree frog, *Leptopelis palmatus*, and the ranid *Phrynobatrachus feae*. *P. dispar* is endemic to the two islands.

Conservation Areas

There are no protected areas in either São Tomé or Príncipe. However, on both islands an ecological zone (*zona ecológica*) has been proposed. The land within the proposed zones belongs to the state and is the responsibility of the Ministério de Agricultura e Pescas. However, the boundaries of these two areas remain to be formally delineated and there is no management or legal protection for them other than the prohibition of tree felling.

On São Tomé the proposed ecological zone is located between 0°05'–0°21'N latitude and 6°28'–6°39'E longitude and covers an area of 245 sq. km. It comprises all the remaining untouched lowland and montane rain forest at all altitudes in the catchments of the major rivers draining the central massif of São Tomé. The canopy is closed except on the steepest slopes and summits, but there are small, scattered areas of shrub and bamboo thicket in the region.

Príncipe's proposed ecological zone, an area of approximately 45 sq. km, is located between 1°32'–1°37'N latitude and 7°20'–7°26'E longitude. It covers all the remaining untouched rain forest south of and including Pico Papagaio, and consists mainly of lowland and submontane rain forest with small areas of bamboo and shrub thicket.

Initiatives for Conservation

Although formal conservation legislation has not yet been promulgated, forest protection and wildlife conservation are the responsibility of the recently created Commission for Forestry Coordination (Comissão de Coordenação Florestal) within the Ministry of Agriculture and Fisheries. São Tomé and Príncipe will receive EEC funding over five years to support the Commission in establishing fuelwood and timber plantations to avoid exploitation of primary forest and for conservation education. Similar financial resources for forest conservation were pledged in 1990 by the Gesellschaft für Technische Zusammenarbeit (GTZ, German technical aid). The International Council for Bird Preservation with the International Centre for Conservation Education (Harrison and Steele, 1989) have, with USAID and EEC funding, already provided wildlife conservation publicity and educational materials to the Association of Friends of Nature (Associação dos Amigos da Natureza), a Sãotomean NGO constituted in 1988 to promote wildlife conservation. IUCN has recently published a detailed account of the forest conservation issues affecting the islands (UICN, 1991).

References

BDPA (1985) *Potencialidades agricolas: República Democrática de São Tomé e Príncipe.* Bureau pour le Développement de la Production Agricole, Paris, France.

Bredero, J. T., Heemskerk, W. and Toxopeus, H. (1977) *Agriculture and Livestock Production in São Tomé and Príncipe (West Africa)*. Unpublished report, foundation for Agricultural Plant Breeding, Wageningen, The Netherlands. 43 pp.

Collar, N. J. and Stuart, S. N. (1985) *Threatened Birds of Africa and Related Islands. The ICBP/IUCN Red Data Book Part 1.* ICBP/IUCN, Cambridge, UK. 761 pp.

Collar, N. J. and Stuart, S. N. (1988) *Key Forests for Threatened Birds in Africa.* ICBP Monograph No. 3. ICBP, Cambridge, UK. 102 pp.

Exell, A. W. (1944) *Catalogue of the Vascular Plants of São Tomé (with Príncipe and Annobon).* British Museum (Natural History), London, UK. xi + 428 pp.

Exell, A. W. (1956) *Supplement to the Catalogue of the Vascular Plants of São Tomé (with Príncipe and Annobon)*. British Museum (Natural History), London, UK. 58 pp.

Exell, A. W. (1973) Angiosperms of the Islands of the Gulf of Guinea (Fernando Po, Príncipe, São Tomé and Annobon). *Bulletin of the British Museum (Natural History) London* **4**(8): 327–411.

FAO (1988) *An Interim Report on the State of Forest Resources in the Developing Countries*. FAO, Rome, Italy. 18 pp.

FAO (1991) *FAO Yearbook of Forest Products 1978–1989.* FAO Forestry Series No. 24 and FAO Statistics Series No. 97. FAO, Rome, Italy.

Harrison, M. J. S. and Steele, P. (1989) *ICBP/EEC Forest Conservation Mission to São Tomé and Príncipe, January–March 1989.* Report on Conservation Education and Training. Unpublished report to ICBP. 13 pp.

Interforest AB (1990) *Democratic Republic of São Tomé and Príncipe. Results of National Forest Inventory and Study of Supply and Demand of Primary Forest Products*: Q3770-Ejpl-1 and -2.

Jones, P. J. and Tye, A. (1988) *A Survey of the Avifauna of São Tomé and Príncipe.* ICBP Study Report No. 24. ICBP, Cambridge, UK.

UICN (1991) *Conservação dos Ecossistemas Florestais na República Democrática de São Tomé e Príncipe.* Report by P. J. Jones, J. P. Burlison and A. Tye. UICN, Gland, Switzerland and Cambridge, UK.

Authorship

Peter Jones of the Department of Natural Resources and Forestry, Edinburgh University, UK.

29 Sierra Leone

Land area 71,620 sq. km
Population (mid-1990) 4.2 million
Population growth rate in 1990 2.5 per cent
Population projected to 2020 8.9 million
Gross national product per capita (1988) US$370
Rain forest (see map) 5064 sq. km
Closed broadleaved forest (end 1980)* 7400 sq. km
Annual deforestation rate (1981–5)* 60 sq. km
Industrial roundwood production† 140,000 cu. m
Industrial roundwood exports† nd
Fuelwood and charcoal production† 2,874,000 cu. m
Processed wood production† 12,000 cu. m
Processed wood exports† 3000 cu. m
* FAO (1988)
† 1989 data from FAO (1991)

Sierra Leone is located on the Atlantic coast of West Africa. It lies at the western end of the Upper Guinea Forest Block and is one of the more severely deforested of the countries in this region. Over 50 per cent of the country has climatic conditions suitable for tropical rain forest, but less than 5 per cent is still covered with mature, dryland closed forest. Deforestation is mainly a result of the rapidly increasing human population requiring more agricultural land and fuelwood.

Most of the remaining closed forest is in government-controlled forest reserves. However, timber extraction usually takes place without adequate forest management and some of the forest has been over-exploited. In addition, there has been little effort to replant or encourage regeneration once an area has been logged. Although there are several areas proposed for protection, the only gazetted conservation area at present is the tiny Tiwai Island Game Reserve.

Introduction

Sierra Leone, one of the smallest countries in Africa, shares borders with Guinea and Liberia. It extends approximately 352 km from north to south between latitudes 6°55'N and 10°00'N and 323 km from east to west between longitudes 10°14'W and 13°17'W. The coastal half of the country lies below the 200 m contour, while the interior half is mostly above 500 m. Much of the coastline is fringed by mangroves extending 30–50 km inland along estuaries. Thereafter, the land rises to a gently undulating plateau which continues inland to the foothills of the Guinean Dorsale crossing the interior from north-west to south-east. Mt Bintumani in the Loma Mountains, at 1946 m the highest point in sub-Saharan Africa west of Mt Cameroon, and Sankan Biriwa Peak (1715 m) in the Tingi Mountains, are the highest points in the country.

Mean annual rainfall of 3000–5000 mm is experienced at the coast, decreasing inland to 2000 mm on the northern border. There is a long wet season from May to November, but the dry season is severe with virtually no rain and cool, desiccating Harmattan winds. The mean annual temperature fluctuates little from 27°C; August is the coolest month and March/April the hottest.

Sierra Leone has a dense and rapidly growing human population, estimated to have increased by 78 per cent from 1963 to 1988. Density is about 58 individuals per sq. km with concentrations in the south-east and north-west. About 72 per cent of the population are rural (PRB, 1990) and it is the clearing and burning of vegetation for subsistence crops that is the major cause of forest loss. The staple crop is rice which tends to be planted in upland areas, while the main cash crops grown are cacao, coffee and oil palm. Forest areas are selectively thinned and under-planted with cacao and coffee bushes; cola trees are also grown in these plantations.

Three main vegetation zones occur, coastal mangroves, evergreen high forest in the east and drier guinea savanna in the north and north-west. Almost everywhere the vegetation has been altered by human pressures. Indeed, land formerly covered in forest can often be identified by the presence of oil palms, which have economic value so have been retained by the farmers. Now only a very limited area of the country supports climax vegetation. However, small patches of closed forest (generally under five hectares) are often maintained near settlements as sacred sites for traditional religious uses.

The Forests

Cole (1978) distinguished three main moist forest types. These are evergreen, semi-deciduous and montane forest. In addition, edaphic climax forests occur in freshwater swamps (with *Raphia* palms), gallery and fringing forest and mangroves.

The moist evergreen or semi-deciduous forests are mostly located in forest reserves on hill slopes: Gola, Kambui, Nimini, Dodo Hills, Freetown Peninsula, Tama-Tonkoli, Kasewe, Loma and Tingi Hills (FAO/UNEP, 1981). They have a closed canopy with trees over 30 m high and those now in existence are mostly secondary forests.

The moist evergreen forest in the Gola Reserve, which contains the last large remnant of lowland, closed canopy rain forest in Sierra Leone (see case study), is typical of climax Upper Guinean rain forest. The main canopy, with a height of 30–35 m, is dominated by *Heritiera utilis* and *Cryptosepalum tetraphyllum* with some *Erythrophleum ivorense* and *Lophira alata*. There are emergents of up to 60 m (Fox, 1968; Davies, 1987). Other typical trees of the moist evergreen forests are *Klainedoxa gabonensis*, *Uapaca guineensis*, *Oldfieldia africana*, *Brachystegia leonensis* and *Piptadeniastrum africanum* (FAO/UNEP, 1981). Tree species diversity is comparatively low in the Golas: only 60 species or so (with a diameter of at least 10 cm) per hectare were recorded there (Davies, 1987). One tree, *Didelotia idae*, is endemic to Gola North.

The moist semi-deciduous forests, such as those found on the Loma Mountains and Tingi Hills, commonly contain tree species such as *Daniellia thurifera*, *Terminalia ivorensis*, *T. superba*, *Parkia bicolor* and *Anthonotha fragrans*. These species are associated with evergreen trees such as *Parinari excelsa*, *Bridelia grandis*, *Treculia africana* and *Pycnanthus angolensis* (FAO/UNEP, 1981). The forests of Loma and Tingi gradually change into submontane gallery forests reaching 1700 m. In these areas, tree ferns are found along river courses at higher altitudes. Most other areas of semi-deciduous forest have now been logged.

Young secondary forests, derived from former clearing of the forests for agriculture, are found mainly in the south-eastern part of the country on hill slopes. Trees range in height from 10–30 m. Species characteristic of secondary forest are *Funtumia africana*, *Holarrhena floribunda* and *Pycnanthus angolensis* (Savill and Fox, 1967). Near villages, these secondary forests provide a suitable habitat for an undercropping of cacao and coffee.

Fringing swamp forests, which are subject to seasonal flooding, are mainly present in the northern part of the coastal area and situated along valley bottoms, shallow drainage ways and lake margins. There is usually a closed canopy and trees can be up to 30 m high. Typical tree species include *Pterocarpus santalinoides*, *Napoleonaea vogelii*, *Uapaca heudelotii*, *Newtonia elliotii*, *Myrianthus arboreus*, and *Cynometra vogelii*. *Mitragyna stipulosa* is an important tree in the swamps. It remains uncut even in intensively farmed inland valley swamps because the leaves are useful for wrapping goods, especially cola nuts. *Raphia* spp. also occur along valley bottoms within both the moist forest and the savanna zones. The canopy is usually closed and may be 20 m in height.

Mangroves

Mangrove swamps used to occupy most of Sierra Leone's 825 km of highly indented coast. The northern coast, in particular, was entirely cloaked with mangrove forest that extended far up rivers and then gave way to *Pandanus* and *Raphia* swamps which were, in turn, replaced by freshwater swamp forest along the river sides. The main tree species in the mangroves are *Rhizophora racemosa*, *R. mangle* and *R. harrisonii* and they may reach heights of 20 m. In the late 1970s it was estimated that 1716 sq. km of mangrove remained (Davies and Palmer, 1990 after Cole, 1978 and Gordon *et al.*, 1976). Map 29.1 indicates an area of only 1015 sq. km of mangrove and 43 sq. km of swamp forest left in the country by 1987. A great deal of the mangrove and swamp forests have been cleared for agriculture, particularly rice cultivation. In addition, the mangroves are used extensively for fuelwood and building materials and for materials with which to make fish traps. Many of them are over-fished and none is protected.

Table 29.1 Estimates of forest extent based in Sierra Leone

Rain forests	*Area (sq. km)*	*% of land area*
Lowland	3,925	5.5
Montane	81	0.1
Mangrove	1,015	1.4
Swamp	43	0.06
Totals	5,064	7.1

(Based on analysis of Map 29.1. See Map Legend on p. 250 for details of sources.)

The mangrove, palm and swamp forests are important bird habitats. The threatened West African manatee *Trichechus senegalensis* occurs in the mangroves, but it is heavily hunted for its meat and because it is considered to be a pest by rice growers and fishermen (Reeves *et al.*, 1988).

Forest Resources and Management

The most thorough survey of vegetation in the country is based on aerial photographs taken in 1976 (Gordon *et al.*, 1979). At that time it was estimated that 3652 sq. km of the southern and eastern part of Sierra Leone was covered in closed-canopy high (over 30 m) forest. Another 2610 sq. km was covered in secondary forest (canopy between 10 and 30 m). Map 29.1, based on satellite images taken in 1987 by UNEP/GRID (see Map Legend on p. 250), shows 3925 sq. km of lowland rain forest and 81 sq. km of montane forest (in the Loma and Tingi Hills) in the country (Table 29.1). This gives a total dryland forest cover of 4006 sq. km, a reduction of over 2000 sq. km from the 1976 figures calculated by Gordon *et al.* (1979). Primary and secondary forest have not been differentiated on the UNEP/GRID dataset and it is possible that some of the more open, lower, secondary forest was not identified as forest on the satellite images. FAO (1988) estimated that there were 7400 sq. km of closed forest remaining in 1980, but this figure presumably includes both the secondary and primary dryland forest and the mangroves.

Extensive opencast excavations at Baomahun gold mine near Kangari Hills Forest Reserve. Most of the visible hills are gold ore-bearing and are due to be mined in the future.
G. Drucker

Sierra Leone

Most of the remaining closed canopy forest in Sierra Leone is in the government gazetted forest reserves, which were established to serve either protection or timber production functions. Local communities were compensated for giving up management of the land that was incorporated into the reserves. They did, however, retain ownership of the area. There are 29 forest reserves in Sierra Leone, covering a total area of 2850 sq. km (3.9 per cent of the country) and forest within these covers 2.3 per cent of the land area (Davies, 1987). FAO/UNEP (1981) estimated that, in 1980, there was 2190 sq. km of productive closed forest remaining in the country and of this, an estimated 1000 sq. km was unlogged.

The most important remaining forest areas are the highlands of the southern and eastern provinces, on the axial mountain chain of the Freetown Peninsula and in the lowland Gola Reserves on the eastern border of the country (Davies, 1987). Manpower shortages and lack of equipment and financial support have severely limited the effectiveness of the Forest Department in protecting reserve boundaries against agricultural encroachment. This is a particular problem in the Tama-Tonkoli Forest Reserve but there is, as yet, little encroachment in the Gola Reserves.

The Forestry Division has established plantations using the taungya system, organising farmers to tend tree seedlings alongside their agricultural crops. Once the farmers have harvested their crops and moved on to farm other pieces of land, the Forest Department takes over the care of the remaining tree crops. In the past this has worked well, with some of the forest reserves being planted with indigenous trees, and many roadside 'protected forests' being planted to supply rural communities with timber. Unfortunately, in recent times, it has proved impossible to control farmers who return to fell and burn timber plantations when the soil has recovered sufficiently to give a good yield of food crops, but before the trees have reached commercial size.

Most of the timber extraction in Sierra Leone takes place without any form of forest management. There is supposedly a limit on the size of trees that can be cut – at least 60 cm dbh – but this regulation is rarely enforced by the under-resourced and undermanned forest department. More than 60 species of tree are logged but the nine most important account for over 70 per cent of the production: *Mimusops heckelii*, *Terminalia ivorensis*, *Brachystegia leonensis*, *Oldfieldia africana*, *Berlinia confusa*, *Piptadeniastrum africanum*, *Didelotia idae*, *Heritiera utilis* and *Ceiba pentandra* (FAO/UNEP, 1981).

The two major logging companies, Forest Industries Sierra Leone Ltd (formerly Forest Industries Corporation) and Panguna sawmills, and three other small ones, operate under licences and not under concession agreements. Logged areas are usually left unguarded once the logging concerns leave the area and are frequently invaded by farmers.

The timber trade in Sierra Leone began as early as 1816. This may have been because the country was an important outpost of the slave trade and many of the slaving outposts converted to forest exploitation when the British government suppressed slavery. The trees were initially cut for shipbuilding and were then exported to England. By 1827 about 10,742 tons of timber had been shipped to England, and between then and 1835 a further 82,911 tons were exported. By 1840 the supply of timber near the coast was more or less exhausted and the exploiters progressed inland. A forestry department was set up in 1911 with a Chief Conservator of Forests and the first forest reserve was created.

In 1980, some 50,000 cu. m of timber was felled by five sawmills over an area of about 17 sq. km, but by 1984 industrial production had fallen to 17,748 cu. m with only three mills in production (Davies and Palmer, 1990). The permitted cut for Kambui Hills, Gola East and West and Gola North before 1975 was 24,440 cu. m per annum but maximum production never reached even 50 per cent of the allowed volume. The weak economic situation, combined with poor industrial management has kept log production low. However, pitsawyers' production has increased from 2000 cu. m in 1975 to 20,000 cu. m in 1985 (Davies and Palmer, 1990).

Forest resource depletion through controlled felling is not excessive since areas are not clearfelled and timber stocks are not destroyed, although some rare species which are favoured timber trees (e.g. *Entandrophragma* spp.) do appear to be severely depleted. However, in areas where over-logging has occurred such as Kambui Hills, Dodo Hills, Nimini South, Gboi Hills and parts of Gola East and Gola West, the forest is severely damaged, regeneration is impaired and potential for future yields is reduced.

Estimates of fuelwood consumption vary enormously. FAO (1991) estimated annual consumption of almost 3 million cu. m while the World Bank/UNDP Energy Commission (1987) estimated 7 million cu. m for 1988 and an annual increase of 14–20 per cent. The two principal industrial uses for fuelwood are for fish drying along the coast and tobacco drying in the central parts of the country. In rural areas, firewood is collected from farmland for domestic use and sold along the roadside for use by urban dwellers. In the drier north and west, where trees are scarcer, villagers will cut live trees for firewood, thus causing degradation of the savanna.

Three areas are being over-exploited for fuelwood: the Freetown Peninsula, *Lophira* savanna and the mangroves. In some of the forested areas, collecting of fuelwood has degraded tree species composition. For instance, in the John Obey area, where charcoal burners have been active for decades, there is an absence of *Uapaca guineensis*, while the preference for high-heat producing *Rhizophora* has led to the destruction of many mangroves along the Sierra Leone River, Sherbo River and Yawri Bay (Davies and Palmer, 1990).

Rural communities depend on numerous other forest products including cooking oils, medicines, wild fruits and vegetables, rattan palms and bamboos. Little is known of the rates of exploitation and depletion of these resources, but most of them are declining through conversion of the forest to farmland rather than through over-exploitation (Davies and Palmer, 1990).

Bushmeat is a traditional source of protein in West Africa and, as a result, hunting is a major cause of the decline of mammals in Sierra Leone. Monkeys and antelopes are most commonly shot or trapped, but few species are safe; even large birds are killed. In the 1940s and 1950s, the Ministry of Agriculture in Sierra Leone organised monkey drives with a bounty paid for each head. Jones (1951) reported that between 1948 and 1950 alone, head bounties were paid for over 60,000 monkeys. Liberian hunters used to kill the animals, take the meat across the border and leave the heads with the local farmers so they could claim the bounty money (Davies, 1987). Commercial hunting by Liberians occurred on a large scale until 1985 since when, with the devaluation of the Liberian dollar and border closures, the trade has subsided. However, most of the local people now hunt for economic gain, to supply a rapidly expanding domestic market, rather than for their own subsistence. The logging roads through the forests provide the hunters with improved access to their quarry.

The Forestry Act of 1988 enables the Chief Conservator of Forests to classify the forests as either national forests or community forests. National forests may be constituted as production forests where the primary objective is the extraction of timber and wood products, or as protection forests with the primary objective of preserving the entire forest environment including the flora, fauna, soil and water-catchment values. Classified forests may already be under state ownership or may be purchased or leased under the act. There are provisions for reforestation following timber extraction and for the preparation and authorisation of forest management and working plans.

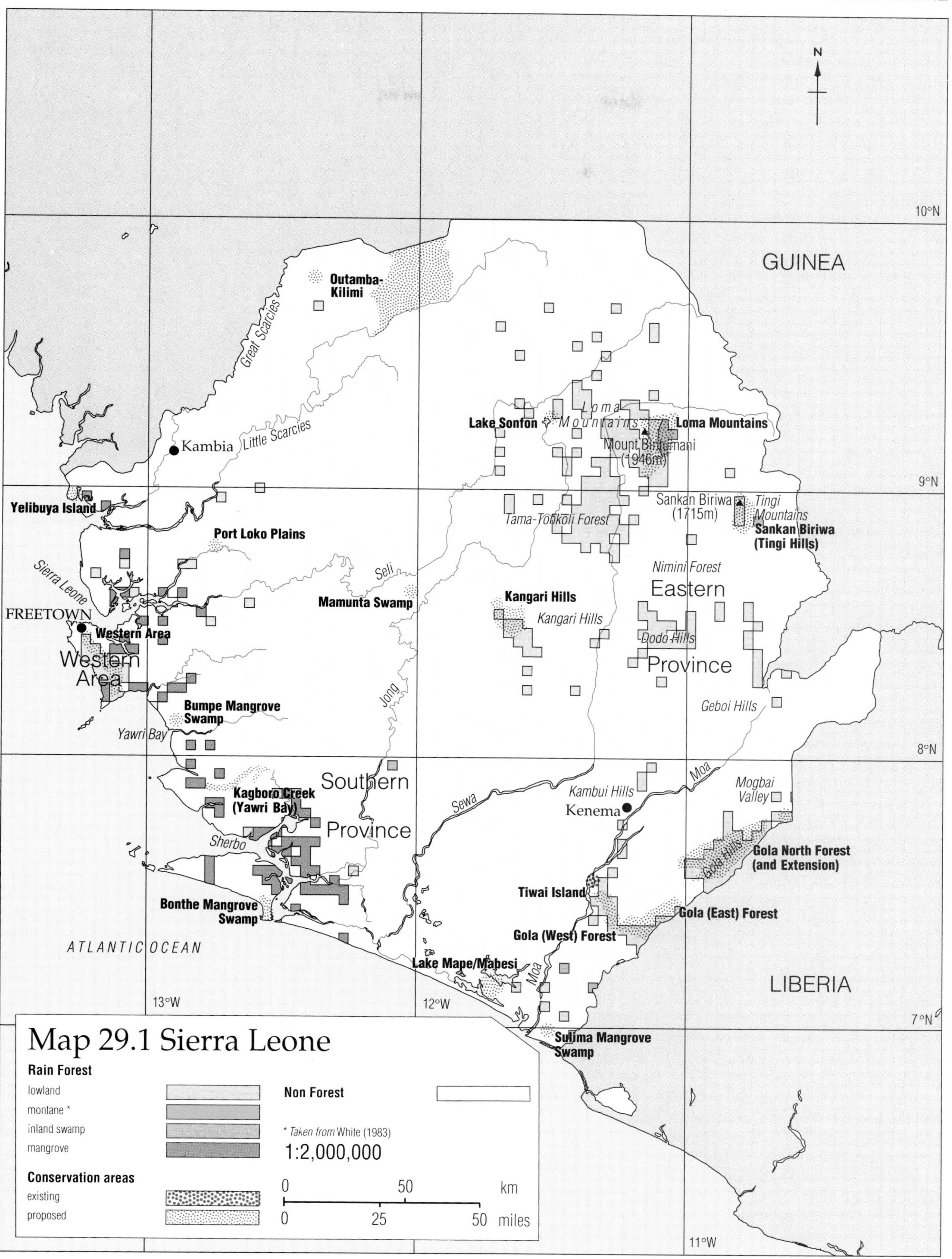

GUINEA
LIBERIA
ATLANTIC OCEAN
10°N
9°N
8°N
7°N
13°W
12°W
11°W
N
Outamba-Kilimi
Great Scarcies
Little Scarcies
Kambia
Lake Sonfon
Loma Mountains
Loma Mountains
Mount Bintumani (1945m)
Yelibuya Island
Port Loko Plains
Sierra Leone
FREETOWN
Western Area
Western Area
Mamunta Swamp
Seli
Tama-Tonkoli Forest
Sankan Biriwa (1715m)
Tingi Mountains
Sankan Biriwa (Tingi Hills)
Nimini Forest
Eastern Province
Kangari Hills
Kangari Hills
Dodo Hills
Geboi Hills
Bumpe Mangrove Swamp
Yawri Bay
Jong
Southern Province
Kagboro Creek (Yawri Bay)
Sherbo
Sewa
Moa
Kambui Hills
Kenema
Mogbai Valley
Gola Hills
Gola North Forest (and Extension)
Tiwai Island
Gola (East) Forest
Gola (West) Forest
Bonthe Mangrove Swamp
Lake Mape/Mabesi
Moa
Sulima Mangrove Swamp
Map 29.1 Sierra Leone
Rain Forest
lowland
montane *
inland swamp
mangrove
Non Forest
* Taken from White (1983)
1:2,000,000
Conservation areas
existing
proposed
0 50 km
0 25 50 miles

Sierra Leone

Deforestation

At least 50 per cent of Sierra Leone has climatic conditions suitable for tropical evergreen forest. However, agriculture and timber exploitation have greatly depleted the area under forest and by 1976 only 5 per cent of the country was covered in primary forest; another 3.5 per cent was under secondary forest (Gordon *et al.*, 1979; Davies, 1987). FAO (1988) estimated that 60 sq. km of forest is lost each year but the most serious problem is the progressive deterioration that is occurring in all the forests.

Slash and burn agriculture has been the main cause of forest loss in Sierra Leone. Most cleared areas are used for rice production – hill rice is grown in preference to swamp rice – and crops are cultivated for one or two years before the area is left fallow for eight to ten years. There is not enough time for the forest to regenerate before the land is recleared, so large areas of scrubby bush develop instead. In addition, there is little effort to replant or encourage regeneration once an area of forest reserve has been logged; moreover, the timber roads open up the forest to further exploitation by subsistence farmers. Kingston (1986) estimated that at least 2340 sq. km of woody vegatation is burned annually for farms, of which 44 sq. km is in closed forest and 14 sq. km in mangroves.

Biodiversity

More than 2000 species of plants occur in the country, of which a minimum of 74 species and one genus are endemic (Davis *et al.*, 1986). The Gola Forests and the hill forests of the Loma Mountains and Tingi Hills are centres of plant diversity and endemism.

Numbers of invertebrates are unknown but two endemic species of dragonfly, *Argiagrion leoninum* and *Allorhizucha campioni*, have been recorded in the country (Stuart *et al.*, 1990). The rare African giant swallowtail butterfly *Papilio antimachus* reaches its westernmost limit in Sierra Leone.

There are two endemic species of amphibian in the country, *Cardioglossa aureoli* in the Western Area Forest Reserve and the toad, *Bufo cristiglans*, which is known only from Tingi Hills. Four other species are near-endemics, three (*Hyperolius occidentalis*, *Phrynobatrachus alticola*, *Ptychadena retropunctata*) being found also in Guinea and the fourth (*Pseudhymenochirus merlini*) in Guinea and Guinea-Bissau (WCMC, 1987).

Some 614 bird species have been recorded in Sierra Leone (Stuart *et al.*, 1990). The six threatened species reported by Collar and Stuart (1985) are all forest birds. The white-breasted guineafowl *Agelastes meleagrides*, which was recently rediscovered in Sierra Leone (Davies, 1987; Allport *et al.*, 1989), is considered one of the most threatened birds in continental Africa (Collar and Stuart, 1985). The western wattled cuckoo-shrike *Campephaga lobata*, the yellow-throated olive greenbul *Criniger olivaceus* and the white-necked rockfowl *Picathartes gymnocephalus* are all found only in western Africa and are all vulnerable to extinction as a result of forest destruction. The rare rufous fishing owl *Scotopelia ussheri* is restricted to the rain forest zone west of the Dahomey Gap and is probably threatened by the destruction of the forest and mangroves. Little is known of the Gola malimbe *Malimbus ballmanni* which was described as recently as 1974; it has been found in only Sierra Leone, Liberia and Côte d'Ivoire. Two other threatened species have recently been recorded (Allport *et al.*, 1989) from Gola Forest: the Nimba flycatcher *Melaenornis annamarulae* and the yellow-footed honeyguide *Melignomon eisentrauti*.

The two small populations of forest elephants *Loxodonta africana cyclotis* in the Gola Forest are probably the last viable populations in the country (Roth and Merz, 1983). However, by 1989 those in Gola North had dispersed from their core area in the Mogbai valley and their future is very uncertain (A. G. Davies, *in litt.*). There are some elephants in the Kangari Hills, Outamba-Kilimi National Park and Tama-Tonkali Forest Reserve, but these areas are small and under heavy hunting pressure (Davies, 1987). The Moa River around Tiwai island and the Mahoi River in the Gola forest are important sites for the rare pygmy hippopotamus *Choeropsis liberiensis*.

Most of Sierra Leone's 18 species of antelope are considered threatened to a greater or lesser extent (Teleki *et al.*, 1990). Populations are depleted by over-hunting and habitat destruction. Of particular concern are the very rare Jentink's (see Davies and Birkenhager, 1990) and zebra duikers (*Cephalophus jentinki* and *C. zebra*). There are 15 species of primate in the country of which 11 are forest species. Of these the chimpanzee, diana monkey, red and olive colobus (*Pan troglodytes*, *Cercopithecus diana*, *Procolobus [badius] badius* and *Procolobus verus* respectively) are listed as threatened (Lee *et al.*, 1988). Temminck's squirrel *Epixerus ebii* appears to be restricted to primary forest and is probably the squirrel most in need of protection in Sierra Leone (Davies, 1987).

Table 29.2 Conservation areas of Sierra Leone

Existing and proposed conservation areas are listed below. Forest reserves are not included or mapped.

	Existing area (sq. km)	*Proposed area (sq. km)*
National Parks		
Lake Mape/Mabesi		75
Lake Sonfon*		52
Loma Mountains*		332
Outamba-Kilimi		808
Western Area*		177
Strict Nature Reserves		
Bonthe Mangrove Swamp		101
Gola (East) Forest*		35
Gola (West) Forest*		62
Gola North Forest }*		40
Gola North Extension }*		39
Mamunta Swamp		21
Port Loko Plains		26
Sulima Mangrove Swamp		26
Yelibuya Island		39
Game Reserves		
Bagru-Moteva Creek†		50
Kagboro Creek (Yawri Bay)*		50
Kangari Hills*		86
Kpaka-Pujehun†		25
Sankan Biriwa (Tingi Hills)*		119
Sewa-Waanje†		100
Tiwai Island*	12	
Game Sanctuaries		
Bo Plains†		26
Bumpe Mangrove Swamp		49
Totals	12	2,338

(*Sources:* Teleki, 1986; FAO, 1990)

* Area with moist forest within its boundaries according to Map 29.1. Note that the forested area on Tiwai Island is too small to show on map.

† Location data unavailable – not mapped.

Conservation Areas

The Ministry of Agriculture, Natural Resources and Forestry (MANRF) has overall responsibility for the environment; forest resources are under the authority of the Forestry Division, the Wildlife Conservation Branch of which is responsible for protecting the flora and fauna.

The Wildlife Conservation Act (1972) allows for several categories of legally protected land: game sanctuaries, strict nature reserves, game reserves and national parks. There is no restriction on land use in game sanctuaries but hunting and trapping within them are prohibited. The nature reserves are supposed to protect land, and the flora and fauna therein, from any kind of injury or destruction, while the purpose of the national parks is to propagate, conserve and manage wildlife and wild vegetation and to protect sites, landscapes or geological formations of scientific or aesthetic value for the benefit and enjoyment of the public. There are two categories of forest reserve, those established for the management of forest resources and those established to restrict hunting. In addition, there are protected forests, which can be defined as managed plantation areas.

In his review of wildlife conservation in Sierra Leone, Phillipson (1978) made recommendations for a network of conservation areas to ensure the preservation of all types of vegetation and faunal communities. Additions and changes to the 18 key areas originally proposed by Phillipson have been made by Oates (1980), Teleki (1986), Davies (1987) and Chong (1987). Although the government endorsed Phillipson's report, little, as yet, has come of the recommendations (Davies and Palmer, 1990).

To date, only the small Tiwai Island Game Reserve, has been gazetted (see Table 29.2). This sanctuary contains a field research station, operated by Njala University College in collaboration with foreign universities. The proposed Outamba-Kilimi National Park has been notified but not yet gazetted although wildlife surveys, park development and conservation education are all being carried out in and around the area.

There are four non-hunting forest reserves: Western Area and Loma Mountains (proposed national parks), while Kangari Hills and Sankan Biriwa (Tingi Hills) are proposed game reserves, but these are intensively hunted and are being encroached by agriculturalists (see Table 29.2).

The proposed protected areas account for only 3.2 per cent of the country. At present, no major vegetation type is secure from farming, timber extraction or hunting. The absence of legal protection is a major obstacle to conservation management (Davies and Palmer, 1990).

Initiatives for Conservation

The Sierra Leone Council for the Protection of Nature (formerly the Sierra Leone Environment and Nature Conservation Association – SLENCA) has been active for many years in conservation education and wildlife surveys. More recently, in 1988, the Conservation Society of Sierra Leone was formed. Its objectives are to promote the efficient use and management of natural resources through education and public awareness, to support and encourage research on conservation issues and to act as a forum for the exchange of information both domestically and internationally. The Society has been supported by WWF and the MacArthur Fund.

The Royal Society for the Protection of Birds is working with the Conservation Society of Sierra Leone and the Forestry Division to initiate better management of the Gola Forest Reserves (see case study). This includes sustained yield logging and education programmes. Wildlife Conservation International and the US Peace Corps are also involved with this effort and they are trying to link Tiwai Island Game Reserve with the Gola Reserves via a chain of islands in the Moa River. The whole of the Gola/Tiwai area is being proposed as a Biosphere reserve.

The Gola Rain Forest Conservation Programme

The Gola Rain Forest Conservation Programme was launched on 14 May 1990 with the signing, in Freetown, of an agreement between the government of Sierra Leone, the Conservation Society of Sierra Leone, the Royal Society for the Protection of Birds (RSPB) and the International Council for Bird Preservation (ICBP). The aim of the programme is to ensure the sustainable management and conservation of the Gola Forest Reserves. Covering an area of 748 sq. km they have been proposed as strict nature reserves and contain the only substantial areas of lowland primary rain forest remaining in Sierra Leone. They represent one of the most important tracts of Upper Guinean rain forest for the conservation of threatened and endemic birds and mammals, ranking alongside Taï forest in Côte d'Ivoire as one of the two most significant forests for threatened bird species in West Africa.

The conservation programme is administered by the Forestry Division of the Ministry of Agriculture, Natural Resources and Forestry, with finance and other inputs supplied by RSPB and ICBP. The programme is multi-disciplinary, involving research, training of Forestry Division staff, education and awareness programmes and forest management. The research carried out, both on the ecological requirements of threatened species and on sustainable agricultural and silvicultural practices, will be used to design management plans for the forest which ensure long-term conservation while meeting the needs of local people and allowing some timber production to occur. The plans will include the establishment of strict nature reserve areas where no logging or hunting will be permitted.

As a first stage in the education and awareness part of the programme, two Sierra Leonean education officers have been appointed. They, together with programme coordinators, have undertaken an extensive tour of the Gola area, visiting all the paramount chiefs to hear their views and to inform people of the aims and activities of the scheme. The education office in Kenema arranges conservation teaching and activities in local schools and forms and supports conservation clubs.

Other activities in the early phase of the programme include re-marking of the old forest reserve boundaries which have become overgrown, participation in an environmental conference at Njala University College, and continuing research on threatened species of birds in the forest, including a PhD study on white-necked rockfowl by a lecturer from Freetown University. A small field station has been established in Gola forest and university students from within and outside Sierra Leone are being encouraged to make use of this and to carry out research which will contribute to the conservation of this precious resource.

Source: Nonie Coulthard, RSPB

Sierra Leone

WWF supported Peace Corps volunteers in the establishment of Outamba-Kilimi National Park and this project has been included in the Wildlands and Human Needs Program which has focused on rural development projects around the proposed park. In 1988, the United Nations Development Programme and FAO undertook a forestry sector review as part of the Tropical Forestry Action Plan process. This review (FAO, 1990) produced recommendations for a programme to address shortcomings in management of Sierra Leone's forest resources. However, the grave economic situation of Sierra Leone provides little room for optimism that these plans will be realised before the remaining forest disappears entirely.

References

Allport, G., Ausden, M., Hayman, P. V., Robertson, P. and Wood, P. (1989) *The Conservation of the Birds of Gola Forest, Sierra Leone.* Report of the UEA/ICBP Gola Forest Project (Bird Survey), October 1988–February 1989. ICBP Study Report No. 38. ICBP, Cambridge, UK.

Chong, P. W. (1987) *Proposed Management and Integrated Utilisation of Mangrove Resources in Sierra Leone.* FAO, Rome, Italy.

Cole, N. H. A. (1978) The Gola Forest in Sierra Leone: a remnant tropical primary forest in need of conservation. *Environmental Conservation* 7(1): 33–40.

Collar, N. J. and Stuart, S. N. (1985) *Threatened Birds of Africa and Related Islands. The ICBP/IUCN Red Data Book Part 1.* ICBP/IUCN, Cambridge, UK. 761 pp.

Davies, A. G. (1987) *The Gola Forest Reserves, Sierra Leone. Wildlife Conservation and Forest Management.* IUCN, Gland, Switzerland and Cambridge, UK. 126 pp.

Davies, G. and Birkenhager, B. (1990) Jentink's duiker in Sierra Leone: evidence from the Freetown Peninsula. *Oryx* **24**(3): 143–6.

Davies, G. and Palmer, P. (1990) Conservation of Forest Resources in Sierra Leone. In: *Tropical Forestry Action Plan. Inter-Agency Forestry Sector Review: Sierra Leone.* UNDP/FAO, Rome, Italy.

Davis, S. D., Droop, S. J. M., Gregerson, P., Henson, L., Leon, C. J., Villa-Lobos, J. L., Synge, H. and Zantovska, J. (1986) *Plants in Danger: What do we know?* IUCN, Gland, Switzerland and Cambridge, UK.

FAO (1988) *An Interim Report on the State of Forest Resources in the Developing Countries.* FAO, Rome, Italy. 18 pp.

FAO (1990) *Tropical Forestry Action Plan. Inter-Agency Forestry Sector Review: Sierra Leone.* UNDP/FAO, Rome, Italy.

FAO (1991) *FAO Yearbook of Forest Products 1978–1989.* FAO Forestry Series No. 24 and FAO Statistics Series No. 97. FAO, Rome, Italy.

FAO/UNEP (1981) *Tropical Forest Resources Assessment Project. Forest Resources of Tropical Africa. Part II: Country Briefs.* FAO, Rome, Italy.

Fox, J. E. D. (1968) Exploitation of the Gola forest. *Journal of West African Science Association* **13**: 185–210.

Gordon, O. L. A., Kater, G. and Schwaar, D. C. (1979) *Vegetation and Land Use in Sierra Leone.* UNDP/FAO Technical Report No. 2 AG: DP/SIL/73/002.

Jones, T. S. (1951) *Notes on the Monkeys of Sierra Leone.* Sierra Leone Agricultural Notes No. 22. GPO, Freetown.

Kingston, B. (1986) *A Forestry Action Plan for Sierra Leone.* FAO, Rome, Italy.

Lee, P. C., Thornback, J. and Bennett, E. L. (1988) *Threatened Primates of Africa. The IUCN Red Data Book.* IUCN, Gland, Switzerland and Cambridge, UK.

Oates, J. F. (1980) *Report on the Survey of* Colobus verus *and Other Forest Primates in Southern Sierra Leone with Comments on Conservation Problems.* Unpublished report. 25 pp.

Phillipson, J. A. (1978) *Wildlife Conservation and Management in Sierra Leone.* Special report to MANRF, Freetown.

PRB (1990) *1990 World Population Data Sheet.* Population Reference Bureau, Inc., Washington, DC, USA.

Reeves, R. R., Tuboku-Metzger, D. and Kapindi, R. A. (1988) Distribution and exploitation of manatees in Sierra Leone. *Oryx* **22**(2): 75–84.

Roth, H. H. and Merz, G. (1983) *Conservation of Elephants in Sierra Leone, with Special Reference to the Management of the Gola Forest Complex.* Unpublished final report to IUCN, Project 3039.

Savill, P. S. and Fox, J. E. D. (1967) *Trees of Sierra Leone.* Government Printers, Freetown.

Stuart, S. N., Adams, R. J. and Jenkins, M. D. (1990) *Biodiversity in Sub-Saharan Africa and its Islands: Conservation, Management and Sustainable Use.* Occasional Papers of the IUCN Species Survival Commission No. 6. IUCN, Gland, Switzerland.

Teleki, G. (1986) *Outamba-Kilimi National Park. A Provisional Plan for Management and Development.* Prepared for the Government of Sierra Leone, IUCN and WWF. George Washington University, Washington, DC, USA.

Teleki, G., Davies, A. G. and Oates, J. F. (1990) Sierra Leone. In: *Antelopes Global Survey and Regional Action Plans. Part 3: West and Central Africa.* East, R. (ed.). IUCN, Gland, Switzerland.

WCMC (1987) *The Amphibians of Africa.* Unpublished report.

World Bank/UNDP (1987) *Sierra Leone: issues and options in the energy sector.* Mission Report.

Authorship

Caroline Harcourt, WCMC with contributions from Glyn Davies, University College, London, John Waugh, WWF-US, John Oates, Hunter College, New York, Nonie Coulthard, Neil Burgess and Pete Wood, RSPB, UK, and Prince Palmer, Sierra Leone Forestry Division.

Map 29.1 Forest cover in Sierra Leone

Information on the remaining rain forests of Sierra Leone was taken from 1989–90 UNEP/GRID data which accompanies an unpublished report *The Methodology Development Project for Tropical Forest Cover Assessment in West Africa* (Päivinen and Witt, 1989). UNEP/GEMS/GRID together with the EEC and FINNIDA have developed a methodology to map and use 1 km resolution NOAA/AVHRR-LAC satellite data to delimit forest/non-forest boundaries in West Africa. These data have been generalised for this Atlas to show 2 ×2 km squares which are predominantly covered in forest. The study has also made use of higher resolution satellite data (Landsat MSS and TM, SPOT) and field data from Ghana, Côte d'Ivoire and Nigeria. Forest and non-forest data have been categorised into five vegetation types: forest (closed, defined as greater than 40 per cent canopy closure); fallow (mixed agriculture, clear-cut and degraded forest); savanna (includes open forests in the savanna zone and urban areas); mangrove and water. In addition this dataset portrays areas obscured by cloud. The forest and mangrove types and cloud obscured areas have been mapped in this Atlas. As shown on Map 29.1 montane, lowland rain forest, mangrove and swamp forest have been delimited by overlaying White's (1983) vegetation map on to the UNEP/GRID 'forest' and 'mangrove' categories.

Existing and proposed protected areas are mapped from a 1:950,000 (*c.* scale) map *The Forest Estate at 31st March 1961*, compiled by the Directorate of Overseas Surveys (n.d.), showing forest reserves and protected forests with hand drawn additions and updates from the Forestry Division, Freetown, and from spatial data held within files at the WCMC.

30 Southern Africa

ANGOLA
Land area 1,246,700 sq. km
Population (mid-1990) 8.5 million
Population growth rate in 1990 2.7 per cent
Population projected to 2020 18.5 million
Gross national product per capita (1988) nd
Closed broadleaved forest (end 1980)* 29,000 sq. km
Annual deforestation rate (1981–5)* 440 sq. km
Industrial roundwood production† 1,067,000 cu. m
Industrial roundwood exports† nd
Fuelwood and charcoal production† 4,335,000 cu. m
Processed wood production† 7000 cu. m
Processed wood exports† nd

MALAWI
Land area 94,080 sq. km
Population (mid-1990) 9.2 million
Population growth rate in 1990 3.4 per cent
Population projected to 2020 22.0 million
Gross national product per capita (1988) US$160
Rain forest 320 sq. km‡
Closed broadleaved forest (end 1980)* 1860 sq. km
Annual deforestation rate (1981–5)* nd
Industrial roundwood production† 337,000 cu. m
Industrial roundwood exports† nd
Fuelwood and charcoal production† 7,070,000 cu. m
Processed wood production† 37,000 cu. m
Processed wood exports† nd

‡ Dowsett-Lemaire (1989a)

MOZAMBIQUE
Land area 781,880 sq. km
Population (mid-1990) 15.7 million
Population growth rate in 1990 2.7 per cent
Population projected to 2020 31.9 million
Gross national product per capita (1988) US$100
Closed broadleaved forest (end 1980)* 9350 sq. km
Annual deforestation rate (1981–5)* 100 sq. km
Industrial roundwood production† 1,005,000 cu. m
Industrial roundwood exports† 1000 cu. m
Fuelwood and charcoal production† 15,022,000 cu. m
Processed wood production† 39,000 cu. m
Processed wood exports† 1000 cu. m

ZIMBABWE
Land area 386,670 sq. km
Population (mid-1990) 9.7 million
Population growth rate in 1990 3.2 per cent
Population projected to 2020 20.9 million
Gross national product per capita (1988) US$660
Rain forest 80 sq. km**
Closed broadleaved forest (end 1980)* 2000 sq. km
Annual deforestation rate (1981–5)* 0
Industrial roundwood production† 1,593,000 cu. m
Industrial roundwood exports† 2000 cu. m
Fuelwood and charcoal production† 6,269,000 cu. m
Processed wood production† 216,000 cu. m
Processed wood exports† 10,000 cu. m

* FAO (1988)
† 1989 data from FAO (1991)
** See Figure 30.7

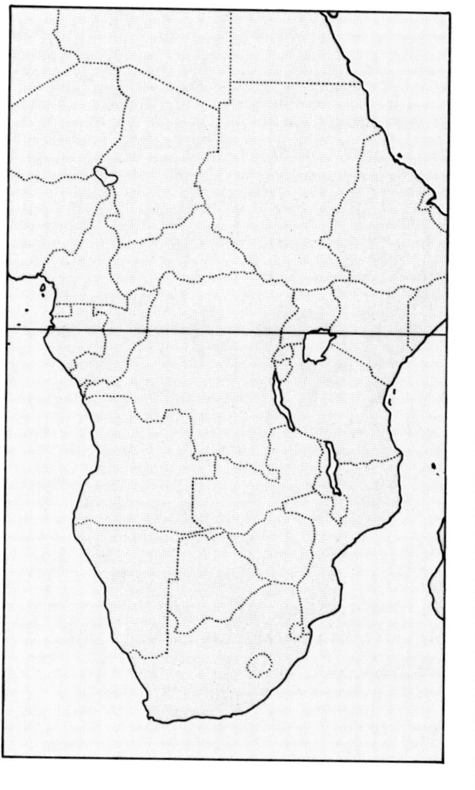

The countries included in this chapter are Angola, Malawi, Mozambique and Zimbabwe. Zambia has been excluded as the only rain forests within the country are small areas of montane forest on the border with Malawi. These countries cannot even begin to match the forest riches contained within West or Central Africa. No single country in southern Africa has more than relict patches of natural forest. The many reasons for this include a habitat generally unsuitable for forests, the intensity of past and present-day human activities and the prevalence of bush fires. Civil war, too, has taken its toll on the environment, with complete disintegration of protected area networks in Angola and Mozambique and widescale poaching and burning of the forests. Nonetheless, the fragmented relict forests harbour impressive levels of species diversity, both faunal and floral, frequently at a highly localised level.

Angola

The Forests

The Guinea forest biome in Angola comprises the evergreen and semi-deciduous forests, which are restricted to the interior of Cabinda, and the large but discontinuous patches of forest in the districts of Zaïre, Uige, Cuanza Norte and Cuanza Sul (Huntley, 1974, Figure 30.1). The evergreen forest in Cabinda is multi-storeyed with a canopy of between 40 and 60 m in height, and is dominated by genera such as *Gilletiodendron*, *Librevillea*, *Tetraberlinia* and *Julbernardia*. Much greater areas are covered by forests with a lower canopy of 30–40 m where some of the dominant species are deciduous. In Cabinda, genera in this forest type include *Gossweilerodendron*, *Pentaclethra* and *Oxystigma*, while further south the common genera are *Celtis*, *Morus*, *Albizia*, *Bombax* and *Pterocarpus* (Huntley, 1974).

An extended but fragmented series of forests and forest patches exists along the Angolan escarpment from Dondo, south to Quilenges. These forests range in size from just a few hectares to several thousand hectares and form a continuum from dry scrub forest or thicket to tall moist rain forest, following a gradient of altitude and the availability of moisture. The largest and most important is the so-called 'Amboim forest' immediately north of Gabela. At the dry extreme, in which there is only 400–600 mm of rainfall, a dense thicket of trees 10 to 20 m tall, occurs, including *Ceiba pentandra*, *Bombax reflexum*, *Pteleopsis myrtifolia*, *Adansonia digitata*, *Lannea welwitschii*, *Albizia glabrescens* and numerous species of climbers. Inland of the dry thicket, on the higher (400–1200 m), moister (600–1200 mm per annum) slopes, an increasingly luxuriant and tall forest occurs, growing up to 40 m in height. Species such as *Bombax reflexum*, *Khaya anthotheca*, *Blighia unijugata*, *Zanha golungensis*, *Piptadeniastrum africanum*, *Celtis mildbraedii* and *Spathodea campanulata* dominate a moist cloud-forest type with abundant epiphytes, but fewer lianas than the low-altitude thickets (Collar and Stuart, 1988 from Airy Shaw, 1947 and B. Huntley, *in litt.*).

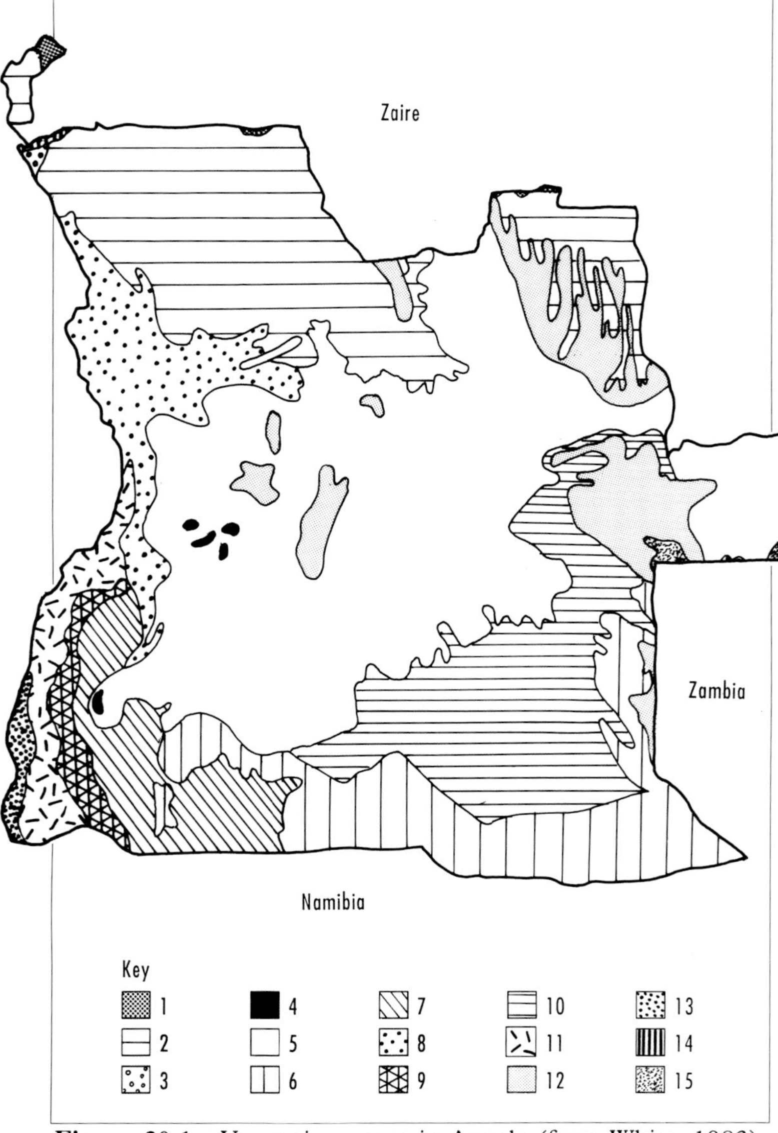

Figure 30.1 Vegetation zones in Angola (from White, 1983). The present distribution of moist forest is not known

1 Guinea-Congolian rain forest: drier types. 2 Mosaic of Guinea-Congolian lowland rain forest and secondary grassland (with some Zambezian dry evergreen forest). 3 West African coastal mosaic. 4 Undifferentiated Afromontane vegetation. 5 Wetter Zambezian miombo woodland (with some Zambezian dry evergreen forest). 6 Mosaic of Zambezian dry deciduous forest and secondary grassland. 7 *Colophospermum mopane* woodland and scrub woodland. 8 Undifferentiated Sudanian woodland. 9 Transition from *Colophospermum mopane* scrub woodland to Karoo-Namib shrubland. 10 Mosaic of *Brachystegia bakerana* thicket and edaphic grassland. 11 Bushy Karoo-Namib shrubland. 12 Edaphic and secondary grassland on Kalahari sand. 13 The Namib desert. 14 Mangroves. 15 Zambezian dry evergreen forest.

Montane forests are today represented by just a few isolated patches on protected slopes in the mountains of Huambo, Benguela, Cuanza Sul and Huila districts on the Bailundu Highlands. The combined area of the relicts is probably less than 2 sq. km, yet they provide sufficient habitat for the maintenance of faunal and floral communities separated by over 2000 km from their closest allies and are thus of considerable biogeographical interest. The best examples of this forest type are to be found in the Luimbale area, in particular on Mount Môco where at least 15 patches from 1 to 20 ha survived in the early 1970s (Huntley, 1974). The forests are mostly in deep ravines between 2000 m and 2500 m in altitude. The dominant trees, 10–15 m in height, include species from the genera *Podocarpus*, *Pittosporum*, *Olea* and *Ilex*.

The mangrove flora in Angola is richest in the northern part of the country, particularly in Cabinda, and decreases in density and diversity towards the south, a reflection of the strong influence of the cold Benguela Current. The dominant tree of this ecosystem is *Rhizophora racemosa*, which often reaches 20 m in height, backed by *R. harrisonii* which, at least in the north, grades into *Pandanus-Raphia* swamps, dominated by *R. hookeri* and *R. palma-pinus* (Huntley, 1974). Estimates of the remaining surface area of mangroves range from 700 sq. km (Hughes and Hughes, 1991) to 1250 sq. km (FAO/UNEP, 1981). During the past 50 years, a substantial proportion has either been cleared or severely degraded, being regularly cut for firewood.

Forest Resources and Management

It has not been possible to obtain recent information on the extent or the management of the forests in Angola for this Atlas. In 1980, FAO estimated that there were 29,000 sq. km of closed broadleaved forest and that the 18 forest reserves in the country covered some 18,560 sq. km (FAO/UNEP, 1981).

Logging output steadily increased in Angola during the 1960s, from 258,000 cu. m in 1961 to 740,000 cu. m in 1969, but with the onset of civil war in the 1970s, decreased again to less than 200,000 cu. m. According to FAO (1991), there are no exports of logs (unprocessed or processed) from the country at present. Logging is selective, the main species exploited including *Guibourtia coleosperma*, *Marquesia macroura*, *Berlinia* spp., *Baikiaea plurijuga* and *Brachystegia spiciformis*.

Forestry plantations in the country date back to the 1930s when *Eucalyptus* plantations were established to provide fuel for the Benguela railway line. Fifty years later, more than 1570 sq. km were under plantation.

Threats to the forest patches in the Angolan escarpment are currently unknown other than that undergrowth in some regions has been cleared for coffee cultivation (but much of the formerly cultivated area has been abandoned and is now regenerating) and that all the sites are so small that they are permanently vulnerable to exploitation, modification and clearance (Collar and Stuart, 1988). The tiny patches of montane forest are severely threatened by the continued extraction of timber and fuelwood by the local people. Estimated rate of deforestation for closed broadleaved forest was 440 sq. km per annum (FAO/UNEP, 1981).

Biodiversity

An estimated 5000 plant species are thought to occur in Angola, excluding Cabinda (Airy Shaw, 1947; Davis *et al.*, 1986). The country's high level of species endemism – 1260 plant species – ranks it second in Africa after Zaïre. Most of the endemics are in the highland and escarpment zones (Stuart *et al.*, 1990).

With 275 mammal species recorded in the country, Angola is one of the richest on the continent in this respect (Stuart *et al.*, 1990). However, most of the large mammals are threatened by poaching. These include the black rhinoceros *Diceros bicornis*, elephant *Loxodonta africana* and 15 of the 26 species of antelope in the country (Estes, 1990; Stuart *et al.*, 1990). Angola is the only country in which the giant sable antelope *Hippotragus niger variani* occurs. Forest primates recorded in Angola include the chimpanzee *Pan troglodytes*, gorilla *Gorilla gorilla*, the endemic black-nosed monkey *Cercopithecus ascanius atrinasus* and a rare and distinctive subspecies of the black mangabey *Cercocebus aterrimus opdenboschi* which is known only from the gallery forests of northern Angola and an adjacent part of Zaïre (Stuart *et al.*, 1990).

Angola has a diverse avifauna, with 872 species of birds listed for the country (Stuart *et al.*, 1990). The forests of the Angolan

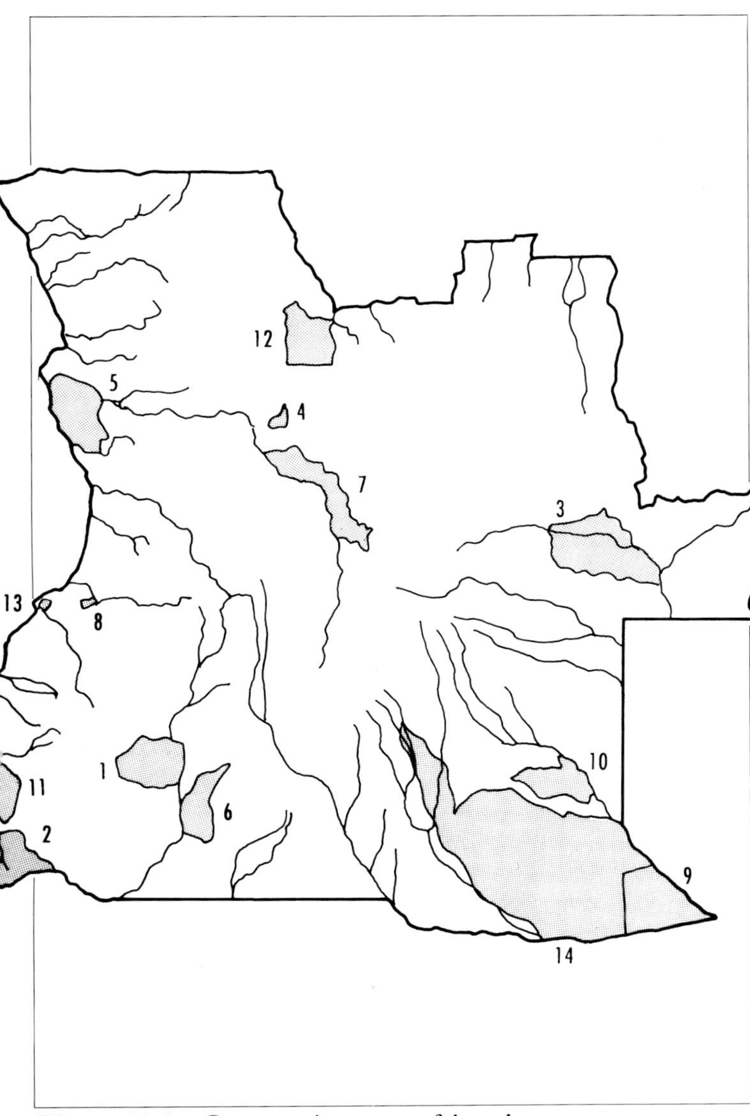

Figure 30.2 Conservation areas of Angola

(*Sources:* IUCN, 1990; WCMC, *in litt.*)

Table 30.1 Conservation areas of Angola

Conservation areas as shown on Figure 30.2

	Area (sq. km)	*Number*
National Parks		
Bikuar	7,900	1
Iona	15,150	2
Kameia	14,450	3
Kangandala	630	4
Kisama	9,960	5
Mupa	6,600	6
Integral Nature Reserves		
Ilheu dos Passaros†	2	
Luando	8,280	7
Partial Reserves		
Bufalo	400	8
Luiana	8,400	9
Mavinga	5,950	10
Mocamedes	4,684	11
Special Reserves		
Milando	nd	12
Regional Nature Parks		
Chimalavera	150	13
Controlled Hunting Areas		
Cuando-Cabango Coutadas	nd	14
Total	82,556	

(*Sources:* IUCN, 1990; WCMC, *in litt.*)

† Not shown on Figure 30.2.
nd no data

escarpment are particularly rich in species: indeed, they are ranked eleventh in importance in Collar and Stuart's (1988) list of 75 key forests in Africa. Rare bird species found here include the Gabela helmet-shrike *Prionops gabela* and Gabela akalat *Sheppardia gabela* (both restricted to the Gabela region), as well as Monteiro's bush-shrike *Malaconotus monteiri* and Pulitzer's longbill *Macrosphenus pulitzeri* which are endemic to the area (Collar and Stuart, 1988). No data exist on other fauna or flora, although the factors that explain the ornithological importance of the Angolan escarpment (see Hall, 1960) seem likely to have influenced the diversity of species in other forms of life there as well.

In the montane forests, again only the avifauna has been studied in detail. Rare birds that occur there include the Swierstra's francolin *Francolinus swierstrai*, the Fernando Po swift *Apus sladeniae* and the black-chinned weaver *Ploceus nigrimentum*. However, only the francolin is endemic to the area and it is the only forest species (Collar and Stuart, 1988). Of the 30 bird species collected in the montane forests of Mount Môco, seven are also found in the relic montane forests of Bioko (Fernando Poo), the Cameroons, Rwenzori, Tanzania, Ethiopia and Malawi but nowhere in between. Information on other vertebrate or invertebrate groups is generally sparse for this area and, indeed, for the entire country.

Conservation Areas and Initiatives

In Angola, responsibility for the environment resides with the Ministry of Agriculture and, in particular, the National Directorate for Conservation of Nature and the National Directorate for Fisheries and Agriculture.

Although a few vegetation types are well represented in the current reserve system in Angola, which accounts for almost 7 per cent of the country (Table 30.1, Figure 30.2), several biomes are not represented at all. These include the forest habitats in the mountains, along the western escarpment and in the north. Small areas of mangrove forest are protected within the Kisama National Park.

The minimal protection currently being afforded Angola's conservation areas, due to the lack of security in parts of the country, means that many of the country's most critical sites are in serious danger. In addition, because Angola is so poorly known biologically, there must be many critical sites that could easily be lost before their true value is known. Although the country desperately needs outside assistance to help protect its unique flora, fauna and spectacular scenery, there are at present few international agencies operational in Angola and none that is concerned with forest or nature conservation.

MALAWI

The Forests

The floristic composition of the forests of Malawi has been described in detail by Chapman and White (1970) and by Dowsett-Lemaire (1985, 1988, 1989a, 1990). Three main categories of rain forest can be recognised in the country: lowland, mid-altitude and Afromontane.

The lowland forests are characterised by 30–40 m high emergents with buttresses and high spreading crowns such as *Burttdavya nyasica*, *Khaya nyasica* and *Newtonia buchananii*. Other trees, forming a canopy around 20 m, include *Milicia excelsa*, *Ficus* spp. and *Terminalia sambesiaca*. There is generally a closed middle layer of trees 6–15 m high with species such as *Afrosersalisia emarginatum* and *Margaritaria discoidea* (FAO/UNEP, 1981).

The mid-altitude forests, from 1100 or 1200 m (or 900 m on Mt Mulanje) to 1500 or 1600 m above sea level, are secondary in structure and species composition. The forests south of 14°S have a more diverse flora than those further north, with 195 species in the former and 169 in the latter (Dowsett-Lemaire, 1989a). No extensive areas of this forest type remain and species composition is very variable in the fragmented patches. Details are given by Dowsett-Lemaire (1989a). The largest patch remaining is on Mt Mulanje and it is dominated by *Newtonia buchananii*, with *Chrysophyllum gorungosanum*, *Parinari excelsa* and *Strombosia scheffleri* as common tall trees. According to Dowsett (1985) this is probably the most vulnerable single habitat in this part of Africa.

Most of the forest remaining in the country is Afromontane. Only that on the Nyika Plateau (above 2250 m), Dedza Mts (above

Figure 30.3 Areas of Afromontane, mid-altitude and lowland forest in Malawi (*Source:* Dowsett-Lemaire, 1989a, 1990)

Nos 1–31 are Afromontane and mid-altitude forests; positions a–i are lowland rain forest. **1** Misuku Hills. **2** Mafinga Mts. **3** Jembya Plateau. **4** Musisi Hill. **5** Nyika Plateau. **6-7-8** Uzumara. Chimaliro and Choma on the North Viphya Plateau. **9** Kaningina Hills. **10-11-12-13** Nthungwa, Chamambo, Kawandama and Iwanjati-Champila on the South Viphya Plateau. **14** Chipata Mt. **15** Ntchisi Mt. **16** Dzalanyama Range. **17** Chongoni Mt. **18** Mlunduni Mt. **19** Dedza Mt. **20** Chirobwe Mt. **21** Kirk Range. **22** Phirilongwe Hill. **23** Namizimu Hills. **24** Mangochi Mt. **25** Chikala Hill. **26** Malosa and Zomba Mts. **27** Chiradzulu Mt (and Lisau). **28** Hills around Blantyre (Ndirande, Soche, Bangwe and Malabe in the east). **29** Thyolo Mt. **30** Manze Hill. **31** Mt Mulanje. **a** Igembe Hills. **b** Kalwe(Nkhata Bay). **c** Nkuwadzi. **d** Mzuma (Chintheche). **e** Kuwilwe Hil.l **f** Thambani and Zobue Hills. **g** Machemba Hil.l **h** Thyolo tea estates. **i** Malawi Hills. Dotted lines represent provincial boundaries between northern, central and southern regions.

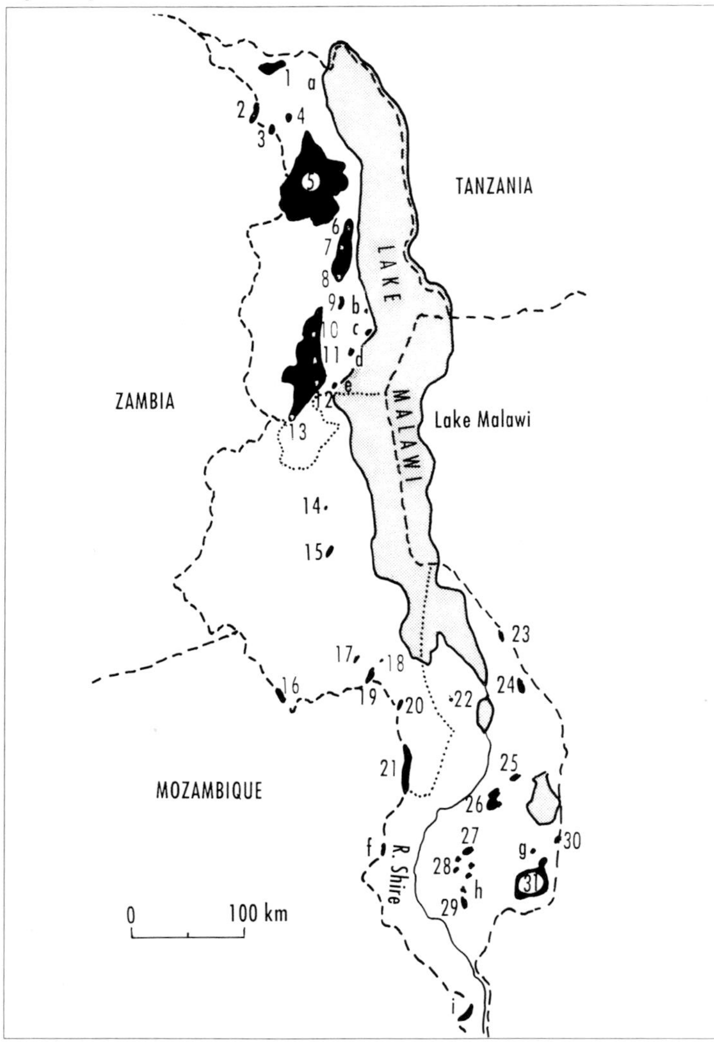

Table 30.2 Sizes and locality of the main patches of evergreen forest remaining in Malawi (sizes were measured from aerial photographs)

Forest types and locality	*Size (sq. km)*
Afromontane forest	
Misuku Hills	31
Nyika Plateau	*c.* 63
North Viphya Plateau	8
South Viphya Plateau	*c.* 28
Mafinga, Jembya Muisisi, Ntchisi	9.3
Dedza, Chongoni, Mlunduni Mts	4.1
Chirobwe Mt	6.1
Kirk Range	1.1
Namizimu, Mangochi Mts	2.7
Shire Highlands (Malosa, Zomba, Chiradzulu)	14.8
Mulanje Mt	50
Total	218.1
Mid-altitude forest	
Kaningina Hills	6.7
South Viphya (E escarpment)	*c.* 25
Elsewhere (Chipata, Ntchisi, etc)	1
Dzalanyama and Kirk Range	1.1
Liwonde Hills (Chikala, Chaone)	3
Shire Highlands (Zomba and Lisau to Thyolo)	14.8
Mauze Hill	2
Mulanje Mt	18
Total	71.6
Lowland forest	
Lake shore (Nkhata bay to Kuwilwe)	15
Mulanje	*c.* 2
Thyolo tea estates	6
Malawi Hills	4
Elsewhere (Thambani, Machemba, etc)	1.4
Total	28.4

(*Source:* Dowsett-Lemaire, 1989a)

Table 30.3 Conservation areas of Malawi

Existing and proposed conservation areas as shown on Figure 30.4. For data on World Heritage sites see chapter 9.

	Existing arean (sq. km)	*Proposed area (sq. km)*	*Number*
National Parks			
Kasungu	2,316		1
Lake Malawi	94		2
Lengwe	887		3
Liwonde	548		4
Nyika	3,134		5
Game Reserves			
Majete	784		6
Mwabvi	104		7
Nkohotakota	1,802		8
Vwaza Marsh	1,000		9
Conservation Area			
Michiru†		46	
Totals	10,669	46	

(*Source:* IUCN, 1990)

† Not shown on Figure 30.4.

2050 m) and Mulanje Mt (above 1850 m) is strictly montane; the majority is submontane. As for the lowland forests, detailed descriptions of each forest are given by Dowsett-Lemaire (1989a); however, a more generalised account is reported by FAO/UNEP (1981). The broadleaved montane forest usually has a dense, continuous canopy not more than 15 m high. Common tree species include *Kiggelaria africana*, *Olinia usambarensis* and *Rapanea melanophloeos*, while *Podocarpus milanjianus* and *Pygeum africanum* are rather more scarce. In the submontane forests, *Entandrophragma excelsum*, *Ficalhoa laurifolia* and *Ocotea usambarensis* are emergents of between 25 m and 50 m tall. The upper canopy at 18–30 m is composed of *Cassipourea congoensis*, *Chrysophyllum gorungosanum*, *Cola greenwayi*, *Diospyros abyssinica*, *Drypetes gerrardii* and *Parinari excelsa*. There are also poorly differentiated middle and lower strata of trees, with prominent ferns and lianes among them.

Forest Resources and Management

Malawi has many small patches of evergreen rain forest scattered over 40 major sites, but they total only 320 sq. km or thereabouts. Of this, around 218 sq. km is Afromontane forest, 72 sq. km is mid-altitude forest, while lowland forest makes up the remaining 30 sq. km (Table 30.2, Figure 30.3; Dowsett-Lemaire, 1989a, 1990). This is considerably less than the closed broadleaved forest cover of 1860 sq. km estimated by FAO (1988) for 1980. Single blocks of forest in excess of 5 sq. km can now be found in only seven localities: Misuku Hills, the eastern scarp of Nyika Plateau, Uzumara on the North Viphya Plateau, on the northern shore of Lake Malawi (at Nkuwadzi and Mzuma), Chirobwe, Mulanje and Thyolo (Dowsett-Lemaire, 1989b). Details of locality, altitude, size and forest type of most of the patches can be found in Dowsett-Lemaire (1989b).

Most of the surviving evergreen forests in Malawi were included in reserves in the 1920s. Overall, they have received adequate protection and have mostly not been exploited for timber (Dowsett-Lemaire, 1989a). They are not now being managed for production purposes.

Plantations were begun at the end of the 19th century and, since then, small areas have been planted with a wide range of species, including *Pinus*, *Cupressus*, *Eucalyptus* and other exotics. Fuelwood and pole plantations have also been established in many areas to alleviate possible shortages of these commodities in the future. An estimated 750 sq. km had been planted by the end of 1978 (FAO/UNEP, 1981).

Deforestation

Afromontane forests have suffered least from direct destruction – there has been comparatively little agricultural settlement at that altitude – but dry season bushfires have taken their toll. Much of the montane grassland and secondary growth on high plateaux such as South Viphya and Nyika is derived from forest (Dowsett-Lemaire, 1989a).

Early this century large areas of lowland forest were cleared round Thyolo and Mulanje for the establishment of farms. Similarly most of the lowland forest around Nkhata Bay gave way to tea and rubber plantations. Elsewhere, for instance in the Misuku Hills and in central Malawi, human population was already dense by 1900 and all that remains of the mid-altitude forest is relict patches in graveyards (Dowsett-Lemaire, 1989a). In the past two decades, increasing human population in the south of the country has posed a growing threat to the mid-altitude forests; some have been lost to subsistence cultivation (for example, on Mt

Figure 30.4 Conservation areas of Malawi (*Source:* IUCN, 1987)

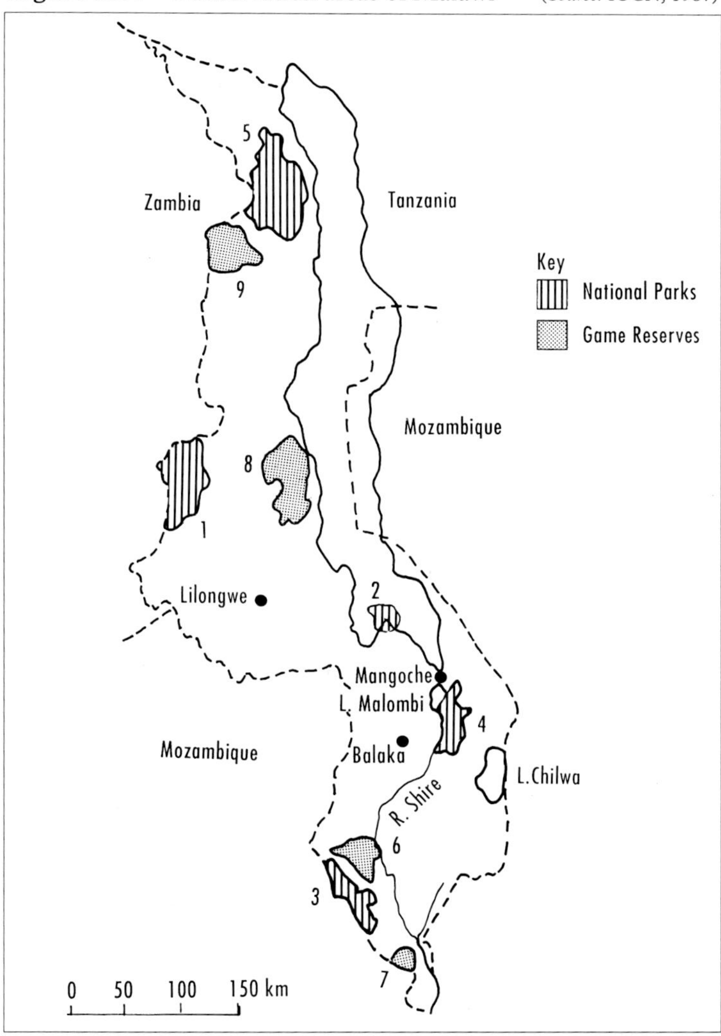

Coffee plantations in the Misuku Hills, Malawi. There is now very little forest remaining in this area. C. Harcourt

Mulanje) while others, particularly those on the Shire Highlands, suffer from illegal felling (Dowsett-Lemaire, 1989a).

Biodiversity

Malawi has an impressive range of species, particularly in birds (with 630 species) and fish. No bird species is endemic to Malawi, but there are several of limited distribution, shared only with neighbouring Zambia, Tanzania or Mozambique. Important bird species on Mt Mulanje include the endangered Thyolo alethe *Alethe choloensis*, the rare spotted ground-thrush *Turdus fischeri* and the white-winged apalis *Apalis chariessa*. Four species and four subspecies of reptile are endemic to Mt Mulanje (Stuart *et al.*, 1990). This mountin also harbours the largest number of forest butterfly species in Malawi (118), of which three, *Baliochila woodi*, *Charaxes margaretae* and *Cymothoe melanjae* are entirely endemic and may be threatened by forest destruction. There are about 30 plants endemic to this mountain but only a few are forest species. Its fern flora is the richest in Malawi, with at least 100 species having been recorded.

Mt Thyolo in southern Malawi is another important site for biological diversity. Its mid-altitude evergreen forests harbour species such as Delegorgue's pigeon *Columba delegorguei* and the green-headed oriole *Oriolus chlorocephalus* (Collar and Stuart, 1988).

Lake Malawi probably contains the largest number of fish species of any lake in the world, over 500 species from ten families. Endemism is high – thought to be over 90 per cent – and adaptive radiation and speciation within the lake is remarkable. Particularly noteworthy are the Cichlids, of which all but five of more than 400 species are endemic to Lake Malawi.

There are 187 mammal species recorded from the country but most of the larger species, outside protected areas at least, are at risk from habitat clearance and illegal hunting. A number of species depend on the forest for survival such as the tiny Zanzibar bushbaby *Galagoides zanzibaricus*, Syke's monkey *Cercopithecus albogularis*, the chequered elephant shrew *Rhynchocyon cirnei hendersoni* and the greater hamster rat *Beamys major* (Stuart *et al.*, 1990).

Conservation Areas and Initiatives

The national parks and game reserves in Malawi have been managed by the Department of National Parks and Wildlife since 1973. The forestry estate is the responsibility of the Department of Forestry and Natural Resources. At present, natural resource management is governed by six separate Acts of Parliament, but these are currently being rationalised into a single new act.

Malawi has an extensive protected area network covering about 11.5 per cent of the country (Table 30.3, Figure 30.4) and including representative samples of all major vegetation types. Overall, Malawi probably has one of the better-run protected area systems in Africa's tropical regions. The main threats to reserves are poaching, wild fires and encroachment from the expanding human population. Some game and timber exploitation is allowed in the game reserves on a sustained yield basis. Although Malawi's forests are protected by law, many are at risk because of their small size: some may not be large enough to hold viable populations of certain important species. Montane evergreen forests are already fairly well represented in the protected area system, but priority should be given to extending coverage of this biome, as these forests contain large numbers of endangered plants and animals. Priority forests include Chikala, Malawi Hills, Mulanje, Mzuma, Nkuwadzi, Soche and Thyolo.

Although a number of projects are taking place in Malawi, none is directly concerned with conservation of the forests. There is, however, a WWF project run in conjunction with the Wildlife Society of Malawi (WSM) for the support of environmental education centres at Lengwe and Nyika National Parks and Nkhotakota Game Reserve, while FAO/UNDP are working with the government on the development of wildlife management strategies, including training, around Kasungu and Liwonde National Parks and developing soil conservation measures with small scale farmers. WSM is also active in the provision of student accommodation in some of the national parks, the production of environmental education literature and the development of several of the conservation areas.

MOZAMBIQUE

The Forests

Like that of its neighbours, the flora of Mozambique is predominantly Zambezian, with Afromontane elements on high ground. Abundant mangroves on the coastline grade into non-tidal swamps and riverine forests and patches of coastal forest. Montane forests are mostly confined to the border with eastern Zimbabwe (Figure 30.5).

The montane forests in Mozambique are composed of trees up to 20 m in height. They are mainly confined to areas of the Gorongosa and Chimanimani mountains. Dominant trees are *Aphloia theiformis*, *Maesa lanceolata*, *Curtisia dentata*, *Tabernaemontana stapfiana*, *Celtis africana*, *Widdringtonia cupressoides* and *Podocarpus latifolius*.

Moist evergreen or 'sub-hygrophilous' forest occupies small areas dispersed at the base of mountainous areas where moisture is higher. It is, for example, found on the southern and eastern slopes of Namuli, Milange, Tamasse, Gorongosa and Chimanimani mountains and on the Mueda plateau as well as in patches throughout the most humid zone of the country. The most representative species are *Cordyla africana*, *Chrysophyllum gorungosanum*, *Bombax schumannianum*, *Diospyros mespiliformis*, *Manilkara discolor*, *Cussonia spicata*, *Milicia excelsa*, *Kigelia africana*, *Morus mesozygia*, *Newtonia buchananii*, *Berchemia zeyheri* and *Sideroxylon inerme*.

There are also a number of small forest patches along the coastline, south of the Zambezi River. Some of these, such as the moist evergreen forest patches at Cheringoma, have been superficially surveyed and found to consist predominantly of *Pteleopsis myrtifolia*, *Erythrophleum suaveolens*, *Hirtella zanguebarica*, *Pachystela brevipes*, *Milicia excelsa* and others (Tinley, n.d., cited in Collar and Stuart, 1988).

Mangrove stands occur at all river mouths in Mozambique, as well as in many sheltered bays and lagoons. Their area has been estimated at 4550 sq. km by interpretation of 1972–3 satellite images (FAO/UNEP, 1981). Along the northern section of the coastline the mangrove association is continuous with that in Tanzania. Generally *Sonneratia alba* provides a dense outer coastal fringe of up to 20 m in height, backed by *Rhizophora mucronata* and *Bruguiera gymnorrhiza*, the latter occurring in areas where flooding is less deep and less frequent. *Xylocarpus granatum* is a common associate of both *Rhizophora* and *Bruguiera*, especially near streams. In drier sites, patches of *Ceriops tagal* occur. In the Zambezi Delta and along other accessible parts of the coastline, the mangrove swamps have been heavily exploited for firewood, charcoal and building materials. Many local people are dependent on the extensive mangroves for fishing.

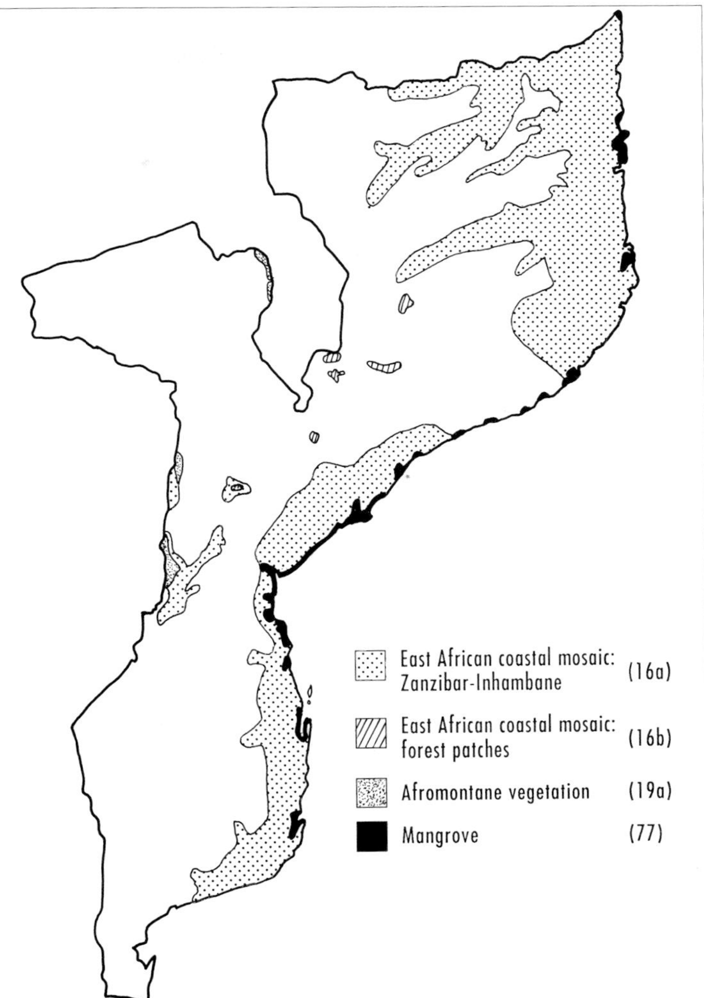

Figure 30.5 Moist forest zones in Mozambique (simplified from White, 1983). Present day forest cover within these zones is not known

Forest Resources and Management

No recent information has been obtained on the present extent of the forests or their management in Mozambique. In 1973, virgin closed forest covered 925 sq. km and logging in these forests was then very limited or non-existent (FAO, 1980). The figure of 9350 sq. km estimated by FAO for the area of closed broadleaved forests at the end of 1980 includes gallery forests and mangroves and is based on data collected from satellite imagery in 1973. Figure 30.5 shows the forest zones in the country.

A decade ago five species accounted for 90 per cent of the sawn timber produced in Mozambique: *Millettia stuhlmannii* (panga-panga), *Pterocarpus globuliflora* and *P. angolensis* (both known locally as umbila) and *Brachystegia spiciformis* and *Julbernardia globiflora* (both called messassa). Logging is permitted through concession of areas or cutting licences granted by the Department of Agriculture but there is no control within the forests (FAO/UNEP, 1981). Extracted volume is reported to have been as high as 15 cu. m per ha in closed forests. As long ago as 1978, it was reported that most of the valuable timbers had been cut in the easily accessible areas (FAO, 1978). There is, however, still a small quantity of timber being exported from the country both in the form of unprocessed logs and processed wood (FAO, 1990).

Casuarina equisetifolia plantations were first established in Mozambique to stabilise sand dunes. By 1980, plantations of this species and a number of other exotics (*Eucalyptus* spp., *Pinus* spp. and *Cupressus lusitanica*) covered an area of about 402 sq. km (J. Timberlake, *in litt.*).

The annual deforestation rate of 100 sq. km for 1981–5 given by FAO (1988) is, essentially, a best guess. The virgin forest is not expected to suffer any significant deforestation, but agricultural encroachment, fires and over-exploitation of other areas will probably occur.

Biodiversity

The flora of Mozambique is believed to consist of more than 5500 species (Lebrun, 1960), with an estimated 216 of these being endemic (Brenan, 1978). The northern part of the coastline is especially rich in endemics because of the extension of the coastal vegetation mosaic southwards from Tanzania. The fauna of Mozambique has been little studied and it has not been possible to find information on the biodiversity within the forests except for some bird species reported on by Collar and Stuart (1988).

The rare dappled mountain robin *Modulatrix orostruthus* was first described from Mt Namuli, which also holds the endangered Thyolo alethe *Alethe choloensis*, the Thyolo green barbet *Stactolaema olivacea belcheri* and the Namuli apalis *Apalis lynesi*. To the west, Gorongosa Mountain is the home of the rare Swynnerton's forest robin *Swynnertonia swynnertoni* and the Chirinda apalis *Apalis chirindensis*.

In the coastal forest blocks, some notable species are the rare east coast akalat *Sheppardia gunningi*, the southern banded snake eagle *Circaetus fasciolatus*, Rudd's apalis *Apalis ruddi*, Neergaard's sunbird *Nectarinia neergaardi*, lemon-breasted canary *Serinus citrinipectus* and the pink-throated twinspot *Hypargos margaritatus*.

Conservation Areas and Initiatives

The Ministry of Mineral Resources is charged with overseeing environmental matters. Responsibility for wildlife and forestry rests with the Ministry of Agriculture, especially the National Directorate for Forestry and Wildlife.

More than 15,000 sq. km of land have been set aside as national parks (IUCN, 1987). About 6 per cent of the country is protected (Table 30.4, Figure 30.6) but this does not include any moist forest areas. However, Mozambique's civil war has led to a breakdown in protective measures being applied in the country's parks and reserves. As a result, most of its important sites are suffering from encroachment and severe illegal hunting of large mammals. Most wildlife populations are now seriously at risk. The Wildlife and Forestry Department suffers from extreme shortages of manpower and equipment and only a few reserves have permanent staff. A report by Tello (1986) reviews the protection and management in Mozambique's conservation areas.

Protection in all current reserves urgently needs great improvement but this cannot occur until the security situation permits. The establishment of additional protected areas in both highland and coastal forests is needed as soon as possible, as well as surveys of coastal forest areas to identify key conservation sites. Few international conservation projects are in progress in Mozambique at present and none is concerned with forest protection.

Figure 30.6 Protected areas in Mozambique

(*Source:* Mozambique, Economic Map: Scale 1:7,500,000; Main Administration of Geodesy and Cartography under the Council of Ministers of the USSR, Moscow, 1987)

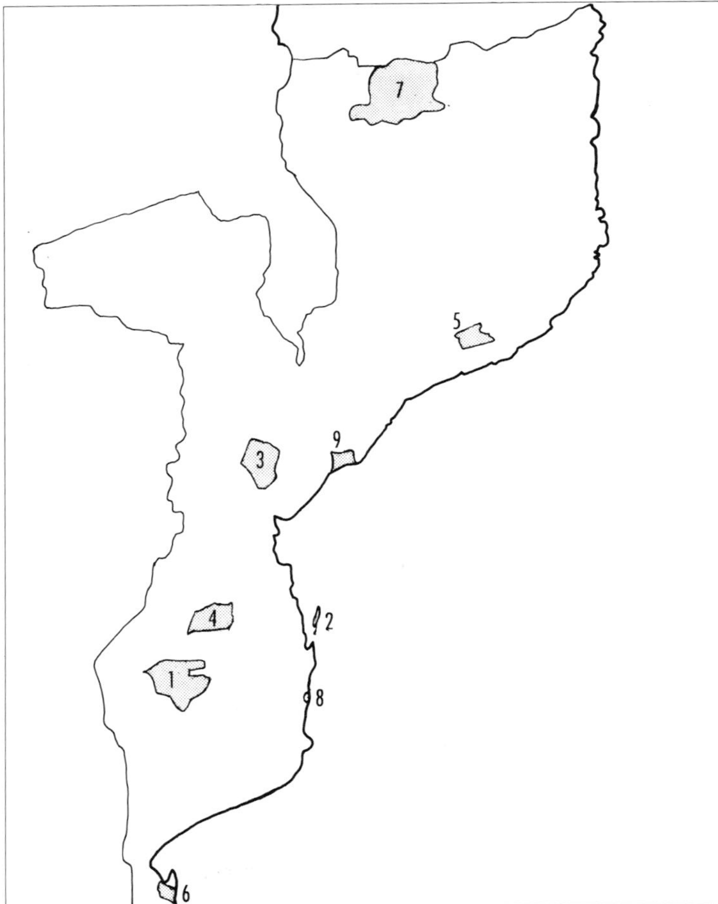

Table 30.4 Conservation areas of Mozambique

Existing and proposed conservation areas are listed below and see Figure 30.6. Marine national parks, wildlife utilisation areas and forest reserves are not listed.

	Existing area (sq. km)	*Proposed area (sq. km)*	*Number*
National Parks			
Banhine	7,000		1
Bazaruto	150		2
Gorongosa	3,750		3
Primeira and Segunda Islands		nd	
Quirimba Islands		nd	
San Sebastian Peninsula		nd	
Zinave	5,000		4
Faunal Reserves			
Ilhas da Inhaca e dos Portugeses	20		
Game Reserves			
Gile	2,100		5
Maputo	900		6
Niassa†	15,000		7
Pomene	100		8
Reserves			
Marromeu	10,000		9
Totals	44,020		

(*Source:* IUCN, 1990)

nd no data
† proposed as NP, to be called Rovuma

ZIMBABWE

The Forests

The vegetation in Zimbabwe is predominantly dry woodland, but there is some montane rain forest and very small areas of moist forest at lower levels within the country. The highest rain forest zone is above 1800 m and here three types can be distinguished. First the *Syzygium masukuense* dominated forest which is found only on the Nyanga Mountains, where it covers most of the areas where forest can occur up to an altitude of about 2100 m. This species can account for two-thirds or more of the total number of trees where the forest is least disturbed. Other common canopy trees are *Rapanea melanophloeos*, *Aphloia theiformis*, *Podocarpus latifolius*, *Prunus africana*, *Ilex mitis* and *Schefflera umbellifera*. Common trees of the sapling layer include *Peddiea africana*, *Diospyros whyteana* and *Pavetta umtalensis*. The height of the canopy is 10–12 m with a few emergents up to 15 m in height. No sub-canopy can be distinguished but there is a layer of scattered small trees at about 4–5 m and a dense shrub layer of 1.5–2 m.

The second distinct type is the *Afrocrania volkensii* forest. This has a limited distribution in the Nyanga area but is more extensive on the hills of the south-west of Chimanimani District. *Afrocrania* is dominant or co-dominant with *Ilex mitis* and *Olea europaea capensis*. Other common trees are *Prunus africana*, *Kiggelaria africana* and *Ekebergia capensis*. The canopy height of these relatively undisturbed forests is 30 m.

There is also some *Widdringtonia cupressoides* dominated forest but it is limited in extent and is confined to fringing rain forest and streams in rain-shadow areas. The height of the canopy rarely exceeds 10 m and stratification is poorly developed.

The high to medium altitude rain forest zone lies between 1400 and 1800 m in the Nyanga Mountains but can reach up to 1900 m elsewhere. There has been widespread human disturbance over most parts of this zone; indeed, shifting cultivators have strongly influenced the species composition. Overall the most common canopy tree is *Syzygium guineense* but others include *Cassipourea malosana*, *Rapanea melanophloeos*, *Pterocelastrus echinatus* and *Faurea racemosa*. The canopy height in this altitudinal belt increases from about 25–30 m in the upper part to nearly 50 m in the *Albizia*-dominated forests below 1600 m. The different layers are generally well developed, especially where no recent disturbance has taken place. Forests above 1500 m have not been significantly reduced in modern times but those below this altitude have been mostly lost to agriculture. In the more mature forests of the lower portion of this zone, species diversity reaches its peak for rain forests in Zimbabwe.

The medium altitude rain forest zone lies between 1000 m and 1400 m. In the Chimanimani and Chipinge districts of the south-east this forest type probably covered 150 sq. km but elsewhere it has always been very limited in extent. Since the beginning of this century, however, it has been reduced to less than 10 sq. km by advancing agriculture. In the canopy the dominant species are *Craibia brevicaudata* and *Chrysophyllum gorungosanum*, while other common canopy species are *Strychnos mitis*, *Ficus chirindensis*, *Lovoa swynnertonii* and *Diospyros abyssinica*. The canopy height is 50–55 m with some emergent trees reaching over 60 m, and stratification is well developed.

Chirinda Forest is the best developed and preserved example of medium altitude rain forest, its only disturbance being the removal of *Khaya nyasica* trees from some areas earlier this century. Just over 6 sq. km in extent, it covers the plateau-like top of a broad mountain. Apart from Chirinda, only small fragments of medium altitude forest still exist and these are either the inaccessible remains of once larger forests or have always been restricted in size by an unsuitable environment.

The low altitude rain forest zone occurs between 350 and 750 m, but there are only two valleys in the country low enough and with a sufficiently high rainfall to support this type of vegetation. These are the Pungwe valley at the foot of the Nyanga Mountains and the Rusitu valley at the south-eastern base of the Chimanimani mountain range. In the Pungwe valley, practically all of the 40 sq. km of lowland rain forest which existed in the early part of this century has disappeared, with the last few hectares being destroyed before 1980. Apart from a pure *Newtonia buchananii* forest on an isolated hill, all that remains are occasional groups of trees and some small fragments of regenerating forest which indicate where rain forest could occur.

In the lowest portion of the Rusitu valley, adjacent to the border with Mozambique, there is one continuous block of lowland rain forest of about 2 sq. km in extent. This is all that remains of what was probably once the largest area of low altitude rain forest in Zimbabwe. Canopy height is in the region of 50 m. The dominant tree species in the canopy is *Newtonia buchananii*, while other common trees are *Maranthes goetzeniana* and *Xylopia aethiopica* with *Khaya nyasica* and *Erythrophleum suaveolens* locally frequent. *Funtumia africana* often forms a high sub-canopy and species such as *Aporrhiza nitida*, *Blighia unijugata* and *Uapaca guineensis* are also found there. The shrublayer is often varied and dominated by young lianes.

Small outliers of rain forest occur on windward gullies on a number of mountains scattered over a wide area south of the central watershed. Examples include Wedza Mountain, the mountains around Bikita, Mt Buhwa and Nyoni Range. They contain a number of rare species, among them the endangered *Bivinia jalbertii* on Nyoni Range and *Oreobambos buchwaldii* on Mt Buhwa.

Forest Resources and Management

Rain forest in Zimbabwe is extremely localised and now covers only 80 sq. km or thereabouts (Figure 30.7). It is found mainly on the windward slopes of mountains in the eastern part of the country. The moist forests occur between altitudes of 350 m and 2100 m, but those which have survived have done so mainly because they are in inaccessible places or on land which is not suitable for farming. Consequently most of the forests existing today are found above 1500 m, while those at lower altitudes have been cleared to make way for agriculture. The three main centres of rain forest development are, from north to south, the Nyanga, the Vumba and the Chimanimani mountain ranges, along the eastern border of the country. A few smaller areas occur north and south of the Vumba and to the south of the Chimanimani District.

Practically all forest above 1500 m is at present well protected, with much of it falling under the Department of National Parks and Wildlife Management. Some smaller forests are looked after by the Forestry Commission and the rest are on private land, including the *Afrocrania* forest in the Chimanimani District. Most of the medium altitude forest was destroyed by early settlers and much of what remained has been severely reduced over the past 50 years by shifting cultivation and for tea or coffee plantations. Chirinda Forest, now the most significant area of medium altitude forest, is protected by the Forestry Commission. Many of the forests below 1400 m occur on private land, either forestry estates, commercial farms or smallholdings, or on communal land. Most of those on private land are, at least for the time being, well protected, whereas the few forests which still exist on communal land are greatly endangered and most of them are in the process of being destroyed. The communal land is under multiple ownership and therefore there is little control over land use, which makes conservation of natural vegetation particularly difficult.

The moist forests are now generally too small for commercial exploitation and timber extraction is taking place on only an insignificant scale. Local people are using the forests to obtain natural products and wood for construction, but the impact of this has not been fully investigated. Only the very small fragments are vulnerable to fire, particularly those occurring in narrow strips such as the *Widdringtonia cupressoides* forests.

Figure 30.7 Major areas of rain forest occurrence in Zimbabwe

(*Source:* T. Muller, *in litt.*)

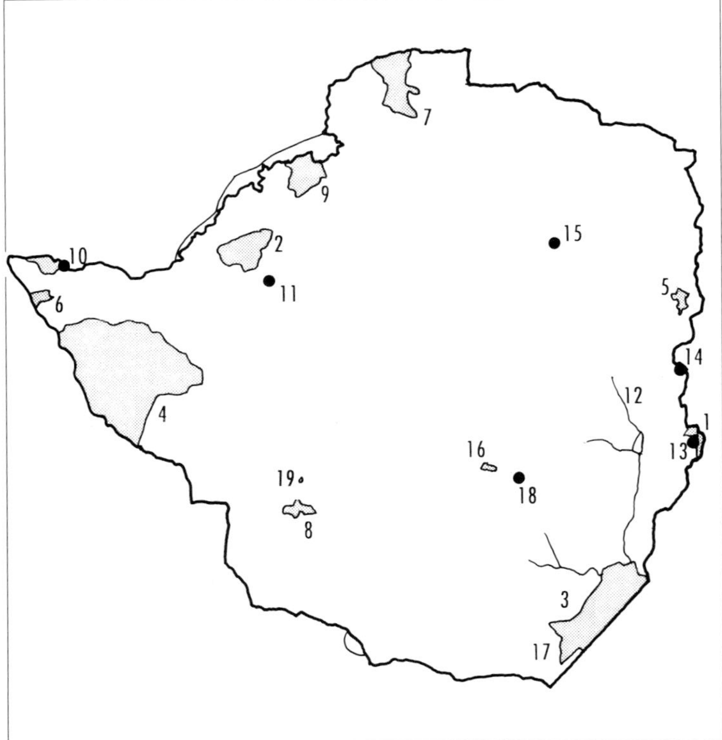

Figure 30.8 Protected areas in Zimbabwe

(*Source:* Government of Zimbabwe, 1973)

Biodiversity

Out of the nearly 6000 vascular plants recorded in Zimbabwe, about 740 occur in rain forest. However, the rain forests in the country have a much lower biodiversity than those of the equatorial belt. Very few endemic or endangered plant species occur and some of the species which are rare here are common elsewhere.

There are about 190 species of mammal listed for Zimbabwe, 37 of which have been recorded in rain forest and ten of which are confined to it (Smithers and Wilson, 1979). Two species, the golden mole *Chlorotalpa arendsi* and the Selinda rat *Aethomys silindensis* have so far been found only in Zimbabwe (Smithers, 1979). Of the 634 birds listed for Zimbabwe, 46 have been recorded in rain forest and two are confined to it (Irwin, 1981). The number of endemics has shrunk to two in recent years: Roberts' prinia *Prinia robertsi* and the Chirinda apalis *Apalis chirindensis*. Of the five species considered to be threatened only one, Swynnerton's forest robin *Swynnertonia swynnertoni*, is found in forests (Collar and Stuart, 1985).

There are 156 reptile species and 54 amphibians in the country. Of these, seven reptiles and five amphibians are confined to rain forest, among them one endemic chameleon *Rhampholeon marshalii* and one endemic frog *Probreviceps rhodesianus*. There is also another frog and one toad that are endemic to the forests of Zimbabwe and Mt Gorongosa in Mozambique (Broadley, 1988). The rain forests are also rich in Lepidoptera; for example, 182 butterfly species have been recorded in Chirinda Forest.

Conservation Areas

The protected areas in Zimbabwe cover over 28,289 sq. km or at least 7.3 per cent of the country (Table 30.5, Figure 30.8). (This figure does not include state forests, safari areas, protected forests or recreation parks. When these are included, protected areas cover 49,792 sq. km or 12.7 per cent of the country.) Rain forest is protected in two national parks and five botanical reserves. There are also 19 state forest areas totalling 8250 sq. km, some of which contain rain forest.

Table 30.5 Conservation areas of Zimbabwe

State forests, safari areas, protected forests and recreation parks are not listed here or mapped in this Atlas. For data on World Heritage sites see chapter 9.

	Area (sq. km)	*Number*
National Parks		
Chimanimani	171	1
Chizarira	1,910	2
Gonarezhou	5,053	3
Hwange (Wankie)	14,651	4
Inyanga	289	5
Kazuma Pan	313	6
Mana Pools	2,196	7
Matobo (Matopos)	432	8
Matusadona	1,370	9
Mtarazi Falls	25	5
Victoria Falls	23	10
Zambezi	564	10
Wildlife Research Areas		
Sengwa	373	11
Nature Reserves		
Cecil Kop†	17	
Game Parks		
Imire†	10	
Iwabe†	200	
Botanical Reserves		
Bunga Forest	16	12
Chirinda†	9	
Rusitu Forest	2	13
Vumba	2	14
Ewanrigg	3	15
Sanctuaries		
Chimanimani Eland	12	1
Mushandike	129	16
Tshabala†	nd	
Manjnji Pan	nd	17
Rhodes Bulawayo	nd	18
Natural Monuments		
Great Zimbabwe	7	19
Victoria Falls	nd	10
Wilderness Areas		
Mavuradona†	500	
Parks		
Gweru Antelope†	12	
Total	28,289	

(*Sources:* IUCN, 1990; WCMC, *in litt.*)

† Not mapped as location data are not available.
nd no data

The Department of National Parks and Wildlife Management within the Ministry of Natural Resources and Tourism is responsible for all aspects of wildlife and protected areas management, though management of wildlife has been delegated to villagers on communal lands and to private land-owners in many places. The conservation and management of forests comes under the Forestry Commission.

Zimbabwe's parks and protected areas are some of the best managed in the world, with a highly trained and efficient National Parks and Wildlife Department. Considerable emphasis is placed on the role of protected areas in raising rural living standards and generating material revenue through domestic and foreign tourism and use of wildlife (Child, 1984). The greatest threats to wildlife and the parks system are poaching and fires.

The National Parks Department recently acquired 136 sq. km of land in the Nyanga Mountains, including the rugged eastern portion of the Nyangani massif where, on the lower slopes, most of the remaining patches of medium altitude forest occur. There is also some *ex situ* conservation of rain forest species taking place. For instance, at the National Botanic Garden in Harare, an artificial rain forest, 5 ha in extent, has been created over the past 28 years.

While much of the moist forest is now adequately conserved, better laws are required to give it total protection and rain forests on private land should be given legal protection by declaring them botanical reserves.

Conclusion

It is apparent that much more information is needed on the extent of the forests within Angola and Mozambique where the political situation has meant that such data are not readily available. In these countries, although conservation areas exist on paper, it is impossible to protect them in reality and, anyway, none of them includes forested areas. In both Malawi and Zimbabwe, comparatively little moist forest remains, most of it is montane forest and most is comparatively well protected. In none of the countries is logging a problem at the moment, but all have a growing human population and the attendant problem of an ever increasing need for more agricultural land. It is very important that the remaining areas of moist forest do not suffer as a result.

References

Airy Shaw, J. K. (1947) The vegetation of Angola. *Journal of Ecology* **35**: 23–48.

Brenan, J. P. M. (1978) Some aspects of the phytogeography of tropical Africa. *Annals Missouri Botanical Garden* **65**(2): 437–78.

Broadley, D. G. (1988) Checklist of the reptiles of Zimbabwe with synoptic keys. *Arnoldia Rhodesia* **9**(30): 369–430.

Child, G. (1984) Managing wildlife for people in Zimbabwe. In: *National Parks, Conservation and Development. The role of protected areas in sustaining society*. McNeely, J. A. and Miller, K. R. (eds). Smithsonian Institution Press, Washington, DC, USA.

Chapman, J. D. and White, F. (1970) *The Evergreen Forests of Malawi*. Oxford Commonwealth Forestry Institute, University of Oxford, UK.

Collar, N. J. and Stuart, S. N. (1985) *Threatened Birds of Africa and Related Islands. The ICBP/IUCN Red Data Book Part 1*. ICBP/IUCN, Cambridge, UK. 761 pp.

Collar, N. J. and Stuart, S. N. (1988) *Key Forests for Threatened Birds in Africa*. ICBP Monograph No. 3. ICBP, Cambridge, UK.

Davis, S. D., Droop, S. J. M., Gregerson, P., Henson, L., Leon, C. J., Villa-Lobos, J. L., Synge, H. and Zantovska, J. (1986) *Plants in Danger: What do we know?* IUCN, Gland, Switzerland and Cambridge, UK.

Dowsett, R. J. (1985) The conservation of tropical forest birds in central and southern Africa. In: *Conservation of Tropical Forest Birds*, pp. 197–212. Diamond, A. W. and Lovejoy, T. E. (eds). ICBP, Cambridge, UK.

Dowsett-Lemaire, F. (1985) The forest vegetation of the Nyika Plateau (Malawi-Zambia): ecological and phenological studies. *Bulletin du Jardin Botanique National de Belgique* **55**: 301–92.

Dowsett-Lemaire, F. (1988) The forest vegetation of Mt Mulanje (Malawi): a floristic and chronological study along an altitudinal gradient (650–1950 m). *Bulletin du Jardin Botanique National de Belgique* **58**: 77–107.

Dowsett-Lemaire, F. (1989a) The flora and phytogeography of the evergreen forests of Malawi. I: afromontane and mid-altitude forests. *Bulletin du Jardin Botanique National de Belgique* **59**: 3–131.

Dowsett-Lemaire, F. (1989b) Ecological and biological aspects of forest bird communities in Malawi. *Scopus* **13**: 1–80.

Dowsett-Lemaire, F. (1990) The flora and phytogeography of the evergreen forests of Malawi. II: lowland forests. *Bulletin du Jardin Botanique National de Belgique* **60**: 9–71.

Estes, R. D. (1989) Angola. In: *Antelopes Global Survey and Regional Action Plans. Part 2: Southern and South-Central Africa*. East, R. (ed.). IUCN/SSC Antelope Specialist Group, IUCN, Gland, Switzerland.

FAO (1978) *Forest Resources in Mozambique and their Rational Use: a preliminary evaluation*. Report by J. H. Ferreira de Castro. Project Working Document FO:MOZ/76/00, Rome, Italy.

FAO (1980) *Evaluacion de los Recursos Forestales de la Republica Popular de Mozambique, Mayo de 1979 a Junio de 1980*. Draft report by J. Malleux. MOZ/76/007, Maputo, Mozambique.

FAO (1988) *An Interim Report on the State of Forest Resources in the Developing Countries*. FAO, Rome, Italy. 18 pp.

FAO (1991) *FAO Yearbook of Forest Products 1978–1989*. FAO Forestry Series No. 24 and FAO Statistics Series No. 97. FAO, Rome, Italy.

FAO/UNEP (1981) *Tropical Forest Resources Assessment Project. Forest Resources of Tropical Africa. Part II: Country Briefs*. FAO, Rome, Italy.

Government of Zimbabwe (1973) *Zimbabwe: Relief*. Map (1:1,000,000) compiled and printed by Government of Zimbabwe, Harare.

Hall, B. P. (1960) The faunistic importance of the scarp of Angola. *Ibis* **102**: 420–42.

Hughes, R. H. and Hughes, J. S. (1991) *A Directory of Afrotropical Wetlands*. IUCN, Gland, Switzerland and Cambridge, UK/UNEP, Nairobi, Kenya/WCMC, Cambridge, UK.

Huntley, B. J. (1974) Ecosystem Conservation Priorities in Angola. *Journal Southern Africa Wildlife Management Association* **4**(3): 157–66.

Irwin, M. P. S. (1981) *The Birds of Zimbabwe*. Quest Publishing, Salisbury, Rhodesia.

IUCN (1987) *IUCN Directory of Afrotropical Protected Areas*. IUCN, Gland, Switzerland and Cambridge, UK. xix + 1043 pp.

IUCN (1990) *1990 United Nations List of National Parks and Protected Areas*. IUCN, Gland, Switzerland and Cambridge, UK. 275 pp.

Lebrun, J.-P. (1960) Sur la richesse de la flore de divers territoires africains. *Bulletin Séances Academie Royale des Sciences d'Outre-mer* **6**(2): 669–90.

Smithers, R. H. N. and Wilson, V. J. (1979) Checklist and atlas of the mammals of Rhodesia. *Museum Memoir No. 9*. Salisbury, Rhodesia.

Stuart, S. N., Adams, R. J. and Jenkins, M. D. (1990) *Biodiversity in Sub-Saharan Africa and its Islands: Conservation, Management and Sustainable Use*. Occasional Papers of the IUCN Species Survival Commission No. 6. IUCN, Gland, Switzerland.

Tello, J. (1986) *Survey of Protected Areas and Wildlife Species in Mozambique, with recommendations for strengthening their conservation*. Report to WWF International, Gland, Switzerland.

Authorship

David Stone, Switzerland, and Caroline Harcourt, with contributions from Th. Muller, National Herbarium and National Botanic Gardens, Causeway, Harare, Zimbabwe.

31 Uganda

Land area 199,550 sq. km
Population (mid-1990) 18 million
Population growth rate in 1990 3.6 per cent
Population projected to 2020 42.2 million
Gross national product per capita (1988) US$280
Rain forest (see map) 8795 sq. km‡
Closed broadleaved forest (end 1980)* 7500 sq. km
Annual deforestation rate (1981–5)* 100 sq. km
Industrial roundwood production† 1,858,000 cu. m
Industrial roundwood exports† nd
Fuelwood and charcoal production† 12,507,000 cu. m
Processed wood production† 31,000 cu. m
Processed wood exports† nd

* FAO (1988)
† 1989 data from FAO (1991)
‡ Figure derived from an old source map; Howard's (1991) figure of 7,400 sq. km is probably a closer reflection of the true situation.

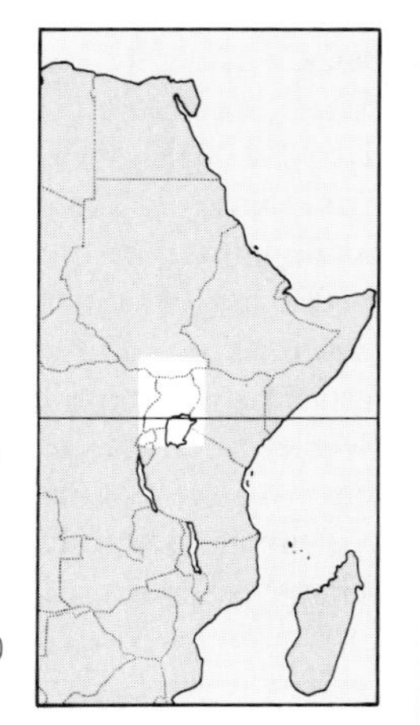

At the turn of the century, Uganda's forests extended over an estimated 12.7 per cent of the country; today they cover less than 3 per cent. This depletion can be attributed to the demand from the rapidly growing human population for agricultural land and for wood.

The issues of deforestation in Uganda revolve round the classic conflict between short-term, non-sustainable exploitation on the one hand and long-term sustainable management on the other. Ineffective law enforcement during times of political and economic turmoil and high population growth tilted the balance of this conflict towards deforestation. Recently, however, the forests have come under stricter control. Several large-scale international projects have been initiated to strengthen national capacity to conserve and manage forests. The fate of the forests of Uganda is likely to be decided within the next five years.

Introduction

Uganda is a comparatively small, landlocked country straddling the equator in East Africa. It lies between 4°07'N and 1°30'S latitude and between 29°33'W and 35°20'E longitude. It is bordered to the north by Sudan, by Kenya to the east, Zaïre to the west and Rwanda and Tanzania to the south. The total area of the country is 235,880 sq. km, but about one-sixth of that is made up of lakes, the biggest of which is Lake Victoria. Most of Uganda is situated on the east-central African plateau at an altitude of between 900 and 1500 m. Towards the Sudanese border in the north, the plateau slopes downwards, forming gently rolling plains interrupted by occasional mountains and hills (Hamilton, 1984). In the south, near Lake Victoria, the topography consists of flat-topped hills and broad, frequently swampy valleys (Hamilton, 1984). The rift valley, with its associated mountains and lakes, runs through the west of the country and it is here that the Rwenzori Range rises to its highest peak at 5119 m. In the east, along the border with Kenya, three high volcanic mountains dominate: Elgon (4321 m), Kadam (3068 m) and Moroto (3083 m).

Most of Uganda receives between 1015 mm and 1525 mm of rain in the year (Department of Lands and Surveys, 1967). There are two distinct rainy seasons in the south, one in April and May and the other in October and November. Reliability of rainfall declines towards the north, where there is only a single rainy season, July being the wettest month. The driest area is the north-east, where Karamoja receives less than 750 mm of rain per annum (Department of Lands and Surveys, 1967). Temperatures are related principally to altitude; at around 1000 m, mean maximum temperatures are between 25° and 30°C, while mean minimum temperatures are in the range of 16–18°C. The climate has, however, been somewhat erratic and unpredictable since about 1970.

The population of Uganda is growing rapidly; already it is mainland Africa's fourth most densely populated country, after Rwanda, Burundi and Nigeria (PRB, 1990). In 1920 there were three million people in the country; the last official census, in 1980, suggested that there were almost 13 million people but now there are estimated to be around 18 million. Distribution of the population within the country is very uneven. Ethnically, it is exceptionally diverse, with 36 recognised tribal divisions (Department of Lands and Surveys, 1967). The southern half of the country is occupied by Bantu, who make up about two-thirds of the population, while in the north most groups are pastoralists of Sudanic, Nilotic and Nilotic-Hamitic origin (Howard, 1991).

The population is 93 per cent rural (PRB, 1990) and it is the pressing need for more agricultural land that is the principal agent of forest destruction. The economy of the country is heavily reliant upon agriculture which engages 83 per cent of its labour force (Howard, 1991). Agricultural products, particularly coffee and cotton but also tea, tobacco, beans, groundnuts, simsim and other crops account for nearly all the country's export earnings and contribute 60 per cent to GDP (World Bank, 1986).

Uganda was a British Protectorate from 1894 until 1962 when it became independent. It enjoyed several years of prosperity until Idi Amin's military coup in 1971, which heralded more than a decade of turmoil. After the National Resistance Movement government of Yoweri Museveni took power in 1986, Uganda began to regain its economic and social stability and the government has now given conservation of the nation's resources a high priority.

The Forests

Uganda's tropical high forest occurs in three distinct geographical zones where favourable rainfall regimes occur (Howard, 1991). One of these is along the eastern rim of the rift valley escarpment in the west, another is a broad belt, up to 80 km wide, around the north-western shores of Lake Victoria and on numerous islands in the lake and the third on mountains scattered throughout the country (Hamilton, 1984).

The forests vary widely in structure and composition. This can mostly be related to altitude, but the climate, soil and historical factors also play a part. During the last dry climatic period (21,000–12,000 BP) there was a major forest refugium in east Zaïre. Some lowland forest may have survived very locally in south-west Uganda. On the mountains, vegetation zones were lowered by 1000 m and the higher parts of some mountains, especially Elgon and to a lesser extent Rwenzori, were dry with little forest cover. As the climate became wetter, starting 12,000 years ago, the forests expanded eastwards across the country. The western forests contain a greater number of tree species because they are nearer the Central African forest refugium.

Hamilton (1984) describes the different forest types following Langdale-Brown *et al.* (1964). The lowland forests are tall, evergreen or semi-deciduous rain forests with canopies of over 30 m. There are four different types, each named after characteristic trees:

1 The *Parinari* zone, found above 1400 m, is virtually confined to western Uganda. It used to be present on Mt Elgon (Namatale Forest), but this has now mostly been cleared.

2 The *Celtis-Chrysophyllum* zone, represented in both western Uganda and in the Lake Victoria region, occurs between 1000 and 1400 m.

3 The *Cynometra-Celtis* zone, which is found between 700 and 1200 m, is confined to western Uganda.

4 The *Piptadeniastrum* zone, found only close to Lake Victoria, replacing the *Celtis-Chrysophyllum* zone within 15 km or so of the lake shore.

Hamilton (1984) also describes four montane forest zones in Uganda:

1 The Upper Montane forest zone at 3050–3300 m is found on all mountains of sufficient height and degree of soil maturity (i.e. Rwenzori, Elgon, Sabinio and Mgahinga). Canopy height is only 15 m and plant species diversity is low.

2 The Bamboo (*Arundinaria alpina*) zone is found between approximately 2450 and 3050 m. It occurs on wet mountains. It is also present at lower altitudes (2260–2450 m) in Echuya Forest in the south-west, but it may have colonised this area after clearance by man.

3 The Moist Lower Montane Forest occurs at 1500–2450 m in areas of high rainfall. Structurally similar to lowland forest, this type occurs in Bwindi (Impenetrable) Forest, Rwenzori and on the western and southern slopes of Mt Elgon.

4 The Dry Lower Montane Forest Zone contains abundant evergreen sclerophyllous trees and is found between 2000 and 3050 m in drier areas, principally on north-east Elgon and the mountains of Karamoja.

Forest Resources and Management

Were it not for extensive disturbance by man, about 20 per cent of Uganda would be covered with moist broadleaved forest (Langdale-Brown, 1960; Hamilton, 1984). As it is, less than 3 per cent of the country's land surface is occupied by closed forest and most of this is within government forest reserves (Struhsaker, 1987). These cover an area of approximately 14,900 sq. km (Aluma, 1987) of which 5900 sq. km are tropical high forest and 1500 sq. km are montane catchment forest (Howard, 1991). The remainder is savanna woodland and plantations. In approximate agreement with these figures, FAO estimated that, in the early 1980s, there were 7500 sq. km of closed broadleaved forest in the country (FAO, 1988). Map 31.1 and Table 31.1 indicate rather more forest (8795 sq. km) than suggested by Howard (1991) or FAO (1988), but as the source map is over 25 years old, it is likely that Howard's figures are more accurate. The forests remaining are now widely separated from one another, forming ecological islands surrounded by other vegetation types (Hamilton, 1984).

Table 31.1 Estimates of forest extent in Uganda

Rain forests	*Area (sq. km)*	*% of land area*
Lowland	6318	3.2
Montane	2212	1.1
Swamp	265	0.1
Totals	8795	4.4

(Based on analysis of Map 31.1. See Map Legend on p. 269 for details of sources.)

The establishment of forest reserves dates back to around the turn of the century when the British Protectorate government signed agreements with some of the ancient kingdoms, making all forest land the property of the Protectorate government (Howard, 1991). However, no demarcation of the reserves was made until 1932, but then 3657 sq. km of forest reserves were formally gazetted (Forest Department, 1951). Over the next two decades, additional smaller areas, under district administration, were added as Local Forest Reserves, and were brought under centralised Forest Department control in 1967 (Hamilton, 1984).

There are now some 679 forest reserves (Pomeroy, 1990). The status of the 12 principal reserves, which cover approximately 4000 sq. km of tropical high forest and 1500 sq. km of montane forest, has been studied in detail by Howard (1991). The degree of encroachment, the level of pitsawing activity and the extent of mechanical harvesting in each of these is shown in Figure 31.1. The primary function of the forest reserves is to satisfy domestic

Figure 31.1 The status of Uganda's twelve principal forest reserves

Much of the undisturbed forest occupies steep mountainous terrain that is unsuitable for timber harvesting activities.
Pitsawing activity was not assessed in the Kibale and Budongo forests, but is thought to be widespread, at least in Kibale (J. Kasenene, pers. comm.). (*Source:* Howard, 1991)

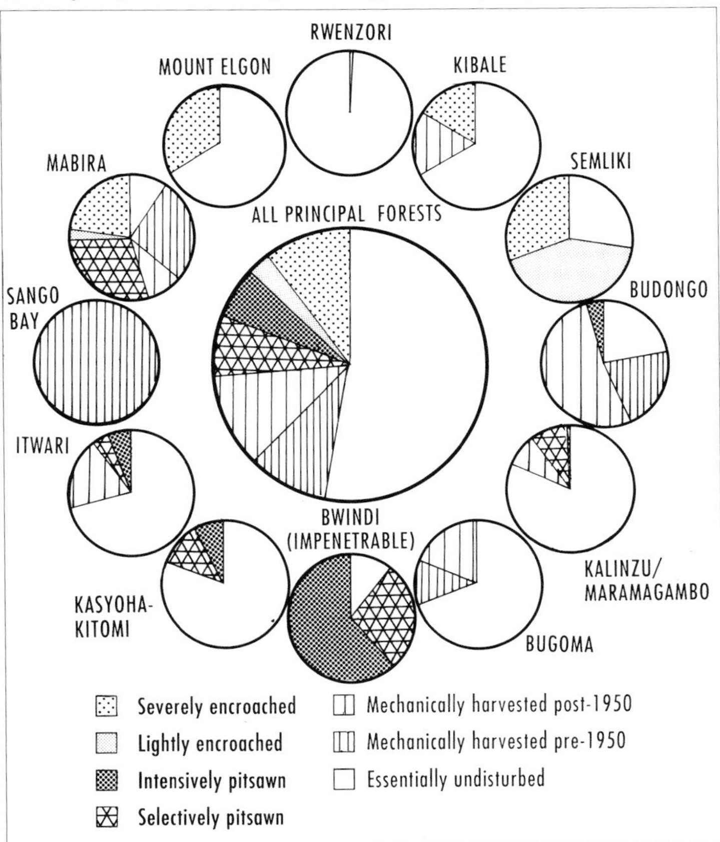

wood requirements. People may live, graze their livestock or cultivate the land in these reserves but only with the written authority of a senior forest officer. Removal of any forest products also requires a permit from a senior forest officer and usually the payment of a fee. Certain local people retain the right to take most forest products for their own domestic use without permission or payment.

The government of Uganda first adopted a forestry policy in 1929 and this placed emphasis on the role of forestry in the protection of the environment (Forest Department, 1951, 1955). Initially the Forest Department was an effective organisation with far-sighted policies and a high degree of control over its land. Forest areas were divided into management compartments each of which was harvested in an orderly manner. No felling was allowed in any area prior to systematic stock mapping by the Forest Department and only marked trees could be cut. Minimum girth limits were also endorsed. None of the timber could be removed before the royalties payable were assessed (Howard, 1991). However, this changed in the early 1970s under the Amin regime, when there was a general breakdown of forest control. As a result, there was greater emphasis on short-term profit from timber extraction while protective forestry was ignored (Hamilton, 1984). In addition, many of the management practices designed to control activities in the forest reserves had become in-effective, while maintenance and expansion of the tree plantations had almost ceased (Hamilton, 1984; Struhsaker, 1987). The timber industry also collapsed soon after Amin came to power. Before then Uganda had developed a thriving forest industry which employed around 3000 people, had sawmills in most of the principal reserves and, in 1971, processed 170,000 cu. m of timber (World Bank, 1986). However, the majority of the industrial plants have since broken down and the country's sawn timber requirements are today satisfied largely by an estimated 3000 pitsawyers, many of whom are probably cutting illegally in forest reserves (Howard, 1991).

Approximately 21 per cent of the forests within the principal reserves have been mechanically harvested (Figure 31.1; Howard, 1991). Until 1960, a polycyclic felling system was used: trees exceeding a specified minimum girth were harvested about every 30 years on a 60–90 year rotation (Howard, 1991). This system interfered relatively little with the natural state of the forest and regeneration was natural rather than by enrichment planting. In 1960, the polycyclic system was replaced by a monocyclic one and there was no lower limit on the size of trees cut (Howard, 1991). This probably had a far more deleterious impact on the wildlife and the ecology of the forests (Struhsaker, 1987 and see Kibale Forest case study).

In the early years, various silvicultural treatments were applied after the timber harvest. These included enrichment planting, and arboricide treatment to remove undesirable trees which were felled and together with waste timber were made into charcoal. This helped to open up the canopy and stimulate growth of timber tree seedlings. From around the mid-1970s post-harvest silviculture techniques have rarely been applied, so any potential regeneration is usually suppressed (Howard, 1991).

Effects of Logging on the Primates in Kibale Forest

In general, small species of primates are better able to adapt to moderate levels of habitat disturbance than larger ones and fruit-eating species suffer more than leaf-eaters. Skorupa (1987) found that population densities of five out of seven primate species were significantly lower in an area that had been heavily logged (i.e. about 50 per cent of the original stand had been destroyed) 12–14 years previously compared with an adjacent area of undisturbed forest (see Figure 31.2). However, the black-and-white colobus is favoured by intensive logging; severely disturbed areas were found to support populations up to five times as high as those in nearby undisturbed areas. Where logging had been comparatively light, only L'Hoest's monkey showed significantly reduced population densities. Forest that had been heavily logged and subsequently treated with arboricides exhibited the lowest primate densities.

Primate populations have been monitored in three heavily logged compartments of Kibale forest over a period of about 17 years (Struhsaker, 1975; Skorupa, 1987), from two years post-felling to 19 years afterwards. The results of these studies indicate that all but one (red-tail monkey) of the seven resident primates have remained at post-logging population densities without any significant return to densities characteristic of undisturbed habitat. In other words, in an area where there was no post-harvest enrichment planting, there has been no recovery of the primate community 19 years after logging. These findings in an African rain forest are very different from the results of studies in Malaysia where it was found that there were no long-term detrimental effects on the primates after heavy selective logging (Johns, 1983).

It can be concluded that, in Africa, a lightly logged forest will support primate populations comparable to unlogged forests, while a heavily logged forest will have its conservation value substantially compromised. Selective logging in Africa can, therefore, be compatible with primate conservation in particular and biological conservation in general, but only if levels of damage are strictly limited.

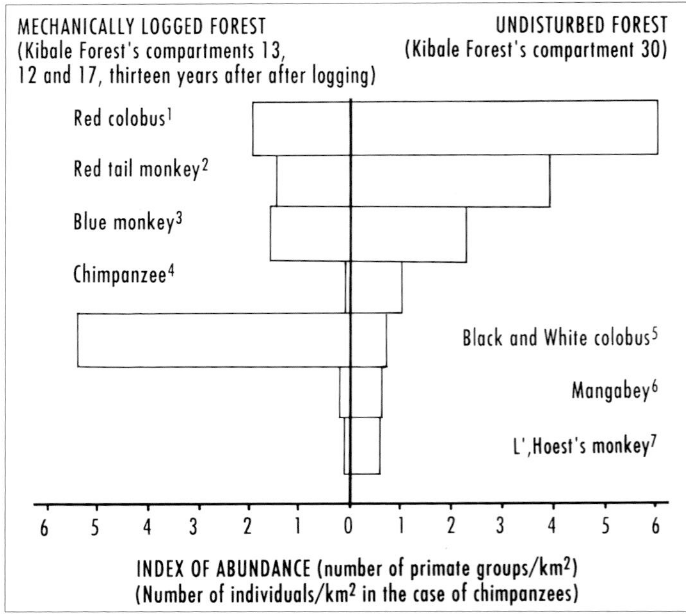

Figure 31.2 The relative abundance of seven primates in adjacent areas of mechanically logged and undisturbed forest at Kibale

(*Sources:* Data from Skorupa, 1987; figure from Howard, 1991)

[1] *Procolobus [badius] rufomitratus tephrosceles* [2] *Cercopithecus ascanius* [3] *Cercopithecus mitis stuhlmanni* [4] *Pan troglodytes* [5] *Colobus angolensis ruwenzorii* [6] *Cercocebus albigena johnstoni* [7] *Cercopithecus lhoesti*

The demand for fuelwood in Uganda is enormous, much greater than that for sawn-wood. Various estimates indicate that 90–95 per cent of the wood consumed in the country is for fuel. At present, most of this is taken from public land outside the forest reserves – from fuelwood plantations, small remnant patches of natural forest and woodland – but demand from an increasing population and a growing industrial sector is already outstripping the supply available in these areas (Howard, 1991). The burgeoning brick-making industry consumes large quantities of fuel during firing while the recent trend of rural people to migrate to the larger towns and cities also greatly increases the demand for charcoal (Struhsaker, 1987). The demand for building poles is also substantial, estimated by the World Bank (1986) to be between four and five million poles or 300,000 to 400,000 cu. m of wood. These come mostly from small private plantations, natural forest remnants on public land and trees on agricultural land, but poles are also taken from all the forest reserves. Furthermore, the exploitation of bamboos, palm nuts and rattan canes has a locally significant impact on the forests (Howard, 1991).

In 1987, Uganda issued a revised forest policy which emphasised the importance of protective forestry. In theory, enough forest land is to be maintained and safeguarded to ensure sufficient supplies of timber, fuel and other forest products for the long-term needs of the country and, where feasible, for export. Water supplies and soils are to be protected, plants and animals will be conserved in natural ecosystems, but the forests will also be available for amenity and recreation. The forests will be managed to optimise economic and environmental benefits by ensuring that conversion of the forest resources to timber, charcoal and the like is carried out efficiently; that the forest estate is protected against encroachment, illegal tree cutting, fires, diseases and pests; and that sustainable methods are used for harvesting. Biodiversity will be preserved and research undertaken to improve forest growth and yield. In addition, the new policy will encourage agroforestry, by providing services to help farmers grow and protect their own trees and by promoting scientific research.

Hunting is common within the Ugandan forests and it is having a major impact on populations of many of the larger mammals. Ungulates are favourite targets, particularly the larger ones such as buffalo *Syncerus caffer*, giant forest hog *Hylochoerus meinertzhageni*, bushbuck *Tragelaphus scriptus* and bushpig *Potamochoerus porcus*, but also various species of duiker *Cephalophus* spp. and other antelopes (Howard, 1991). Primates are hunted only by particular ethnic groups, the Batwa, Baamba, Bakonjo and Bagisu and only in Semliki, Rwenzori and Mt Elgon forests which are in the traditional areas of these groups. Ground-dwelling birds are also sometimes caught. Hunting methods include snares, deadfall and pitfall traps, poison-tipped arrows and the use of dogs to drive animals into nets. Guns are rarely used (Howard, 1991).

Deforestation

Ever since the introduction of agriculture some 2500 years ago, Uganda's forests have been cleared to make way for crops and pasture. This process continues, although most of the forest loss is now from the little remaining on public lands (A. Johns, *in litt.*). It was estimated by FAO (1988) that 100 sq. km of forest were lost annually between 1981 and 1985, while Hamilton (1984) estimated that 110 sq. km (2 per cent) of Uganda's tropical high forest were lost each year, which meant that the remaining patches of forest were shrinking fast. Recently, though, the losses from the forest reserves have mostly been curtailed (A. Johns, *in litt.*) and it is likely that the annual deforestation rate is now quite low.

Most of the accelerating destruction that occurred between 1971 and 1986, was due both to agricultural encroachment and to increased demands for fuel, especially charcoal (Struhsaker, 1987). Howard (1991) estimated that 12 per cent of the forested land within the country's principal reserves had been affected by agricultural encroachment. Four reserves were badly affected (Kibale, Semliki, Mabira and Mt Elgon) while eight had suffered little encroachment (Figure 31.1). The encroachment of Kibale and Semliki was generally for subsistence farming while that of Mabira and Mt Elgon involved cash-crop production.

Buhoma Forest, near Bwindi in the west of Uganda. C. Harcourt

The increased demand for agricultural land and forest products can be attributed primarily to the high population growth rate of more than 3 per cent. The population has doubled since 1960 (Hamilton, 1984). Immigrants from overcrowded districts, such as Kabale and Rukungiri in the south-west, move to less populated regions, often forested areas on public land (Struhsaker, 1987). Immigrants used to move into forest reserves as well, but, since 1987, protection of the reserves has improved and in some cases encroachers have left (D. Pomeroy, *in litt.*). Indeed, around 350 sq. km of reserved forest have been cleared of settlers since 1987 and only 10 sq. km or so are still affected by their presence.

Biodiversity

Uganda is exceptionally rich in species of both plants and animals. Its location in east-central Africa means that communities characteristic of East Africa's savannas overlap there with those of West Africa's rain forests (Howard, 1991). Indeed, White (1983) shows in his vegetation map of Africa that Uganda contains seven of mainland Africa's 18 phytochoria (biogeographic divisions based on plant distribution), more than any other single nation (Howard, 1991). The lowland and montane forests of western Uganda are of particular importance for the diversity of species they contain (Stuart *et al.*, 1990). The forest with the greatest biological diversity is Bwindi Forest, not only because it was one of the forest refugia but because it is the only forest in East Africa with contiguous lowland and montane communities (see case study).

There are approximately 5000 flowering plant species in the country but the vegetation regions are shared with other countries and only about 30 species are endemic (Davis *et al.*, 1986), compared with 265 in Kenya and 1100 in Tanzania.

Uganda is one of the top three or four countries in Africa for number of mammal species with an estimated 311 (Stuart *et al.*, 1990). Of these, nine are listed by IUCN (1990) as threatened: mountain gorilla *Gorilla gorilla berengei*, chimpanzee *Pan troglodytes*, L'Hoest's monkey *Cercopithecus lhoesti*, Uganda red colobus *Procolobus [badius] rufomitratus tephrosceles*, elephant *Loxodonta africana*, leopard *Panthera pardus*, cheetah *Acinonyx jubatus*, black rhinoceros *Diceros bicornis* and Jackson's mongoose *Bdeogale jacksoni*, but none of these is endemic to the country. The black rhino may now be extinct in Uganda, while the white rhino *Ceratotherium simum* certainly is.

Uganda has more species of bird for its size than any other country in Africa (Pomeroy and Lewis, 1987). There are 989 bird species listed for the country (Stuart *et al.*, 1990), but new studies suggest that there are at least 1000 species present (D. Pomeroy, *in litt.*). Ten are considered to be threatened (Collar and Stuart, 1985) and six of these are forest species. Five are listed as rare: Nahan's francolin *Francolinus nahani* which is found in Kibale, Semliki, Budongo, Bugoma and Mabira forests; the forest ground thrush *Turdus oberlaenderi* found in Semliki; the African green broadbill *Pseudocalyptomena graueri* from Bwindi; Chapin's flycatcher *Muscicapa lendu*, which also occurs in Bwindi, and Turner's eremomela *Eremomela turneri* found in Nyondo Forest in the extreme south-west. The threatened Kibale ground thrush *Turdus kibalensis* is endemic to Kibale, but its status is 'indeterminate' as it is known only from two individuals collected in 1966 (Collar and Stuart, 1985).

Total numbers of reptiles and amphibians are unknown. In the former group, only the dwarf, Nile and slender-snouted crocodiles (*Osteolaemus tetraspis*, *Crocodylus niloticus* and *C. cataphractus*) are listed as threatened by IUCN (1990) and it is possible that the slender-snouted species is extinct in Uganda. None of the ten amphibians considered to be of conservation concern is endemic to Uganda (Stuart *et al.*, 1990).

Numbers of fish and invertebrates are not known. However, there are important populations of endemic fish, particularly cichlids, in Lake Victoria and other Ugandan lakes. Those in Lake Victoria are threatened by the introduction of Nile perch and some are listed as endangered (IUCN, 1990). Two threatened swallowtail butterflies, the African giant *Papilio antimachus* and the cream-banded *Papilio leucotaenia* are found in Uganda's forests (Collins and Morris, 1985).

Conservation Areas

There are six categories of protected areas in Uganda. National parks have the highest conservation status. They can be created or abolished only by an act of parliament and their purpose is to preserve wild animal life and natural vegetation. There are six national parks (Table 31.2) with a total area of 8336 sq. km, and three of these contain forest. Two, Mt Rwenzori and Gorilla, are very recent additions to the network. Queen Elizabeth, Uganda's only Biosphere reserve, contains the northern part of Maramagambo Forest.

THE IMPENETRABLE FOREST CONSERVATION PROJECT

The Impenetrable Forest Conservation Project, funded by WWF and USAID, was set up in 1986 to assist in preventing further degradation of the once extensive rain forests of south-west Uganda. The welfare and protection of three of the remaining natural forest remnants, namely the Bwindi (Impenetrable), Echuya and Mgahinga Forests are the project's main concern. These areas, managed as forest reserves by the Forest Department, are among the most important and most biologically diverse in Africa. The terrain over all three forests is rugged and the dense vegetation protects vital water catchments and provides important sources of bamboo, timber, fuelwood, honey, medicines and other natural products. They are among Uganda's most precious natural resources, yet prior to the initiation of the Impenetrable Forest Conservation Project, the forests were poorly managed and protected.

Bwindi Forest Reserve (321 sq. km) in the Kigezi Highlands is one of the largest forests in East Africa and one of the few containing montane and lowland forest in a continuum. As a result of its size and altitudinal range (1160 m–2600 m asl), and its likely role as a Pleistocene refugium, it is probably the richest forest in East Africa in terms of plant, mammal, bird and butterfly species. It has possibly the highest number of montane bird and butterfly species in Africa, many endemic to the forest. Three species of bird found there are listed as threatened (Collar and Stuart, 1985), namely the African green broadbill, Grauer's swamp warbler *Bradypterus graueri* and Chapin's flycatcher. One of the most notable endangered mammals is the mountain gorilla. Approximately 300 individuals, half the world's population, live in Bwindi forest.

Mgahinga Forest Reserve, situated on the Uganda, Rwanda and Zaïre borders, forms part of the Virunga Volcanoes ecosystem with Zaïre's Virungas National Park and Rwanda's Volcanoes National Park. It comprises the northern slopes of three of the Virunga volcanoes – Sabinio, Mgahinga and Muhavura – and covers an area of 24 sq. km (probably soon to increase to 33 sq. km). The forest, with an altitudinal range of 2400 m–4300 m and largely consisting of bamboo and alpine vegetation, is one of the few sanctuaries for the golden monkey *Cercopithecus mitis kandti*.

Echuya Forest, one of the least studied forests in Uganda, is located between Bwindi and Mgahinga forests, and covers an area of 34 sq. km primarily consisting of bamboo. Muchuya swamp, in the centre of the forest, is one of the highest and largest upland swamps remaining in East Africa.

Beyond the borders of these three forests there is virtually no natural forest remaining in south-west Uganda, the land being intensively cultivated in what is one of the most densely populated parts of Africa. Massive deforestation has led to wood and water shortages, soil and watershed damage, flooding and siltation. In short, the environment of south-western Uganda has been badly degraded.

At the end of the first four years (1986–90) the project has managed to achieve a number of its objectives, the most notable being the training of guards. In Impenetrable and Mgahinga this has resulted in the virtual elimination of poaching, grazing, timber theft and gold mining. The project is working on the establishment of a national park in the area. CARE and USAID are collaborating in rural development activities around the forest in an attempt to reconcile conflicts with local people.

Source: Tom Butynski

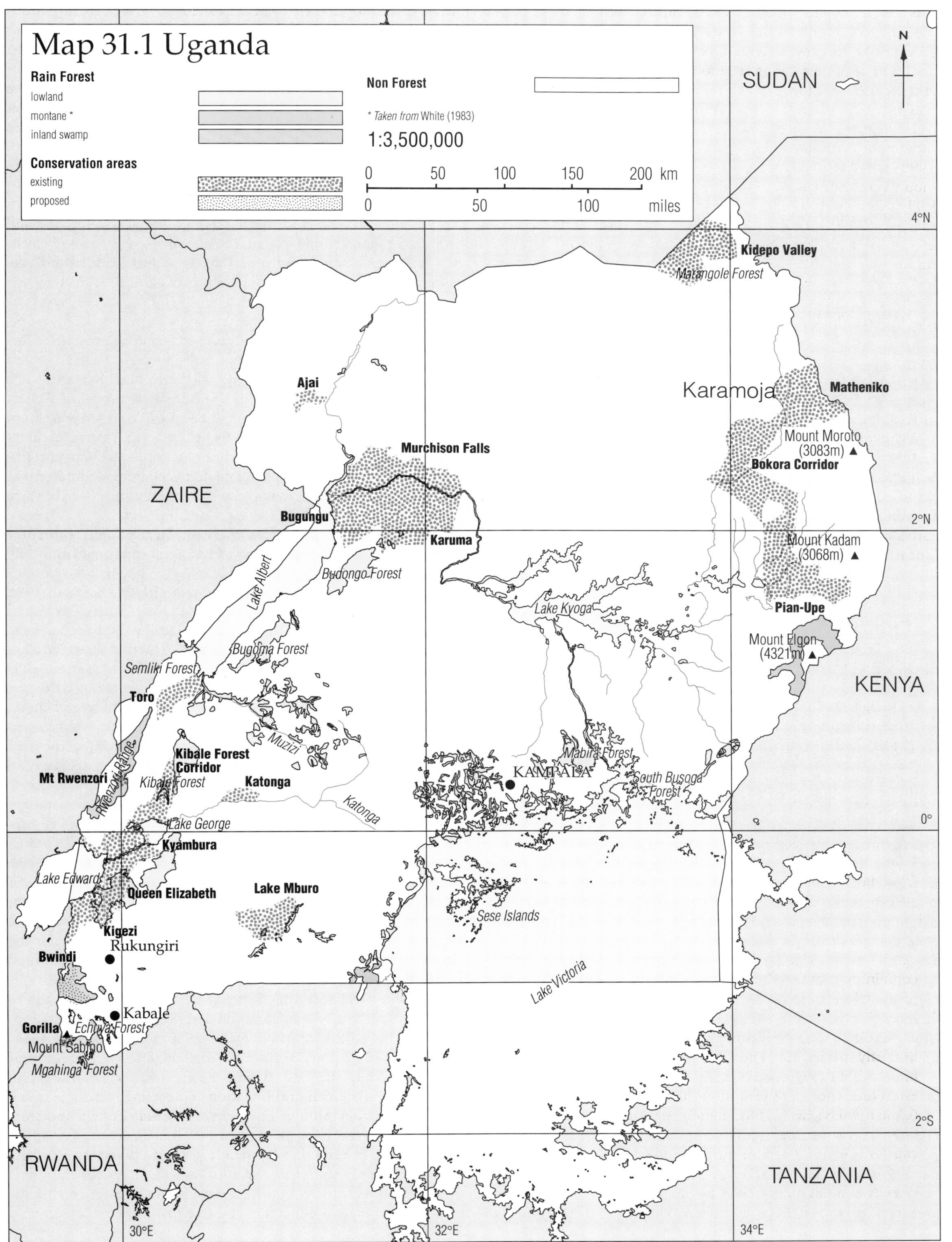
Map 31.1 Uganda
Rain Forest
lowland
montane *
inland swamp
Conservation areas
existing
proposed
Non Forest
* Taken from White (1983)
1:3,500,000
0 50 100 150 200 km
0 50 100 miles
SUDAN
ZAIRE
KENYA
RWANDA
TANZANIA
4°N
2°N
0°
2°S
30°E
32°E
34°E
Kidepo Valley
Marangole Forest
Ajai
Karamoja
Matheniko
Mount Moroto (3083m)
Bokora Corridor
Murchison Falls
Bugungu
Karuma
Mount Kadam (3068m)
Budongo Forest
Lake Albert
Lake Kyoga
Pian-Upe
Mount Elgon (4321m)
Bugoma Forest
Semliki Forest
Toro
Muzizi
Kibale Forest Corridor
Kibale Forest
Katonga
Mt Rwenzori
Rwenzori Range
Mabira Forest
KAMPALA
South Busoga Forest
Lake George
Kyambura
Lake Edward
Queen Elizabeth
Lake Mburo
Sese Islands
Kigezi
Rukungiri
Bwindi
Lake Victoria
Kabale
Gorilla
Echuya Forest
Mount Sabino
Mgahinga Forest

Table 31.2 Conservation areas of Uganda

Forest reserves, game sanctuaries and controlled hunting areas are not listed or mapped. For information on Biosphere reserves and Ramsar sites see chapter 9.

	Existing area (sq. km)	*Proposed area (sq. km)*
National Parks		
Bwindi *		310
Gorilla (Mgahinga)	24	
Kidepo Valley	1,344	
Lake Mburo	536	
Mt Rwenzori*		
Murchison Falls (Kabalega)	3,840	
Queen Elizabeth (Rwenzori)*	1,978	
Game Reserves		
Ajai	158	
Bokora Corridor	2,056	
Bugungu	520	
Karuma*	820	
Katonga	208	
Kibale Forest Corridor*	560	
Kigezi*	330	
Kyambura*	157	
Matheniko	1,600	
Pian-Upe	2,314	
Toro	555	
Nature Reserves		
Igwe Luvunya†	10	
Kasagala†	21	
Kisanju†	12	
Maruzi Hills†	68	
Ngogo†	72	
North Mabira†	34	
Ntendwe Hill†	3	
Nyakafunjo†	7	
Waibira†	32	
Wambabya†	34	
Zoka†	61	
Totals	17,968	310

(*Source:* WCMC, *in litt.*)
* Area with moist forest within its boundaries as shown on Map 31.1..
† Not mapped – location data unavailable to this project.

In game reserves, animals were allowed to be captured or killed only with the permission of the Chief Game Warden but hunting in Uganda has now been forbidden for more than a decade. None of the vegetation within a game reserve is protected to any greater extent than in a forest reserve, but grazing of livestock, cultivation or settlement of an area within the reserve requires permission from the Chief Game Warden (Howard, 1991). The Bwindi Forest, Mgahinga and part of Kibale Forest Reserve carry dual status as game and forest reserves. There are now 11 game reserves in the country, but one of these (the Gorilla Game Reserve) has recently been promoted to the status of national park; the remainder cover an area of 9278 sq. km (Table 31.2).

There are 14 controlled hunting areas and seven game sanctuaries; these are the responsibility of the Game Department and can be gazetted or degazetted by the minister responsible for wildlife (IUCN, 1987). Human settlement, livestock and cultivation are permitted in these reserves and, therefore, neither category has been shown on Map 31.1 or in Table 31.2. Some of the forest reserves contain nature reserves within their boundaries and these areas are protected from all forms of extractive use. Most of Uganda's major forest types are covered by the existing forest reserves, but not all are represented within nature reserves (Pomeroy, 1990).

The Forest Department has recently proposed the establishment of a new category of multiple-use conservation areas known provisionally as forest parks. These would fall under the control of a Forest Parks Commission and be managed by the Forest Department under the following conditions:

- At least 50 per cent of the area would be protected against extractive use.
- No mechanised exploitation would be permitted.
- Manual harvesting of forest products would be licensed and strictly regulated within designated areas.
- Park guards would be empowered to enforce regulations protecting both plant and animal life (Kigenyi, 1988).

Initiatives for Conservation

The Ugandan Government is concerned that the demand for forest products is rapidly outstripping the supply. As a result, several important programmes of external aid have been initiated. The largest of these is a World Bank Forestry Rehabilitation Project with a budget of US$38.3 million to be spread over six years from March 1988. The project aims to improve the management of Uganda's forests to meet domestic needs for wood on a sustainable basis. It will extend and rehabilitate plantations and, at the same time, improve the management of natural forests (Howard, 1991).

The EEC is managing the conservation component of the project and the cost of this is approximately a quarter of the total finance (IUCN, 1990; Tabor *et al.*, 1990). This includes boundary marking around the forest estate and already, between 1988 and 1990, 2600 km of boundary has been made with a cut line or a line of planted marker trees. There is also a reforestation programme, under which so far 43 sq. km of degraded forest land has been planted with tree seedlings. In addition, the value of each forest reserve for nature conservation is being documented. Further plans exist to establish more nature reserves with buffer zones around them (Kigenyi, 1990), so that the two categories eventually will cover 20 per cent and 30 per cent respectively of the total forested area. The EEC will supervise the surveying, delimitation and management in these areas (IUCN, 1990).

A number of smaller forest projects are being undertaken. For instance, since 1984 CARE-International has been helping to establish village tree nurseries and to train Forest Department staff. CARE-International and WWF also have a programme to assist with the integration of rural developments and the conservation of forests in south-western Uganda. IUCN and the Ministry of Environment Protection, with funding from the Norwegian International Development Agency, have started a conservation project for the degraded forest of Mt Elgon. Development of the rural areas around this region is being integrated with protection of the forests.

FAO has a Wildlife and National Parks Project in Uganda to review the current system of national parks and game reserves, to examine the institutional arrangements and legislation for management of these areas and to increase the effectiveness of their protection (IUCN, 1990). Bwindi Forest has had a conservation project since 1986 (see case study on p. 266) and there is also a conservation management programme in Kibale Forest. Makerere University has a biological field station in Kibale supported by WCI/NYZS, USAID and the EEC, to research and monitor the forest and train Ugandan students. In addition, these organisations are involved with several projects outside the forest involving the local people, such as education and tree planting.

There have been numerous recommendations that large areas of forest be set aside as protected areas. This is most urgent for the important water catchment areas of Mt Elgon, Mt Kadam and the Bwindi Forest (see Howard, 1991). Semliki, Kibale, Kaniyo-Pabidi, Marangole and South Busoga forests have also all been proposed as worthy of national park status (see Howard, 1991). In recognition of this, the Forest Department is considering upgrading some of these areas to forest parks. There is a general consensus that no simple, direct solution can be applied to solving the problems of deforestation in Uganda and that wide-ranging integrated programmes of activities are needed. These programmes are now being initiated and the next five years will be critical in deciding the fate of the entire forest ecosystem in Uganda (Tabor *et al.*, 1990).

References

Aluma, J. R. (1987) *Uganda Forestry Resources and Action Plan.* Draft report to UNEP, Nairobi, Kenya.

Collar, N. J. and Stuart, S. N. (1985) *Threatened Birds of Africa and Related Islands. The ICBP/IUCN Red Data Book Part 1.* ICBP/IUCN, Cambridge, UK. 761 pp.

Collins, N. M. and Morris, M. G. (1985) *Threatened Swallowtail Butterflies of the World. The IUCN Red Data Book.* IUCN, Gland and Cambridge, UK. vii + 401 pp.

Davis, S. D., Droop, S. J. M., Gregerson, P., Henson, L., Leon, C. J., Villa-Lobos, J. L., Synge, H. and Zantovska, J. (1986) *Plants in Danger: What do we know?* IUCN, Gland, Switzerland and Cambridge, UK.

Department of Lands and Surveys (1967) *Atlas of Uganda.* Government Printer, Entebbe, Uganda.

FAO (1988) *An Interim Report on the State of Forest Resources in the Developing Countries.* FAO, Rome, Italy. 18 pp.

FAO (1990) *FAO Yearbook of Forest Products 1977–1988.* FAO Forestry Series No. 23 and FAO Statistics Series No. 90. FAO, Rome, Italy.

Forest Department (1951) *A History of the Uganda Forest Department 1898–1929.* Government Printer, Entebbe, Uganda.

Forest Department (1955) *A History of the Uganda Forest Department 1930–1950.* Government Printer, Entebbe, Uganda.

Hamilton, A. C. (1984) *Deforestation in Uganda.* Oxford University Press, Nairobi, Kenya.

Howard, P. C. (1991) *Nature Conservation in Uganda's Tropical Forest Reserves.* IUCN, Gland, Switzerland and Cambridge, UK. 330 pp.

IUCN (1987) *IUCN Directory of Afrotropical Protected Areas.* IUCN, Gland, Switzerland and Cambridge, UK. xix + 1043 pp.

IUCN (1990) *1990 IUCN Red List of Threatened Animals.* IUCN, Gland, Switzerland and Cambridge, UK.

Johns, A. D. (1983) *Ecological Effects of Selective Logging in a West Malaysian Rain-forest.* PhD thesis. University of Cambridge, UK.

Kigenyi, F. (1990) Forest Reserves. In: *Conservation of Biodiversity in Uganda.* Pomeroy, D. (ed.). Proceedings of the Second Conservation Forum, pp. 15–16.

Langdale-Brown, I. (1960) *The Vegetation of Uganda (excluding Karamoja).* Series 2, No. 6, Memoirs Research Division, Department of Agriculture, Uganda.

Langdale-Brown, I., Osmaston, H. A. and Wilson, J. G. (1964) *The Vegetation of Uganda and its Bearing on Land Use.* Government Printer, Entebbe, Uganda.

Pomeroy, D. (ed.) (1990) *Conservation of Biodiversity in Uganda.* Proceedings of the Second Conservation Forum. 63 pp.

Pomeroy, D. E. and Lewis, A. D. (1987) Bird species richness in tropical Africa: some comparisons. *Biological Conservation* **40**: 11–28.

PRB (1990) *1990 World Population Data Sheet.* Population Reference Bureau, Inc., Washington, DC, USA.

Skorupa, J. P. (1987) *The effects of habitat disturbance on primate populations in the Kibale Forest, Uganda.* PhD thesis. University of California, USA.

Struhsaker, T. T. (1975) *The Red Colobus Monkey.* University of Chicago Press, Chicago, USA.

Struhsaker, T. T. (1987) Forestry issues and conservation in Uganda. *Biological Conservation* **39**: 209–34.

Stuart, S. N., Adams, R. J. and Jenkins, M. D. (1990) *Biodiversity in Sub-Saharan Africa and its Islands: Conservation, Management and Sustainable Use.* Occasional Papers of the IUCN Species Survival Commission No. 6. IUCN, Gland, Switzerland.

Tabor, G. M., Johns, A. D. and Kasenene, J. M. (1990) Deciding the future of Uganda's tropical forests. *Oryx* **24**(4): 208–14.

White, F. (1983) *The Vegetation of Africa: a descriptive memoir to accompany the Unesco/AETFAT/UNSO vegetation map of Africa.* Unesco, Paris, France. 356 pp.

World Bank (1986) *UNDP/World Bank Energy Sector Management Assistance Program.* World Bank, Washington, DC, USA.

Authorship

Caroline Harcourt with contributions from Peter Howard, WWF-Uganda, Alan Hamilton, WWF-UK, Derek Pomeroy, Makerere University, Uganda, Tom Butynski, Impenetrable Forest Conservation Project, Uganda and Andy Johns, Kibale, Uganda.

Map 31.1 Forest cover in Uganda

Information on forest distribution is extracted from the 1:500,000 map *Uganda Vegetation* (1964, reprinted 1972) published in four sheets, and drawn and printed by the Department of Lands and Surveys, Uganda. The map was compiled by I. Langdale-Brown, H. A. Osmaston and J. G. Wilson and is one of five which accompany a book entitled *The Vegetation of Uganda*, published by the Government of Uganda.

Of the 86 vegetation classes shown, ranging from high altitude types, through to forest, savanna, woodlands, shrub steppes, grass steppes, bushland, thickets and wetter formations, 13 types have been digitised and harmonised to produce Map 31.1. The 13 classes on the source map are: B: High Altitude Forest – B1, B2, B3, B4; C: Medium Altitude Moist Evergreen Forests – C1, C2, C3; D: Medium Altitude Moist Semi-Deciduous Forests – D1, D2, D3, D4, and Y: Swamp Forests – Y1 and Y2. The swamp forests are small seasonal formations located on the west side of Lake Victoria. White's (1983) vegetation map has been overlaid on to the source map to distinguish between lowland and montane forests. Forest types on the Sese Islands in Lake Victoria have not been mapped, as the source map does not distinguish between rain forest and grass savanna types and no further data have been made available. The reader must be cautious when using the forest statistics produced from Map 31.1 as the source map is more than 25 years old.

Conservation areas data have been digitised from a 1:1,500,000 map *Uganda Game Conservation* (1969), printed by the Department of Lands and Surveys, Entebbe, which is an extract from the *Atlas of Uganda* (1967).

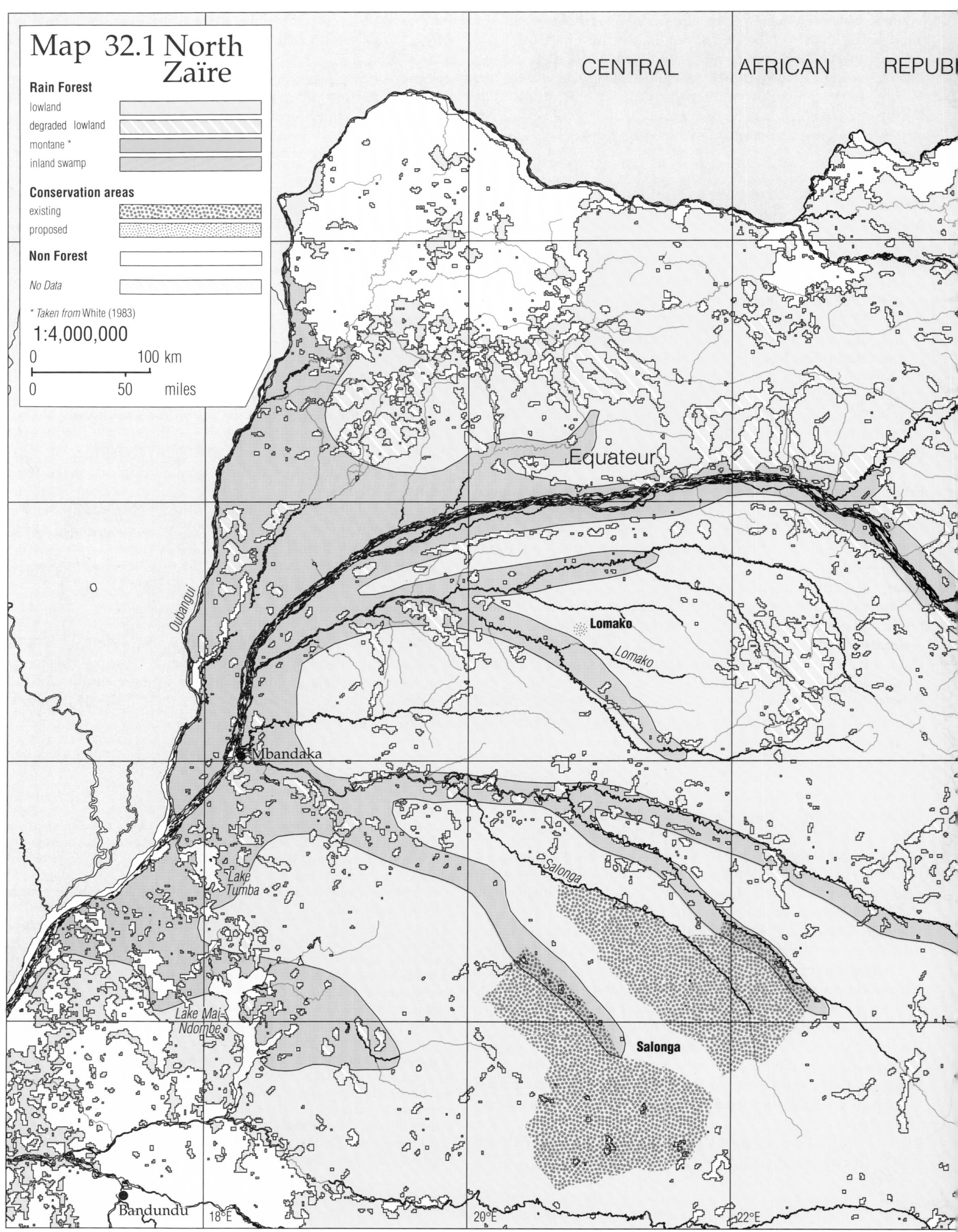
Map 32.1 North Zaïre
Rain Forest
lowland
degraded lowland
montane *
inland swamp
Conservation areas
existing
proposed
Non Forest
No Data
* Taken from White (1983)
1:4,000,000
0
100 km
0
50
miles
CENTRAL
AFRICAN
REPUB
Equateur
Oubangui
Lomako
Lomako
Mbandaka
Salonga
Lake Tumba
Salonga
Lake Mai-Ndombe
Bandundu
18°E
20°E
22°E

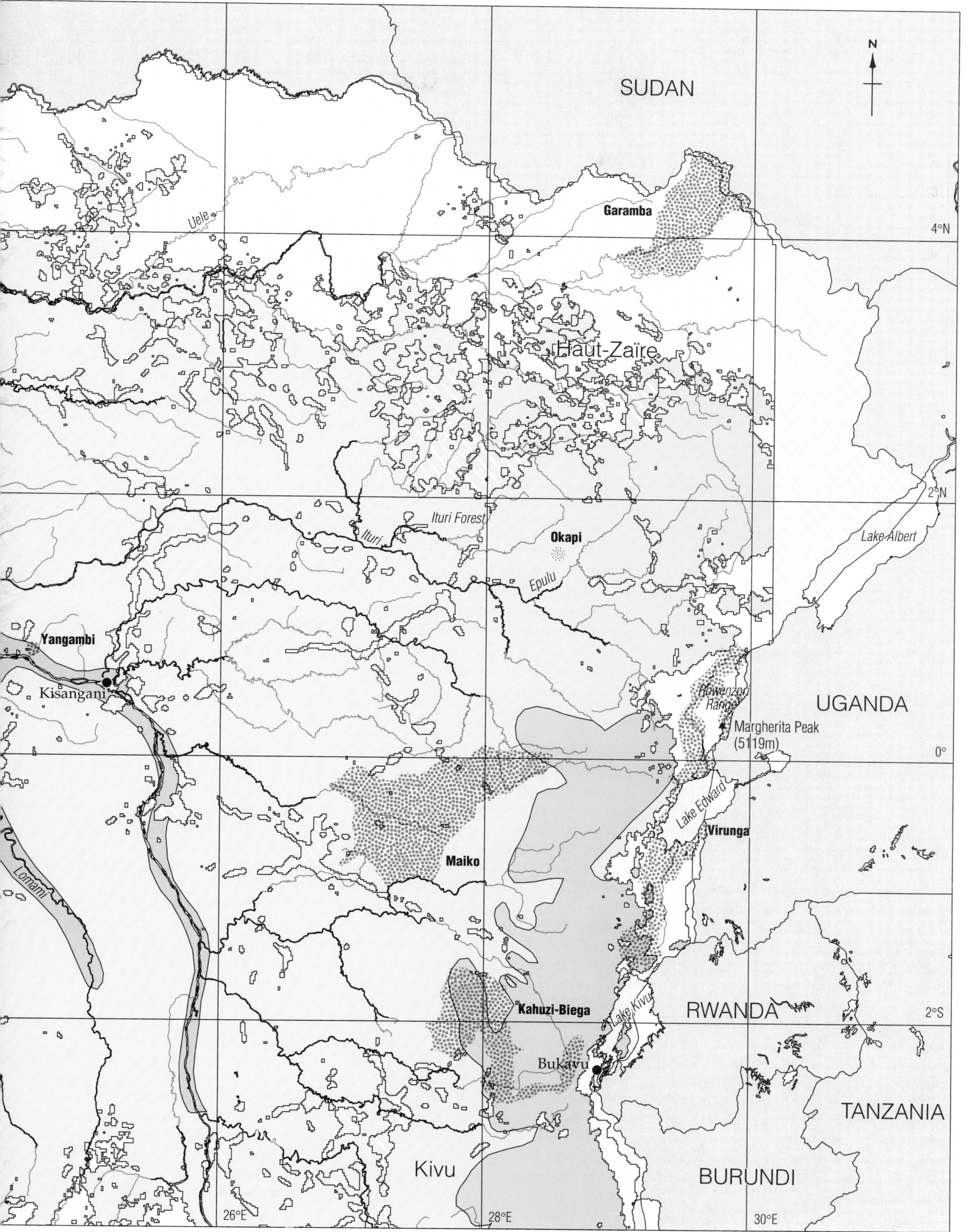
SUDAN
Garamba
4°N
Uele
Haut-Zaïre
2°N
Ituri Forest
Ituri
Okapi
Epulu
Lake Albert
Yangambi
Kisangani
Rowenzori Range
UGANDA
Margherita Peak
(5119m)
0°
Lake Edward
Virunga
Maiko
Lomami
Kahuzi-Biega
Lake Kivu
RWANDA
2°S
Bukavu
TANZANIA
Kivu
BURUNDI
26°E
28°E
30°E

32 Zaïre

Land area 2,267,290 sq. km
Population (mid-1990) 36.6 million
Population growth rate in 1990 3.3 per cent
Population projected to 2020 90 million
Gross national product per capita (1988) US$170
Rain forest (see maps) 1,190,737
Closed broadleaved forest (end 1980)* 1,056,500 sq. km
Annual deforestation rate (1981–5)* 1800 sq. km
Industrial roundwood production† 2,791,000 cu. m
Industrial roundwood exports† 117,000 cu. m
Fuelwood and charcoal production† 32,557,000 cu. m
Processed wood production† 174,000 cu. m
Processed wood exports† 27,000 cu. m

* FAO (1988)
† 1989 data from FAO (1991)

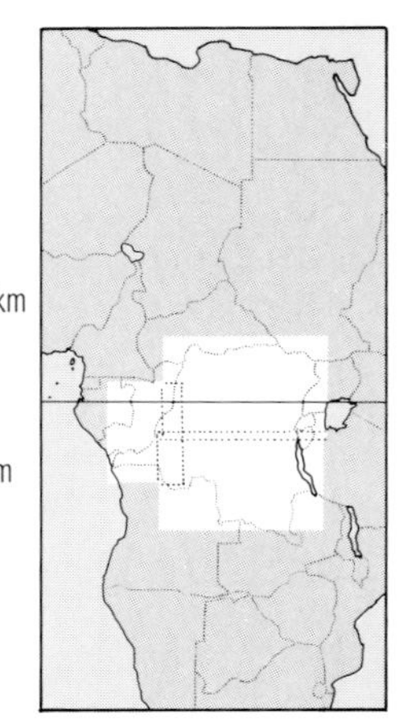

Zaïre is Africa's third largest country, after Algeria and Sudan, and it contains over half of the continent's tropical moist forests. The Salonga National Park is the largest rain forest park in the world (36,000 sq. km) and the Virunga National Park certainly supports a greater biological diversity than any other single protected area in Africa. The extent of the forest is such that it is still possible to fly in a jet for two hours over virtually undisturbed forest, from Bandundu in the west to Bukavu in the east.

Zaïre is renowned for its biological richness, with more species of birds and mammals than any other African country. This diversity is due in part to the country's vastness, but more especially to the range of geomorphological and climatic conditions that it encompasses. The forests are of great importance to local people – for firewood, bushmeat and crops – as well as for the production of export commodities such as timber and palm oil.

Despite having serious conservation problems Zaïre also has a considerable record of conservation achievement. There is already a comprehensive system of protected areas within the country, and a commitment has been made by the government to increase this to cover 12–15 per cent of national territory. Conservation programmes have received considerable support from the country's charismatic leader, President Mobutu Sésé Séko. In his 25 years of absolute power, President Mobutu amassed a considerable personal fortune and acquired a degree of international notoriety. Yet this controversial figure spent his free time fishing and observing wildlife in the country's national parks. He drew on his personal wealth to equip and pay the guards of the parks and, in his opening address to the IUCN General Assembly in Kinshasa in 1975, he referred to Zaïre's parks as the 'cathedrals of my country'.

Introduction

With a total country area of 2,344,510 sq. km, Zaïre dominates central Africa; it extends from the savanna woodlands of the southern fringes of the Sahara in the north, to the woodlands and grasslands of Zambia and Angola bordering the Kalahari, in the south. From west to east it encompasses mangrove forests on the Atlantic coast and glaciers in the Rwenzori, the summit of which, Margherita peak, is Africa's third highest mountain (5119 m). The vast expanse of the equatorial forest lies in the central basin – the Cuvette Centrale – at an altitude of just 300 m above sea level.

Zaïre has a population of almost 37 million people, of which approximately 40 per cent live in cities. These are located around the periphery of the Cuvette, along two main axes: north–south, along the eastern border, and west–east, from Bas-Zaïre to Kasaï. The economy of the country is based on a mixture of mining and agricultural activities, with copper, cobalt, diamonds and crude oil being the main mineral exports, and coffee the main agricultural export. There is some industrial development including brewing, cement manufacture and oil refining. The generation of hydro-electricity is also important.

The central forested areas have always been sparsely populated. In recent years, the poor international market for the few agricultural commodities that the region produces has further weakened the economy of this zone and, as a result, there is a gradual outward movement of the population towards the more prosperous areas to the south, east and west.

The sheer size and ethnic diversity of Zaïre have been significant factors in determining the present state of its forests. Formerly known as the Belgian Congo, Zaïre gained its independence in 1960, since which time it has been riven with conflicts inspired by the hunger for regional autonomy. These conflicts, together with an inadequate communication system, have seriously weakened the country's infrastructure, so that large areas are cut off from government control and many parts of the country suffer major depopulation. It is now more difficult to travel by river or road in Zaïre than it was 30 years ago and, in some remote parts of the east and north of the country, the only mail, health and educational services available are those provided by foreign missionary groups. The isolated mission stations are kept supplied by light aircraft operating out of the adjoining East African countries.

The history of Zaïre would undoubtedly have been very different but for an accident of geography. The Zaïre River is navigable for most of its length and could have provided a major highway for human penetration of the centre of the continent. But this highway is broken by two sets of impassable rapids, the first below Kinshasa and the second above Kisangani. Railways now bypass both rapids but the time and cost involved in transhipment of goods have always prevented exploitation and development of the interior. This factor has been of special significance for forest preservation. The cost of transporting timber from the forests of the Cuvette to European markets has always been so high that only the most valuable species could be exploited profitably. This has meant that the central forest zone has only experienced a low volume of very selective logging which has had little impact on the conservation status of the forests.

The Forests

Few of the forests of Zaïre are true evergreen rain forests. Even in the central basin, they are mostly semi-evergreen, becoming more deciduous towards the north and south of the forest zone. Most regions experience pronounced dry seasons and, in much of the central Cuvette, the total rainfall is less than 2000 mm and is therefore at the lower limit of the amount required to support moist evergreen forest. Around the Cuvette, this low rainfall and the poor soils make the forests extremely sensitive to disturbance, particularly to fire. Thus, even the relatively small scale disturbance caused by shifting cultivation in the forest zone has a severe impact: it allows fire to enter the forest, thus creating large patches of secondary vegetation and grassland in areas which were forested until recently. This poses a particularly serious threat to the forest area around the periphery of the Cuvette.

The original vegetation of the areas to the north and south of the closed forest block were Sudanese and Zambezi dry forests, respectively. The former have now been degraded by fire, forming open *Isoberlinia* woodlands, with species such as *Lophira lanceolata*, *Daniellia oliveri*, and *Parkia biglobosa*. The Zambezi dry forest exists only as relict stands of dense forest dominated by *Entandrophragma delevoyi*, *Parinari excelsa* and *Cryptosepalum pseudotaxus*. Most of the Zambezi forests have suffered the effects of fire, and vast areas are now covered by secondary Miombo woodland dominated by *Brachystegia* spp. and *Julbernardia globiflora*.

The closed evergreen forests of Zaïre are generally grouped into three major categories (Lebrun and Gilbert, 1954; Ipalaka, 1988; UICN, 1990). These are: swamp and riverine forests; the Guineo-Congolian lowland rain forests of the central Cuvette and Bas-Zaïre; and the various Afromontane forest communities of the highlands on the eastern borders of the country.

These forest types can be further subdivided into numerous more localised communities, the characteristic species of which are given in UICN (1990). Within the semi-evergreen forests, there is a major distinction between the forests of the Bas-Zaïre region, and those of the Cuvette and its environs. In the Bas-Zaïre, the only remaining fragments are found on the hills of the Mayombe, between Kinshasa and the Atlantic coast. These contain several tree species not found elsewhere in Zaïre, such as safukala *Dacryodes pubescens*, mutényé *Guibourtia arnoldiana* and *Daniellia klainei*. These forests are also rich in species such as tola *Gossweilerodendron balsamiferum*, and limba *Terminalia superba*.

One particularly interesting feature of the forests of the Cuvette is the tendency for very large areas to be dominated by a single species of the family Caesalpiniaceae. For example, in the Ituri Forest, lower slope forests are often composed of almost pure stands of limbali *Gilbertiodendron dewevrei*, whereas communities on the higher plateaux are dominated by bomanga *Brachystegia laurentii*. In other areas, mufimbi *Cynometra alexandri* dominates.

The montane forests of the so-called Albertine rift mountains, which form the watershed between the Nile and the Zaïre rivers, do not have a high diversity of tree species but are, nonetheless, important as they exhibit an interesting pattern of altitudinal zonation. Species of *Podocarpus*, *Prunus* and *Ocotea* occur at intermediate elevations, giving way, with increasing altitude, to elfin forests and communities dominated by bamboos, tree ferns, giant groundsels and lobelias. At the highest altitudes, Afroalpine moorlands occur, with a variety of ericaceous shrubs and grasses. Much of the mid-elevation forest has suffered from human disturbance and is now covered with open *Hagenia abyssinica* woodland and other secondary plant communities.

Mangroves

A small area, 226 sq. km according to Map 32.1, of mangrove forest exists in the estuary of the Zaïre River (see UICN, 1990). The principal species represented are *Rhizophora racemosa* and *Avicennia nitida*, while other, less widely distributed species are *Conocarpus erectus* and *Laguncularia racemosa*. This site is of particular interest as it provides refuge for a small and highly endangered population of the West African manatee *Trichechus senegalensis*.

The mangroves have suffered from pollution from the oil terminals in the adjacent Cabinda enclave of Angola. In recent years, some logging has taken place at this site. Unesco has contributed towards a proposal to establish a mangrove national park, but sufficient financial resources have not yet been found to implement the project. However, a warden and guards have now been installed.

Forest Resources and Management

FAO (1988) estimated that in 1980 closed broadleaved forests in Zaire covered approximately 1,056,500 sq km, and that a further 718,400 sq km of open forests also existed. Other studies have been published (USAID, 1981; Ipalaka, 1990; MacKinnon and MacKinnon, 1986) which give a variety of figures for either original or present forest cover or both. Ipalaka (1990) gives estimates of forest cover produced by Zaire's forest department, the Service Permanent d'Inventaire et d'Amenagement Forestier (SPIAF). Having undertaken Landsat studies of forest cover throughout the country, SPIAF has produced unpublished maps at scales ranging from 1:100,000 to 1:400,000. Their estimates of forest cover for each region are given in Table 32.1. SPIAF also presents estimates of original coverage of different forest formations (Table 32.2).

Table 32.1 Official estimates of the total forest cover in Zaïre by region

Forest Region	*Area of Region (sq. km)*	*Area of Forest (sq. km)*	*% Forest cover*
Equateur	403,292	402,000	99 .7
Haut-Zaïre	503,239	370,000	73 .5
Kivu	256,662	180,000	70.1
Bandundu	295,658	120,000	40.6
Kasai-oriental	168,216	100,000	59 .4
Kasai-occidental	156,967	40,000	25.5
Bas-Zaïre	53,855	10,000	18 .6
Shaba	496,965	10,000	2 .0
Ville de Kinshasa	9,965	–	–
Totals	2,344,819	1,232,000	48.7

Source: Ipalaka (1990)

Table 32.2 Original area of main forest types

Forest Formation	*Area of coverage (sq. km)*
Dense evergreen forest	900,000
Dense forest on hydromorphic soils (swamp forest)	100,000
Montane forest	53,148
Total closed broadleaved forest	1,053,148
Mosaic of open forest and savanna	1,291,852
Total	2,345,000

Source: Ipalaka (1990)

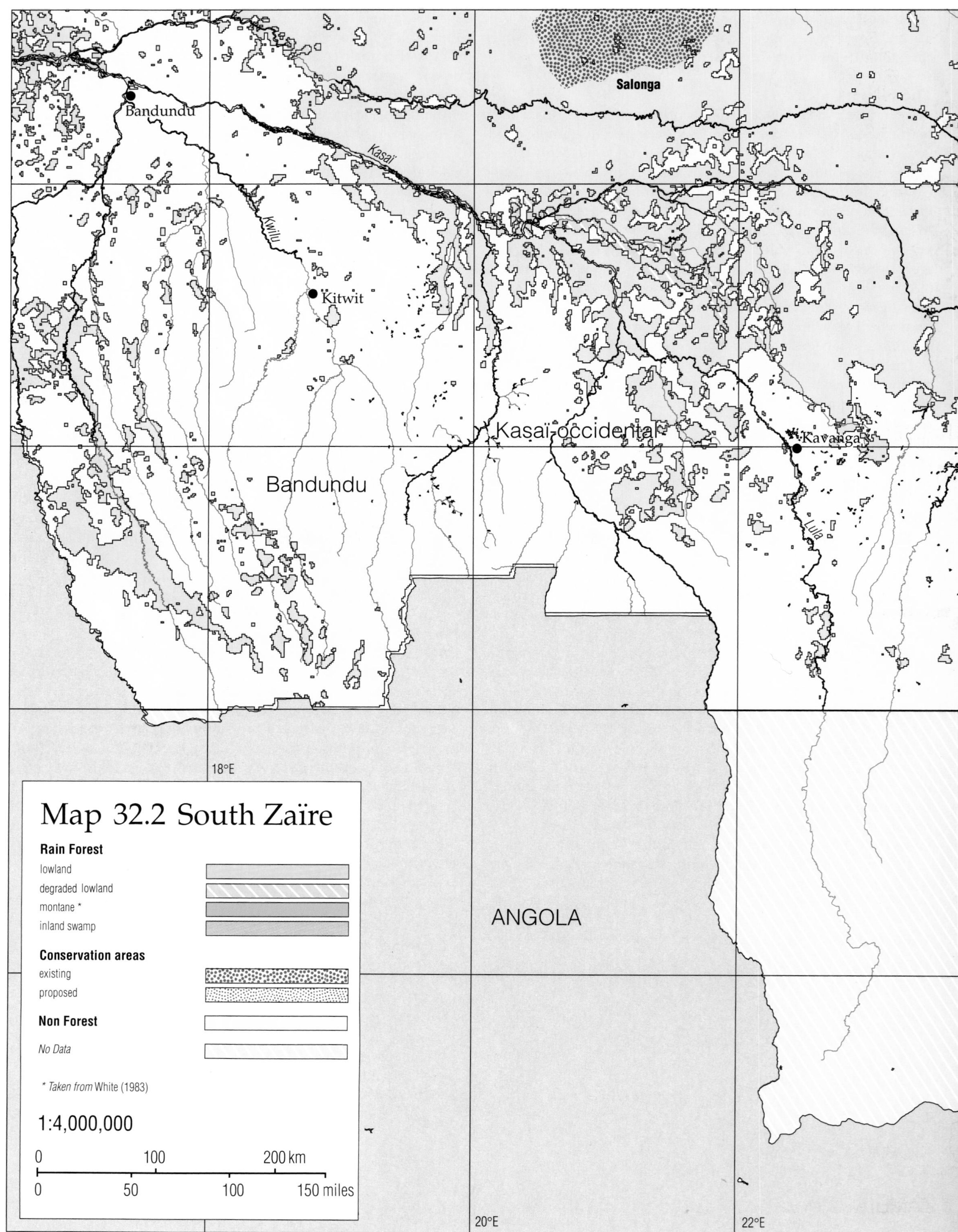

Map 32.2 South Zaïre

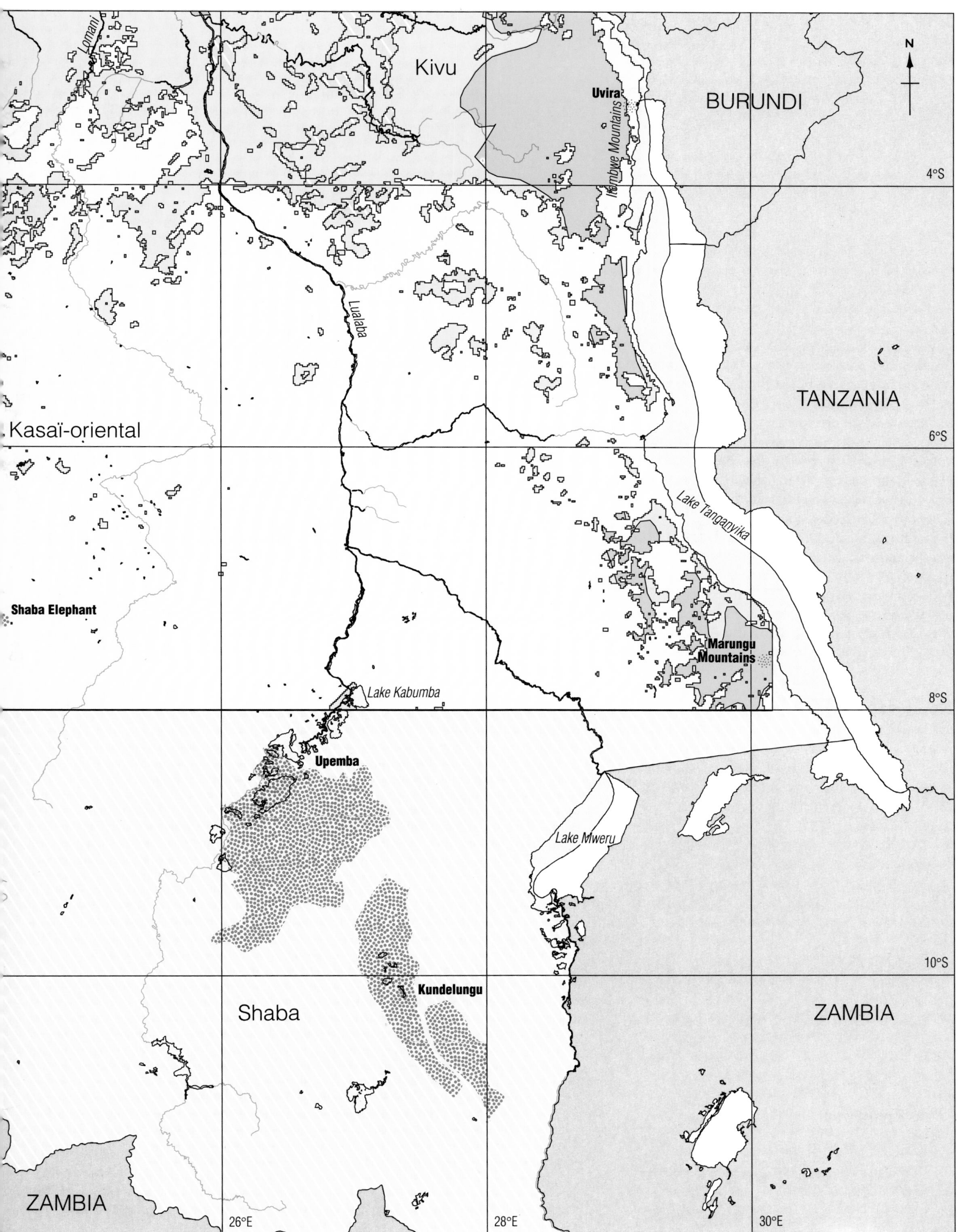

Lomani
Kivu
Uvira
Itombwe Mountains
BURUNDI
N
4°S
Lualaba
TANZANIA
Kasaï-oriental
6°S
Lake Tanganyika
Shaba Elephant
Marungu Mountains
Lake Kabumba
8°S
Upemba
Lake Mweru
10°S
Kundelungu
Shaba
ZAMBIA
ZAMBIA
26°E
28°E
30°E

Table 32.3 Estimates of forest extent

	Area (sq. km)	*% of land area*
Rain forests		
Lowland (closed)	919,736	40.6
Lowland (degraded)	86,547	3.8
Montane	59,487	2.6
Swamp	124,741	5.5
Mangrove	226	<0.01
Totals	1,190,737	52.5

(Based on analysis of Map 32.1. See Map Legend on p. 282 for details of sources.)

Table 32.4 Timber production and exports from Zaïre in 1985

Veneer and sawlog production by region 1985		*Timber exports 1985*	
Bas-Zaïre	94,300 cu. m		
Bandunda	82,300 cu. m	Logs	110,000 cu. m
Equateur	165,050 cu. m	Sawnwood	26,000 cu. m
Haut-Zaïre	56,070 cu. m	Plywood/veneer	28,000 cu. m
Other regions	22,400 cu. m		
Total	420,120 cu. m	Total	164,000 cu. m

(*Source:* UICN, 1990)

The figures for veneer and sawlog production in Table 32.4 differ from those cited for industrial roundwood at the head of this chapter as the latter includes unprocessed timber used by the local communities. The figures in the table are for roundwood delivered to mills or ports for processing or export.

As the original area of closed broadleaved forest in Table 32.2 is less than that given as the present extent of 'forest' in Table 32.1, it would appear that both open and closed formations are included in the earlier table.

The forest cover shown on Map 32.1 on p. 282 has been derived from satellite imagery interpreted and produced by NASA in collaboration with FAO (see Map Legend for details). White's (1983) map has been used to classify the forest into the types used throughout this Atlas, but his mosaic of swamp and lowland forest (category 9) has been shown here as lowland forest as it is impossible to distinguish between the two classes of forest using NASA's data. Figure 32.1 depicts the position of this mosaic. NASA's category of '*Degraded Forest including shifting agriculture and palm*' has been shown as degraded forest on Map 32.1. The extent of the different forest types as shown on Map 32.1 is given in Table 32.3.

According to MacKinnon and MacKinnon (1986) there were originally 1,784,000 sq km of closed forest in Zaire (see Chapter 10 and Table 10.1). Our figure of 1,190,737 sq. km of moist forest present in Zaïre is, therefore, 67 per cent of what is estimated to have been the original forest cover. However, it must be noted that 86,547 sq. km of this is degraded. According to the figures in Table 32.3, the intact forest present today covers 62 per cent of its original area and extends over 49 per cent of the land area of Zaire.

Figure 32.1 Map of Zaïre showing the mosaic of swamp and lowland forest (category 9 in White, 1983), which has been depicted as lowland forest in Map 32.1

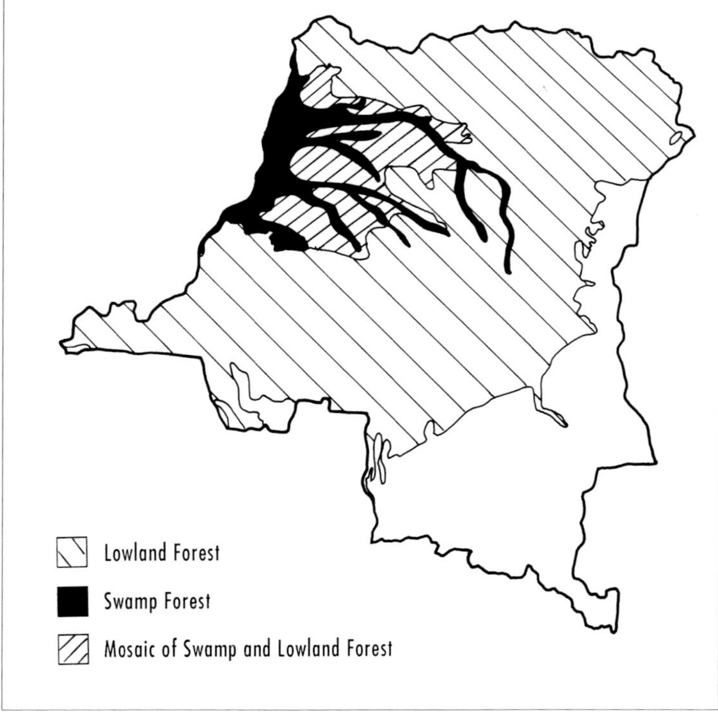

Differences between the estimates given here and those produced by SPIAF (Ipalaka, 1990) and other authorities are undoubtedly due to different criteria for defining and measuring forests.

Few attempts have been made to apply silvicultural techniques to the forests of Zaïre. In the past, the timber industry focused largely on the forests in the Mayombe region in Bas-Zaïre. These forests were readily accessible from the sea and have produced more than half of all the timber ever exported from Zaïre. Virtually all of the commercially valuable timber in the Mayombe region has now been exhausted and much of the forest has been encroached by farmers. The focus of forest exploitation is now moving inland from Bas-Zaïre to Bandundu and the Cuvette.

At present the exploitation of timber in Zaïre is inhibited by transport problems and low market prices. Total exploitation is estimated at roughly 500,000 cu. m per year and, of this, only 164,000 cu. m are thought to have been exported in 1985 (Table 32.4). Thus Zaïre has over 50 per cent of Africa's forests, yet produced only 3.4 per cent of Africa's entire timber exports.

Recent studies by SPIAF have estimated that the million sq. km forest in the Cuvette has a standing volume of 200 cu. m of timber per hectare. On the basis of a postulated 100-year cutting cycle, and assuming that 40 per cent of the forest is accessible and that half of this area should be totally protected in parks and reserves, SPIAF concludes that the forests could sustain a yield of 40 million cu. m per year. Even if only 10 per cent of this volume was of trees of species and dimensions suitable for industrial use it would give an annual yield of 4 million cu. m, or eight times the present estimated crop. The Tropical Forestry Action Plan for Zaïre established this as a target for the Zaïrean timber industry. This proposed eight-fold increase in logging provoked criticism from many conservation groups. However, it is worth noting that this yield would still constitute a much lower offtake per area of forest than that of other African and Asian timber producers.

Few forests in Zaïre are managed but management plans have now been prepared for several pilot areas. These propose a degree of post-logging silvicultural treatment to produce a more uniform stand of commercially valuable timber. However, these plans are not being implemented and logging practices are almost entirely determined by market forces. Large trees, of approximately ten species, account for most of the crop and yields rarely exceed 5–6 cu. m per hectare.

Kahuzi-Biega National Park is included on the Unesco World Heritage list as being of outstanding natural value. Dr M. Merker/WWF

On the periphery of the forest areas, and especially in the Bas-Zaïre region, agricultural encroachment often destroyed the residual forest once logging was completed. In the areas of the Cuvette that are now being logged, there are few inhabitants; biologists have observed that, without population pressures, the forests quickly recover from logging and wildlife remains abundant.

Logging for domestic needs poses a particular, localised threat to the forests. In the Shaba and Kivu provinces, small-scale enterprises are now taking timber from the southern and eastern fringes of the forest. Some of this timber is exported to Rwanda, Uganda, Kenya and possibly Burundi. The increasing demand for charcoal and fuelwood in urban areas is also posing a major threat to the forests. Areas surrounding, or within easy reach of, main cities are now totally deforested and traders travel further and further into the forest in search of fuelwood.

Until 1980 the total area of plantations in Zaïre was only 235 sq. km. The rate of establishment of plantations has since increased but most comprise fast growing fuelwood species in peri-urban areas. Attempts to plant high value timber species are restricted to pilot projects in the Bas-Zaïre where trials are being conducted with *Terminalia superba*. There is a very large EEC-funded project to plant *Acacia auriculiformis* for fuelwood on the deforested Batéké plateau near Kinshasa.

For most Zaïreans the non-timber products of the forest are of far greater significance than wood. People collect honey and fibres as well as plants for food and medicinal purposes. Widespread hunting takes place and large quantities of bushmeat are consumed both in rural areas and in the cities. With the significant exception of ivory, the non-timber products are not traded internationally on any great scale. Despite the combined efforts of Zaïrean and international conservationists to control the illegal trade of ivory, Zaïre continues to be the origin of a large proportion of the ivory exported from Africa and, as a result, elephant populations in the country are declining rapidly.

Deforestation

There are no accurate studies of the rate of deforestation in Zaïre. FAO/UNEP (1981) estimated that 1650 sq. km of forest were destroyed every year from 1967 to 1980, of which 1350 sq. km (81.8 per cent) were apparently cleared for shifting agriculture, much of it in logged-over areas. FAO's (1988) estimate of deforestation for the period 1981–5 was 1800 sq. km per year. Another 2000 sq. km of gallery and open forest are thought to be lost through fire. Apart from agriculture, it appears that exploitation of the forests for fuelwood, over-grazing and fire continue to be the main causes of forest loss.

Deforestation is most severe in Kivu, Shaba, Kasaï and Bas-Zaïre and is particularly acute around Kinshasa, where virtually all of the forest has been cleared within a radius of 60–100 km of the city (USAID, 1981).

Biodiversity

Zaïre has an extraordinary diversity of both animal and plant species. This diversity is due in part to the huge size of the country and, in part, to the range of climates, topography and geology that it contains. The extent and generally good condition of its forests also contribute towards maintaining such levels of bio-diversity. During past periods of drier climates the forests of Zaïre became fragmented and

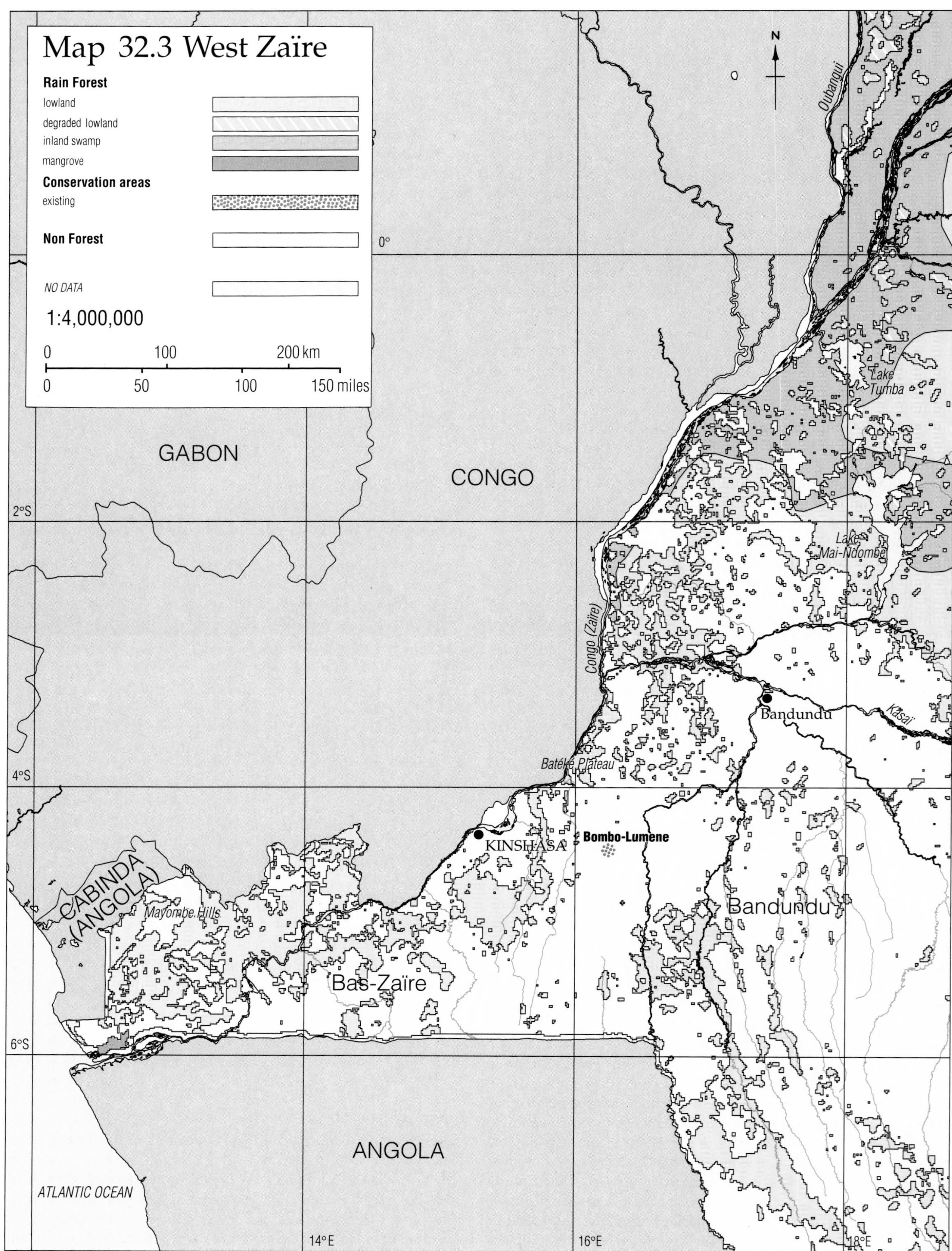

Map 32.3 West Zaïre
Rain Forest
lowland
degraded lowland
inland swamp
mangrove
Conservation areas
existing
Non Forest
NO DATA
1:4,000,000
0 100 200 km
0 50 100 150 miles
GABON
CONGO
ANGOLA
CABINDA (ANGOLA)
ATLANTIC OCEAN
Oubangui
Lake Tumba
Lake Mai-Ndombe
Congo (Zaïre)
Kasaï
Bandundu
Batéké Plateau
KINSHASA
Bombo-Lumene
Mayombe Hills
Bas-Zaïre
Bandundu
0°
2°S
4°S
6°S
14°E
16°E
18°E
N

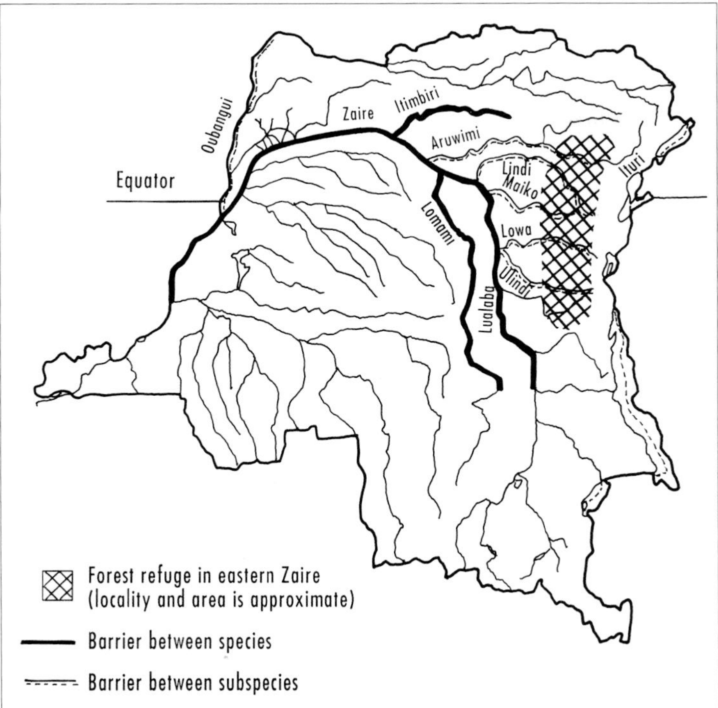

Figure 32.2 The forest refuge in eastern Zaïre and the rivers which have acted as barriers to faunal speciation

(*Source:* after Colyn, 1987, from UICN, 1990)

the subsequent competition between forms of the same species that had diverged during their isolation in forest refuges has resulted in the evolution of numerous endemic species. Many of the country's rivers constitute faunal barriers (Figure 32.2) and there is evidence, particularly for primates, that this has influenced the richness of the mammalian fauna (Colyn, 1987). Similar findings have been reported for many plant communities (Ndjele, 1988).

Over 10,000 species of plants are known from Zaïre, a total exceeded in Africa only by South Africa and, possibly, Madagascar. Of the 3921 species described in the first ten volumes of the *Flore du Congo Belge et Ruanda-Urundi* (now published as the *Flore d'Afrique Centrale)*, 1280 (32.6 per cent) were considered to be endemic (Brenan, 1978). Many of these species will undoubtedly be found eventually in other countries, and there is now evidence that Gabon and Cameroon have higher floral diversity and endemism. Nonetheless, Zaïre remains among the most floristically rich countries in Africa.

Zaïre has 409 species of mammals, 1086 species of birds, 80 species of amphibians (of which 51 are endemic) and 400 species of fish. There are local endemics in all groups and many species are considered to be in danger of extinction (IUCN, 1990; UICN, 1990).

Thirty species of primate – of which 19 species or subspecies are endemic – are known from Zaïre, the highest specific diversity in continental Africa. A notable example is the pygmy chimpanzee *Pan paniscus*, one of man's closest relatives, occurring in forests in the Lomako and Salonga region of the central Cuvette. Other rarities include the mountain gorilla *Gorilla gorilla berengei* and the eastern lowland gorilla *G. g. graueri*. The grey-cheeked mangabey *Cercocebus albigena*, the Salongo guenon *Cercopithecus salongo* (a local endemic), the owl-faced guenon *C. hamlyni*, and L'Hoest's guenon *C. lhoesti* all occur in the forests of Zaïre. There are around 30 species of antelopes including seven or eight species of duiker as well as the spectacular endemic okapi *Okapia johnstoni*, a 'forest giraffe' (see Ituri Forest case study). Some 18 species of fruit bats are found in Zaïre (Stuart *et al.*, 1990), four of which are of conservation concern: short-palate fruit bat *Casinycteris argynnis* (only known from the lowland forests of Zaïre and Cameroon); Hayman's epauletted fruit bat *Micropteropus intermedius;* Anchieta's fruit bat *Plerotes anchietae* and Rwenzori long-haired rousette *Rousettus lanosus*.

An estimated 25 bird species are thought to be endangered, 11 of which are endemic (Collar and Stuart, 1985). The threatened species include the shoebill *Balaeniceps rex*, the wattled crane *Bugeranus carunculatus*, the remarkable endemic Congo peacock *Afropavo congensis*, and the endemic Itombwe owl *Phodilus prigoginei*. The mountain forests on the eastern borders of the country are particularly rich in birds. The forests on Itombwe Mountain are home to at least 564 species of birds, and this site is thought to be the single richest area for birds in Africa (see Itombwe case study).

Conservation Areas

Zaïre has some of the world's most extensive and spectacular national parks (see Table 32.5). Four of the seven parks are included on the Unesco World Heritage list as being of outstanding natural value. These are: Garamba, which is open grassland; Kahuzi-Biega, largely comprising forest; Salonga (entirely forest), and Virunga (extensive forests). Collectively, the parks encompass 84,880 sq. km and four include extensive areas of moist forest.

Apart from these national parks, several new protected areas have recently been proposed. Of these, the most significant area for forest conservation is in the Ituri Forest (see case study), where an extensive park is planned to conserve the okapi and other species characteristic of this otherwise unprotected forest. A second area in the north-western

Table 32.5 Conservation areas of Zaïre

Conservation areas are listed below. Forest reserves, hunting reserves and marine parks are not listed or mapped. For data on World Heritage sites and Biosphere reserves see chapter 9.

	Existing area (sq. km)	Proposed area (sq. km)
National Parks		
Garamba	4,920	
Kahuzi-Biega*	6,000	
Kundelungu	7,600	
Lomako*		nd
Maïko*	10,830	
Marungu Mountains*		nd
Okapi*		nd
Salonga*	36,000	
Upemba	11,730	
Uvira†		nd
Virunga*	7,800	
Nature Reserves		
Bomu†	nd	
Eaux Delcommune†	nd	
Lac Fwa†	nd	
Game Reserves		
Bombo-Luméné	2,400	
Flora Reserves		
Yangambi*	2,500	
Reserves		
Shaba Elephant	nd	
Totals	89,780	nd

(*Source:* WCMC, *in litt.*)

* Area with moist forest within its boundaries according to Maps 32.1–3.
† Unmapped – no location data available.
nd no data.

ITOMBWE

The Itombwe massif is on the site of the largest of the Central African forest refugia, one of those believed to have persisted through the driest periods of the Pleistocene. It is of exceptional importance for the conservation of biological diversity. Indeed, the avifauna is the most species-rich for any single forest in Africa and it is considered to be the most important forest for bird conservation on the continent (Collar and Stuart, 1988). The massif, lying above the north-western shore of Lake Tanganyika, is a relict of the ancient relief which predates the formation of the western rift system. It rises over a distance of some 100 km, from around 700 m in the west to the scarp summit of 3475 m and then falls away abruptly to the level of the lake at 770 m.

The main feature of Itombwe is a 5000 sq. km forest block ranging from below 1200 m to above 2700 m. With 1600 sq. km of montane forest, this block constitutes the largest single tract of such forest in the Central African highland region. It may be contiguous with extensive lowland forest to the west. A belt of forest/grassland mosaic separates the main forest block from a further 1840 sq. km expanse of montane and bamboo forest, rough grassland, heath and marsh which run the 180 km length of the scarp top.

This forest harbours 50 per cent of all the African montane bird species, 94 per cent of those bird species characteristic of the Central African highlands and 89 per cent of the birds endemic to this region. The status of other faunal groups is less well known, but there are probably many endemics. Indeed, both shrew species discovered in the area are found only there and there is a high level of endemicity among the amphibians. A number of mammals of conservation concern occur in Itombwe, including eastern lowland gorilla, elephant *Loxodonta africana* and leopard *Panthera pardus*.

There is little permanent human settlement in the main forest block, but mineral prospecting and traditional hunting are widespread. The forest edge is being cut back by farmers along the entire length of the northern and eastern escarpments. In addition, the lowland forest area to the west is marked by a zone of slash and burn agriculture, although this does not yet appear to be encroaching seriously on the main forest. On the highlands to the south of the forest, cattle-raising is the main activity and here a characteristic mosaic of forest patches and grassland is found. The high altitude area of the scarp summit is also used extensively for grazing. However, although the effects of human activity are locally marked and even intense in places, they do not yet threaten the main forest area.

Low human population density and the forest's isolation have proved, so far, to be Itombwe's best defences, but this state of affairs cannot continue for ever. Despite recognition of its great value, no area of the massif is protected. A joint IZCN/WWF/FFPS preliminary survey of the region was undertaken in 1989 and it is hoped that a conservation programme will now be developed for this important forest. *Source:* Roger Wilson

section of Ituri – the Rubi-Télé hunting area – which also contains okapi, is now being investigated. This area is deemed to be important as it includes some drier forest types than those found in the Ituri. Other proposals have existed for several years for the development of protected areas in the Lomako region of the Cuvette, home of the pygmy chimpanzee, and for the mangrove forests in the Zaïre River estuary.

There are three Biosphere reserves in Zaïre: Lufira (147 sq. km), Luki (330 sq. km) and Yangambi (2500 sq. km). The last two regions are forested. Though these areas are the sites of research initiatives supported by the Unesco Man and Biosphere Programme, they are not effectively protected.

Fifty-six areas have been accorded the status of Hunting Reserves, although only 21 of these receive any sort of protection. Together these sites total 104,831 sq. km. The only activity that is subject to restrictions in these areas is big-game hunting. In general, their value for ecosystem conservation is relatively limited, and for this reason they are not listed in Table 32.5 or shown on Map 32.1.

The Institut Zaïrois pour la Conservation de la Nature (IZCN) is responsible for managing national parks and hunting areas. It is part of the Ministry of Environment and Nature Conservation, but enjoys a degree of autonomy that is unusual for African conservation organisations. The IZCN has a long and distinguished conservation record and has survived periods when civil disturbances and lack of resources made its operations very difficult. It is now receiving support from many international sources including the World Bank, the European Community, German bilateral aid (GTZ) and WWF.

A study by Doumenge (UICN, 1990) identified a number of other forest sites of critical importance for the conservation of biological diversity (Figure 32.3) and recommended additional protection measures for four of the existing national parks and six hunting areas and Biosphere reserves which contain important forest ecosystems. This report also identified numerous other forest areas of local importance for conservation.

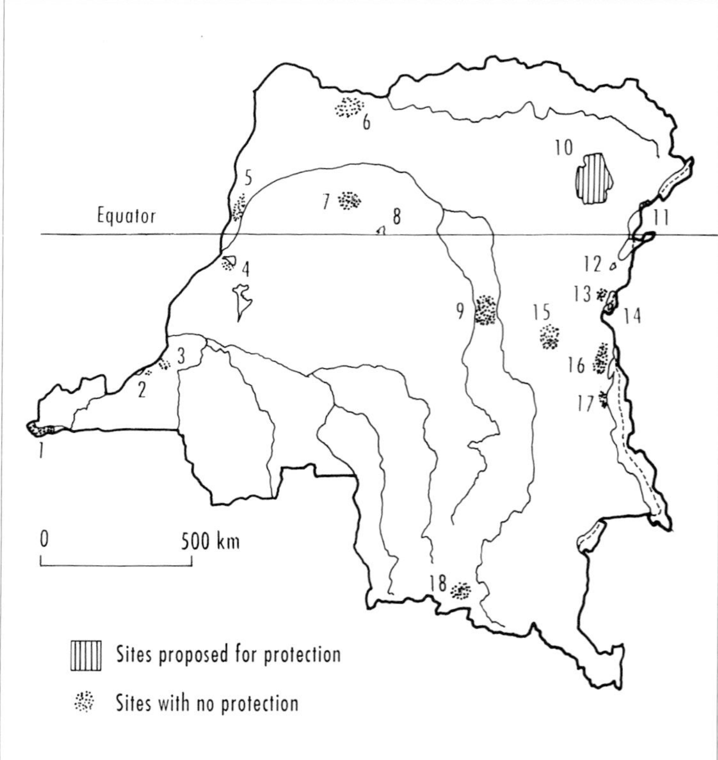

Figure 32.3 Critical sites for conservation in Zaïre's dense forest

(*Source:* UICN, 1990)

1 **Mangroves**. 2 Lake Ngaenké. 3 Maï Mpili-Lidji. 4 Lac Tumba. 5 Ngiri. 6 Abumonbazi-Mobaye. 7 Lomako-Yekokora. 8 **Luo**. 9 Lomami-Lualaba. 10 **Okapi**. 11 **Semliki**. 12 **Tonga**. 13 Ouest lac Kivu. 14 Iles Shushu. 15 Maniéma. 16 Itombwe. 17 Kabobo. 18 Kyamasumba-Kolwezi. (Those in bold type are the sites presently proposed for protection according to UICN, 1990.)

THE ITURI FOREST

The 60,000 sq. km watershed of the upper Ituri river in Zaïre contains one of the biologically and culturally most important forests of Africa. From an elevation of about 600 m in the west, where the rolling plateaux of the Ituri drop on to the central Zaïre river basin, the forest rises to more than 1000 m in the east, giving way abruptly to the savanna hills of the Albert Rift. Various forest types, including monospecific forests, mixed forests of varying composition and unique xerophyllic communities on the bare granite tops of its central massif, make up the Ituri Forest's great botanical wealth.

Nearly 15 per cent of the Ituri Forest's mammals are endemic to the region. It has more anthropoid primate species than any other African forest. A number of its birds occur nowhere else. Among the exceptional animals found here is the okapi, a short-necked rainforest giraffe, the size of a small horse, that was first discovered at the turn of the century in the eastern Ituri. The entire range of the okapi is within Zaïre's borders, and the largest population of this species occurs in the Ituri Forest.

Hunter-gatherers (see case study in chapter 5) and shifting cultivators have occupied the Ituri Forest for centuries, probably for millennia. The ancestries of today's forest peoples can be traced back to both Sudanic and Bantu migrations as well as to more ancient pygmoid stocks. Occupying small, transitory villages and camps, these communities never threatened the integrity of the forest. Rather the patches of secondary forest of varying age that they left behind in their migrations enriched the overall composition of the forest by providing suitable habitats for an array of successional plant and animal species.

Ituri's richness attracted attention even during colonial times. In the early 1950s, a government station was established at Epulu in the central Ituri Forest to capture okapi. The station adapted indigenous capture techniques and employed local forest peoples, notably the Mbuti and Efe pygmies, to capture, habituate and feed the okapi. A unique system of locally controlled forest reserves was established along the road for 70 km on either side of Epulu to serve as okapi capture zones. Each reserve was associated with a specific village whose residents undertook okapi captures. The economic benefits and the status these villagers realised by providing okapi to the station served as an impetus for the protection of the capture zones.

The reserves were exploited for a variety of purposes in addition to capturing okapi; they were used as hunting areas and for the gathering of forest products such as house building poles and wild foods. Their central purpose as capture zones was never forgotten, however. Village tradition protected them from agricultural incursions and excessive hunting likely to kill okapi. As a result, significant stands of mature forest remain intact along the road crossing the central Ituri. Village settlements and agricultural clearings are correspondingly dispersed and the destructive impact of local exploitation is reduced.

This traditional system of reserve management has assured forest conservation in the Ituri Forest up to the present. But the traditional conditions are now changing. Thanks in part to the opportunities provided by available land, timber and gold, the forest has become an economic and demographic frontier, with growing numbers of immigrant peasants, prospectors and entrepreneurs descending on the region. This swelling tide has led to fundamental social and economic changes for the less numerous and politically less organised indigenous forest peoples. Some of the traditional forest reserves have now been opened to settlement, clearing and degradation.

In recognition of its biological significance, and in response to the increasing threats to its integrity, there has been a growing demand to create a national park and forest reserve in the central Ituri. The first park proposals were put forward by the Zaïrean national parks service (IZCN) in the early 1970s, but funds to effect this were lacking. IZCN initiated a park project in 1986 in collaboration with WWF, and funded by Tabazaïre, a private Zaïrean firm. The traditional system of okapi capture reserves is seen as a key component for effective forest conservation. The current proposal calls for a reserve of about 13,500 sq. km, with multiple-use buffer zones centred on the capture areas along the road and with the remote interior being assured the more complete protection of a national park.

The success of the Ituri conservation proposal depends upon adequate information to define the limits of forest degradation and faunal decimation in the buffer zones. A research project to provide these guidelines has been based at Epulu since the early 1980s, and has been supported since 1986 by Wildlife Conservation International (WCI). This initiative has included long-term studies of natural and selectively logged forest, research into the socio-economic impact of human migration and the development of a Landsat map. The project has been unique in the degree to which it has incorporated local field assistants, in particular the expertise of the pygmies into its programmes. Where possible the emphasis is on the training of Zaïrean students in forest ecology and conservation biology.

It is doubtful whether these developments would have been possible had it not been for the clear link of this forest with the okapi. The WCI radio telemetry study of the okapi, the first study of the species in the wild, has provided estimates of 0.33 to 0.5 okapi per sq. km for the proposed reserve, indicating a total population of 4500 to 6500, the most important protected okapi population in existence. Despite its striking coat pattern, the okapi's solitary and secretive life has made viewing of the species in its natural habitat very difficult. Nevertheless, a significant tourist attraction has developed around the viewing of captive okapi in semi-natural enclosures at the capture station. More than half the visitors are Zaïrean nationals and the captive animals, along with troops of free ranging primates and abundant bird life, provide an exceptional opportunity for conservation education.

The okapi has become a national symbol of the forest and forest conservation. Although few Zaïreans may recognise the significance of their country's unique fauna, many know that the spectacular and widely esteemed okapi occurs only in Zaïre and is dependent on its forests. Government authorities may fail to grasp the importance of a forest for its abstract biodiversity value, but the okapi presents them with a concrete conservation target. By preserving the forest for the free-ranging okapi, the project will help conserve the entire variety of plants and animals that occur with it.

Source: John A. Hart and Terese B. Hart

Initiatives for Conservation

The IZCN has for many years struggled to extend the protected area system of the country and to maintain its integrity. The award of the order of the Golden Ark by Prince Bernhard of the Netherlands and the IUCN Packard Valour award to its current president, Mankoto ma Mbaelele, are the latest of many international tributes that have been paid to the competence and commitment of its workforce. In recent years 21 IZCN staff have been killed while trying to protect the country's parks against poachers.

WWF is now supporting projects in the Ituri Forest, the Virunga National Park and the Garamba National Park. Frankfurt Zoological Society is contributing to a project in Garamba National Park, while Wildlife Conservation International is supporting research on the okapi in the Ituri Forest (see case study). The British-based Fauna and Flora Preservation Society (FFPS) has a small but significant project in the Itombwe Mountains.

Official development assistance agencies are now recognising the immense value of Zaïre's parks system. German aid has a project in Kahuzi-Biega, the EEC is making major contributions to the Virunga National Park and Unesco to the Garamba National Park project. A planned World Bank loan will support IZCN's field staff in the Salonga National Park, Maïko National Park, and in proposed conservation areas in the Itombwe Mountains, the Ituri Forest and Rubi-Télé.

The EEC has a major project which is intended to improve management of the Salonga National Park and its buffer zones. This activity will be part of a major regional programme including seven central African countries (UICN, 1989).

The Tropical Forestry Action Plan for Zaïre, when complete, will give high priority to supporting conservation programmes in several existing and proposed protected areas. It supports the government's declared policy of eventually giving total protection to 12–15 per cent of the national territory and of maintaining natural forest cover or plantations on 35 per cent of the land surface.

References

Brenan, J. P. M. (1978) Some aspects of the phytogeography of tropical Africa. *Annals Missouri Botanical Garden* **65**: 437–78.

Collar, N. J. and Stuart, S. N. (1985) *Threatened Birds of Africa and Related Islands. The ICBP/IUCN Red Data Book Part 1.* ICBP/IUCN, Cambridge, UK.

Collar, N. J. and Stuart, S. N. (1988) *Key Forests for Threatened Birds in Africa.* ICBP Monograph No. 3. ICBP, Cambridge, UK.

Colyn, M. (1987) Les primates de la forêt ombrophile de la Cuvette du Zaïre: interprétations zoogéographiques des modèles de distribution. *Revue Zoologie Afrique* **101**: 183–96.

FAO (1988) *An Interim Report on the State of Forest Resources in the Developing Countries.* FAO, Rome, Italy. 18 pp.

FAO (1991) *FAO Yearbook of Forest Products 1978–1989.* FAO Forestry Series No. 24 and FAO Statistics Series No. 97. FAO, Rome, Italy.

FAO/UNEP (1981) *Tropical Forest Resources Assessment Project. Forest Resources of Tropical Africa. Part II: Country Briefs.* FAO, Rome, Italy.

Ipalaka Yobwa (1988) Proposition de zonage du territoire forestier. In: Département des Affaires foncières, Environnement et Conservation de la Nature et IIED. *Séminaire sur la politique forestière au Zaïre, Kinshasa, 11–13 mai 1988.* 18 pp.

Ipalaka Yobwa (1990) *L'Aménagement et la Gestion des Forêts du Zaïre.* Paper presented at the Conference sur la Conservation et Utilisation Rationelle de la Forêt Dense d'Afrique Centrale et de l'Ouest. African Development Bank/World Bank/IUCN, Abidjan, Côte d'Ivoire. November 3–9 1990.

IUCN (1990) *1990 IUCN Red List of Threatened Animals.* IUCN, Gland, Switzerland and Cambridge, UK. 228 pp.

Lebrun, J. and Gilbert, G. (1954) Une classification écologique des forêts du Congo. *Publ. Inst. Natl. Etude Agron. Congo Belge Sér. Sci.* **63**: 1–89.

MacKinnon, J. and MacKinnon, K. (1986) *Review of the Protected Areas System in the Afrotropical Realm.* IUCN/UNEP, Gland, Switzerland. 259 pp.

Ndjele, M. (1988) Principales distributions obtenues par l'analyse factorielle des éléments phytogéographiques présumés endémiques dans la flore du Zaïre. *Monograph Syst. Bot. Missouri Botanical Garden* **25**: 631–8.

Stuart, S. N., Adams, R. J. and Jenkins, M. (1990) *Biodiversity in Sub-Saharan Africa and its Islands. Conservation, Management and Sustainable Use.* IUCN, Gland, Switzerland and Cambridge, UK.

UICN (1989) *La Conservation des Ecosystèmes forestiers d'Afrique Centrale.* UICN, Gland, Switzerland and Cambridge, UK. 124 pp.

UICN (1990) *La Conservation des Ecosystèmes forestier du Zaïre.* Basé sur le travail de Charles Dommenge. UICN, Gland, Switzerland and Cambridge, UK. 242 pp.

USAID (1981) *Le Zaïre. Profil écologique du pays.* Harza Engineering Company/USAID, Kinshasa.

Authorship

Jeffrey Sayer with contributions from Ipalaka Yobwa, Mankoto ma Mbaelélé and Hadelin Mertens in Kinshasa, Jean-Pierre d'Huart in Brussels, Charles Doumenge, IUCN, Gland, Roger Wilson, FFPS and John and Terese Hart, Wildlife Conservation International, Project Okapi, Epulu, Zaïre.

Map 32.1–3 Forest cover for Zaire

The spatial data for Zaire's forest cover were provided solely for the use of this project by NASA/GSFC and the University of Maryland, USA. The vegetation map, derived from lkm resolution NOAA/AVHRR 1988 data, was produced at NASA, Goddard Space Flight Center. The vegetation classification encompasses those forests which fall north of 8.01°S and west of 30.23°E. This is a preliminary version of the map which will later be published by NASA with a detailed accompanying map legend, and is at present being evaluated by FAO and the Service Permanent d'Inventaire et d'Amenagement Forestier (SPIAF) in Zaire.

The original data have been categorised by NASA into six land use/vegetation types. These are: *Non-Zaire and/or continuous cloud regions; Water; Primary Forest including swamp forest; Degraded Forest including shifting agriculture and palm; Mixed savanna, degraded forest, and mature secondary forest occurring in the forest/savanna transition zone; and Savanna.* For the forest cover shown on Map 32.1 two of these data types have been mapped, namely, *Primary Forest including swamp forest and Degraded Forest including shifting agriculture and palm.* These have been harmonised into the rain forest types discussed in this atlas by overlaying White's map (1983). The distribution of swamp forest on Map 32.1 has been demarcated by White's type no. 8 *'Swamp forest'*. Type no. 9 *'Mosaic of 8 and la'* (swamp and Guineo-Congolian lowland rain forest) has been amalgamated into the lowland rain forest category, as there is no distinction between swamp and lowland rain forest in the original NASA dataset (but see Figure 32.1). The NASA Degraded Forest class is mapped here as degraded lowland rain forest.

Conservation areas spatial data have been taken from a published Russian map, *Zaire* (1987), at a scale of 1:2,500,000, compiled by the Main Administration of Geodesy Under the Council of Ministers of the USSR and from data within WCMC files.

ACRONYMS

AECCG African Elephant Conservation Coordination Group
AETFAT Association pour l'Etude de la Flore d'Afrique Tropicale
AIDS Acquired Immune Deficiency Syndrome
APN Amélioration des Peuplements Naturels
ASEAN Association of South East Asian Nations
AVHRR Advanced Very High Resolution Radiometry
AWF African Wildlife Foundation
BP British Petroleun
CAMPFIRE Communal Area Management Plan for Indigenous Resources
CAR Central African Republic
CARE Care and Relief Everywhere
CCFI Collaborative Community Forestry Initiative
CDC Commonwealth Development Corporation
CECI Centre Canadien d'Etude et de Coopération
CENAREST Centre National de la Recherche Scientifique et Technique
CENPAF Centre National pour la Protection et l'Aménagement de la Faune
CERGEC Centre de Recherche Géographique et de production Cartographique
CIDA Canadian International Development Agency
CIEO Centre International pour l'Exploitation des Océans
CILSS Comité Inter-états de Lutte contre la Sécheresse au Sahel
CITES Convention on International Trade in Endangered Species of Wild Fauna and Flora
CML Centre for Environmental Studies, University of Leiden
CNPPA Commission on National Parks and Protected Areas (IUCN)
CNRS Centre National de la Recherche Scientifique
CRNP Cross River National Park
CTFT Centre Technique Forestier Tropical (Paris)
DANIDA Ministry of Foreign Affairs, Department of International Development
DGFC Direction Générale des Forêts et Chasse
DNFC Direction Nationale des Forêts et Chasse
DPIAF Direction du Projet Inventaire et Aménagement de la Faune
EAP Environmental Action Plan
EC European Community
EDWIN Environment and Development Wetland Information Network
EMP Environmental Management Plan
EPC Environmental Protection Council
EROS Earth Resources Orbital Satellite
ERP Economic Recovery Programme
FAO Food and Agriculture Organisation of the United Nations
FDA Forest Development Authority
FFPS Flora and Fauna Preservation Society
FINNIDA Finnish International Development Agency
FoE Friends of the Earth
FPF Forest Products Fee
FRIN Forestry Research Institute of Nigeria
GACON Ghana Association for the Conservation of Nature
GDP Gross Domestic Product
GEF Global Environment Facility
GEMS Global Environment Monitoring System
GFM German Forestry Mission
GHAFOSIM Ghana Forest Simulation Model
GIS Geographic Information System
GPF Gestion et Protection des Forêts
GREEN Ghana Rainforest Expedition Eighty-Nine
GRID Global Resources Information Database (UNEP/GEMS)
GSFC Goddard Space Flight Center
GTZ Deutsche Gesellschaft für Technische Zusammenarbeit
HIV Human Immunodeficiency Virus
ICBP International Council for Bird Preservation
IDA International Development Association
IFAN Institut Fondamental d'Afrique Noire
IGADD Inter-Governmental Authority on Drought and Development in Eastern Africa
IIED International Institute for Environment and Development
IIF Industrialisation Incentive Fee
IMF International Monetary Fund
INEP Institut National d'Etudes et des Recherches
IRNR Institute of Renewable Natural Resources
ITTO International Tropical Timber Organisation
IUCN International Union for Conservation of Nature and Natural Resources – The World Conservation Union
IWGIA International Working Group on Indigenous Affairs
IZCN Institut Zaïrois pour la Conservation de la Nature
JWPT Jersey Wildlife Preservation Trust
KfW Kredit anstalt für Wiederaufban (German Development Bank)
KREMU Kenya Rangelands Ecological Monitoring Unit
LAC Local Area Coverage
MAB Man and the Biosphere Programme (Unesco)
MAFNR Ministry of Agriculture, Forestry and Natural Resources
MANRF Ministry of Agriculture, Natural Resources and Forestry
MGP Mountain Gorilla Project
MPAEF Ministère de la Production Animal et Eaux et Forêts
MPEA Ministry of Planning and Economic Affairs
MSS Multispectral Scanner
MVP Minimum Viable Population
NASA National Aeronautic and Space Administration
NCS National Conservation Strategy
NCF Nigerian Conservation Foundation
NGO Non-governmental Organisation
NMK National Museums of Kenya
NOAA National Oceanic and Atmospheric Administration
NYZS New York Zoological Society
ODA Overseas Development Administration (UK)
ONAREF Office National de Régéneration des Forêts
ONADEF Office National de Développement des Forêts
ONC Operational Navigation Charts
ONF Office National des Forêts
ORSTOM Institut Français de Recherche Scientifique pour le Developpement en Cooperation
ORTPN Office Rwandais de Tourism et Parcs Nationaux
PCFN Projet de Conservation de la Forêt de Nyungwe
PNUD Programme des Nations Unies pour le Développement
PRB Population Reference Bureau
PSG Primate Specialist Group
PVA Population Vulnerability Analysis
RSPB Royal Society for the Protection of Birds
SADCC Southern African Development Coordinating Conference
SA.FA.FI Sampan'Asa Fambolena Fiompiana
SCN Société pour la Conservation de la Nature
SECA Société d'Eco-Aménagement
SIDA Swedish International Development Authority
SNBG Société Nationale des Bois du Gabon
SODEFOR Société du Developpement Forestier
SPIAF Service Permanent d'Inventaire et Aménagement Forestier
SPOT Systeme Probatoire d'Observation de la Terre
SSC Species Survival Commission
SZDP Support Zone Development Programme
TFAP Tropical Forestry Action Plan
TM Thematic mapper
TMF Tropical Moist Forest
TSS Tropical Shelterwood System
UAIC Unite d'Afforestation Industrielle du Congo
UEA University of East Anglia
UFA Unité Forestière d'Aménagement
UICN Union Internationale pour la Conservation de la Nature
UNDP United Nations Development Programme
UNECE United Nations Economic Commission for Europe
UNEP United Nations Environment Programme
Unesco United Nations Educational, Scientific and Cultural Organisation
UNSO United Nations Sundano-Sahelian Office
US-AID US Agency for International Development
WCI Wildlife Conservation International
WCED World Commission on Environment and Development
WCMC World Conservation Monitoring Centre
WHO World Health Organisation
WIWO Werkgroep Internationaal Wed-En Watervogelonderzoek
WRI World Resources Institute
WSM Wildlife Society of Malawi
WWF World Wide Fund for Nature

GLOSSARY

Afromontane Montane areas of tropical Africa.
Afrotropical Africa and its islands between the Tropics of Capricorn and Cancer.
Afroalpine vegetation Vegetation which is confined to the highest mountains of tropical Africa.
Agroforestry Interplanting of farm crops and trees.
Anthropic *see* ANTHROPOGENIC.
Anthropogenic Created by man.
Avifauna Birdlife of a region or period of time.
Bimodal Frequency distribution with two peaks.
Biodiversity Richness of plant and animal species and in ecosystem complexity.
Biogeographic Area defined by fauna and flora it contains.
Biomass Amount of living matter in a defined area.
Biome A large naturally occurring community of flora and fauna adapted to the particular conditions in which they are found.
Biosphere Reserve Concept introduced by Unesco's Man and Biosphere Programme. A reserve including zones with different degrees of land use.
Biota The flora and fauna of an area.
Biotic Relating to living things.
Blanket bog A bog that forms in areas of high rainfall and low evaporation.
broadleaved (tree) Any tree belonging to the subclass Dicotyledonae of the class Angiospermae (flowering plants).
Buffer zone An area peripheral to a national park or reserve which has restrictions placed on its use to give an added layer of protection to the nature reserve itself.
Bush fallow Agricultural fallow with trees and shrubs growing on the land.
Bushmeat Meat from wild animals intended for human consumption.
Cay A low insular bank or reef of coral, sand, etc.
Clear felling Complete clearance of a forest, as opposed to selective fellings. *See also* monocyclic/polycyclic systems.
Congeneric Of the same genus.
Coupe An area of forest concession to be cut in a year.
Creaming Light exploitation of a forest (removal of the most valuable trees).
Cuticle The waxy or corky layer on the epidermis in plants.
Deciduous (of a tree) One which sheds its leaves annually.
Desertification Expansion of deserts by climatic change or by overgrazing and clearing of vegetation in adjacent areas.
Dipterocarp Member of the Dipterocarpacae, a family of old-world tropical trees valuable for timber and resin.
Ecosystem A natural unit consisting of organisms and their environment.
Ecotourism Tourism for people interested in the ecology of an area.
Edaphic Produced or influenced by the soil.
Emergents Trees whose crowns are conspicuously taller than the surrounding canopy.
Endemic Native or confined to a particular area.
Endemism Noun from endemic.
Eolian (also aeolian) Wind-borne.
Entrophication Over-enrichment of a water body with nutrients, resulting in excessive growth of organisms and depletion of oxygen.
Epiphyte A plant which uses another for support, not for nutrients.
Ericaceous Belonging to the genus *Erica* or its family Ericaceae, a shrub or heath having small leathery leaves and bell-like flowers.
Ericoid With heather-like leaves.
Escarpment Long cliff or slope separating two more or less level slopes, resulting from erosion or faults.
Ethnotourism Tourism for those interested in the culture and lifestyle of the local people in the region.
Eutrophication Destruction of animal life in water bodies that are rich in nutrients and thereby support a dense plant life that uses up all the available oxygen.
Floristics The plant species composition of an ecosystem.
Forest outliers Small patches of forest away from the main forested areas.
frugivorous Fruit-eating.
Gallery forests Forests along rivers.
Gazetted Legally established.
Geomorphological Pertaining to geomorphology.
Geomorphology The study of the physical features of the surface of the earth and their relation to its geological structures.

GLOSSARY

Guenon African monkey of the genus *Cercopithecus.*
Herbaceous Of or like herbs.
Herpetofauna Reptile fauna.
Hydromorphic Having an affinity for water.
Hypergyny Having the outer parts of a flower (petals, stamens and sepals) attached below the pistil (stigma, style and ovary).
Inselberg An isolated hill or mountain rising abruptly from its surroundings.
Intercropping Raising a crop among plants of a different kind, usually in the space between rows.
Interglacial Between glacial periods.
Isoenzyme One of two or more enzymes with identical function but different structure.
Isohyet A line on a map connecting places having the same amount of rainfall in a given period.
Isotope One of two or more forms of an element differing from each other in relative atomic mass and in nuclear but not chemical properties.
Jurassic The geological period that began about 190 million years ago and ended around 136 million years ago.
Lianes Climbing and twining plants in tropical forests.
Limnology The study of the physical phenomena of lakes and other fresh waters.
Littoral Situated near a (sea) shore.
Macrofossil A fossil visible to the naked eye.
Mangrove A salt-adapted evergreen tree of the intertidal zone of the tropical and subtropical latitudes.
Manioc Any plant of the genus *Manihot,* especially the cultivated varieties, having starchy tuberous roots or the starch or flour made from these roots. Also called cassava or tapioca.
Massif Large mountain mass.
Mesic Pertaining to or adapted to life with a moderate supply of moisture.
Microclimate The climate of a small local area.
Microflora Extremely small plants, usually those invisible to the naked eye.
Miocene The geological epoch that began about 26 million years ago and ended around 5 million years ago.
Miombo A distinctive dry woodland dominated by the genera *Brachystegia* and *Julbernardia.*
Monoculture Cultivation of a single crop.
Monotypic Having a single representative (used of a boilogical group).
Monsoon forest Closed canopy forests in seasonal tropical climates (*see* CHAPTER 1).
Montane Growing or living in mountainous areas.
Mulching Spreading a mixture of wet straw, leaves etc. around or over a plant to enrich or insulate the soil.
Neogene Includes the geological epochs of the Miocene and Pliocene, i.e. began about 26 million years ago and ended 1.8 million years ago.
Oligocene The geological epoch that began some 37 million years ago and ended around 26 million years ago.
Onchocerciasis A disease of man also known as river blindness, common in tropical regions of Africa (and America), caused by infestation of a filarial worm *Onchocerca volvulus* which is transmitted by various species of black fly.
Palaearctic Of the Arctic and temperate parts of the Old World.
Palaeobotany The study of fossil plants.
Palaeoclimatology The study of the climate in geologically past times.
Palaeoecology The study of the ecology of fossil plants and animals.
Palaeogeography The study of the geographical features at periods in the geological past.
Palaeolimnology The study of lakes of past ages.
Palynology The study of pollen, spores etc. for the interpretation of past environments.
Pandemic Prevalent over a whole country or the world.
Parastatal Partially controlled by the state.
Passerine Any bird of the order Passeriformes; they are perching birds, sparrow-like in form.
Phytochorion Floristic region; pl. phytochoria.
Phytogeography The geography of plant distribution.
Pitsawing In situ cutting up of felled trees using a large manually-operated saw, generally with one person standing below the tree (in a pit) and one above it.
Pleistocene A geological epoch that commenced about 2 million years ago and ended about 10,000 years ago.
poison girdling Poisoning of unwanted trees and climbers to enable nearby trees to develop.
Pooid grass A member of the subfamily Pooideae.
Postglacial The period after the end of the last Ice Age.
Precambrian The earliest geological era that commenced with the consolidation of the earth's crust about 4600 million years ago and ended about 600 million years ago.
Primary forest *See* PRISTINE FOREST.
Pristine forest Forest in a primary, virgin or undisturbed state.
Prosimian A primate of the suborder Prosimii (e.g. lemurs, lorises, galagos and tarsiers), related more closely to the ancestral primate than are the simians (monkeys and apes).
Quaternary The present geological period that began around 2 million years ago.
Rain forest Closed canopy forests in aseasonal climates; may be found in tropical and temperate latitudes.
Refugium, plur. refugia Region where biological communities have remained relatively undisturbed over long periods.
Relic stand A group of trees left after the loss of the major proportion of a forest.
riparian Land bordering water.
Riverine Living or growing on a river bank.
Roundwood Wood in its natural state as felled or otherwise harvested.
Sapropel Slimy sediment laid down in water, largely organic in origin.
Savanna A tall grass plain in tropical or subtropical regions with widely spaced trees.
Scarp A steep slope at the edge of an upland area.
Sclerophyll Plant with tough evergreen leaves.
Secondary forest Forest which is regenerating after a greater or lesser degree of disturbance, often by selective logging or agriculture. It is characterised by a lack of large trees and a significant proportion of coloniser species.
Selective logging The logging or felling of only particular species, size or type (e.g. those with straight, solid trunks) of tree within an area (as opposed to clearfelling where all trees are cut). Selective logging is the more common method of felling though the number of trees taken from an area varies considerably.
Shifting cultivation System of agriculture that depends on clearing and burning an area of forest for farming over a temporary period. *See* SWIDDEN AGRICULTURE.
Skidder path Bulldozer or tractor track through a forest along which logs are pulled to the main road during selective logging.
Silviculture Treatment often involving removal in a natural forest of unwanted climbers, damaged trees or uncommercial species. Replanting is rare.
Slash and burn *See* SHIFTING CULTIVATION.
Softwood Wood from a conifer, technically recognised by the absence of vessels. Softwoods have abundant fibres and make good paper.
Speciation Formation of a new biological species.
Stand dynamics The age structure and growth rate of communities of trees.
Stratification (in geology) The arrangement of horizontal layers of sediment that comprise sedimentary rock.
Stratigraphy The geological study of strata and their succession.
Subsistence farming The production of food and other resources to satisfy the needs of the household, rather than for cash.
Sub-hygrophilous A plant growing in a damp environment.
Superspecies A group of closely related species; this taxonomic term is usually used by ornithologist or primatologists.
Swidden agriculture Shifting agriculture carried out in the traditional, sustainable way, i.e. with periods of fallow to restore soil fertility.
Symbiosis An interaction between two different organisms living in close physical association, usually to the advantage of both.
Sympatric Two or more species that occupy the same geographic area.
Taungya system The planting of trees in mixture with agricultural crops with the farmer agreeing to tend the trees at the same time as his own crops. When the crops are harvested, further tending of the trees is usually carried out by the Forestry Service.
Tavy A Malagasy term for shifting cultivation.
Taxon Any of the taxonomic groups into which an Order is divided and which contains one or more related genera.
Tectonic Relating to the deformation and movement of rock in the earth's crust.
Tertiary The geological period that followed the Cretaceous period around 65 million years ago and that ended 2 million years ago.
Tillage The process by which soil is prepared to form a seedbed favourable to crop growth.
Transhumant pastoralism The seasonal or periodic movement of pastoral farmers and their livestock in search of grazing, usually between areas that have distinctly different climatic and ecological conditions.
Transpiration The act of transpiring, i.e. (in plants) giving off water vapour.
Trophic chain Food chain.
Tropical shelterwood system A form of forest management most commonly used in West Africa, especially Nigeria. The objective was to enhance the natural regeneration of valuable tree species by gradually opening up the canopy by such methods as poisoning undesirable trees and cutting climbers.
Trypanosomiasis A disease caused by a trypanosome (a flagellate protozoon), especially sleeping sickness.
Ungulate A hoofed mammal.
Vascular Used to describe channels carrying fluids in plants (and animals).
Watershed The line separating two river basins.
Xerophyllic A plant tolerant of a very arid habitat.

INDEX OF SPECIES: FAUNA

INDEX

INDEX OF SPECIES: FLORA

General Index

INDEX